ELEMENT OF WATER POLLUTION CONTROL ENGINEERING

Om Prakash Gupta
B. Tech. (Chemical Engg.)
Ex DGM, HRD
Bokaro Steel Plant
Steel Authority of India Limited (SAIL)

KHANNA BOOK PUBLISHING CO. (P) LTD.
Publisher of Engineering and Computer Books
4C/4344, Ansari Road, Darya Ganj, New Delhi-110002
Phone : 011-23244447-48 **Mobile:** +91-99109 09320
E-mail : contact@khannabooks.com
Website : www.khannabooks.com

Price: ₹ 450.00

ELEMENT OF WATER POLLUTION CONTROL ENGINEERING

OM PRAKASH GUPTA

ISBN: 978-93-86173-22-5

Edition: 2019

Published by:

Khanna Book Publishing Co. (P) Ltd.
4C/4344, Ansari Road, Darya Ganj, New Delhi-110 002
Phone: 011-23244447-48 **Mobile:** +91-9910909320
E-mail: contact@khannabooks.com

Printed in India by:

Vaibhav Products Tronica City

PREFACE

It gives me immense pleasure to present this book titled 'Elements of Water Pollution Control Engineering'. This book is meant for the students, teachers, practicing engineers. To prepare a book on this subject in its true sense would be very exhaustive requiring many volumes to do real justice to the subject . However, attempt has been made to assimilate the essentials of the subject in lucid manner in a single volume .

The book contains twelve chapters followed by appendices (meant for specific target reader groups) pertaining to complete domain of water pollution control engineering. Besides , it also contains two chapters devoted to short questions & answers and multiple choice questions & answers drawn from the examination papers of various engineering colleges for the benefits of the students. Latest and classified informations are given in the appendices.

This book is targeted to benefit the following:

- Diploma in engineering students
- Degree in engineering students
- AMIE, AMIIM, AMIICHE Students.
- Faculty members and teaching staff
- Practicing engineers / professionals / Consultants
- M.Sc.& B.Sc. students

The scope and coverage of the subject matter presented can be gauged from glancing the contents portion of the book to judge its suitability and relevance for a particular target group. The subject matter has been presented in a simple, lucid and easy to understand manner for easy grasping of the subject.

Latest and updated informations/data/statistics pertaining to the subject matter has been included in this edition for the benefit of the readers.

Constructive criticisms and suggestion for further improvement of the book are solicited.

Om Prakash Gupta

Contents

Chapter

Introduction to Water Pollution

Water pollution occurs when a body of water is adversely affected due to the addition of large amounts of materials to the water. When it is unfit for its intended use, water is considered polluted. Two types of water pollutants exist; point source and non-point source. Point sources of pollution occur when harmful substances are emitted directly into a body of water.

Many causes of pollution including sewage and fertilizers contain nutrients such as nitrates and phosphates. In excess levels, nutrients overstimulate the growth of aquatic plants and algae. Excessive growth of these types of organisms consequently clogs our waterways, use up dissolved oxygen as they decompose, and block light to deeper waters.

TYPES OF WATER POLLUTION

It can be classified into three types as domestic pollution, industrial pollution and agricultural pollution.

Domestic pollution

Important source of water pollution is domestic sewage system, which pollutes well and rivers, which are important source of drinking water. The problem of excreta disposal is clearly as old as mankind itself and the need for careful disposal is highlighted in a number of religious books including Hindu, Islamic and Christian texts. The proper management of excreta acts as the primary barrier to prevent the spread of pathogens(ie. disease causing bacteria) in the environment. It directly influences disease transmission through person to person contact, water and the food chain. At least 2 billion people in developing countries lack an adequate system for disposing of their faeces. For the poor in many developing countries, faeces deposited near their homes constantly threaten household hygiene, by direct contact with people or by being inadvertently carried into homes and kitchens by children, domestic animals or insects. Domestic neighbourhood or district water supplies such as wells, tanks and reservoirs may be contaminated by poorly designed or maintained sewage disposal systems. The potable water contaminated with faeces is the chief cause of some important diseases of human beings. About 98% of the people of rural areas use open field for defecation. They deem that latrines are meant for city dwellers, where there are no fields for defecation. Faeces are the most common pollutant of potable water. They consist primarily of intestinal bacteria.

An increase in the quantity of water has a greater health impact than improved water quality, because an adequate water supply makes it possible or at least more feasible for people to adopt safe hygiene behaviors. Since a primary cause of contamination of water is inadequate or improper disposal of human & animal excreta, better water quality only improves health when sanitation is improved as well and when the quantity of water is sufficient. Deteriorating water quality is a particular threat in developing countries, where hundreds of millions of people lack access to clean drinking water and the vast majority of sewage are discharged into surface waters without wastewater treatment. When untreated or inadequately treated wastewater or excreta (faecal sludge) is applied to soil and crops, disease transmission can occur. The persons at risk are the farmers, farm workers and their families as well as consumers of crops produced in such a way.

Industrial Pollution

Sewage is not the only cause of water pollution, industrial waste is also a significant polluter giving rise to contamination with heavy metals. The chemicals discharged by the factories are more harmful than the sewerage that flows into the river. The effluent discharged by the factories contains detergents, non-biodegradable materials and toxic chemicals hazardous to health & hygiene. Other sources of water pollution are distillery potassic fertilizers, electroplating plant, which contains harmful heavy metals, and cyanides which causes excessive acidity in water, which is harmful for aquatic life. Intensive use of pesticides and fertilisers has increased the level of nitrates in water. This can lead to brain damage and even death among children. Water pollution ranges from moderate to very high in case of textiles, jute, leather, vegetables fats and brewery. Moderate pollution was noted from dairy and canning establishments.

Agricultural Pollution

In addition to problems of waterlogging, desertification, salinization, erosion, etc., that affect irrigated areas; the problem of downstream degradation of water quality by salts, agrochemicals and toxic leachates is a serious environmental problem. 'It is of relatively recent recognition that salinisation of water resources is a major and widespread phenomenon of possibly even greater concern to the sustainability of irrigation than is that of the salinisation of soils. Indeed, only in the past few years has it become apparent that trace toxic constituents, such as Se, Mo and As in agricultural drainage waters may cause pollution problems that threaten the continuation of irrigation in some projects

Water pollution parameters

pH value of water

pH is a measurement of how **acidic** or how **basic (alkaline)** a solution is. When substances dissolve in water they produce charged molecules called ions. Acidic water contains extra hydrogen ions (H+) and basic water contains extra hydroxyl (OH-) ions.

Pollution can change pH in water, which in turn can harm animals and plants living in the water. For instance, water coming out of an abandoned coal mine can have a pH of 2, which is very acidic and would definitely affect any fish crazy enough to try to live in it. By using the logarithm scale, this mine-drainage water would be 100,000 times more acidic than neutral water - so stay out of abandoned mines.

TOTAL SUSPENDED SOLIDS (TSS)

The TSS is solid materials, including organic and inorganic, that are suspended in the water. These would include silt, plankton and industrial wastes.

High concentrations of suspended solids can lower water quality by absorbing light. Waters then become warmer and lessen the ability of the water to hold oxygen necessary for aquatic life. Because aquatic plants also receive less light, photosynthesis decreases and less oxygen is produced. The combination of warmer water, less light and less oxygen makes it impossible for some forms of life to exist.

Suspended solids affect life in other ways. They can clog fish gills, reduce growth rates, decrease resistance to disease, and prevent egg & larval development. Particles that settle out can smother fish eggs and those of aquatic insects, as well as suffocate newly-hatched larvae. The material that settles also fills the spaces between rocks and makes these microhabitats unsuitable for various aquatic insects, such as mayfly nymphs, stonefly nymphs & caddisfly larva.

Suspended solids can result from erosion from urban runoff and agricultural land, industrial wastes, river bank erosion, bottom feeders (such as carp), algae growth or wastewater discharges.

Electrical Conductivity of Water

Conductivity is a measurement of the ability of an aqueous solution to carry an electrical current. An ion is an atom of an element that has gained or lost an electron which will create a negative or positive state. For example, sodium chloride (table salt) consists of sodium ions (Na+) and chloride ions (Cl-) held together in a crystal. In water it breaks apart into an aqueous solution of sodium and chloride ions. This solution will conduct an electrical current. Unit (mho/cm or mg/L) An equation which shows this is:

Na (atom) + Cl (atom) Na + Cl - (ionic crystal)

Na+Cl - (in a water solution) = Na + (ion) + Cl - (ion)

There are several factors that determine the degree to which water will carry an electrical current. These include:

- The concentration or number of ions;
- Mobility of the ion;
- Oxidation state (valence) and;
- Temperature of the water.

Resistance, which is an electrical measurement expressed in ohms, is the opposite of conductivity. Conductivity is then expressed in reciprocal ohms. A more convenient unit of measurement in the chemical analysis of water is micromhos. The specific conductance or conductivity measurement is related to ionic strength and does not indicate what specific ions are present.

Conductivity is a measurement used to determine a number of applications related to water quality. These are as follows:

- Determining mineralization: This is commonly called total dissolved solids. Total dissolved solids information is used to determine the overall ionic effect in a water source. Certain physiological effects on plants and animals are often affected by the number of available ions in the water.
- Noting variation or changes in natural water and waste waters quickly.
- Estimating the sample size necessary for other chemical analyses.
- Determining amounts of chemical reagents or treatment chemicals to be added to water sample.

Elevated dissolved solids can cause 'mineral tastes' in drinking water. Corrosion or encrustation of metallic surfaces by waters high in dissolved solids causes problems with industrial equipment and boilers as well as domestic plumbing, hot water heaters, toilet flushing mechanisms & washing machines.

Indirect effects of excess dissolved solids are primarily the elimination of desirable food plants and habitat forming plant species. Agricultural uses of water for livestock watering are limited by excessive dissolved solids and high dissolved solids can be a problem in water used for irrigation.

Hardness of Water

Hardness is due to the presence of multivalent metal ions which come from minerals dissolved in the water. Hardness is based on the ability of these ions to react with soap to form a precipitate or soap scum.

In fresh water the primary ions are calcium and magnesium; however iron and manganese may also contribute. Carbonate hardness is equal to alkalinity but a non-carbonate fraction may include nitrates and chlorides.

This is an electrochemical procedure. The technique for analysis uses potentiometric titration on the computer aided titrimeter (CAT) with a copper ion-specific electrode. A reference substance, EDTA, is used as a titrant. Hardness is expressed in mg/l of $CaCO_3$ (even though all the hardness may not be due to $CaCO_3$).

Table 1.1: Classification of Water by Hardness Content (Hardness Unit: mg/l)

Concentration mg. $CaCO_3$/l	Water type
0 - 75	Soft
75 - 150	Moderately hard
150 - 300	Hard
300 and up	Very hard

The most important impact of hardness on fish and other aquatic life appears to be the affect the presence of these ions has on the other more toxic metals such as lead, cadmium, chromium and zinc. Generally, the harder the water, the lower the toxicity of other metals to aquatic life. In hard water some of the metal ions form insoluble precipitates and drop out of solution and are not available to be taken in by the organism. Large amounts of hardness are undesirable mostly for economic or aesthetic reasons. If a stream or river is a drinking water source, hardness can present problems in the water treatment process. Hardness must also be removed before certain industries can use the water. For this reason, the hardness test is one of the most frequent analyses done by industrial units that use water.

TOTAL DISSOLVED SOLIDS (TDS)

The TDS comprise inorganic salts (principally calcium, magnesium, potassium, sodium, bicarbonates, chlorides & sulphates) and small amounts of organic matter that are dissolved in water. The TDS in drinking water originate from natural sources, sewage, urban runoff and industrial wastewater. Concentrations of TDS in water vary considerably in different geological regions owing to differences in the solubilities of minerals.

Reliable data on possible health effects associated with the ingestion of TDS in drinking water are not available, and no health-based guideline value is proposed. However, the presence of high levels of TDS in drinking water may be objectionable to consumers.

BIOLOGICAL OXYGEN DEMAND (BOD)

Biochemical Oxygen Demand (BOD) refers to the amount of oxygen that would be consumed if all the organics in one litre of water were oxidised by bacteria and protozoa. The first step in measureing BOD is to obtain equal volumes of water from the area to be tested and dilute each specimen with a known volume of distilled water which has been thoroughly shaken to ensure oxygen saturation.

After this, an oxygen meter is used to determine the concentration of oxygen within one of the vials. The remaining vial is than sealed and placed in darkness and tested five days later. BOD is then determined by subtracting the second meter reading from the first. The range of possible readings can vary considerably: water from an exceptionally clear lake might show a BOD of less than 2 ml/l of water. Raw sewage may give readings in the hundreds and food processing wastes may be in the thousands.

Generally, when BOD levels are high, there is a decline in DO (Disolved Oxygen) levels. This is because the demand for oxygen by the bacteria is high and they are taking that oxygen from the oxygen dissolved in the water. If there is no organic waste present in the water, there would not be as many bacteria present to decompose it and thus the BOD will tend to be lower and the dissolve oxygen level will tend to be higher.

Table 1.2: BOD and Water Quality

BOD (in ppm)	Water Quality
1 - 2	**Very Good** There will not be much organic waste present in the water supply.
3 - 5	**Fair: Moderately Clean**
6 - 9	**Poor: Somewhat Polluted** Usually indicates organic matter is present and bacteria are decomposing this waste.
100 or greater	**Very Poor: Highly Polluted** Contains organic waste.

CHEMICAL OXYGEN DEMAND (COD)

The COD test will give a good estimate of the first stage oxygen demand for most wastewaters. An advantage of the COD test over the biochemical oxygen demand (BOD) test is 2 to 3 hours versus 5 days. The COD test also is used to measure the strength of wastes that are too toxic for the BOD test. The COD test should be considered an independent measurement and not a quick substitute for the BOD test. The COD is usually higher than the BOD, but the amount will vary from waste to waste. The COD test should be considered an independent measurement of organic matter in a sample rather than a substitute for the BOD test.

Chemical Oxygen Demand measures the ability of hot chromic acid solution to oxidize organic matter. This analyzes both biodegradable and non-biodegradable (refractory) organic matter expressed as O_2.

The results of the COD (chemical oxygen demand) tests are usually higher than the corresponding BOD test for several reasons. Many organic compounds which are dichromate oxidizable are not biochemically oxidizable; Certain inorganic substances, such as sulphides, sulphates, thiosulphates, nitrites and ferrous iron are oxidized by dichromate, creating an inorganic COD, which is misleading when estimating the organic content of the wastewater. If this is true, a BOD value for a glucose/glutamic acid standard should be 60-70% of the COD value for the same sample.

The BOD to COD ratio is nothing more than the BOD concentration divided by the COD concentration for the same sample (e.g., if BOD is 60 mg/l, and COD is 100 mg/l for a sample, the ratio is 60/100, or 0.60). There are three fairly reliable 'rules of thumb' correlations between COD/BOD. 1. Ratios for COD to BOD of 0.5 to 2 are usually found in potable water or exceptionally clean surface or groundwater. 2. Ratios of COD to BOD of 2 to 4 are usually seen in routine domestic/municipal sewage wastes. 3. Ratios of COD to BOD of 4 to 6 are usually indicative of industrial type wastes. Of course, each specific treatment system may be checked for its own particular ratios.

In some industrial effluents (pretreatment program), BODs *can* be higher than CODs (for example, some effluents which are high in sugars, as can be found in the bakery industry, or soda bottling. Some industrial effluents will have higher demand because of the higher quantities of chemicals that demand oxygen.

Most applications of COD determine the amount of organic pollutants found in surface water (e.g. lakes and rivers), making COD a useful measure of water quality. It is expressed in milligrams per liter (mg/l), which indicates the mass of oxygen consumed per liter of solution. Older references may express the units as parts per million (ppm).

The following formula is used to calculate COD:

$$COD = \frac{8000(b - s)n}{sample\ volume}$$

where b is the volume of FAS (Ferrous Ammonium Sulphate) used in the blank sample, s is the volume of FAS in the original sample, and n is the normality of FAS. If milliliters are used consistently for volume measurements, the result of the COD calculation is given in mg/l.

CHLORIDE

Chloride in drinking water originates from natural sources, sewage & industrial effluents, urban runoff containing deicing salt and saline intrusion. The main source of human exposure to chloride is the addition of salt to food, and the intake from this source is usually greatly in excess of that from drinking water. Excessive chloride concentrations increase rates of corrosion of metals in the distribution system, depending on the alkalinity of the water. This can lead to increased concentrations of metals in the supply. No health based guideline value is proposed for chloride in drinking water. However, chloride concentrations in excess of about 250 mg/litre can give rise to detectable taste in water.

SULPHATE

Sulphates occur naturally in numerous minerals and are used commercially, principally in the chemical industry. They are discharged into water in industrial wastes and through atmospheric deposition; however, the highest levels usually occur in groundwater and are from natural sources. In general, the average daily intake of sulphate from drinking water, air and food is approximately 500 mg, food being the major source. However, in areas with drinking water supplies containing high levels of sulphate, drinking water may constitute the principal source of intake.

COLIFORMS

Coliforms are bacteria that are always present in the digestive tracts of animals, including humans, and are found in their wastes. They are also found in plant and soil material.

Total coliforms, Faecal coliforms, and E. coli

The most basic test for bacterial contamination of a water supply is the test for total coliform bacteria. Total coliform counts give a general indication of the sanitary condition of a water supply. Total coliforms include bacteria that are found in the soil, in water that has been influenced by surface water, and in human or animal waste.

Fecal coliforms are the group of the total coliforms that are considered to be present specifically in the gut and feces of warm blooded animals. Because the origins of fecal coliforms are more specific than the origins of the more general total coliform group of bacteria, fecal coliforms are considered a better indication of animal or human waste than the total coliforms.

The major species in the fecal coliform group is *Escherichia coli (E. coli)*. Of the five general groups of bacteria that comprise the total coliforms, only *E. coli* cannot be found growing and reproducing in the environment. Consequently, *E. coli* is considered to be the species of coliform bacteria that is the best indicator of fecal pollution and the possible presence of pathogens.

HARMFULNESS OF COLIFORM BACTERIA

Coliform bacteria may harm us by causing diseases and food spoilage. Of environmental concern are the many types of coliform bacteria. Fecal coliform bacteria are a group of bacteria that are passed through the fecal excrement of humans, livestock and wildlife. They aid in the digestion of food. A specific subgroup of this collection is the fecal coliform bacteria, the most common member being Eschericia coli. These organisms may be separated from the total coliform group by their ability to grow at elevated temperatures and are associated only with the fecal material of warm-blooded animals. Bacteria reproduce rapidly if conditions are right for growth. Most bacteria grow best in dark, warm, moist environments with food. Some bacteria form colonies as they multiply which may grow large enough to be seen. By growing and counting colonies of fecal coliform bacteria from a sample of stream water, we can determine approximately how many bacteria were originally present.

Most coliform bacteria do not cause disease. However, some rare strains of *E. coli*, particularly the strain 0l57:H7, are pathogenic and can cause serious illness. Recent outbreaks of disease caused by E. coli 0157:H7 have generated much public concern about this organism.

E. coli 0157:H7 has been isolated from cattle, chickens, swine, and sheep. Most of the reported cases have been attributed to the consumption of partially cooked hamburger. Cases of E. coli 0157:H7 caused by contaminated drinking water supplies are extremely rare.

The Standards

- With these bacteria in mind the first standard is a mandatory one that must not be exceeded; there should be no more than 10,000 total coliforms per 100ml of water or 2,000 faecal coliforms per 100ml of water.
- The second standard is a guideline that should be achieved where possible; within 100ml of water there should be no more than 500 total coliforms, 100 faecal coliforms and 100 faecal streptococci.

MAJOR TYPES OF WATER POLLUTANTS AND THEIR EFFECTS

Human activities give rise to water pollution by introducing various categories of substances or waste heat into a water body. The more common types of polluting substances include pathogenic organisms, oxygen demanding organic substances, plant nutrients which stimulate algal blooms, inorganic & organic toxic substances and oil & grease.

Pathogenic Organisms: Many serious human diseases such as cholera, typhoid, bacterial & amoebic dysentery, enteritis, polio and infectious hepatitis are caused by water-borne pathogens. In addition, malaria, yellow fever and filariasis are transmitted by insects that have aquatic larvae. Faecal pollution of water resources by untreated or improperly treated sewage is a major cause for the spread of water borne diseases. To a lesser extent, disease causing organisms may also be derived from animal rearing operations & food processing factories with inadequate waste water treatment facilities.

The spread of water borne infectious diseases can be controlled through the introduction of water & sewage treatment facilities and through improved hygiene.

Biodegradable Organic Substances: Human & animal wastes as well as effluents from industries processing plant or animal products contain a mixture of complex organic substances such as carbohydrates, proteins & fats as their major pollution load. These substances are readily biodegradable and when introduced into the environment are quickly decomposed through the action of natural microbial populations. Some of the organic matter is oxidised to carbon dioxide & water while the rest is assimilated and used for the synthesis of new microbial cells. In due course, these organisms will also die and become food for other decomposers. Eventually virtually all of the organic carbon will be oxidised.

When a biodegradable organic waste is discharged into an aquatic ecosystem such as a stream, estuary or lake, oxygen dissolved in the water is consumed due to respiration of micro-organisms that oxidise the organic matter. The more biodegradable a waste, the more rapid is the rate of its oxidation & the corresponding consumption of oxygen. In an aquatic ecosystem, a greater number of species of organisms are supported when the dissolved oxygen concentration is high. Oxygen depletion due to waste discharge has the effect of increasing the numbers of decomposer organisms at the expense of others.

When oxygen demand of a waste is so high as to eliminate all or most of the dissolved oxygen from a stretch of a water body, organic matter degradation occurs through the activities of anaerobic organisms which do not require oxygen. Not only does the water then become devoid of aerobic organisms, but anaerobic decomposition also results in the formation of a variety of foul smelling volatile organic acids & gases such as hydrogen sulfide & mercaptans. The stench from these is quite unpleasant.

Plant Nutrients: The availability of plant nutrients, *e.g.*, nitrogen & phosphorus are important determinants of the biological productivity of aquatic ecosystems. Nutrient deficient aquatic environments are called "oligotrophic" and those rich in nutrients, "eutrophic". Young lakes are generally oligotrophic, but they naturally accumulate nutrients over time, derived from drainage and sediment run off from its catchment. When human activities greatly accelerate nutrient enrichment of water bodies, the process is called "cultural eutrophication". Sewage, animal wastes and many industrial effluents contain high levels of nitrogen & phosphorus. Another major source is fertiliser run off from urban and agricultural catchments.

While in the long term, cultural eutrophication accelerates the natural successional progress of aquatic ecosystems towards a terrestrial system,in the short term problems arise due to cyclic occurrences of algal blooms & decay. In warm weather, nutrients stimulate rapid growth of algae and floating aquatic weeds. The water often becomes opaque and has unpleasant tastes & odours. When these organisms die they become food for decomposer bacteria. Depletion of dissolved oxygen leads to anaerobic conditions and a general decline in the ecological & aesthetic qualities of the water body. Algal blooms also reduce light penetration into the water making it impossible for sea grasses and other bottom anchored plants to survive.

Toxic Inorganic Pollutants. Many inorganic substances are released by natural weathering of rocks and are washed into water courses. However, human activities such as mining and mineral processing as well as wastage are responsible for far greater quantities of toxic inorganic pollutants entering water supplies & aquatic ecosystems. Of particular concern among these are arsenic, an ingredient of some pesticides, and heavy metals such as mercury, lead, tin & cadmium as they tend to accumulate in tissues. Mine drainage and leaching of mine tailings as well as metal finishing & inorganic chemical industries are major sources of metal pollution in the water environment.

MERCURY poisoning causes birth defects and permanent brain damage. The worst case of mercury poisoning of a community to date occurred in Minamata in Japan, due to the consumption of contaminated fish & shell fish from the bay which received discharges of chemical industry effluents over a long period of time. Insoluble metallic mercury is converted by bacteria in the marine sediments into water soluble methyl mercury. It is then concentrated through the trophic levels of the food chain due to selective retention in tissues.

LEAD causes damage to the nerve system, and experts recommend a tolerance limit of less than 10 parts per billion in drinking water. Prior to the introduction of PVC water pipes, lead piping and cast iron pipes using lead solder were widely used for public water supplies. The main source of environmental lead is automobile exhausts. Lead has long been an additive to petrol to boost its octane rating. Exhaust fumes from cars & other vehicles are eventually washed down by rain and enter watercourses. Some inorganic industrial effluents also contain lead.

Other heavy metals such as **Chromium** and **Zinc** are present in effluents from metal finishing industries. Discharge of waste waters from many small metal finishing industries causes ground water pollution. Chromium salts are added to cooling waters of air conditioning systems for corrosion prevention. Significant contribution of this metal therefore comes from this source as well.

Production of alumina from bauxite gives rise to highly caustic effluents which also contain dissolved aluminium and other metals. **Aluminium** has recently been implicated in the Alzhiemers disease, although there is still uncertainty about the link. **Caustic Effluents** can raise the pH of receiving waters to levels unsuitable for many organisms. **Acidic Effluents** are produced from mine drainage and by many industries such as ore smelting, metal finishing, leather tanning and petroleum processing. Many lakes are acidified due to acid precipitation. Increased levels of sulfur & nitrogen oxides in the atmosphere from the combustion of fossil fuels such as coal and petroleum are the principal causes of acid rain.

Acidification lowers the pH of the water, especially when there is little buffering capacity in the form of alkalinity to neutralise it. Fish, amphibians & many insects will be killed by increased acid levels and in severe cases only a few resistant species such as fungi may survive.

Toxic Organic Chemicals (TOC). Many thousands of natural and synthetic organic chemicals are in use today for the manufacture of a variety of products ranging pesticides, pigments, pharmaceuticals & plastics. Several of these are known to cause birth abnormalities, genetic defects and cancer. Some chemicals like DDT and PCB's (Poly Chloro Biphenyls) are concentrated in tissues to dangerous levels. Many are only very slowly biodegradable and persist in the environment for long periods of time.

The major causes of organic chemical pollution are improper disposal of domestic & industrial wastes and herbicide & pesticide run off from farming areas, where they are used in substantial quantities. In addition, large quantities of hazardous wastes have, in the past, been disposed of in landfills with inadequate containment from where they are slowly leached into surface & ground water supplies, eventually finding their way into the food chain.

Oil Pollution. Petroleum is one of the major energy sources today and huge volumes of oil are transported between points of production & consumption around the globe. All along these major transportation routes oil spills happen regularly and oil slicks are ever present. With serious spills, many marine birds and other animals are choked to death by the oil slick. Even when dispersed, many hydrocarbons in the oil are toxic to aquatic organisms. Some are thought to be carcinogenic. Oil being lighter than water floats on the surface as a thin film which can interfere with the transfer of gases such as oxygen & carbon dioxide, as well as heat, between the water and the atmosphere.

Routine petroleum refining, storage & use also results in pollution by leaking oil, oily waste water and sludge. Underground fuel storage tanks in service also leak oil into the ground water.

Thermal Pollution. Tremendous quantities of waste heat is produced by power plants and to a lesser extent by a broad spectrum of other industries. Cooling water drawn from the ocean, river, lake or aquifer is passed through heat exchangers where it absorbs the waste heat and is subsequently discharged back into the water environment.

Significant rises in water temperature can be caused in the receiving water in the vicinity of cooling water disposal sites. Such increases in temperature alter the species composition in ecosystems as organisms normally tolerate temperature variations only over a very small range. At higher temperatures, oxygen solubility is reduced, but bacterial respiration rate will increase, making the water more prone to deoxygenation. Temperature variations will also cause alterations in pH due to changes in the degree of ionisation and increased solubility or precipitation of bottom deposits.

VARIOUS METHODS OF CONTROL OF WATER POLLUTION.

With increasing urbanisation and expanding agricultural & industrial production, water pollution problems have become serious necessitating the adoption of suitable control measures for reducing water pollution.

Approaches to controlling sources of water pollution may be grouped into three broad categories: (1) minimisation of waste or pollutant generation, (2) Treatment prior to disposal of waste water streams at source, and (3) "in-situ" reduction or elimination of pollution.

Minimisation of Pollutant Generation: Reduction of quantity of waste /pollutants generated is the most desirable approach to pollution control. Since it conserves resources that would otherwise be wasted, and at the same time eliminates the cost of removing pollutant after they are produced, it is the cheapest & the most effective alternative. For non-point pollution sources, this is the only practicable method of pollution control. Yet, this approach has not been exploited to its fullest extent.

As a general rule, a resource becomes a waste when it can no longer be economically utilised or recovered. It is then disposed off in the environment in the cheapest manner possible. Availability of economical technology for resource processing & usage has been a main determinant of when the resource is discarded as waste.

Minimising soil erosion by improved agricultural practices (e.g. by minimising surface runoff & leaving crop residues in the ground), more efficient use of nutrients (e.g., though the use of slow release fertilisers) and the development & use of biological pest control techniques in preference to the use of non-biodegradable toxic chemicals are some of the measures for minimising water pollution from agriculture.

Considerable potential also exists in many industries to reduce waste generation. Development and use of non-polluting technology to modify or replace existing manufacturing processes, and recycling or recovering materials that would otherwise be wasted are two approaches which not only reduce pollutant generation, but can even result in a saving for the industry by minimising or eliminating the need for waste treatment for pollutant removal.

In some cases, it may be more practical to segregate strong & weak waste streams to facilitate materials or energy recovery. Good house keeping practices, like minimising spillage & materials wastage, also leads to waste reduction and savings in production cost.

Wastewater Treatment at Source: In nature, a variety of different mechanisms operate to degrade & transform waste materials into stable, harmless end products such as carbon dioxide. This cleansing ability is referred to as the "self-purification" or "assimilative" capacity. When the quantities of wastes to be disposed off are large, however, the natural purification processes become overloaded and can no longer assimilate the wastes without adversely affecting environmental quality. Man made treatment systems are then needed to reduce pollutant loads to acceptable levels for discharge. For the most part, these purification systems make use of the same mechanisms as in the natural environment to bringing about waste stabilisation.

The wastewater treatment technologies are classified as physical, chemical & biological processes, depending on the nature of the purification mechanism employed. The character of the pollutants and the form (suspended or dissolved) in which they are present usually determine the most suitable process for their removal. For example, total suspended solids and floatable materials such as oil & fat are readily removed by physical processes such as sedimentation or flotation respectively.

Biological Methods are effective & economical when the waste water contains mostly biodegradable pollutants such as organic matter. An advantage of biological processes is that the microorganisms involved in waste stabilisation are themselves produced in the process. For dilute wastes including general domestic wastewaters, "aerobic" biological processes (activated sludge, oxidation ponds and aerobic biofilter) are usually favoured since they are capable of producing an effuent with very low residual pollutant concentrations. These processes, require oxygen, in proportion to the pollutant load present. Oxygen is supplied through aeration, which is a significant cost component. For strong wastes, "anaerobic" biological treatment in enclosed vessels is generally preferred as they proceed in the absence of oxygen, and in addition produce a useful, energy-rich by-product in the form of methane. The effuent from anaerobic processes, however, contain higher levels of residual organic materials and may require further polishing treatment (often in aerobic processes).

Chemical Treatment is used when the pollutant is non biodegradable and is not amenable to removal by simple physical means (e.g. when it occurs in dissolved form). Heavy metals are typically removed by chemical precipitation, while toxic substances such as cyanide may be chemically oxidised. An important disadvantage of chemical treatment methods is that they generally require dosing with a chemical which can be quite expensive. In addition, disposal of the chemical sludge produced in these processes may also pose some problems.

When a community based treatment system is impractical, it is possible to provide a degree of treatment prior to discharging sewage into the environment. A popular method used for individual homes & small groups of residences is the Septic Tank. It consists of a simple baffled tank which traps most of the solids in the waste water and also affords some decomposition of soluble organic matter. The effluent is disposed off into the ground through a system of leach drains. As solids progressively accumulate in the tank, it is necessary to periodically desludge the system, typically every 3 to 7 years.As deep sewering in built up areas is very expensive, other more efficient alternatives to the septic tank are also desirable for on site use.

In-situ Pollution Control: Waste minimisation & treatment help prevent pollution from occurring and should be the principal approaches to water quality maintenance. Occasionally, however, when a water body is already adversely affected, it will be necessary to consider action aimed at helping the ecosystem recover from the impact of pollution. Methods to facilitate this are collectively grouped under in-situ control techniques.

Aeration of lakes & reservoirs, especially when they are thermally stratified (in summer), has been used to prevent anaerobic conditions from occurring. Forced circulation of water in stratified lakes is an alternative method. Dredging nutrient rich superficial sediments from highly eutrophic lakes, while very expensive, helps reduce occurrence of severe algal blooms. Addition of aluminium or iron salts to assist the precipitation of phosphorus has also been practiced in some lakes to control dissolved phosphorus levels in the water.

APPENDIX 1.1

Relative bio degradability of certain organic compounds present in polluted water

S.No	Biodegradable organic compounds	S.No.	Compounds generally resistant to biological degradation
1	Acrylic	1	Ethers
2	Aliphatic acids	2	Ethylene chlorohydrin
3	Aliphatic alcohols (normal,iso,secondary)	3	Isoprene
4	Aliphatic aldehydes	4	Methyl vinyl ketone
5	Aliphatic esters	5	Morpholine
6	Alkyl benzene sulphonates	6	Oil
7	Propylene based Benzaldehyde	7	Polymeric compounds
8	Aromatic amines	8	Polypropylene benzene sulphonate
9	Dichlorophenols	9	Selected hydrocarbons
10	Ethanolamine	10	Aliphatics `
11	Glycols	11	Aromatics
12	Ketones	12	Alkyl-aryl groups
13	Methacrylic	13	Tertiary aliphatic alcohols
14	Methyl methacrylate	14	Tertiary benzene sulphonates
15	Mono chloro phenols	15	Trichloro phenols
16	Nitriles		
17	Phenols		
18	Primary aliphatic amines		
19	Styrene		
20	Vinyl acetate		

Note: Water pollution has been seriously affecting the life of humans, Plants as well as animals The eco-system of rivers, streams, lakes, seas and oceans is also getting deteriorated due to the contamination of water, through various sources. This condition also leads to the outbreak of numerous diseases, majority of them being lethal and contagious.

Causes of water pollution

Dumping of industrial wastes, containing heavy metals, harmful chemicals, by-products organic toxins and oils, into the nearby source of water is one of the visible causes of water pollution. Cause for the contamination of water is the improper effluents disposal of human & animal wastes from factories, refineries, injection wells and sewage treatment plants are dumped into urban water supplies, leading to water pollution. Number of pollutants, both harmful and poisonous, enter the ground water system through rain water. The residue of agricultural practices, including fertilizers and pesticides are some of the major sources of water pollution.

- Untreated pollutants are drained into the nearest water body, such as stream, lake or harbor, causing water pollution.
- Another major sources of water pollution comprises of organic farm wastes. When farm land, treated with pesticides & fertilizers, is irrigated, the excess nitrogen and poisons get mixed into the water supply, thereby contaminating it.
- Pathogens, sediments and chemical pollutants are other sources of water pollution.

The corporeal wastes produced by humans and farmed organisms pollute rivers. lakes, oceans and other surface waters.

Harmful Effects of water pollution

- A number of waterborne diseases are produced by the pathogens present in polluted water, affecting humans and animals alike.
- Pollution affects the chemistry of water. The pollutants, including toxic chemicals, can alter the acidity , conductivity and temperature of water.
- Polluted municipal water supplies are found to pose a threat to the health of people using them.
- People perish or incur various communicable diseases due to the consumption of contaminated drinking water.
- The concentration of bacteria and viruses in polluted water causes increases in solids suspended in the water body, which, in turn, leads to health problems.
- Marine life become deteriorated due to water pollution. Lethal killing of fish and aquatic plants in rivers oceans and seas is an aftereffect of water contamination only.
- Diseases affecting the heart, poor circulation of blood & the nervous system and aliments like skin lesion cholera & diarrhea are often linked to the harmful effects of water pollution.
- Carcinogenic pollutants found in polluted water causes cancer.
- Alteration in the chromosomal makeup of the further generation is foreseen, as a result of water pollution.
- Discharges from power station reduce the availably of oxygen in the water body, in which they are dumped.
- The flora and fauna of rivers, sea and oceans is adversely affected by water pollution.

APPENDIX 1.2

Characteristic of domestic waste water pollutants in Bombay entering sea water

S.No	Parameters	mg/litre
1	Suspended solids	245×10^3
2	Dissolved solids	1450×10^3
3	BOD	258×10^3
4	Sulphate	75×10^3
5	Chloride	587×10^3
6	Phosphorous	9×10^3
7	Nitrogen	35×10^3
8	Manganese	507×10^3
9	Iron	2529
10	Cobalt	30
11	Nickel	81
12	Copper	110
13	Zinc	251
14	Lead	11

APPENDIX 1.3

Level of pollution parameters in the Bombay Harbour water

S. No	Parameters	Concentration,mg/litre	Safe limit,mg/litre
1	BOD	110-225	4
2	COD	212-448	180
3	Oil and grease	4-15.6	1
4	Ammoniacal nitrogen	0.7-3.5	1.2
5	Nickel	30-240	0.3
6	Chromium	ND-20	0.2
7	Lead	50-100	0.1
8	Cadmium	ND-45	0.3

APPENDIX 1.4

Upper permissible limit of effluents discharged into surface water

S No.	Parameters	Upper permissible limit
1	B O D (5 days, 20 °C)	20 mg / litre
2	Suspended solids	30 mg / litre
3	Particulates	15 mg / litre
4	Oil & Grease	0.1 mg / litre
5	Cyanide	0.1 ppm
6	Phenols	0.5 ppm
7	Sulphate	1 ppm
8	Toxic Metals	0.5 ppm
9	pH value	5-9

APPENDIX 1.5

Problems caused by various water contaminants

S No.	Contaminants in water	Problems / difficulties caused by their presence
1	Alkalinity	Caused by the presence of carbonates, bi-carbonates, & various caustics ; its presence generates & scales particularly in boilers
2	Hardness	Caused primarily by calcium & magnesium salts, other salts (eg, iron salt) are also minor contributors, it causes scales particularly in boilers
3	Sodium salts	Occurs often as sulfates, chlorides, nitrates & bicarbonates, it is very bad for some industries like cellulose and drug industry
4	Silica	Results in scale formation in boilers
5	Iron & manganese	Tends to stain, its presence in water used in paper, textile & tanning industry is very objectionable, upper, limit of iron is 0.3 ppm & that of manganese is 0.05 ppm
6	Aluminum	Usually not a problem for industrial purpose water
7	Fluorides	Usually not a problem, expected in the production of baby foods concentration is < 0.1 ppm, higher concentration (> 1.2ppm) causes mottling of teeth or enamel fluorosis
8	Carbon dioxide	It increases the alkalinity and enhances the corrosive behavior of dissolved oxygen
9	Oxygen	Corrosive to iron, Steel, Galvanised iron and brass
10	Nitrogen	It helps retain dissolved oxygen in water, Thus increased corrosiveness
11	Hydrogen sulfide	Rotten egg odor, corrosion of iron pipes, fitting & equipments
12	Methane	Poses fire & explosion hazard
13	Micro-organisms	Forms coating in pipes, stains. Tastes and odours, decompose organic substances such as cellulose
14	Organic matter	Causes taste & odours, often absorbed by various processes, forms colored colloidal suspensions
15	Cyanide	High concentration in water may prove fatal, concentration above 0.2 ppm is not good for health
16	Lead	Lead should be < 0.05 ppm, higher concentration is inferior to health or even lethal
17	Nitrate & nitrites	Fatal poisoning of infants by ingestion of well water having high level of nitrates (<45ppm) , it also causes irritation of mucous membranes of stomach in adults & increased dieresis
18	Sulfates	Higher sulfate content in water (> 250 ppm) causes diarrhea It also affects taste of water.
19	Total dissolved solids (TDS)	High TDS is associated with correspondingly high levels of sulfates and/or chlorides. Higher concentration (> 500ppm) causes undesirable taste and diarrheic

Note: Major categories of water pollutants and their sources are tabulated below

Category of water pollutants	Examples	Sources
A. Causes health problems Infectious agents Organic Chemicals Inorganic Chemicals Radioactive materials	Bacteria, Viruses, parasites Pesticides, plastics, oil, gas, Detergents Acids, caustics, salts, metals, Uranium, thorium, cesium, iodine, radon	Human and animal excreta Industrial, household, and farm use Industrial effluents, household cleansers, surface runoff. Mining/ processing ores, power plants, weapons, natural sources
B. Causes ecosystem disruption: Sediment O_2-demanding wastes Thermal	Soil, silt Animal manure and plant residues Heat	Land erosion Sewage, agricultural runoff, paper mills, food processing Power plants, industrial cooling

Table 1.5T_1: Common causes of Water Related problems

S. No	Observed Problems	Causes
1	Water turns black, smell	Waste water
2	Acidic taste	Low pH
3	Alkaline taste	High pH
4	Boiled Rice hard and yellow	High Alkalinity
5	White deposits on boiling	Hardness
6	Iron taste, change in colour after exposure to atmosphere, change in colour of cloths, utensils Oily appearance on top of water body	Presence of Iron compounds
7	Soap not lathering	Hardness
8	Brownish black steaks on teeth	Fluoride
9	Growth of Algae	Nitrate, phosphate
10	Fish kills	Low pH , less DO
11	Salty taste	Chloride

APPENDIX 1.6

List of ecologically sensitive/ protected areas

S No.	Ecologically protected Areas
1	Religious and historical Place
2	Archaeological monuments
3	Scenic areas
4	Hill resorts
5	Beach resorts
6	Health resorts
7	Coastal areas rich in mangroves, corals, breading grounds of specific species
8	Estuaries
9	Gulf areas
10	Biosphere reserves
11	National parks and sanctuaries
12	Natural lakes, swamps
13	Seismic zones
14	Tribal settlements
15	Areas of scientific and geological interest
16	Defense installations especially those of security importance and sensitive to pollution
17	Border areas (international)
18	Airports

APPENDIX 1.7

List of pollution control monitoring equipments for excise duty exemption (INDIA)

S No.	Description of goods qualifying for excise duty exemption
1	Electrostatic precipitations system including auto mining spray nozzle, CO analyzers, emitting electrode
2	Felt filter system including pulse, valve, filter bag
3	Dust collector system
4	Scrubber-counter current/ venturi/ packed bed/ cyclonic scrubber
5	Air sampler for CO, SO_2, NOx, H_2, SO_3
6	Continuous analyser for CO, NO_2, SO_3
7	Halogen detector
8	Mechanically skimmed oil and grease removal system
9	Mechanical flocculators
10	Diffused air/surface aeration system for activated sludge system
11	Surface aerated system for aerated lagoon system
12	Bio filters rooting arms
13	Methane recovery anaerobic digestor systems
14	Air floatation systems
15	Marine outfall systems
16	Centrifuge and vacuum filter for dewatering sludge
17	pH meter (digital as well as analog type)
18	Conductivity meter
19	Turbidity meter
20	Water pollution Test kit
21	Water flow meter
22	BOD incubator
23	Mercury analyser
24	Spectrophotometers (UV-VIS and VIS-IR digital)
25	Continuous flow recorder with integrator
26	Continuous PH recorder
27	Fly ash handling system

Ref. Notification no. **78/90 central excise date 20.3.92 as amended on 25.7.92**

APPENDIX 1.8

Major effects of waste water pollutant materials on water receiving systems

S No.	Water receiving systems	Effects of waste water pollutants
1	Natural bodies water	(a) oxygen depletion of water (microbiological uptake depressed reiteration) (b) Interference with aquatic life (toxicity, fish –flesh tainting) (c) Danger to public health safety (pathogens, toxic chemicals) (d) Miscellaneous economic damage (Water potability, recreational losses)
2	Municipal Sewerage systems	(a) Hazards to maintenance personnels (explosion, toxic gases) (b) Damage to sewers (corrosion, deposition, overload) (c) Decrease in efficiency of sewerage treatments system (overload of units capacity, micro biological toxicity)
3	Underground strata	(a) Contaminations of potable water supplies (b) Contamination of brines and minerals
4	Land	(a) Irrigation: spray, overflow (Vegetation damage) (b) Burial (Seepage to ground or surface water)

Note:- Consequences of human activities in various economic sectors and their effects on aquatic ecosystems:

Consequence	Sedimentation	Eutrophication	Thermal Pollution	Dissolved Oxygen	Acidification	Microbial Contamination	Stabilization	Trace metals Contamination	Mercury	Non-metallic Tox-ins	pesticides	Hydrocarbons	Micronutrient Depl-etion
Sector:													
Agriculture	✓	✓	✓	-	-	✓	✓	✓	-	-	✓	-	-
Urban use	✓	✓	✓	✓	-	✓	✓	✓	-	-	✓	✓	-
Forestry	✓	✓	✓	-	-	-	-	-	-	-	✓	-	-
Hydroelectric power generation and water storage	✓	✓	✓	✓	-	-	-	-	-	-	-	-	✓
Mining	✓	✓	✓	✓	✓	-	-	✓	-	✓	-	-	-
Industries	✓	✓	✓	✓	✓	-	✓	✓	✓	✓	-	✓	-

APPENDIX 1.9

Water pollutants generated by various industrial, agricultural & urban activities

S.N.	Origin of activities classifications	Classifications	Activities/operations	Water pollutants generated
1.	Industrial activities	(a) Basic metal industries	Iron & steel, non-ferrous industries	SS, cyanides, sulfides, greases, oils, phenols
		(b) Chemical industries	Fine chemicals, pharmaceuticals, plastic industries	SS, COD, BOD, ammonia, metals, pH, nitrates, solvents, surfactants, phenol,
		(c) Mechanical engineering	Machinery, electronics, electrical equipments	SS, COD, BOD, pH,oil, surfactants, solvents
		(d) Mining & quarrying	Mineral extraction & processing induding mineral oil	SS, COD, BOD, pH , oils, odour, phenols, sulfides
		(e) Food industry	Sugar refining, milk& oilseed processing, canning, wine production, brewing and distilling	SS, COD, BOD, ammonia, nitrates, dyes
		(f) Non-metallic mineral products industries	pottery, cement, glass, building materials	SS, pH, dyes
		(g) Power generation	Thermal power plants, coking operations, others	SS,COD,BOD, ammonia, phenols, cyanides, pH
		(h) Textile industry	Spinning fibre, processing, dyeing, clothing industry	SS,. COD, BOD, surfactants, dye, ammonia, solvents
		(i) Wood and paper industries	Cellulose fibres, paper & board manufacture	SS, COD, BOD, sulfides,
2.	Agricultural activities	(a) Crop farming (b) Livestock rearing	Fertilisers use Poultry, pigs, cattles, horses	SS, COD, BOD, Phosphates, nitrates, ammonia
3.	Urban activities	(a) Human activities (b) Domestic activities (c) Handicrafts	Cooking, cleaning, electroplating, laundries, vehicle repairing shops. haspitals	SS, COD, BOD, oil, solvents, metals, ammonia, phosphates

Note: SS-suspended solids

Areas of Concern in Water Pollution

There are seven areas of concern which have preoccupied individuals with water pollution & contamination and ecological mainfestations.

(i) They are Oxygen demanding wastes, (ii) Pathogens (iii) Nutrients, (iv) Salts and minerals (v) Heated water discharges (iv) Trace metals and (vii) Pesticides & other volatile organic compounds.

(i) Oxygen demanding wastes–Oxygen demanding wastes are biodegradable and require oxygen from receiving waters. Since the amount of oxygen dissolved is limited,. it get depleted quickly and degradation proceeds under anaerobic (non toxic) conditions. The amount of oxygen required to degrade the organic matter is measured as Biochemical Oxygen Demand and/or Chemical Oxygen Demand (BOD/COD). A quantitative relationship exists between the amount of Oxygen required to convert a definite amount of Organic matter to Carbon dioxide, water ammonia.

$$C_n H_a O_b N_c + O_2 \rightarrow CO_2 + H_2O + CNH_3$$

For a fully biodegradable waste which is also chemically oxidisable , BOD: COD ratio would be 0.6 to 1.0

(ii) Pathogens–Domestic waste/sewage carry a plethora of micro organisms of intestinal origin which may be bacteria/ virus/protozoa. They produce disease by growing and multiplying in the body of host. Some of these pathogens are Salmonella typhi (Typhoid), Salmonella paratyphi (Paratyphoid), Salmonella enteritis (gastro enteritis), Shigella (bacillary dysentery), Cholera vibro (Cholera), & giardiasis, worms (helminthes) cause significant disorders. The intestinal discharges of persons suffering, carriers, contain billions of these pathogens.

A group of bacteria which are not pathogenic but are always associated with pathogens and unsanitary & unhygienic conditions are non pathogens (Coliform, faecal Coliform, Streptococcus). In routine water and waste water examination the number of non pathogens (also known as indicator bacteria) is estimated.

(iii) Nutrients–Nutrients are essential plant requirements such as nitrogen , phosphates, carbon, sulfates, potassium, iron manganese, boron and cobalt. Some of these chemicals present in water accelerate the growth of plants (hydrophytes) and algae. Sometimes they can lead to blooms. The decay of blooms when nutrients are exhausted depletes dissolved oxygen and add color & odor to water body (termed as Eutrophication).

In water pollution control strategy, Nitrogen & Phosphorus are important and need to be controlled to combat eutrophication and / or after growth. Nitrogen as NO_3 in excess pf 45 mg/l are toxic to infants (methemoglobinemia-blue babies). Such concentrations are more frequent in ground waters.

(iv) Salts and minerals–Water gathers a variety of dissolved salts during flow on geologic strata. These include cations viz. calcium (Ca^{++}), Sodium (Na^{+}), Magnesium (Mg^{++}), Potassium (k^{+}) & anions ie, chlorides (CI^{-}),carbonates (CO^{-}_3), bicarbonates (HCO_3^{-}) & sulfates ($SO^{=}_4$). The normal method of their measurement is as total dissolved solids (TDS) or conductance. The fresh water normally have TDS < 1500 mg/ l ; brackish waters upto 5000 mg/ I and waste water normally in the range of 30,000-35000 mg/1. TDS is an important criterion to determine the usefulness of water. In public water supplies the recommended value is 500 mg/1.

Acid mine drainage is another mineral pollution problem. The major ions as pollutants in it are Fe^{++}, $SO_4 =$, H^+.

(v) Heated water discharges–Thermal power generation is the main source of pollution due to heat, as it uses enormous amounts of water for cooling. The heat from fossil fuels and nuclear fuels is not fully converted into power and reaches cooling waters. For each kilowatt-hour energy produced in an efficient coal fired plant , 2/3rd heat is dissipated in cooling water.

Consequent to the discharge of heated water is water bodies, the ecology is altered.

(iv) Trace metals–There are 80 ≈ elements in nature which are classed as metals. (heavy metals have sp. Gravity > 5). More often heavy metals are those which are toxic. The list includes:

Chromium (Cr), Cobalt (Co), Copper (Cu), Iron (Fe), Lead (Pb), Manganese (Mn), Mercury (Hg), Nickel (Ni) Selenium (Se) and Zinc (Zn).

The source of these metals in waters is through industrial discharges. They may undergo change to produce more toxic substances (Hg Salts). Excess of metal in body through ingestion and accumulation damages, liver & kidneys.

(vii) Pesticides and other volatile organic compounds–There are large majority of synthetic compounds which are manufactured to kill pests, (animals / plants harmful to man & plants), (insecticides, herbicides, rodenticides & fungicides).

The commonly used pesticides are washed out in waste waters and may be found in trace quantities in water bodies (organo chlorines, organophosphates, carbamates)

Other volatile organic compounds include Vinyl Chloride (used in manufacture of resins), Chloroethylene (Solvent used in heat transfer), Chlorofluorocarbons (air conditioning), dichloromethane (metal degreaser), Carbon tetra chloride (domestic degreaser). Some of them are carcinogenic and can damage nervous system.

Table 1.9:T_1: Major Water Pollutants and Their Sources & examples are listed below in table.

S. No.	Type/Effects	Examples	Major Sources
1.	Infectious agents (pathogens): Cause diseases	Bacteria, viruses, Protozoa, parasites	Human and animal wastes
2.	Oxygen-demanding wastes: Deplete dissolved oxygen needed by aquatic species	Biodegradable animal wastes and plant debris	Sewage, animal feedlots, food-processing facilities, paper mills
3.	Plant nutrients: Cause excessive growth of algae and other species	Nitrates (NO_3^-)	Sewage, animal wastes, inorganic fertilizers
4.	Organic chemicals: Add toxins to aquatic systems	Oil, gasoline, plastics, pesticides, fertilizers, cleaning solvents	Industry, farms, households, mining sites, runoff from streets and parking lots

S. No.	Type/Effects	Examples	Major Sources
5.	Inorganic chemicals: Add toxins to aquatic systems	Acids, bases, salts, metal compounds	Industry, households, mining sites, runoff, from streets and parking lots
6.	Sediments: Disrupt photosynthesis, food webs, other processes	Soil, silt	Land erosion from farms and construction & mining sites
7.	Heavy metals: Cause cancer, disrupt immune and endocrine systems	Lead, mercury, arsenic	Unlined landfills, household chemicals. mining refuse, industrial discharges
8.	Thermal pollution of water: Make some species vulnerable to disease	Heat	Electric power and industrial plants

APPENDIX 1.10

Major physical, chemical, and biological treatment unit operations & processes used for water/wastewater treatment

S. No.	Unit Operations/Unit Processes	Principal Applications & objectives
1.	Screening	Racks or bar screens are the first step in wastewater treatment. They are used to remove large objects.
2.	Flow equalization	Used to dampen the flow rate and mass loading.
3.	Skimming	Used for oil separation where free oil is floated to the surface of a tank and then skimmed off. The oil separator specified by American Petroleum Institute is commonly used.
4.	Dissolved Air Flotation (DAF)	Process used to remove suspended or refractory solids and oil & grease from the waste stream. Uses dissolved air to produce line bubble that float the solids.
5.	Foam separation	Process used to remove minute concentrations or refractory organics and heavy metal ions. Contaminants are carried up by rising bubbles to the poor surface where they are deposited as the bubbles exit.
6.	Sedimentation	Used to remove settleable solids.
7.	Neutralization	Process used to neutralize acidic or alkaline water streams prior to chemical and/or biological treatment.
8.	Precipitation	Dissolved soiled in solution are chemically transformed into insoluble form. Process is extensively used to precipitate, phosphorous and heavy metals. Heavy metals are generally precipitated as dioxide through the addition of lime or caustic to a pH of minimum solubility
9.	Chemical Oxidation	Used to oxidize pollutants to terminal end products, Common oxidants are Cl_2, O_3, H_2 and $KMNO_4$.Oxidation of iron and manganese is used for precipitation. Cyanides, sulfides and many organics are destroyed.
10.	Chemical Reduction	Used to precipitate certain ions from solutions, e.g., hexavalent chromium. Also many oxidizing agents are destroyed by reduction, Common reducing agents are ferrous sulfate, sodium metabisulfite, and sulfur dioxide.
11.	Coagulation	Used to agglomerate suspended materials so that efficient settling can take place. Commonly used chemicals are alum, iron salts, and polymers. The process consists of chemical feed. Flash mix and flocculation basins.

S. No.	Unit Operations/Unit Processes	Principal Applications & objectives
12.	Suspended growth biological reactor	Biological process using suspended biomass to remove dissolved organics. Principal variations in an activated sludge process.
13.	Attached growth biological reactor	Biological process using attached biomass to remove dissolved organics, Principal variations are a trickling filter and a rotating biological contactor (RBC).
14.	Anaerobic Processes	Attached or suspended microbial process operated in the absence of oxygen. Normally used to treat high strength organic wastes. Methane is produced as an energy source. Sludge (biomass) production is small.
15.	Oxidation Pond	Large basins (normally earthen) used for waste storage and treatment. Treatment is achieved by natural processes of settling and biological decomposition.
16.	Aerated Lagoon	Basins (normally earthen with plastic liners) with aeration equipment where waste are aerated over long periods of time. The waste characteristics are altered by biological oxidation.
17.	Land Treatment	Organic waste is applied over land for treatment. Common application methods are slow rate infiltration, rapid infiltration and overland flow.
18.	Nitrification	Process is used to convert ammonia nitrogen to nitrate nitrogen. It can be achieved in suspended or attached growth biological reactors.
19.	Denitrification	Nitrate nitrogen and nitrite nitrogen are reduced to nitrogen gas by microorganisms. Denitrification is achieved under anaerobic condition in suspended or attached growth reactors.
20.	Disinfection	Process used to reduce the number of water borne pathogens is water, Normally applied to effluents of wastewater treatment plants containing microbes, Chlorination and zonation are the most common methods.
21.	Ammonia stripping	Ammonia gas is air stripped from the wastewater, normally by using stripping tower, High pH values are required.
22.	Breakpoint Chlorination	Ammonia nitrogen is oxidized to nitrogen gas by breakpoint chlorination in a mixing basin.
23.	Filtration	Used to polish the effluent by removing total suspended solids and turbidity. Biological activity in the filter bed may reduce some of the biochemical oxygen demand (BOD).
24.	Carbon adsorption	Used to remove soluble refractory organics from wastewater effluent.

S. No.	Unit Operations/Unit Processes	Principal Applications & objectives
25.	Ion Exchange	A demineralization process in which the cations and anions in wastewater are selectively exchanged for the ions in an insoluble resin bed. When the resin capacity is used up ,it is regenerated by using high concentrations of the original ions that are exchanged for the polluted ions attached to the resin. Flow from regeneration is composed of highly concentrated brine.
26.	Reverse osmosis or ultrafiltration	A demineralization process applicable to production of high-quality water in one stream and a high concentration of mineral ions in another.
27.	Electro dialysis	A demineralization process where electrical potential is used to transfer the ions through ion-selective membranes. Produces two streams; one high in mineral content and the other free of minerals.

APPENDIX 1.11

Sources of urban storm water contaminants

S.No	Storm Water Contaminants	Typical sources of storm water contamination
1.	Sediment/floatables	Streets, lawns, driveways, roads, construction activities, atmospheric deposition, drainage channel erosion
2.	Pesticides/herbicides	Residential lawns and gardens roadsides, utility right-of-ways, commercial and industrial landscaped areas, soil wash-off
3.	Organic materials	Residential lawns and gardens. Commercial landscaping, animal wastes
4.	Metals	Automobiles, bridges, atmospheric deposition, industrial areas, soil erosion, corroding metal surfaces, combustion processes.
5.	Oil/grease/hydrocarbons	Roads, driveways, parking lots, vehicle, maintenance areas, gas stations, illicit dumping to storm drains.
6.	Microbial pathogens	Lawns, roads, leaky sanitary sewer lines, sanitary sewer cross-connections, animal waste, septic systems
7.	Nutrients (N and P)	Lawn fertilizer, atmospheric deposition, automobile exhaust, soil erosion, animal waste, detergents

Water Pollution Sources

- Water pollution is the contamination of water bodies e.g. lakes, rivers, oceans and groundwater.
- Water pollution occurs when pollutants are discharged directly of indirectly into water bodies without adequate treatment to remove harmful compounds.
- Water pollution affects plants and organisms living in these bodies of water.
- The effect is damaging to individual species & populations and natural biological communities.

Water pollution-point sources

- Contaminants that enter a waterway from a single, identifiable source, such as a pipe of ditch.
- Traced to a specific source
- Leaking chemical tank,. effluents coming from a waste treatment of industrial plant, or a manure spill from a hog confinement lagoon

Examples of point sources

- Discharges from a sewage treatment plant; a factory; a city storm drain; municipal storm sewer systems; industrial storm water, such as from construction sites

Water Pollution-non point Sources

- Non-point source pollution (NPS)-contamination that does not originate from a single discrete source.
- NPS Pollution is the cumulative effect of small amounts of contaminants gathered from a large area.
- Pollutants will come from wide spread area
- They can't be tracked to a single point of source
- Examples: Soil erosion, chemical runoff, animal waste
- Leaching out of fertilizers/nutrients from agricultural lands.
- Nutrient runoff in storm water-agricultural field/forest.
- Contaminated storm water washed off of parking lots, roads and highways, called urban runoff.

Water Pollution-Specific Sources

- Septic systems– Use a large tank buried in the ground to contain and break down household sewage; Fats, oils, & grease as well as large waste particles, are stored and later pumped out of the holding tank; source of concern for groundwater pollution & surface water pollution
- Lagoons: Shallow holding pits into which wastes are pumped and treated; Water Quality Problems: Poorly constructed lagoons (leakage); lagoons built on high water table; Nitrates: most often found contaminant
- Waste Disposal: Underground or above ground disposal practices of domestic, municipal, or industrial liquid waste.

  Industrial liquid waste-treated/untreated

  Solid waste-Land disposal of municipal and industrial solid waste-leaching

  Storage and transport of Commercial Materials-storage tanks & spills

  Mining operations: Mines, Oil & gas-Acid mine drainage, Leaching of toxic metals, Wastewater generated

  Agricultural operations: Fertilizers; Pesticides

  Saline water intrusion

APPENDIX 1.12

All about Eutrophication of Water Bodies

Eutrophication is the process by which of water acquires a high concentration of nutrients, especially phosphates and nitrates which promote excessive growth of algae which is called "As they die & decompose, high levels of organic matter and the decomposing organisms deplete the water of available oxygen, causing the death of aquatic organisms, such as fish.

Eutrophication can occur naturally. It is considered as a natural ageing process of the water body but human interference has paced up the degradation process. The natural process is called **"Natural Eutrophication"** and Eutrophication by human interference where large amounts of pollutants are released into water because of societies are termed as **"Cultural Eutrophication"**

Eutrophication:

Is Too Much of a Good Thing.

Eutrophication is the enrichment of a shallow lake, estuary, or slow-moving water stream and is caused by runoff into lake that contains nitrates and phosphates. Causes & effects of eutrophication is shown given in figure.

Eutrophication, naturally is both prominent in fresh water bodies like lakes & rives and large salt water bodies like oceans.

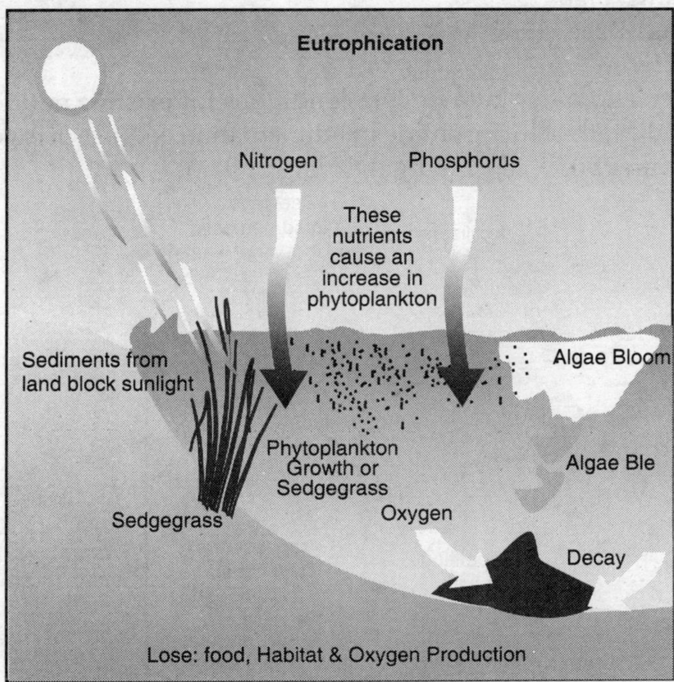

Fig.1.12F_1: Causes & effects of eutrophication

But because of Human interference and the extent of pollution now , it would take less than a century for even a large water body to get succumbed to Eutrophication.

Cultural Eutrophication is caused by release of pollutants into water by human activities Human activities causing cultural eutrophication include:

- Sewage disposal in water.
- Industrial waste disposal in water.
- Fertilizers and pesticides dissolved in water being channelized into water bodies.
- Use of organic fertilizers.
- Dissolution of pollutant on surface in rain water and flowing into water bodies.

Sources of Cultural eutrophication are:

- Nitrates and phosphates from human sources.
- Farms, feedlots, streets, parking lots
- Fertilized lawns, mining sites, sewage plants.

During hot weather or droughts , it results in:

- Algal blooms
- Increased bacteria
- More nutrients
- Anaerobic bacteria

Process of eutrophication is shown below in figure.

The process of Eutrophication can be roughly divided into the following steps based on the progress of process:

- Addition of nutrients
- Growth of plants/algae
- Death of plants/algae
- Growth of bacteria
- Lack of Oxygen and other unfavorable conditions for existing entities.
- Suffocation and death of living entities, Sedimentation because of excess waste (Detritus)
- Death of the water body and eventual drying of land.

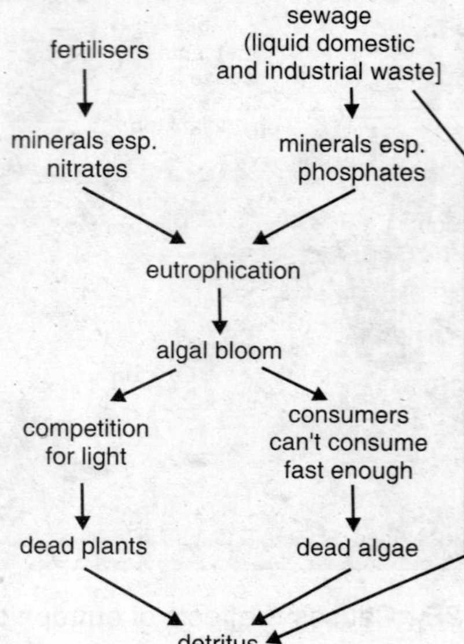

Fig.1.12F$_2$: Sequence of events in the process of eutrophication

Various Sources & causes of Eutrophication is shown below in figure.

Effects of eutrophication

Because of algal bloom sunlight cannot enter into the depths of the water body which makes plants unable to live. Decrease in O2 levels occurs resulting in the eventual death of living organisms inside the water body (rivers, lakes, oceans etc.)

During this process of death of living organisms inside, more oxygen is consumed during the decomposition process. Eutrophication not only effects the living organisms inside the water but also degrades the water quality: Water taste & odor degrade because of tike Geosmin, MIB etc. being released during algal bloom

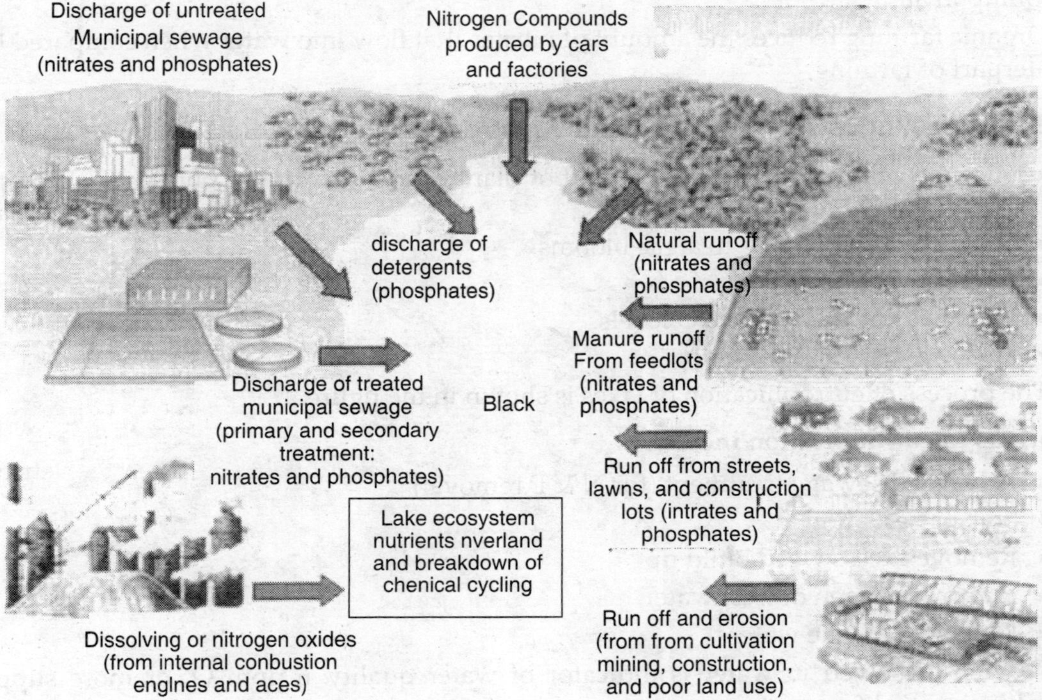

Fig.1.12F_3: Various Sources & Causes of Eutrophication

As water bodies go on dying, the chain breaks. Only those species which take high nutrients and can grab them fast survive in the habitat. Tubificid worms larvae and bacteria replace the vertebrates in water. This broken food chain can lead to extinction of many species and many adverse conditions.

More effects of eutrophication include

- Increased biomass like phytoplankton in water.
- Decrease in harvestable fish and shell fish.
- Increase in turbidity of water.
- Loss of aesthetic value of water body.

Biological Oxygen Demand (BOD) serves as a measure for finding out about Eutrophication in water. A high BOD indicates occurrence of eutrophication in the water body

Prevention and Reversal of Eutrophication

If organic material addition stops, the bacteria use up all the nutrients available and die off after sometime. This can let the left off plants and vertebrates grow their numbers again. Concentration of dissolved O2 increases through natural means like atmospheric replenishment and photosynthesis.

Cleaning up of water bodies will help in removing excess weeds sedimentation and allowing plants & fish grow again. For cleaning of lakes, use herbicides algaecides down side and pump in air.

Organic farming reduces the amount of nitrites that flow into water when compared to its counterpart of farming.

Controlling pollution and flow of polluting substances into water bodies

Eutrophication of lakes mostly from runoff of plant nutrients (nitrates phosphates) comprise of following sequential processes:

- Hot dry weather leads to algae blooms.
- Decrease of photosynthesis.
- Dying algae then drops DO levels.
- Fishes die, bad taste & odor.

The process of eutrophication of lakes is shown in the figure

Solutions to eutrophication include:

- Advanced sewage treatment (for N & P removal)
- Soil conservation
- Remove excess weed build up
- Pump in oxygen of freshwater.
- Diversion of lake water

Oxygen dissolved in water is indicator of water quality. 6 ppm O_2 or more supports desirable aquatic life.

BOD (Biochemical oxygen demand) measures the amount of dissolved oxygen consumed by aquatic microorganisms. Sewage, paper pulp, or food wastes can cause an oxygen sag, where few fish survive.

Oxygen sag curve in polluted surface water stream is shown below in figure.

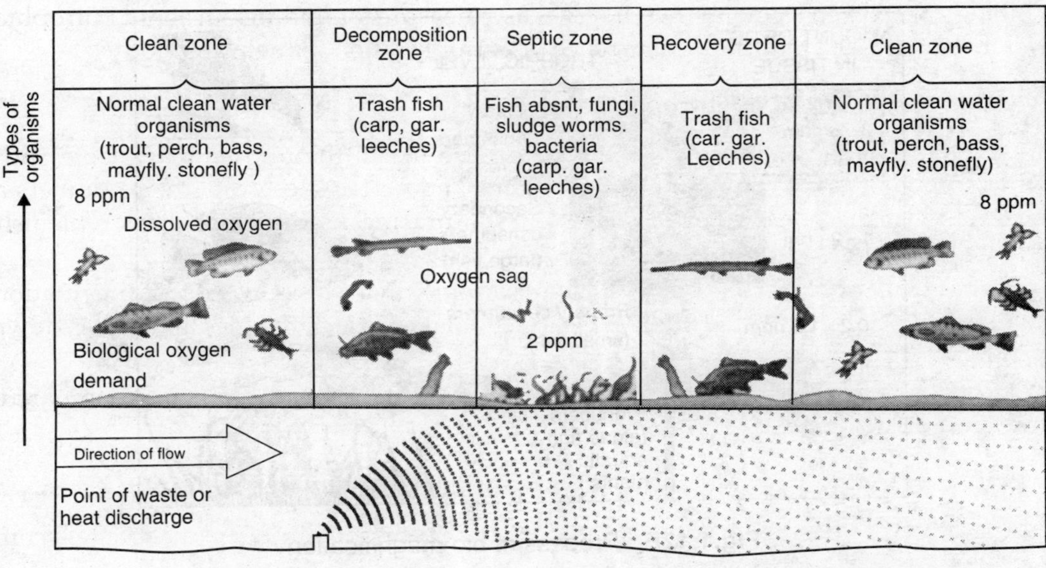

Fig.1.12F$_4$: Oxygen sag curve in polluted surface water

BOD and Eutrophication: Rapid succession in a body of water because of an increase in biological productivity. (Oligotrophic lakes & rivers have clear water and low biological productivity).

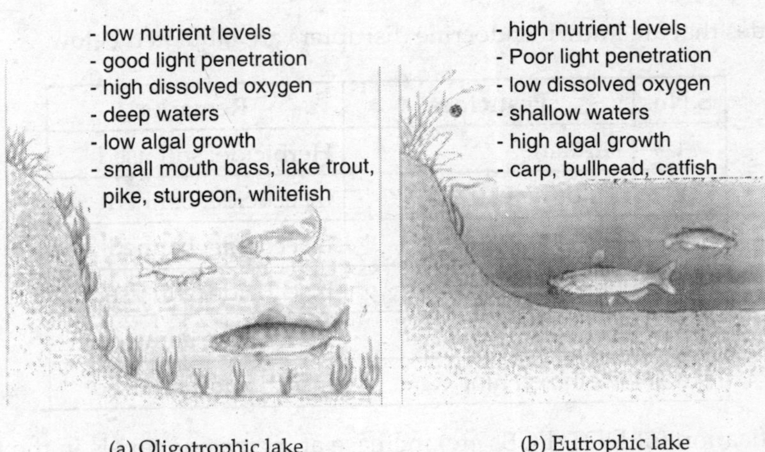

(a) Oligotrophic lake (b) Eutrophic lake

Problems with DDT: DDT is not metabolized very rapidly by animals; instead it is deposited and stored in the fatty tissues–Bio magnification.

DDT is stable and slowly degrades. It causes:

- Toxicity to Fish
- Increased mortality in birds: calcium decreased in egg shells
- Estrogen mimic in Vertebrates: feminizes males-lower sperm count; alters behavior

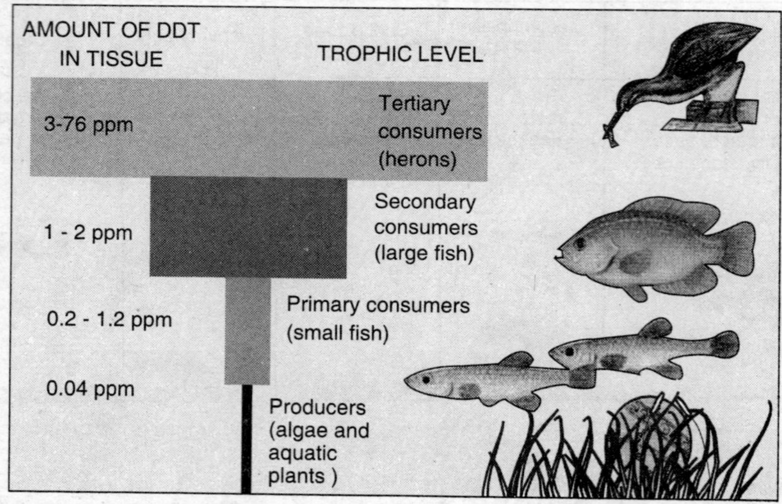

Fig.1.12F$_5$: Process of bio magnification

- Human Health problems:
  - Decreased mental function
  - Male infertility
  - Cancer

Some pesticides that are known endocrine disrupters are tabulated below.

S.No.	Pesticides	Remarks
1	Atrazine	Herbicide; still used
2	Chlordane	Insecticide; banned
3	DDT	Insecticide; banned
4	Endosulfan	Insecticide; still used
5	Kepone	Insecticide; banned
6	Methoxychlor	Insecticide; still used

In biomagnification (of DDT. PCBs etc).increase at increasing levels in the food chain as shown below diagrammatically.

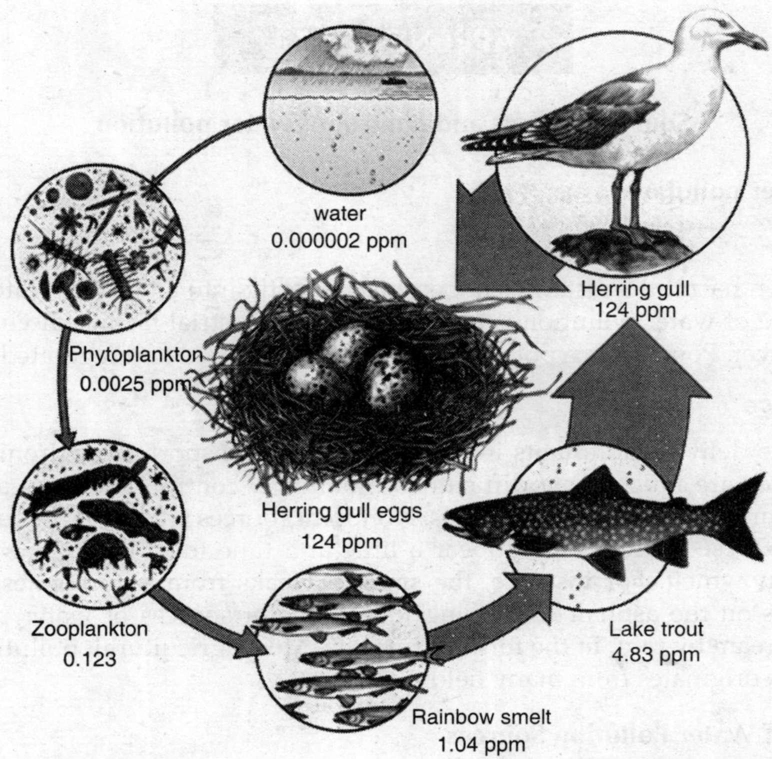

Fig.1.12F_6: Biomagnification of DDT , PCBs etc.

APPENDIX 1.13

Sources, effects and control of water pollution

Sources of Water pollutants

Point Source

This occurs when harmful substances are emitted directly into a body of water, An example of a point source of water pollution is a pipe from an industrial facility discharging effluent directly into a river. Point-source pollution is usually monitored and regulated.

Nonpoint Source

Nonpoint source delivers pollutants indirectly through transport or environmental change. Nonpoint sources are much more difficult to monitor & control, and they account for the majority of contaminants in streams & lakes. Nonpoint sources are more difficult to recognize, Pollutants from these sources may appear a little at a time form large areas, carried along by rainfall or snowmelt. For instance, the small oil leaks from automobiles that produce discolored spots on the asphalt of parking lots nonpoint sources of water pollution when carried into a stream by rain, in the form of run-offs. Most agricultural pollution is nonpoint since it typically originates from many fields.

Classification of Water Pollution Sources

The major sources of water pollution can be classified as municipal. Industrial, and agricultural.

Municipal Water Pollution

Municipal water pollution consists of waste water from homes and commercial establishments. For many years, the main purpose of treating municipal wastewater was simply to reduce its content of suspended solids, oxygen-demanding materials, dissolved inorganic compounds, and harmful bacteria. In recent years, however, more stress has been placed on improving means of disposal of the solid residues from the municipal treatment processes. The basic methods of treating municipal wastewater fall into three stages: primary treatment , including grit removal, screening, grinding & sedimentation; secondary treatment, which entails oxidation of dissolved organic matter by means of using biologically active sludge, which is then filtered off; and tertiary treatment in which advanced biological methods of nitrogen removal and chemical & physical methods such as granular filtration & activated carbon absorption are employed.

Industrial Water Pollution

The characteristics of industrial waste can differ considerably both within and among industries. The impact of industrial discharges depends not only on their collective characteristics, such as biochemical oxygen demand , the amount of suspended solids, but also on their content of specific inorganic & organic substances. Three options are available in controlling industrial wastewater. Control options are available in controlling industrial wastewater. Control can be at the point of generation in the plant; wastewater can be pretreated for discharge to municipal treatment sources; or wastewater can be treated completely at the plant , either reused or discharged directly into receiving waters.

Agriculture Water Pollution

Agriculture, including commercial livestock & poultry farming, is the source of many organic & inorganic pollutants in surface waters and groundwater. These contaminants include both sediment from erosion cropland and compounds of phosphorus & fertilizers. Animal wastes are high in oxygen demanding material, nitrogen & phosphorus, and they often harbor pathogenic organisms. Wastes from commercial feeders are contained and disposed of on land: their main threat to natural waters, therefore, is from runoffs & leaching. Control may involve settling basins for liquids, limited biological treatment in aerobic or lagoons, and a variety of other methods.

Category of Water Pollutants

The major water pollutants are chemical, biological, or physical materials that degrades water quality. Pollutants can be classed into following eight categories which are described below:

Petroleum Products

Oil & chemicals derived from oil are used for fuel, lubrication, plastics manufacturing, and many other purposes. These petroleum products get into water mainly by means of accidental spills from ships, tanker trucks, pipelines, and leaky underground storage tanks. Many petroleum products are poisonous if ingested by animals, and spilled oil damages the feathers of birds or the fur of animals, often causing death. Land based petroleum pollution is carried into waterways by rainwater runoffs. This includes drips of oil, fuel, and fluid from cars & trucks; dribbles of gasoline spilled onto the ground at the filling station; and drips from industrial machinery.

Pesticides and Herbicides

Pesticides that get applied to farm fields and roadsides & homeowners' lawns goes off into local streams & rivers or drain down into groundwater, contaminating the fresh water. Farming is a big contributor to this problem.

Chemicals used to kill unwanted animals & plants on farms or in suburban yards, may be collected by rainwater runoffs and carried into streams. Some of these chemicals are biodegradable and quickly decay into harmless or less harmful forms, while others are nonbiodegradable and remain dangerous for a long time.

When animals consume plants that have been treated with certain nonbiodegradable chemicals, such as chlordane and dichlorodiphenyltrichloroethane (DDT), these chemicals are absorbed into the tissues of organs of the animals. When other animals feed on these contaminated animals, the chemicals are passed up the food chain. With each step up the food chain, the concentration of the pollutant increases. This process is called bio-magnification. Animals at the top of food chain may, as a result of these chemical concentrations, suffer cancers, reproductive problems, and death.

Many drinking water supplies are contaminated with pesticides from widespread agricultural use.

Heavy Metals

Heavy metals, such as copper, lead , mercury, and selenium, get into water from many sources, including industries, automobile exhaust. Mines and even natural soil can pollute water resources. Like pesticides, heavy metals become more concentrated as animals feed on plants and are consumed in turn by other animals. When they reach high levels in the body, heavy metals can be immediately poisonous, or can result in long-term health problems similar to those caused from sewage sludge that can be absorbed by crops. If these crops are eaten by humans in sufficient amounts, the metal can cause diarrhea and, over time liver & kidney damage. Lead can get into water from lead pipes and solder in older water systems; children exposed to lead in water can suffer mental retardation.

Hazardous Wastes

Hazardous wastes are chemical wastes that are either toxic (poisonous)] reactive (capable of producing explosive or toxic gases), corrosive (capable of corroding steel), or ignitable (flammable). If improperly treated or stored, hazardous wastes can pollute water supplies, PCBs, a class of chemicals once widely used in electrical equipment such as transformers, can get into the environment through oil spills and can reach toxic as organisms eat one another.

Excess Organic Matter

Fertilizers and other nutrients used to promote plant growth on farms and in gardens may find their way into water. At first, these nutrients encourage the growth of plants and algae in water. However, when the plant matter die and settle underwater, microorganisms decompose them. In the process of decomposition, these microorganisms consume oxygen that is dissolved into the water. Oxygen levels in the water may drop to such dangerously low levels that oxygen-dependent animals in the water, such as fish, die, This process of depleting oxygen to deadly levels is called eutrophication.

Sediment

Sediment, soil particles carried to a stream bed, lake, or ocean, can also be a pollutant if it is present in large enough amounts. Soil erosion produced by the removal of soil-trapping tress near waterways, or carried by rainwater & floodwater from croplands, strip mines, and roads, can damage a stream of lake by introducing too much nutrient matter. This leads to eutrophication. Sedimentation can also cover stream bed gravel in which fishers, such as salmon and trout, lay their eggs.

Infectious Organisms

Many disease-causing organisms that are present in small numbers in most natural waters are considered pollutants when found in drinking water. Such parasites as Giardia lamblia and *Cryptosporidium parvum occasionally* turn up in urban water supplies. These parasites can cause illness, especially in people who are very young, and in people who are already suffering from other diseases.

Thermal Pollution

Water is often drawn from rivers, lakes, or the ocean for use as a coolant in factories and power plants, The water is usually returned to the source warmer than when it was taken. Even small temperature changes in a body of water can drive away the fishes and other species in them. Thermal pollution can accelerate biological processes in plants and animals or deplete oxygen levels in water. The result may be fishes and other wildlife deaths near the discharge source. Thermal pollution can also be caused by the removal of vegetation that shade and cool streams.

Effects of Water Pollution

Water pollution effects can be categorized as described below:

Waterborne Infectious Diseases

Human Infectious are among the most serious effects of water pollution Waterborne diseases occur when parasites or other disease-causing microorganisms are transmitted via contaminated water, particularly water contaminated by pathogens originating from excreta. These include typhoid, intestinal parasites, and most of the enteric and diseases caused by bacteria, parasites, & viruses. Amongst the most serious parasitic diseases are amoebiasis, giardiasis, ascariasis, and hookworm. Every year there are thousands of beach closing worldwide, and outdated monitoring methods may in some cases leave beach goers vulnerable to a range of illnesses. Polluted beach water can cause rashes, ear aches, pink eye, respiratory infections, hepatitis, encephalitis, diarrhea, vomiting, and stomach aches.

Nutrient Pollution

Nutrient pollution is the most widespread, chronic environmental problem in the coastal ocean. The discharges of nitrogen, phosphorus, & other nutrients comes from agriculture , waste disposal, coastal development and fossil fuel use. Once nutrient pollution reaches the coastal zone, it stimulates harmful overgrowths of algae, which can have direct toxic effects and ultimately result in low-oxygen conditions. Monitoring for toxic algal blooms and closing fisheries has reduced the incidence of related human illness but has had the obvious economic cost of lost related to human illness , lost income for fishermen and related businesses. Nutrient-pollution blooms of non-toxic algae & sea weed can also cause problems by reducing water clarity, making it hard for marine animals to find food and blocking the sunlight needed by sea grasses, which serve as nurseries for many important fish species.

When the algal overgrowths finally die, they sink to the bottom and begin decomposing. This process uses oxygen from the surrounding water. The decomposition process takes enough oxygen out of the water such that the level falls too low to support normal aquatic life and the region becomes a coastal dead zone. Finally, nutrient pollution can trigger unusual outbreaks of fish diseases.

Chemical Contamination

Over the years, many of chemicals have entered into our waterways and they continue to do so today. Chemical water pollution typically occurs because:

- The chemicals were dumped into the water intentionally;
- The chemicals seeped into groundwater, streams, or rivers because of failing pipes or storage tanks;
- The chemicals catastrophically contaminated waterways because of industrial accidents;

- The pollution settled out of polluted air (or was precipitated out of polluted air)
- Chemicals were leached out of contaminated soil.

  The above types of chemical contamination are considered 'point sources' of water pollution. Non-point-source chemical pollution also occurs via pesticide runoffs from farm fields & homeowners' lawns as well as runoffs of automotive fluids and other chemicals from roads, parking lots, driveways,& other surfaces.

  The effect of chemical that has polluted water, can be as follows:

- Severe chemical spills and leaks into surface waters usually have an immediate effect on aquatic life (kills fishes, etc).
- Chronic lower-level chemical pollution has more subtle effects, with problems manifesting over a long period of time and sometimes being difficult to tie directly to the water pollution.
- The human effects of chemical pollution in water can generally be viewed the same way as any other form of chemical contamination–water is just the delivery mechanism.

There are a few broad categories of water pollution effects related to chemicals that are described below.

Effects of Pesticides pollution

Pesticides are carried in rainwater runoffs from fields, suburban lawns, or roadside embankments into the nearest creeks and streams. Occasionally they are even intentionally sprayed into waterways as part of a pest-control effort.

Below given are some noteworthy examples of the effects of pesticide water pollution:

- Atrazine (Commonly used herbicide) causes feminization of male frogs even at concentrations in water as low as 0.1 parts per billion(ppb). Atrazine water pollution has been linked to a number of human cancers, including prostate cancer and non-Hodgkin's lymphoma, as well as hormonal problems that could disrupt reproductive & developmental processes.
- Glyphosate another of the common herbicides, causes a decline in frog biodiversity & in the total mass of tadpoles when the glyphosate gets into water.
- Pesticides have been found in well water in countries such as India, The Netherlands, Italy, Israel, Japan, Canada and Australia. Pesticide contamination of drinking water is a peculiar problem in rural agricultural areas where pesticide use is heavy and drinking water supplies come directly from groundwater or surface water.
- Pesticides can migrate via water into the food chain as well, ultimately consumed by humans or animals in food.
- In the most infamous case of pesticide pollution, widespread use of the insecticide DDT polluted waterways, contaminating fish, and ultimately poisoning bale eagles (and other animals) that ate the fish. DDE, the principal breakdown product of DDT built up in the fatty tissues of female eagles and prevented sufficient calcium being released to produce strong egg shells. The thin shells would break when the parents sat on the eggs to keep them warm. DDT affected many other species as well.
- In terms of general human health effects, pesticides can:
  - affect and damage the nervous system
  - cause liver damage

- damage DNA and cause a variety of cancers
- cause reproductive and endocrine damage
- Cause other acutely toxic or chronic effects.

Effects of Oil and Petroleum Chemicals Pollution

When oil pollution gets in water, some of the components are degraded and dispersed by evaporation, photochemical reactions. or bacterial degradation, while others are more resistant any may persist for many years, especially in shallow waters with muddy sediments.

- Exposure to oil or its constituent chemicals alter the ecology of aquatic habitats and physiology of marine organisms.
- Oil (or chemical components of oil) can seep into marsh and sub-tidal sediments and lurk there for decades, negatively affecting marsh grasses, marine worms, and other aquatic life forms that live in. on, or near the sediment.

The components of crude oil, called polycyclic aromatic hydrocarbons (PAHs), persist in the marine environment for years and are toxic to marine life at concentration in the low parts per billion (ppb) range. Chronic exposure to PAHs can affect development of marine organisms, increased susceptibility to disease, and jeopardize normal reproductive cycles in many marine species.

Mercury

Mercury finds its way into water primarily through air pollution from coal-fired power plants and some other industrial processes. In the water, the elemental mercury is converted to methylmercury by certain bacteria, after which it moves up the food chain of fish gobbling each other up.

Exposure to mercury in the womb can cause neurological problems, including slower reflexes, learning deficits, delayed or incomplete mental development, autism, and damage, Mercury in adults is also a problem, causing:

- Central nervous system effects like Parkinson's disease, multiple sclerosis, and Alzheimer's disease.
- Heart disease..
- Causing death or irreversibly damaging areas of the brain.

Animals in any part of the food chain affected by the bioaccumulation of mercury can also suffer the effects of mercury pollution. Possible effects include death, reduced reproduction, slower growth & development, and abnormal behavior.

Effects of other Chemicals

Tens of thousands of chemicals are used in industrial processes and are found in car-maintenance products.

The effects of hormone disrupting chemicals include interrupted sexual development; thyroid system disorders; inability to breed; reduced immune response; and abnormal mating & parenting behavior. In humans,. Endocrine disruptors lead to degraded immune function, mental impairment, decreased fertility, and increases in some types of cancers.

Effects of Mining Operations

There are a number of negative water-pollution effects form mining operations as described below:

- **Acid mine drainage:** When rain or surface water flows over exposed rock and soil, it can combine naturally causing sulfur to form sulfuric acid. The acidified rain water eventually finds its way to streams & groundwater, polluting them and impacting local aquatic life. Some streams can become so acidic (more acidic than car-battery acid) that the aquatic ecosystem is completely destroyed. The same leaching process that causes acid mine drainage can impart heavy metal pollutants from the soil and rock as well.
- **Spills and leaks:** Whether it's a leak in the containment system of a cyanide leach heap or a breach in a coal-slurry impoundment dam, the result is the same-pollution of streams, rivers & groundwater, killing aquatic life and poisoning drinking water.
- **Mountain top Removal Mining:** In this technique, the tops of coal-rich mountains are removed and the overburden is dumped into nearby valleys, burying habitats altogether, with the obvious catastrophic effect on whatever life form lived in or around the stream.

Effects of Marine Debris

Marine debris is basically trash in the ocean. Trash fouls inland waterways too, for sure, but trash seems to be a particular problem in our seas, Marine debris is one of the world's most pervasive marine pollution problems. The debris includes escaped inland trash & garbage thrown overboard by ships & boaters–plastic bottles and bags, six-pack rings, cigarette butts, Styrofoam, etc. Marine animals can swallow the trash items, which often look similar to prey they would normally eat. Or the trash item may have barnacles or other delectable attached and is inadvertently ingested with the food. For instance, sea turtles will eat a plastic bag believing it to be a jellyfish. The bag can cause an intestinal blockage and sometimes death.

A new devastating effect of marine debris is emerging. After years of degradation at sea, plastic breaks up. The plastic has not biodegraded but rather has disintegrated into very small pieces. Marine animals near the bottom of the food chain are now ingesting these tiny little pieces of plastic. How far up the food chain, the stuff will go is unknown.

Discarded or lost fishing gear line, rope, nets and certain trash items can get wrapped around marine animals fins or flippers, causing drowning or amputation. Marine debris can also degrade coral reefs, sea grass beds, and other aquatic habitats.

Effects of Thermal Pollution of Water

It's easy to see how discharging the heated-up from a power plant into a river could cause problems for aquatic organisms used to having their water home stay at a fairly specific temperature. Industrial thermal pollution is a problem for our waterways fish and other organisms adapted to a particular temperature range can be killed from thermal shock, and the extra heat may disrupt spawning of kill young fish.

- Additionally, warmer water temperatures lower the dissolved oxygen content of the water. That's a double-whammy to aquatic organisms, since the warmer water also causes them to increase their respiration rates and consume oxygen faster. All this increases aquatic organisms' susceptibility to disease & the effects of toxic chemicals.

- Global warming is imparting extra heat to our oceans, which have absorbed about 20 times as much heat as the atmosphere over the past half-century. Sea levels will rise (because of thermal expansion & melting ice), increasing coastal flooding and inundation.
- There will be more intense hurricanes as they gather additional strength from warmer surface waters.
- Temperature sensitive species like corals will see tougher times. An increase in the mean sea surface temperature of only 2 ^{0}C could cause the global destruction of coral reef ecosystems.

The effects of thermal pollution of water has been shown below in figure

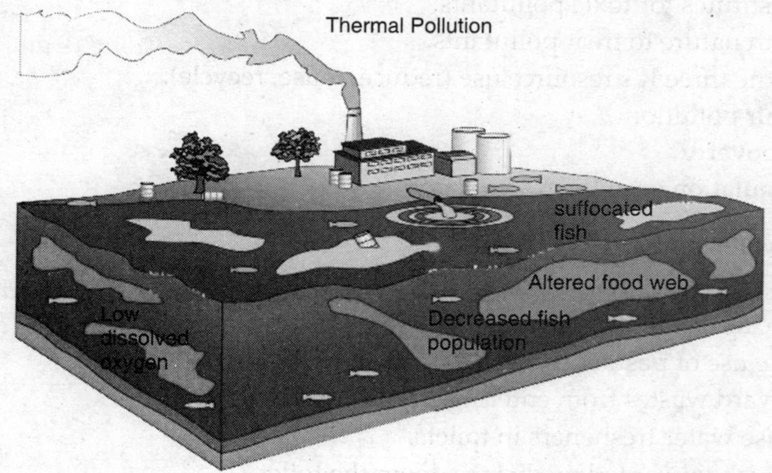

Fig.1.13F$_1$: Effects of thermal pollution of water

Controlling water Pollution

Because of the many environmental tragedies in the past, many nations instituted comprehensive regulation designed to repair the past damage of uncontrolled pollution and prevent future environmental contamination. International agreement have played a role in reducing global pollution. The Montreal Protocol on substances the Deplete the Ozone Layer (1987) set international target dates for reducing the manufacture and emission of the chemicals, such as CFCs, known to deplete the ozone layer. The Basel Convention on the Control of Transboundary Movements of Hazardous Wastes and their Disposal (1989) serves as a framework for the international regulation of hazardous waste transport & disposal.

Since 1992 representatives from more than 160 national have met regularly to discuss method to reduce greenhouse gas emission. In 1997 the Kyoto Protocol was devised, calling for industrialized countries to reduce their gas emission by 2012 to an average 5 percent below 1990 levels. The Kyoto Protocol went into force in February 2005 with more than 130 countries having ratified it. Regulations and legislation have led to considerable progress in reducing air and water pollution in developed countries. Vehicles in the 21st century emit fewer nitrogen oxides than house in the 1970s did; power plants now burn low sulfur fuels; industrial stacks have scrubbers to reduce emissions; and lead has been removed from gasoline.

Developing countries, however, continue to struggle with pollution control because they lack clean technologies and desperately need to improve economic strength, often at the cost of environmental quality. The problem is compounded by developing countries attracting foreign investments and industry by offering cheaper raw materials, and fewer environmental restrictions.

Solutions: Methods for preventing and Reducing Water Pollution

For preventing water pollution:

- Prevent groundwater contamination.
- Reduce non-point runoff.
- Reuse treated wastewater for drinking and irrigation.
- Find substitutes for toxic pollutants.
- Work with nature to treat pollutants.
- Practice the three R's resource use (reduce, reuse, recycle).
- Reduce air pollution
- Reduce poverty
- Slow population growth.

For reducing water pollution

- Fertilize garden and yard plants with manure or compost instead of commercial inorganic fertilizer.
- Minimize use of pesticides, especially near bodies of water.
- Prevent yard wastes from entering storm drains.
- Do not use water fresheners in toilets.
- Do not flush unwanted medicines down the toilet.
- Do not pour pesticides, paints, solvents, oil, antifreeze, or other products containing harmful chemicals or onto the ground.

Water Quality

Water quality is closed linked to water use and to the state of economic development. In industrialized countries, bacterial contamination of surface water caused serious health problems in major cities throughout the mid 1800s.

While waterborne diseases have been eliminated in the developed world, outbreaks of cholera and other similar diseases still occur with alarming in frequency in the developing countries. Water quality has been heavily impacted worldwide by industrial and agricultural chemicals.

Eutrophication of surface waters from human & agricultural wastes and nitrification of groundwater from agricultural practices has greatly affected large parts of the world.

Main points of sources, effects and control of water pollution is summarised below:

- Water Pollution is the contamination of streams, lakes, underground water, bays, or oceans by substances harmful to living things.
- Point source pollution occurs when harmful substances are emitted directly into a body of water.

- Nonpoint source pollution occurs when pollutants are indirectly introduced into the environment through transport or environment change.
- Major water pollutants are chemical, biological, or physical materials.
- Pollution may not damage our health immediately but can be harmful after long term exposure.
- Human infectious diseases are among the most serious effects of water pollution, especially in developing countries, where sanitation may be inadequate or non-existent.
- Discharges of nitrogen, phosphorus, & other nutrients come from agriculture, waste disposal, coastal development, and fossil fuel use.
- Pesticides are carried in rainwater runoffs from farm fields, suburban lawns, or roadside embankments into the nearest creeks and streams.
- When oil gets in water, some of the components are degraded and dispersed by evaporation, photochemical reactions, or bacterial degradation.
- Exposure to mercury in the womb can cause neurological problems, including slower reflexes, learning deficits, delayed or incomplete mental development, autism and brain damage.
- Industrial thermal pollution of water is a problem for our waterways—fish and other organisms adapted to a particular temperature range can be killed from thermal shock, and the extra heat may disrupt spawning or kill young fish.
- Environmental tragedies of the mid-20th century, has made many nations to institute comprehensive regulations, designed to repair the past damage of uncontrolled pollution and prevent future environmental contamination.
- Regulations and legislation have led to considerable progress in cleaning up some air and water pollution in developed countries.
- Developing countries, however, continue to struggle with pollution control because they lack clean technologies and desperately need to improve economic strength, often at the cost of environmental quality.
- Water quality is closely linked to water use and to the state of economic development.
- While waterborne diseases have been eliminated in the developed world, outbreaks of cholera and other similar diseases still occur with alarming frequency in the developing countries.
- Acidification of surface waters by air pollution (acid rain) is a recent phenomenon and threatens aquatic life in many areas of the world.

The problems associated with water pollution have the capabilities of disrupting life on our planet. We must become familiar with our local water resources and learn about ways for disposing harmful household wastes so they do not end up in sewage treatment plants that can't handle them or landfills not designed to receive hazardous materials. In our yards, we must determine whether additional nutrients are needed before fertilizers are applied, and look for alternatives where fertilizers might run off into surface waters. We have to preserve existing trees & plant, new trees & shrubs to help prevent soil erosion and promote infiltration of water into the soil. Around our houses, we must keep litter, pet waste, leaves, and grass clippings out of gutters & storm drains. These are just a few of the many ways in which we, have the ability to combat water pollution. If these measures are not taken and water pollution continues, life on earth will suffer severely.

Chapter

Characteristics and Properties of Water

We live on a planet that is dominated by water. More than 70% of the Earth's surface is covered with the water. Hydrosphere contains about 14 billion cubic kilometres of this substance mostly in the form of a liquid (water) which mostly occupies topographic depressions on the Earth. The second most common form of the water on our planet is ice. If the entire planet's ice is melted, then the sea level would rise by about 70 metres.

Most of the earth's water sources get their water supplies from precipitation, which may fall in various forms, such as, rain, snow, hail, dew etc. Rains form the principal and the major part of the resultant supplies. When rain starts falling, it is first of all intercepted by buildings and other objects. When the rainfall rate exceeds the interception rate, water starts reaching the ground and infiltration starts. This is the source of groundwater storage.

SOURCES OF WATER

The primary sources of water include: rainwater, surface water (stored in lakes, streams, & ponds), and groundwater. The distribution of water, however, is quite varied; many locations have plenty of it while others have very little. Water exists on earth in three forms solid (ice), liquid or gas (water vapour). Oceans, rivers, clouds, and rain, all of which contain water, are in a frequent state of change (surface water evaporates, cloud water precipitates, rainfall infiltrates the ground, etc.). However, the total amount of the earth's water does not change owing to glaciers, rivers and ground water flow.

Water is essential to life. Without it, the biosphere that exists on the surface of the earth would not be possible. The earth is called as the 'water' planet. Water's molecular arrangement is very simple, two hydrogen atoms attached to each oxygen atom. One special characteristic of water is its ability to change state very easily under earth conditions. It can be found readily on the planet in all of its three forms, solid, liquid, and gas.

The average annual rainfall in India is 1170 mm, which corresponds to annual precipitation, including snowfall of 4000 Billion Cubic Metres (BCM). Out of this volume of precipitation, only 1869 BCM appears as average annual potential flow in rivers. Due to various constraints, only 1123 BCM is assessed as the average annual utilisable water – 690 BCM from surface water and 433 BCM from groundwater.

The present(year 2015) total water use is 900 BCM of which 80% is for irrigation. This is projected to grow to 1100 BCM by 2025 and 1500 BCM by 2050, against utilizable quantum of 1100 BCM. Thus the demand will outstrip availability in another 35 years.

HYDROLOGIC CYCLE

The movement of water on the earth's surface and through the atmosphere is known as the hydrologic cycle. Water is taken up by the atmosphere from the earth's surface in vapour form through evaporation. It may then be moved from place to place by the wind until it is condensed back to its liquid phase to form clouds. Water then returns to the surface of the earth in the form of either liquid (rain) or solid (snow, sleet, etc.) precipitation. Water transport can also take place on or below the earth's surface by flow.

The hydrologic cycle is used to model the storage and movement of water between the biosphere, atmosphere, lithosphere and hydrosphere. Water is stored in the following reservoirs: atmosphere, oceans, lakes, rivers, glaciers, soils, snowfields, and groundwater. It moves from one reservoir to another by processes like: evaporation, condensation, precipitation, deposition, runoff, infiltration, sublimation, transpiration, and groundwater flow.

Water is stored in the atmosphere in all three states of matter. Water vapour in the atmosphere is referred to as humidity. If liquid and solid forms of water can overcome atmospheric updrafts they can fall to the Earth's surface as precipitation. The formation of ice crystals and water droplets occurs when the atmosphere is cooled to a temperature that causes condensation or deposition.

Precipitation can be defined as any aqueous deposit, in liquid or solid form, that develops in a saturated atmospheric environment and generally falls from clouds and includes rain, freezing rain, snow, ice pellets, snow pellets, & hail. Fog represents the saturation of air near the ground surface.

In certain locations on the Earth, acid pollutants from the atmosphere are being deposited in dry and wet forms to the Earth's surface.This process is called acid deposition. If the deposit is wet it can also be called acid precipitation. Normally, rain is slightly acidic. Acid precipitation, however, can have a pH as low as 2.5. Evaporation and transpiration are the two processes that move water from the Earth's surface to its atmosphere. Evaporation is movement of free water to the atmosphere as a gas. It requires large amounts of energy. Transpiration is the movement of water through a plant to the atmosphere. The term evapotranspiration is used to describe both processes.

In general, the following four factors control the amount of water entering the atmosphere via these two processes: energy availability; the humidity gradient away from the evaporating surface; the wind speed immediately above the surface; and water availability. The growth of crops is a function of water supply. If crops experience drought, yields are reduced. Irrigation can supply crops with supplemental water.

The distribution of precipitation falling on the ground surface can be modified by the presence of vegetation. Vegetation changes this distribution because of the fact that it intercepts some of the falling rain. How much is intercepted is a function of the branching structure and leaf density of the vegetation. Some of the water that is intercepted never makes it to the ground surface. Instead, it evaporates from the vegetation surface directly back to the atmosphere. A portion of the intercepted water can travel from the leaves to the branches and then flow down to the ground via the plant's stem. This phenomenon is called stem flow.

Another portion of the precipitation may flow along the edge of the plant canopy to cause canopy drip. Both of the processes described above can increase the concentration of the water added to the soil at the base of the stem and around the edge of the plant's canopy. Rain that falls through the vegetation, without being intercepted, is called through fall.

Infiltration is the movement of water from precipitation into the soil layer. After a rain, infiltration can create a condition where the soil is completely full of water. This condition is, however, only short lived as a portion of this water quickly drains (gravitational water) via the force exerted on the water by gravity. The portion that remains is called the field capacity. In the soil, field capacity represents a film of water coating all individual soil particles to a thickness of 0.06 mm. The soil water from 0.0002 to 0.06 mm (known as capillary water) can be removed from the soil through the processes of evaporation and transpiration. Both of these processes operate at the surface. Capillary action moves water from one area in the soil to replace losses in another area (biggest losses tend to be at the surface because of plant consumption and evaporation). This movement of water by capillary action generally creates a homogeneous concentration of water throughout the soil profile. Losses of water stop when the film of water around soil particles reaches 0.0002 mm. Water held from the surface of the soil particles to 0.0002 mm is essentially immobile and can only be completely removed with high temperatures (> 100 ^{0}C). Within the soil system, several different forces influence the storage of water.

Runoff is the surface flow of water to areas of lower elevation. On the microscale, runoff can be seen as a series of related events. At the global scale runoff flows from the land masses to the oceans. The Earth's continents experience runoff because of the imbalance between precipitation and evaporation.

Through flow is the horizontal subsurface movement of water on continents. Rates of through flow vary with soil type, slope gradient, and the concentration of water in the soil. Groundwater is the zone in the ground that is permanently saturated with water. The top of groundwater is known as the water table. Groundwater also flows because of gravity to surface basins of water (oceans) located at lower elevations.

The flow of water through a stream channel is commonly called stream flow or stream discharge. On many streams humans gauge stream flow because of the hazards that can result from too little or too much flow. Mechanical gauging devices record this information on a graph known as a hydrograph.

Oceans cover most of the Earth's surface. On average, the depth of the world's oceans is about 3.9 kilometers. However, maximum depths can be greater than 11 kilometers. The distribution of land and ocean surfaces on the earth is not homogeneous. In the Southern Hemisphere there is 4 times more ocean than land. Ratio between land and ocean is almost equal in the Northern Hemisphere.

The water found in the ocean is primarily a by product of the lithospheric solidification of rock that occurred early in the Earth's history. A second source of water is volcanic eruptions. The dissolved constituents found in the ocean come from the transport of terrestrial salts in weathered sediments by leaching and stream runoff. Seawater is a mixture of water and various salts. Chlorine, sodium, magnesium, calcium, potassium, and sulfur account for 99 % of the salts in seawater. The presence of salt in seawater allows ice to float on top of it. Seawater also contains small quantities of dissolved gases including: carbon dioxide, oxygen, and nitrogen. These gases enter the ocean from the atmosphere and from a variety of organic processes. Sea water changes its density with variations in temperature, salinity, and ocean depth. Sea water is least dense when it is frozen at the ocean surface and contains no salts. Highest sea water densities occur at the ocean floor.

Atmospheric circulation drives the movement of ocean currents. Within each of the ocean, the patterns of these currents are very similar. In each basin, the ocean currents form several closed circulation patterns known as gyres. Ocean currents differ from each other by direction of flow, by speed of flow, and by relative temperature.

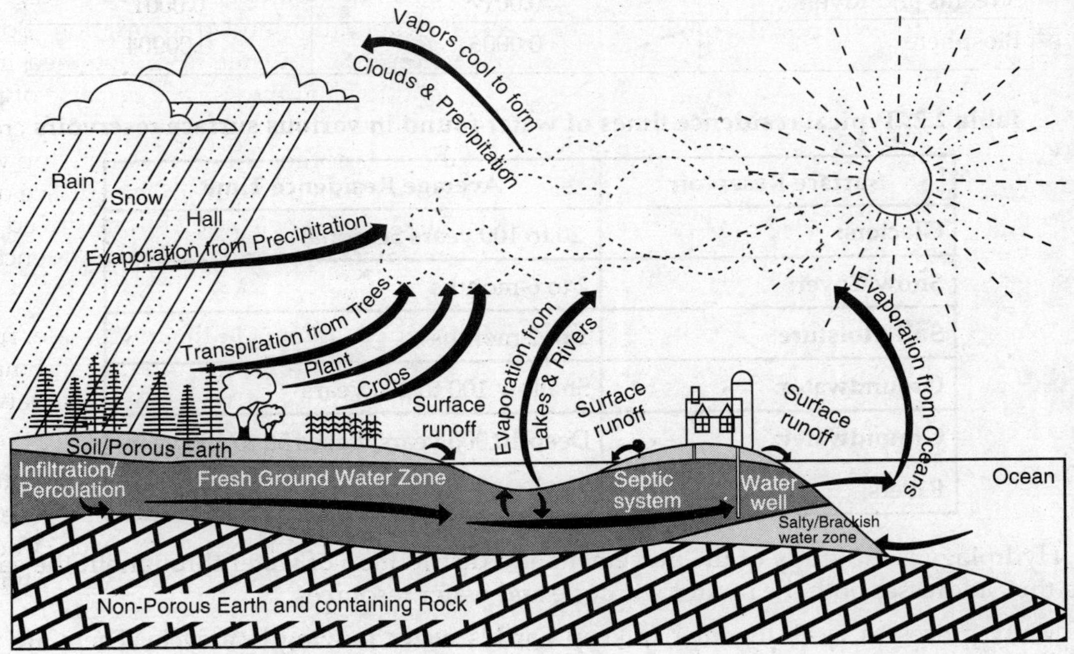

Fig.2.1: Hydrological Cycle

The planetary water supply is dominated by the oceans. Approximately 97 % of all the water on the earth is in the oceans. The other 3 % is held as freshwater in glaciers & ice caps, groundwater, lakes, soil, the atmosphere, and within life.

Water is continually cycled between its various reservoirs. This cycling occurs through the processes of evaporation, condensation, precipitation, deposition, runoff, infiltration, sublimation, transpiration, melting, and groundwater flow.

Hydrological cycle is shown above in figure.

Table below shows the water volumes in various reservoirs and residence times of water in the reservoirs.

Table 2.1: Water volume at the Earth's various reservoirs

Water Reservoir	Water Volume (Cubic km x 10^6)	Percent of Total
Oceans	1370	97.25
Ice Caps and Glaciers	29	2.05
Groundwater	9.5	0.68
Lakes	0.125	0.01
Soil Moisture	0.065	0.005
Atmosphere	0.013	0.001
Streams and Rivers	0.0017	0.0001
Biosphere	0.0006	0.00004

Table 2.2: Typical residence times of water found in various surface reservoirs

Surface Reservoir	Average Residence Time
Glaciers:	20 to 100 years Seasonal
Snow Cover:	2 to 6 months
Soil Moisture:	1 to 2 months
Groundwater:	Shallow 100 to 200 years
Groundwater:	Deep 10,000 years, Lakes 50 to 100 years
Rivers:	2 to 6 months

Hydrology is the study of the movement and distribution of water throughout the earth, and thus addresses both the hydrologic cycle and water resources.

Many processes work together to keep Earth's water moving in a cycle. There are five processes at work in the hydrologic cycle: condensation, precipitation, infiltration, runoff, and evapotranspiration. These occur simultaneously and except for precipitation, continuously. Water vapour condenses to form clouds, which result in precipitation when the conditions are suitable. Precipitation falls to the surface and infiltrates the soil or flows to the ocean as runoff. Surface water (e.g., lakes, streams, oceans, etc.), evaporates, returning moisture to the atmosphere, while plants return water to the atmosphere by transpiration.

The water cycle technically known as the hydrologic cycle , is the circulation of water within the earth's hydrosphere, involving changes in the physical state of water between liquid, solid, and gas phases. The hydrologic cycle refers to the continuous exchange of water between atmosphere, land, surface & subsurface waters, and organisms. In addition to storage in various compartments (the ocean is one suchcompartment); the multiple cycles that make up the earth's water cycle involve five main physical actions: ***evaporation, precipitation, infiltration, runoff, and subsurface flow.***

Evaporation – It occurs when radiant energy from the sun heats water, causing the water molecules to become so active that some of them rise into the atmosphere as vapour. It is the transfer of water from bodies of surface water into the atmosphere. This transfer entails a change in the physical nature of water from liquid to gaseous phases. Along with evaporation can be counted transpiration from plants. Thus, this transfer is sometimes referred to as evapotranspiration. About 90% of atmospheric water comes from evaporation, while the remaining 10% is from transpiration. Transpiration occurs when plants take in water through the roots and release it through the leaves, a process that can clean water by removing contaminants & pollution. Evapotranspiration is water *evaporating* from the ground and *transpiration* by plants. Evapotranspiration is also the way water vapour re-enters the atmosphere.

Precipitation

In cold air way up in the sky, rain clouds will often form. Rising warm air carries water vapor high into the sky where it cools, forming water droplets around tiny bits of dust in the air. Some vapor freezes into tiny ice crystals which attract cooled water drops. The drops freeze to the ice crystals, forming larger crystals we call snow flakes. When the snowflakes become heavy, they fall. When the snow flakes meet warmer air on the way down, they melt into raindrops. In tropical climates, cloud droplets combine together around dust or sea salt particles. They bang together and grow in size until they're heavy enough to fall.

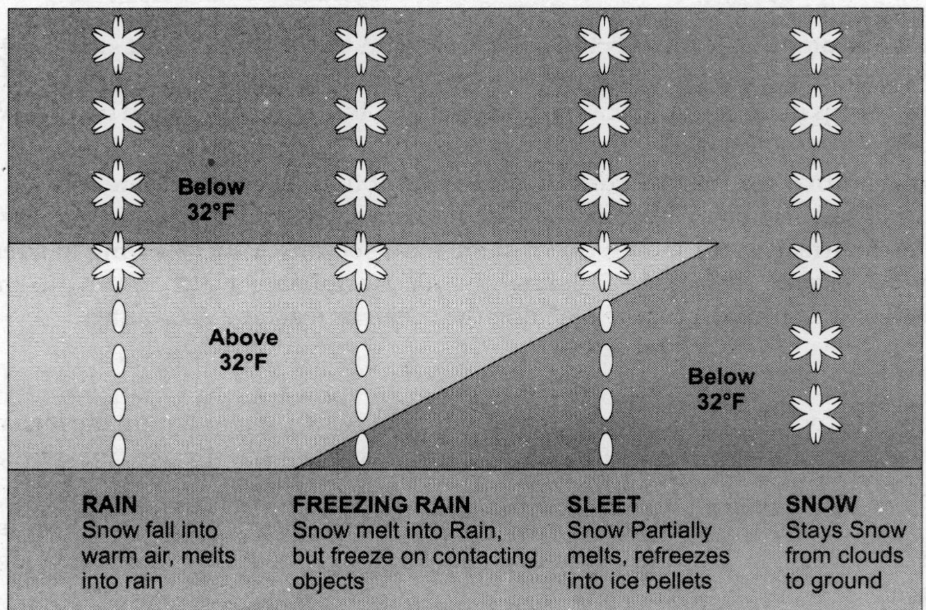

Fig.2.2: Precipitation

Sometimes there is a layer of air in the clouds that is above freezing point ie, 32° F. Then closer to the ground, the air temperature is once again below freezing. Snow flakes partially melt in the layer of warmer air, but then freeze again in the cold air near the ground. This kind of precipitation is called sleet. It bounces when it hits the ground. If snow flakes completely melt in the warmer air, but temperatures are below freezing near the ground, rain may freeze on contact with the ground or the streets. This is called freezing rain, and a significant freezing rain is called an ice storm. Ice storms are extremely dangerous because the layer of ice on the streets can cause traffic accidents. Ice can also build up on tree branches and power lines, causing them to break and our lights to go out. There is another kind of precipitation that comes from thunder storms called hail.

Infiltration

Under some circumstances precipitation actually evaporates before it reaches the surface. More often, though, precipitation reaches the earth's surface, adding to the surface water in streams and lakes, or infiltrating a portion of the precipitation that reaches the earth's surface seeps into the ground through the process called infiltration.

Infiltration into the ground is the transition from surface water to groundwater. The infiltration rate will depend upon soil or rock permeability as well as other factors. Infiltrated water may reach another compartment known as groundwater (i.e., an aquifer). Groundwater tend to move slowly, so the water may return as surface water after storage within an aquifer for a period of time that can amount to thousands of years in some cases. Water returns to the land surface at lower elevation than where it infiltrated, under the force of gravity or gravity induced pressures.

Runoff

The amount of water that infiltrates the soil varies with the degree of land slope, the amount & type of vegetation, soil type & rock type, and whether the soil is already saturated by water. The more openings in the surface (cracks, pores, joints), the more infiltration occurs. Water that doesn't infiltrate the soil flows on the surface as runoff.

Precipitation that reaches the surface of the earth but does not infiltrate the soil is called **runoff**. Runoff can also come from melted snow and ice. Also it includes the variety of ways by which land surface water moves down slope to the oceans. Water flowing in streams and rivers may be delayed for a time in lakes. Not all precipitated water returns to the sea as runoff; much of it evaporates before reaching the ocean or reaching an aquifer.

Subsurface Flow

Surface flow incorporates movement of water within the earth, either within the recharge zone or aquifers. After infiltrating, subsurface water may return to the surface or eventually seep into the ocean.

Water is the major constituent of almost all life forms. Most animals and plants contain more than 60% water by volume. Without water life would probably never have developed on our planet.

GROUND WATER RESOURCES

About 70% of the rainfall is lost to evaporation. Most of the rest runs off the surface into streams & lakes, and eventually returns to the sea. Some, however, slowly percolates down through the soil until it reaches a layer of impermeable rock. The water which builds up on the top of the impermeable rock fills all the openings, crevices, & pores in the overlying sand and rock, to a level called the *water table.* It then very slowly flows to the oceans. When the ground level drops below the water table, the ground water helps supply streams and lakes giving a continuity of discharge they would not otherwise have. Groundwater resources are extremely important because they supply to water withdrawal. Difficulties exist with this in that the rate of replenishment of these supplies is so slow that they must be considered to be essentially nonrenewable resources.

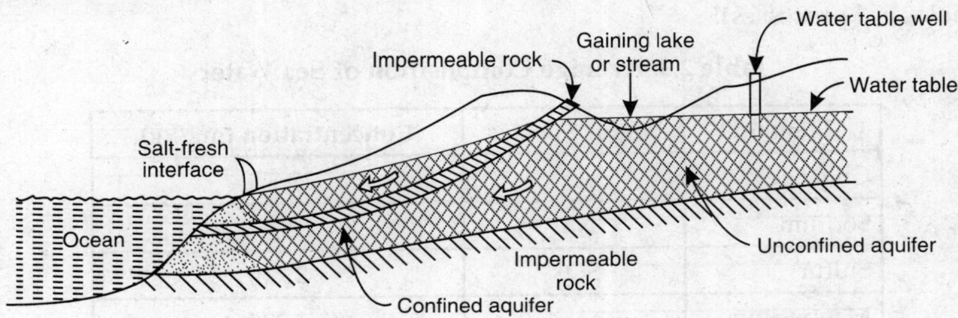

Fig.2.3: Groundwater resources

One problem that can result from a drop in the water table is a corresponding sinking of the ground level in a process called *subsidence.* Another problem that can result from over exploitation of groundwater is that the water table can drop so low that it becomes uneconomical to pump it out. When this happens, the area involved may lose its principal water supply. Coastal areas face another sort of groundwater problem because of the close proximity of saline and fresh water. Under natural conditions there is a balance between the inflow of salt water and the outflow of fresh water, which establishes a boundary between the two regimes. However, as freshwater is pumped out of the area, the water table drops and the boundary moves inland. After a period of time, a coastal freshwater well may need to be abandoned because of this intrusion.

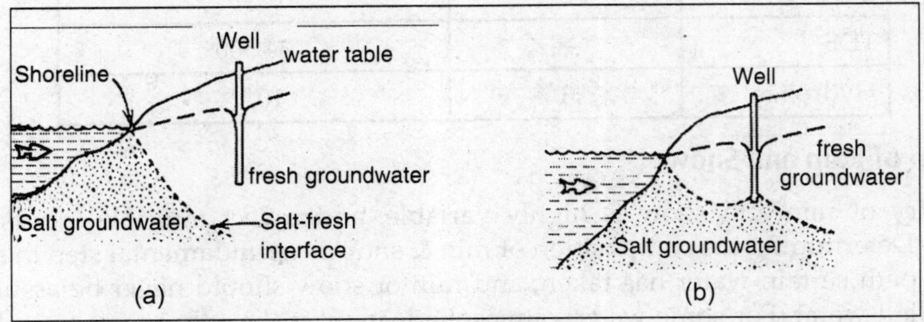

Fig.2.4: Saltwater intrusion into a freshwater well due to over-pumping. (a) before, (b) after.

Water is called the 'universal solvent' because it dissolves more substances than any other liquid. This means that wherever water goes, either through the ground or through our bodies, it takes along valuable chemicals, minerals, and nutrients.

COMPOSITION OF VARIOUS WATER

Sea Water

The oceans account for 97 % of the world's water, and can be generalized as a 1.1 molar solution of solutes with the average composition as given below in the table.

(The composition of seawater varies with location and depth, with higher total solutes found in colder polar waters, and with large changes in non-conservative elements with depth due to biological processes).

Table 2.3: Average Composition of Sea Water

Element	Major Species	Concentration (mg/kg)
Chloride	Cl^-	19,350
Sodium	Na^+	10,760
Sulfur	SO_4^{2-}	2,710
Magnesium	Mg^{2+}	1,290
Calcium	Ca^{2+}	411
Potassium	K^+	399
Carbon	HCO^{3-}	142
Bromide	Br^-	67
Strontium	Sr	8
Boron	BO_4^{3-}	4.5
Fluorine	F^-	1.3
Silicon	H_4SiO_4	0.5-10
DOC	C	0.3-2
TDS	—	34,500
Hydrogen	H^+	$10^{-8.35}$

Composition of Rain and Snow

The chemistry of rain and snow is highly variable, and reflect multiple inputs from the atmosphere. Determining the composition of rain & snow is a fundamental step in evaluating the reaction path certain water has taken, and rain or snow should never be assumed to be simply distilled water. For some waters, atmospheric input is the only source for sulphate and chloride.

In some areas, such as the New England and Norway, atmospheric contamination by acid gases results in extremely low pH rain water, significant input of sulphate, nitrate, & chloride, and the secondary mobilisation of aluminum in receiving waters, are resulting in Al toxicity in fish.

Table 2.4: Typical composition of some low pH acidic rainwater

Species	Sierra. Snow	N.C VA Rain	Menlo Park Rain
Ca^{2+}	O	0.65	0.8
Mg^{3+}	0.2	0.14	1.2
Na^{+}	0.6	0.56	9.4
K^{+}	0.6	0.56	9.4
NH_4^{+}	0	-	-
HCO_3^{-}	3.0	-	4.0
SO_4^{2-}	1.6	2.18	7.6
Cl^{-}	0.2	0.57	17
NO_2^{-}	0.02	-	0.02
NO_3^{-}	0.1	0.62	0
pH	5.6	-	5.5

Composition of Rivers and Lakes

The composition of a river, and eventually a lake, is a reflection of four inputs: atmospheric input of gases & solutes, biologic processes, discharge of groundwater, and local interactions with mineral components in the soil or stream bed.

At base flow, the water in a gaining river is derived from groundwater, so the composition of the river is really a reflection of the aquifer, while for a losing stream the reverse holds - the composition of the shallow groundwater reflects the river, for example an alluvial aquifer.

The composition of a river is a dynamic thing, with concentrations generally (but not always) increasing down stream, but pH generally increasing.

WATER HARDNESS

Hard water occurs when excess minerals in the water create certain problems. While these water problems can be frustrating, water hardness is not a safety issue. Hard water is safe for drinking, cooking, and other household uses.

Hard water can cause several problems for consumers including decreased life of household plumbing & water-using appliances, increased difficulty in cleaning & laundering tasks, decreased efficiency of water heaters, and white/chalky deposits on items such as plumbing, tubs, sinks, pots & pans.

Causes of Hard Water

Approximately 22 percent of the earth's fresh water is groundwater, and naturally, as it flows through soil & rock, it picks up minerals. Hard water results when an excessive amount of calcium and magnesium are present. Total hardness is measured in milligrams per litre (mg/l). Sometimes hardness is measured in parts per million (ppm). Parts per million (ppm) measures the unit(s) of a substance for every one million units of water. Milligrams per litre (mg/l) and parts per million (ppm) are roughly equal in water analysis.

Milligrams Per Liter (mg/1) or parts Per Million (ppm)	Water Rating
less than 17.1	Soft
17.1-60	Slightly Hard
60-120	Moderately Hard
120-180	Hard
Over 180	Very Hard

Identifying Hard Water

- Decreased cleaning capabilities of soaps & detergents, resulting in dingy laundry and reduced life of fabrics.
- Increased buildup of scale on plumbing fixtures and cooking utensils.
- Film left on the body resulting in dry skin and dull, limp hair.
- Soap scum on bathtubs, shower tiles, and basins.
- Increased water heating costs due to scale buildup & mineral deposits, and more frequent replacement of hot water heating elements.
- Clogged pipes or appliances resulting in reduced water flow and increased repairs.

TYPES OF WATER HARDNESS

Temporary Hardness

This refers to hardness whose effects can be removed by boiling the water in an open container. Such waters have usually percolated though limestone formations and contain bicarbonate HCO_3^- along with small amounts of carbonate CO_3^{2-} as the principal negative ions. Boiling the water promotes the reaction:

$$-2\, HCO_3 \rightarrow CO_3 + CO_2$$

by driving off the carbon dioxide gas. The CO_3^{2-} reacts with Ca^{2+} or Mg^{2+} ions, to form insoluble calcium and magnesium carbonates which precipitate out. By tying up the metal ions in this way, the amounts available to form soap scum are greatly reduced.

Permanent Hardness

Waters that contain other anions such as chloride or sulphate cannot be remediated by boiling, and are said to be 'permanently' hard. The only practical treatment is to remove all the ions.

Most conventional water-softening devices depend on a process known as ion-exchange in which 'hardness' ions trade places with sodium and chloride ions that are loosely bound to an ion-exchange resin or a zeolite.

FEATURES & PROPERTIES OF WATER

Features & properties of water are described below:

- **Water has a high specific heat** -Since water has a high specific heat, it can absorb large amounts of heat energy before it begins to get hot. It also means that water releases heat energy slowly when situations cause it to cool. Water's high specific heat allows for the moderation of the Earth's climate and helps organisms regulate their body temperature more effectively.
- **Water in a pure state has a neutral pH**.- Pure water is neither acidic nor basic. Water changes its pH when substances are dissolved in it. Rain has a naturally acidic pH of about 5.6 because it contains natural derived carbon dioxide ,nitrogen oxides and sulfur dioxide.
- **Water conducts heat more easily than any other liquid except mercury. -**This feature causes large bodies of liquid water like lakes and oceans to have a uniform vertical temperature profile.
- **Water molecules exist in liquid form over a wide range of temperature from 0 to 100 ^{0}C -** This range allows water molecules to exist as a liquid in most places on our planet.
- **Water is a universal solvent.-** It is able to dissolve a large number of different chemical compounds. This feature also enables water to carry solvent nutrients in runoff, infiltration, groundwater flow, and also in living organisms.
- **Water has a high surface tension-** Water is adhesive & elastic, and tends to aggregate in drops rather than spread out over a surface as a thin film. This phenomenon also causes water to stick to the sides of vertical structures despite gravity's downward pull. Water's high surface tension allows for the formation of water droplets & waves, allows plants to move water (and dissolved nutrients) from their roots to their leaves, and the movement of blood through tiny vessels in the bodies of some animals.
- **Water molecules are the only substance on Earth that exists in all three physical states of matter: solid, liquid, and gas-** This feature plays an important role in the redistribution of heat energy in the earth's atmosphere. In terms of heat being transferred into the atmosphere, approximately 75 % of this process is accomplished by the evaporation and condensation of water.
- **The freezing of water molecules causes their mass to occupy a larger volume-** When water freezes it expands rapidly adding about 9% by volume. Fresh water has a maximum density at around 4 ^{0}C. Water is the only substance on this planet where the maximum density of its mass does not occur when it becomes solidified.

Key physical properties of water are tabulated below.

Table 2.5: Physical properties of water

Property	Value
Molar mass	18.015;
Molar Volume	55.5 mole/liter
Boiling Point (BP)	100°C at 1 atm

Freezing point (FP)	0°C at 1 atm
Triple point	273.16 K at 4.6 torr
Surface Tension	73 dynes at 20°C
Vapor pressure	0.0212 atm at 20°C
ΔH of Vaporization	40.63kJ/mol
ΔH of Fusion	6.013 kJ/mol
Heat Capacity(cp),	4.22 kJ/kg.K
Dielectric Constant(ε)	78.54 at 25°C
Viscosity	1.002 centipoise at 20°C
Density	1 gm/cc
Density maxima at	4°C

APPENDIX 2.1

Typical composition of domestic waste water

S No.	Constituents of domestic water	Concentration of constituents in various grade of domestic waste water, mg/litre		
		Strong waste water	Medium waste water	Weak waste water
1.	Total solids (suspended + dissolved)	1200	700	350
2.	Total suspended solids	350	200	100
	-Fixed	75	50	30
	-volatile	275	150	70
3.	Total dissolved solids	850	500	250
	-Fixed	525	300	145
	-volatile	325	200	105
4.	Settleable solids (mg/litre)	20	10	5
5.	Biological oxygen demands (BOD), 5 days, 20°C	300	200	100
6.	Chemical oxygen demand (COD)	1000	500	250
7.	Nitrogen (total as N)	85	40	20
	– Organic	35	15	8
	– Free ammonia	50	25	12
	– Nitrites	0	0	0
	– Nitrates	0	0	0
8.	Phosphorous (as total p)	20	10	6
	– organic	5	3	2
	– Inorganic	15	7	4
9.	Total organic carbon (TOC)	200	135	65
10.	Chlorides	100	50	30

S No.	Constituents of domestic water	Concentration of constituents in various grade of domestic waste water, mg/litre		
11.	Grease	150	100	50
12.	Alkalinity	200	100	50

APPENDIX 2.2

Effect of temperature on water properties

Temperature, °C	Density, gm/c. c	Absolute viscosity, centipoise	Vapor pressure, mm Hg	Dissolved oxygen at saturation, mg/litre
0	0.99987	1.7921	4.58	1.46
4	1.00000	-	-	-
5	0.99999	1.5188	6.54	12.80
10	0.99973	1.3077	9.21	11.30
15	0.99913	1.1404	12.80	10.20
20	0.99823	1.0050	17.50	9.20
25	0.99707	0.8937	31.80	7.60
30	0.99567	0.8007	31.80	7.60
35	0.99406	0.7225	42.20	7.10
40	0.99224	0.6560	55.30	6.60

APPENDIX 2.3

Solubility of various gases (mg/litre) in pure water in contact with gas at a pressure of 1 atm

Temperature, °C	N_2	0_2	H_2	CH_4	CO_2	H_2S	SO_2	NH_3
0	29.40	69.80	1.93	39.80	3,360	7,100	186,700	471,000
5	26.10	61.20	1.84	34.30	2790	6040	1,62,300	438,000
10	23.20	54.30	1.76	29.90	2345	5160	140,600	406,000
15	21.10	48.70	1.70	26.40	2000	4475	170500	375000
20	19.30	44.30	1.64	23.60	1720	3925	103,300	344,000
25	17.90	40.40	1.58	21.50	1495	3470	88200	322,000
30	16.80	37.20	1.53	19.70	1305	3090	74700	288,000
40	14.80	32.90	1.48	16.90	1040	2520	54800	252000
60	12.77	27.80	1.44	14.00	704	1810	-	192,000
80	11.96	25.10	1.44	12.70	-	1394	-	134000
100	11.85	24.20	1.44	12.20	-	1230	-	-

* Total solubility including ionization products

APPENDIX 2.4

Some physical characteristics/constant of water

S. No	Physical characteristics/constant	Unit	Value
1	Absorption co-efficient	At 0°C	1.713 for 0_2 0.049 for CO_2 0.024 for N_2
2	Boiling point	°C	100
3	Constituents	At 0°C , 1 atm	N_2=2%, 0_2=1 % & CO_2=0.05%
4	Color	-	Colorless
5	Compressibility	At 20°C and 1 atm	1.0016
6	Conductance	Ohm^{-1} at 25°C	5.54×10^{-5}
7	Dipole moment	Debye	1.84
8	Density	at 20°C	0.99823
9	Dielectric constant	at 20°C	78.54
10	Electrical conductivity	Mhos/cm at 18°C	4×10^{-8}
11	Freezing point	°C	0
12	Heat of fusion (latent heat of melting)	Kcal/kg at 20°C	79.67
13	Ionic product of water	At 25°C	10^{-14}
14	Latent heat of vaporization	Kcal/kg at 20°C	585
15	Molecular depression of freezing point	In 1000 gm of solvent	1.86
16	Molecular elevation of boiling point	-do-	0.512
17	PH value	-	7
18	Ratio of Cp/Cv	At 100°C	1.324
19	Refractive index	At 20°C	1.33
20	Specific heat	At 20°C	0.99859 (It is between 0 and 1°C)
21	Specific conductance	$Ohm^{-1} cm^{-1}$ at 25°C	5.54×10^{-8}
22	Surface tension	Dynes/cm at 20°C	73.75
23	Thermal conductivity	at 4.1°C	1.29×10^{3}
24	Viscosity	Centipoise at 20°C	1.005

Note- The impurities in water may be: dissolved, suspended and colloidal.

Dissolved Impurities

All natural water supplies, either in the raw state or after treatment, contain dissolved mineral matter that differ greatly in the amounts in different water supplies. The most abundant are the bicarbonates, sulphates & chlorides of calcium , magnesium and sodium, present as mixtures of their respective cations & anions. Nitrates of calcium, magnesium & sodium, iron , manganese compounds , silica and alumma may also be present. Iron and manganese occur in water in amounts ranging from a fraction to several mg/l. They are present in many ground waters as ferrous and manganese bicarbonates, in combination with organic matter in some surface waters, and as ferrous & manganese sulphates in acid waters such as those contaminated with mine drainage.

The dissolved gases that occur in various water supplies are: oxygen , carbon dioxide and hydrogen sulphide. Dissolved oxygen is usually not present in deep well waters but it is found in other waters in amounts varying upto the limit of solubility, which is about 14 mg/1 at 0 °C and about 10 mg/1 at 16 °C. Free carbon dioxide, in varying amounts, is found in most natural water supplies. Most surface waters, sampled at their surface, have low contents of free carbon dioxide: zero to about 5 mg/1. In the case OF SURFACE samples will usually range from zero to 2 mg/1 while depth samples may show higher contents. Carbon dioxide is found in deep well water in quantities upto 40 mg/1 or more in unusual cases. In most of the river samples, carbon dioxide usually varies from zero to 5 mg/1. However , since oxidation of organic matter furnishes carbon dioxide, rivers containing considerable organic matter may at times show much higher quantities of carbon dioxide. Waters which contain sulphides are commonly known as 'sulphur waters'. Most sulphur waters are ground waters. Sulphide content of some of the well waters vary from zero to 50 mg/I. Waters having higher amounts of sulphides than this maximum are occasionally encountered but nearly all of the usable sulphur waters will be found in the range below 10 mg/1 and of these the great majority will range below 5 mg/1.

Suspended Impurities

The suspended impurities present in water include mud , vegetable matter , bacteria and industrial wastes sewage. These suspended impurities may be inorganic in nature, such as clay, silt, sand, silica and calcium carbonate or they may be organic matter , oils, fats, greases, microorganisms, etc. The quantities of clay and sand carried by 'muddy rivers are very large, ranging upto several g/1. Microorganisms are common in surface waters but are usually either absent or present only in small amounts in deep well waters. Shallow wells, some springs and waters from infiltration galleries may also contain appreciable amounts of microorganisms. With iron or manganese bearing waters, iron or manganese , bacteria, popularly known as 'Crenothrix' may become exceedingly troublesome. Dissolved, suspended and colloidal impurities introduced by industrial wastes into some surface waters may be so great that it may be economically impossible to purify the water for boiler feed or other industrial purposes.

Colloidal Impurities

Impurities may exist in water in colloidal form. They may be so finely divided that they cannot be seen under microscope ; they exist in a state intermediate between true suspension and solution. Colour in water is usually due to colloidal suspension of ultramicroscopic particles.

Sometimes other impurities such as silica and clay may be present in colloidal form. Matter in colloidal suspension does not settle out easily whereas suspended material settles out rapidly. Some shallow well waters, a few springs and an occasional deep well water may contain noticeable amounts of colour. The colours noted in water range from a very light straw through a yellowish brown, upto dark brown. Swampy waters may appear black because of ink formed due to tannates and gallates. Since colour can be removed through adsorption by colloids, and since turbidity is caused principally by colloidal matter, turbid waters are less likely to be as highly coloured as clear waters subjected to similar conditions.

Tastes and odours are associated with the presence in water of: (1) decaying organic matter,(2) living algae (3) other microscopic organisms that contain essential oils (4) other odourous, compounds- iron , manganese and metallic products of corrosion, industrial wastes, particularly phenolic substances and (5) chlorine and its substitution compounds used for disinfection of water. An increasing cause of disagreeable tastes in water is pollution by industrial wastes like coke oven effluents. These tastes are characteristically medicinal or phenolic, are detectable in extremely low concentration of the order of one ppm, are intensified by chlorination, and are difficult to remove. The odour of water changes with temperature, sometimes not being noticeable when the water is cold. To be palatable, water should be significantly free from colour, turbidity, taste , odour and of moderate temperature in summer & winter. To be wholesome, water must be uncontaminated and hence unable to infect its user with a water borne disease, free from poisonous substances and free from excessive amounts of mineral organic matter.

Characteristics of Water From Different Sources

When rain water or the water formed from melting snow comes in contact with the ground, some of it evaporates, some collects or flows over the surface and some sinks into the ground. Those waters which collect or flow over the surface to form lakes, ponds, reservoirs, rivers, canals or creeks are termed surface waters. Those waters which sink into the ground to be tapped as springs, or by means of wells, shafts or infiltration galleries, are termed groundwaters. As water slowly percolates through the ground,-the strata through which it passes exert a filtering, absorbent and purifying action on it. Therefore, deep well water is usually clear, colourless and low in bacterial count.

Surface Waters

Streams and Rivers: Owing to the inadequacy of underground sources to meet the growing demand for water, rivers and streams are increasing in importance as sources of supply for drinking & industry. Water supplies obtained from rivers streams usually require chemical coagulation because of high turbidity. Many surface waters, particularly those emanating from swampy areas, are often coloured to the extent that they are not sometimes acceptable for domestic or some industrial uses without treatment to remove the colour. The colouring material results from contact of water with organic debris, such as leaves, needles of conifers-wood, all in various stages of decomposition. The quality of the river water is subject to the widest variation, according to the character of the gathering ground, the land over which the river flows, the discharge of sewage, the industrial wastes it receives, the season of the year and climatic conditions.

Ponds, Lakes and Reservoirs: In general the water from lakes, ponds reservoirs are less subject to variations in composition than a river water and many large lakes are remarkably constant in composition even over long periods of times. In the case of small lakes, ponds and-reservoirs, some are fairly constant in composition while others show considerably seasonal variations. Owing *to* their relative quiescence, lakes serve as very efficient settling basins. Very large lakes usually constitute a very dependable water supply. Some of these are extremely low in mineral content while others contain a high content of hardness and other mineral constituents.

Underground Waters

Springs: The yield from springs is inadequate, except for small supplies. It is often supposed that spring water is pure. This is sometimes true, but more often it is not. Spring water may vary considerably in quality, depending on geological and topographical conditions. The water may be hard or soft, pure or polluted, sometimes saline, sulphurous, ferruginous etc. The content of free carbon dioxide is sometimes high and the water may possess corrosive & plumbo solvent properties. As far as potability is concerned, no spring water should be judged safe until it is tested.

Wells: Well water supplies are very widely used in industry. They have a number of advantages over surface supplies, among which usually are clarity, freedom from colour, taste odour, relatively constant composition and freedom of pollution from sewage & industrial wastes. Shallow wells yield waters of widely varying character and as a class they are of a low grade of bacterial purity. Shallow wells are seldom used for public water supplies.

Deep wells provide water supplies to many communities & industries and often yield waters of the highest degree of organic & bacterial purity. Such waters are therefore some times supplied to the consumers in their natural state. When treatment is necessary it is usually for such purposes as softening or the removal of iron. While deep well waters are almost invariably clear and quite free from turbidity & colour, some may contain iron, some hydrogen sulphide and a relatively small number may contain manganese. In general, it might be expected that in a given area, the deeper well waters would be more highly mineralised than those drawn from higher levels.

APPENDIX 2.5

Glossary of terms related to water

A

Acid neutralizing capacity

Measure of the buffering capacity of water; the ability of water to resist changes in pH.

Acid rain

Rain that has a flamboyantly low pH, due to contact with atmospheric pollutants such as sulphur oxides.

Activated coal

This is the most commonly used adsorption medium, produced by heating carbonaceous substances or cellulose basses in the absence of air. It has a very porous structure and is commonly used to remove organic matter and dissolved gasses from water. Its appearance is similar to coal of peat. Available in granular, powder or block form it has the highest adsorption capacity.

Activated sludge

Oxygen dependent biological process that serves to convert soluble mater to solid biomass, that is removable by gravity of filtration.

Adsorption

Separation of liquids, gases, colloids or suspended matter from a medium by adherence to the surface of pores of a solid.

Advanced oxidation process

One of several combination oxidation processes, advanced chemical oxidation processes use chemical oxidants to reduce COD/BOD, and to remove both organic & oxidisable inorganic components. The process can completely oxidize organic materials to carbon dioxide and water, although it is often not necessary to operate the process to this level of treatment.

A wide variety of advanced oxidation processes are available:

Chemical oxidation process using hydrogen peroxide, ozone, combined ozone & peroxide, hypochlorite, Fenton's reagent, etc.

Ultra-violet (UV) enhanced oxidation such as UV/ hydrogen, UV /air

wet air oxidation and catalytic wet air oxidation (where air is used as the oxidant)

Advanced water treatment

The level of water treatment that requires an 85 percent reduction in pollutant concentration, also known as tertiary treatment.

Advanced Wastewater Treatment

Any treatment of sewage water that includes the removal of nutrients such as phosphorous & nitrogen and a high percentage of suspended solid.

Aerated lagoon

A water treatment pond that speeds up biological decomposition of organic waste by stimulating the growth and activity of bacteria, which are responsible for the degradation.

Aeration

Technique that is used with water treatment that demands oxygen supply, commonly known as aerobic biological water purification. Either water is brought into contact with droplets by spraying or air is brought into contact with water by meads of aeration facilities. Air is pressed through a body of water by bubbling.

Aeration tank

A tank is used to inject air into water.

Aerobic

A process that takes place in the presence of oxygen, such as the digestion of organic matter by bacteria in an oxidation pond.

Aggressive water

Water that is soft & acidic and can corrode plumbing, pipes & appliances.

Algae

Single or multi-celled organisms that are commonly found in surface water, such as duckweed. They produce their own food through photosynthesis. The algae population is divided up into green algae and blue algae, of which the blue algae are very damageable to human health. Excessive algae growth may cause the water to have undesirable odours or tastes. Decay of algae diminishes oxygen supplies in the water.

Algal blooms

Periods of enlarged algal growths that affect quality. Algal blooms indicate potentially hazardous changes in the chemistry of water.

Alkalinity

Alkalinity means the buffering capacity of water; the capacity of the water to neutralize itself, It prevents the water pH levels from basic or acid. It also adds carbon to water. Alkalinity stabilizes water at harmful conditions for aquatic life.

In water chemistry , alkalinity is expressed in ppm or mg/ l of equivalent calcium carbonate. Total alkalinity of water is the sum of all three sorts of alkalinity; carbonate and hydroxide alkalinity.

Alluvium

Sediments deposited by erosion usually by streams.

Anaerobic

A process that takes place in absence of oxygen, such as the digestion of organic matter by bacteria in a reactor.

Anion

A negatively charged ion that results from the dissociation of salts, acids or alkali's solution.

Anode

A site in electrolysis where goes into solution as a cation leaving an equivalent of electrons to be transferred to an opposite electrode, called a cathode.

Aquatic

Growing in water, living in water, or frequenting water.

Aqueous solubility

The maximum concentration of a chemical that dissolves in a given amount of water,

Aquifer

A layer in the soil that is capable a significant volume of groundwater.

Assimilation

The ability of water to purify itself of pollutants.

Assimilative Capacity

The capacity of natural water to receive wastewaters of toxic materials without negative effects and without damage to aquatic life or humans who consume the water.

Attenuation

The process of reduction of a compound's concentration over time. This can be through absorption, adsorption, degradation, dilution or transformation.

Attrition

The action of the particles rubbing against the other in the filter media or ion exchange bed that can in time cause breakdown of the particles.

Available chlorine

A measure of the amount of chlorinated lime , hypochlorite compounds, and other materials.

B

Backflow

The flow of water in a medium in a direction opposite to normal flow. Flow is often returned into the system by backflow, if the wastewater in a purification system is severely contaminated.

Back sphonage

Reverse seepage of water in a distribution system.

Bacteria

Microscopiacally small single-cell organisms, that reproduce by fission of spores.

Backflow

The flow of water in a medium in a direction opposite to normal flow. Flow is often returned into the system by backflow, if the wastewater in a purification system is severely contaminated.

Back espionage

Reverse seepage of water in a distribution system.

Backwashing

Reversing the flow of water back through the filter media to remove entrapped solids.

Bacteria

Microcopically small single-cell organisms, that reproduce by fission of spores.

Bacterial water contamination

The introduction of unwanted bacteria into a water body.

Bed Load

Sediment particles resting on or near the channel bottom that are pushed or rolled along by the flow of water.

Benthic zone

Sediment particles resign on or near the channel bottom that are pushed or rolled along by the flow of water.

Bioaccumulation

The increase in concentration of a substance in living organisms, as they take in contaminated air, water, or food, due to slow metabolization and excretion.

Biocide

A chemical that is toxic to microorganisms, Biocides are often used to eliminate bacteria and other single-cell organisms from water.

Biodegradable pollutants

Pollutants that are capable of decomposing under natural conditions.

Biofilm

Population of various microorganisms. Trapped in a layer of slime and excretion products, attached to a surface.

Biological contaminants

Living organisms such as viruses, bacteria, fungi and mammal & bird antigens that can case harmful heath effects to humans.

Biologically activated carbon

Activated carbon that supports active microbial growth, in order to aid in the degradation of organics that have been absorbed on its surface and in its pores.

Biological oxidation

Decomposition of complex organic materials by microorganisms through oxidation.

Bio monitoring

The use of living organisms to test the suitability of effluents for discharge into receiving waters and to test the quality of such waters downstream from the discharge.

Bio mediation

The biological treatment of wastewater & sludge, by inducing the breakdown of organics and hydrocarbons to carbon dioxide & water.

Biota

All living organisms in a region or ecosystem.

Biotransformation

Conversion of a substance into other compounds by organisms; including biodegration,

Black water

Water that contains waste of humus ,animals or food.

Blinding

A bled-up of particles in a fitter medium, that prevents fluids from flowing through.

BOD_5

The amount of dissolved oxygen consumed in five days by bacteria that perform biological degradation of organic matter.

Brackish water

Water that is neither falls in the category of salt water, nor in the category of fresh water. It holds the middle between either one of the categories.

Breakpoint chlorination

Addition of chlorine to water until there is enough chlorine present in floc particulate matter through a filter.

Brine

Highly salty & heavily mineralized water, containing heavy metal and organic contaminants,

C

Calcium hypo chlorite

A Chemical that is widely used for water disinfection, for instance in swimming pools or water purification plants. It is especially useful because it is stable , dry and can be made into tablets.

Candle filter

A relatively coarse aperture filter, designed to retain a coat filter medium on an extended surface.

Cappillary membranes

Membranes about the thickness of a human hair, used for Reverse Osmosis, nanofiltration, ultrafiltration and microfiltration.

Carcinogen

Any dissolved pollutant that can induce cancer.

Carbonate hardness

Hardness of water caused by carbonate and bicarbonate by-products of calcium & magnesium.

Catch basin

A sedimentation area designed to remove pollutants from runoff before being discharged into a stream or pond.

Cathode

A site in electrolysis where cations in solution are neutralized by electrons that plate on the surface or produce a secondary reaction with water.

Cation

A negatively charged ion, resulting from dissociation of molecules in solution.

Centrifugation

A separation process, which uses the action of centrifugal force to promote accelerated settling of particles in a solid-fluid mixture.

Check valve

A valve that allows water to one direction and will then close to prevent development of a back flow.

Chelating agents

Organic compounds that have the ability to draw ion from their water solutions into soluble complexes.

Chemical Oxgen Demand (COD)

The amount of oxygen (measured in mg/l) that is consumed in the organic and oxidisable inorganic matter, under test conditions. It is used to measure the total amount of organic and inorganic pollution in wastewater. Contrary to BOD, with COD practically all compounds are fully oxidized.

Chemical pollution

Introduction of chemical contaminants into a water body.

Chloramines

A chemical complex that consists of chorine and ammonia. It serves as a water disinfectant in public water supplies in place of chorine because chlorine can combine with organics to form dangerous reaction products. In which forms chloramines exist depends on the physical/chemical properties of the water source.

Chlorinated hydrocarbons

Hydrocarbons the contain chlorine. These include a class of persistent hydrocarbons that accumulate in the aquatic food chain. Among them are DDT , aldrin, dieldrin, heptachlor, chlordane, lindane, endrin, Mirex, hexachloride, and toxaphene.

Chlorination

A water purification process in which chlorine is added to water for disinfection, for the control of present microorganisms. It is also used in the oxidation of compound impurities in water.

Coagulation

Destabilisation of colloidal particles by addition of a reactive chemical, called a coagulant. This happens through neutaliztaion of the charges.

Coalescence

Liquid particles in suspension that unite to create particles of a greater volume.

Coastal zone

Lands and waters near the coast, whose uses & ecology are affected by the sea.

Coliform bacteria

Bacteria that serve indicators of pollution and pathogens when found in water. These are usually found in the intestinal tract of humans and other warm-blooded animals.

Collector sewers

Pipes top collect and carry wastewater from individual sources to an interceptor sewer that will carry it to a treatment facility.

Combined sewer

A sewer system that carries both sewage and rain water runoff.

Composite sample

Series of water samples taken over a given period of time and weighted by flow rate.

Conductivity

The amount of electricity the water can conduct. It is expressed in a chemical magnitude.

Conduit

A natural or artificial track through which fiulds may be transported.

Consumptive water use

Water removed from available supplies without return to a water resources system. Water used in manufacturing, agriculture, and food preparation.

Contact time

The length of time a substance is in contact with a liquid, before it is removed by filtration or the occurrence of the chemical change.

Conventional sewer systems

Systems that were traditionally used to collect municipal wastewater in gravity sewers and convey it to a central primary or secondary treatment plant, before discharge on receiving surface waters.

Conveyance loss

Water loss in pipes and channels by leakage or evaporation.

Cooling tower

Large tower used to transfer the heat in cooling water from a power or industrial plant to the atmosphere either by direct evaporation or by convection & conduction.

Cross flow filtration

A process that uses opposite flows across a membrane surface to minimize particle build-up.

Cryptosporidium

A microorganism in water that causes gastrointstestinal illness in humans. It is commonly found in untreated surface water and can be removed by filtration. It is resistant to disinfectants such as chlorine.

Cultural eutrophication

Decline of the oxygen rate in water, which has serious consequences for aqutic life, caused by humans.

D

Dealkalinisation

Any process that serves to reduce the alkalinity of water.

Decarbonation

The process of removing carbon dioxide from water, using towers or scrubbers.

Decomposition

The break down of organic matter by bacteria and fungi, to change the chemical structure & physical appearance of matter.

Defoaming agents

Chemicals that are added to wastewater discharges to prevent the water from foaming when it is discharged into a receiving water body.

Degasification

The process of removing dissolved gasses from water, using vacuum or heat.

Demineralistaion

Processees to remove minerals from water, usually the term is restricted to ion exchange process.

Dentirification

Removal of nitrate and nitrate product from water to produce a quality that answers common water standards.

Depression storage

The storage water in low areas, such as ponds, and wetlands.

Depth filtration

The actual time that a small amount of water is in a settling basin or flocculating basin. In storage reservoirs, It means the length of time water will be stored.

Detergent

A water-soluble cleansing agent, other than soap.

Dewater

The separation of water from sludge, to produce a solid cake.

Digester

A closed tank for wastewater treatment, in which bacterial action is induced to break down organic matter.

Diluting water

Distilled water that has been stabilized, buffered, and aerated. It is often applied in the BOD tests.

Direct run-off

Water that flows from the ground surface directly into streams, rivers, and lakes.

Disinfection

The decontamination of fluids and surfaces. To disinfect a fluid or surface a variety of techniques are used, such as ozone disinfection. Often disinfection means eliminating the present micorganisms with a biocide.

Dissolved air flotation (DAF)

A procedure of induced flotation with very fine air bubbles or 'micro bubbles' of 40 to 70 microns.

Dissolved oxygen

The amount of oxygen dissolved in water at a certain time. Expressed in ppm or mg/l.

Dissolved solids

Solids material that totally dissolves in water and can be remove by means of filtration.

Dredging

Cleaning, deepening, or widening a waterway, using a machine (dredge) that removes materials by means of a scoop or a suction device.

Dystrophic lakes

Acidic bodies of water that contain many plants but few fish due to the presence of great amounts of organic matter.

E

Effluent

The outlet or outflow of any system that deals with water flows, for an oxidation pond for biological water purification. It is the product water of the given system.

Ejector

A device used to inject a chemical solution into wastewater during water treatment.

Electrolyte

Substance that dissociates when it dissolves in water.

Electrodialysis

Process where electrical energy will change in chemical energy. The process happens in a electrolyte, a watery solution or a salt melting which gives the ions a possibility to transfer between two electrodes. The electrolyte is the connection between the two electrodes, which are also connected to a direct current. At the electrodes, the cations will be reduced and the anions will be oxidated.

Elutriation

Freeing sludge of its mother liquor by washing it with water.

Emulsifier

A chemical that helps suspending one liquid in another.

Emulsion

Dispersion of one liquid in another liquid, occurs when a liquid is insoluble.

End-of-pipe techniques

Techniques for water purification that save the reduction pollutants after they have formed.

Enrichment

When the addition of nutrients, such as pathogen and phosphorus, from effluent or agricultural runoff to surface water, greatly increases algal growth.

Erosion

The wearing away of the land surface by wind, waster, ice or other geological agents. Erosion occurs naturally from weather or runoff but is often intensified by human land use practices.

Escherichia coli (E. Coli)

Coliform bacterium that is often associated with human & animal waste and is found in the intestinal court. It is used by health departments and private laboratories to measure the purity of water.

Esturary

Region of interaction between rivers and near-shore waters, where tidal action and river flow mix fresh & salt water. Therefore estuaries mainly consist of brackish water.

Eutrophic

Eutrophic refers to water that is rich in nutrients such as nitrogen and phosphrous.

Eutrophication

Enrichment of water, which causes excessive growth of aquatic plants and increasing activity of anaerobic microrganisms. As a result the oxygen levels in the water quickly decline and the water chokes, making life impossible for aerobic water organisms.

Evapotranspiration

The loss of water from the soil through vaporizing, both by direct evaporation and by transpiration from plants.

F

Facultative bacteria

Bacteria that can live under aerobic or anaerobic conditions.

Fermentation

The conversion of organic matter to methane, carbon dioxide and other molecules by anaerobic bacteria..

Fitter medium

The permeable material that separates solids from passing through it.

Filtrate

A liquid that has passed through the filter medium.

First draw

The water that comes out when a tap is first opened. It is likely that is has the highest level of lead.

Floc

A flocculent mass that is formed in the accumulation of suspended particles. It can naturally, but is usually induced in order to be able to remove certain particles from wastewater.

Flocculation

The accumulation of destabilized particles and micro flakes, and subsequently the formation of sizeable flakes. One must add another substance called flocculent in order to facilitate the formation of flakes called flocs.

Floodplain

The flat of nearly flat land along a river or stream that is covered by water during a flood.

Floatation

A solid-liquid or liquid-liquid separation procedure, which is applied to particles of which the density is lower than that of the liquid they are in, natural, aided and induces flotation.

Flow augmentation

The addition of water to meet flow needs.

Flux

The rate at which a Reverse Osmosis Membrane allows water to pass through it.

Fouling

The deposition of organic matter on the membrane surface.

Freshwater

Water containing less than 1 mg/l of dissolved solids of any type.

G

Giardia

A microorganism that is commonly found in untreated surface water and can be removed by filtration. It is resistant to disinfectants such as chlorine.

Gray water

Domestic wastewater composed of wash water from kitchen, bathroom & laundry sinks and from tubs & washers.

Groundwater

Water that can be found in the saturated zone of the soil; a zone that consists merely of water. It slowly moves from places with high elevation & pressure to places with low elevation & pressure, such as rivers and lakes.

Groundwater discharge

Ground water entering that deals with the occurrence, movements, replenishment & depletion, properties and methods or investigation and ultillsation of groundwater.

Gully

A deeply eroded channel created by the concentrated flow of water.

H

Hard water

Water that contains a great number of positive ions. The hardness is determined by the number of calcium and magnesium atoms. Soap usually dissolves badly in hard water.

Heavy metals

Metals that have a density of 5.0 or higher and a high elemental weight. Most are toxic to humans, even in low concentrations.

Heavy water

Water is which all the hydrogen atoms have been replaced by deuterium.

Henry's Law

A way of calculating the solubility of a gas in a liquid, based on temperature and partial pressure, by means of constants.

Holding Pond

A pond or reservoir, usually made of earth, built to store polluted runoff.

Hydraulic conductivity

The rate at which water can move through a permeable medium.

Hydraulic gradient

In general, the direction of groundwater flow due to changes in the depth of the water table.

Hydrogeology

The science of chemistry and movement of groundwater.

Hydrograph

A chart measures the amount of water flowing past a point as a function of time.

Hydrologic cycle

The natural cycle of water passing through the environment, including evaporation, contestation, retention and run off.

Hydrophilic

Having an affinity for water.

Hydrophobic

Having an aversion for water

Hydrosphere

Region that includes all the earth's liquid water, frozen water, floating ice, frozen upper layer of soil, and the small amounts of water vapour in the atmosphere.

Hypochlorite

An anion that forms products such as calcium and sodium hypo chlorite. These products are often used for disinfection and bleaching.

Hypoxic waters

Waters with dissolved oxygen concentrations of less than 2 mg/1, the level generally accepted as the minimum required for life and reproduction of aquatic organisms.

Imhoff cone

A clear, cone-shaped container used to measure the volume of settleable solids in a specific volume of water.

Impermeable

Not easily penetrated by water.

Impurities

Particles or other impurities that cause cause water to be unclear.

Indicator

Any biological entity or process, or community whose characteristics show the presence of specifics.

Indicator organisms

microorganisms, such as coliforms, whose presence is indicative for pollution or for the presence of more harmful microorganisms.

Indirect discharge

Introduction of pollutants from a non-domestic source into a publicly owned wastewater treatment system. Indirect dischargers can be commercial or industrial whose wastes enter local sewers.

Infiltration

Penetration of water into a medium, for Instance the soil.

Injection

The introduction of a chemical or medium into process water to alter its chemistry of specific compounds.

Ion exchange

The replacement of undesirable ions with a certain charge by desirable ions of the same charge in a solution, by an ion-permeable absorbent.

Inorganic chemicals

Chemical substances of mineral origin, not of basically carbon structure.

Irrigation

Applying water or wastewater to land to supply the water and nutrient needs of plants.

J

Jar test

A laboratory test procedure with differing chemical doses. Mix speeds, and settling times, to estimate the minimum or ideal coagulant dose required to achieve water quality goals.

L

Lagoon

A shallow pond where sunlight, action, and oxygen work to purify wastewater.

Lake

An inland body of water, usually fresh water, formed by glaciers, river drainage, etc. It is usually larger then a pool or pond.

Land Application

Discharge of wastewater onto ground for treatment of reuse.

Leachate

Water that contains solute substances in solution after percolation through a filter or soll.

Leaching

The process by which soluble constituents are dissolved and through the soil by a percolating fluid.

Light absorption

The amount of light water can absorb over time.

Limnology

The study of the physical, chemical, hydrological, and mineralogical aspects of fresh water.

M

Maximum Contaminant Level (MCL)

The Maximum level of a contaminant allowed based on health effects and currently available treatment methods.

Mechanical aeration

Use of mechanical energy to inject air / water to cause a waste stream to absorb oxygen.

Mechanical floatation

A term used in the mineral industry to dispersed air to produces bubbles that measure 0.2 to 2 mm in diameter.

Membrane

A thin barrier that allows some compounds or liquids to pass through. It is a semipermeable skin of which the pass-through is determined by size or special nature of the particles. Membranes are commonly used to separate substances.

Mesotrophic

Reservoirs and lakes which contain moderate quantites of nutrients and are moderately productive in terms of aquatic animal & plant life.

Metabolise

Conversion of food, for instance soluble organic matter, to cellular matter and gaseous by-products through a biological process.

Microbial growth

The multiplication of microrganisms such as bacteria, algae, diatoms plankton, and fungi.

Microorganisms

Organisms that are so small that they can only be observed through a microscope, for instance bacteria, fungi or yeasts.

Mineral Water

Contains large amounts of dissolved minerals such as calcium, sodium, magnesium, and iron. Some tap waters contain as many or more minerals than commercial mineral waters. There is no scientific evidence that either high or low mineral content water is beneficial to humans.

Municipal discharge

Discharge of effluent from wastewater treatment plants, which receive wastewater from households, commercial establishments, and industries in the coastal drainage basin.

Municipal sewage

Semi liquid residue that remains from the treatment of municipal water and wastewater.

N

Neutralization

The addition of substances to neutralize water, so that it is neither acid, nor basic. Neutralization does not specifically mean a pH 7.0, it just means the equivalent point of an acid base reaction.

Nitrification

A biological process, during which nitrifying bacteria convert toxic ammonia to less harmful. It is commonly used to remove nitrogen / nitrogen substances from wastewater, but in lakes and ponds it occurs naturally.

Non-point sources

Diffuse water pollution sources without a specific point of origin. The pollutants are generally carried off the land by storm water. Common non-point sources are agriculture and atmospheric disposal.

Non-potable

Water that is unsafe or unpalatable to drink because it contains pollutants, contaminants, minerals or infective agents.

Nuisance Contaminant

Constituents in water, which are not normally harmful to health but may cause offensive taste, odor, color, corrosion, foaming, or staining.

Nutrient

Any substance that promotes growth with living organisms. The term is generally applied to nitrogen and phosphorus in wastewater, but is also applied to other essential and trace elements.

O

Oligotrophic lakes

Deep clear lakes with few nutrients, little organic matter and a high dissolved-oxygen level.

Osmosis

Water molecules passing through membranes naturally, to the side with the highest concentration of dissolved impurities.

Outfall

The place where a wastewater treatment plant discharges treated water into the environment.

Overflow rate

One of the guidelines for design of the settling tanks and clarifiers in a treatment plant to determine if tanks and clarfiers in a treatment plant are used enough.

Oxidation pond

A man-made body of water in which waste is consumed by Bacteria.

Oxidation-reduction potential

The electric potential required to transfer electrons from the oxidant to the reductant, used as a qualitative measure of the state of oxidation in water treatment systems.

Oxygen depletion potential

The reduction of the dissolved oxygen level in a water body.

Ozone

An unstable oxidizing agent, that consists of three oxygen atoms and can be found in the ozone layer in the atmosphere. It is produced by electrical discharge through oxygen or by specifically designed UV-lamps.

Ozone generator

A device that generates ozone by passing a voltage through oxygen or by specifically designed UV lamp -used as a disinfection system.

P

Particulate loading

The mass of particulates per unit volume of water.

Parts per billion

Expressed as ppb; a unit of concentration equivalent to the ug/1.

Parts per million

Expressed as ppm; a measure of concentration, One ppm is one unit weight of solute per million unit weights of solution. In water analysis the ppm is equivalent to mg/1.

Pasteurisation

The elimination of microorganisms by heat applied for a certain period of time.

Pathogens

Disease-producing microorganisms.

Percolating water

Water that passes through rocks or soil under the force of gravity.

Permeability

The ability of a medium to pass a fluid under pressure.

Photosynthesis

The process of conversion of water and carbon dioxide to carbohydrates. It takes place in the presence of chlorophyll and is activated by sunlight. During the process oxygen is released. Only plants and a limited number of microorganisms can perform photosynthesis.

Physical and chemical treatment

Processes generally used in wastewater treatment facilities. Physical processes are for instance filtration. Chemical treatment can be coagulation, chlorination, or ozon treatment.

Physical weathering

Breaking down of rock into bits and pieces by exposure to temperature and changes & the physical action of moving ice & water, growing roots, and human activities such as farming & construction.

Phytoplankton

Free-floating, mostly, microscopic aquatic plants.

Point source

A stationary location from which pollutants are discharged. It is a single identifiable source of pollution, such as a pipeline or a factory.

Polar substance

A substance that carries a positive or negative charge, for instance water.

POP's

Persistent Organic pollutants are complex compounds that are very persistent and difficult to degrade biologically.

Potable water

Water that is safe for drinking and cooking.

Potentiometric surface

The surface to which water in an aquifer can rise by hydrostatic pressure.

Pressure sewers

A system of pipes in which water, wastewater, or other liquid is at higher elevation.

Pre-treatment

Processes used to reduce or eliminate wastewater pollutants from before they are discharged.

Primary wastewater treatment

The removal of suspended, floating and precipitated solids from untreated wastewater.

Process water

Water that serves in any level of the manufacturing process of certain products.

Product water

Water that has passed through a water treatment plant and is ready to be delivered to consumers.

Protozoa

Large micrорganisms, which consume bacteria.

Putrefaction

Biological that is produced by bacteria and is fairly stable. It causes fever in mammals.

Q

Qualitative water assessment

Analyses of water used to assess water properties and concentrations of compounds & contaminants in order to define its suitability for a particular purpose.

R

Raw sewage

Untreated wastewater and its contents.

Raw water

Intake water before any treatment or use.

Reaeration

Renewing air supplies in the lower layers of a reservoir in order to lower the pH.

Receiving waters

A river, lake, ocean, stream or other, watercourse into which wastewater or treated effluent is discharged.

Recharge Area

An area where rainwater soaks through the ground to reach an aquifer.

Recirculation

Recycling water after it is used. Often it has to pass a wastewater purification system before it can be reused.

Redox

Shortened term for reduction/oxidation reactions. Reactions are a series of reactions of substances in which electron transfer takes place. The substance that gains electrons is called oxidising agent.

Reserve Capacity

Extra treatment capacity built into wastewater treatment plants and sewers to be able to catch up with future.

Reservoir

A natural or artificial holding area used to store water.

Reverse Osmosis process

The Reverse Osmosis (RO) Process uses a semi-permeable membrane to separate and remove dissolved solids, organics, submicron colloidal matter, viruses & bacteria from water. The process is called 'reverse' osmosis since it requires pressure to force pure water across a membrane, leaving impurities behind.

Run-off

The part of precipitation water that runs off the land into streams or other surface water.

S

Safe water

Water that does not contain harmful bacteria, toxic materials, or chemicals, and is considered safe for drinking.

Safe yield

The annual amount of water that can be taken from a source of supply over a period of years without depleting that source beyond its ability to be naturally refilled.

Salinity filtration

Sand filtration is a frequently used and very robust method to remove suspended solids from water. The filtration medium consists of a multiple layer of a multiple layer of sand with a variety in size and specific gravity. Sand filters can be supplied in different sizes and materials both hand operated and fully automatically.

Saturated zone

The area below the water table where all open spaces are filled with water.

Screening

Use of screens to remove coarse floating and suspended solids from sewage.

Sedimentation

Settling of solid particles in a liquid system due to gravity.

Sediments

Soil, sand, and minerals washed form land into water, usually after rain.

Semi-confined aquifer

An aquifer partially confined by soil layers of low permeability through which recharge and discharge can still occur.

Semipermeable

A medium that allows water to pass through, but rejects dissolved solids, so that it can be used to separate solids from water.

Separate sewer

A sewer system that carries only sanitary sewage; no storm-water runoff. When a sewer is constructed this way, wastewater treatment plant can be sized to treat sanitary wastes only and all of the water entering the plant receives complete treatment at all times.

Septic tank

An underground storage tank for waster from homes not connected to a sewer line. Waste goes directly from the home to the tank.

Settleable solids

Those suspended solids in wastewater that will settle over a certain period of time and are removed in that way.

Sewage

Waste fluid in a sewer system.

Sewage contamination

The introduction of untreated sewage into a water body.

Sewage sludge

Sludge produced in a public sewer.

Sewerage

The entire system of sewage collection, reatment, and disposal.

Sludge

A semi-solid residue, containing microoroganisms and their products, from any water treatment process.

Softening

The removal of calcium and magnesium from water to reduce hardness.

Soft water

Any water that does not contain large concentrations of the dissolved mineral calcium or magnesium.

Sparger

A device that introduces compressed air into a liquid.

Sparging

Injection of air below the water table to strip dissolved volatile organic compounds and to facilitate aerobic biodegradation of organic compounds.

Specific conductance

Method to estimate the dissolved solid content of a water supply by testing its conductivity.

Spring

Ground water seeping out of the earth where the water table exceeds the ground surface.

Surface tension

The elastic like force in a body, especially a liquid , tending to minimize, or constrict, the area of the surface.

Suspended solid

Solid organic or inorganic particles that are held in suspension in a solution.

T

Tertiary treatment

Advanced cleaning of wastewater that goes beyond the secondary or biological stage, removing nutrients such as phosphorus, pathogen, and most BOD & suspended solid.

Thermal pollution

Discharge of heated water from industrial processes in receiving surface water, causing death or injury of aquatic organisms.

TDS

Total Dissolved Solid. The weight per unit volume of water of suspended solids in a filter media after filtration or evaporation.

TH

Total Hardness. The sum of calcium and magnesium hardness, expressed as a calcium carbonate equivalent.

TS

Total Solids. The weight of all present solids per unit volume of water. It is usually determined by evaporation. The total weight concerns both dissolved and suspended organic & inorganic matter.

Total solids

All the solids in wastewater or sewage water, including suspended solids and filterable solids.

Toxic water pollutants

Compounds that are not naturally found in water at the given concentrations and filterable solids.

Transmission lines

Compounds that are not naturally found in water at the given concentrations and that cause death, disease, or birth defects in organisms that ingest or absorb them.

Transpiration

The process by which water vapors is released into the atmosphere after transpiring of living plants.

Treatment plant

A structure built to treat wastewater before discharging it into the environment.

Trickling filter

A wastewater treatment unit that contains medium material with bacteria. The stream of wastewater is trickled over the medium and the bacteria break down the organic waster. Bacteria are collected on the filter medium.

Tube settler

Device using bundles of tubes to let solids in water settle to the bottom for removal by sludge.

Turbidity

A measure of non-transparency of water due to the presence of suspended mater.

U

Ultra violet oxidation

A process using extremely short wave length that can kill micro-organisms (disinfection) or cleave organic molecules (photo oxidation) rendering them polarized or ionized and thus more easily removed from the water.

Unloading

The release of the contaminant that was captured by a filter medium.

Up-flow

The area above the water table where soil pores are not fully saturated with water.

Urban run-off

Water from city streets , domestic properties that carries pollutant into the sewer system and receiving waters.

UV

Ultra Violet. Radiation that has a wavelength shorter than light. It is often used to kill bacteria and destroy pathogen.

V

Venturi

A channel that serves the measurement of water flows.

Viruses

The smallest life forms, that are not cellular in nature. They live inside the cells of animals, plants & bacteria and cause disease. They are made up of a chromosome by a protein shell.

W

Wastewater

The spent or used water from a home, community, farm, or industry that contains dissolved or suspended matter.

Wastewater Infrastructure

The plan or network for the collection, treatment and disposal of sewage in a community.

Water monitoring

The process of constant control of a body of water by means of sampling and analyses.

Water quality

The condition of water with respect to the amount of impurities in it.

Water recycling

Using water again for the same or another process step, after a small form of purification is applied.

Watershed

A land area from which water drains to a particular water body.

Water storage pond

An impound for liquid water designed to accomplish some degree of biochemical treatment.

Water supply system

The collection, treatment , storage and distribution of water from source to consumer.

Water table

The surface of groundwater in the soil.

Weir

A spill over devices used to measure or control water flows.

Well

A deep hole with the purpose to reach underground water supplies.

Wetland

An area that is saturated by surface water or groundwater, with vegetation adapted for life under those soil.

X

Xenobiotic

Any biological substance, displaced habitat; a chemical foreign to a biological system.

Z

Zero discharge water

The principle of ''Zero discharge' is recycling off all industrial wastewater. This means that wastewater will be treated and again put in the process. Because of the water reuse , wastewater will not be released on the sewer system or surface water.

Zeta potential

An electro kinetic measurement which can be used for the control of coagulation processes.

Zone of saturation

The space in the soil below the water table in which all the pores are filled with water. The water in the zone of saturation is groundwater.

Zooplankton

Tiny aquatic animals eaten by fish.

APPENDIX 2.6

Water Quality Criteria for various usage

Water is colorless, tasteless, and odorless. It is an excellent solvent that can dissolve most minerals that come in contact with it. Therdefore, in nature, water always contains chemicals & biological impurities i.e. suspended and dissolved inorganic & organic compounds and micro organisms. These compounds may come from natural sources and leaching of waste deposits. However. Municipal and Industrial water also contribute to a wide spectrum of both organic & inorganic impurities. Inorganic compounds, in general, originate from weathering & leaching of rocks, soils, and sediments, which principally are calcium and potassium salts of bicarbonate, chloride, sulfate, nitrate & phosphate. Besides, lead, copper, arsenic, iron and manganese may also be present in trace amounts. Organic compounds originate from decaying plants & animal matters and from agricultural runoffs, which constitute natural humic material to synthetic organics used as detergents, pesticides, herbicides & solvents. These constituents and their concentration influence the quality & use of the natural water resource.

Primary water quality criteria for designated best classes (for drinking water, outdoor bathing, propagation of wildlife & fisheries, irrigation, industrial cooling) as per the Central Pollution Control Board are tabulated below. (CPCB, India)

Table 2.6T$_1$: Primary Water Quality Criteria for Designated Best Use Classes

S.N	Designated best use	Class	Water Quality Criteria
1	Drinking Water Source without conventional treatment but after disinfection	A	Total Coliform organism MPN/100 ml shall be 50 or less. pH between 6.6 and 8.5 Dissolved Oxygen 6 mg/1 or more. Biochemical Oxygen Demand 5 days 20°C, 2 mg/1 or less
2	Outdoor bathing (organized)	B	Total Coliform organism MPN/100 ml shall be 500 or less pH between 6.6 and 8.5 Dissolved Oxygen 6 mg/1 or more Biochemical Oxygen Demand 5 days 20°C, 3 mg/1 or less
3	Drinking water source after conventional treatment and disinfection	C	Total Coliform organism MPN/100 ml shall be 5000 or less pH between 6 and 9 Dissolved Oxygen 4 mg/1 or more Biochemical Oxygen Demand 5 days 20°C, 3 mg/1 or less
4	Propagation of wild life and fisheries	D	Total Coliform organism MPN/100 ml shall be 5000 or less pH between 6 and 9 Dissolved Oxygen 4 mg/1 or more Biochemical Oxygen Demand 5 days 20°C, 3 mg/1 or less

S.N	Designated best use	Class	Water Quality Criteria
5	Irrigation, industrial cooling, controlled waste disposal	E	pH between 6.5 and 8.5 Electrical Conductivity at 25°C micro mhos/ Cm Max. 2250 Sodium absorption ratio max 26 Boron max. 2 mg/1

The water quality criteria developed for raw waters used for organized community supplies is being reworked and the proposed criterion for the organized community supplied is given below in table.

Table 2.6T$_2$: General Quality Criteria for Raw water for organized Community Water Supplies (Surface water and Ground water)

A. Primary Parameters (frequency of monitoring may be daily or even continuous using even continuous using even automatic for few parameters like pH. DO and Conductivity)

S.N.	Parameters	Range/Limiting Value of Water Quality			Remarks
		High	Medium	Poor	
1	pH	6.5-8.5	6-9	6-9	To ensure prevention of corrosion in treatment plant & distribution system and interference in coagulation & chlorination.
2	Colour, Pt Scale, HZ units	< 10	< 50	< 500	Colour may not get totally removed during treatment,
3	Total Suspended solids, mg/Litre.	< 1000	< 1000	< 2000	High suspended solids may increased the cost of treatment
4	Odour dilution factor	< 3	< 10	< 20	May not be easily tackled during treatment to render water acceptable.
5	Nitrate, mg/1	< 50	< 50	< 50	High nitrate/ nitrite may cause methamoglobinemia
6	Sulphates, mg/1	< 150	< 250	< 250	May cause digestive abnormality on prolonged consumption
7	Chloride, mg/1	< 200	< 300	< 400	May cause physiological impact and unpalatable mineral taste.
8	Fluoride, mg/1	< 1	< 1.5	< 1.5	Prolonged consumption of water containing high fluorosis may cause fluorosis.
9	Surfactants, mg/1	< 0.2	< 0.2	< 0.2	May impair treatability and coagulation.
10	Phosphates mg/1	< 0.4	< 0.7	< 0.7	May interfere with coagulation.
11	DO(%Saturation)	60-110	80-120	90-140	May imply with higher chlorine demand.

S.N.	Parameters	Range/Limiting Value of Water Quality			Remarks
12	Biochemical oxygen demand, mg/1	< 3	< 5	< 7	Could cause problems in treatment, larger chlorine demands and residual taste & odour problem
13	Total Kjeldahl Nitrogen, mg/1	< 1	< 2	< 3	Same as above
14	Ammonia, mg/1	< 0.05	< 1	< 2	Same as above
15	Total Coliform, MPN/100 ml	< 500	< 5000	< 50000	The criteria would be satisfied if during a period not more than 5% samples show greater than 50000 MPN/100 ml, and not more than 20% of sample show greater than prescribed limit.
16	Faecal Coliform ,MPN/100 ml	< 200	< 2000	< 20000	The criteria would be satisfied during a period not more than 5% samples show greater than 20000 MPN/100 ml, sample show greater than prescribed limit.
17	Faecal Streptococci	200	1000	10000	Same as above

Note: There should not be any visible discharge in the upstream (up to 5 kms) of the water intake point.

- High Quality water: Raw water simple disinfection
- Medium Quality Water: Normal Conventional treatment i.e. pre-chlorination, coagulation, flocculation, settling and disinfections
- Poor Quality of Water: Intensive physical and chemical treatment i.e. chlorination, aeration, chemical precipitation , coagulation, flocculation, settling, filtration, adsorption (activated carbon), disinfections, epidemiological surveys needs to be carried out frequently to ensure that the supplied water quality is not resulting in any health problems.

B. Table 2.6T_3: Additional Parameters for periodic (say monthly/seasonal) monitoring

S. N	Parameters	Range/Limiting Value of Water Quality			Remarks
		High	Medium	Poor	
1	Dissolved iron, mg/1	< 0.3	< 1	< 1	Higher iron affects the taste of beverages and causes stains.
2	Copper, mg/1	< 1	< 5	< 1	May result in damage of liver.
3	Zinc, mg/1	< 5	< 5	< 0.05	May cause bitter stringent taste.
4	Arsenic, mg/1	< 0.01	< 0.05	< 0.05	May cause hyperkeratosis and skin cancer in human beings
5	Cadmium, mg/1	< 0.001	<0.005	< 0.005	Toxic to man

S. N	Parameters	Range/Limiting Value of Water Quality			Remarks
6	Total-Cr mg/1	< 0.05	< 0.05	< 0.05	Toxic at high doses
7	Lead, mg/1	< 0.05	< 0.05	< 0.05	Irreversible damage to the brain in children, anemia, neurological dysfunction and renal impairment.
8	Selenium, mg/1	< 0.01	< 0.01	< 0.01	Toxic symptoms similar to arsenic.
9	Mercury, mg/1	< 0.0005	< 0.0005	<0.0005	Deadly poisonous and carcinogenic.
10	Phenols, mg/1	< 0.001	< 0.001	< 0.001	Toxic and carcinogenic, may also cause major problem of taste and odour.
11	Cyanides, mg/1	< 0.05	< 0.05	< 0.05	Larger consumption may lead to abnormality
12	Polycyclic	< 0.0002	< 0.0002	< 0.002	Carcinogenic.
13	Total Pesticides, mg/1	< 0.001	0.0025	< 0.0025	Tend to bio accumulation and bio magnify in the environment, toxic.

C. Table 2.6T$_4$: Quality criteria for water of mass bathing

S.N.	Parameter	Desirable	Acceptable	Remarks
1	Total coliform, MPN/100 ml	< 500	< 5000	If MPN is noticed to be more than 500/100 ml, then regular tests should be carried out. The criteria would be satisfied if during a period not more than 5% samples show greater then 10000 MPN/100 ml and not more than 20% of samples show greater than 5000 ml.
2	Faecal Coliform, MPN/100 ml	< 100	< 1000	If MPN is noticed to be more than 100/ 100 ml, than regular tests should be carried out, The criteria would be satisfied if during a period not more than 5% samples show greater than 5000 MPN/100 ml and not more than 20% of samples show greater than 1000/100 ml.
3	Faecal streptococci, MPN/100 ml	< 100	< 1000	Same as above
4	pH	6-9	6-9	-
5	Colour	No abnormal colour		-

S.N.	Parameter	Desirable	Acceptable	Remarks
6	Mineral oil, mg/1	No film visible, <0.3	No film visible	-
7	Surface active substances, mg/1	< 0.3	-	Skin problem likely
8	Phenols, mg/1	< 0.005	-	Skin problem and odour problem
9	Transparency (Sechhi depth)	< 2m	> 0.5 m	-
10	BOD, mg/1	< 5	-	High organic matter may be associated with coliform /pathogens.
11 Dissolved oxygen (% saturation)	80-120	-	-	May be associated with coliform/ pathogens
12	Floating matter of any type	Absent	Absent	-

Note: No direct or indirect visible discharge of untreated domestic/ industrial Wastewater.

D. Table 2.6T$_5$: Water quality criteria for irrigation waters (for selected suitable soil crop combinational only)

S.N	Parameter	General	Relaxation for special planned (exceptional notified cases)	Remarks
1	Conductivity , μmohs/cm	< 2250	< 4000	The irrigation water having conductivity more than 2250 μ mhos/ cm at 20°C may reduce vegetative growth and yield of the crops. It may also increase soil salinity, which may affect its fertility.
2	Total Coliform, MPN/100 ml	< 10000	-	No limit for irrigating crops not eaten raw.
3	Faecal coliform, MPN/100 ml	< 5000	-	No limit for irrigating crops. Not eaten raw.
4	Faecal streptococci, MPN/100 ml	< 1000	-	No limit for irrigating crops. Not eaten raw.
5	pH	6-9	-	Soil characteristics are important
6	BOD, mg/1	< 100	-	Land can adsorb organic matter faster than water

S.N	Parameter	General	Relaxation for special planned (exceptional notified cases)	Remarks
7	Floating materials such as wood, plastic, rubber etc.	Absent	-	May inhibit water percolation
8	Boron	< 2	-	Boron is an essential nutrient for plant growth, however, it becomes toxic beyond 2 mg/1.
9	SAR	< 26	-	SAR beyond 26 may cause salinity and sodicity in the soil, When it exceeds the limit, method of irrigation and salt tolerance of crops should be kept in mind
10	Total heavy metals	< 0.5 mg/1	< 5 mg/1	-

Note: Significance of anions and Cations in Natural Water:

The principal constituents of ionic species and their distribution in natural waters very greatly depending on the geographical formations and soil type. Important ionic species (Cation & Anion) in all natural waters that influence water quality and represent the principal constituents, which are listed below:

Cation	Anions
Calcium (Ca^{+2})	Bicarbonate (HCO_3 -) and
Magnesium (Mg^{2+})	Carbonate (CO_3^{2-})
Sodium (Na^+)	Chloride (Cl^-)
Potassium (K^+)	Sulfate (SO_4^{2-})
Iron (Fe^{2+})	Nitrate (NO_3^-)
Manganese (Mn^{2+})	Fluoride (F)

Calcium: It is derived mostly from rocks, and maximum concentrations come form lime stone, gypsum & gypsiferrous shale. Calcium is the second major constituent, after bicarbonate, present in most natural waters, with a concentration range between 10 and 100 mg/1. Calcium is a primary constituent of water hardness & calcium level between 40 & 100 mg/1 are generally considered as hard to very hard.

Magnesium: Source of magnesium includes ferro magnesium minerals in igneous & metamorphic rocks and magnesium salts are more soluble than calcium, but they are less abundant in geological formations. At high concentration in drinking water, magnesium salts may have laxative effects. They may also cause unpleasant taste at concentrations above 500 mg/l. For irrigation purposes, magnesium is a necessary plant as well as a necessary soil conditioner. Magnesium is associated with hardness of water, and is undesirable, in several industrial processes.

Sodium: The major source of sodium in natural waters is from weathering of feldspars, evaporates, and clay. Sodium salts are very soluble and remain in solution. Typical sodium concentrations in natural waters range between 50 mg/1. Excessive sodium intake is linked to hypertension in humans. A deficiency may result in hypernatremia and muscle fatigue. The recommended USEPA limit of sodium in drinking water supply is 20 mg/1.

Potassium: Potassium is less abundant than sodium in natural waters. Its concentration rarely exceeds 10 mg/1 in natural waters. In highly cultivated areas, runoff may contribute to temporarily high concentrations as plants take up potassium and release it on decay. From the point of view of domestic water supply, potassium is of little importance and creates no adverse effects. There is presently no recommended limit in drinking water supply.

Iron: Iron is present in soils and rocks as ferric oxides (Fe_2O_3) and ferric hydroxide [Fe (OH_3)]. In natural waters, iron may be present as ferrous bicarbonate [Fe (HCO_3)], ferrous hydroxide, ferrous sulfate ($FeSO_4$) and organic (chelated) iron. The USEPA secondary water regulations limit for iron is 0.3 mg/1, for reasons of aesthetics and taste.

Manganese: Manganese is present in rocks and soils. In natural waters, it appears with iron. Common manganese compounds in natural waters are manganous bicarbonate [Mn $(HCO_3)_2$], manganous chloride ($MnCl_2$), and manganous sulfate (MnSO4). The toxicity of Mn include neuro behavioral changes. The USEPA secondary standard for Mn is 0.05 mg/1

Bicarbonate-Carbonate: Bicarbonate is the major constituent of natural water. It comes from the action of water containing carbon dioxide on limestone, marble, chalk, calcite, dolomite and other minerals containing calcium & magnesium carbonate. The carbonate-bicarbonate system in natural waters controls the pH and the natural buffer system. The typical concentration of bicarbonate in surface waters is less than 200 mg/l as HCO_3. In groundwater, the bicarbonate concentration is significantly higher.

Chloride: Chloride in natural waters is derived from chloride-rich sedimentary rock. In typical surface waters, the chloride concentration is less than 10 mg/I.

Drinking water standards have been formulated and updated time to time, Drinking water standards are, formulated by Bureau on Indian Standards (BIS) based on guidelines of central Public Health and Environmental Engineering Organization (CPHEEO), and World Health Organization (WHO).

Global total water and freshwater reserves are shown below.

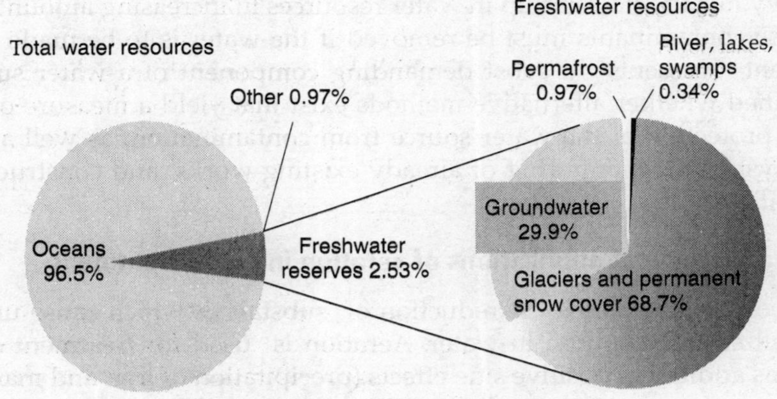

Fig.2.6F$_1$: Global total water and fresh water reserves

Chapter

Purification of Water/Water Treatment

Any physical, biological, or chemical change in water quality that adversely affects living organisms or makes water unsuitable for desired uses is considered pollution. The most serious water pollutants for human health are pathogenic organisms from human and animal wastes. We have traditionally taken advantage of the capacity of ecosystems to destroy these organisms, but as the population density has increased , these systems have become overloaded & ineffective necessitating water treatment to make it usable.

In industrialized nations, toxic chemical wastes have become an increasing problem. Agricultural & industrial chemicals have been released or spilled into surface waters and are seeping into groundwater supplies.

The water treatment processes are classified as:

- Aeration;
- Sedimentation;
- Coagulation and flocculation;
- Filtration;
- Disinfection.

Potential industrial and agricultural contaminants (chemicals such as oil, phosphates, sulphates, heavy metals, etc.) end up in water resources in increasing amounts causing water pollution. These contaminants must be removed if the water is to be made potable. Since water treatment presents the most demanding component of a water supply system, it must be examined whether alternative methods exist that yield a measure of water quality improvement: protection of the water source from contamination, as well as rehabilitation, upgrading & systematic monitoring of already existing works, and construction of efficient sanitation facilities.

Basic purpose and range of applications of aeration in water treatment

The basic purpose of aeration is the reduction of substances which cause unpleasant tastes & odors as well as discoloration in water. Aeration is used for treatment of groundwater where it also has additional positive side effects (precipitation of iron and manganese). When treating surface water aeration is useful in adding oxygen to the raw water. Aeration always precedes some other treatment process.

Aeration equipment is used to intensively mix air and water to facilitate the transfer of gases into or out of the water. The following effects can be achieved by aeration of water:

- **Addition of oxygen**; this may be necessary for surface water where the natural oxygen content was depleted due to the presence of large amounts of organic substances. Aeration contributes positively to subsequent biological treatment (e.g. slow sand filtration).
- **Removal of dissolved iron & manganese**. -These are oxidized and form nearly insoluble hydroxide sludges. They can be removed in a settling tank or by means of a coarse filter.
- **Removal of excess carbon dioxide** (CO_2) to prevent corrosion of metal and concrete surfaces.
- **Reduction of H_2S, CH_4 and other volatile compounds** which produce objectionable taste & odor.
- **Temperature reduction.**

Aeration Filter

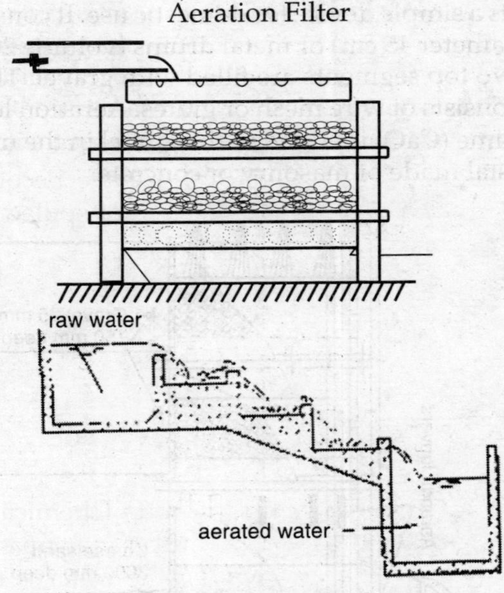

Cascade aeration

Fig.3.1: Aeration Filter and Cascade aeration

Aeration can be done in various ways. Open aeration is possible by means of spraying the water or running it over surfaces multiple tray aerators or trickling aerators consisting of a series of vertical trays with wire mesh bottoms over which water is distributed and made to fall into a collection basin at the base. The water is dispersed in fine droplets of spray which efficiently take in oxygen from the atmosphere. If the trays are filled with coarse material, such as gravel, the efficiency can be increased.

A cascade aerator is another possible aeration device. A simple cascade consists of a lateral sequence of basins (masonry, concrete or timber) at various levels, the water spilling over from one basin to the next lower one. Total height of the cascade may be between 1 and 6 meters.

The large water surface thus created allows simple and fast aeration. Baffles obstructing the flow of the water increase the effect.

If there are only small amounts of iron and manganese to be removed, or if the purpose of aeration is the addition of oxygen, it is sufficient to install a small weir just above the downstream clarifying tank so as to feed the water into that tank through a perforated pipe.

A third method of aeration which is the most efficient of all and the most expensive & complicated is based on the principle of diffusion. Water is forced into the air through fixed nozzles. Large contact surfaces for gas transfer are commonly set up above a settling tank or a filter.

Aeration of water usually requires an interruption of the gravity flow of water through a treatment plant. This means that downstream from the aeration, the water must be lifted once more. Exceptions may be possible in cases of gravity flow with significant differences in altitude (hills).

Small Aerators for Removal of Iron and Manganese

The figure below exhibits a simple device for domestic use. It consists of four vertically stacked round concrete pipes (diameter 45 cm) or metal drums (volume 200 litres) which are protected against corrosion. The two top segments are filled with gravel. The third from the top is filled with sand. The bottom consists of wire mesh or grates. Aeration louvres are placed around the device. A low pH value lime (CaO) is added to the gravel in the upper segments. The device is mounted on a low pedestal made of masonry or concrete.

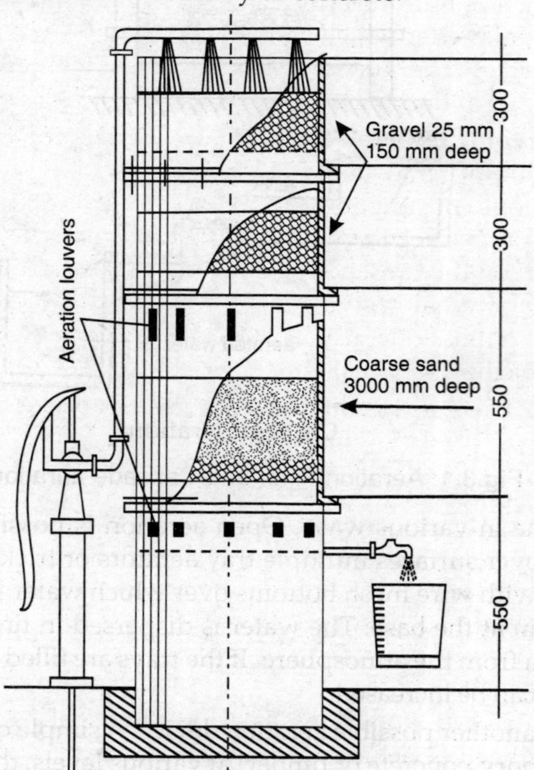

Fig.3.2: Manual device for removal of iron and manganese; capacity 200 litres/hr.

A handpump lifts the raw water, forcing it through nozzles on to the gravel. The water then trickles through layers of stones and trays. It is collected in the bottom segment and can be drawn off by means of a faucet. Particles precipitated from the water due to aeration accumulate on the lower sand layer. The latter is to be exchanged once or twice a month.

An aerator of this size is capable of treating some 200 l/h, i.e., some 1400 l/h m^2.It can be easily modified in size in accordance with the actual needs.

SEDIMENTATION

Sedimentation occurs in nature perpetually and aids the natural purification of lakes & rivers. Use is made of this physical process in the treatment of water by passing it through settling basins or storage tanks at low and uniform velocities. This is a simple means of reducing the contents of suspended matter and partially of bacteria. It can be combined by preceeding it with coagulation & flocculation, and succeeding it with slow sand filtration. Following these procedures, disinfection is required for high bacteria contents.

Areas of Application of sedimentation

Turbidity control: Under the influence of gravity, suspended matter in rain water settles out if it as it has a density greater than that of the water itself. The efficiency of a settling basin depends on the nature (shape, size, density) of the particles that are accountable for the turbidity; gravity, sand & silt, which pollute surface waters heavily and settle easily, especially during the rainy season. Colloidal matter which contributes much to turbidity is held in suspension mainly by electrostatic forces & because of its low density. Colloidal particles,when brought in contact with coagulants, form flocculent material that can be settled or filtered out.

Pathogenic Organisms removal: Simple sedimentation by passing water through a settling tank does not achieve a significant removal of pathogens. Two to four weeks storage, though, can reduce bacteria populations considerably (50- 90%) by means of biological processes. Storage in excess of one month can reduce the viral count. The degree of purification depends on the severity of pollution and on the presence of other pollutants. Storage induced contamination (mosquito breeding due to algal growth) must be avoided by covering tanks. Schistosoma larvae, infectious agents of Bilharzia, usually cannot survive more than two days of protected storage, provided suitable hosts (snails) are not present.

Color removal. Removal of color without the use of chemical procedures can only be achieved by very long storage times.

SIMPLE SETTLING BASINS

Settling basins can be operated either continuously or in batch mode. The choice of method may depend on whether water is readily available and/or must be supplied continuously. Simple methods are available for either mode.

Batch Mode

Batch operation is mainly used if only small amounts of water are to be treated and stored. A settling tank is filled with water, which is retained for between two days and several months, depending on water availability, demand and desired level of purification. When used, the water is drawn off the top, down to a depth which just covers the layer of deposits. This sludge layer at the bottom of the tank is to be removed from time to time. This can be done manually after the tank has been emptied. A tank floor sloping towards the drain greatly simplifies sludge extraction. Tanks can be constructed simply by raising earth embankments, which have to be sealed to prevent seepage. On the household level, clay vessels or other locally available jars can be used. It is important to protect receptacles from contamination: the water must not be taken out with soiled jars. Instead, an outlet spout should be provided. Layout and design of settling & storage tanks are determined by the desired retention time and the water demand of the consumers.

Continuous Mode

For larger amounts of water, it is more economical to operate a settling tank continuously. The rain water is slowly and uniformly passed through the tank either horizontally or vertically. The through flow velocity must be kept smaller than the settling velocity of the suspended matter. Horizontal flow tanks generally achieve higher rates of removal for high solids concentrations.

The most common geometric form of sedimentation tanks are circular, square or rectangular. Triangular shapes are possible. Water inlet and outlet are to be positioned such that short circuiting is prevented, and the detention time of the water is long enough to allow complete settling of particles.

Circular tanks have radial flow patterns. The water can be introduced either in the center or around the periphery. The clarified liquid is then drawn off in a trough either at the rim or in the center. Rectangular tanks have horizontal flow patterns. Inlet & outlet troughs are provided at the head and tail ends of the tank.

Ideally, the tank may be divided into four distinct zones, each of which acts characteristically different (see Figure below).

- **Inlet Zone**: In this zone, the entering water is spread out uniformly and at low turbulence over the entire cross-section of the tank.
- **Settling zone**: Portion of the tank where sedimentation occurs.
- **Outlet zone**: Slow uniform draw-off of the clarified water from the settling zone. The outward progression of the flow shall not disturb the settling process.
- **Sludge zone**: Collection of the deposits. If the sludge is to slide down by itself, the floor of the tank should be sloped at 45°. The draw off occurs at a sludge drain.

The tanks may be built above ground with sealed masonry, concrete, or reinforced concrete. Alternatively, earth basins may be used with vertical or inwardly sloping water tight walls.

Settlement basin impounded by earth embankments is shown below in figure.

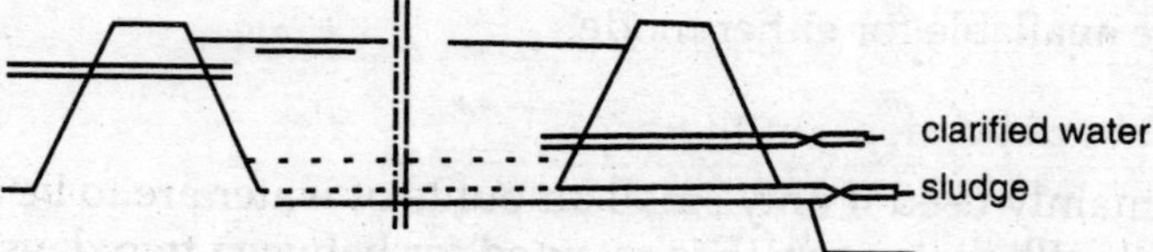

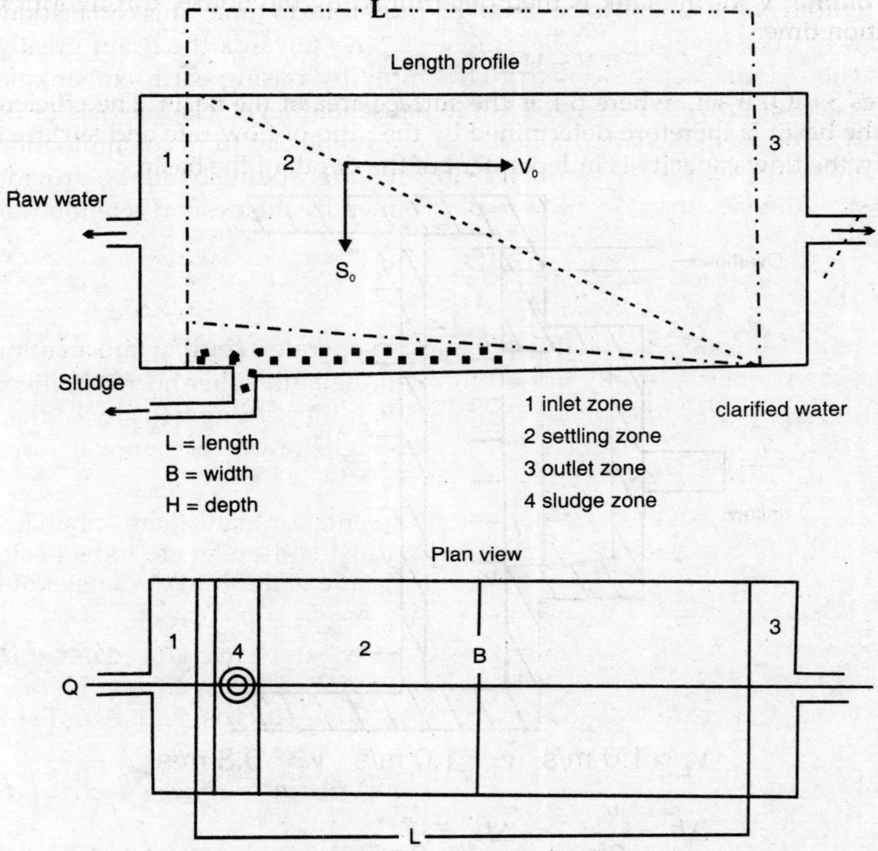

Fig.3.3: Sketch of a rectangular settling basin with horizontal flow,

Design of a Rectangular Settling Tank with Horizontal Flow

Settling tanks are designed such that the reduced flow velocity of the water allows suspended particles to settle out within the settling zone. Generally, the smaller the particles, the smaller their settling velocity(s), i.e., the lower the horizontal flow velocity of the water must be.

The necessary design parameters are determined as follows:

- Decide on the hourly throughput Q (m^3 /h).
- In a laboratory test, determine the settling velocity s, also called surface loading rate, of the suspended matter in the raw water. The settling velocity is obtained by measuring the time T (detention time) it takes a particle to drop from the surface to the bottom of the tank at depth H.

$$s = H/T$$

s and T are both dependent on the nature of the particles to be removed; s normally ranges between 0.1 and 1 m/hr; for particles with diameter = 0.01 mm, the settling velocity is approximately. s = 0.6 m/hr. If flocculation preceded settling, the aggregated particles settle at a velocity 'S' between 1 and 3 m/hr. The detention time 'T' may range between 4 to 12 hours.

- The volume 'V' of the tank is then determined by the hourly throughput Q, and the detention time:

$$V = H.\ B.\ L = Q.\ T$$

This gives S = Q/B + L, where B L is the surface area of the basin. The efficiency or flow capacity of the basin is therefore determined by the ratio of flow rate and surface area of the basin. Ideally, the flow capacity is independent of the depth of the basin.

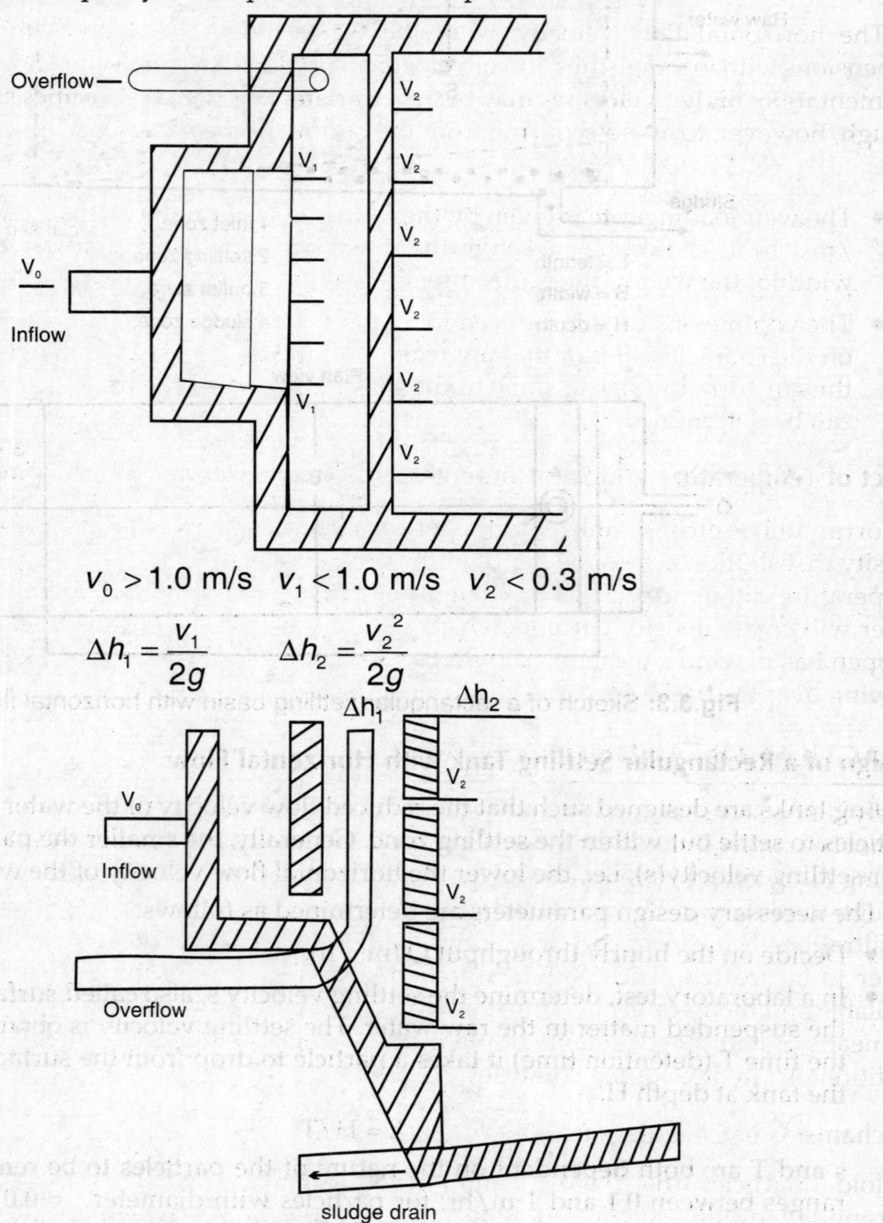

Fig.3.4: Inlet zone of a settling basin (example). The entering water first hits a baffle. It is then passed through a perforated partition wall.

- The required geometry of the tank can now be calculated. The following ranges should not be exceeded:

Depth of the tank	1:5 m < H < 2.5 m
Ratio H/L	1:5 < H/L < 1:10
Ratio B/H	1:4 < B/H < 1:8

The horizontal flow velocity, v_0 of the water ranges between 3 and 36 m/min. For suspensions with low densities, lower velocities should be chosen. When flocculation precedes sedimentation, higher velocities may be appropriate. Horizontal velocities should be kept low enough, however, to avoid scouring from the bottom of the basin.

$$v_0 = Q/B.\ H$$

- The weir loading rate is given by the flow rate Q per unit width of the weir, Q/R (m^3 /m x h). It should be chosen in the range between 3 and 10 m^2/h. An increase in the width of the weir reduces the effluent velocity.
- The volume of sludge produced in m^3 per m^2 of tank area and per unit of time depends on the characteristics of the raw water and the design, i.e., efficiency of the tank. From this, in turn, the size & slope of the settling zone and the frequency of sludge removal can be determined.

Effect of Temperature and Salt Content of the Raw Water and Wind Conditions

Unfortunately, settling tanks seldom perform in accordance with the theory. A non uniform density distribution across the depth of the tank may disturb the settling process. Even small temperature differences (1° C) or changes in the salt content(1 g/l. hour) of the entering raw water will create density currents which reduce the efficiency of the plant. When designing an open basin, wind conditions should be examined, since surface currents induced by wind blowing over the basin affect the basin performance.

COAGULATION AND FLOCCULATION

Finely dispersed , suspended and colloidal particles causing turbidity & color of the water cannot be removed sufficiently by the ordinary sedimentation process. Adding a coagulant and mixing & stirring the water causes the formation of settleable particles. These flocs are large enough to settle rapidly under the influence of gravity, and may be removed from suspension by filtration. The choice and dose rates of coagulants depend on the characteristics of the water to be treated. The coagulation chemicals like aluminium sulfate should be readily available and their application must be closely monitored.On the household level, coagulation by means of natural coagulants of plant & soil origin and simple devices has been practiced traditionally in developing countries.

Mechanisms of Coagulation

Colloidal particles (dia:10^{-4} to 10^{-6} mm) generally carry a negative electrical charge and are surrounded by an electrical double layer (due to attachment of positively charged ions from the ambient solution) and thus inhibit the close approach of each other. They remain finely divided & don't agglomerate. Due to their low specific gravity, they don't settle out. A coagulant (generally positively charged) causes compression of the double layer and thus the neutralization of the electrostatic surface potential of the particles. The resulting destabilized particles stick sufficiently together when contact is made. Rapid mixing (a few seconds) is important at this stage to obtain uniform dispersion of the chemical and to increase the opportunity for particle to particle contact. Subsequent gentle and prolonged (several minutes) mixing cements the still microscopic coagulated particles into larger flocs. These flocs then are able to aggregate with suspended polluting matter. When increased sufficiently in size and weight, the particles settle to the bottom.

Various coagulants commonly used in water treatment

Coagulants used include chemicals , materials of soil origin, materials of plant origins & natural coagulants.

Chemicals. It is common practice to use aluminium and iron salts. Both salts hydrolyze when added to water. They form insoluble material -aluminium & ferric hydroxides -when reacting with calcium and mangenese hydrogen carbonates, which are almost always present in water (alkalinity and hardness of the water). If those carbonates are not present in sufficient concentration (soft water) hydrated lime $Ca(OH)_2$ or sodium carbonate Na_2CO_3 may be added also. In the case of aluminium sulphate, these reactions can be represented as follows:

$$Al_2(SO4)_3 + 3\,Ca(HCO_3)_2 = 2\,Al(OH)_3 + 3\,CaSO_4 + 6\,CO_2\ Al_2(SO_4)_3 + 3\,Ca(OH)_2 = 2\,Al(OH)_3 + 3\,CaSO_4$$

$$Al_2(SO_4)_3 + 3\,Na_2CO_3 + 3\,H_2O = 2\,Al(OH)_3 + 3\,Na_2SO_4 + 3\ CO_2$$

The formation of the insoluble hydroxides depends on the pH: it has been shown that aluminium sulphate coagulates best in a pH range between 4.4 and 6. At higher pH values, higher rates of soluble aluminate ions form.

Sodium aluminate is generally used at medium pH values (6.5 to 8). Irons salts have the advantage of being effective over a wide range of pH values (except for values between 7 and 8.5).

Whereas turbidity is best removed within a pH range of 5.7 to 8.0, color removal is generally obtained at acid pH's of about 4.4 to 6.0. To improve the coagulation & flocculation process and to reduce the dose of coagulants, flocculation aids may be used. The most commonly used material is activated silica. Yet diatoms (kieselgur), activated carbon in powder form, bentonite & certain other types of adsorbtive clays, organic substances and cellulose derived materials are also used.

Materials of Soil Origin. Mineral substances are used as flocculation aids in modern water treatment. A dose of 10 mg/l of bentonite, for instance, together with 10 mg/l of aluminium sulphate , yield significantly better results than a higher dose of aluminum sulphate alone. In rural households in developing countries, however, various naturally occurring materials are traditionally used as coagulants: e.g., fluvial clays from rivers , clarifying rock material from desert regions, and earth from termite hills. Their main constituents are quartz, montmorillonite, kaolinite, calcite and feldspar. Their coagulating mechanisms differ greatly from those of metal salts.

Application of clay as a coagulant yields the following results:

- Reduction of turbidity;
- No effect on pH value;
- An initial mineral taste, later on normal;
- No effect on bacteria count.
- Clays contain traces of heavy metals (mostly chromium and manganese). High intakes of these metals may have toxic effects;
- Viruses survive in the settled sludge.

Coagulants of Plant Origin. Such substances are widely used in developing countries to purify water. Usually the plants are not cultivated. Rather, according to passed on experience, certain substances are gathered, prepared and added to the water that is to be purified. Some examples of traditionally used coagulants and coagulant aids are:

- Seeds from the Indian Nirmali tree (strychnos potatorum);
- Seeds of the trees of the family of the Moringaceae: Moringa Olifeira, occurring in India, Senegal, Sudan (Behenus tree) and Moringa Stenopetala, Kenya;
- Sap from the stem of the tuna cactus (opuntia ficus indica) occurring in Peru and Chile.
- The bark of the south American tree Schinopsis Quebracho-Colorado which contains tannin: it is known commercially as "Floccatan ;"
- Potato starch. For most of these plant materials, it is not known which particular substance actually triggers the coagulation. Neither is it known whether there are toxic side effects from frequent use. To obtain the optimal dose for various substances and raw water qualities, coagulation experiments must be carried out; generally this dose is smaller than that of aluminum sulphate.

For coagulation with Moringa Olifeira seeds, the following effects have been achieved.

- Significant reduction of turbidity;
- Pleasant taste;
- Unchanged pH value;
- Initial reduction of the bacteria count, followed by a secondary rise after only 24 hours, reaching or even surpassing the initial concentration;
- Antibiotic effect on various bacteria and fungi.

Nirmali seeds and Tunaflex as natural coagulants and aid substances combined with alum salts have been successfully used in municipal water treatment. Substantial savings in primary coagulants can be achieved which, reduces the overall cost considerably.

Other Natural Coagulants include

- Algae-derived substances;
- Chitosan, acting faster than any known coagulant from plant materials (produced from the shells of shrimp and lobster).
- Dough from millet bread (Sudan) or curds (thin layers).

Coagulation and flocculation processes are dependent on a multitude of variable interrelated factors: temperature turbidity, color, pH value, alkalinity, nature of coagulant & intensity and duration of stirring during mixing &flocculation.

Application procedure for Alum and Iron Salts

- In a test the required dose of the coagulant is determined. The pH value is adjusted.
- The coagulant solution is prepared. Usually the coagulant is introduced in a solution or suspension of known concentration (3-7%). Jars made of resistant material are to be used. Addition of the coagulant in solid form is also possible.
- **Constant dosing of the coagulant by means of an adequate closer.** A dosing apparatus should be used (such as those for chlorine dosing) which delivers a constant yet adjustable dose rate. A simple example is exhibited in the Figure above.

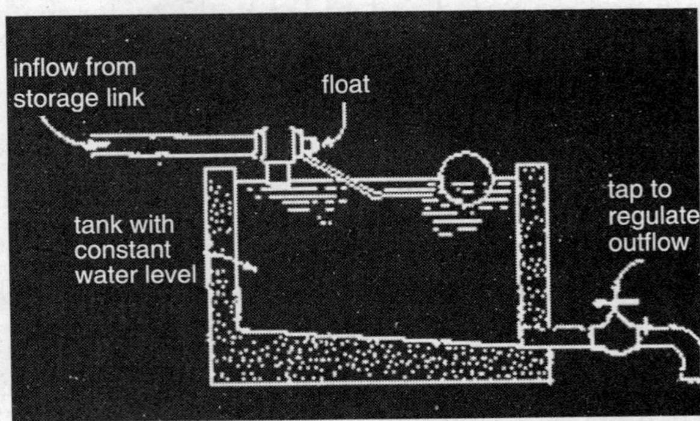

Fig.3.5: Dosing device for continuous feeding of coagulant solution.

- **Immediate rapid mixing:** Upon the addition of the coagulant, rapid mixing & dispersion must be provided for between 1 and 5 minutes. In fact, hydrolysis and polymerisation occurs almost instantly. Also, the destabilization of the colloids takes very little time. Principally, there are two practical methods:
  - **Hydraulic mixing:** channels, weirs or hydraulic jumps are used to create turbulence. At appropriate points the coagulant is introduced (Fig below);
  - **Mechanical mixing:** electrically driven mixers create a uniform dispersion. This requires a reliable supply of electricity and maintenance. Also, mechanical parts are susceptible to wear. This is why the former method is preferable.
- **Flocculation:** slow and even mixing allows the particles to collide and contact so as to form flocs (30-60 minutes). The efficiency of floc formations is contingent on the frequency of particle to particle contact.
- **Hydraulic mixing:** this can be done by routing water through a vertically or horizontally baffled flocculation basin. The resulting turbulence has a mixing effect (see Fig below). This method does not allow any adjustment or control in case of changing characteristics of the water quality.
- **Mechanical mixing:** flocculation takes place in tanks equipped with an electrically driven stirring system. This stirring system consists of screws, paddles or blades mounted on vertically or horizontally rotating shafts.

- In the sedimentation tank, the particles are allowed to settle. Or, alternatively, they are removed by filtration.

  In order to obtain optimal coagulation and flocculation performances, a number of design considerations must be followed: after finding the required dose of coagulant through experimentation, the flocculation and mixing chamber must be hydraulically designed. Approximate speed and duration of mixing, flow velocity, hydraulic profile and detention time of the particles in the tank must be determined. This procedure, however, is more suited to the design of larger scale plants. Smaller plants usually operate without these more sophisticated engineering solutions, e.g.:

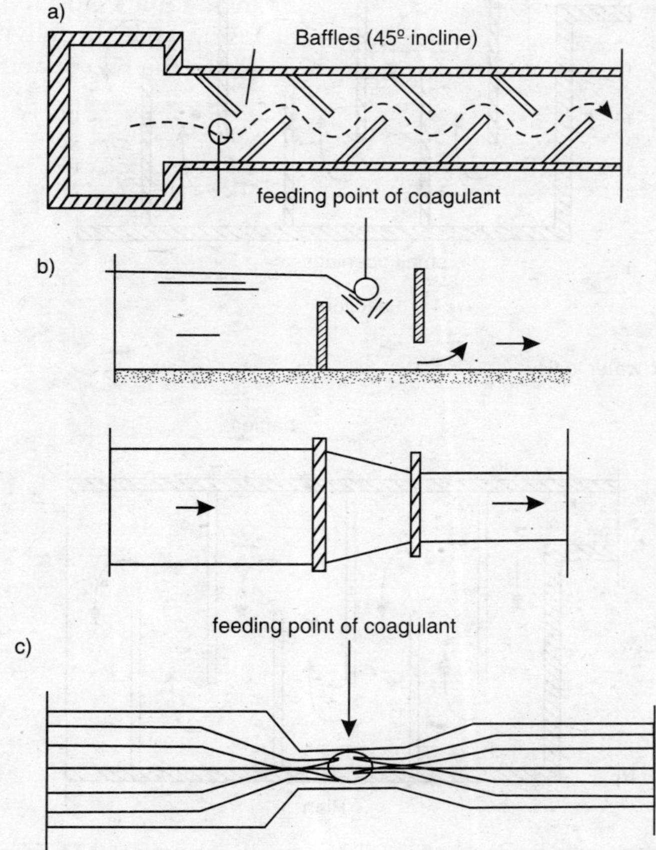

Fig.3.6: Hydraulic mixing in water flow. a) channel with baffles, b) overflow weir, c) hydraulic jump.

- The introduction of the coagulant may be made in the feeder pipe preceding a rapid filter;
- The addition of the coagulant may be made at the point of the inlet weir to the sedimentation basin.

Coagulation on the Household Level with Materials of Plant and Mineral Origin

The following are standard recipes for coagulation with locally available materials which may be modified according to the specific conditions.

The doses for the coagulant are best determined by experiments. Locally used jars, e.g., clay vessels, may be used for the purification process For mixing, wooden twirling sticks would be appropriate.

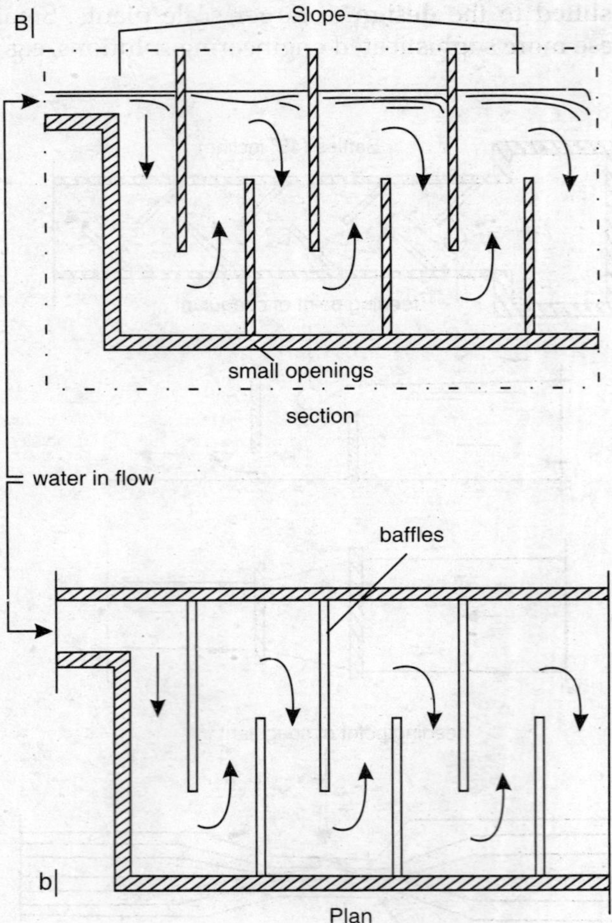

Fig.3.7: Hydraulic mixing in flocculation tank. a) vertical, b) horizontal flow.

After the flocs are settled, the supernatant water is to be transferred carefully into a clean jar. To avoid secondary pollution by unhygienic contact with a jug, the water could be scopped out with a ladle or syphoned off into a nearby vessel. The purified water should be consumed within a few hours, so as to avoid renewed contamination due to temperature induced growth of bacteria. Even better, is to boil the water prior to consumption or to disinfect it by some other method.

The settled mud on the bottom of tile jar is to be collected carefully. It should be exposed to the sun for some time (several days) to assure that potentially existing pathogens be destroyed completely.

Coagulation with fluvial clay

Dose of coagulant: 3.5 gm/l. For a 40 l capacity jar, this translates into 140 gm (1 teaspoon of the pulverized clay corresponds to 2.5 -3 gm).

- Dried clay is pounded to powder and added to water (possibly clarified) in a small bowl.
- The suspension is added to the turbid water.
- Very slow stirring of the water for about 5 min.
- Jar is covered and the water left to settle.

Coagulation with seeds of Moringa Olifeira

Dose: 150 -200 mg/l. For a jar of 40 l capacity this translates into 30 seeds.

- After removing the seed husks, the white kernel material is crushed in a clean mortar or a stone covered with a piece of clean cloth. The powder must be prepared fresh before every use. Humidity causes deterioration.
- The power is then dissolved in a small amount of clarified water and a suspension is prepared.
- The suspension is added to the raw water under short and rapid mixing (coagulating).
- Gentle and slow stirring follows (for flocculation, 10 to 15 min).
- Finally, the water is left covered in the jar to allow the floes to settle.

FILTRATION

Filtration is the deliberate passage of polluted water through a porous medium, thus utilizing the principle of natural cleansing of the soil. This widely used technique in water treatment is based on several simultaneously occurring phenomena:

- Mechanical straining of undissolved suspended particles (screening effect);
- Charge exchange, flocculation adsorption of colloidal matter (boundary layer processes);
- Bacteriological-biological processes within the filter.

Filters may be divided into two different types

- Slow sand (or biological) filtration (v = 0.1 to 0.3 m/h),
- Rapid filtration (v = 4 to 15 m/h).

  In-between types also exist. Depending on the filtration rate, different mechanisms are operative within the filter. Resulting from this is a variety of possible applications of the various types of filters.

  Generally, a filter consists of the following components:

- Filter medium (inert medium: quartz sand; or chemically activated medium: burnt material),
- Support bed (gravel) and under-drain system,
- Influent and effluent pipes, wash and drain lines, control and monitoring appurtenances.

RAPID FILTRATION

Rapid filtration is mainly based on the principle of mechanical straining of suspended matter due to the screening effect of the filter bed (sand, gravel, etc.). The particles in the water pass into the filter bed and lodge in the voids between grains of the medium. It is because of this phenomenon that rapid filters are sometimes called space filters. The cleaning of the rapid filter is facilitated by backwashing i.e., by reversing the flow direction; a backwash may be conducted simply with water or by use of a water-air mix (upward air scour). The impurities are thus dislodged and removed from the filter bed. Also operative to some degree in rapid filters are boundary layer and biological mechanisms - their extent largely depends on the filtration rate, filter medium, depth of the filter bed, and quality of the raw water.

The performance of a rapid filter regarding the removal of suspended matter is determined by the following filtration process variables and parameters:

- Filtration rate (v),
- Influent characteristics, i.e., particle size, distribution, etc.,
- Filter medium characteristics which control the removal of the particles and their release upon backwashing, respectively.

The treatment effect can be improved by:

- Reduced filtration rates,
- Smaller granulation size of the filter medium,
- Increasing depth of the filter bed, - increasing size of the flocs,
- Decreasing concentration of particles to be retained.

Table below shows the purification effects of rapid filters for various water quality.

Table 3.1: Purification effects of rapid filters for various water quality

Water Quality Parameters	Purification Effect
Coarse particles of organic origin up to 250 mg/l	Removal at high filtration rates, using coarse filter material (backwashing is simple).
High turbidity due to gravel, sand or mud.	Removal by rapid filtration, preceding sedimentation is recommended.
Low turbidity up to max. 100 NTU	Direct rapid filtration.
Colloids	Difficult to remove;
• low concentration	Addition of coagulant to inflowing water prior to sedimentation; flocs are retained by the filter; backwashing is difficult.
• high concentration	Preceding coagulation/flocculation and sedimentation in separate tank, rapid filtration
Bacteria of fecal origin, eggs of parasites	Removal of some 50 % at low filtration rate and fine material, subsequent disinfection is required.
Iron and manganese contents upto 25 mg/l	Precipitated compounds are removed upon aeration.

TYPES OF RAPID FILTERS

a. Down flow Filters

Rapid filtration is demanding and expensive in design & operation. This is due to the need for frequent filter washing which requires elaborate backwashing systems. Additional complexities associated with the generation of pressure arise for pressure filters. Monitoring, operation and maintenance of these filtration plants require well-trained personnel. Combined with coagulation, flocculation and sedimentation, rapid filtration is a very efficient treatment process for the removal of impurities. However, it should only be used in larger plants and at well equipped sites.

For smaller plants in rural areas, simple rapid filters without backwashing capabilities are recommended. A number of filter types operating at filtration rates lower than those for conventional filers are discussed hereinafter. Generally, they serve as pretreatment units to reduce the turbidity of the water. The removal of pathogens requires, in addition, either slow sand filtration and/or disinfection.

b. Up flow Filter

In up flow filters, the direction of flow of the raw water is upwards through the filter bed. Backwashing is done by abrupt reversal of the flow direction. The effect of the filter depends on the type of the filter medium, the filtration rate, and possible preceding aeration or addition of a coagulant. For coarse organic and inorganic substances, the filter may act as a simple screen. Or else it may retain precipitated iron compounds. At low filtration rates and sufficient oxygen content of the raw water, biological activity can be observed.

The advantages of up flow filters as compared with gravity rapid filters are

- Can be constructed from locally available materials,
- Quality requirements (uniformity and gradation) and volume of the filter medium are lower. Instead of sand, gravel, crushed bricks, coconut and other type fibers can be used,
- Longer filter runs,
- Better turbidity removal.

Up flow filters can be constructed at a variety of degrees of complexity. A rather simple type can be built from a 200 litres drum. It can be equipped with a raw water inlet pipe, a somewhat larger size drain at the bottom, and an outlet pipe for the clarified water near the top of the drum (see Fig below).

Filtration effect: Reduction of 50- 70% of organic & inorganic coarse and fine particles, slight reduction of bacteria.

Filter output: up to 230 l/h. Filtration rate: 0.5 to 1.5 m/h.

Filter medium: Coarse sand, grain size between 3 and 4 mm diameter. Filter bed depth: 0.3 m.

Support layer and underdrain: gravel covered by perforated metal tray.

Cleaning: Shut off of the inlet. Quick removal of drain stopper so that supernatant as well as water in the filter bed drain out together with retained particles.

Cost: for drain, sand, pipes, tap and stopper.

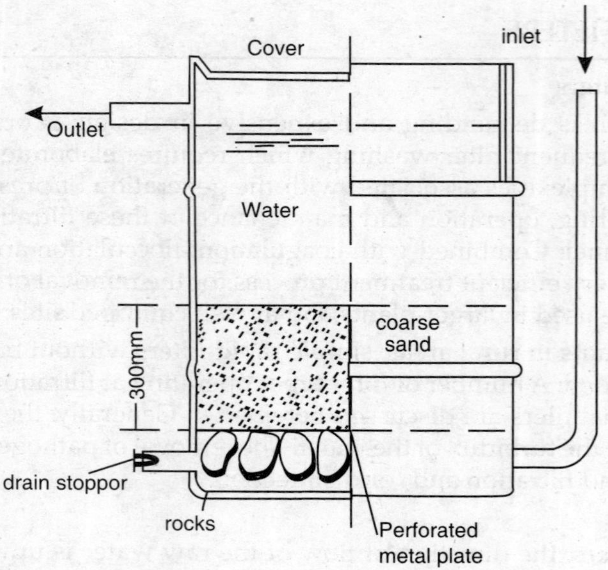

Fig.3.8: Up flow filter made from a 200 litres drum.

Cleaning of the filter (which takes ten minutes) should be done every day to prevent the filter bed from clogging. The 200 litres drum has a capacity to filter up to 230 litres/hour. As bacteria cannot be sufficiently removed, subsequent disinfection is indispensible in case of bacterial water contamination.

Higher filtration rates result in higher buoyancy forces on the filter medium. The top layer of the sand may be spewn up. This can be avoided by covering the filter bed with a metal grate or by raising the depth of the filter bed. In the latter case' though, backwashing by means of simply draining the water in a reversed direction may become increasingly impossible. Conventional backwashing capability may have to be added.

Better results may be obtained by using smaller grains and stratified filter beds with decreasing grain size from bottom to top (e.g., 0.7 to 2 mm over a depth of 1 to 1.5 m).

COARSE FILTERS

Rapid filters preceding slow sand filters are frequently used to retain coarse particles and to sufficiently reduce turbidity. Coarse sand, gravel or plant fibres are used as a filter medium. It can be replaced upon cleaning.

Such pre filtration can be done either horizontally or vertically. The filtration rates for a coarse filter are lower than those for a conventional rapid filter.

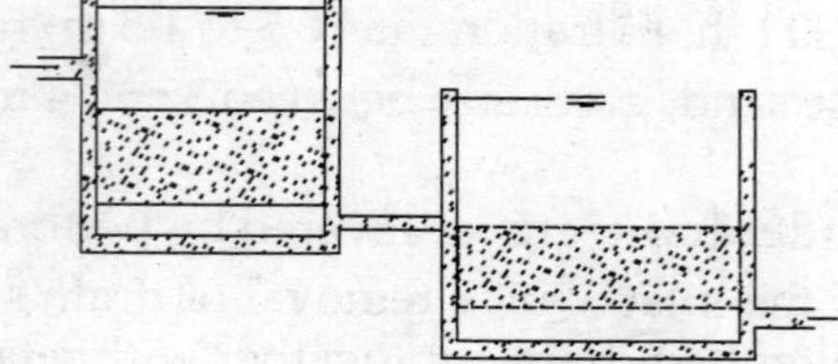

Fig.3.9: Coarse filtration followed by slow sand filtration

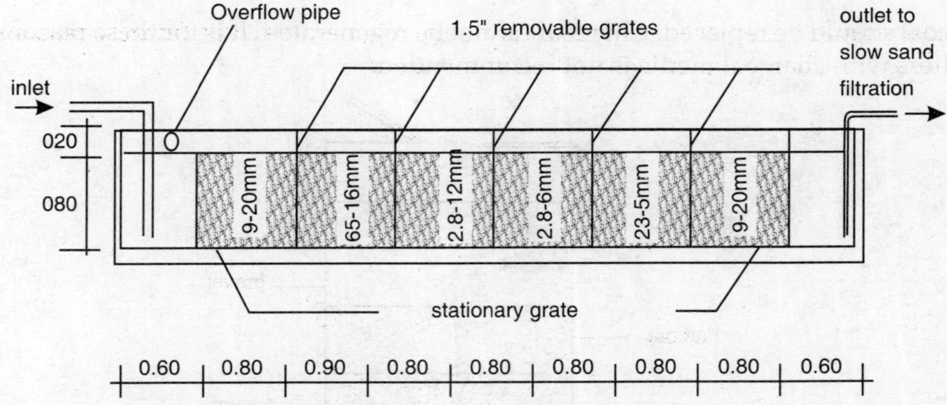

Fig.3.9: Coarse filter with horizontal flow.

Gravity rapid filter as coarse filter (Fig above)

Filtration effect: Reduction of turbidity by 50 - 80% (max. load 250 NTU)

Filtration rate: 0.5 to 1.0 m/h

Filter box: same as slow sand filter Operating head: 1-1.5m.

Filter medium: coarse sand, gravel, shredded coconut fibres.

Two or more layers of different material possible (coarser material up top and finer material below).

Filter bed depth: 0.8 to 1.4 m.

Drainage system: same as slow sand filter.

Cleaning: replace medium completely when head loss exceeds certain value, i.e., when too big (once every 3-4 months).

Household Size Rapid Filter

Household filters can be made from sand or gravel of different grain sizes, from ceramics, porcelain or other fine porosity materials. They basically operate on the principle of mechanical straining of the particles contained in the water. The filter performance depends on the porosity of the filter medium. Through additives in the filter material, additional effects can be obtained (adsorption, disinfection).

Multiple layer filter

Using metal drums, plastic containers or clay vessels and filling them with several layers of sand, gravel or charcoal, simple household filters can be put together. They do not perform well at removing pathogens, though. After filtration, the water therefore needs to be disinfected.

Charcoal adsorbs organic substances which cause disagreeable color and taste. This effect can only be sustained, however, if the charcoal is frequently renewed. If this is not possible, for whatever reason, or if the filter (empty or filled with water) is left unused for some time, the charcoal can become a breeding ground for bacteria. The result is that the filtered water exhibits a higher bacteria count than the raw water. Monitoring of the filter condition is rendered more difficult by the fact that there is no visual indication given for the point when the charcoal should be replaced. Charcoal cannot be regenerated. It is for these reasons that the use of filters with charcoal media is not recommended.

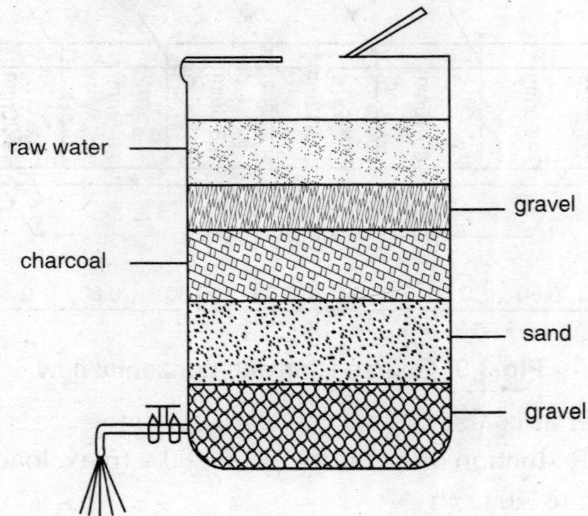

Fig.3.10: Multiple layer filter

Ceramics filter

On the household level ceramics filters may be used for the purification of drinking water. If there are native potters, the filter can be manufactured locally. Otherwise they can be readily obtained from various commercial manufacturers.

The purifying agent is a filter element, also called candle, through which the water is passed. Suspended particles are thus mechanically retained, and, depending on the size of the pores, also pathogens. Ceramics filters should only be used if the water is not too turbid, as the pores clog rather quickly.

Ceramics filter elements can be made from various different material compositions (e g., diatomaceous earth, porcelain); they have pore sizes of between 0.3 and 50 μ. If the pore size is smaller than or equal to 1.5 μ all, pathogens get removed with certainty. Post treatment of the water prior to consumption is rendered unnecessary.

Filters with larger pores only retain macro organisms such as cysts and worm eggs. The filtered water must be boiled subsequently or otherwise disinfected.

The impurities held back by the candle deposit on the candle's surface. At regular intervals, this coating can be brushed off under running water. After the cleaning, the candle should be boiled. Candles made from diatomaceous earth which contain silver, have the advantage that recontamination of purified water due to infestation of the filter material with bacteria laden washing water can be avoided.

Depending on their type, ceramics filters can be operated in the following ways:

- Gravity filter (Fig given below),
- Siphon filter (Fig given below),
- Pump filter,
- Pressure filter.

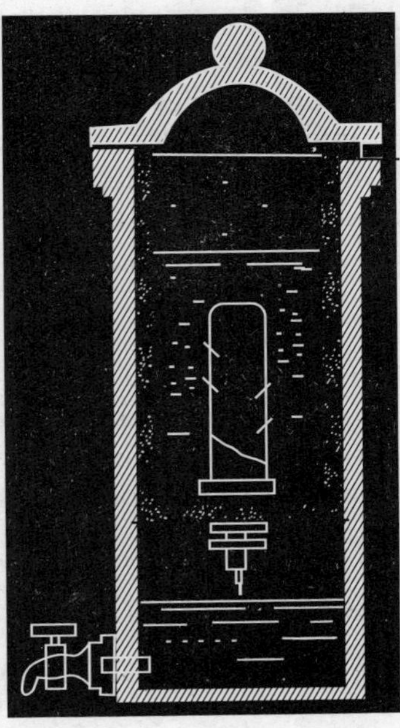

Fig.3.11: Household filter with candle (gravity filter)

The filtration rate depends on the filter material, the pore size and the nature of the particles to be retained.

Filters operating at atmospheric pressure exhibit a very slow rate of percolation. This can be increased considerably by forcing the water through the medium. Ceramics filters must be handled with care. From time to time they must be checked for fissures so as to prevent the water from passing through the medium without being filtered.

Fig.3.12: Siphon filter(Filtration is started by sucking the water by mouth into the siphon system)

Cartridge microfilter

Besides ceramics filters, other microfilters made from fine porosity materials are also available: synthetics, paper, felt like material (pore size between 25 and 50 μ). They are inserted into a bell like filter device, which is mounted on the top of a water pipe. When the filter material becomes clogged, i.e., used up, it must be discarded and replaced new. Even though these filters are cheaper to purchase than ceramics filters, their use is more expensive, since the filter material cannot be regenerated.

Slow Sand Filtration

Slow sand filtration is accomplished by passing raw water slowly - driven by gravity through a medium of fine sand. On the surface of the sand bed, a thin biological film develops after some time of ripening (different from the rapid filter). This film consists of active microorganisms and is called "Schmutzdecke", or filter skin. It is responsible for the bacteriological purification effect. The slow sand filter is therefore also called "surface filter" or biological filter.

Mechanisms of slow sand filtration

The principle purification processes taking place during slow sand filtration are:

Sedimentation

The water body sitting on top of the filter bed acts as a settling reservoir. Settleable particles sink to the sand surface.

Mechanical straining

The sand acts as a strainer. Particles too big to pass through the interstices between the sand grains are retained.

Adsorption

The suspended particles and colloids that come in contact with the surface of the sand grains by following the passage of the water are retained by:

- Adhesion to the biological layer ,
- Physical mass attraction (Van der Waals force), and
- Electrostatic and electrokinetic attractive forces (Coulomb forces).

  On account of these forces, an agglomerate of opposite charged particles forms within the top layer of sand. This process needs some time of ripening to fully develop.

  Biochemical processes in the biological layer:

- Partial oxidation and breakdown of organic substances forming water, CO_2 and inorganic salts,
- Conversion of soluble iron and manganese compounds into insoluble hydroxides which attach themselves to the grain surfaces,
- Killing of E. Coli and of pathogens.

Organic substances are deposited on the upper layer of sand, where they serve as a breeding ground & food for bacteria and other types of microorganisms (assimilation and dissimilation). These produce a slimy, sticky, gelatinous film which consists of active bacteria, their wastes & dead cells and partly assimilated organic materials. The dissimilation products are carried away by the water to greater depth. Similar processes occur there. The bacterial activity gradually decreases with depth. Different types of bacteria are normally found at various depths.

Algae can contribute to the breakdown of organic material and bacteria. They can improve the formation of the biological layer (filter skin). In uncovered filters, growth of algae is driven by photosynthesis. The presence of large amounts of algae in the supernatant reservoir of a filter generally impedes the functioning of the filter. Dead cell material may clog the filter. Increased consumption of oxygen due to the presence of dead cell material increases the possibility that anaerobic conditions will occur. There is always a diurnal variation in the oxygen content due to growth and decay of the algae mass. When algae growth is strong, the algae must be either removed regularly or the filter must be covered.

The conditions necessary for those biochemical processes are:

- Sufficient ripening of the biological layers,
- Uniform and slow flow of water through the filter, approx. 0.1 to 0.3 m/hr
- A depth of the filter bed of 1 m (0.5 m is needed solely for the biochemical process) of specific grain sizes,
- Sufficient oxygen in the raw water (at least 3 mg/l) to induce biological activity.

Purification effects of slow sand filtration for various water quality is tabulated below.

Table 3.2: Range of Application of Slow Sand Filters According to Raw Water Quality

Water Quality Parameters	Purification Effect
Bacteria	Pathogenic bacteria and E. Coli removed at 99 -99.9 %*; cysts,
helminth	eggs and Schistosoma larvae removed completely.
Viruses	Complete removal.
Organic substances	Complete removal.
Color	Partial removal
Turbidity	Significant reduction; average turbidity of raw water should not.
be greater than 10 NTU.	At higher turbidity, pretreatment necessary to prevent clogging of filter.
Substances difficult to degrade biologically e.g., detergents, phenols	Only minor degradation possible.

* At MPN-Contents Greater than 1000 E. Coli/100 ml, Raw Water Should Subsequently Be Disinfected. It is worth noting that microbiological processes and chemical activity are very sensitive to changes in temperature. Both slow down under conditions of low temperature. A reduction in filtration rate can compensate for this effect. Under prolonged cold conditions, the filter should be covered to prevent heat loss, and subsequent disinfection should be provided.

OPERATION AND MAINTENANCE OF SLOW SAND FILTER

A major advantage of slow sand filters is that operation and maintenance of a well designed & constructed filter is rather simple.

Normal operation

- **Normal through flow**: The filtration rate is controlled jointly by valves E and F. Initially, F is all but closed. It is opened gradually as the filter head loss increases so as to maintain a constant rate of filtration. The increase in bed resistance is due to a gradual accumulation of retained impurities in the interstices of the filter bed.
- **Operation at decreasing through flow**: This mode of operation' which is well suited for overnights, reduces the required number of personnel and related costs. The raw water inlet is closed, and the outlet remains open. Consequently, the head of the supernatant drops and the filtration rate decreases. The effluent weir should be fixed at such a height as to prevent the supernatant from dropping below a certain minimum depth (e.g., 0.2 m) above the filter skin. When this period is terminated, raw water should be admitted quickly.
- **Temporary shutdown**: Close both inlet and outlet valves. (The necessary quick- closing valves must be provided.) It is preferable to continue filtration and divert the effluent to waste or other use since a shutdown of the filter causes a deterioration of the quality of the biological agents (filter skin, etc.).

Filter Cleaning

- When the filtration rate starts to drop at fully opened regulating valve F, it is time to clean the filter bed.
- A, I, F valves are closed, C opened to allow the supernatant to drain off. Alternatively, the foregoing mode of operation for decreasing through flow could be chosen.
- By opening valves F and particularly D (waste valve) the water within the bed is lowered still further until it is some 0.2 m below the surface.
- The filter skin and the surface sand adhering to it (top 1.5 to 2 cm of filter) are stripped off quickly and carefully so as not to pollute or disturb the filter to a greater depth.
- Refilling the filter box follows the pattern described for initial commissioning. Only a day or two will be necessary for reripening (water analysis).

Resanding

Since for each cleaning, the top layer of the filter is removed, the depth of the filter material drops until the minimum design level is reached. This is typically about 0.6 m above the supporting gravel. The filter must then be resanded. The sand is to be washed thoroughly to remove all impurities (especially organic coating). This can be rather difficult (use of washing machine). If readily available, new sand may be better used instead. Also, the reuse of the old sand replenished by new material has its economic merits.

Modifications

The procedures and characteristics discussed represent a complete scheme necessary to achieve the best possible purification effects. There is room, however, to modify this scheme sufficiently to scale it down to the household level. Examples are:

- Substitution of sand by alternative filter material ,
- Reduction of the depth of the supernatant reservoir,
- Effluent discharge via rising pipe rather than by a weir.

Mounted on the effluent pipe is a stop cock to regulate the filtration rate and to shut off the outflow during cleaning. Too drastic a simplification of the full scale scheme may reduce the filter efficiency. It may give rise to the danger of insufficient biological effectiveness, necessary conditions for which are slow inflow and uniform through flow. A pure and clean appearance of filtered water is no assurance of sufficient bacteriological quality.

HORIZONTAL FLOW SLOW SAND FILTER

This type of filter (Figure given below) is constructed by excavation of an earth basin which is subsequently filled with sand. A biological skin develops at the surface of the sand around the inlet point. The filtration rate of the water percolating through the sand body is controlled by the filter resistance and the head differential between inflow and outflow. The retention time in such filters is between 36 hours and 30 days.

Filtration rate: 0.2 to 0.4m/hr

Filtration effect: reduction of bacteria count, turbidity, organic content

Filter basin: excavation, water tight lining (e.g., with plastic sheets); depth between 0.5 m and1 m; length 5 m; bottom slope 1: l0 to 1:20

Cleaning: When the filter starts clogging, the point of inflow is simply switched. As soon as the water has drained from the clogged inflow trough, the top sand layer is scraped off. The point of inflow can then be switched back. This technique offers the possibility of uninterrupted operation.

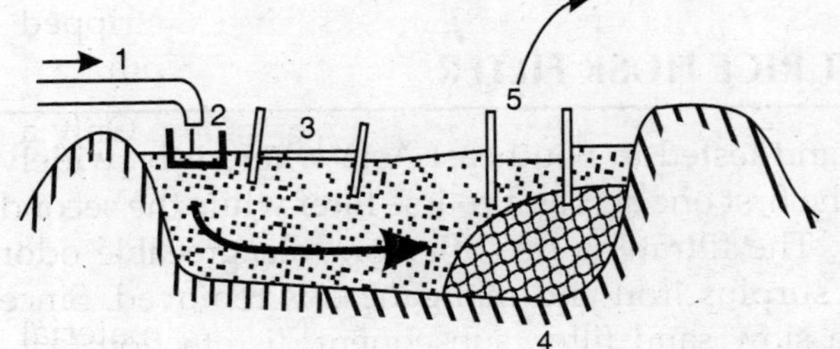

1 Inlet pipe,
2 Inlet trough to prevent scouring,
3 barriers,
4 gravel 50 mm,
5 outlet trough,
6 flow direction

Fig.3.13: Horizontal Flow Slow sand filter of household size

Household filter can be simply made from a used metal drum. A thorough cleaning and disinfection e.g., with Sodium hypochlorite (NaOCl) is necessary prior to its use as a filter casing. A drum previously filled with oil or chemicals should not be used.

Filter casing: 200 litres metal drum, 0.5 m diameter

Depth of supernatant: 0.1 to 0.3 m so as to facilitate steady flow conditions

Filter medium: sand

Filter bed depth: at least 0.6 m, better 0.75 m

Support layer and outlet: Collection of the filtered water in a gravel layer. Effluent discharge via riser pipe, which is partly perforated. The effluent pipe mounted with a stop cock rises just above the level of the filter bed so as to prevent the filter from running dry.

Filter output: 60 l/hr (as compared to up to 230 l/hr for the rapid version).

Operation: setting of the filtration rate through effluent stop cock.

Cleaning: necessary whenever filtration rate below certain specified value (at fully open valve).

In case of high turbidity, pretreating the water is recommended, by means of an up flow rapid filter.

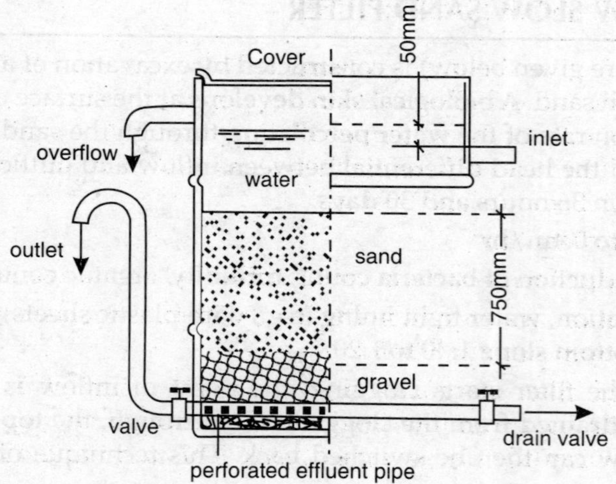

Fig.3.14: Slow sand filter in household size, capacity 60 litres/hour

TWO-STAGE COCONUT FIBER/BURNT RICE HUSK FILTER

This type of filtration plant was developed and tested in Southeast Asia where it is widely used. Two filters are operated sequentially. The first one acts as a coarse filter while the second one operates similarly to a slow sand filter. The filtrate is free of color, disagreeable odor and taste. The turbidity is greatly reduced, surplus iron and manganese is removed. Since pathogen removal is not as high as using a slow sand filter, subsequent disinfection (e.g., chlorination in the storage tank) is recommended.

The fact that the plant is mostly made from locally available materials & residues keeps the initial capital cost and the operating cost low. For filter vessels, clay jars or containers made of concrete, metal or zinc-plated sheet metal can be used. Feasible operating capacities range between 1 and 15 m^3/hr, depending mainly on the size of the system.

Coarse filter (dispersible if raw water turbidity is low)

Filter medium: shredded fibers of coconut shells (washed).

Filtration effect: Reduction of turbidity by 60 to 70%. Removal of dissolved particles; due to certain superficial phenomenon coagulation like effects are achieved by the medium. At high concentrations of colloidal particles (turbidity > 300 NTU) , the addition of a coagulant is recommended.

Depth of filter bed: 0.6 m to 0.8 m; depth of supernatant water 1 m above filter bed

Cleaning: Replacement of entire medium, when the supernatant reaches the rim of the tank (every 3 to 4 months).

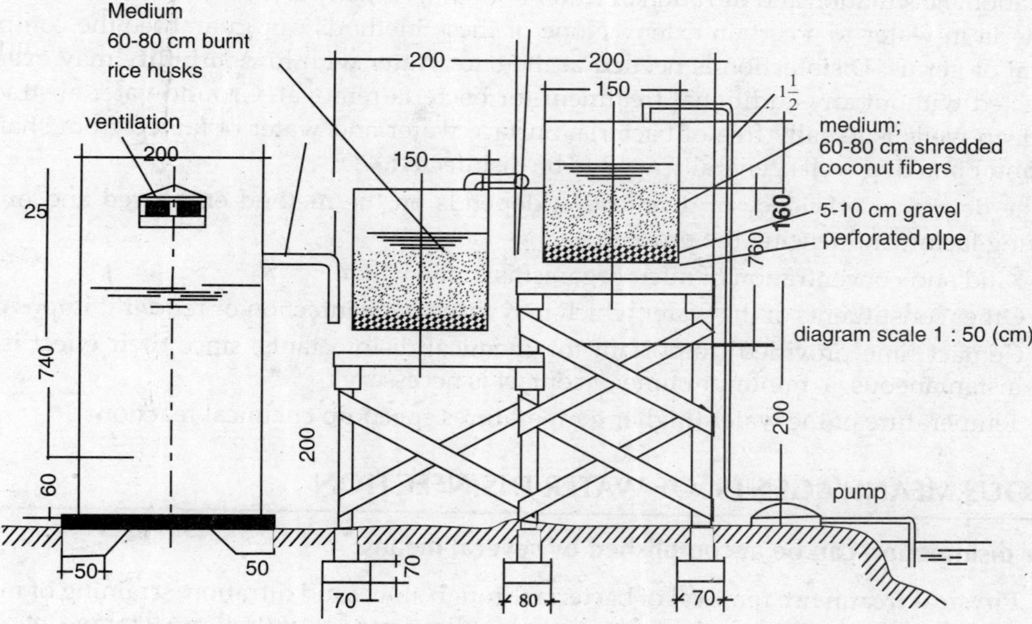

Fig.3.15: Two stage filter.

Slow filter

Filtration rate: 1.25 to 1.5 m/hr

Filter medium: burnt rice husks (washed, difference between 0.3 and 0.5 mm; UC between 2.3 and 2.6

Filtration effect: Removal of residual turbidity up to 95%, reduction of coliform bacteria by 60 -90%, removal of iron & manganese up to 90%, removal of color, odor & objectionable taste through adsorptive effect of the activated carbon of the burnt medium.

Depth of filter bed: 0.6 to 0.8 m; depth of supernatant 1 m.

Supporting layer: 0.05 to 0.1 m of gravel.

Drainage: perforated drain pipe.

Cleaning: Necessary when supernatant reaches rim of tank (approx. every 3-4 months).

After draining of the tank, a layer of 5-10 cm of the filter medium is removed from the top. A refill of the medium is called for when the depth of the filter bed has dropped to a minimum of 0.6 m.

DISINFECTION

Purpose of disinfection of water in water treatment

Water disinfection processes are designed to destroy disease producing organisms by means of disinfectants. Drinking water should be free of pathogenic organisms. Storage, sedimentation, coagulation, flocculation and filtration of water both individually & jointly reduce the contents of bacteria in water to a certain extent. None of these methods can guarantee the complete removal of germs. Disinfection is needed at the end. Water with low turbidity may even be disinfected without any additional treatment for bacteria removal. Groundwater abstracted from deep wells is usually free of bacteria. Surface water and water obtained from shallow wells and open dug wells generally need to be disinfected.

The degree or efficiency of disinfection depends on the method employed and on the following factors influencing the process:

- Kind and concentration of microorganisms in the water,
- Other constituents of the water which may impede disinfection or render it impossible,
- Contact time provided (important for chemical disinfectants, since their effect is not instantaneous, a minimum time of contact is necessary),
- Temperature of the water (higher temperatures speed up chemical reactions).

VARIOUS MEANS/AGENTS OF WATER DISINFECTION

Water disinfection can be accomplished by several means:

- **Physical treatment**: removal of bacteria through slow sand filtration, straining of macro organisms by means of micro screening , application of heat (boiling), storage, etc.
- **Irradiation**, such as UV (Ultra Violet) light,
- **Metal ions**, such as silver (and copper),
- **Chemical treatment**, use of oxidants (halogens & halogen compounds -chlorine, Iodine, bromine , ozone, potassium permanganate, hydrogen peroxide, etc.).

REQUISITES OF A GOOD CHEMICAL DISINFECTANT

A good chemical disinfectant should have the following capabilities:

- Destroy all organisms present in the water within reasonable contact time, the range of water temperature encountered, and the fluctuation in composition, concentration & condition of the water to be treated;
- Accomplish disinfection without rendering the water toxic or carcinogenic;
- Permit simple and quick measurement of strength & concentration in the water,
- Persist in residual concentration as a safeguard against recontamination;

- Allow safe & simple handling, application and monitoring;
- Ready and dependable availability at reasonable cost.

Though there are potential hazards for the human organism associated with prolonged ingestion of chemicals , the application of chlorine & its compounds for the purpose of water disinfection is the best and most tested compromise when evaluated according to the aforementioned criteria. The other methods differ significantly from each other in terms of their effect, the technological level and particularly in their applicability.

CHLORINATION AS A MEANS OF WATER DISINFECTION

Chlorination is the addition of chlorine gas or some other oxidizing chlorine compound (sodium or calcium hypochlorite, chlorinated lime, chlorine dioxide etc.) to the water to be treated. The actual disinfection agent is hypochlorous acid (HOCl) which forms when chlorine is added to water: Hypochlorous acid also forms subsequent to dissociation, when chlorinated lime or hypochlorites are added. At pH levels between 3 and 6, hypochlorous acid dissociates poorly. Chlorination is most effective in that range of pH. At pH levels greater than 8, hypochlorite ions predominate or exist almost exclusively. Hence the disinfecting effect drops off rapidly as the pH level increases. Simultaneously with the dissociation, hypochlorous acid partly breaks up, forming monatomic oxygen, which contributes to the oxidizing effect.

Chlorination is the most widely used method for drinking water disinfection. It is effective and economical. The fraction that becomes effective as an oxidizing agent when chlorine or some of its compounds is added to raw water is called "free available" or "active" chlorine. Small amounts of chlorine, due to its ability to penetrate cells of microorganisms, are sufficient to destroy many different strains of bacteria. Many types of viruses & macro-organisms like schistosoma larvae can be killed. A contact time of at least 30 minutes is required, at the end of which the residual chlorine concentration in the water must still be between 0.1 and 0.5 mg/l (= ppm). Amoebic cysts and spores with resistant cell membranes require higher doses & longer contact times.

Chlorine also reacts with many other oxidizable water constituents such as iron & manganese compounds, ammonia, and compounds thereof (forming chloramines), as well as numerous types of organic particles. The presence of these substances reduces the germicidal effect considerably. Sufficient chlorine must be added to the water to make sure that there is a residual concentration to prevent recontamination.

It is advisable to remove or reduce prior to chlorination, those substances by sedimentation and/or filtration which would impede disinfection. Through such pretreatment,helminth eggs (parasitic worms) can be removed which are insensitive to chlorination. Through chlorination, certain undesirable side effects may occur. Particularly in industrialized areas, synthetic organic compounds may enter the hydrologic cycle in high concentrations. The presence of chlorine enhances the danger of the formation of carcinogenic compounds (e.g., chloroform and other trihalomethanes).

Several chlorine compounds which have various active chlorine contents (Table below) are more easily applicable. In some form or another they are available virtually anywhere.

Table 3.3: Strength of Various Chlorine Preparations

Name	% Active Chlorine	Amount for Preparation of 1 litre of 1% Solution
Sodium Hypochlorite	14 (10-15)	71 g
Household Bleach	5 (3-5)	200 g
Javelle Water	ca. 1	1000 g
Chlorinated Lime	30 (25-37)	40 g
HTH	70 (60-70)	15 g

Merits and demerits of various chemicals used for chlorination of water for its disinfection

Chlorine gas &chlorine dioxide are the most widely used in water treatment on account of their high efficiency and ease of application. Handling & transport, however, are considered too demanding and hazardous (explosive, toxic). Several other chlorine compounds which have various active chlorine contents are more easily applicable. In some form or another they are available virtually anywhere.

Sodium hypochlorite (NaOCl), commonly known as bleach is generally available in dissolved form. Its commercial strength in terms of active chlorine is between 1 and 15%. It is stored in dark glass or plastic bottles. The solution loses some of its strength during storage. Prior to use, active chlorine content should be tested. Sunlight and high temperatures accelerate deterioration of the solution. The containers therefore should be stored in cool darkened areas. The stability of the solution decreases with increasing contents of available chlorine. A 1% solution is relatively stable. But it is not economical to store. Even though hypochlorite solutions are less hazardous than chlorine gas, precaution should be taken to avoid skin contact and to protect containers against physical damage.

Chlorinated Lime or Bleaching Powder ($CaO \cdot 3\ CaOCl_2 \cdot 3\ H_2O$) in the powder form is readily available and inexpensive. It is stored in corrosion resistant cans. When fresh, it contains 35% active chlorine. When exposed to air, it quickly loses its effectiveness. It is usually applied in solution form which is prepared by adding the powder to a small amount of water to form a soft cream. Stirring prevents lumping when more water is added. When the desired volume of the solution has been prepared, it is allowed to settle before decanting. Solutions should have concentrations between 5 and 1% of free chlorine, the latter being the most stable solution. Some 10% of the chlorine remains in the settled sludge.

High Test Hypochlorite (HTH) is a stabilized version of calcium hypochlorite $Ca(OCl)_2$ containing between 60 and 70% available chlorine. Under normal storage conditions, commercial preparations will maintain their initial strength with little loss. Even though HTH is expensive, it may be economical, due to its properties. It is available in tablet or granular form.

These chemicals must be handled with great caution. They are caustic, corrosive and sensitive to light. They should be stored in tightly closed containers and in darkened spaces. When handling the material, contacts with skin, eyes and other body tissues must be avoided. Chlorine corrodes metal and to a less extent, wood and some synthetic materials. Metal parts which come in contact with the chemicals should be resistant.

Factors for Determination of Chlorine Dose during disinfection by chlorination

Chlorine of any type must be added to water in closely controlled concentrations which depend on the characteristics of the water. As the use of dry chemicals doesn't always permit sufficient accuracy of dosing, solutions are preferred. Chlorine is usually added to the water for disinfection at the end of the treatment process. This allows the most effective treatment at the lowest level of chlorine application. Measurements of the chlorine demand & residual chlorine must be taken to assure that sufficient free chlorine is available to accomplish disinfection. Water characteristics and, hence, the chlorine demand may vary due to external influences (e.g., rainy season, etc.). It is therefore necessary to monitor the water quality from time to time, at the points of consumption in cases where the chlorine dosage is fixed. The objective of disinfection via chlorination can only be obtained if the chlorine dosage is adjusted to the changed water characteristics.

The chlorine demand of water of a given quality can be determined as follows:

One litre samples of the water are taken. Chlorine solution of a known concentration is added and mixed with the water. After 30 minutes of contact time, the residual chlorine content is measured. The difference to the amount added then yields the chlorine consumption.

Chlorine demand = chlorine consumption + desired residual chlorine

Usually 1% chlorine solutions are applied. The chlorine flow is set such that a chlorine residual level of between 0.1 and 0.3 mg/l is obtained. Higher levels are used if rapid recontamination is likely.

Colorimetric tests are employed to determine total chlorine residuals. Chemical agents are used which are oxidized by chlorine to produce a colored complex, intensity of which is proportional to the amount of chlorine present. Reading the colors and matching color standards by means of a comparator and disks, gives the amount of free, available & residual chlorine. Various simple test kits are available, using permanent glass and containing DPD reagents in liquid /compressed tablet form.

Various factors to be considered for Practical Application of chlorination for disinfection of water

Chlorination should never be performed prior to slow sand filtration (residual chlorine destroys biological agents). Sedimentation and filtration preceding chlorination enhance the disinfection effect. The lower the turbidity, the smaller the amount of chlorine necessary for effective disinfection. The chlorine solution can either be added to a batch of water (non-continuous or diffusion chlorination) or alternatively, it can be fed continuously to a constant flow of water.

Batch Chlorination. Where tanks are used for storage of drinking water, the required amount of chlorine can be added to the tank periodically. It is advantageous to alternate between two tanks. While one tank is in use, the other one is refilled and treated with chlorine. The water can be used after a minimum of 30 minutes contact time. This procedure allows uninterrupted supply.

If the water quality of a given source varies, the chlorine demand must be reevaluated from time to time. Before a tank is used the first time for storing water, it must be cleaned carefully and disinfected (application of between 50 and 100 ppm active chlorine). Once the water has been disinfected, recontamination must be carefully prevented. Tanks should be covered. A tap should be used to release the water so as to avoid scooping out the water with unclean jars and the like. If the water is not used immediately but left in the house for awhile, only well cleaned and covered jars should be used.

Chlorine Tablets: In certain situations, e.g., while travelling, chlorine tablets can be used. They are used for periodic chlorination of small batches of water.

Diffusion Chlorination. Open wells are often bacteriologically contaminated because of non hygienic methods for lifting the water, or due to careless use of the surroundings of the well.

Various devices for diffusion chlorination include

Type I which is a clay jar (12 to 15 l volume), filled nearly half-way with a mix of 1.5 kg bleach powder and 3 kg coarse sand (grain size 1.4 to 1.6 mm). It has two holes above the sand surface. The jar is covered with a plastic foil. The jar is suspended approximately 1 m below the water surface in the well. The chlorine can thus diffuse through the two holes into the well water.

Type II also consists of a clay jar (volume 7 to 10 l). It has 6 to 8 holes in the bottom. These are covered with stones on top of which a layer of gravel is placed. On top of that is put a mix of 1.5 kg bleaching powder and 3 kg of coarse sand. Stones are filled to the rim of the jar, which is then lowered into the water.For larger wells & higher rates of water use, two jars should be used which are interchangeably refilled.

In **Type III** used for small household wells, a double jar is recommended which releases less chlorine per time unit. The inner jar contains a mix of 1 kg bleaching powder and 2 kg coarse sand. The diffusion openings are provided. As these devices are not fit for large variations in water use, insufficient chlorination may occur at higher rates of water use.

Continuous Chlorination. Simple chlorine dosing instruments can be installed in piped water supply systems. Chlorine is fed to the water in proportion to the flow rate. A pipeline transmits the water from the source to the reservoir, passing through some sort of pretreatment (e.g., coagulation/flocculation and settling). Before entering the reservoir, the water is passed through a mixing chamber where a dosing apparatus introduces droplets of chlorine into the water.

A variety of types of chlorine dosing instruments are commercially available. They range from manually controlled types to fully automated ones. Usually a unit consists of a storage tank and a diaphragm pump for feeding the hypochlorite solution. The feed rate is proportional to the water flow rate and, thus depends on the consumption. The use of these devices is limited to piped water supply systems.

USE OF IODINE AS A WATER DISINFECTANT

Iodine is an excellent water disinfectant, effective against bacteria, amoeba cysts, cercerea and some viruses. It is added to the water mostly in the form of an aqueous solution @ 2 droplets per liter of water of a 2% iodine tincture. Iodine preparations are also available in tablet form.

In comparison to chlorination, the use of iodine has the following advantages:

- Effectiveness over a wider range of pH values (up to pH 10), except at very low temperatures;
- Ammonia and organic nitrogenous compounds have little effect on germicidal efficiency because they do not form substitution compounds with iodine;

- Action depends less on contact time and temperature;
- Effectiveness against more pathogenic organisms within short times;
- Use and handling is simpler.

Since operating costs are too high, the use of iodine is not expected to ever become an important widely applied disinfectant. The applicability of iodine as a disinfectant is limited due to the following disadvantages:

- Higher concentrations than chlorine (on a ppm basis) are necessary for effective action; -muddy or turbid water substantially affect germicidal action;
- Iodine is about 20 times as expensive as chlorine per unit of germicidal effectiveness;
- Taste & slight color produced by the iodine affect palatability and aesthetic quality;
- Physiological effect of prolonged use of iodine (especially in children) is suspected. Allergies were ascertained.

In view of these economic and health implications the use of iodine for disinfection is recommended only for occasional application (e.g., in case of catastrophe or while traveling). Aside from that, iodine is a highly effective and technically widely applicable disinfectant.

Ozonation (application of ozone) as a disinfectant in water treatment

Ozone (O_3) is one of the most effective water disinfectants. As a powerful oxidant, it reduces the contents of iron, manganese, & lead, and eliminates most of the objectionable taste & odor present in water. Its effectiveness does not depend on pH value, temperature or ammonia content of the water. Since ozone is relatively unstable, it is generated almost invariably at the point of use. Ozone is obtained by passing a current of dried and filtered air (or oxygen) through between two electrodes (plates or tubes) subjected to an alternating current potential difference. A portion of the oxygen is then converted into ozone.

This principle of ozone production has been used in Europe for a long time, since it has the advantage of being applicable under a wide range of conditions. It leaves no chemical residuals behind in the treated water. On the other hand, no lasting protection against recontamination is provided either. Capital costs for the instrumentation of ozone production & feeding, as well as operating costs due to the electrical energy requirements, are very high. Moreover, operation of ozonizers requires continuous and skilled monitoring.

Use of Potassium Permanganate as a disinfectant for water

Potassium permanganate ($KMnO_4$), a powerful oxidant, is rarely applied in water treatment for disinfection. It is effective against cholera bacteria, but not against other pathogenic germs. A dose of 1 to 5 ppm $KMnO_4$ is recommended for application. It creates a purple brown precipitate which coats the walls of the storage tank. It cannot be removed easily. In recent years, potassium permanganate has gained steadily in the application in pretreatment since it has proved effective at:

- Removing objectionable odor & taste by means of oxidation of organic material,& hydrogen sulfide;
- Preventing algal growth;
- Removing iron & manganese compounds by means of oxidation and subsequent separation by filtration.

Use of silver in disinfection of water

Preparations containing silver (available as liquid/powder) may be used to reduce the germ count of water. They are readily soluble in water and can be dosed easily. Effectiveness of silver can be explained by the oligo-dynamic properties of silver ions (silver nitrate or salt compounds). Even minute concentrations (0.03 to 0.04 ppm) are notably effective. The silver ions curb the growth of germs. After contact of between 30 minutes and 6 hours, depending on the level of bacteriological contamination, water of a very low germ content may be obtained. Odor and taste of the water are not affected by the application of silver. Disinfection by silver is a simple and very effective method. Its major advantage is that it provides already treated water with long lasting protection against recontamination by germs.

The effect of silver and other metals has been known to many peoples for a long time. The tradition of storing drinking water in silver vessels is still maintained in wealthier families. Although there is a tendency at present to exchange the metallic containers for plastic ones, even Indian villagers can still be seen fetching water from the well in brass or copper vessels. The metallic vessels are believed to have antiseptic qualities. Silver preparations are also used in ceramic filters. Major disadvantages of silver for the purpose of disinfection are the costs of treatment (about 200 times higher than gaseous chlorine), relatively long contact periods are required, organic substances and iron, sulphur, etc., inhibit action, thus limiting the applicability of the technique. If combined with chlorine,silver preparations are more widely applicable (direct disinfection and protection against recontamination).

Water boiling as a means of water disinfection

Boiling water is a very effective though energy consuming method to destroy pathogenic germs: bacteria, viruses, spores, cercerea & amoeba cysts, worm eggs, etc. The presence of turbidity or other impurities has little effect on germicidal effectiveness. If boiling is the only type of treatment available, it is recommended to let the water settle before, and decant it or filter it through a fine-meshed cloth so as to remove coarse impurities and suspended particles. The water is then brought to a strong boil which is maintained for at least five, preferably twenty minutes. For storing, it must not be transferred to a different vessel, but left in the former one and covered, so as to protect it from recontamination.

Boiling, together with the associated release of gases, especially CO_2, alters the taste of water. But through stirring while boiling and by letting the water sit in the partially filled vessel for a few hours afterward, the water picks up air and loses its bland taste. To improve the taste of the water, flavoring plant materials may be added during boiling. If done properly, boiling is a very effective and simple disinfection method. Since it requires a significant amount of energy, this method is only recommended in exceptional cases.

Use of Ultra Violet (UV) Radiation for disinfection of water

The germicidal effect of UV rays is well known to harness it for water disinfection. The effect of sunlight on surface water is initated in a more intense and controllable way. The most commonly used source of UV radiation is a low pressure quartz mercury vapor lamp which emits invisible light at a wavelength in the range of 200 to 300 nm (nanometer).The germicidal effect depends on the electric power of the lamp and on the time of exposure of the water to the UV radiation. It decreases with increasing distance between water and lamp. Also, many substances present even in pretreated water (e.g., small amounts of dissolved iron) absorb UV light. Other constituents (turbidity, suspended matter) inhibit or prevent the transmission of radiation. A disinfection unit is built such that the water is made to flow through a pipe in a thin film around the lamp, which is located at the pipe's center, emitting radiation. The flow rate is adjusted as required. The water must be pre-filtered. Disinfection by UV radiation is a "clean" process, since no chemical additives are used. Residual matter doesn't occur, and tastes & odors are neither produced in the water nor altered. Automatic devices are available which indicate when the lamp's output is not sufficient.

Some of disadvantages as given below, forbid its widespread use for water disinfection:

- Commercially available devices are relatively expensive,
- There is a dependence on steady power supply ,
- The lamp's powers of penetration are limited; thin water films are necessary,
- Turbidity, and impurities reduce the effectiveness notably,
- The lamps gradually lose their radiation power, accelerated by a coating of dirt. The lamp's average life is 1000 to 5000 hours,
- Disinfection occurs rather quickly and effectively (up to 99.9%), though no protection for recontamination is provided.

ION EXCHANGE METHOD OF WATER PURIFICATION

Ion exchange is a process for water purification in which ions are exchanged between a solution and an **ion exchanger**, a non-aqueous solid or gel. Typical ion exchangers are ion exchange resins, zeolite, montmorillonite, clay, and humus. Ion exchangers are either **cation exchangers** for positively charged cations or **anion exchangers** for negatively charged anions. There are also ion exchangers called mixed beds which have both anion & cation exchanging resins within them. Ion exchange is a reversible process and the ion exchanger can be *regenerated* or *loaded* by washing with an excess of the ions to be exchanged.

Ion exchange method is widely used in household and industrial water purifications to produce soft water. This is accomplished by exchanging calcium Ca^{2+} and magnesium Mg^{2+} cations against sodium Na^{+} or hydrogen H^{+} cations. Ion exchange is also applied in water softeners.Ion Exchange can be used in the food & beverage, hydrometallurgical, metals finishing, chemical & petrochemical, pharmaceutical, sugar & sweeteners, ground & potable water, nuclear, softening & industrial water, semiconducter, power, and a host of other industries.

FEATURES OF ION EXCHANGE RESINS USED IN WATER PURIFICATION

An ion exchange resin is an insoluble matrix (on support structure) normally in the form of small (1-2 mm diameter) beads, usually white or yellowish, fabricated from an organic polymer substrate on the surface of which are sites with easily trapped and released ions in a process called ion exchange. There are different types of ion exchange resin which are fabricated to selectively prefer one or several different types of ions. This resin is used extensively for water softening during water purification. They were developed as a more flexible alternative to the use of natural or artificial zeolites.

Most ion exchange resins are based on crosslinked polystyrene. The required active groups can be introduced after polymeration, or substituted monomers can be used. The crosslinking is usually achieved by adding a small proportion of divinyl benzene to styrene. Non-crosslinked polymers are used only rarely because of their tendency to change dimensions in dependence on the ions bonded. However, crosslinking somewhat decreases the capacity of the resin and prolong the time for reaching the equilibrium for the ions in solution and in the resin. Particle size also influences the resin parameters; smaller particles have larger surface, but cause larger head loss in the column. Membranes made of highly cross-linked ion exchange resins that allow passage of ions but not of water are used for electrodialysis.

There are four main types of ion exchange resins differing in their functional groups:

- Strongly acidic (sulfonic acid groups, eg. sodium polystyrene sulfonate or polyAMPS)
- Strongly basic, (trimethylammonium groups, eg. polyAPTAC)
- Weakly acidic (carboxylic acid groups)
- Weakly basic (amino groups, eg. polyethylene amine)

There are also specialised types:

- Chelating resin

DIFFERENCE BETWEEN NORMAL OSMOSIS PROCESS AND REVERSE OSMOSIS PROCESS

In the normal osmosis process, more dilution of concentrated solution occurs by the movement of molecules from the less concentrated side. But incase of reverse osmosis process, due to the application of high external pressure, the molecules of solution move from the more concentrated side to the less concentrated side through the semi-permeable membrane.

REVERSE OSMOSIS PROCESS FOR DESALINATION OF BRACKISH WATER & SEA WATER

To desalinate water, we must create a flow through a membrane, causing the water to leave the salty side of the membrane, flowing into the unsalted side. To achieve this, pressure must be created upon the water column on the salt side of the membrane; firstly, to remove the natural osmotic pressure and secondly, to create extra pressure on the water column, in order to push the water through the membrane. For the desalination of seawater, the pressure must be about 50-60 bars.

There are several different techniques that can be applied for water desalination. Examples are reverse osmosis, electro dialysis, distillation and ion exchange: Reverse osmosis is the most economic process for the desalination of brackish water and seawater. When we compare this process to the traditional thermic process of distillation, the capital investments and the energy use are much lower. Membranes occupy through a selective separation wall. Certain substances can pass through the membrane, while other substances are caught.

Membrane filtration

This can be used as an alternative for flocculation, sediment purification techniques, adsorption (sand filters & active carbon filters, ion exchangers), extraction and distillation. Membrane filtration can be divided up between micro & ultra filtration on the one hand and nano filtration

& Reverse Osmosis (RO or hyper filtration) on the other hand. When membrane filtration is used for the removal of larger particles, micro filtration and ultra filtration are applied. Because of the open character of the membranes the productivity is high while the pressure differences are low. When salts need to be removed from water, nano filtration and Reverse Osmosis are applied. Nano filtration and RO membranes do not work according to the principle of pores; separation takes place by diffusion through the membrane. The pressure that is required to perform nano filtration and Reverse Osmosis is much higher than the pressure required for micro & ultra filtration, while productivity is much lower.

USE OF ELECTRODIALYSIS IN TREATING WASTEWATER

Electrodialysis is used to transport salt ions from one solution through ion exchange membranes to another solution under the influence of an applied electric potential difference. This is done in an electrodialysis cell. The cell consists of a feed (diluate) compartment and a concentrate (brine) compartment formed by an anion exchange membrane and a cation exchange membrane placed between two electrodes. In practical electrodialysis processes, multiple electrodialysis cells are arranged into a configuration called an electrodialysis stack, with alternating anion and cation exchange membranes forming the multiple electrodialysis cells.

Electrodialysis processes are unique compared to distillation techniques and other membrane based processes (such as reverse osmosis) in that dissolved species are moved away from the feed stream rather than the reverse. Because the quantity of dissolved species in the feed stream is far less than that of the fluid, electrodialysis offers the practical advantage of much higher feed recovery in many applications.

WORKING PRINCIPLE OF ELECTRODILYSIS CELL USED FOR WATER PURIFICATION

In an electrodialysis stack, the diluate (D) feed stream, brine or concentrate (C) stream, and electrode (E) stream are allowed to flow through the appropriate cell compartments formed by the ion exchange membranes. Under the influence of an electrical potential difference, the negatively charged ions (e.g., chloride) in the diluate stream migrate toward the positively charged anode. These ions pass through the positively charged anion exchange membrane, but are prevented from further migration toward the anode by the negatively charged cation exchange membrane and therefore stay in the C stream, which becomes concentrated with the anions. The positively charged species (e.g., sodium) in the D stream migrate toward the negatively charged cathode and pass through the negatively charged cation exchange membrane. These cations also stay in the C stream, prevented from further migration toward the cathode by the positively charged anion exchange membrane. As a result of the anion and cation migration, electric current flows between the cathode and anode. Only an equal number of anion and cation charge equivalents are transferred from the D stream into the C stream and so the charge balance is maintained in each stream.

The result of the electrodialysis process is an ion concentration increase in the concentrate stream with a depletion of ions in the diluate solution feed stream.

The E stream is the electrode stream that flows past each electrode in the stack. This stream may consist of the same composition as the feed stream (eg,sodium chloride) or may be a separate solution containing a different species (e.g., sodium sulfate). Depending on stack configuration, anions & cations from the electrode stream may be transported into the C stream, or anions & cations from the D stream may be transported into the E stream. In each case, this transport is necessary to carry current across the stack and maintain electrically neutral stack solutions.

Anode and Cathode Reactions

Reactions take place at each electrode.

At the cathode,

$$2e- + 2\,H_2O \rightarrow H_2\,(g) + 2\,OH$$

While at the anode,

$$H_2O \rightarrow 2\,H+ + ½\,O_2\,(g) + 2e- \text{ or, } 2\,Cl- \rightarrow Cl_2\,(g) + 2e-$$

Small amounts of hydrogen gas are generated at the cathode and small amounts of either oxygen or chlorine gas (depending on composition of the E stream and end ion exchange membrane arrangement) at the anode. These gases subsequently dissipated as the E stream effluent from each electrode compartment is combined to maintain a neutral pH and discharged or recirculated to a separate tank.

VARIOUS PRACTICAL APPLICATIONS OF ELECTRODIALYSIS

The major application of electrodialysis is for desalination of brackish water / seawater as an alternative to RO for potable water production & seawater concentration for salt production. In normal potable water production without the requirement of high recoveries, RO is generally believed to be more cost effective when total dissolved solids (TDS) are 3,000 parts per million (ppm) or greater, while electrodialysis is more cost-effective for TDS feed concentrations less that 3,000 ppm or when high recoveries of the feed are required.

Another important application for electrodialysis is the production of pure water and ultrapure water by electro deionization (EDI). In EDI, the compartments (diluate, concentrate, or both) of the electrodialysis stack are filled with ion exchange resin. When fed with low TDS feed (e.g., feed purified by RO), the diluate can reach very low levels (e.g., 18 Megaohms). The ion exchange resins act to retain the ions, allowing these to be transported across the ion exchange membranes. The main usage of EDI systems are in electronics, pharmaceutical, and cooling tower applications.

Some practical applications of electrodialysis include:

- Large scale brackish & seawater desalination and salt production.
- Small and medium scale drinking water production (e.g., towns & villages, construction & military camps, nitrate reduction, hotels and hospitals).
- Water reuse (e.g., industrial laundry wastewater, produced water from oil/gas production, cooling tower makeup & blowdown, metals industry fluids, wash-rack water).
- Pre-demineralization (e.g ,boiler makeup and pre treatment, ultrapure water pre treatment, process water desalination, power generation, semiconductor, chemical manufacturing, food & beverage).

- Food processing.
- Agricultural water (e.g., water for greenhouses, hydroponics, irrigation, livestock).
- Glycol desalting (e.g., antifreeze or engine-coolants, capacitor electrolyte fluids, oil & gas dehydration, conditioning & processing solutions, industrial heat transfer fluids, secondary coolants from heating, venting,& air conditioning (HVAC).
- Glycerene Purification.

In application, electrodialysis systems can be operated as continuous production or batch production processes. In a continuous process, feed is passed through a sufficient number of stacks placed in series to produce the final desired product quality. In batch processes, the diluate or concentrate streams are re-circulated through the electrodialysis systems until the final product or concentrate quality is achieved. Electrodialysis is usually applied to deionization of aqueous solutions. However, desalting of sparingly conductive aqueous organic and organic solutions is also possible.

LIMITATIONS OF ELECTRODIALYSIS

Electrodialysis has inherent limitations, working best at removing low molecular weight ionic components from a feed stream. Non-charged, higher molecular weight, and less mobile ionic species will not typically be significantly removed. Also, in contrast to RO, electrodialysis becomes less economical when extremely low salt concentrations in the product are required and with sparingly conductive feeds: current density becomes limited & current utilization efficiency typically decreases as the feed salt concentration becomes lower, and with fewer ions in solution to carry current, both ion transport & energy efficiency greatly declines. Hence, comparatively large membrane areas are required to satisfy capacity requirements for low concentration (and sparingly conductive) feed solutions.

As with RO, electrodialysis systems require feed pretreatment to remove species that coat, precipitate onto, or otherwise "foul" the surface of the ion exchange membranes. This fouling decreases the efficiency of the electrodialysis system. Species of concern include calcium & magnesium hardness, suspended solids, silica, and organic compounds. Water softening can be used to remove hardness, and micron or multimedia filtration can be used to remove suspended solids. Hardness in particular is a concern since scaling can build up on the membranes. Various chemicals are also available to help prevent scaling. Also, electrodialysis reversal systems seek to minimize scaling by periodically reversing the flows of diluate & concentrate and polarity of the electrodes.

APPENDIX 3.1

Various chemicals used in treatment and control of aqueous water pollutants (√ means used for)

S.No.	Chemicals used	Various water treatment processes						
		Adsorption	Coagulation/ Flocculation	Organic control	Oxidation	Reduction	Emulsion breaking	Neutralisation
1	Activated Carbon	√	—	√	—	—	—	—
2	Alum and Aluminium Sulphate	—	√	√	—	—	√	—
3	Aluminium chlorohydrate	—	√	√	—	—	√	—
4	Ammonium Phosphate	—	—	√	—	—	√	—
5	Calcium Carbonate	√	—	—	—	—	—	√
6	Calcium hydroxide or oxide	—	√	√	—	—	√	√
7	Calcium hydroxide	—	—	—	√	—	—	—
8	Calcium chloride	—	√	√	—	—	—	—
9	Chlorine	—	√	√	√	—	—	—
10	Ferric chloride	—	√	—	—	—	√	—
11	Ferrous sulphate	—	√	—	—	—	—	—
12	Ferric sulphate	—	√	—	—	—	—	—
13	Hydrochloric acid	—	√	—	—	—	√	√
14	Polyelectrolytes	—	√	—	—	—	—	—
15	Sodium aluminate	—	√	—	—	√	—	—
16	Sodium bisulphite	—	—	—	—	√	—	—
17	Sodium corbonate	—	—	—	—	—	—	—
18	Sodium Phosphates	—	√	—	—	—	—	—

S.No.	Chemicals used	Various water treatment processes						
		Adsorption	Coagulation/ Flocculation	Organic control	Oxidation	Reduction	Emulsion breaking	Neutralisation
19	Sodium hydroxide	—	—	—	—	—	—	√
20	Sodium sulphite	—	—	—	—	√	—	—
21	Sulphur dioxide	—	—	—	—	√	—	—
22	Sulphuric acid	—	—	—	—	—	—	√

Note:

Water Pollution-Causes

- Wide spectrum of chemicals, pathogens, and physical or sensory changes such as elevated temperature and discoloration.
- Natural occurring – eg. Salts, fluoride, arsenic, Ca, Mg, Na etc.
- Artificial – disposed by humans from various sources
- Water's chemistry – physical chemistry includes acidity (change in pH), electrical conductivity, temperature, and eutrophication

Water Pollutants Types

- Detergents; Disinfection by-products
- Food processing waste, fats and grease
- Insecticides & herbicides, organohalides and other chemical compounds
- Petroleum hydrocarbons, including fuels (motor oil), & fuel combustion byproducts
- Tree and bush debris from logging operations
- Volatile organic compounds (VOCs), such as industrial solvents,
- Chlorinated solvents, dense non-aqueous phase liquids (DNAPLs), Polychlorinated biphenyl (PCBs) – Trichloroethylene, Perchlorate.
- Various chemical compounds found in personal hygiene & cosmetic products
- Acidity caused by industrial discharges (sulfur dioxide from power plants)
- Ammonia from food processing waste
- Chemical waste as industrial by-products
- Fertilizers containing nutrients-nitrates and phosphates
- Heavy metals from motor vehicles (via urban storm water runoff) and acid mine drainage
- Silt (sediment) in runoff from construction sites, logging, slash and burn practices or land clearing sites

Pollution-macroscopic/Micro pollutants

Macroscopic-Large visible items polluting the water may be termed 'floatables' in a urban storm water , or marine debris when found on the open seas like:

- Trash of garbage (e.g- paper, plastic, or food waste) discarded by people on the ground
- Dumping of rubbish, that are washed by rainfall into storm drains and eventually discharged into surface waters
- Hurdles, small ubiquitous waterborne plastic pellets
- Shipwrecks, large derelict ships

  Microscopic pollutants-micro organisms, dissolved/dispersed pollutants

Thermal pollution of water

- The rise or fall in the temperature of a natural body of water caused by human influence.
- Thermal pollution results in a change in the physical properties of water.
- A common cause of thermal pollution is the use of cooling water in power plants.
- Elevated water temperatures decrease oxygen levels (which can kill fish) and affects ecosystem.
- Urban runoff may elevate temperature in surface waters.
- Thermal pollution can also be caused by the release of very cold water from the base of reservoirs into warmer rivers.

APPENDIX 3.2

Classification of waters by concentration of coliform bacteria and treatment required to render the water of safe Quality

Classification Group No.	Maximum permissible average MPN coliform bacteria per month	Treatment required
1	Not more 10% of all 10 ml or 60% of 100 ml portions positive; not more than 1.0 coliform tactician per 100 ml	No treatment required of underground water, but a minimum of chlorination required to surface water, advised for ground water
2	Not more than 50 per 100 ml	Simple chlorination or equivalent.
3	Not more than 5000 per 100 ml and this MPN exceeding in not more than 20% of sample	Rapid sand filtration (including coagulation) or its equivalent plus continuous chlorination
4	MPN greater than 5,000 per 100 ml in more than 20% of samples ; 20,000 per 100 ml in more than 5% of the samples.	Auxiliary treatment like 30 & 90 days storage, pre settling , pre-chlorination, or equivalent plus complete filtration and chlorination.
5	MPN exceed group 4	Prolonged storage or equivalent to bring within group 1-4

(a) Physical/inorganic/organic chemicals and radioactive concentrations in the raw water and ease of removal by proposed treatment must also to be taken into consideration.

(b) Fecal coliforms not to exceed 20% of total coliform organisms. The monthly geometric mean of the MPN for group 2 may be < 100 and for 3&4 < 20000 per 100 milliliter with the indicated treatment.

Water Quality (WQ) –Tolerance & Classification

As per IS 2296-1982, the tolerance limits of parameters are specified as per classified use of water (Table below) depending on various uses of water.

Table 3.2T$_1$: Classification of inland surface water

Classification of inland surface water	Type of use
Class A	Drinking water source without conventional treatment but after disinfection
Class B	Outdoor bathing
Class C	Drinking water source with conventional treatment followed by disinfection
Class D	Fish culture and wild life propagation
Class E	Irrigation, Industrial cooling or controlled waste disposal

Table 3.2T_2: Tolerance Limits for inland Surface Water-Class A

S. No.	Characteristics	Tolerance
1	pH	6.5 to 8.5
2	Dissolved Oxygen. Mg/1	6.0
3	Bio-chemical Oxygen Demand	2.0
4	Total Coliform Organisms, MPN/100 ml, Max	50
5	Colour, Hazen units, Max	10
6	Odour	Unobjectionable
7	Taste	Agreeable taste
8	Total Dissolved Solids, Mg/1. Max	500
9	Total Hardness (as $CaCO_3$), mg/1. Max	300
10	Calcium Hardness (as $CaCO_3$), Mg/1 , Max	200
11	Magnesium (as $CaCO_3$), mg/1, Max	100
12	Copper (as Cu), mg/ 1 , Max	1.5
13	Iron (as Fe), mg/1 , Max	0.3
14	Manganese (as Mn),. Mg/1, Max	0.5
15	Chlorides (as Cl), Mg/1, Max	250
16	Sulphate (as SO_4) Mg/1, Max	400
17	Nitrates (as NO_4), Mg/1, Max	20
18	Fluorides (as F), Mg/1, Max	1.5
19	Phenolic compounds (as C_6H_5OH) , Mg/1, Max	0.002
20	Mercury (as Hg), Mg/1, Max	0.001
21	Cadmium (as Cd) ,Mg/1, Max	0.01
22	Selenium (as Se), Mg/1, Max	0.01
23	Arsenic (as As), Mg/1, Max	0.05
24	Cyanides (as CN), Mg/1, Max	0.05
25	Lead (as Pb), Mg/1, Max	0.01
26	Zinc (as Zn), Mg/1, Max	15
27	Chromium (as Cr^{6+}), Mg/1, Max	0.05
28	Anionic detergents, (as MBAS), Mg/1, Max	0.2
29	Poly-nuclear aromatic hydrocarbons (PAH)	0.2
30	Mineral oil, Mg/1, Max	0.01
31	Barium (as Ba), Mg/1, Max	1.0
32	Silver (as Ag), Mg/1, Max	0.05
33	Pesticides	Absent
34	Alpha emitters, μc/ml, Max	10^{-9}
35	Beta emitters, μc/ml, Max	10^{-8}

Table 3.2T$_3$: Tolerance Limits for Inland Surface Water- Class B

S.No.	Characteristics	Tolerance Limit
1	pH Value	6.5 to 8.5
2	Dissolved Oxygen, mg/1, Max	5.0
3	Biochemical Oxygen Demand (5 days at 20 °C)	3.0
4	Total Coliform Organisms, MPN/100 ml. Max	500
5	Fluorides (as F) , Mg/1, Max	1.5
6	Colour, Hazen units, Max	300
7	Cyanides (as CN), Mg/1, Max	0.05
8	Arsenic (as As), mg/1. Max	0.2
9	Phenolic Compounds (as C_6H_5OH) mg/I,Max	0.005
10	Chromium (as Cr^{6+}) , Mg/1, Max	1.0
11	Anionic detergents (as MBAS), mg/l, Max	1.0
12	Alpha emitters, μc/ml. Max	10^{-8}

Table 3.2T$_4$: Tolerance Limits for Inland Surface Water- Class C

S.No	Characteristics	Tolerance limit
1	pH	6.5 to 8.5
2	Dissolved Oxygen, mg/l , Minimum	4.0
3	Biochemical Oxygen Demand	3.0
4	Total coliform organisms, MPN/100 ml, Max	5000
5	Colour, Hazen, units , Max	300
6	Fluorides (as F), mg/I, Max	1.5
7	Cadmium (as Cd), mg/I, Max	0.01
8	Chlorides (as CI), mg/I, Max	600
9	Chromium (as Cr^{6+}), mg/I, Max	0.05
10	Cyanides (as CN), mg/I, Max	0.05
11	Total Dissolved Solids, mg/I, Max	1500
12	Selenium (as Se), mg/I, Max	0.05
13	Sulphates (as SO_4), mg/I, Max	400
14	Lead (as Pb), mg/I, Max	0.1
15	Copper(as Cu), mg/I, Max	1.5
16	Arsenic (as As), mg/I, Max	0.2
17	Iron (as Fe), mg/I, Max	50
18	Phenolic compounds (as C_6H_5OH), mg/I, Max	0.005
19	Zinc (as Zn), mg/I, Max	Absent

S.No	Characteristics	Tolerance limit
20	Insecticides (as F), mg/I, Max	1.90
21	Anionic detergents (as MBAS), mg/I, Max	1.0
22	Oils and grease , mg/I, Max	0.1
23	Nitrates (as NO_3), mg/I, Max	50
24	Alpha emitters (as NO_3), mg/I, Max	10^{-8}
25	Bata emitters , µc/mI, Max	10^{-8}

Table 3.2T_5: Tolerance Limits for Inland Surface Water- Class D

S.No	Characteristics	Tolerance limit
1	pH value	6.5 to 8.5
2	Dissolved Oxygen, mg/l , Min.	4.0
3	Free Ammonia (as N) mg/I, Max	1.2
4	Electrical Conductance at 25° C, µS, Max	1000
5	Free Carbon Dioxide (as CO_2) , mg/I, Max	6.0
6	Oils and Grease mg/I, Max	0.1
7	Alpha emitters µc/mI, Max	10^{-9}
8	Beta emitters µc/mI, Max	10^{-9}

Table 3.2T_6: Tolerance Limits for Inland Surface Water- Class E

S.No	Characteristics	Tolerance limit
1	pH value	6.5 to 8.5
2	Electrical Conductance at 25°C, µS,Max	2250
3	Sodium Adsorption Ratio , Max	26
4	Boron (as B), mg/I, Max	2.0
5	Total Dissolved Solids (inorganic), mg/I, Max	2100
6	Sulphates(as SO_4), mg/I, Max	1000
7	Chlorides (as CI), mg/I, Max	600
8	Sodium Parcentage , Max	60
9	Alpha emitters µc/mI, Max	10^{-9}
10	Beta emitters µc/mI, Max	10^{-8}

Water Quality (WQ) issues in India

- Water resources are over-exploited resulting in major WQ problems
- Common issues of Surface & Ground water: Pathogenic (Bacteriological) Pollution; Salinity; Toxicity (micro pollutants and other industrial pollutants)
- Surface Water: Eutrophication; Oxygen depletion;
- Ground Water:

- Fluoride
- Nitrate
- Arsenic
- Iron
- Sea water intrusion

Water Quality Degradation-Major factors

- **Domestic:** About 430 class I cities and 500 class II towns harboring population of 30 Crore generate about 3000 mld of wastewater of which only about 25% is treated.
- **Industrial:** About 60,000 polluting industries in India generate about 15000 mld of wastewater out of which nearly 60% (generated from large & medium industries) is treated.
- Non-point sources also contribute significant pollution loads mainly in rainy season. Pesticides consumption in India in surface water
- Sewage along with agricultural run-off and industrial effluents also contributes large amount of nutrients.
- A large part of the domestic sewage is not even collected.

APPENDIX 3.3

Water treatment systems used for removal of major contaminants found in domestic & industrial waste water

S.No.	Water Contaminants	Treatment systems	Treatment method
1	Suspended solids	Screening and comminution	P
		Sedimentation	P
		Floatation	P
		Filtration	P
		Coagulation/sedimentation	C/P
		Land treatment	P
2	Biodegradable	Activated sludge	B
	Organics	Trickling filters	B
		Rotating biological Contactors	B
		Aerated lagoons	B
		Oxidation ponds	B
		Intermittent sand filtration	P/B
		Land treatment	B/C/P
		Physical/Chemical	P/C
3	Pathogens	Chlorination	C
		Ozonation	C
		Land treatment	P
4	Nutrients	Suspended-growth	B
5	Nitrogen	Nitrification and denitrification	
		Fixed-Film nitrification and denitrification	B
		Ammonia stripping	C/P
		Ion exchange	C
		Breakpoint Chlorination	C
		Land treatment	B/C/P
6	Phosphorus	Metal salt coagulation/sedimentation	C/P

S.No.	Water Contaminants	Treatment systems	Treatment method
		Lime coagulation/sedimentation	C/P
		Biological/chemical phosphorus removal	B/P
		Land treatment	C/P
7	Refractory	Carbon adsorption	P
		Tertiary ozonation	C
		Land treatment system	P
8		Chemical precipitation	C
		Ion exchange	C
		Land treatment	C/P
9	Dissolved inorganics Solids	Ion exchange	C
		Reverse osmosis	P
		Electrodialysis	C

B = Biological, C = Chemical, P = Physical

Note:- For domestic water supply it is desired to make the water wholesome and pleasant to drink.The object of purification therefore is to remove (I) bacteria that cause diseases and (2) turbidity, colour, taste , odour which cause the water to be distasteful. The treatment usually comprises of:

- Sedimentation
- Chemical coagulation
- Filtration
- Chlorination

In preparing the water for industrial uses, softening occupies the most prominent place, although sedimentation, filtration, aeration etc. are frequently required as adjuncts to the softening process. Water treatment processes are described below.

Sedimentation

Sedimentation is one of the most widely used processes in the treatment of water, second perhaps to chlorination. The simplest method of removing suspended impurities is by plain sedimentation. The water is allowed to stand quiescent or move very slowly through natural or artificial basins until the suspended impurities settle to the bottom and the relatively clear water is drawn off the top. The degree of removal of suspended impurities depends upon the length of retention period, the size of the suspended particles and the temperature of the water. For certain purposes, such as condensing water where complete clarification is not necessary, sedimentation alone may suffice; but generally sedimentation is preliminary to filtration. It reduces the load on the filters, being especially necessary with very turbid waters, by preventing them from becoming so rapidly clogged with dirt that they would require frequent washing and would thereby consume excessively large amounts of wash water. Sedimentation is used in two stages: (I) sedimentation of raw water and (2) sedimentation following coagulation, flocculation or softening.

In Sedimentation basins water flows through the basin continuously in a horizontal direction, entering and leaving at the same rates. Horizontal flow tanks may be either rectangular or circular. Circular, horizontal flow tanks may have (1) centre feed with radial flow, (2) peripheral feed with radial flow, or (3) peripheral feed with spatial flow; however, in current practice only the first is in common use. Sludge may be removed from the basin by emptying the basin and flushing or scraping the sludge from it, or it may be removed intermittently.

In general, circular horizontal flow tanks are designed to admit the influent at the centre of the tank and to withdraw the effluent in a peripheral trough at the outer edge of the tank or at some predetermined distance between the centre and outer edge.

Important considerations in the design of rectangular sedimentation basins include (1) period of retention, (2) number.& capacity of basins, (3) velocity of flow, (4) depth of the basin, (5) relation of the length to the width of the basin, (6) inlet & outlet devices, (7) the cover or roof, (8) the "provisions for cleaning and (9) sludge storage capacity

Many of the impurities suspended in water have relative density greater than one and are held in suspension by virtue of the turbulence or currents maintained in the water. When these currents are retarded, the suspended matter generally settles to the bottom of the body of water. The particles tend to move down wards in water by gravity until the frictional resistance of water or the drag approaches the impelling force due to gravity. Thereafter the particles travel with a constant vertical velocity called the settling velocity.

The settling velocity of the particle depends upon (1) the horizontal flow velocity of water, (2) shape & size of the particle, (3) relative density of the particle and (4) temperature of the water.

Chemical Coagulation

Chemical coagulation consists of adding to the water small amounts of certain chemicals which form flocculent precipitates which coalesce the suspended impurities and cause them to sink rapidly. This allows the process to be carried out in a much smaller basin than if sedimentation alone is used. Sedimentation and chemical coagulation are both necessary as preliminaries to filtration. Coagulation and sedimentation remove only the suspended & colloidal impurities in the water and do not remove dissolved mineral salts which constitute. the hardness.

Coagulation consists of mixing the coagulating chemicals with the water in sufficient strength to produce the required precipitate or floc and then allowing the water to stand quietly or to flow slowly either horizontally or vertically up ward through a sedimentation tank.

Coagulation of turbid water consists of a series of operations that result in the destabilisation of the colloids leading to its removal from the suspending fluid by four sequential steps in coagulation viz. (1) dispersion of the coagulant, (2) nucleation,' (3) flocculation and (4) phase separation. As the coagulant is distributed through the liquid (dispersion), particles are formed by nucleation. At the beginning, the particles are small. They populate the initial suspension of flocs and they later grow or flocculate (flocculation) until they are large enough for efficient separation from the water. By sedimentation or filtration, or both, the flocs are eventually taken away from the water carrying with them the suspended or dissolved materials intended to be removed.

The most common coagulants in use are aluminium sulphate, (usually called alum) and ferrous sulphate, (also known as copperas) in combination with lime. Ferric chloride, ferric sulphate, sodium aluminate and chlorinated copperas are also used to a lesser extent.

Alum is the most commonly used chemical for the coagulation of water because of its excellent floc formation, its relative economy, its stability and its ease in handling. In order that alum may react to form a precipitate, it is necessary that the water should have some alkalinity, usually as calcium carbonate or bicarbonate. For waters having low natural alkalinity, it is necessary to add soda ash or lime.

The dose of coagulant required depends upon a number of factors such as turbidity, pH value and temperature of the water. It is found that greater the turbidity of raw water, more coagulant is required. It is found that alum coagulates best in the pH range 6.0 to 8.0 and the optimum pH range for iron salts lies between 8.0 to 10.0. The amount of coagulant increases with decrease in temperature.

Coagulants can be added in powdered form or in the form of a solution. When large amounts of water are being treated the chemicals may best be applied with dry feeds to avoid the large tanks and pumps required for feeding solutions. With smaller quantities of water and when very accurate control is necessary, the chemicals may be fed in the form of a solution

After dosing the chemical, it must be mixed with water. Tanks in which water and the coagulants are mixed together are called mixing basins. Mixing basins are usually of two types: (1) the baffle type basins and (2) the mechanically agitated type basins.

A mechanically operated coagulator is shown below in figure.

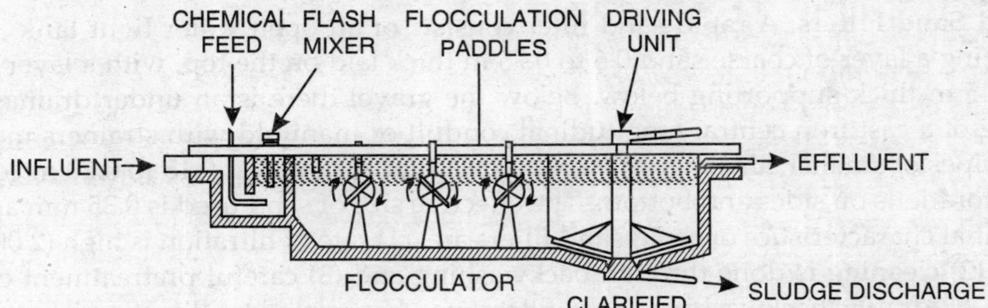

Fig.3.3F$_1$: Mechanically Operated Coagulator

The use of activated silica has spectacular effects upon floc formation as it increases the rate of coagulation in addition to widening the optimum pH range of alum.

In a flocculator, the slow stirring is mechanically brought about inside a circular tank equipped with paddles revolving on a vertical shaft.

Filtration

Filtration consists of passing water through some porous material to remove suspended and colloidal impurities. For large quantities of water, sand is generally used for the filter medium, as it is inexpensive and effective. Passing water through a bed of the proper sized sand at the correct rate and with suitable pretreatment will remove not only coarse particles of suspended matter but also material in colloidal suspension. Some of the particles removed by filtration are smaller than the interstices of the sand bed but are prevented from passing between the sand grains by the gelatinous layer of organic and colloidal matter on the top of the sand. This matter on the sand bed is deposited by the natural colloids in the water and the colloidal precipitates formed by the addition of coagulants. Water filters may be classified on the basis of;

- Rate of flow per unit of surface area - Slow filter, rapid filter
- Kind and type of filter media employed - Sand, coal (or anthracite) or diatomaceous earth
- Direction of flow through the bed - Downflow, upflow or biflow , Pressure or gravity filters.

Gravity Filters

Slow Sand Filters: Slow sand filters consists of a water tight tank 2.5 to 3.5 m in depth, having a sand bed 0.6-0.9 m thick, supported on a bed of gravel 0.30 to 0.45 m thick laid in 5 to 6 layers, beneath which the under drainage system is laid over a concrete bed sloping towards central longitudinal drain. The essential characteristics of a slow sand filter are: (1) rate of filtration is low (100 to 200 1/m2 h), (2) cleaning of sand is by scraping removal and (3) -Removal of bacteria from water. The effective size (i.e. the sieve size in mm that permits 10% of the sand by mass to pass through) of sand used is 0.25 to 0.35 mm.

Slow sand filters find their greatest usefulness in municipal plants but are being supplemented by other types which require less space and are less expensive to construct. Slow sand filters are unsuitable for waters having turbidity greater than 50 mg/1.

Rapid Sand Filters: A rapid sand filter consists of an open water tight tank 3 to 3-5 m deep, having a layer of coarse sand 0.6 to 0.75 m thick laid on the top, with a layer of graded gravel 0.45 m thick supporting below. Below the gravel there is an under drainage system consisting of a cast iron central longitudinal conduit or manifold with strainers mounted on top and pipes of smaller diameter called laternais that are fixed at 0.15 to 0.20 m centres and carry perforations on sides and bottom. The effective size of sand used is 0.35 mm and higher. The essential characteristics of rapid sand filters are: (1) rate of filtration is high (2 000 to 6 000 1/m2 h), (2) cleaning is done through backwashing and (3) careful pretreatment of water is necessary. Hydraulic backwashing of the filters may be assisted by directing jet of water into the expanded media, by stirring with mechanical rakes or injecting air into the bed before or during back washing.

Pressure Filters

Pressure filters are essentially the same as rapid filters, except that the water is filtered through the filter bed under suitable pressure. This form of filtration is used in industries, especially for the treatment of boiler feed water, since the filter may be installed on pressure lines, thereby avoiding repumping. Such units are built as vertical or horizontal units. The former is adaptable primarily for filtering relatively smaller amount of water and the latter for greater volumes. Ordinarily, the condition of filter media in a pressure filter cannot be checked visually on a routine basis. It is possible for the filter media in a pressure filter operating at high pressure to be violently disturbed by sudden loss of pressure on the effluent side of unit. Pressure filtration is often preceded by coagulation and sedimentation; if not, the coagulants are introduced to the unfiltered water pipe ahead of the filters.

Vertical pressure filters range in diameter from 0.5 to 25 m and height from 2 to 2.5 m. The horizontal units are usually 2.5 m diameter and any desired length upto 7.5 m. The pressure filters occupy less space than gravity filters of the same capacity.

For best filtration results, it is advantageous to filter water first through coarse grains to improve water quality and to use the entire depth of the filter bed.. In up-flow filtration, the water is passed from bottom to top. As the water percolates upwards, it is progressively clarified till it emerges in a fully filtered condition at the top of the bed. During backwash, compressed air scouring is desirable to dislodge the arrested suspension

Up-flow filters have the disadvantage of expanding as head loss increases, with resulting loss of efficiency and safety. Expansion is overcome in biflow filter by placing the filter outlet at about one fourth bed depth and filtering simultaneously downwards through the short depth of fine sand and upwards through the greater depth of coarse sand.

Radial flow filters are usually operated in batch, because the filtration is to be stopped for back washing.

A radial flow filter is shown below in figure.

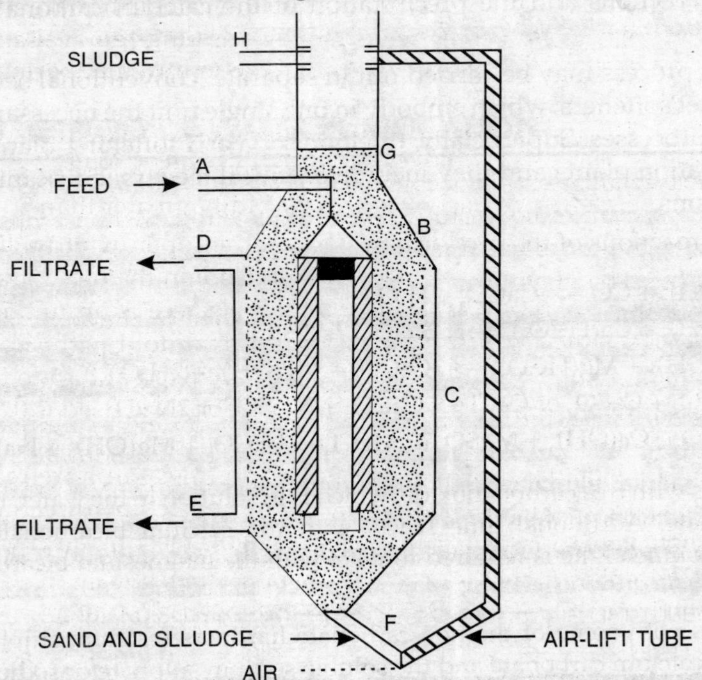

Fig.3.3F$_2$: Radial Flow Filter

Softening of Water

In preparing water for industrial uses, softening occupies the most prominent place. Many industries such as brewing, distilleries, rayon, paper and ice need soft water. In canning, hard water causes toughening of some vegetables, notably beans & peas and in textile finishing, the precipitates formed by the lime in water combining with the soap leave spots on the fabrics or react unfavourably with dyes. Hard water deposits scale in pipe lines & boilers and is the cause of vast economic waste.

The principal compounds that give the hardness to water are the salts of calcium magnesium and more particularly, the bicarbonates & sulphates of these metals. The purpose of softening is to reduce its contents of calcium / magnesium so as to remove most of its soap consuming properties and to minimise its scale forming tendencies in hot water & cooling system. The salts that compose the hardness are present in solution and not in suspension.which makes it impossible to remove them by ordinary filtration, and therefore a chemical process is required to eliminate them. The oldest used processes is the lime-soda process. It depends on the low solubility of calcium carbonate and magnesium hydroxide.

Lime Soda Process

If the proper quantities of lime and soda ash are added to a given quantity of hard water in a reaction tank and if the water is thoroughly mixed and allowed to settle, the calcium/ magnesium in the water will form calcium carbonate / magnesium hydroxide, which will precipitate and settle to the bottom of the tank, allowing the softened water to be decanted off the top. In addition to lime and soda, a coagulant such as alum or sodium aluminate is added to accelerate the reactions and the precipitation of the calcium carbonate and magnesium hydroxide.

The softening process may be carried out in separate, conventional sedimentation basins or in sludge blanket softeners, which embody in one single unit the necessary mixing, floccula-ting and settling processes. Superficially, the lime soda ash softening plants are quite similar to rapid sand filtration plants and they include chemical feeders, quick mix, flocculation and sedimentation basins.

The principal reactions of the process are:

$$Ca(OH)_2 + CO_2 \text{->} CaCO_3 + H_2O$$

$$Ca(OH)_2 + Ca(HCO_3)2 \rightarrow \text{à.}\ 2CaCO_3 + 2H_2O$$

$$2Ca(OH)_2 + Mg(HCO_3)_2 \text{->} 2CaCO_3 + Mg(OH)_2 + 2H_2O$$

$$Na_2CO_3 + CaSO_4 \text{ or } CaCl_2 \text{->} CaCO_3 + Na_2SO_4 \text{ or } 2NaCl$$

$$Na_2CO_3 + Ca(OH)_2 + MgSO_4 \text{ or } MgCl_2 \text{->} CaCO_3 + Mg(OH)_2 + Na_2SO_4 \text{ or } 2NaCl$$

Lime first reacts with free carbon dioxide, which is in solution in the water, to form a precipitate of calcium carbonate. Additional lime reacts with the calcium bicarbonate to form calcium carbonate. Twice as much lime is required to react with the magnesium bicarbonate compared to calcium bicarbonate.

Soda ash reacts with some of the non-carbonate hardness (calcium sulphate or chloride) to form the insoluble calcium carbonate and the soluble sodium sulphate or sodium chloride.

It is apparent that waters higher in carbonate than in non-carbonate hardness will require relatively more lime than soda ash for their treatment and conversely, waters with a high non-carbonate hardness as compared to the carbonate hardness, will require more soda ash and less lime.

Complete elimination of calcium magnesium cannot be effected by lime soda softening, because calcium carbonate and magnesium hydroxide are slightly soluble. Theoretically, the hardness of the water can be reduced by this process to about 25 mg/1 (as $CaCO_3$), but in actual practice it will range upto 50 to 60 mg/1 or higher.

Cold Lime Soda Softeners

There are four basic types of cold lime soda softeners namely (1) the sludge blanket type, (2) the conventional type, (3) the catalyst or spirator type and (4) the intermittent type. The first three are continuous in operation. Of these, sludge blanket type, owing to its higher efficiency, shorter detention period and smaller space requirements, is rapidly displacing the conventional type. The catalyst or spirator type is of interest as it gives a granular sludge which can be handled easily. The intermittent type because of its larger space requirements and greater amount of labour required to operate it, has practically gone out of existence.

The sludge blanket type of softeners differ from the older conventional type in that the treated water is filtered upwardly through a suspended sludge blanket composed of previously formed precipitates. In a single unit, all the processes such as mixing, softening and clarification are carried out. in the conventional type of equipment. some of the added lime suspension is carried down in the sludge formed by the precipitates, before it has time to dissolve and react with the hardness constituents of the raw water and therefore some of the lime is wasted. In the sludge blanket type, this does not occur because the upward filtration through the suspended sludge blanket results in the complete use of the added lime. Secondly, it is a common occurrence with the conventional type of equipment to have 'after-precipitates' or after-deposits form on the granules or the filter medium in the filters and in pipe lines or distribution systems carrying the filtered effluents. This has led to the practice of recarbonation with carbon dioxide to avoid formation of these after deposits. In the sludge blanket type of equipment, the intimate contact of the treated water with a large mass of solid phase largely prevents super saturation or the formation of after-deposits. Absence of after-precipitates in the sludge blanket type of equipment results in the production of an effluent which is clear enough (turbidity usually under 10 mg/l)for many industrial applications, so that subsequent filtration is often unnecessary. With sludge blanket type, detention required is one hour as against four hours in the conventional units. Silica is removed better in sludge blanket units.

The spirator consists of a conical tank which is about two-thirds filled with finely divided, granular catalyst. The tank may be a closed one to operate under pressure, or it may be of the open type for gravity operation. In either case, the raw water and chemicals enter tangentially near the bottom of the cone and spiral upwards through the suspended catalyst bed.

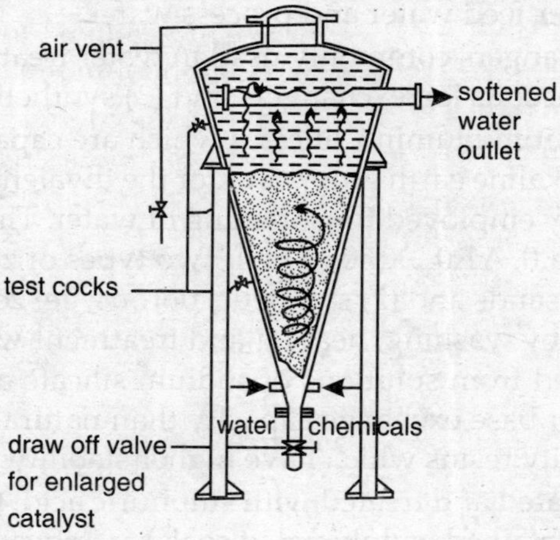

Fig.3.3F$_4$: Spirator for Rapid Lime Soda Softening

In this equipment, catalyst may be any finely granular (0.3 to 0.6 mm diameter) insoluble mineral substance. Crushed and graded calcite was originally employed, but sand or green sand fines are now often used in place of the calcite.

The retention time is only 8 to 12 minutes. Since the byproduct is a granular material which drains rapidly to less than 5 percent moisture, its disposal is easy.

The calcium carbonate formed by the reaction deposits on the catalyst grains in an adherent form so that these granules grow in size.'

Hot Lime Soda Process

The hot lime soda process is used principally for boiler feed water treatment and makes use of the same principle that is used in the cold continuous treatment, viz. the application of lime and soda ash dosage to water passing slowly through a reaction tank. The precipitates formed in the water settle to the bottom of the tank and the soft water is drawn off and filtered to remove residual hardness & other residual suspended matter. Various means are used, as with the cold lime soda softener, to proportion the flow of chemical solution to the flow of water.

The outstanding difference between the two processes is in the temperature of the water being treated, which in the cold process is usually the temperature at which it comes from the city mains or the lake or well, whereas in the hot process the raw water is heated to 94 to 100 °C before entering the reaction tank, utilising exhaust steam where possible, supplemented with live steam if necessary. In this temperature range, the reactions take place quickly, and the decreased viscosity of the water permits rapid settling of the precipitates. A greater degree of softening is accomplished than in the cold process, the effluent in a carefully controlled plant having a hardness of 17 to 34 mg/1 (as $CaCO_3$). A smaller excess of chemicals is needed than with the cold treatment and no coagulant is required.

Ion Exchange

Ion exchange is the displacement of one ion by another. As applied to water treatment, it may be defined as a reversible exchange of ions between a liquid and a solid. This exchange does not involve any radical change in the physical structure of the solid. Both cations and anions may be exchanged. Ion exchanger, must have ions of its own to exchange for others. Cation or base exchange is used for softening water. In industry, both cation and anion exchangers find use in the preparation of boiler feed water and process water.

Ion Exchangers: Ion exchangers commonly used in water treatment include (1) natural & synthetic zeolites, (2) carbonaceous ion exchangers and (3) synthetic resins.

Zeolites are insoluble sodium-alumino silicates which are capable of exchanging sodium ions for multivalent ions of alkaline earth group and for the divalent ions of some of the metals in water. They are particularly employed for softening of water. The chemical structure of the sodium zeolite is given by Na_20, $A1_20_3$, $XSi0_2$, YH_20. Two types of zeolites are in common use: (1) natural non-porous green sands and (2) synthetic, porous, gel zeolites. The natural zeolites are derived from green sand by washing, heating and treatment with sodium hydroxide.The synthetic zeolites are prepared from solutions of sodium silicate and aluminium hydroxide. Synthetic zeolites have higher base exchange capacity than natural zeolites. Zeolites are now being replaced by high capacity resins which have higher stability.

If certain coals are granulated and treated with sulphuric acid, they develop ion-exchange properties and the products obtained (sulphonated coals) are physically and chemically stable towards acids and to some extent towards alkalis.

New resins available for water treatment include cation-exchange resins of the weakly & strongly acidic types, anion-exchange resins of the weakly & strongly basic types, and highly porous modifications of strong acid cation exchangers & strong base anion exchangers. These resins are produced in a granular or bead like form and may be obtained in an effective size.

In order to be effective in the treatment of water, ion exchangers should possess the following properties: (1) they should have large surface area since ion exchange is a surface phenomenon and at the same time, their resistance to flow must be compatible with hydraulic requirements, (2) they should be physically durable & resistant to chemical attack and possess a high exchange capacity, (3) they must be relatively inexpensive and capable of being easily regenerated & backwashed and (4) they should be non-toxic & should not discolour the waters.

In India, National Chemical Laboratory, Poona has developed polystyrene based cation exchangers and melamine based anion exchangers.

Softening of Water Using ION Exchange: Water can be softened using ion exchange materials either in sodium or hydrogen form. In the first method, hard water is passed through bed of an ion exchanger that has been placed in sodium form.

$$\begin{matrix} Ca \\ Mg \end{matrix} \left\{ \begin{matrix} (HCO_3)_2 \\ SO_4 \\ Cl_2 \end{matrix} \right. + 2NaR \rightarrow \left. \begin{matrix} Ca \\ Mg \end{matrix} \right\} R_2 + Na_2 \left\{ \begin{matrix} (HCO_3)_2 \\ SO_4 \\ Cl_2 \end{matrix} \right.$$

When the ability of the cation exchanger bed to produce a completely softened water is exhausted, the softener unit is temporarily put out of service, back-washed to cleanse & hydraulically regrade the bed; regenerated with a solution of sodium chloride, rinsed free from salt, and then the unit is returned to service ready to soften another equal volume of hard water. The regeneration reaction is:

$$\left. \begin{matrix} Ca \\ Mg \end{matrix} \right\} R_2 + NaCl \rightarrow 2NaR + \left. \begin{matrix} Ca \\ Mg \end{matrix} \right\} Cl_2$$

A typical ion exchanger is shown in figure.

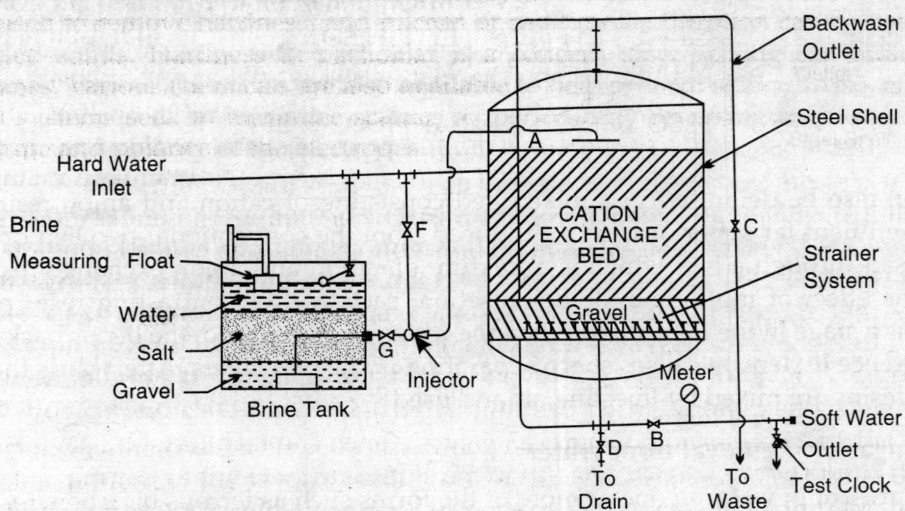

Fig.3.3F$_5$: A typical ion exchanger

For many industrial water supplies, chemical softening (lime soda ash process) is utilised for removal of the larger fraction of the total hardness and the remainder is softened by means of ion exchange. For softening water, usually sodium cycle is preferred.

The ion exchange softener closely resembles a filter, the exchanger taking the place of sand. Either the pressure or the open gravity type may be used depending upon local conditions and they are obtainable for either upflow or downflow. Usually pressure type is used. Turbid waters must be filtered before softening by this method; otherwise the suspended matter will foul the exchanger bed and renders it inactive. Raw waters with a hardness above 1 000 mg/1 cannot be softened economically in cation exchangers partly because of the size of the softener required and the volume of salt and wash water involved. In general, waters with temperatures above 35 °C will have an injurious effect on exchange materials. Raw waters containing dissolved iron or manganese, suspended matter or those high in certain salts (particularly chlorides or other substances deleterious to the exchange material) should be treated before softening.

The removal of calcium and magnesium compounds with the reduction of hardness to extremely low amounts sometimes creates a corrosive water that requires subsequent pH adjustments to avoid corrosion.

Demineralisation or Deionisation

For many years, the only means available for complete removal of soluble salts from water was distillation. Since the development of cation and anion exchange resins, it has been possible to produce demineralised or deionised water of highest purity by a two step ion exchange process.

The first step in the process is hydrogen cation exchange treatment. The effluent from this step is then passed through an anion exchanger bed. Strongly basic anion exchanger is used when silica is to be removed. Weakly basic anion exchangers are generally used when silica removal is not required. Forced draft or vacuum degasifier unit is provided when carbon dioxide removal is also required. Chemical reactions taking place are:

$$\left.\begin{matrix} Ca \\ Mg \end{matrix}\right\} \begin{matrix} SO_4 \\ Cl_2 \end{matrix} + H_2R \rightarrow R \left\{\begin{matrix} Ca \\ Mg \end{matrix}\right. + \left\{\begin{matrix} H_2SO_4 \\ 2HCl \end{matrix}\right.$$

$$H_2SO_4 \text{ or } HCl + ROH \rightarrow R \left\{\begin{matrix} SO_4 \\ Cl_2 \end{matrix}\right. + H_2O$$

Water can also be deionised in a mixed bed consisting of cation and anion resins. This produces an effluent far superior to that produced from the conventional two bed operation Water is passed through a bed which is a mixture of cation and anion exchange resins and which has the effect of multiple-cycle-deionisation, each pair of contrasting resin particles functioning as a stage in the treatment. When the resins are exhausted, the bed is backwashed and the difference in density causes sharp separation in the two layers. Afterwards the bed is washed; the resins are mixed by injecting air and used.

Iron and Manganese Removal from Water

Iron may be present in water in one or more of the forms such as ferrous bicarbonate, ferrous sulphate, ferric hydroxide or organic chelated iron. The commonest form in which iron is found in water-supplies in troublesome amounts is as ferrous bicarbonate. Manganese is a much rarer constituent in water supplies than iron, but where found, occurs in much the same forms as iron. The commonest form in which it is found in ground waters is as manganese bicarbonate.

The presence of iron or manganese in water in quantities as small as 0.1 to 0.3 mg/1 may be seriously objectionable. These elements cause difficulties in various industrial processes such as textile dyeing finishing, paper making, rayon processes and manufacture of beverages. They cause discolouration of bathroom fixtures & laundry work and they promote troublesome bacterial (iron bacteria) growths in water mains which impede the flow and give rise to offensive tastes & odours.

A number of different processes can be employed for the precipitation or removal of iron and manganese from water. Choice & economy of the treatment method depend upon the characteristics of water and the nature of the iron & manganese

Iron and manganese in water are generally removed by (1) aeration followed by settling and filtration, (2) ion exchange process, (3) lime soda process or (4) manganese zeolite process.

Aeration

The easiest method for removing iron and manganous compounds into insoluble ferric & manganic compounds which are then removed by settling and filtration. The oxidation which simultaneously expels carbon dioxide and admits air to water.

Hard water respond well to this treatment, but when the water is soft or has a low pH or when the iron / manganese are combined with organic matter, the aeration must be supplemented with chemical treatment. Lime or soda ash is used to raise the alkalnity to the point necessary for oxidation reaction and coagulants are used to expedite settling. chemicals like chlorine and permanganate may be utilised to release iron from organic combinations oxidising the organic matter. Instead of using dissolved oxygen for manganous compounds in water, it is possible to use chlorine as the oxidising agent.

Passage of the water after settling in sedimentation basins through a bed of filter media is necessary in all cases of iron and manganese removal. These filters may be either gravity pressure type. Contact filters or beds of coke, gravel, or manganese ore may also be employed to take some of the load and they have been placed either and operated at higher rates of flow.

The Type of aeration equipment that is most widely used in iron & manganese removal is the coke-tray type aerator. To a lesser extent, wood slot aeration have been employed.

A typical Indian iron removal unit consists of four dismantleable cylinders. The top two cylindrical vessels each contain 150 mm layers of 20 to 50 mm assorted stones. The third cylinder contains a 50 mm layer of 10 or 20 mm stones and 300 mm coarse sand. The bottom cylinder is a collection trough to which is attached a 12 mm tap through which the iron free water is collected. Ventilator holes for aeration are provided in between the cylinders.

The raw water containing iron is sprayed over the gravel in the top cylinder and trickles down. The gravel gradually gets coated with oxides of iron which helps further oxidation of iron. A portion of the precipitate is retained on the gravel. Although the gravel needs no washing, the sand requires washing once every month. This can be done by removing the sand and washing it manually to free it from sediment.

A unit of 450 mm dia is capable of handling 180 litres per hour.

Ion Exchange

Iron and manganese can be removed from water simultaneously with calcium and-magnesium in cation exchange process. It is necessary in the process that the raw water is not aerated before entering the ion exchange bed; otherwise the precipitate of hydrated metallic oxides may clog the bed and coat the exchanger. For iron and manganese removal, cation exchangers can be used either in sodium form or hydrogen form. With the high capacity resins, it is the usual practice to limit the application of this process to waters containing not more than 0.5 mg/1 of iron (as Fe) for each mg/1 of hardness present upto a maximum of 50 mg/l. With the older zeolites and carbonaceous exchangers, the usual limit is 0.5 mg/1 of iron per 17 mg/1 of hardness with a maximum of 10 mg/1 of iron. The operation of the ion exchange unit is similar to that used for softening. The production of a low alkalinity, iron-free water is possible by passing a portion of the water through a cation exchanger in hydrogen form & the rest through a cation exchanger in sodium form and mixing the effluents in suitable proportions.

Lime Soda Process

Lime soda process is also effective in removing iron and manganese from water. For reasons of economy, the use of lime soda process for iron and manganese removal is restricted to hard waters which are also to be softened. Precipitation of iron manganese from hard waters by lime is similar and effective as precipitation of calcium / magnesium. As the pH of the lime treated water is raised to around 10.0 by the chemicals used to precipitate hardness, the oxidation of iron / manganese is rapid and complete. If oxygen is kept out of system, iron precipitates as ferrous oxide or ferrous carbonate and manganese as hydrous manganous oxide. If iron and manganese removal is effected in a sludgv blanket type of softener, filtration of the treated water may not be necessary. But if water is required to be free from all traces of turbidity and iron, then filtration will be necessary.

Manganese Zeolite

For removal of iron and manganese by zeolite without softening, a special 'manganese zeolite' is used, to oxidise these impurities. It is regenerated or replenished with a solution of potassium or sodium permanganate. The amount and cost of regenerant required may be reduced by aeration of the water, which may be accomplished by introducing a limited amount of air into the water under pressure, before the manganese zeolite treatment. Manganese zeolite is prepared from zeolite by alternate treatments with manganese sulphate potassium permanganate. This results in the precipitation of the higher oxides of manganese in and around the granules of zeolites. In this process, iron manganese are mostly removed by contact oxidation.

The manganese zeolite is used as a filter medium in pressure type filter. If free carbon dioxide content is high in water, it should be neutralised to prevent manganese pick up from the exhausted bed. In large industrial plants, the use of this process is limited to water containing less than 1.0 mg/1 of iron and manganese.

Boiler Feed Water Treatment

The prevention of hard scale is the primary object of boiler feed water treatment but modern treatment is also designed to prevent foaming, corrosion and caustic embrittlement. Scale formation is caused by carbonates & sulphates of calcium & magnesium and silicates.

Foaming may be caused by oil and grease or other organic impurities in the feed water or result from a high concentration of dissolved & suspended mineral salts. A high rate of steam liberation from a small water surface has a greater foam building effect than the same amount of steam liberation distributed over a greater area. Embrittlement results in the development of intergranular cracks in boiler plates, which are caused by excessive stress and chemical attack. In boilers, the chemicals that are responsible for embrittlement are caustic soda and silica. Caustic soda is not generally present in water, but some of the waters contain sodium bicarbonate which breaks down in the boiler to form caustic soda. Lime-soda softened water also contains sodium carbonate. Corrosion in boilers is due to dissolved oxygen and the rate at which it proceeds is a function of the hydrogen ion concentration.

The deposition of scale in boilers can be prevented by (1) external treatment of make up water to remove hardness causing substances and (2) internal treatment with certain chemicals which react inside the boiler with scale-forming salts & form soft, easily removable deposits instead of hard scale. Two classes of materials are used for internal treatment: (1) organic materials such as tannin extract, seaweed extract & corn starch and (2) inorganic compounds such as soda ash & the sodium phosphates. Among the organic materials, the most commonly used is tannin which renders the scale nonadherent, inhibits embrittlement and reduces foaming. Internal treatment chemicals are directly added to the water in the boilers in the proportion required, depending upon the composition of water.

Silica in boiler feed water is troublesome, especially at high pressures. combines with calcium & magnesium to form dense scales. Silica in suspension can be removed by coagulation and filtration but soluble silica is difficult to be removed. Silica can be removed in a demineralising unit using strong base anion exchanger. Magnesium compounds (magnesium oxide, magnesium hydroxide and special dolomitic lime) can be used where softening the water is to accompany the silica removal. Silica removal by magnesium compounds is usually done simultaneously with coagulation.

By maintaining a minimum ratio of sulphate to carbonate in water, caustic embrittlement can be minimised in a boiler. Higher the pressure, higher should be the ratio of sulphate to alkalinity. The ratios suggested are: Na_2S0_4 / Na_2CO_3 = 1 for boilers pressure upto 1 MN/m^2 (150 psi), 2 for 1 - 1.7 MN/m^2 (250 psi) and 3 for pressure over 1.7 MN/m^2. The presence of phosphates, tannates, chromates, nitrates, acetates etc in feed water inhibits the embrittlement action of caustic soda, if these salts are present in sufficient quantities. When feeding hard waters to boiler, the problem of maintaining certain ratio of sodium sulphate to sodium alkalinity does not usually arise because there is seldom any sodium alkalinity in a natural water. With ion exchange softened water or lime-soda ash softened water, this will be a problem and a modification of the treatment of the water is necessary. The treatment may consist of (l) installation of a hydrogen-ion exchanger to handle part of the water, the effluent being mixed with the sodium ion exchanger effluent to bring down the alkalinity, (2) lime treatment preceding the sodium ion exchanger to precipitate carbonate hardness, (3) sulphuric acid treatment either preceding or following sodium-ion exchanger to convert a part of the alkalinity into sulphate, (4) direct addition of sodium sulphate to water and (5) using trisodium phosphate in chemical softening process.

Corrosion in boilers can be prevented considerably by (1) deaeration of feed water, (2) increasing the pH of feed water by adding alkali and (3) feeding to the boilers chemicals like sodium sulphite or hydrazine phosphate which combine with oxygen and prevent its entry into the corrosion reaction. Deaerating heaters or deaerators are used to reduce dissolved oxygen content of feedwater. Deaerating heater brings steam into immediate contact with the water as it is being broken up into thin films or drops by splashing over trays or being sprayed into the heating chamber. Deaerators contain certain refinements over deaerating heaters, such as more tray area for bringing the steam and water into contact.

The remedies to reduce foaming from boilers are to remove oil from the feed water using oil filters, increase the blow down, operate more boilers at lower ratings and treat the feed water to reduce the content of dissolved solids.

Cooling Water Treatment

Large quantities of water are needed for various cooling purposes in industry. The source of the cooling water in an industry may be (1) ground water, (2) surface water or (3) sea water. The specifications for cooling waters vary so widely with different classes of water for various industrial purposes that the only general specifications that can be given are that the water should not form heat insulating scales or other clogging deposits, either inorganic or organic and that it should not be unduly corrosive under the particular conditions of use or reuse. Cooling waters have therefore to be treated to prevent scale formation, corrosion and formation of slimes in heat exchangers, condensers and other cooling surfaces. The extent of treatment depends on the type of cooling water system employed like once-through system, open recirculation system and closed-recirculation system.

Once-through system is used where a plentiful supply of water is available. Since the water is to be used once and then wasted, the treatment should be cheap. In the industry, the commonest recirculating system is one in which the water circulates through the cooling system, is cooled through a cooling tower or spray pond and is then recycled. The treatment required should be based not only on the composition of the raw water but also on what the composition of the treated water will be after it has been concentrated after reuse. In closed cooling systems usually employed for cooling diesel engines , other types of internal combustion engines and for various other cooling purposes, there is but little make up water. The recirculated water should, however, be treated for the initial start and for whatever small amount of make-up is needed.

In treating cooling waters, one of the main considerations is the elimination of carbonate hardness which deposits on heating. For this, cold lime treatment may be used to advantage. The chemicals used are usually lime to reduce the bicarbonate hardness, a small dosage of alum as a coagulant, frequently followed by a small dosage of sulphuric acid to adjust the alkalinity of the effluent to the desired amount. In cooling water systems, sodium hexametaphosphate in the range of 2 to 6 mg /I is used for precipitation of calcium carbonate. Demineralisation using ion exchange process is used for closed recirculation systems only where a very high quality cooling water is required.

Turbid waters containing either coarse or colloidal suspensions are objectionable in most of the cooling systems except surface condensers for steam as they are very troublesome and form clogging deposits. The removal of turbidity is usually accomplished by coagulation, settling and filtration.

Slime and algae growths in condensers and heat exchangers cause serious operational problems in cooling water systems. It is generally preferable to add chemicals like copper sulphate, chlorine or chloramine that will either prevent the formation of the growths or destroy existing growths. Ordinarily the chemicals are added intermittently in heavy doses. It is not practicable to clean periodically condensers and heat exchangers when slime growth develops because it is an expensive operation as the equipment has to be dismantled.

Disinfection of Water

The main aim of disinfection is to destroy pathogenic bacteria and thus prevent water-borne diseases. Disinfection of water can be achieved by irradiation of water by ultraviolet light of suitable wavelength, by use of ozone, iodine, chlorine or chlorine compounds. Irradiation of water by ultraviolet light is commonly used for disinfection in food industry. The use of chlorine has become practically universal in the disinfection of water supplies

In the early years of water chlorination, the commercial sources of chlorinated lime (bleaching powder) and sodium hypochlorite solution. In 1910 liquid chlorine was produced commercially. From 1928, a high test calcium hypochlorite which is relatively stable was available. In recent years chlorine dioxide is coming to use. Chlorine may be applied to water with equal effectiveness in the form of bleaching powder or gas.

When chlorine (solution, gas or compounds) is added to water, a mixture of hypochlorous (HOC1) and hydrochloric acids is formed.

$$Cl_2 + H_2O = HOCl + H^+ + Cl^-$$

The hypochlorous acid formed is a weak acid and this may further dissociate according to the reaction:

$$HOCl = H^+ + OCl^-$$

The extent of dissociation depends upon the pH of the solution. In pure water containing chlorine, both HOC1 and OCl– are present. Of these, HOCl is far more active than the latter. The disinfection efficiency of chlorine decreases with an increase in pH. This can be explained on the assumption that HOCl is the effective agent in disinfection with chlorine and HOCl decreases rapidly in a solution with increasing pH.

If solid or liquid hypochlorite or bleaching powder is used, the following reactions take place:

$$Ca(OCl)_2 + H_2O = Ca^{++} + H_2O + 2OCl^-$$

$$OCl^- + H^+ = HOCl$$

The relative amounts of HOCl and OC1– present depend upon the ionisation constant of HOCl and the acidity of the solution.

The relative distribution of HOCl and OCl– will depend on pH and temperature. At 20°C and pH 8.0, the percentage distribution of HOCl is 29%. This equilibrium relationship, therefore, permits the identification of the species of chlorine that constitute ' free available chlorine ' which is important in virus & bacterial removal.

The factors that influence the efficiency of disinfection include (1) temperature, (2) time of contact, (3) concentration of reacting substances & disinfectant bacterial protoplasm, (4) presence of extraneous matter, particularly of a colloidal & organic nature, (5) surface tension, (6) pH and (7) age of bacterial cells spores & the resistivity of the individual cells.

The earliest water chlorination practice known as plain chlorination, simple chlorination or marginal chlorination is still practiced in water supplies. The chlorine ammonia treatment was thereafter introduced to limit the development of objectionable tastes & odours, sometimes associated with marginal chlorination. Subsequently superchlorination and dechlorination were developed for the additional purposes of destroying objectionable taste and odour producing substances associated with chlorine containing organic materials. The introduction of breakpoint chlorination and the recognition that residual chlorine can exist as combined or free chlorine, established the modern chlorination practices.

The application of ammonia to water prior to the addition of chlorine has many advantages like prevention of tastes particularly those due to phenols. It is important that the ammonia is applied at the right point and that adequate mixing and time of contact are provided before the application of the chlorine. The ammonia and chlorine combine together to form the mono-, di- and tri-chloramines (NH_2C1, $NHC1_2$, NCl_3) respectively.

The break point in the chlorination of the water may be defined as the point on the applied chlorine - residual chlorine curve at which all or nearly all the residual is free chlorine. The dose of chlorination necessary to reach break-point will vary widely even for a given water with time. The application of chlorine at or slightly higher than break-point concentration will remove tastes & odours, will have adequate bactericidal effect and will leave a desired chlorine residual.

Superchlorination is the application to water of an excess amount of chlorine in order to insure disinfection in the time available and destroy odours & tastes. Resulting high residual chlorine is removed by dechlorination using chemicals like sodium thiosulphate, sodium bisulphite, sulphur dioxide or active carbon. Intensive use of chlorine in this manner without the break-point reaction is called 'superchlorination and dechlorination.

Chlorine is generally applied after all other treatments have been given to the water supply. This is known as 'post chlorination' and is generally applied in water works. In some cases , chlorine is used preceding filtration by adding into the suction pipes of raw water pumps or to the water as it enters the mixing basin and this process is known as 'prechlorination'. Prechlorination reduces bacterial load on filters resulting in increased filter runs and oxidises excessive organic matter thus reducing tastes and odours. In the process of 'double chlorination chlorine is applied at two points in the treatment process. It is essentially prechlorination with an added treatment to the final effluent from the filters. Advantages of this method include (1) decrease in the load on filters, (2) greater removal of bacteria and (3) control of algae & slimy growths in coagulating basins filters.

The dose of chlorine required for effective disinfection would depend on chlorine demand of water. The dose applied is usually less than 1 mg/1 so as to produce a residue of 0.1 to 0.2 mg/1.

The organisms which are to be eliminated by disinfection of water are varied in character and include (1) the enteric bacteria, comprising members of the groups salmonella, shigella and vibrio, (2) the intestinal protozoa, of which Entamoeba histolytica is most common (3) a few worms for example, the Schistosomes, (4) viruses, for example, that of infectious hepatitis and (5) coliform organisms which although not pathogens are the primary indices of pollution of water. These organisms vary widely in resistance to disinfection ; coliforms and enteric bacteria are easily destroyed. Viruses and cysts of E.histolytica are very resistant.

Desalination of Water

As a part of his quest for water, man has been trying to get fresh water from sea or other saline sources. The techniques and economic aspects of this have been subjects of increased worldwide attention.

Good progress has been made in reducing the cost of desalination.

In India, acute water shortage exists in some states like Gujarat, Rajasthan, Bihar , Orissa and Maharashtra.

Several processes are known for desalting water but only a few have been developed to the point of actual use. Desalination is effected either by removing water from salt or salt from water. Processes such as distillation, freezing and reverse osmosis fall under the first category. The methods using ion-exchange materials which include ion exchange resins and membranes fall under the second category.

IMPORTANT WATER DESALINATION PROCESSES ARE AS FOLLOWS

A. Processes which separate water from the solute

1. Distillation (or evaporation)
   a. Multistage flash distillation
   b. Evaporation in long tube vertical evaporator
   c. Vapour reheat distillation
   d. Solar distillation
2. Freezing (or crystallisation)
3. Reverse osmosis

B. Process that separate it from the solution

1. Electrodialysis
2. Piezodialysis.

Desalination involves transport of salt molecules, water, ions and ion complexes in aqueous solutions, organic liquids, gases and solids. These processes are discussed below.

Distillation

Most of the desalination plants now being constructed utilise one of the distillation processes. Distillation of water on a major scale is a problem more of economics than of technology. Distillation has been in successful use for many years. Costs are continually decreasing with improvements in process design, materials of construction and increase in operating temperatures.

Various power sources (from solar to nuclear) have been used for distillation. Installing a dual purpose plant (i.e. for distillation of water and electricity generation) is found to be advantageous.

Nuclear power needs high purity water to make up its water requirement. Desalination plants need energy to move the large mass flows required from the production of desalted water

Distillation process basically consists of (1) vaporisation of some of the saline water at suitable temperature & pressure and (2) condensation of the vapour to provide distilled water. Operational problems include scale formation and corrosion. Single effect distillation is inefficient because of the heat lost with outgoing condensed water and brine. Multiple effect distillation plants.reduce the heat losses. The vapour from the first effect is condensed during the passage through the coils in second effect, which operates at a lower temperature. Saline water in the second effect serves as a coolant and absorbs heat from decondensing vapours. This causes the water in the second effect to boil and in a similar fashion in the third effect giving up heat to saline water in that unit. Vapour from the last effect is condensed in a separate heat exchanger. Vacuum is applied to one or more effects to allow distillation at a relatively low temperature and to minimise scaling. Distillation is the most economical process today for the production of large quantities of fresh water, specially when salinity is high. This process requires large amount of low pressure steam which can be obtained at a cheap rate by combining a power plant with a desalination plant, The production of electrical power can be more efficiently accomplished With high temperature and high pressure steam. On the other hand, desalination plants cannot use steam above 125 °C because of scale formation on heat transfer surfaces. Thus bv combining the two processes, that is, producing power from the higher temperature section and placing a distillation plant in the lower temperature portion, thermal energy can be used more effectively. This is the concept of the dual purpose plant

Multistage Flash Distillation (MSF)

Until now, multistage flash distillation has been the most popular and economical process for desalination. The heat supplied provides only the temperature differential required for heat transfer in stages rather than the latent heat of evaporation. The brine reaches its highest temperature after passing through the brine heater, enters the first flash chamber in which pressure is reduced, and boils. The steam released is condensed on the tubes carrying the incoming brine. The desired pressure is maintained in each chamber by the condensing action of the heat exchangers. Only a small fraction of the water entering a given chamber can be evaporated in that chamber unless temperature differential across the chamber is large. The rest of the brine passes on to the next stage at a lower temperature.

In the present plants, the flash evaporation is carried out by admitting heated sea water to a series of chambers arranged in sequence. The over-all process becomes more nearly reversible, and hence more efficient as the number of stages is increased for a given temperature range. But in practice, this number is sharply limited by (1) the high cost of constructing the individual, sealed compartments & their connecting piping and (2) the necessity of maintaining sufficient pressure drop between stages to overcome the pressure losses & maintain flow without an excessive depth of liquid.

Evaporation In Long Tube Vertical Evaporator (Ltve)

The multistage flash distillation has some built in limitations, thereby requiring large surface areas and consumption of energy. This has led to the development of multi effect long tube vertical evaporators which are potentially more efficient. Some of the advantages of this process are (1) reduction in formation of mineral scales on heat transfer surfaces, (2) increased efficiency due to better thermodynamic reversibility, (3) reduced size of equipment and (4) less consumption of pumping energy due to low volume of circulated brine. Innovations like the falling film / rising film techniques and the use of extended heat transfer surfaces make the LTVE process more favourable economically than MSF process.

The major difference between the LTVE and MSF processes is that the tubes in the former system are fixed in vertical position. The evaporation occurs as sea water flows down the inside of each tube in a thin film, collecting heat from the condensation of vapour outside. The sea water inside of tube and the steam generated passes to the next stage at lower pressure, to heat the incoming sea water and to condense on the outside of the tube, giving fresh water.

To improve the LTVE process further and to make it feasible for large scale applications by reducing large investment and operational & maintenance requirements, the vertical evaporators can be used in conjunction with the multistage flash process, the latter acting as preheaters for incoming sea water feed. Such a combination, popularly known as LTVE-MSF, would considerably improve the thermodynamic efficiency of the process.

Vapour Reheat Distillation

This system of desalination is similar, in several respects, to multistage flash distillation and is being currently used in many successful sea water conversion plants. In this system deaerated sea water passes through a heat exchanger counter-current to hot fresh water. The temperature is then raised with heat from an external source. The hot sea water then cascades through a series of flash chambers countercurrent to a stream of fresh water flowing in open channels. In each stage, some sea water flashes to form steam, which condenses in the stream of fresh water. As a result , sea water is cooled and fresh water is heated. Hot fresh water leaving the highest pressure stage is used to heat incoming sea water. Part of the cooled fresh water is recycled to the lowest pressure stage; the rest is the product. In this system scaling problems are reduced to a minimum and heat transfer coefficients are high. But heat must be transferred three times.

A multiple effect evaporator with vapour compression and a secondary heat transfer medium is another method of distillation which has been used in desalination plants.

Solar Distillation

Solar energy may be used to generate heat or power which can then be used for operating any desalination process or it may be employed directly for distillation of saline water in equipment which serves both as a solar energy absorber and as a distiller. Utilisation of free solar energy for distillation of sea water was an old idea. As of fresh water— Solar distillation occupies a favoured position among all desalination processes in capacity ranges of 200 to 400 m3 /day.

The latest development includes the use of polyvinyl fluoride films as a covering material with a metaled finish on the underside. The transmitting sun's rays are reflected and bounce around till they are reflected back. The brine temperature goes as high as 85 °C.

Freezing

If sea water is reduced in temperature to a value very close to that of the freezing point, ice crystals will begin to form within the solution. The resulting crystals are then physically separated from the mother liquid, washed to remove any entrapped brine and remelted into salt-free fresh water. Apart from the successful development of an ice-brine separation method, the freezing system must operate in a closed cycle integrating the ice making and ice melting steps so that minimum of energy is consumed. In order to induce the inherently low energy requirement of the system, direct heat interchange is employed in the freezing and melting steps thus eliminating costly metallic surface. This facilitates heat removal at the highest possible temperature and heat rejection at the lowest possible temperature. The power requirement for the direct contact freezing process has been estimated to be in the range of 18 to 28 MJ / m3 (ca. 5-8 kWh /m3,) of potable water produced.

In the indirect refrigeration method, the incoming sea water is first pumped through a heat exchanger which reduces its temperature, if necessary. It is then admitted to a freezing chamber, which is cooled by means of the refrigeration coils of a separate refrigerating system. The sea water is sufficiently reduced in temperature to a level where the ice crystals are formed. The ice and brine slurry then flow to a separation unit, where the ice and brine are separated. The brine is returned to the supply source and the ice is transferred to a melting unit, where the heat originally removed by the refrigerant is used to melt the ice as the refrigerant is condensed. The melted ice is then taken from the melting unit as product water, part of which is by-passed to the separating unit, where it is used for washing the crystals.

In the direct refrigeration method developed in Israel , water itself is used as the refrigerant. By introducing the sea water to be purified into a vacuum chamber which is maintained at a pressure equal to the vapour pressure of sea water at or below the freezing point, some of the water immediately flashes into vapour. The evaporation of the water removes heat from the sea water. The heat removed is equal to the total latent heat of vaporisation for the mass of evaporated water. With proper design, this will reduce the sea water to a temperature where nucleation of ice crystals will begin.

One of the more recent approaches to saline water conversion is that of chemically separating pure water from salt solution by the use of hydrate forming materials. Certain low molecular mass hydrocarbons like propane and their derivatives combine with water to form solid hydrate compounds which are complexes in which molecules of one substance are contained or enclosed within the crystal lattice of another component like ice crystals. These hydrates reject ionic constituents and accept only pure water molecules in their lattice. Water is released by melting. The liquid hydrating agent is recycled.

Membrane Processes

Reverse Osmosis: In reverse osmosis, desalination is achieved by forcing salt solution under pressure through a membrane which generally passes water more readily than salts. Reverse osmosis differs from electrodialysis, the other important membrane method, in that the water is removed from salt rather than salt from water as in electrodialysis and in that the driving force is pressure rather than electrical potential. The equipment used in the two methods is very different as are the membranes. A cellulose acetate membrane is commonly used to separate salt and water at 5.5 to 10.0 MN/m^2 (ca. 800 to 1 500 psi) pressure. The membrane is very thin and has to be supported on a physically strong backing material. Flow channels have to be provided for feed and product water. The membrane support assembly must be housed in a pressure vessel to permit high pressure operation required for reverse osmosis. There are different modules of membrane unit, each directed towards achieving better performance of the system. The advantages of the process are low cost, no operational problems, no corrosion and no scale formation.

Electrodialysis: Electrodialysis is the transport of ions through ion, exchange membranes as a result of an electrical driving force. The cations & anions of the saline feed migrate towards cathode & anode through cation exchange & anion exchange membranes respectively, leaving the feed depleted in salt. A number of such membranes are kept alternately in a stack making multiple cell pairs. The concentrated brine and the depleted product are collected from alternate cell pairs. Ion exchange membranes are of heterogeneous and homogeneous types. The former is generally prepared by mixing finely powdered ion exchange resins with a suitable binder to form thin sheets while the latter has a homogeneous phase of ion exchange material with or without a base By packing cation and anion exchange membranes alternately in between two electrodes, thin chambers are formed with the help of gaskets of non-conducting plastic materials. Any saline water to be desalted is split into two streams and fed into alternate sets of compartments in the stack. Electrode chambers are separately flushed with feed water to remove the products of electrodialysis.

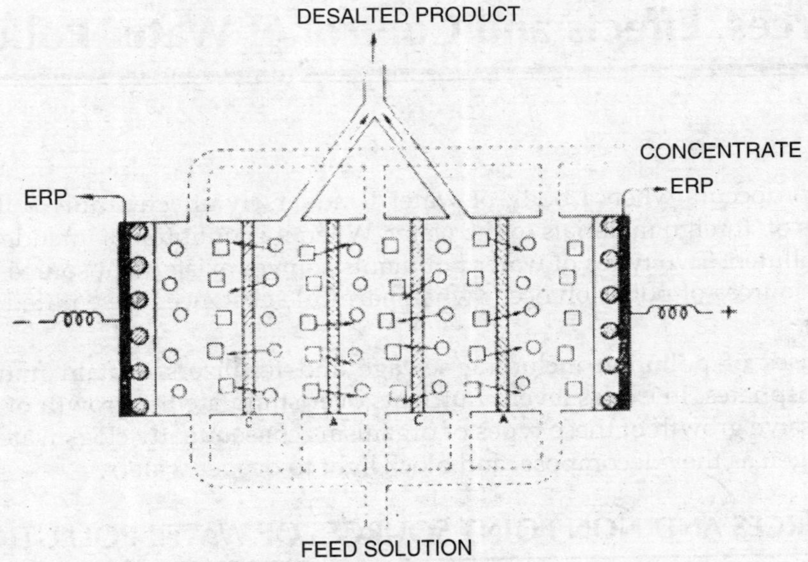

Fig.3.3F$_6$: Electrodialysis Unit

When a direct electric current is passed through this unit, the alternate chambers in the stack get depleted in salt content while the remaining chambers get concentrated with salt. The process is made continuous as the concentrate and dilute solutions are drawn out continuously from the systems. When high salinity waters are encountered, stage-wise and batch-wise operations are employed.

Piezodalysis: It is the flow of salt species through mosaic membrane (anionic and cationic in one matrix) under a hydrostatic pressure gradient. Like electrodialysis, salt is transported through the membrane; like reverse osmosis, the driving force is the pressure gradient. Cellulose acetate membranes suitable for desalting brackish water (having 90-95% salt rejection and 0.5-1.0 m^3/m^2 per day water flow) are now made in India.

Chapter

Sources, Effects and Control of Water Pollution

Water pollution occurs when a body of water is adversely affected due to the addition of large amounts of foreign materials to the water. When it is unfit for its intended use, water is considered polluted. Two types of water pollutants sources exist; point source and non-point source. Point sources of pollution occur when harmful substances are emitted directly into a body of water.

Many causes of pollution including sewage and fertilizers contain nutrients such as nitrates & phosphates. In excess levels, nutrients over stimulate the growth of aquatic plants & algae. Excessive growth of these types of organisms consequently clogs waterways, use up dissolved oxygen as they decompose, and block light to deeper waters.

POINT SOURCES AND NON POINT SOURCES OF WATER POLLUTION

Point Sources of water Pollution–When pollutants enter the river though a specific entry point, such as a drainpipe draining directly into a river, it is called point source pollution. Industrial water discharges and sewage treatment plants are the main sources of this type of pollution. Point source pollutants can include many different organic & inorganic substances, including human waste and toxic metals.

They include transfer of pollutants from municipal & industrial liquid waste disposal sites and from municipal & household hazardous waste and refuse disposal sites.

Point source pollution can be traced to a specific discharge point and therefore, it is the easiest source of pollution to control & regulate.

Non-Point Sources of water Pollution–In contrast to point source pollution, nonpoint source (NPS) of water pollution comes from many different diffuse sources and is extremely difficult to regulate and control. NPS pollution is mainly caused by runoff, when rain water and snow melt move over the land, picking up pollutants along the way & dumping the pollutants into rivers, tanks and lakes. Some common NPS pollutants include fertilizers & pesticides from agricultural lands and home owners; oil, grease & salt from highways;sediment from construction sites & eroding shorelines; and animal & human waste.

Non Point Sources include

Diffuse agricultural sources

Wash off and soil erosion from agricultural lands carrying materials applied during agricultural use, mainly fertilizers, herbicides & pesticides.

Diffuse urban sources:

Run off from city streets, from horticultural, gardening & commercial activities in the urban environment and from industrial sites & storage areas.

CATEGORIES OF WATER POLLUTANTS

Various categories of water Pollutants are as follows

- ***Oxygen-demanding wastes*** are biodegradable organic compounds contained in domestic sewage or certain industrial effluents. When these compounds are decomposed by bacteria, oxygen is removed from the water. If the oxygen level drops low enough, the fish will die.
- ***Disease-causing agents*** are various pathogenic microorganisms which usually enter the water with human sewage. Contact with these microbes can be made by drinking the water or through various water-contact activities.
- ***Synthetic organic compounds*** include detergents & other household aids, pesticides,and various synthetic industrial chemicals. Many of these compounds are toxic to aquatic life and may be harmful to humans.
- ***Plant nutrients*** such as the nitrogen & phosphorus which drains from fertilized lands, as well as the effluent from most sewage treatment plants, stimulate the growth of algae and water weeds.
- ***Inorganic chemicals and mineral substances*** includes the acids which form when water drains from abandoned mines, as well as the heavy metals such as mercury & cadmium.
- ***Sediments*** are particles of soils, sands, and minerals washed from the land. They can smother bottom life such as shellfish & coral, as well as fill in reservoirs and harbors. Improper soil management leading to erosion is a major contributor.
- ***Radioactive substances*** can enter the water from the mining & processing of radioactive ores, from various nuclear power operations, from medical facilities, and from nuclear weapons testing.
- ***Thermal discharges of hot water*** from steam-electric power plants raise the temperature of the receiving water by 7-10^{0}C resulting in various changes in local eco systems.

IMPORTANCE & IMPLICATION OF DISSOLVED OXYGEN IN WATER

The amount of dissolved oxygen (DO) in water is an important parameter of water quality. Fish, for example, require certain minimum amounts of DO for their survival depending upon their species, stage of development, level of activity, and the water temperature. For example, for a well-rounded, warm water fish population, it is recommended that the DO remains above 5 ppm for at least 16 hours of the day and during the other 8 hours it should not drop below 3 ppm.

There are four processes which affect the amount of oxygen in the water: **Reaeration, photosynthesis, respiration, and the oxidation of wastes.**

Reaeration is the process wherein oxygen enters the water through the contact the water surface makes with the atmosphere. The solubility of oxygen in fresh water, at a pressure of 1 atmosphere, decreases with temperature as shown in Figure below.

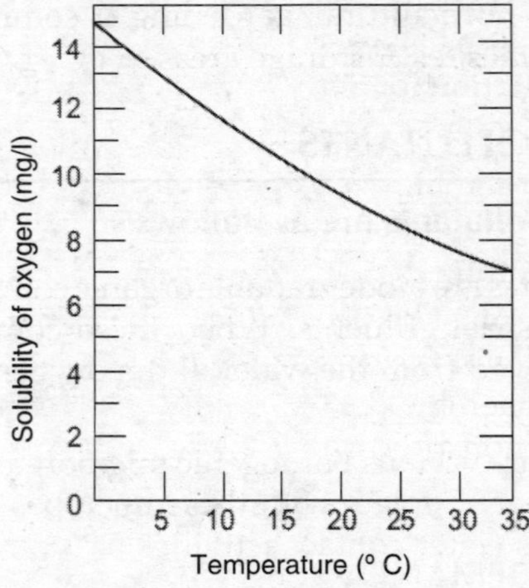

Fig.4.1: Solubility of oxygen in fresh water exposed to water saturated air at 1 atmosphere pressure

When the actual amount of oxygen in the water is less than the saturation value given in the figure, atmospheric oxygen passes into the water at a rate which is proportional to the deficit. By increasing the surface area in contact with the atmosphere, the transfer of oxygen in contact with the atmosphere is increased, so a bubbling stream will pick up oxygen easier than a stagnant pond.

During photosynthesis, which of course occurs only during the daylight hours, oxygen is liberated, thereby increasing the DO level in the water. Respiration, however, is a process which continuously removes oxygen from the water. Combining the three effects of photosynthesis, respiration, and reaeration produces a diurnal variation of DO as shown in Figure below. In this figure, photosynthesis is assumed to occur from 6 AM to 6 PM, bringing the DO level to slightly above saturation in the afternoon. While the water is supersaturated, oxygen diffuses out of the water instead of into it.

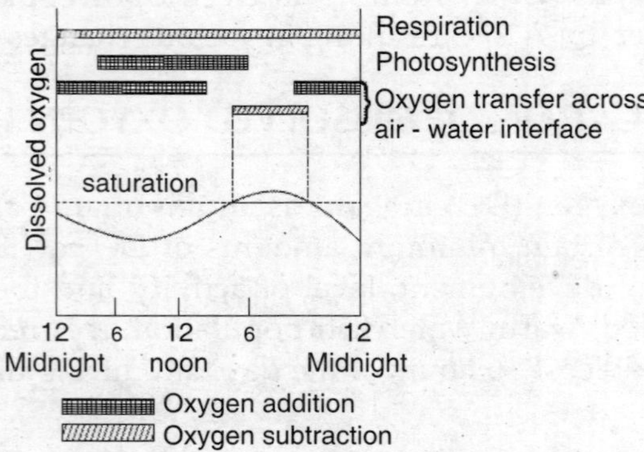

Fig.4.2: Diurnal variation of DO assuming photosynthesis from 6 AM to 6 PM.

The fourth process which affects the amount of oxygen in the water is the oxidation of wastes. Microorganisms, especially bacteria, use organic wastes as food and in the process break down the complex organics into simple organic & inorganic materials. This decomposition may occur in the presence of oxygen, in which case it is called *aerobic* decomposition, or it may occur in the absence of oxygen in which case it is called *anaerobic* decomposition.

The general form of the equation for aerobic decomposition of organic material is:

Organic matter + bacteria + $O_2 \rightarrow CO_2 + H_2O$ + new bacterial cells

The reaction products from aerobic decomposition are not obnoxious, being simply carbon dioxide & water along with some sulfates and nitrates. During aerobic decomposition oxygen is removed from the water, resulting in a drop in DO.

If too much organic matter must be decomposed, the DO supply in the water may fall to zero. If this happens, not only will the oxygen-dependant aquatic life be killed, but the end products from the resulting anaerobic decomposition will be toxic and malodorous. Typical anaerobic decomposition reaction products include ammonia, methane, hydrogen sulfide, carbon dioxide, and water.

BIOCHEMICAL OXYGEN DEMAND (BOD) AND ITS IMPORTANCE IN WATER POLLUTION

One of the most widely used measurements for estimating the water pollution potential of a given amount of organic waste is called the *biochemical oxygen demand*, or BOD. The BOD is a measure of the amount of oxygen required by bacteria to oxidize waste aerobically to carbon dioxide and water.

If we were to introduce an amount of organic matter into some water and monitor its decomposition, we would observe that the amount of organic material remaining would decrease exponentially with time. The amount of BOD which remains after some period of time is proportional to the remaining organic material, which results in a curve as shown in Figure below.

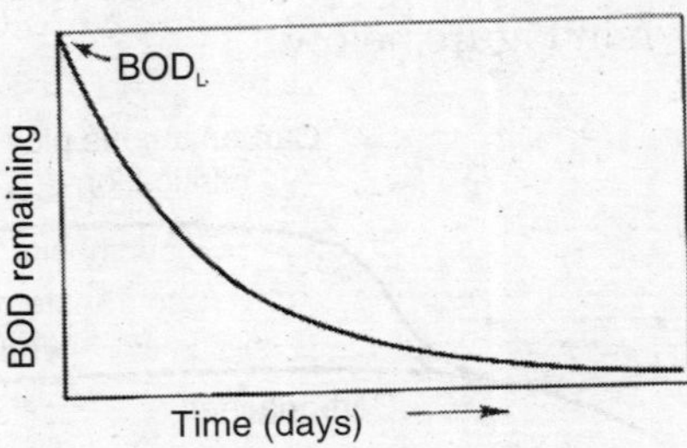

Fig.4.3: BOD remaining as a function of time with initial value BOD_L

The initial value of BOD will of course be the total oxygen requirement to oxidize the organic material. This quantity is called the *ultimate BOD*, or BOD_L.

Usually when curves of BOD versus time are drawn, they are actually made for BOD utilized and not BOD remaining. BOD utilized will be proportional to the amount of organic matter which has been oxidized and hence will have the shape as shown in Figure below.

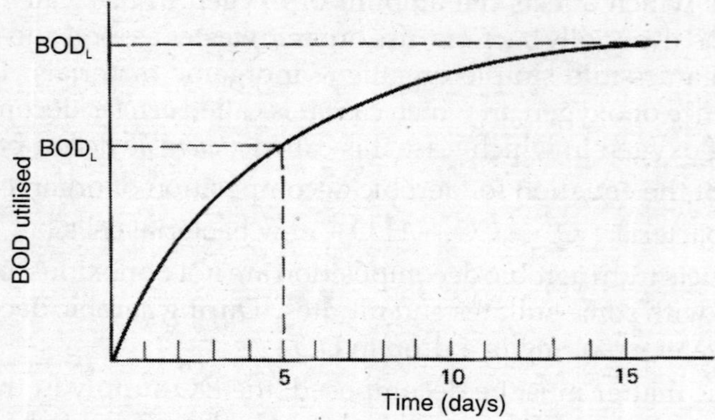

Fig.4.4: This is the usual curve for BOD, and is actually BOD utilized. Also shown is the 5 day BOD.

A standard way to measure BOD is to determine the amount of oxygen required by the bacteria during the first 5 days of decomposition (at 20°C). This is the 5-day BOD called BOD_5, which is shown in Figure above.

The importance of the concept of BOD in water quality management must be emphasized. Consider that normal domestic sewage may have a 5-day BOD of around 200 mg/1 and that industrial sewage may have a BOD of several thousands of milligrams per liter. Compare these requirements for oxygen to the amounts actually dissolved in water.

It is thus easy to anticipate the quick depletion of oxygen in any receiving water unless the dilution facter is quite high. Depleting the water of oxygen, will kill the fish and the anaerobic decay products will be extremely objectionable. There is a secondary effect which causes the demand for oxygen to suddenly increase after about 8-10 days. The organic nitrogen in the wastes is converted to ammonia during decomposition and the subsequent oxidation of the ammonia to nitrite (NO_2) and then nitrate (NO_3) requires oxygen.

The total oxygen demand curve showing both the carbonaceous demand and the nitrification requirement is shown in Figure below.

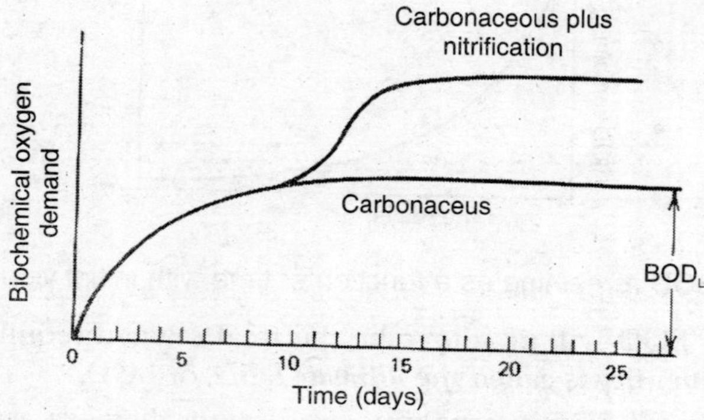

Fig.4.5: The complete BOD curve includes the carbonaceous and nitrification demands

OXYGEN SAG CURVE AND ITS IMPORTANCE IN WATER POLLUTION

When wastes are discharged into a body of water, the amount of dissolved oxygen will decrease due to oxidation by bacteria. Opposing this drop in dissolved oxygen is reaeration which replaces oxygen through the surface, at a rate which is proportional to the depletion of oxygen below the saturation value. The simultaneous action of deoxygenation and reaeration produces what is called the oxygen sag curve, as shown in Figure below.

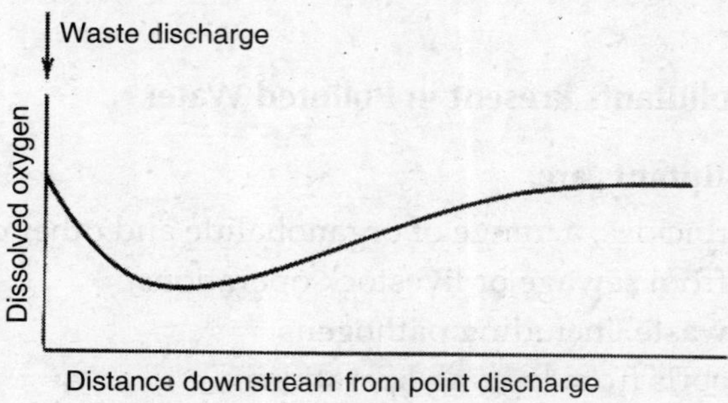

Fig.4.6: The oxygen sag curve

The DO curve initially drops as the wastes deplete the oxygen faster that it can be replaced. At the point where the DO is a minimum, the rate of reaeration becomes equal to the rate of oxygen utilization. Beyond that point the rate of reaeration exceeds the rate of utilization and the DO level eventually returns to normal. This sequence is referred to as the natural self-purification ability of water.

The horizontal axis of the oxygen sag curve may be either time or distance. If for example, a certain amount of waste is released all at one time into some impounded water, the DO level will be a function of time. If, however, there is a continuous discharge of wastes into a stream, then the oxygen sag curve will be a function of the distance downstream from the point of discharge. As we move downstream from the point of discharge, and the dissolved oxygen begins to drop, there will be a corresponding change in the biota. Normal game fish will be replaced by fishes tolerant of the turbid, low DO waters. At the point downstream where the DO reaches its lowest value, conditions will be at their worst. If the stream goes anaerobic, there will be no fish and the only organisms present will be those able to obtain their oxygen from the surface, or those which are tolerant of low oxygen conditions.

MAJOR SOURCES OF WATER POLLUTION

Principal sources of water pollution are

- Industrial discharge of chemical wastes and byproducts.
- Discharge of poorly-treated or untreated sewage.
- Surface runoff containing pesticides.
- Slash and burn farming practice, which is an element within shifting cultivation agricultural systems.
- Surface runoff containing spilled petroleum products.

- Surface runoff from construction sites, farms, or paved and other impervious surfaces e.g. silt.
- Discharge of contaminated and/or heated water used for industrial processes.
- Acid rain caused by industrial discharge of sulfur dioxide (by burning high-sulfur fossil fuels).
- Excess nutrients added by runoff containing detergents or fertilizers.
- Underground storage tank leakage, leading to soil contamination, thence aquifer contamination.

Organic & Inorganic Pollutants Present in Polluted Water

Some organic water pollutants are:

- Insecticides & herbicides, a range of organohalide and other chemicals.
- Bacteria, often is from sewage or livestock operations;
- Food processing waste, including pathogens.
- Tree and brush debris from logging operations.
- VOCs (Volatile Organic Compounds, industrial solvents) from improper storage.

Some inorganic water pollutants are:

- Heavy metals including acid mine drainage.
- Acidity caused by industrial discharges (especially sulfur dioxide from power plants).
- Chemical waste as industrial by products.
- Fertilizers, in runoff from agriculture including nitrates & phosphates.
- Silt in surface runoff from construction sites, logging, slash and burn practices or land clearing sites.

CAUSES, EFFECTS AND CONTROL MEASURES OF WATER POLLUTION

Water pollution is defined as alteration in physical, chemical or biological characteristics of water making it unsuitable for designated use in its natural state.

Causes of water pollution: Water is an essential commodity needed for drinking, cooking, bathing, washing, irrigation, and for industrial operations. Most water for such uses comes from rivers, lakes or ground water sources. Water dissolves many substances in it , therefore, it can easily get polluted. Pollution of water can be caused by point sources or non-point sources. Point sources are specific sites near water bodies which directly discharge effluents into them. Major point sources of water pollution are industries, power plants, underground coal mines, offshore oil wells etc. The discharge from non-point sources is not at any particular site, rather, these sources are scattered, which individually or collectively pollute water. Surface run-off from agriculture fields, overflowing small drains, rain water sweeping roads & fields, atmospheric deposition etc. are the non-point sources of water pollution.

Sources of water pollution

Ground water pollution: Ground water forms about 6 % of the total water available on earth and is about 30 times more than surface water. Ground water seems to be less prone to pollution as the soil mantle through which water passes helps to retain various contaminants due to its caption exchange capacity. However, there are a number of potential sources of ground water pollution. Septic tanks, industry, deep well injection, mining etc. are mainly responsible for ground water pollution, which is irreversible. Ground water pollution with arsenic, fluoride & nitrate are posing serious health hazard.

Surface water pollution: The major sources of surface water pollution are:

Sewage: Pouring the drains and sewers in fresh water bodies causes water pollution. The problem is severe in cities.

Industrial effluents: Industrial wastes containing toxic chemicals, acids, alkalis, metallic salts, phenols, cyanides, ammonia, radioactive substances, etc. are sources of water pollution. They also cause thermal pollution of water.

Synthetic detergents: Synthetic detergents used in washing & cleaning products foam and pollute water.

Agrochemicals: Agrochemicals like fertilizers and pesticides washed by rain water and surface run off pollute water.

Oil: Oil spillage into sea-water during drilling and shipment pollute it.

Waste heat: Waste heat from industrial discharges increases the temperature of water bodies and affects distribution & survival of sensitive species.

Effects of Water Pollution

Organic matter which reaches water bodies is decomposed by micro-organisms present in water. For this degradation oxygen dissolved in water is consumed. Dissolved oxygen (DO) is the amount of oxygen dissolved in a given quantity of water at a particular temperature and atmospheric pressure. Amount of dissolved oxygen depends on aeration, photosynthetic activity in water, respiration of animals & plants and ambient temperature.

The saturation value of DO varies from 8-15 mg/l. For active fish species 5-8 mg/l of DO is required whereas less desirable species like carp can survive at 3.0 mg/l of DO.

Lower DO may be harmful to animals especially fish population. Oxygen depletion helps in release of phosphates from bottom sediments and causes eutrophication.

Water pollution affects the health of the waterway, the health of the organisms living in & around the water bodies, and, eventually, the health of humans. The effects of water pollution can range from aquatic deformities to contaminated fish to 'dead' lakes. Toxic pollutants can also alter the genetic makeup of an organism, resulting in either death or extreme deformities.

Persistent Organic Pollutants (POPs), such as dioxin, PCBs and DDT, are chemical substances that persist in the environment & bioaccumulate through the food web.Therefore, POPs can also cause sickness and disease in humans, who are at the end of the food chain. People who regularly consume a lot of fish will have larger levels of toxic chemicals in their bodies than those who only eat fish occasionally. Toxic chemicals can lead to reproductive problems, cancer and neurological disorders. People who are most at risk of health problems due to contaminated fish consumption are those with weakened immune systems, including children, pregnant women & the elderly.

Water-borne diseases are caused by drinking water contaminated by human or animal faeces, which contain pathogenic microorganisms.There are many types of water-borne diseases, like Cholera, Dengue, Diarrhoea, Leptospirosis, Malaria, Hepatatis, Scabies, Typoid and Fluorosis. **Fluorosis** is a dental problem which attacks the children by making the brownish tinch on their teeth.The amount of nutrients entering the water bodies has intensified due to increased urbanization & agriculture, leading to increased biological growth, or eutrophication. Under eutrophic conditions, nutrient loading (more nutrients than the water body can handle) stimulates excessive plant growth, which in turn decreases the amount of oxygen in the water and kills off certain species of animal life.

Freshwater resources all over the world are threatened not only by over exploitation and poor management but also by ecological degradation. The main source of freshwater pollution can be attributed to discharge of untreated waste, dumping of industrial effluent, and run-off from agricultural fields. Industrial growth, urbanization and the increasing use of synthetic organic substances have serious & adverse impacts on freshwater bodies, The developed countries suffer from problems of chemical discharge into water sources mainly groundwater,while developing countries face problems of agricultural run-off in water sources.

EFFECTS OF WATER POLLUTION ON AQUATIC SYSTEM

Water pollution affects the health of the waterway, the health of the organisms living in & around the waterway, and, eventually, the health of humans. The effects of water pollution on aquatic systems can range from aquatic deformities to contaminated fish to 'dead' lakes. Heavy metals like mercury & lead, and human-made organic chemicals such as pesticides, biomagnify as they move up the food chain, resulting in tumors and death for predatory animals, such as lake trout, herring gulls, & even humans.

Toxic pollutants can also alter the genetic makeup of an organism, resulting in either death or extreme deformities. Cormorants suffer from cross-billed syndrome , while terns exhibit birth defects from dioxin, PCBs (Poly Chlorinated Biphenyls) and furan exposure. Other examples of deformities include large fish tumors and three-legged frogs.

Eutrophication and Its Effects

Before the introduction of chemical fertilizers the water contained little plant nutrients and were clear due to their size & depth. These lakes support high levels of animal life and receive proper amounts of nutrients, mainly phosphorous & nitrogen, from natural sources, such as decomposing plant matter. Urbanisation , agricultural revolution and industrialization has changed the situation. The amount of nutrients entering the water bodies has increased greatly, mainly due to increased urbanization and agriculture, leading to increased **biological growth, or eutrophication**.

Under eutrophic conditions, nutrient loading (more nutrients than the water body can handle) stimulates excessive plant growth, which in turn decreases the amount of dissolved oxygen in the water and eventually kills off certain species of animal life. Other pollution-tolerant species, such as worms and carp, grow more rapidly; thus, the ecological balance of the lake is altered.

Transport and Chemical Reactions of Water Pollutants

Most water pollutants are eventually carried by the rivers into the oceans. There are areas of oxygen depletion, caused by chemicals using up oxygen and by algae blooms, caused by excess nutrients from algal cell death & decomposition. Fish and shellfish kills have been reported, because toxins climb the food chain after small fish consume copepods, then large fish eat smaller fish, etc. Each step up the food chain concentrates certain toxins like heavy metals and DDT by approximately a factor of ten.

Many chemicals undergo reactive decay / change over long periods of time in groundwater reservoirs. Such chemicals include chlorinated hydrocarbons such as trichloroethylene (used in industrial metal degreasing) and tetrachloroethylene used in the dry cleaning industry. Both of these chemicals, which are carcinogens , undergo partial decomposition reactions leading to new hazardous chemicals.

Groundwater pollution is much more difficult to abate than surface pollution because ground water can move great distances through unseen aquifers. Non-porous aquifers such as clays partially purify water of bacteria by simple filtration (adsorption & absorption), dilution, and, in some cases, chemical reactions & biological activity.However, in some cases, the pollutants merely transform to soil contaminants. Ground water that moves through cracks and caverns is not filtered and can be transported as easily as surface water.

There are a variety of secondary effects stemming not from the original pollutant, but a derivative condition. Some of these secondary impacts are:

- Silt bearing surface runoff can inhibit the penetration of sunlight through the water column, hampering photosynthesis in aquatic plants.
- Thermal pollution of water can induce fish kills and invasion by new thermophyllic species.

Water Pollutants	Effects
Detergent	Complex formation between DNA and mercury or cadmium results in birth defects. Although ,detergents are not highly toxic to fishes ,they do cause damage to gills and remove the protective mucous from the skin & the intestine.
Domestic waste	Water borne diseases like cholera (Vibrio cholera) typhoid, dysentery and various health problems, depletion of dissolved oxygen, objectionable odour.
Industrial effluents	It causes deleterious effects on living things and may bring death or sub lethal pathology of kidneys, liver, lungs, brain & reproductive system.
Agricultural waste	Excessive fertilizer leads to accumulation of nitrates in children called methemoglobinaemia. Richness of nutrients results in eutrophication.
Eutrophication	During eutrophification, algal bloom release toxic chemicals into the aquatic system. Algal Bloom leads to oxygen depletion and an increase in CO_2 level. Thus aquatic organisms begin to die which leads to succession.

Bioaccumulation (or) Biological magnification	Aquatic plants and animals can accumulate certain pesticides in their body tissues in greater concentration than in water. This phenomenon is commonly referred to as biological magnification or biological amplification eg.,DDT. It is more threatening as its concentration continuosly increases in successive trophic levels in a food chain which results in many health hazards. DDT absorbed by fish eating birds 25 ppm ↑ DDT in large fish 2 ppm ↑ DDT in small fish 0.5 ppm ↑ DDT in zooplankton 0.003 ppm ↑ DDT in water 0.000003 ppm or 0.003 ppb
Lead (Pb)	Anaemia, vomiting, damage of liver, brain and kidney
Arsenic(A)	Mental disturbance, lung cancer, ulcer, kidney damage
Mercury(Hg)	Abdominal pain, headache, diarrhea, chest pain
Cadmium(Cd)	Growth retardation, diarrhea, bone deformation, kidney damage, anemia, damage to lever
Barium(Ba) Chromium(Cr)	Excessive salvation, diarrhea, paralysis Gastro intestinal ulceration, diseases of central nervous system, cancer, nephritis
Zinc(Zn)	Vomiting, renal damage
Copper(Cu)	Hypertension, uremia, coma
Temperature:	Reduction of dissolved oxygen , Increase in Biological Oxygen Demand Early hatching of fish eggs and fish mortality, Mitigation of aquatic biota
Radioactivity	Serious skin cancer, carcinoma, melanoma, breast cancer, leukemia, DNA breakage and cataract
Siltation	Reduced visibility, Reduction in direct light penetration, Decrease in photosynthetic rate, Chances of anaerobic digestion in the benthos zone

CONTROL OF WATER POLLUTION

Following steps will help reduce water pollution:

(i) Prevent run-off of manure. Divert such run off to basin for settlement. The nutrient rich water can be used as fertilizer in the fields.

(ii) Separate drainage of sewage and rain water should be provided to prevent overflow of sewage with rain water.

(iii) Planting trees would reduce water pollution by sediments and will also prevent soil erosion.

For controlling water pollution from point sources, treatment of waste waters is essential before being discharged. Parameters which are considered for reduction in such water are- total solids, biological oxygen demand (BOD), chemical oxygen demand (COD), nitrates & phosphates, oil & grease, toxic metals etc.

Wastewaters should be properly treated by primary and secondary treatments to reduce the BOD, COD levels up to the permissible levels for discharge.

Advanced treatment for removal of nitrates and phosphates will prevent eutrophication. Before the discharge of waste water, it should be disinfected to kill the disease-causing organisms like bacteria.

Proper chlorination should be done to prevent the formation of chlorinated hydrocarbons or disinfection should be done by ozone or ultra violet radiations.

APPENDIX 4.1

Sources, Effects & Control Of Water Pollution

Table: Sources & effects of various constituents / component groups present in waste water

S. No	Component present in wastewater	Effects	Typical sources of wastewater constitutes
1	Biooxidizables expressed as BOD_5	Deoxygenation, anaerobic conditions fish kills, odors	Large amounts of soluble carbohydrates: sugar refining, canning, distilleries, breweries, milk processing, pulping, and paper making
2	Primary toxicants: As, CN, Cr, Cd, Cu, F, Hg, Pb, Zn	Fish kills, cattle poisoning, plankton kills, accumulations in flesh of fish and mollusks	Metal cleaning plating, pickling; phosphate and bauxite refining; chlorine generation; battery making; tanning
3	Acids and alkalies	Disruption of pH buffer systems, disordering previous ecological system	Coal-mine drainage, steel pickling, textiles, chemical manufacture, wool scouring laundries
4	Disinfectants: Cl_2, H_2O_2, formalin, phenol	Selective kills of microorganisms, taste and odors	Bleaching of paper and textiles; rocketry, resin synthesis; penicillin preparation; gas coke, and coal tar manufacture
5	Ionic forms: Fe, Ca, Mg, Mn, Cl, SO_4	Changed water characteristics: staining, hardness, salinity, encrustations	Metallurgy, cement making, ceramics, oil well pumpage
6	Oxidizing and reducing agents: NH_3, NO^{-2}, NO^{-3}, S^-, SO_3^{-2}	Altered chemical balances ranging from rapid oxygen depletion to over nutrition, odors, selective microbial	Gas and coke making, fertilizer plants, explosive manufacture, dyeing and synthetic fiber making, wood pulping, bleaching growths
7	Evident to sight and smell	Foaming, floating, & settleable solids; anaerobic bottom deposits; oils fats, and grease; waterfowl and fish injuries	Detergent wastes, tanning, food and meat processing, beet sugar mills, woolen mills, poultry dressing, petroleum refining
8	Pathogenic organisms: B, anthracis, Leptospira fungi viruses	Infections in humans, reinfection of livestock, plant diseases from fungi-contaminated irrigation water; risks to human sight	Abattoir wastes, wool processing fungi growths in waste treatment works, poultry-processing waste waters

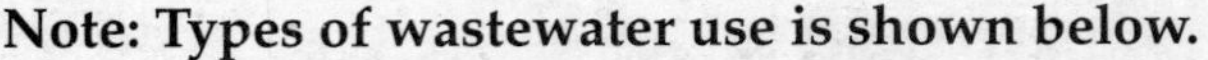

Note: Types of wastewater use is shown below.

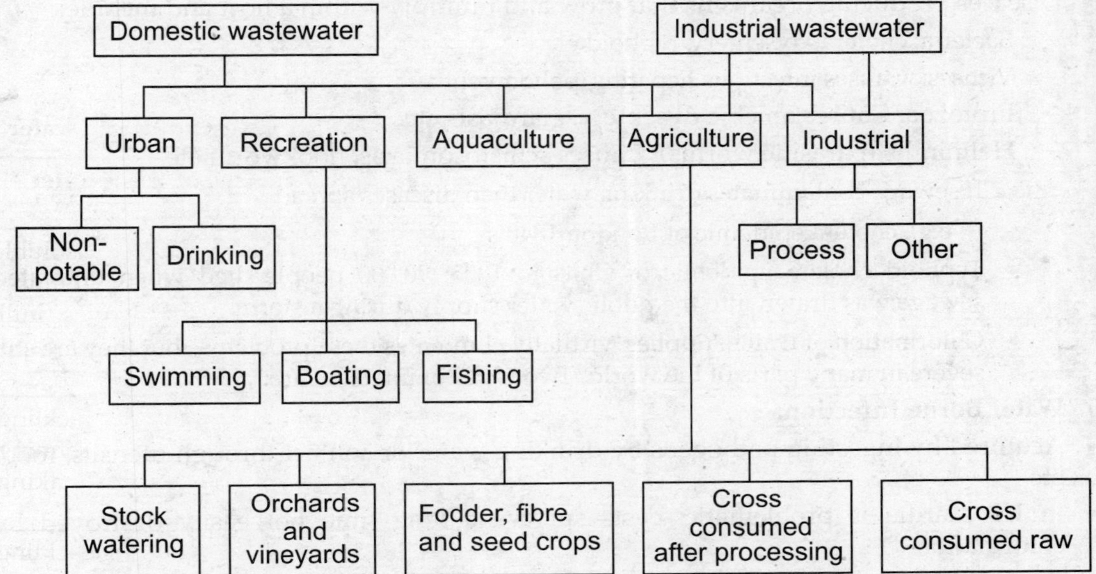

Eight water pollutants of interest are tabulated below:

1	Oxygen demanding wastes
2	Pathogens
3	Nutrients
4	Salts
5	Thermal pollution
6	Heavy metals
7	Pesticides
8	Volatile organic compounds (VOCs)

(1) Oxygen demanding wastes:

- Dissolved oxygen (DO) is important for many forms of aquatic life.
- From Henry's Law we can find the DO concentration in air-saturated water is between 8 &15 mg/l depending on temperature (T) and salinity.
- Higher T Means less DO; Lower T, means more DO.
- Fish need 3 mg to 5-8 mg/l of DO to survive in water.
- The biodegradation of organic wastes from wastewaters (sewage) & industrial effluents create an oxygen demand, so do natural organics and animal wastes.
- As DO falls, odors, tastes, colors make water less acceptable.
- Chemical oxygen demand (COD) means oxygen required for chemical oxidation.
- Biochemical oxygen demand (BOD) means oxygen required by microorganisms to biologically degrade the wastes.

(2) Pathogens

- Diseases producing organisms that grow and multiply within a host and include:
  - Bacteria: cholera, dysentery, typhoid etc
  - Viruses:, Causes infectious hepatitis, poliomyelitis
  - Protozoa: Causes amebic dysentery, giardiasis etc.
  - Helminths (parasitic worms): Causes schistosomiasis, hookworm etc
    - If sewage contaminates drinking water then diseases spread.
      - e.g., cholera epidemic of London 1840's
    - Typhoid cholera epidemic of Chicago 1885: 90,000 people died when untreated sewage was drawn into the public water supply during a storm.
    - Chlorination of water supplies virtually eliminates these problems , but they are still severe in many parts of the world. Two types of infection are:

(i) Water Borne Infection:

- Acquired by ingesting pathogens by drinking water or contact through utensils, food, etc.
- In US giardia is problematic: cysts survive a long time not easily destroyed by chlorination

(ii) Water contact Infection:

- Don't need to drink the water, just be in it.
- Schistosomiasis: eggs develop in snails and mature in liver; snails must have standing water.
- Water breeding insects spread malaria, yellow fever, and the lack of water for hygiene allows skin diseases to develop eg. Leprosy, trachoma.

(3) Nutrients

- Chemicals essential for life
  - Problematic when high concentrations allow for excessive growth of aquatic plants, especially algae.
    - 3 important nutrients are Carbon, Nitrogen & Phosphorous.
    - The one that limits plant growth is called the limiting nutrient
    - Carbon is available from the atmosphere (CO_2).
    - Seawater is generally limited by nitrogen, fresh water by phosphorus. Nitrogen & phosphorus are in wastewater, feedlot and agricultural chemicals.

Nitrate (NO_3) and phosphorus in wastewater is from household detergents containing sodium tri poly phosphate (STPP). Orthophosphate is immediately usable by plants.

(4) Salts

Many salts naturally accumulate in water: e.g. NA^+, Ca^{2+}, Mg^+, K^+, CI^-, SO^{-2}_4, HCO^-_3

Rough measurement of salt content is the total dissolved solids (TDS)

Fresh water has TDS of < 1500 mg/L ;

Saline water > 5000 mg/L

Seawater, 30,000 to 34,000 mg/L TDS

Brackish water up to 5000 mg/L

Drinking water < 500 mg/L (higher: laxative)

Animals can tolerate more than 500 mg/L

Irrigation requires < 500 –2000 mg/L, but Plants would prefer < 500 mg/L of TDS in water.

- Irrigation as well as evaporation from ponds can increase salt concentrations.
- As water flows downstream, there is water added from drainage, further increasing salt concentration.

(5) Thermal pollution of water

The effects of using water in a cooling system where the water discharges are warmer than the input are:

- This can cause stress on aquatic life, which may acclimate to a somewhat different ecosystem but then the plant turns off.
- DO (Dissolved oxygen) Concentration in water as shown below in table is a function of water temperature and chloride

Solubility of Oxygen in Water (mg/l) at 1 atm Pressure

	Chloride Concentration in Water, (mg/l)			
Temperature (°C)	**0**	**5,000**	**10,000**	**15,000**
0	14.62	13.73	12.89	12.10
5	12.77	12.02	11.32	10.66
10	11.29	10.66	10.06	9.49
15	10.08	9.54	9.03	8.54
20	9.09	8.62	8.17	7.75
25	8.26	7.85	7.46	7.08
30	7.56	7.19	6.85	6.51

At the same time, microorganism growth increases.

(6) Heavy Metals

- Metals form cations in solution. Heavy metals have high densities (- >4 of 5 g.cm^3)
  - Most of them are toxic.
- Common Toxics: lead (Pb) , cadmium (Cd), mercury (Hg) and arsenic (As).
  - Toxicity is strongly dependent on the oxidation state of the metal.
- High doses of essential minerals, chromium (Cr^{6+}), iron (Fe), etc. are also toxic.
  - Damages the brain, kidneys or nervous system.
  - Inhaled chromium (Cr^{6+}) clearly causes lung cancer, with very high potency.
  - C^{+6} orally appears to be less toxic because some C^{6+} is reduced to Cr^{3+} in the stomach, however stomach & bladder cancer and liver toxicity rates are elevated in people exposed to Cr^{6+} in drinking water.
- Lead
  - **Causes:** sterility, kidney & neurological damage,. Spontaneous abortions & still births, and more.
  - Lead was used to sweeten wine in ancient Rome, and this combined with lead pipes may have been part of the fall of the roman empire.

- Lead is still used in aviation fuel and in gasoline fuel in many countries like Africa etc.
- Blood, lead is inversely related to Intelligence Quotient (IQ), and the correlation is especially strong at levels between 1 and 10; 10 is the current cut off. A major source of lead in our homes is now batteries.

Metals are not biodegradable.

(7) Pesticides

- Pesticides include insecticides, herbicides, rodenticides & fungicides.
  - Problem is their persistence in the environment.
  - Many are lipophilic (soluble in fat)–accumulate in animal's fatty tissue through chain.
- Organochlorines
  - DDT (para dichloro diphenyl trichloro ethane) is still widely used in the developing world to control malaria.
- Acoustic profiling clearly showed a 20 to 60cm thick low-density sediment layer extending over much of the 3 km by 10 km continental shelf.
- Physical property testing of cores confirmed the presence of a 20 to 60 cm thick low-density sediment layer on the shelf and also showed that the layer extended well down the 3 km by 10 km continental slope.
- Chemical analyses confirmed that virtually the entire sediment is contaminated with DDT and PCBs.
- Significant amounts of DDT and PCBs will remain near the sediment surface and thus remain biologically available well into the next century.
- Acute human toxicities vary, and long term chronic exposure can cause cancer or neurological effects.
- Many organochlorines have since been replaced with organo phosphates and carbamates. The replacement has been motivated by human health concerns, but also developing insect resistance.
- Organophosphates are not as persistent, but have more acute toxicity to humans. They are absorbed through the skin, lungs, intestines.

(8) Volatile Organic Compounds (VOCs)

- Most common groundwater pollutant
  - Mainly solvents
  - Toxicity varies; some are carcinogens.
  - Mutagen: causes cancer
  - Teratogen: causes fetal malformations

Vinyl chloride (chloro ethylene)-Used for making PVC resins , is carcinogen (cancerous)

TCE (trichloroethylene)–solvent to clean from jet engines to septic tanks and CFC production is a Suspected carcinogen.

1,2-dichloroethane (DCA)–used for degreasing; high exposure causes kidney, liver, CNS damage; very soluble

Carbon tetrachloride used for–household cleaning, Fungicide, very toxic; a few mL's can kill.

Perchloroethylene–most widely used dry cleaning solvent. Possible human carcinogen, may cause miscarriages.

Tetrachlorothylene–It is a solvent, causes tumors in animals.

Methyl tertiary butyl ether (MTBE)–formerly used as gasoline oxygenation additive ; currently banned in many parts of the world.

Drinking tap water during the first trimester in Los Angeles (and other places) is associated with increased miscarriage rates.

APPENDIX 4.2

ILL EFFECTS OF WATER POLLUTANTS

S.N.	Pollutants	Effects
1	Zinc	Zinc is essential element for humans, animal and plants. It is also important cell component in several metallo enzymes. Infants need 3-5 mg/day, adult males 15 mg/day, pregnant & lactating females 20-25 mg Zn/day, However, heavy intake causes vomiting , renal damage, cramps, etc.
2	Copper	Excess of Cu in human body (more than 470 mg) is toxic, may cause hypertension, sporadic fever, uremia, coma, Copper also produces pathological changes in brain tissue. However, Cu is an important cell component in several metalloenzymes. Lack of Cu causes anaemia, growth inhibition and blood circulation problem.
3	Barium	Excess of Ba (more than 100 mg) in human body may cause excessive salivation, colic, vomiting, diarrhoea, tremors, paralysis of muscles or nervous.
4	Iron	It is one of the essential mineral for humans and animals. Degree of absorption depends upon solubility and stability of compound. It is a component of blood cells and liveral metalloenzymes. However, more than 10 mg per kg of body weight causes rapid respiration & pulse rates, congestion of blood vessels, hypertension, as many of them require Fe for their growth.
5	Cadmium	Cd is very toxic, 50 mg cause vomiting, diarrhoea, abdominal pains , loss of consciousness. It takes 5-10 years for chronic Cd intoxication, During first phase, discolouration of teeth , loss of smell, mouth dryness occurs. Afterwards it may cause decrease of red blood cells, impairment of bone marrow, lumber pains, disturbance in calcium metabolism, softening of bones, fractures, skeletal deformations, damage of kidney, hypertension , tumor formation, heart disease, impaired reproductive function, genetic mutation, etc.
6	Mercury	Mercury is very toxic. Excess mercury in human body (more than 100 mg) may cause headache, abdominal pain, diarrhea, destruction of hemoglobin, tremors, very bad effects on cerebral functions & central nervous system, paralysis , inactivates functional proteins, damage of renal tissues, hyper coagulability of blood, mama Mata disease, and even coma. It disturbs reproductive and endocrine system. Also causes insomnia , memory loss, gum inflammation, loosening of teeth, loss of appetite, etc.
7	Lead (Pb)	More than 400 mg of lead in human body can cause brain damage, vomiting , loss of appetite, convulsions, uncoordinated body movements, helplessly amazed state, coma. It is retained in liver, kidney, brain, muscle, soft tissues, bones, Leads to high rate of miscarriages, affects skin, & respiratory system, damages kidney, liver & brain cells. Disturbs endocrine system, causes anemia, and long term exposure may cause even death.

S.N.	Pollutants	Effects
8	Arsenic	Poisonous to fishes, animals and humans. Greater than 25 mg of arsenic causes vomiting, diarrhea, nausea, irritation of nose & throat, abdominal pain, skin eruptions , inflammations and even death. It binds globulin of blood haemoglobin in erythrocytes. May cause cancer of skin, lungs and liver, chromosomal aberration & damage, gangrene, loss of hearing, injury to nerve tissue, liver & kidney damage. Minor symptoms of a poisoning, weight loss, hair, nausea, depression, fatigue, white lines across toe nails & finger nails.
9	Vanadium	It is very toxic, may cause paralysis
10.	Silver	Causes pathological change in kidney, liver and may even damage kidney. May cause Argyria (discolouration of skin). Effects mucous membranes and dyes. In high doses, it may be fatal to humans.
11	Radioactive materials/ substances	These generally cause 'Gene' mutation, of body fluids, chromosomal mutations and cancers. Destroy body cell tissue, adversely effects reproductive system. When mother is exposed to radiation during pregnancy, it causes severe mental retardation and leukemia in infants. Radioactive metals are nephrotoxic and damage kidneys.
12	Fluoride	Excess fluoride intake in body results in progressive crippling scourge (sponging)/fluorosis of bones, teeth. May cause metabolic alternations in soft tissues and their mechanism.
13	Selenium	Signs of Se poisoning (more than 4 mg) are fever, vomiting, falling of blood pressure, causes damage to liver, kidney & spleen, loss of nails, causes blindness to animals. Cats are most susceptible. It affects enzyme systems and interfere with sulphur metabolism. It can cause growth inhibition, skin discolouration, bad teeth, psychological problem, gastro intestinal problems, but trace amount of Se is protective against poisoning by Hg Cd, Ag.
14	Chromium	Any chromium compound is toxic but haxavalent Cr greater than 70 mg is very toxic. It causes cancer, anuria, nephritis, gastrointestinal ulceration, perforation in partition of nose. It penetrates cell membrane and badly affects central nervous system. Causes respiratory trouble, lung tumors when inhaled. May cause complication during pregnancy. Has adverse effects on aquatic life. Trace amount of Cr^{III} is essential for normal glucose, protein and fat metabolism and hence it is a essential trace element in diet.
15	Manganese	Mn is essential for mammals but in concentration greater than 100 ppm is toxic,causes retardation fever, sexual impotence, muscles fatigue , eye blindness.
16	Cobalt	High dose (27 mg or above) can cause paralysis, diarrhea, low blood pressure lung irritation, bone defects.
17	Nickel	More than 30 mg may cause changes in muscle, brain, lungs, liver, kidney and can cause canner, tremor*, paralysis and even death.
18	Boron	Essential for plant growth in traces. Harmful to crops and affects metabolic activities of plants in higher concentration, Affects central nervous system.
19	Alkalinity and Acidity	Permissible range of pH value if violated may cause health problems to human & animals and loss of productivity in agriculture.

S.N.	Pollutants	Effects
20	Phosphate and nitrates	Soil nutrient and not toxic in low concentration, Deplete oxygen by excess Algae production-giving bad odour & taste of water and detrimental to aquatic life. They are toxic for human and animal life if concentration is beyond permissible limits. Nitrates also cause cyanosis or blue body disease.
21	Chlorine	Destroys plant & aquatic life and is a biocide.
22	Sulphide	Gives bad odour, toxic to many aquatic organisms and animals
23	Salinity	Very bad for soils which retain salinity, Destroys agricultural land.
24	Oil/Grease/ Oil Sludge	Petroleum product in general are very harmful for soils, aquatic live, animal , human & plant life. They are very toxic. Agricultural land may suffer accumulation of oily waste affecting aeration and fertility. Many constituents of oily sludge are even carcinogenic and potent immune toxicants.
25	Surfactants and detergents	They are toxic and harmful for aquatic life, animals & humans. Inhibit self-purification of water.
26	Phenols	Toxic and impart objectionable odor. Also subdue plant growth generally. Some phenols (nitrophenyl etc) are carcinogens.
27	Cyanides	Cyanide poses a serious health hazard. Apart from acute toxicity and chronic toxicity, it leads to development of iodine deficiency disorders.
28	Pesticides/ Insecticides	Highly poisonous for humans and animals. Also they lower seed germination, plays a role in development of Parkinson's disease, destruction of nerve cells in certain regions of brain resulting in loss of dopamine which is used by nerve cells to communicate with brain. Some of these are physical poisons, some are protoplasmic poisons causing liver damage , some are respiratory poisons and some are nerve poisons.
29	Aluminum	Toxic specially for brain, sometimes may lead to Alzheimer's disease in humans.

Specific Sources of Ground Water Pollution are as follows
(rate of breakdown is extremely slow in ground water)

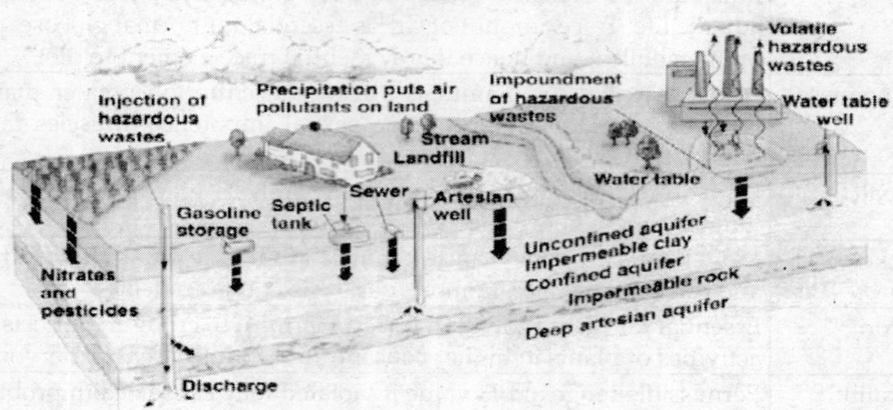

Fig.4.2F$_1$: Specific Sources of Ground Water Pollution

- Industrial waste into aquifer recharge zone
- Surface runoff into abandoned wells-Industry, agriculture, home
- Leaking underground storage tanks of gas stations
- Leaking septic tank into recharge area
- Injection wells for secondary recovery

Impacts of water pollution:

- It causes depletion of oxygen
- It produces foul odour and makes the water oily and brownish.
- Increase the sludge which make the water unfit for recreational and industrial use,
- It induces the growth of algal blooms.

Surface Run-off: The pollutants present on the surface of land and fertilizers are washed down into water bodies.

The nitrites enter the blood and combine with hemoglobin to form methaemoglobin. The letter is unable to transport oxygen and gives rise to disease called as methaemoglobinemia.

Infants are the most affected, showing signs of blueness around the mouth, hands & feet, having trouble breathing as well as vomiting and diarrhoea.

Impacts of Industrial Effluents: They are industrial wastes which are either dumped in the soil or are allowed to pass into cyanides, thiocyanates, chromates, acids, alkalies, organic solvent, etc.

Mercury sources; Coal, smelting of ores, paper/paint industry

Mercury causes Minamata disease. Minamata disease is the name given to mercury toxicosis (poisoning) that developed in people who ate contaminated sea food taken from Minamata Bay in Japan. The disease results in crippling deformity.

Sources & effects of lead pollution: Paints, smelters, chemical and pesticide industries, petrochemicals etc.

Lead poisoning (also known as plumbism, colica pictonium, saturnism, painter's colic) is a medical condition caused by increased levels of the heavy metal lead in the body.

Lead interferes with a variety of body processes and is toxic to many organs & tissues including the heart, bones, intestines, kidneys, reproductive and nervous systems.

Cadmium Pollution Sources include: Smelting and refining of metals, or from the air in plants that make cadmium products such as batteries, coatings, or plastics. Cigarettes are also a significant source of cadmium exposure. Cadmium is used in industry as a protective coat for iron, copper and steel Ex; telephone wires, Paint pigments

Cadmium causes renal damage, emphysema , hypertension and ltai-ltai diseases.

Symptoms of cadmium poisoning: One of the main effects of cadmium poisoning is weak and brittle bones. Spinal and leg pain is common, and a wadding gait often develops due to bone deformities caused by the cadmium. The pain eventually becomes debilitating more common as the bone weakens. Other complications include coughing, anemia, and kidney failure, leading to death

Effects of cadmium poisoning: Organs of toxicity: Central Nervous System. Kidney.

Amoebiasis: Amoebiasis is one those common diseases, caused by a parasite which infects the bowel causing a type of gastroenteritis infection. This disease generally occurs in young to middle aged adults who ingest contaminated food or water containing the Entamoeba histolytica microorganism.

Giardiasis: Giardiasis is an infection of the small intestine that is caused by the parasite, Giardia lamblia The most common manifestations of giardiasis are diarrhoea and abdominal pain, particularly cramping; however, diarrhea is not invariable and occurs in 60 to 90% of patients

Pesticides: They are discharged to the water body through the agricultural run-off.

Chlorinated Hydrocarbon, organic pesticides and inorganic pesticides are poisonous and causes sweating, salication, nausea, vomiting, diarrhea & muscular tremors for the people.

Chromium has carcinogenic properties.

Nickel can cause damage to liver and kidney.

Arsenic can cause hyper-pigmentation, Keratosis and black foot disease

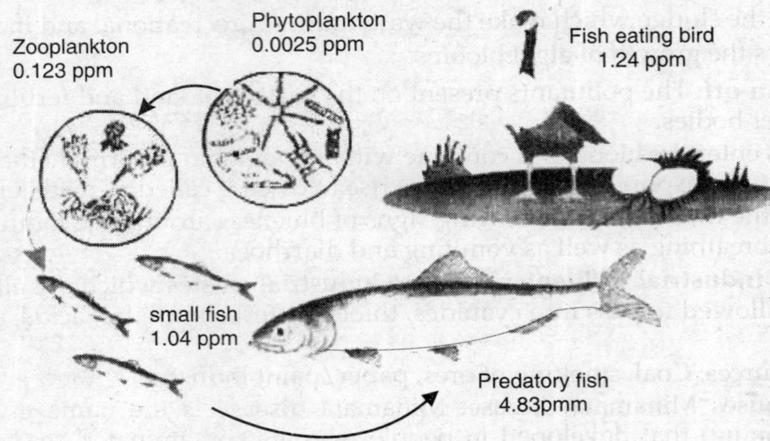

Fig.4.2F$_2$: Biological Magnification Process

Process of Biological Magnification is shown above in figure.

DDT Concentrations increase in organisms along the food chain.

Fluoride Pollution: 96% of fluoride is found in bones and teeth. Fluorine is essential for the normal mineralization of bones and formation of dental enamel.

Water resources near granitic rocks containing more than 2.5 ppm cause Fluorosis The cause of this diseases are mainly due to consumption of fluoride through water, food, cosmetics like fluoridated tooth paste, drugs and inhaling air contaminated with fluoride in industrial environment.

Fluorosis was First detected in India among cattle by the farmers of Andhra Pradesh State during early 1930. The farmers noticed the inability of the bullocks to walk due to painful and stiff joints.This endemic fluorosis had been identified in total 15 state of the Indian union.

Nuclear pollution: The source of nuclear pollution are weapon testing, atomic power plants, recycling plants, nuclear wastes, etc. It causes blood and bone cancer. It disrupts normal functioning of thyroid, and consequently produces abnormal growth and metabolism.

The consequence of two bomb blasts over Hiroshima and Nagasaki (Japan) still send waves of tremor in us. Innumerable persons died. The survivors not only suffered themselves but also passed to their offspring malignant growth, cancer, congenital deformities, mental retardation, etc.

APPENDIX 4.3

Effects of water pollutants on plants/vegetation

The water bodies of the earth are being continuously polluted by a variety of sources. The pollution is occurring in all types of water bodies: both freshwater bodies like ponds, lakes & rivers as well as marine bodies like coastal & deep water seas. Major causes of water pollution are deposition of acid, organic sewage, detergents, agricultural chemicals, industrial effluents, silt , oil and heat into the water bodies.

(i) Effects of acid deposition on plants.

Various acid gases, aerosols and other acidic substances released into the atmosphere from the industrial or domestic sources of combustion of fossil fuels eventually come to the ground. There substances are discharged directly on the water bodies. In addition, these substances also reach the water bodies along with runoff rainwater from the polluted soil. Deposition of these substances causes acidification of water by lowering its pH below 6.0. The sulphates, nitrates and chlorides have been reported to make water bodies like lakes, river & ponds acidic in many countries.

Nutrient deficiency in aquatic ecosystem: The decomposing bacteria and fungi decrease in acidified water. This reduces the rate of decomposition of organic matter and, therefore, the nutrient cycling in the aquatic plants in the affected water body.

Decrease of species diversity: Critical pH for most of the aquatic species is 6.0. The number and variety of aquatic species in the water body generally decreases below this PH.

Change in species composition: The number and abundance of acid tolerant species increases while that of sensitive species decreases.

In the initial phase of water acidification , filamentous algae grow very fast and form thick mats. However, most of the diatoms and green algae disappear below the pH 5.8. Diatoms and small siliceous phytoplankton populations are highly sensitive to pH changes and species composition of their communities shows highly specific changes of the water body. Among green algae, cladophora is highly acid tolerant species and becomes abundant in highly acidic freshwater bodies. Euglena and some other unicellular algae are found up to pH of 1.0.

Macrophysics are generally absent in extremely acidic water. Potamogeton pectinalis is only aquatic saprophyte found in heavily acidified water. At pH lower than 4.0, angiosperm species using dissolved carbon dioxide e.g. Juncos bullous, Juncus effuses, Sporangium emersum, Glycerin fluitans, Eleocharis acicularis, Typhus latifoilia and bryophytes like Polytrichum, Anisothecium, Fantanalis, Catharina become the only survivors. The roots of macrophytes are generally affected adversely in acidic water and result in poor plant growth. Plants with deep roots and rhizomes are less affected while with short root systems are severely affected. Yellowing of plants is common in polluted water.

(ii) Effects of organic matter deposition on Plants:

Large amounts of dead and decaying animal & plant material, fecal material and other organic material is deposited directly from sewage discharges or is washed along with rainwater into the water bodies. Most important consequences of such organic matter deposition are as follows:

Increase in decomposer microbes: Increased addition of organic matter into the water body results in rapid multiplication and increase in decomposer aerobic & anaerobic bacteria. The rapid decomposition of organic matter by these increases nutrient availability in the water.

Eutrophication: Addition of organic matter and its rapid decomposition resulting in increased nutrient supply causes much nutrient enrichment (eutrophication) of water body. In such a condition, planktonic green and blue-green algae grow very rapidly causing water blooms. In addition to these many types of hydrophytes like Salvinia, Azolla, Eicchhornia etc. also become abundant. All this rapid growth of planktonic and free-floating hydrophytese reduces light penetration into deeper layers of water body and submerged flora gradually declines. Abundant flora after death further increases supply of organic matter in the water body.

Oxygen depletion: Rapid decomposition of organic matter by aerobic bacteria eutrophication phase consumes much water dissolved oxygen. On the other hand. Gradual decrease of submerged aquatic flora results in reduced oxygenation of water. Both these phenomena together result in increase in biological oxygen demand (B.O.D.) of the water. Biological oxygen demand (B.O.D) of water is defined as the amount of oxygen needed by a unit volume (usually one litre) of water sample to completely decompose the organic matter present in it by microbial activity. Measured at 20°C and tested at least five days after sampling, The B.O.D. value of fresh, unpolluted water is usually below 1 ppm while in organic matter polluted water, it may be more than 400 mg/litre.

(iii) Effects of detergent deposition on plants:

Various detergents from domestic or industrial use directly released or washed down into the water bodies cause serious effects on plants. Most of the domestic and industrial detergents contain high (up to 40%) phosphate content. Addition of such detergents into water results in phosphate-enrichment of water. Most of the detergents that are toxic, enter the plants through routes of surface absorption. Common effects of detergents on plants are as follows:

- Retardation of plant growth, root elongation, carbon dioxide fixation, photosynthesis, cation uptake, pollen germination and growth of pollen tubes.
- Destruction of the chlorophylls and cell membranes.
- Alteration of the absorption maxima of chlorophylls..
- Binding of membrane lipids and proteins.
- Denaturation of protein and thus causing enzyme inhibition in various metabolic processes.
- Non-degradable alkyl benzene sulphonates and phosphate-rich detergents interfere with gaseous exchange even in very low concentrations.
- Cation-active compounds hinder algal growth between 0.1 and 10.0 ppm while non-ionic compounds hinder algal growth between < 1.0 and around 10,000 concentrations depending upon the species and the compound. Macrophytes are most sensitive to damage by anionic surfactants.

(iv) Effects of agricultural chemical deposition on plants:

Many chemical fertilizers, pesticides, insecticides, herbicides etc. are applied to crops far in excess. Chemicals are washed away by rainwater, first into the soil and then finally into the water bodies. Chemical fertilizers, entering the water bodies result in eutrophication by enriching the water with major plant nutrients. Many of these fertilizers are acidic in nature e.g. ammonium sulphate. These cause acidification of water. Pesticides, herbicides and insecticides also cause pH changes in the water bodies, The effects of these plants on aquatic plants are similar to those of their overdose in foliar application. The herbicides act directly on aquatic flora but insecticides act indirectly by allowing algal blooms to develop in the water body. Different substances have different patterns of their toxic action, decomposition. Some may uncouple oxidative phosphorylation or inhibit nitrate reductase enzyme. The uptake and bioaccumulation of these substances in aquatic plants is great due to their low solubility in water.

(v) Effects of industrial effluent deposition on plants:

Various inorganic & organic wastes from industries and mining activities are directly deposited into the water bodies. A large amount of these substances deposited on the soil, comes to water bodies indirectly along with surface run-off. Fly-ash, various organic/liquid effluents e.g. Hg. Cu. Cd. Pb. Zn. Ni. Ti, Se etc. are important industrial pollutants of water.

Fly ash forms thick, floating covering over the water surface. This reduces the penetration of light into deeper layers of water body. Fly-ash increases the alkalinity of water and thus, causes reduced uptake of essential bases. All these phenomena cause death of aquatic plants.

Organic/liquid effluents disturb the pH of water and depending upon their chemical composition upon their chemical composition, cause specific toxicity effects on the aquatic plants. Change in the floristic composition of the water body is most obvious and direct effect of pollution by such effluents.

Heavy metals usually occur together and with many others of pollutants ; so single metal are usually difficult to interpret. There may be synergistic, additive or antagonistic interactions between metals regarding their effects on plants. Impact of metals is reduced in hard, well-buffered freshwater systems. For example, Cd-uptake by Nitella and Elodea is less in hard water, Zn-toxicity is less with high Ca for Stigeoclouium and Hormidium while is less with high pH for Hormidium, Bioaccumulation of metals is more in mosses than in angiosperms and is usually more in lower plants parts. Chelators decreases while methylated forms increases the metal toxicity to aquatic plants. Reduced oxygen and low temperature also increase metal toxicity. While bryophytes appear to be highly resistant to heavy metal toxicity, in all classes of algae. Stanics vary in tolerance to to Cu^{++} than Scendensmus. Cholrella is retarded more of the algae is inhibited at 1-2 ppm of Cu^{++}. Cholrella is more sensitive ; reduced more by Cd than by Cu or Hg at 0.32 ppm. Lemanea is quite resistant tso Zn and Pb.

(vi) Efects of silt deposition

The top soil removed due to erosion is carried by rainwater or flood water and deposited into the water bodies causing silt deposition in them.The deposition of silt increases the turbidity of water and reduces light-penetration deep into the water, causing decline in submerged flora.Silt deposition, in general, inhibits growth of aquatic plants. Phytoplankton is particularly affected by silt deposition due to reduction in surface exchange of gases and nutrients.

Species tolerant to turbidity (e.g. Ceratophyllum demersum, Lemna minor agg. Nuphar Lutea, Polygonum amphibium, Sagittaria sagittifolia, Scirpus lacutris) becomes highest followed by species of intermediate tolerance (e.g. Callitriche spp. Myriophyllum spicatum potamogeton natans. P, pectinatus, Sparganium emersum. s. erectum) while least tolerant species (e.g. Elodes Canadensis, Poamogeton perfoliatus , Ranunculus spp., Mosses) are much reduced.

(vii) Effects of oil deposition on plants:

Washing of oil tankers and storage containers in many rivers, large lakes and particularly near sea coasts causes deposition of oil slicks on water. Oil spills in water are also common during normal transport operations during accidents involving oil tankers.

- Oil pollution of water body prevents oxygenation of water.
- Oil depletes oxygen of the water body by consuming dissolved oxygen in oil degradation.
- Oil inhibits planktonic growth and photosynthesis in aquatic macrophytes.
- Oil may even cause destruction of aquatic flora , if it catches fire.

(viii) Effects of waste heat deposition:

Many industries, particularly thermal power plants, take water from rivers, lakes or sea to cool the heat-producing boilers and equipment. The heated water is then returned to the water body. The deposition of waste heat into the water body has many consequences for the plants in it.

The solubility of oxygen in water is reduced at higher temperature. Thus, the reduced oxygenation of water adversely affects the aquatic flora. Reduced oxygenation and high temperature of water causes reduction in the activity of aerobic decomposers. The reduced decomposers result in decreased organic matter decomposition and consequently, reduced nutrient availability in the water body.

In high temperature, aquatic plants show increased respiration, reduced photosynthesis and general inhibition of enzyme activity with increasing temperature.The aquatic flora and primary productivity of the aquatic ecosystem declines with temperature. Green algae are mostly replaced by blue-green algae, which have comparatively less primary productivity. With increase in temperature, species diversity of the water declines and heat-tolerant species gradually become dominant.

Note: Some of the viruses, causes different types of disease to plants, man and animals. Viruses affect vast areas of cropland and cause fatal disease to cattle.

The following are the names of some plant viruses, diseases caused by them and the place on which they are attacked.

Table 4.3T_1: Site of infection & disease caused by various plant virus

S.N	Name of plant virus	Name of Disease	Site of Infection
1	Tobacco Mosaic virus (TMV)	Mosaic	Leaves of tobacco Plant
2	Potato Mosaic virus	Mosaic	Leaves of potato
3	Bean Mosaic virus	Mosaic	Leaves of been Plant
4	Pea Mosaic virus	Mosaic	Leaves of pea Plant
5	Sugarcane busy virus	Mosaic	Leaves of sugar beet Plant
6	Tomato busy virus	Leaf curl	Leaves of tomato Plant
7	Tungro virus	Tungro virus infection of vice	Rice plant
8	Wheat yellow mosaic virus	Yellow mosaic of wheat	Wheat leaf

The virus responsible in causing tobacco mosaic , a disease of tobacco plant resulting in chlorosis of leaves and also alternate patches of yellow &green colour in leaves giving mosaic appearance is called Tobacco Mosaic virus or TMV.

Virus infections are transferred from one place to another by the help of certain agencies. Some of the viruses are easily transferable while others are transmitted by artificial means with difficulty.

Transmission of Plant Viruses

- **Transmission through soil:** Cherry leaf roll, Tobacco mosaic viruses, etc. are soil borne diseases. The viruses responsible for these diseases are actually carried by chytrids, nematodes, etc. from soil.
- **Transmission through contact:** Contact between infected and healtly plants may result in the transmission of viruses. For example, TMV may be transmitted in this way through the agency of field workers.
- **Transmission through roots:** Viruses can be transferred through roots which are injured during cultivation.
- **Transmission through vegetatively propagated plant part:** Infected tubers, (e.g. Potato), bulls (e.g. Onion), etc, may transmit viruses during their vegetative propagation by grafting or cutting. Such a process is most common in the infection of sugarcane & potato with mosaic and leaf roll viruses.
- **Transmission through seeds:** Infected seeds carry viruses which are ultimately transferred to the resulting seedlings. The viruses can also be transmitted to healthy seeds which come into contact with infected seeds. Bean mosaic virus is transmitted through seeds.

  Pathogenic Bacteria, for plants Respective Diseases and their site of Infection is tabulated below.

Table 4.3T_2: Site of infection & disease caused by various pathogenic bacteria for plants

Bacteria	Disease	Site of Infection
Pathogens for Plants:		
Streptomyces scabies	Scab disease of potato	Potato tuber
Xanthomonas citri	Citrus canker	Young branches, leaves and fruits of citrus
Erwinia atroseptica	Black rot of potato	Stem and tuber of potato
Pseudomoans maculicola	Cauliflower spot	Cauliflower
Bacterium stewartii	Wilt disease of corn	Vascular tissues of corn
Agrobacterium rhizogenes	Hairy root of apple	Meristemtaic tissue of apple

APPENDIX 4.4

Point and Non Point Sources of Water Pollution

Point Sources	Non point Sources
• Wastewater effluent, both municipal and industrial	• Runoff from agriculture (including return flow from irrigated agriculture)
• Runoff and leachate from waste disposal sites	• Runoff from pasture and range
• Runoff and infiltration from animal feedlots	• Urban runoff from unsewered areas and sewered areas with a population of less than 100,000
• Runoff from mines, oil fileds, and unsewered industrial sites	• Septic leachate and runoff from failed septic systems
• Storm sewer outfalls from cities with a population of greater than 100,000	• Runoff from construction sites smaller than two hectares
• Runoff from construction sites larger than two hecteres	• Runoff from abandoned mines
• Overflows of combined storm and sanitary sewers	• Atmospheric deposition over a water surface
	• Activities on land that generate contaminants, such as logging, wetland conversion, construction and development of land or waterways

Note: The water, pollution may be caused by the addition of physical & chemical impurities, through addition of water from houses, industries or by surface runoff from agriculture & forests.

The sources of pollution are classified into (a) Point sources and (b) non point sources

Point Sources

Point sources are discrete and confined. They emanate either from houses, offices, other commercial buildings or from industries.

(a) Sewage or Domestic Waste: Waste water emanating from house hold activities , other commercial buildings, toilets, bathrooms , kitchen, offices, commercial centers etc. is termed as sewage. This has mostly biodegradable impurities. Some nonbiological chemicals viz. detergents, phenols may also be present. Some of the important properties are:

(i) Physical Properties:

Color – Natural decay of organic materials

Odor- Decomposing waste waters

Solids – Residues of domestic products, salts, soil erosion

(ii) **Chemical properties** (Residues of material used)
Organic-Carbohydrates, Proteins, Hydrocarbons

(iii) **Primary pollutions:**
Organic and inorganic compounds selected on the basis of their suspected Carcinogencity, tetragenicity or high toxicity
Surfactants: Soaps and detergents
Volatile organic compounds

(iv) **Inorganic:**
Alkalinity
Chlorides
Heavy metals (trace quantities)
Nitrogen
pH
Phosphates
Sulfates

(v) **Gases:**
Hydrogen Sulfide
Methane
Oxygen

(vi) **Biological:**
Animals : Protozoans, Nematodes
Plants : Algae
: Pathogenic bacteria
: Indicator bacteria
Virus : Phages and animal virus

Important Pollutants of Concern

- Solids/Residue : Total Solids, Suspended Solids, Dissolved Solids
- Biodegradable (B.O.D./C.O.D.) : Biochemical Oxygen Demand, Chemical Oxygen Demand
- Pathogens : Salmonella, Shigella and Cholera vibro
- Non Pathogens : Indicator bacteria – Coliform/ Faecal Coliform
- Nutrients : Nitrogen saline, Organic, nitrite & nitrate; phosphates
- Priority pollutants : Known to produce cancer and other disabilities
- Refractory organics : Surfactants, Phenols
- Heavy metals : Fe, Ca, Mg, Hg, Pb

(b) **Industrial Wastes:** The volume, strengths of industrial water entirely depends on the type of raw material used and the processes involved. Some components of industrial wastes would be.

Inorganic Salts : Chlorides (Sanlinity), Hardness (Scale forming): Acids/ alkalies

Organic Compounds : Organic materials used
Floating Solids : Suspended , Dissolved , Oil and grease
Toxic Chemicals : Pb, F, As, Se, Cr, Cu, Fe, Mg, Zn, Hg , Phenols, pesticides

Microbes:Tanneries, Slaughter house waste and food industries will have variety of microbes.

Radio active nucleotides: Atomic energy units.

The volume and strength is expressed in terms of kL waste produced, BOD and suspended solids.

Non Point Sources

Non point sources are diffused and intermittent, They are picked up and carried by run off. **The important non-point sources are:**

Dairies, dhobi ghats, motor garages, cattle wallowing, carcass disposal, open defecation, crematorium, river fronts. They have sediment nutrients and decomposable compounds.

Classes0 of non-point source pollution are tabulated below:

Table 4.4T$_1$: Sources and Examples of various Classes of non-point sources of water pollution

S.N	Class	Source	Examples
1	Agriculture	Animal feedlots Irrigation Cultivation Pasture Dairy farming Orchards Aquaculture	Runoff from all categories of agriculture leading to surface and groundwater pollution. In northern climates, runoff from frozen ground is a major problem, especially where manure is spread during the winter. Vegetable handling, especially washing in polluted surface waters in many developing countries, leads to contamination of food supplies. Growth of aquaculture is becoming a major polluting activity in many conutries. Irrigation return flows carry salts, nutrients and pesticides. Tile drainage rapidly carries leachates such as nitrogen to surface waters.
2	Forestry	-	Increased runoff from disturbed land. Most damaging if forest clearing for urbanization.
3	Liquid waste disposal	-	Disposal of liquid wastes from municipal wastewater effluents, sewage sludge, industrial effluents and sludges, wastewater from home septic systems; especially disposal on agricultural land, and legal or illegal dumping in water courses.
4	Urban areas	Residential Commercial Industrial	Urban runoff , roofs, streets, parking lots, etc. leading to overloading of sewage plants from combined sewers, or polluted runoff routed directly to receiving waters; local industries and business may discharge wastes to street gutters and storm drains; street cleaning; road salting contributes to surface and groundwater pollution.
5	Rural sewage systems	-	Overloading and malfunction of septic systems leading to surface runoff and/or direct infiltration to groundwater.
6	Transportation	-	Roads, railways, pipelines, hydro-electric corridors, etc.

S.N	Class	Source	Examples
7	Mineral extraction	-	Runoff from mines wastes, quarries, well sites.
8	Recreational land use	-	Large variety of recreational land used, including ski-resorts, boating and marinas, comp grounds, parks; waste and "grey" water from recreational boats is a major pollutant, especially in small lakes and rivers, Hunting.
9	Solid waste disposal	-	Contamination of surface and groundwater by leachates and gases. Hazardous wastes may be disposed off through underground disposal.
10	Dredging	-	Dispersion of contaminated sediments, leakage from containment areas.
11	Deep well disposal	-	Contamination of groundwater by deep well injection of liquid wastes, especially oilfield brines and liquid industrial wastes.
12	Atmospheric deposition	-	Long-range transport of atmospheric pollutants (LRTAP) and deposition of land water surfaces. Regarded as a significant source of pesticides (from agriculture, etc.), nutrients, metals, etc. especially in pristine environments.

An example of a non-point source is agricultural runoff that enters a stream as overland flow. During precipitation events, runoff from cultivated fields may contain pesticide residues and fertiliser, as well as suspended sediments that absorb contaminants. Obviously, non-point source pollution is much more difficult to identify, measure and control than point sources. **Point and Nonpoint Sources are shown below in figure.**

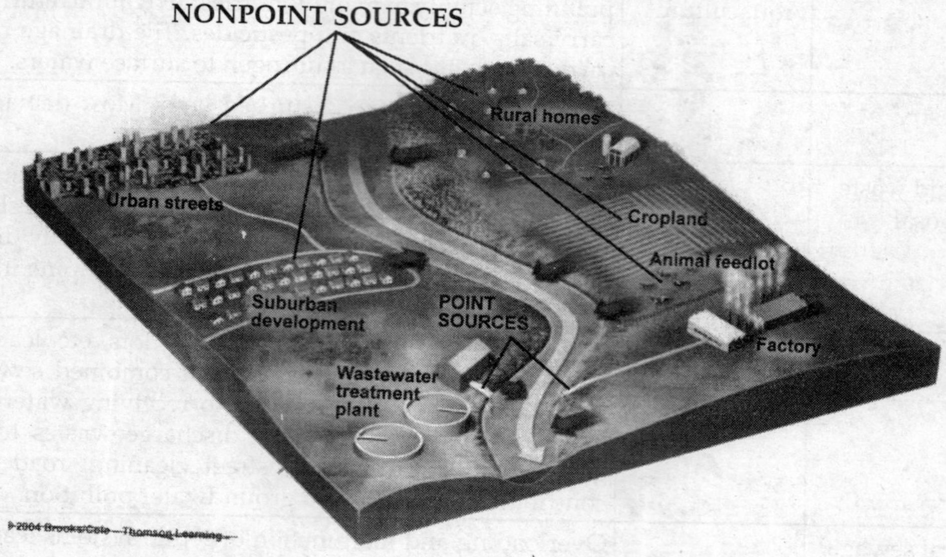

Fig.4.4F$_1$: Point and Non point Sources of water pollution

APPENDIX 4.5

All about the groundwater pollution

Most of the rainfall is lost to evaporation. Most of the rest runs off the surface into streams & lakes, and eventually returns to the sea, Some, however, slowly percolates down through the soil until it reaches a layer of impermeable rock. The water which builds up on the top of the impermeable rock fills all the openings, crevices, slowly flows to the oceans. As shown in Figure below, when the ground level drops below the water table, the groundwater helps supply streams and lakes giving a continuity of discharge they would not otherwise have. Groundwater resources are extremely important because they supply to withdrawal. Difficulties exist with this in that the rate of replenishment of these supplies is so that they must be considered to be essentially non renewable rsesources.

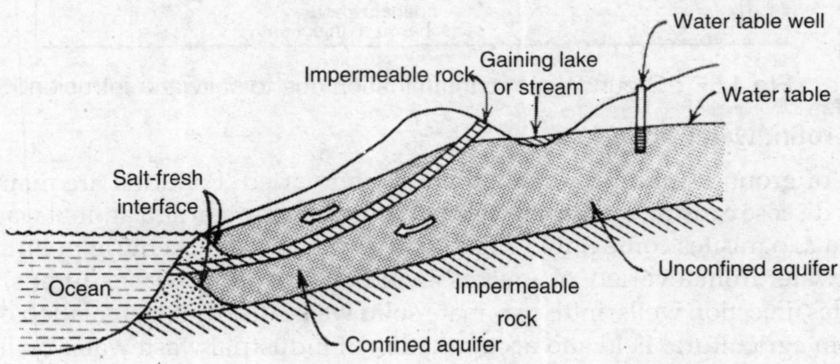

Fig.4.5F$_1$: Groundwater resources

One problem that can result from a drop in the water table is a corresponding sinking of the ground level called subsidence. Another problem that can result from over exploitation of groundwater is that water table can drop so low that it becomes uneconomical to pump it out. When this happens, the area involved may lose its principal water supply. Coastal areas face another sort of groundwater problem because of the close proximity of saline and fresh water. Under natural conditions there is a balance between the inflow of salt water and the outflow of fresh water, which establishes a boundary between the two regimes. However, as freshwater is pumped out of the area, the water table drops and the boundary moves inland. After a period of time a coastal freshwater well may need to be abandoned because of this intrusion.

Ground Water Contamination due to its movement is shown below in fig.

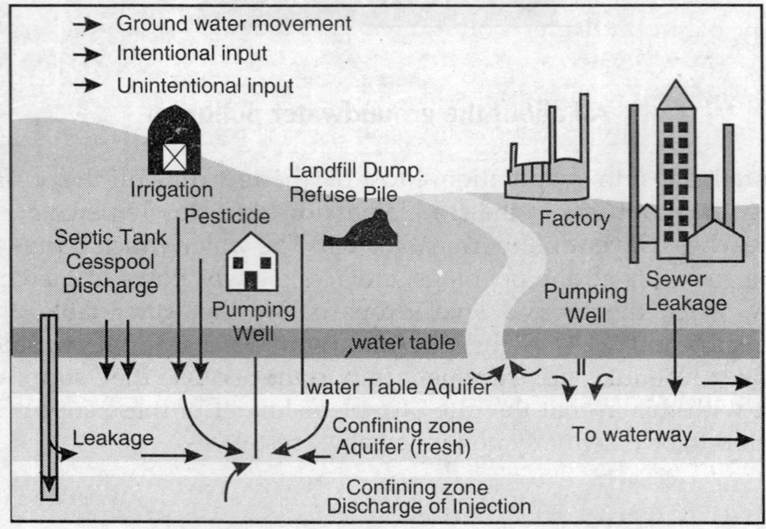

Fig.4.5F$_2$: Ground water contamination due to saltwater intrusion

Source of Groundwater Pollution

The sources of groundwater pollution are many and varied. Pollution are many and varied. Pollution by disease causing micro organisms occur when human and animal waste containing virus, bactria & parasites come into contact with groundwater. Chemical pollutants can leach into groundwater from a variety of sources including hazardous waste dumps, sewage, land treatment sites, injection wells, indiscriminate solid waste disposal, percolation of pesticides & fertilizer from agricultural field and accumulation of industrial wastewater on land.

Abandoned chemical storage tanks and leaking underground storage tanks are potential sources of groundwater pollution.

Sources of Groundwater Pollution is shown in below

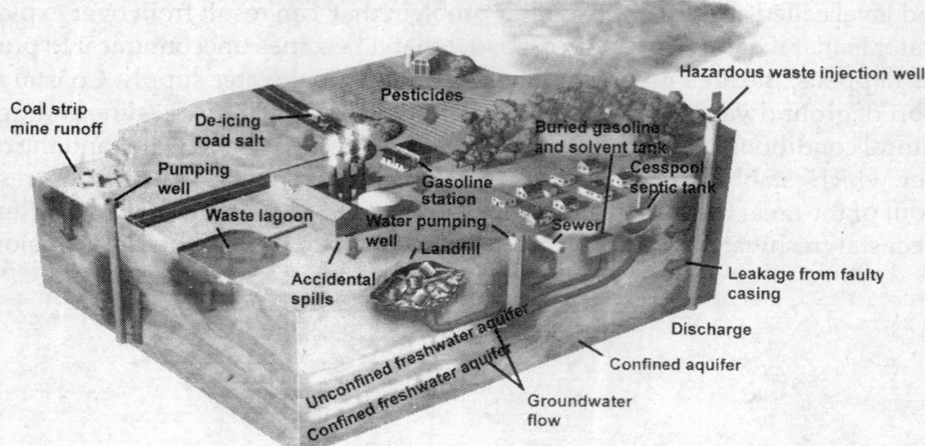

Fig.4.5F$_3$: Various Sources of Groundwater Pollution

Impact of Groundwater Pollution

The severity of groundwater pollution is of particular significance on account of the following:

- Its vulnerability to pollution:
- The associated complexity in pollution source identification
- Limited feasible options for treatment of groundwater, all of which are cost prohibitive;
- Complicated process in fixing geophysical boundaries
- Difficulty in predication of movement
- Insufficient dilution
- Slow movement
- Lack of any natural cleansing capability

A Variety of adverse impacts due to Groundwater contamination is possible including effects on public health, the environment, agricultural productivity (e.g. due to increased salinity in irrigation water) and on the output of industries requiring high quality wa.er

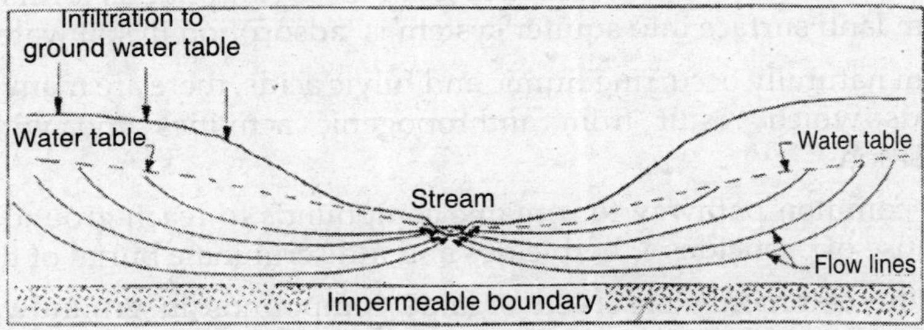

Fig.4.5F$_4$: Groundwater infiltration route

Groundwater infiltration route is shown above in figure.

The type of contaminants usually found in groundwater are discussed below.

A Biological contaminants

Pathogenic biological organisms that are found in groundwater include:

- Bacteria
- Viruses (e.g. Entero-viruses and hepatitis)
- Parasites (e.g. Protozoa, worms and fungi)

  The potential for bacterial contamination of groundwater depends on both the survival rate of the species and characteristics of the sub-surface eg. Moisture content, pH & temperature. Bacterial contamination most commonly result from the introduction of human (or animal) fecal material usually when septic tanks or cess pools leak or over flow. The common disease associated with biological contaminants present in groundwater can be one or more of the following:

- Amoebiasis
- Balantidiasis
- Cholera
- Gastro enteritis

- Hepatitis-A
- Salmonellosis
- Tularemia
- Dracontasis
- Giardiasis
- Paratyphoid
- Typhoid
- Shigellosis

b. **Inorganic Contaminants:** The concentration of most bulk and trace inorganic contaminants found in groundwater varies by area and in many instances, over time.

Some of the common inorganic constituents of groundwater are given in the table.

c. **Organic Contaminants:** The Organic substances that pose the greatest threat to the quality of groundwater are those that are relatively soluble ,non-volatile and refractory. The main mechanism that influences most of these compounds form readily migrating water from land surface into aquifer system is 'adsorption in soil water Zone'.

Apart from naturally occurring humic and fulvic acids, there are many complex organic compounds which result from anthropogenic activities and withstand bacterial decomposition.

The most common pathway for organic compounds to reach groundwater is through excessive use of pesticides & herbicides and indiscriminate burial of their containers.

Some of the commonly observed organic compounds in groundwater are Aldrin,, llinden, methoxychlor, toxaphene, DDT, trichloroethylene, carbon tetrachloride, tetrachloro ethylene, carbon tetrachloride, etc, Almost all the above compounds are known to promote/cause carcinogenicity.

d. **Radionuclides:** These result as waste from nuclear plants, nuclear weapon testing or use of radionuclides in scientific laboratories medicine and industry.

The most commonly found radionuclides in groundwater is Radium-226. It results from specific geological conditions and hence classed as naturally occurring. The abundance of radioactive gas Radium-22 formed by the decay of Radium-226 can also be present in groundwater but has no appreciably adverse impact. However, continuous consumption of water containing Radium-226 at 5 pci/liter may cause between 2 cancer cases per year per million exposed persons.

Unlike surface water, wherein information on direction of movement, Velocity, Sources of contamination are available; monitoring of groundwater involves various investigation prior to the actual assessment of groundwater quality. For groundwater quality the first information to look for is the movement of groundwater.

The most abundant dissolved gases present in groundwater are nitrogen, oxygen, carbon-dioxide, methane, hydrogen sulphide and nitric & nitrous oxide. While the last three gases mostly result from bio geo chemical process , the rest are mostly contributed by atmospheric influence in their relative order or abundance.

Dissolved inorganic constituents in groundwater are tabulated below.

Table 4.5T_1: Dissolved inorganic constituents in groundwater

Major constituents (greater than 5 mg/l)		Copper	Rubidium
Bicarbonate	Silicon	Gallium	Rutenium
Calcium	Sodium	Germanium	Scandium
Chloride	Sulfate	Gold	Selenium
Magnesium		Indium	Silver
Minor constituents (0.01-10.0 mg/1)		Iodide	Thallium
Boron	Nitrate	Lanthanum	Thorium
Carbonate	Potassium	Lead	Tin
Fluoride	Strontium	Lithium	Titanium
Iron		Manganese	Tungsten
Trace constituents (less than 0.1 mg/I)		Molybdenum	Uranium
Aluminium	Bromide	Nickel	Vanadium
Antimony	Cadmium	Niobium	Ytterbium
Arsenic	Cerium	Phosphate	Yttrium
Barium	Cesium	Platinum	Zinc
Beryllium	Chromium	Radium	Zirconium
Bismuth	Cobalt		

Groundwater Pollution - responsible factors and their responses are tabulated below.

Table 4.5T_2: Responsible factors and their responses for Groundwater Pollution

Responsible Factor	**Most Probable Response**
Groundwater pollution originating on land surface:	
Infiltration of contaminated surface water	Contamination of stream side aquifer due to polluted stream
Land disposal of waste	Contamination due to direct disposal of waste
Stock piles (ore) tailings (over burden dumps)	Release of mineralized leachate
Disposal of Sewage/Sludge	Release of Biological mineralized leachate
Salt spreading on road	Pollution due to winter time road salting
Animals feedlots	Biological waste
Fertilizers and Pesticides	Run-off resulting from indiscriminate use of such items
Accidental spills	Spill of in-transit chemicals and contamination due to spray water used during such mishap
Air borne source	Acid/alkali rain particulates as fall out from smoke/flue dust automobile pollutants

Responsible Factor	Most Probable Response
Groundwater pollution originating above the groundwater table:	
Septic tanks Cess pools	Biological contamination of groundwater
Surface impoundment	Leachate from lagoons for storage/treatment of sewage , industrial wastewater , oil , brines , spent acids etc.
Underground storage tanks/pipelines	Corrosion and /or leakage
Artificial Recharge	In case of improper operation the recharge may lead to increased concentration of nitrates, metals, bacteria. Viruses, detergents etc.
Groundwater pollution origination below the **groundwater table:**	
Waste disposal in wet excavations	Contamination through abandoned mines
Agriculture drainage wells	Drainage of agricultural residues from marshes/ponds
Well disposal of waste	Contamination due to direct injection of waste
Secondary recovery of petroleum	Migration and ingress of hydrocarbons
Mines	Percolation of mine water
Exploratory wells and test holes	Inter-linking of aquifers leading to dissemination of pollutants
Abandoned wells	Direct migration of mineralized fluids
Water supply wells	Contamination by surface run off
Excessive groundwater development	Salt water ingress

Once it is established that contamination of groundwater has occurred, some action must be taken to-

- Find and eliminate the sources
- Contain the contaminants in the area already affected
- Restore the water quality of the aquifer.

Wherever feasible identification and elimination of contaminants must be given the first priority. In case of difficulty, the other option source is the aquifer restoration. Leakage is observed in virtually all land-fills, even if various types of plastic liners or clay liners are used to contain the leachate. Capping abandoned land-fill sites with bentonite or other low permeability material prevents rain water from entering the site, thus eliminating the formation of leachate.

A combination of 'containment' and 'abatement' is one way to effect aquifer restoration. Containment usually focuses on some hydraulic means of preventing the spread of the contaminant either through withdrawal of contaminated water or the injection of clean water to create pressure ridge. Withdrawal of groundwater gradient, thereby preventing the advance of the contaminant front.

Slurry walls can also be used to isolate areas of contaminated groundwater. Slurry walls consists of bentonite, water and backfill material placed in deep trenches as deep as 30 metres. Slurry wells last 20 to 40 years, however its life is greatly affected by the groundwater chemistry.

Additional methods of aquifer restoration include:

- Re-oxygenation of groundwater wells by means of air compressors to accelerate the growth of aerobic bacteria that metabolize contaminants.
- Addition of nutrients to wells to stimulate the growth of bacteria.
- Recharge of aquifers to facilitate flushing out contaminants adsorbed onto soil particles.
- Air stripping the contaminated water in an sir-stripping tower to remove volatile organics.
- While monitoring the groundwater level it has been observed that in most of man induced factors leading to over exploitation of groundwater , the effective groundwater resource management stresses on two-pronged strategy, the one relates to 'Static' groundwater level and the other 'Dynamic' groundwater level. While the static water level pertains to the levels not affected by recharge and other seasonal influence, the dynamic water level-has greater, influence of seasonal impacts.

Long term water level decline as a results of injudicious exploitation of groundwater resource may lead to several problems like reduced well yields, land subsidence, intrusion of salty water subsidence, intrusion of salty water especially in coastal areas, leakage into the aquifer of highly mineralized water. In order to overcome these serious environmental implications, the recharge potential of groundwater resource has to be equally or in some cases more important aspect than the abstraction potential.

Groundwater is derived primarily from rain & snow melt that infiltrates the land surface and slowly percolates to the water table. This process of adding water to underground storage is called 'natural' groundwater recharge. Where the materials in the earth's surface are coarse and the slope of the land is gentle, there is generally more groundwater recharge than in areas where the strata consists of fine-grain material such as shale & clay, or where the slope is steep. The purpose of 'artificial recharge' is to increase the rate at which water infiltrates the land surface in order to supplement the quantity of groundwater in storage.

The type of artificial recharge system that can be developed at any specific site, in addition to economic considerations should include at least the following:

- Source of recharge water
- Chemical and physical characteristics of the recharge water
- Chemical characteristics of water in the aquifer
- Availability of an aquifer suitable for artificial recharge
- Thickness and permeability of the aquifer.
- Proximity of the potential recharge site to an appropriate well field cone of depression.
- Water level difference between the aquifer and the recharge site
- Topography.
- Availability of property (land)

Artificial Recharge techniques

Induced filtration

It is basically a method to artificially recharge the aquifers along river, lake or other such surface water bodies. The salient features for this method are as follows:

- Depends to a large extent on the quantity of water that can be diverted from a stream or lake.
- Both horizontal and vertical wells are used.
- Permeability of stream side deposit and its linkage with aquifer to be recharged are the important controlling factor.
- During or after the floods resulting in deposition of low permeable material, dredging is recommended to restore the permeability of river bed medium.
- Chemical characteristics of the recharge source is of prime concern.

Water Spreading

It involves spreading the water mantle over an extensive area with flat/gentle topography. The controlling factors are:

- Water contact time
- Soil permeability
- Area of inundation

In this technique recharge is facilitated through flooding the available area, by excavating ditches, modification in natural channel by building low-head check dams, irrigation during growing season etc. An attractive prospect of irrigation involves treated sewage which further undergoes natural treatment and augments recharge of groundwater simultaneously.

Recharge Pits and Shafts

From a regional point of view, conditions permitting recharge through water spreading technique are rare. In more general cases where aquifers and land surface are separated by low permeable of impermeable material artificial recharge system must penetrate the less permeable strata in order to access the aquifer system to be recharged. Some of the salient issues with Recharge pits and Shafts are:

- They are one of the useful recharge facilities especially in relatively small areas/areas where space or land availability is difficult.
- Construction and maintenance is simple relative inexpensive.
- Low turbidity and low microbial contamination of the source water are some important pre-requisite and are to be seriously looked into especially to avoid plugging of the pits and shafts.

Injection Wells

- In comparison to recharge pits and shafts, the injection wells tap deep aquifers
- Expensive to construct and maintain
- High tendency of plugging
- Introduction of well screen/casing is recommended to prevent caving of aquifer.

Artificial Recharge in Urban areas

Due to space constraints and excessive exploitation of groundwater the most acceptable and feasible mechanisms of artificial recharge in urban areas are Roof top water collection, Storm run-off collection, and Street recharge pits/trenches.

Advantage of Groundwater Recharge

Reduction of land subsidence

Renovation of waste water

Improvement of Ground water quality

Storage of Ground water quality

Storage of stream water during periods of high or excessive flow.

Reduction of flood flows

Increase well yield

Decrease well yield

Decrease the size of areas needed for water supply systems

Reduction of salt water intrusion or leakage of mineralized water

Increase stream flow

Secondary recovery of petroleum (Oil)

Groundwater Treatment Technologies

The clean up of contaminated groundwater is very different from clean up of wastewater on the surface. The most obvious difference is that with groundwater clean ups, the body of water is actually being cleaned. In wastewater (surface) clean ups, we control and treat the waste water that is entering a body of water. The water body river or lake actually cleans itself once we stop putting pollutants into it. Groundwater is not able to clean itself hence we must clean-up the source of pollutants as well as the aquifer itself.

Treatment for Organic Contaminants

Physico-Chemical methods: When a pure organic contaminant is released onto or into the ground the main force on the movement of the compound is gravity. If the ground is porous the spill will move downward. There will also be some lateral spread controlled by soil porosity. Eventually the contaminant may be absorbed on the soil particles, stopped by an impermeable layer of soil or encounters with an aquifer. On reaching the aquifer the contaminant may:

- Float on top of aquifer
- Dissolve into the aquifer
- Sink to the aquifer bottom

For floating contaminants on top of aquifer the best treatment method recommended is Pure compound Recovery. For other possibilities the specific treatment options are discussed below:

a. Pure compound Recovery

This method is applicable for insoluble organic contaminants including most of the straight chain hydrocarbons. The treatment mechanism involves placing a well in the middle of the spill area. Water is removed and the water level in the immediate vicinity decreases. The non salable compounds accumulating in the well, keep floating on top of water and are recovered through oil water separators or through introduction of hydrophobic screens.

Other mechanism to recover pure compounds especially in areas of shallow water table is to dig a trench in the path of ground movement. When Groundwater enters the trench, the floating organic remains on top of water and is later removed by skimmers.

b. Air Stripping

It is a process employed to remove volatile compounds (usually organics) from groundwater. The basic concept of stripping is to bring the contaminated water into intimate contact with air so that the volatile compounds undergo a phase change (from liquid phase to vapor phase). The air will then carry-off the contamination, leaving water free from these compounds. Air striping is accomplished in following ways by employing:

- Aeration tanks
- Cascade aerators
- Spray basins
- Packed towers

c. Carbon Adsorption

This involves use of carbon for removal of odors and taste. The use of activated charcoal is an established technology especially for organic compounds even though these contaminants do not exhibit traditional taste.

Carbon adsorption is particularly suited for low concentration of non-volatile components e.g. phenols, pesticides and high concentration of non-degradable compounds. For better removal efficiency of non-degradable compounds, this method should be combined with Biological 'treatment' or 'Air striping'

d. Ultra Filtration and Synthetic Resin

While ultra filtration method is used for separation of emulsion, the synthetic resin perform similar function as activated carbon, however they can be made more selective and with higher treatment capacity.

e. UV-Ozone Reactors

This technology is used in rapid oxidation of low concentration of chlorinated organics in groundwater and is rated as one of the best treatment technologies.

Biological methods

Once the contaminant has entered the aquifer, biological treatment is the most economical method of treatment. Biological method employing specific bacteria are recommended for 'in-situ' treatment as well as 'above-the-ground' treatment. In both the cases the treatment system is provided with bacteria, nutrient and other favorable condition of pH , availability of oxygen etc. In 'in-situ' treatment applied for both organic and inorganic contaminants selected acids or alkalis are placed in the path of groundwater. In some cases an immobilized agent is introduced underground. The contaminant plume on contact with the acid-base neutralizing agent and in presence of de-composing bio-chemical reaction under controlled environment with excess or oxygen deficient conditions based on aerobic or anaerobic bacteria employed.

'Above-ground' treatment of groundwater requires placement of well in the contaminated aquifer which undergoes biological treatment. The biological treatment above ground are analogous to conventional wastewater treatment technologies wherein, 'Suspended Growth Biological Reactors' or 'Fixed Film Reactors' are employed. The treatment configuration consists of 'Aerated lagoon', and 'Activated Sludge Process'.

Treatment for inorganic Compounds

Unlike organic contaminants, the inorganic contaminants have, a greater chance of removal, transformation and adsorption in the unsaturated zone and hence possess lesser chance of reaching the aquifer zone. As a result of this , it is observed that heavy metals moving through the soil are exchanged with certain material in the soil and later removed. In certain cases of anaerobic zones in the soil, biological transformation of nitrates into nitrogen gas has been observed. The treatment methods usually recommended for inorganic constituents are discussed below. Choice of method and their efficacy may involve bringing the portion of groundwater above ground or in certain cases proved 'in-situ' treatment.

Chemical Addition

This is achieved by way of pH adjustment. It results in precipitation of heavy metals dissolved in groundwater. The pH adjustment (acid-to-neutral-and above) involves passing water through limestone bed , adding caustic soda or soda ash. For adjustment of alkaline groundwater bubbling carbon dioxide or adding strong acid is recommended. In certain, specific cases , introduction of chemical reducing agents like sulfur oxides, sodium bisuluphate or ferrous sulphate are used to precipitate metals, which are soluble at high pH (e.g hexavalent chromium, arsenic etc.)

Ion Exchange

It is basically the exchange of an ion with a high ion exchange selectivity for an ion with a lower selectivity. It is the basis of water softening in which divalent ion (e.g. calcium) usually having higher ion selectivity replaces monovalant ion (e.g. sodium) at an exchange site on an ion exchange bed. This technique though being expensive is recommended for treating wide range of dissolved inorganic e.g. chlorides, nitrates, sulphates and metals.

Reverse Osmosis

Employs semi-permeable membrane and high pressure resins. The membrane rejects inorganic material and allows passage of water through it. It can remove 60-90% of inorganics.

Electrodialysis

The method is a combination of reverse osmosis and ion exchange technology. It is the recommended method for treatment of groundwater with moderate to high conductivity.

Distillation

Involves evaporation of contaminated water by heating followed by recondensation. The inorganics do not evaporate ; hence are left behind. The recondensation brings out clean water free form dissolved inorganics.

At present the groundwater contributes more than 58% drinking water, 42% for agriculture production and for urban & industrial use ; and unscientific management of this resource has led to multiple problems of decline in groundwater level, sea water ingress, in-land, salinity, groundwater, pollution, land subsidence etc.

The measures that need to be adopted to meet the increased water demand would include exploration of deeper aquifers, groundwater recharge, development of aquifers, development of aquifers in flood plains, direct use of groundwater development etc.

Groundwater development has been intensive in alluvial area of Indo-Ganga-Yamuna plains of Punjab, Haryana, Uttar Pradesh, Uttaranchal and in parts of hard rock terrain in southern States. Though over-exploitation of the resource in some parts of the country has created serious problems, a large portion of the davailable resources still remains untapped, particularly in north-eastern areas, where precipitation is high and the demand for irrigation is low and also in eastern States where fragmented nature of land holdings has been a major factor in low development of groundwater.

Solutions for Groundwater Pollution is summed up below:

Table 4.5T_3: Solutions for abatement of Groundwater Pollution

Prevention	Cleanup
Find substitutes for toxic chemicals	Pump to surface clean and return to aquifer (very expensive)
Keep toxic chemicals out of the environment	Inject microorganisms to clean up
Install monitoring wells near landfills and underground tanks.	Pump nanoparticles of inorganic compounds to remove pollutants (may be the cheapest, easiest, and most effective method but is still being developed)
Require leak detectors on underground tanks	-
Ban hazardous waste disposal in landfills and injection wells	-
Store harmful liquids in above ground tanks with leak detection and collection systems.	-

5
Chapter

Waste Water/Sewage Treatment

Wastewater is "used" water that is deposited into drains by residents and businesses. Flushing the toilet, taking a shower, and running the washing machine are all examples of everyday activities that produce wastewater. This wastewater drains down the sink or toilet and travels through a network of underground pipes known as the sanitary sewer system. It includes substances such as human waste, food scraps, oils, soaps and chemicals. In homes, this includes water from sinks, showers, bathtubs, toilets, washing machines and dish washers. Businesses and industries also contribute their share of used water that must be cleaned.

NEED FOR WASTEWATER TREATMENT

Wastewater treatment is needed so that we can use our rivers & streams for fishing, swimming and drinking water. Pollution in the urban waterways results in frequent occurrences of low dissolved oxygen, fish kills, algal blooms and bacterial contamination. Early efforts in water pollution control prevented human waste from reaching water supplies or reduced floating debris that obstructed shipping. Pollution problems and their control were primarily local, not national, concerns.

Since then, population and industrial growth have increased demands on our natural resources, altering the situation dramatically. Progress in abating pollution has barely kept ahead of population growth, changes in industrial processes, technological developments, changes in land use, business innovations, and many other factors.

Increases in both the quantity & variety of goods produced can greatly alter the amount & complexity of industrial wastes and challenge traditional wastewater treatment technology. The application of commercial fertilizers & pesticides, combined with sediment from growing development activities, continues to be a source of significant pollution as runoff washes off the land.

The basic function of the wastewater treatment plant is to speed up the natural processes by which water purifies itself. In earlier years, the natural treatment process in streams & lakes was adequate to perform basic wastewater treatment. As our population and industry grew to their present size, increased levels of treatment prior to discharging domestic wastewater became necessary.

IMPORTANCE OF WASTE WATER TREATMENT

Wastewater treatment aims at removal of unwanted components in waste stream. Its objective is to produce a waste stream (or treated effluent) and a solid waste or sludge also suitable for discharge or reuse back into the environment. This material is contaminated with toxic organic & inorganic compounds. Waste water treatment can be done by physical, chemical and biological methods either alone or in combination. However, most of the treatment processes are not completely effective and also do not offer complete solutions for the removal of water contaminants as it transfers the contaminant to a different component. The wastewater generated by residential, commercial, & industrial sources is collected by sewers and transported to the Wastewater Treatment Plant. A lot of sewage also includes some surface water from roofs or hard standing areas. Municipal wastewater includes residential, commercial, & industrial liquid waste discharges, and may include storm water runoff.

The wastewater contains many of the physical, chemical & biological impurities which may impair quality of the receiving water body as well as interfere with the aquatic ecosystem in many ways. In order to render it safe ,it would be appropriate to treat the wastewater prior to disposal by adopting a systematic logical sequence of treatment.

Sewage systems capable of handling storm water are known as combined systems. Such systems are usually avoided since they complicate and thereby reduce the efficiency of sewage treatment plants owing to their seasonality. Besides, heavy storms may overwhelm the sewage treatment system, causing a spill or overflow. It is preferable to have a separate storm drain system for storm water. Sewerage systems that transport liquid waste discharges and storm water together to a common treatment facility are called combined sewer systems.

As rainfall runs over the surface of roofs and the ground, it may pick up various contaminants including soil particles, (sediment), heavy metals, organic compounds, animal waste, and oil & grease. Some jurisdictions require storm water to receive some level of treatment before being discharged directly into waterways.

EFFECTS OF WASTEWATER POLLUTANTS

If wastewater is not properly treated, then the environment and human health can be negatively impacted. These impacts can include harm to fish & wildlife populations, oxygen depletion, beach closures and other restrictions on recreational water use, restrictions on fish & shellfish harvesting and contamination of drinking water.

Some examples of pollutants found in wastewater and the potentially harmful effects these substances can have on ecosystems and human health are listed below:

- Decaying organic matter and debris can use up the dissolved oxygen in a lake so fish and other aquatic biota cannot survive;
- Excessive nutrients, such as phosphorus & nitrogen (including ammonia), can cause eutrophication, or over-fertilization of receiving waters, which can be toxic to aquatic organisms, promote excessive plant growth, reduce available oxygen, harm spawning grounds, alter habitat and lead to a decline in certain species;
- Chlorine compounds and inorganic chloramines can be toxic to aquatic invertebrates, algae & fish;

- Bacteria, viruses & disease causing pathogens can pollute beaches and contaminate shellfish populations, leading to restrictions on human recreation, drinking water consumption and shellfish consumption;
- Metals, such as mercury, lead, cadmium, chromium & arsenic can have acute and chronic toxic effects on species.
- Other substances such as some pharmaceutical & personal care products, primarily entering the environment in wastewater effluents, may also pose threats to human health, aquatic life and wildlife.

COLLECTION OF WASTEWATER FOR TREATING

The most common form of pollution control consists of a system of sewers and wastewater treatment plants. The sewers collect municipal wastewater from homes, businesses, & industries and deliver it to facilities for treatment before it is discharged to water bodies or land, or reused.

During the early days people living in both the cities & the countryside used cesspools to dispose of domestic wastewater. Cities began to install wastewater collection systems later because of an increasing awareness of water borne disease.

The use of sewage collection systems has brought improvements to public health.

A municipal sewer system is shown below in figure.

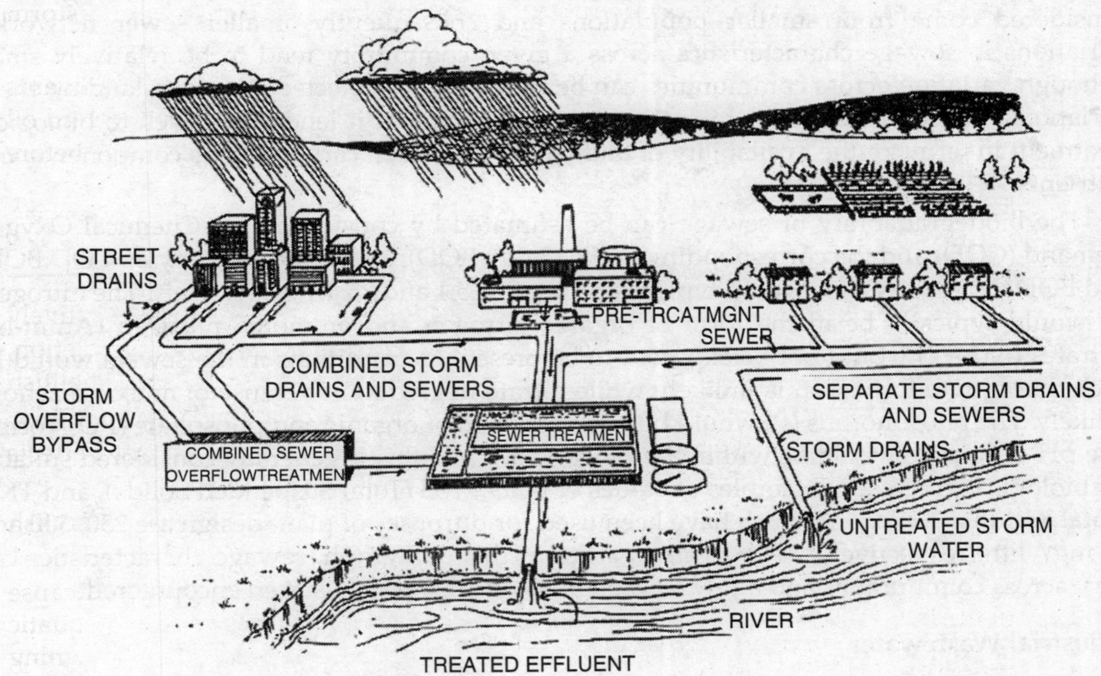

Fig.5.1: Municipal sewer systems

DIFFERENCE BETWEEN DOMESTIC SEWAGE AND INDUSTRIAL WASTEWATER

The term wastewater is commonly used to describe liquid wastes that are collected and transported to a treatment facility through a system of sewers. Wastewater is generally divided into two broad classifications: domestic wastewater and industrial wastewater.

Domestic Wastewater

Domestic sewage is wastewater discharged from sanitary conveniences in residential, office, commercial, factories and various institutional properties. It is a complex mixture containing primarily water (approximately 99%) together with organic and inorganic constituents. These constituents or contaminants comprise of suspended, colloidal and dissolved materials. Domestic sewage, since it contains human wastes, also contains large numbers of micro-organisms and some of these can be pathogenic(disease causing). Waterborne bacterial diseases that can be present in sewage include cholera, typhoid and tuberculosis. Viral diseases can include infectious hepatitis. Inorganic constituents include chlorides & sulphates, various forms of nitrogen & phosphorous, as well as carbonates &bicarbonates. Proteins and carbohydrates constitute about 90% of the organic matter in domestic sewage. These arise from the excreta, urine, food wastes, and wastewater from bathing, washing, & laundering, and because of the latter, soaps, detergents, & other cleaning products can be found as well.

Domestic sewage has a flow pattern which typically shows two peaks — in the morning before the start of working hours and in the evening after the population has returned from work. Typically these hydraulic peaks would also become more distinct as the sewage flows considered come from smaller populations and consequently smaller sewer networks. Variations in sewage characteristics across a given community tend to be relatively small although variation across communities can be more readily detected. Notwithstanding these variations, the composition of domestic sewage is such that it lends itself well to biological treatment in terms of the availability of and balance between carbonaceous components and nutrients.

The biodegradability of sewage can be estimated by considering its Chemical Oxygen Demand (COD) and the corresponding BOD_5 (5 day BOD), and is indicated by its COD:BOD_5 and BOD_5:N:P ratios. This would typically be about 1.5:1 and 25:4:1 respectively. The nitrogen, N, would typically be in the form of organic nitrogen and ammonia nitrogen (Amm-N). Nitrates (NO_3-N) would not be expected to be present as conditions in the sewers would be such that nitrate formation is unlikely while nitrate degradation because of anoxic reactions is likely. The phosphorous (P) would be a combination of organic and phosphate (PO_4) forms. The pH of sewage would be within the range of 6–9 and this is generally considered suitable for biological processes. Examples of values of BOD_5, TSS (Total Suspended Solids), and TKN (Total Kjeldhal Nitrogen) which have been used for purposes of plant design are 250, 300 and 40 mg/ litre respectively. As indicated earlier in this paragraph, sewage characteristics can vary across communities and a raw sewage BOD_5 of 500mg/ l has been encountered.

Industrial Wastewater

Industrial wastewater characteristics vary with each type of discharge. Food processing, dairy operations and meat packing all have wastes that are very high in BOD, sometimes over 1000 mg/l. This BOD is high in dissolved sugars, fats, & proteins and low in suspended material.

It will dramatically increase the loading on the secondary treatment processes. Some food process wastes can also have a high pH, like potato processing and some vegetable canning, or a low pH, like green chili processing or fruit canning.

Toxins, like heavy metals coming from metal plating, battery shops,& heavy equipment manufacturing, and solvents from industries like body shops, dry cleaners, & furniture manufacturing can kill the microorganisms that are needed to treat the waste. Copper, chrome, lead, and cyanide are all chemicals that are commonly found in heavy industrial manufacturing discharges.

Hospitals and other medical facilities have the potential to discharge wastes that are radioactive or represent a biohazard.

In some cases, pretreatment of these industrial discharges may be required. Systems that deal with industrial discharges should adopt a pretreatment ordinance to cover conditions that the system may need to impose on the industrial discharges in order to maintain their NPDES requirements on the treatment plant effluent. If local ordinances are different from specific categorical limits imposed in the Federal Regulations, the most stringent limits of the two must be followed. Pretreatment ordinances can require actual treatment of the industrial waste prior to discharge to the system, or it may levy a monetary surcharge based on the addition loading that results from the discharge.

Any industrial discharger is considered to be a "Significant Industrial User" , if:

- The discharge requires pretreatment.
- The flow is over 25,000 gpd (gallon per day) , or
- The flow is more than 5% of total average plant BOD or hydraulic load.

VARIOUS SEWAGE TREATMENT METHODS

Sewage is created by residences, institutions, and commercial & industrial establishments. It can be treated close to where it is created (in septic tanks, biofilters or aerobic treatment systems), or collected & transported via a network of pipes and pump stations to a municipal treatment plant. Sewage collection and treatment is typically subject to local, state & federal regulations and standards.

Industrial sources of wastewater often require specialized treatment processes. Raw influent (sewage) is the liquid waste from toilets, baths, showers, kitchens, sinks etc. Household waste is that which is disposed of via sewers. Municipal wastewater therefore includes residential, commercial, & industrial liquid waste discharges, and may include storm water runoff. Sewage systems capable of handling storm water are known as combined systems.

Typically , sewage treatment involves three stages, called ***primary, secondary*** **and** ***tertiary*** *treatment*.

Sewage Treatment Methods include

- **Mechanical treatment;**
  Removal of large objects
  Removal of sand and grit
  Pre-precipitation

- **Biological treatment:** Oxidation bed (oxidizing bed) or aeration system
  Post precipitation
  Effluent
- **Chemical treatment** (this step is usually combined with settling and other processes to remove solids, such as filtration.)

Primary treatment. It involves removal of large objects, sand and grit. Primary treatment removes the materials that can be easily collected from the raw wastewater and disposed off. The typical materials that are removed during primary treatment include fats, oils, & greases , sand, gravels & rocks (also referred to as grit), larger settleable solids including human waste and floating materials. This step is done entirely with machinery, hence the name mechanical treatment.

Secondary treatment. It is designed to substantially degrade the biological content of the sewage such as are derived from human waste, food waste, soaps and detergent. The majority of municipal and industrial plants treat the settled sewage liquor using aerobic biological processes. For this to be effective, the biota require both oxygen and a substrate on which to live. There are number of ways in which this is done. In all these methods, the bacteria and protozoa consume bio degradable soluble organic contaminants (e.g. sugars, fats, organic short chain carbon molecules, etc.) and bind much of the less soluble fractions into floc.

A sewage treatment process flow diagram is shown below in figure.

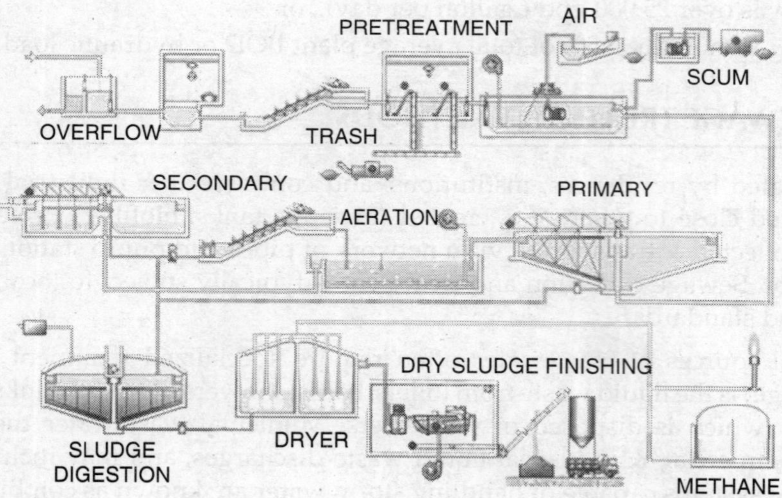

Fig.5.2: A typical sewage treatment process flow diagram

Secondary treatment systems are classified as **fixed film** or suspended growth. Fixed-film treatment process including trickling filter and rotating biological contactors where the biomass grows on media and the sewage passes over its surface. In **suspended growth systems**—such as activated sludge—the biomass is well mixed with the sewage and can be operated in a smaller space than fixed-film systems that treat the same amount of water. However, fixed-film systems are more able to cope with drastic changes in the amount of biological material and can provide higher removal rates for organic material and suspended solids than suspended growth systems.

Tertiary treatment. This treatment provides a final stage to raise the effluent quality before it is discharged to the receiving environment (sea, river, lake, ground, etc.). More than one tertiary treatment process may be used at any treatment plant. If disinfection is practiced, it is always the final process. It is also called "effluent polishing".

CONVENTIONAL WASTEWATER TREATMENT PROCESSES

Conventional wastewater treatment process is described below.

1. **Preliminary treatment:**

   Screening: Upon arrival via the sewer system, the wastewater is sent through a bar screen, which removes large solid objects such as sticks and rags. Leaving the bar screen, the wastewater flow is slowed down entering the grit tank. This allows sand, gravel, and other heavy material that was small enough not to be caught by the bar screen to settle to the bottom. All the collected debris from the grit tank and bar screen is disposed of at a sanitary landfill or recycled. Wastewater entering the treatment plant includes items like wood, rocks, and even dead animals. Unless they are removed, they could cause problems later in the treatment process. Most of these materials are sent to a landfill.

2. **Pumping**

   The wastewater system relies on the force of gravity to move sewage from home to the treatment plant. So wastewater-treatment plants are located on low ground, often near a river into which treated water can be released. If the plant is built above the ground level, the wastewater has to be pumped up to the aeration tanks. From here on, gravity takes over to move the wastewater through the treatment process.

3. **Aerating**

   One of the first steps that a water treatment facility can do is to just shake up the sewage and expose it to air. This causes some of the dissolved gases (such as hydrogen sulfide, which smells like rotten eggs) that taste & smell bad to be released from the water. Wastewater enters a series of long, parallel concrete tanks. Each tank is divided into two sections. In the first section, air is pumped through the water.

4. **Primary treatment**

   Primary treatment is the second step in wastewater treatment. It allows for the physical separation of solids and greases from the wastewater. The screened wastewater flows into a primary settling tank where it is held for several hours. This allows solid particles to settle to the bottom of the tank and oils & greases to float to the top. As organic matter decays, it uses up oxygen. Aeration replenishes the oxygen. Bubbling air through the water also keeps the organic material suspended while it forces 'grit' (sand and other small, dense particles) to settle out. Grit is pumped out of the tanks and taken to landfills.

   Wastewater then enters the second section or sedimentation tanks. Here, the sludge (the organic portion of the sewage) settles out of the wastewater and is pumped out of the tanks. Some of the water is removed in a step called thickening and then the sludge is processed in large tanks called digesters.

Various primary sewage treatment units

Primary treatment removes the materials that can be easily collected from the raw wastewater and disposed off eg; fats, oils, & greases , sand, gravels & rocks , larger settleable solids including human waste & floating materials.

Influx (influent) and removal of large objects

In the mechanical treatment, the influx (influent) of sewage water is strained to remove all large objects that are deposited in the sewer system, such as rags, sticks, condoms, sanitary towels / napkins / tampons, cans, fruit, etc. This is most commonly done with a manual or automated mechanically raked screen. This type of waste is removed because it can damage or clog the equipment in the sewage treatment plant.

Sand & grit removal

Primary treatment typically includes a sand or grit channel or chamber where the velocity of the incoming wastewater is carefully controlled to allow sand , grit & stones to settle, while keeping the majority of the suspended organic material in the water column. This equipment is called a detritor or sand catcher. Sand grit and stones need to be removed early in the process to avoid damage to pumps and other equipment in the remaining sewage treatment stages. Sometimes there is a sand washer (grit classifier) followed by a conveyor that transports the sand to a container for disposal. The contents from the sand catcher may be fed into the incinerator in a sludge processing plant, but in many cases, the sand and grit is sent to a landfill.

Sedimentation

Many plants have a sedimentation stage where the sewage is allowed to pass slowly through large tanks, commonly called "primary clarifiers" or "primary sedimentation tanks". The tanks are large enough that fecal solids can settle and floating material such as grease & oils can rise to the surface and be skimmed off. The main purpose of the primary stage is to produce a generally homogeneous liquid capable of being treated biologically and a sludge that can be separately treated or processed. Primary settlement tanks are usually equipped with mechanically driven scrapers that continually drive the collected sludge towards a hopper in the base of the tank from where it can be pumped to further sludge treatment stages.

5. Secondary treatment

Secondary treatment is a biological treatment process that removes dissolved organic material from wastewater. The partially treated wastewater from the settling tank flows by gravity into an aeration tank. Here it is mixed with solids containing micro-organisms that use oxygen to consume the remaining organic matter in the wastewater as their food supply. The aeration tank uses air bubbles to provide the mixing and the oxygen, both of which are needed for the micro-organisms to multiply.

From here the liquid mixture, composed of solids with micro-organisms and water, is sent to the final clarifier. Here the solids settle to the bottom where some of the material is sent to the solids handling process, and some is recirculated to replenish the population of micro-organisms in the aeration tank to treat incoming wastewater.

As sludge is settling to the bottom of the sedimentation tanks, lighter materials are floating to the surface. This 'scum' includes grease, oils, plastics, and soap. Slow-moving rakes skim the scum off the surface of the wastewater. Scum is thickened and pumped to the digesters along with the sludge.

Many plants also use filtration in sewage treatment. After the solids are removed, the liquid sewage is filtered through a substance, usually sand, by the action of gravity. This method gets rid of almost all bacteria, reduces turbidity & color, removes odors, reduces the amount of iron, and removes most other solid particles that remained in the water. Water is sometimes filtered through carbon particles, which removes organic particles.

Various secondary treatment units

Secondary treatment degrades the biological content of the sewage such as are derived from human waste, food waste, soaps and detergent. The majority of municipal and industrial plants treat the settled sewage liquor using aerobic biological process. For this to be effective, the biota require both oxygen and a substrate on which to live. There are number of ways in which this is done. In all these methods, the bacteria and protozoa consume biodegradable soluble organic contaminants (e.g. sugars, fats, organic short-chain carbon molecules, etc.) and bind much of the less soluble fractions into floc.

Secondary treatment systems are classified as **fixed film** or **suspended growth**. Fixed-film treatment process including trickling filter and rotating biological contactors where the biomass grows on media and the sewage passes over its surface.

In suspended growth such as activated sludge ,the biomass is well mixed with the sewage and can be operated in a smaller space than fixed-film systems that treat the same amount of water.

However, fixed-film systems are more able to cope with drastic changes in the amount of biological material and can provide higher removal rates for organic material and suspended solids than suspended growth systems.

Roughing filters

These are intended to treat particularly strong or variable organic loads, typically industrial, to allow them to then be treated by conventional secondary treatment processes. Characteristics include typically tall, circular filters filled with open synthetic filter media to which wastewater is applied at a relatively high rate. They are designed to allow high hydraulic loading and a high flow-through of air. On larger installations, air is forced through the media using blowers. The resultant wastewater is usually within the normal range for conventional treatment processes.

Activated sludge

Activated sludge plants encompasses a variety of mechanisms and processes that use dissolved oxygen to promote the growth of biological floc that substantially removes organic material. It also traps particulate material and can, under ideal conditions, convert ammonia to nitrite & nitrate ultimately to nitrogen gas.

Fluidized bed reactors

The carbon adsorption following biological treatment are particularly effective in reducing both the BOD and COD to low levels. A fluidized bed reactor is a combination of the most common stirred tank packed bed, continuous flow reactors and has excellent heat & mass transfer characteristics.

In a fluidized bed reactor, the substrate is passed upward through the immobilized enzyme bed at a high velocity to lift the particles. However the velocity must not be so high that the enzymes are swept away from the reactor entirely. This causes low mixing; these type of reactors are suitable for the exothermic reactions. It is most often applied in immobilized enzyme catalysis.

Filter beds (oxidising beds)

In older plants and plants receiving more variable loads, trickling filter beds are used where the settled sewage liquor is spread onto the surface of a deep bed made up of coke (carbonised coal), limestone chips or specially fabricated plastic media. Such media must have high surface areas to support the biofilms to that form.

The liquor is distributed through perforated rotating arms radiating from a central pivot. The distributed liquor trickles through this bed and is collected in drains at the base. These drains also provide a source of air which percolates up through the bed, keeping it aerobic. Biological films of bacteria, protozoa & fungi form on the media's surfaces and eat or reduce the organic content. This biofilm is grazed by insect larvae and worms which help maintain an optimal thickness.

Overloading of beds increases the thickness of the film leading to clogging of the filter media and ponding on the surface.

Biological aerated filters

Biological Aerated (or Anoxic) Filter (BAF) or Biofilters combine filtration with biological carbon reduction, nitrification or denitrification. BAF usually includes a reactor filled with a filter media. The media is either in suspension or supported by a gravel layer at the foot of the filter. The dual purpose of this media is to support highly active biomass that is attached to it and to filter suspended solids. Carbon reduction and ammonia conversion occurs in aerobic mode and sometime achieved in a single reactor while nitrate conversion occurs in anoxic mode. BAF is operated either in upflow or downflow configuration depending on design.

Membrane biological reactors

Membrane biological reactors (MBR) combines activated sludge treatment with a membrane liquid-solid separation process. The membrane component utilizes low pressure microfiltration or ultra filtration membranes and eliminates the need for clarification and tertiary filtration. The membranes are typically immersed in the aeration tank (however, some applications utilize a separate membrane tank). One of the key benefits of a membrane bioreactor system is that it effectively overcomes the limitations associated with poor settling of sludge in conventional activated sludge (CAS) processes.

This technology permits bioreactor operation with considerably higher mixed liquor suspended solids (MLSS) concentration than CAS systems, which are limited by sludge settling.The process is typically operated at MLSS in the range of 8,000–12,000 mg/l, while CAS are operated in the range of 2,000–3,000 mg/l. The elevated biomass concentration in the membrane bioreactor process allows for very effective removal of both soluble and particulate biodegradable materials at higher loading rates. Thus increased Sludge Retention Times,usually exceeding 15 days ensure complete nitrification even under extreme cold weather operating conditions.

The cost of building and operating a MBR is usually higher than conventional wastewater treatment, however, as the technology has become increasingly popular and has gained wider acceptance throughout the industry, the life-cycle costs have been steadily decreasing.

Secondary sedimentation

The final step in the secondary treatment stage is to settle out the biological floc or filter material and produce sewage water containing very low levels of organic material & suspended matter.

Rotating biological contactors

Rotating biological contactors (RBCs) are mechanical secondary treatment systems, which are robust and capable of withstanding surges in organic load. The rotating disks support the growth of bacteria and micro-organisms present in the sewage, which breakdown and stabilise organic pollutants. Microorganisms need both oxygen to live and food to grow. Oxygen is obtained from the atmosphere as the disks rotate. As the micro-organisms grow, they build up on the media until they are sloughed off due to shear forces provided by the rotating discs in the sewage. Effluent from the RBC is then passed through final clarifiers where the micro-organisms in suspension settle as a sludge. The sludge is withdrawn from the clarifier for further treatment.

6. Tertiary treatment of sewage

Tertiary treatment provides a final stage to raise the effluent quality before it is discharged to the receiving environment (sea, river, lake, ground, etc.). More than one tertiary treatment process may be used at any treatment plant. If disinfection is practiced, it is always the final process. It is also called "effluent polishing".

Filtration

Sand filtration removes much of the residual suspended matter. Filtration over activated carbon removes residual toxins.

Lagooning

Lagooning provides settlement and further biological improvement through storage in large man made ponds or lagoons.These lagoons are highly aerobic & colonization by native macrophytes, especially reeds, is often encouraged. Small filter feeding invertebrates such as Daphnia and species of Rotifera greatly assist in treatment by removing fine particulates.

Constructed wetlands

Constructed wetlands include engineered reed beds and a range of similar methodologies, all of which provide a high degree of aerobic biological improvement and can often be used instead of secondary treatment for small communities.

Waste removal

Wastewater may contain high levels of the nutrients ; nitrogen and phosphorus whose excessive release to the environment can lead to a build up of nutrients, called eutrophication, which can in turn encourage the overgrowth of weeds, algae, & cyanobacteria (blue-green algae). This may cause an algal bloom, a rapid growth in the population of algae. The algae numbers are unsustainable and eventually most of them die. The decomposition of the algae by bacteria uses up so much of oxygen in the water that most or all of the animals die,which creates more organic matter for the bacteria to decompose. In addition to causing deoxygenation, some algal species produce toxins that contaminate drinking water supplies.

Different treatment processes are required to remove nitrogen and phosphorus.

Nitrogen removal

The removal of nitrogen is effected through the biological oxidation of nitrogen from ammonia (nitrification) to nitrate, followed by denitrification, the reduction of nitrate to nitrogen gas. Nitrogen gas is released to the atmosphere and thus removed from the water.

Phosphorus removal

Phosphorus can be removed biologically in a process called enhanced biological phosphorus removal. In this process, specific bacteria, called polyphosphate accumulating organisms, are selectively enriched and accumulate large quantities of phosphorus within their cells (up to 20% of their mass). When the biomass enriched in these bacteria is separated from the treated water, these biosolids have a high fertilizer value.

7. **Final treatment: Killing pathogens**

Wastewater that remains is disinfected to kill harmful micro-organisms before being released into receiving waters. Although there are many methods available to kill these micro-organisms, chlorine and ultraviolet disinfection are the most widely used. Dechlorination occurs in the final wastewater treatment step. A solution of sodium bisulfite is added to the chlorinated effluent to remove residual chlorine.

Following disinfection and dechlorination, the treated wastewater (now called final effluent) can be returned to the receiving waters from which it came. The flow is conveyed to an outfall and discharged through a series of diffusers into a surface water body or stream.

Finally, the wastewater flows into a 'chlorine contact' tank, where the chemical chlorine is added to kill bacteria, which could pose a health risk, just as is done in swimming pools. The chlorine is mostly eliminated as the bacteria are destroyed, but sometimes it must be neutralized by adding other chemicals. This protects fish and other marine organisms, which can be harmed by the smallest amounts of chlorine.

The production of water of satisfactory quality is dependent upon the elimination or destruction of two groups of living organisms (1) the pathogenic microorganisms, which may infect humans through the use of contaminated water and (2) the algae and other aquatic growths, which may render water aesthetically unfit for human consumption. The treated water (called effluent) would then be discharged to a local river or the ocean.

Disinfection

The purpose of disinfection in the treatment of wastewater is to substantially reduce the number of microorganisms in the water to be discharged back into the environment. The effectiveness of disinfection depends on the quality of the water being treated (e.g., cloudiness, pH, etc.), type of disinfection being used, the disinfectant dosage (concentration & time), and other environmental variables. Cloudy water will be treated less successfully since solid matter can shield organisms, especially from ultraviolet light or if contact times are low. Generally, short contact times, low doses and high flows all militate against effective disinfection. Common methods of disinfection include ozone, chlorine, or ultraviolet light. Chloramine, which is used for drinking water, is not used in wastewater treatment because of its persistence.

Wastewater Residuals: Solids processing

Primary solids from the primary settling tank and secondary solids from the clarifier are sent to the digester. During this process, micro-organisms use the organic material present in the solids as a food source and convert it to by-products such as methane gas and water. Digestion results in a 90% reduction in pathogens and the production of a wet soil-like material called "biosolids" that contain 95-97% water. To remove some of this water and reduce the volume, mechanical equipment such as filter presses or centrifuges are used to squeeze water from the biosolids. The biosolids are then sent to landfills, incinerated, or beneficially used as a fertilizer or for soil amendment.

Another part of treating wastewater is dealing with the solid-waste material. These solids are kept for 20 to 30 days in large, heated and enclosed tanks called 'digesters.' Here, bacteria break down (digest) the material, reducing its volume, odors, and getting rid of organisms that can cause disease. The finished product is mainly sent to landfills, but sometimes can be used as fertilizer.

Our home is serviced by either a sanitary and a storm sewer or a combined sewer, a single pipe which conveys both sanitary sewage and storm water. It is 99.90 percent water, with only 0.01 percent of the wastewater being dissolved & suspended solid material. The cloudiness of sewage is caused by suspended particles which, in untreated sewage, range from 100 to 350 mg/l. A measure of the strength of the wastewater is biochemical oxygen demand, or BOD_5. The BOD_5 measures the amount of oxygen microorganisms require in five days to break down sewage. Untreated sewage has a BOD_5 ranging from 100 mg/l to 300 mg/l. Pathogens or disease causing organisms are present in sewage. Coliform bacteria are used as an indicator of disease causing organisms. Sewage also contains nutrients (such as ammonia & phosphorus), minerals, and metals. Ammonia can range from 12 to 50 mg/l and phosphorus can range from 6 to 20 mg/l in untreated sewage.

The sewer systems are built to follow the natural slope of the land. This design allows gravity to do most of the work transporting the wastewater to treatment facilities. In some situations, pump stations are constructed in low-lying areas to force the wastewater uphill to the treatment facility.

The organic & inorganic material found in this wastewater is too abundant to be handled by the environment. Therefore, treatment of wastewater is an essential process that prevents contamination & destruction of waterways, drinking water sources, and natural water resources.

Sewage treatment is a multi-stage process to renovate wastewater before it reenters a body of water, is applied to the land, or is reused. The goal is to reduce or remove organic matter, solids, nutrients, disease causing organisms, and other pollutants from wastewater. Each receiving body of water has limits to the amount of pollutants it can receive without degradation. Therefore, each sewage treatment plant must treat it to allowable levels of BOD_5, suspended solids, coliform bacteria, and other pollutants.

A typical waste water treatment facility is shown below in figure.

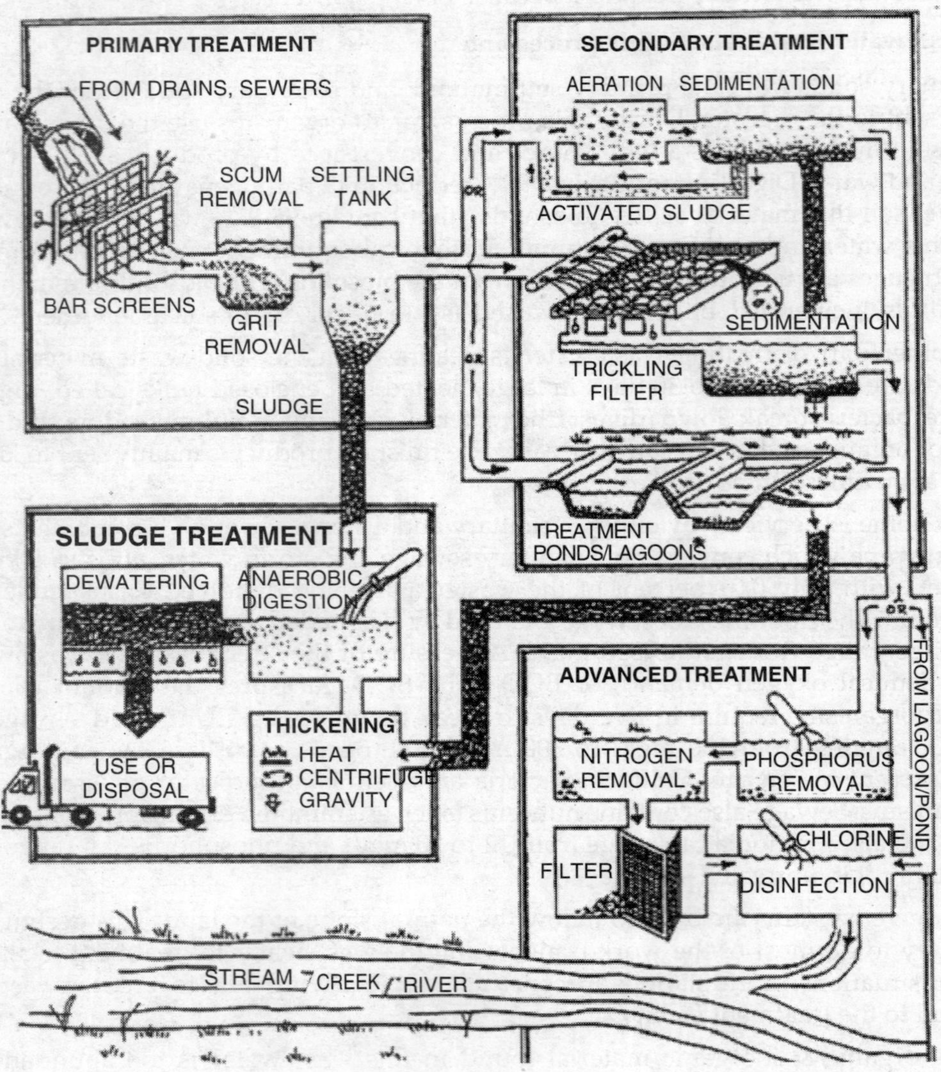

Fig.5.3: Typical wastewater treatment facility

MAJOR CONSTITUENTS OF PRELIMINARY, PRIMARY AND SECONDARY TREATMENT OF WASTE WATER

The wastewater contains many floatable materials that may interfere with the downstream processes or be detrimental to the plant equipments. To increase the waste water treatment efficiency of the secondary treatment units it is essential to have **preliminary and primary treatment** of wastewater.

Materials removed may include rags, plastic, lumber, and grit. This is done by bar screens which remove large objects, rags, and plastic that would cause plugging problems in the downstream lines. The grit basins remove the grit by sedimentation. Grit is abrasive materials such as sand and gravel that must be removed to minimize wear on the downstream equipment. Materials removed by the bar screens and grit removal system are de-watered, discharged to a dump trailer, and transported to the landfill for disposal.

Secondary treatment helps in reducing the organic load present in wastewater. The commonly used units in this treatment are Trickling filter, Activated sludge, Water treatment ponds etc.

The secondary treatment of water is based on the natural self-purification capacity of natural waters. This treatment is designed to substantially degrade the biological content of the sewage such as are derived from human waste, food waste, soaps and detergent. The majority of the municipal & industrial water treatment plants treat the settled sewage liquor using aerobic biological processes. For this to be effective, the biota require nutrients and a substrate on which to live. There are number of ways in which this is done. In all these methods, the bacteria and protozoa consume biodegradable soluble organic contaminants (e.g. sugars, fats, organic short-chain carbon molecules, etc.) and bind much of the less soluble fractions into flocs.

Secondary treatment systems are classified as **fixed film** or suspended growth, such as rock filters where the biomass grows on media and the sewage passes over its surface. In **suspended growth systems** - such as activated sludge - the biomass is well mixed with the sewage and be operated in a smaller space than fixed film systems that treat same amount of water.However, fixed film systems are more able to cope better with drastic changes in the amount of biological material and can provide higher removal rates for organic material & suspended solids than suspended growth systems. This is reflected in two different possibilities for water treatment. The first type is a so-called fixed bed bio-reactor, usually referred to as a trickling filter.

DIFFERENT STAGES IN PRELIMINARY TREATMENT OF WASTE WATER

It is used to reduce Solids(50-60%), and BOD (30- 50%). The commonly used units in the preliminary treatment of water are bar screens, Comminutor, and Grit chamber (Settling tank).

Preliminary treatment, the first treatment process, consists of the removal of substances that may interfere with the downstream processes or be detrimental to the plant equipment. Primary treatment is to reduce oils, grease, fats, sand, grit, and coarse (settleable) solids. This step is done entirely with machinery, hence the name mechanical treatment.

Influx (influent) and removal of large objects

In the mechanical treatment, the influx (influent) of sewage water is strained to remove all large objects that are deposited in the sewer system, such as rags, sticks, condoms, sanitary towels (sanitary napkins) or tampons, cans, fruit, plastic, lumber, grit., etc. This is most commonly done with a manual or automated mechanically raked screen. This type of waste is removed because it can damage the sensitive equipment in the sewage treatment plant.

Sand and grit removal

This stage includes a sand or grit channel where the velocity of the incoming waste water is carefully controlled to allow sand grit and stones to settle but still maintain the majority of the organic material within the flow. This equipment is called a detritor or sand catcher. Sand grit and stones need to be removed early in the process to avoid damage to pumps and other equipment in the remaining treatment stages. Sometimes there is a sand washer (grit classifier) followed by a conveyor that transports the sand to a container for disposal. The contents from the sand catcher may be fed into the incinerator in a sludge processing plant, but in many cases, the sand and grit is sent to a land-fill. Materials removed may include rags, plastic, lumber, and grit. This is done by bar screens which remove large objects, rags, & plastic that may cause plugging problems in the downstream lines. The grit basins remove the grit by sedimentation. Grit is abrasive materials such as sand and gravel that must be removed to minimize wear on the downstream equipments. Materials removed by the bar screens and grit removal system are de-watered, discharged to a dump trailer, and transported to the landfill for disposal.

Grit and Screen Facility

The Grit and Screen Facility is where sewage first enters the plant. The speed of the wastewater is reduced to 0.3 metre per second and air is pumped in to set up a rolling motion.This movement separates the heavier grit and sand materials from the suspended organic matter. The larger sand particles settle to the bottom of the tanks, thus preventing harm to equipment within the Plant or clogging of pipes during the treatment process. The settled sand is removed once a month and buried.Wastewater continues through the grit tanks and passes through bar screens to remove any material larger than 12.5mm in size. The wastewater then flows by gravity to the Primary Settling Basins.

DIFFERENT STAGES IN PRIMARY TREATMENT OF WASTE WATER

The second step in the waste water treatment process is primary treatment. The waste water enters two primary clarifiers (sedimentation basins) which remove suspended and floating materials.The primary clarifiers remove about 60% of the Total Suspended Solids and about 30% of the Biochemical Oxygen Demand (BOD) in the incoming wastewater. BOD is a measure of the amount of oxygen needed to biochemically degrade the organic matter in the wastewater. Primary treatment does not remove colloidal or dissolved solids.

Sedimentation

Many plants have a sedimentation stage where the sewage is allowed to pass slowly through large tanks, commonly called "primary clarifiers" or "primary sedimentation tanks". The tanks are large enough that faecal solids can settle and floating material such as grease & plastics can rise to the surface and be skimmed off. The main purpose of the primary stage is to produce a generally homogeneous liquid capable of being treated biologically and a sludge that can be separately treated or processed. Primary settlement tanks are usually equipped with mechanically driven scrapers that continually drive the collected sludge towards a hopper in the base of the tank from where it can be pumped to further sludge treatment stages.

Primary Settling Basins

This is where suspended solids settle out and floating scum is removed for further treatment. Upon entering the Primary Settling Basins, the velocity of the wastewater is reduced to 3 metres per minute, allowing fine particles to settle to the bottom. Settled sludge on the bottom of the basins is continuously scraped into hoppers at the end of tanks. It takes about 4 hours for waste water to flow through the Primary Settling Basins. Upon completion, the primary effluent is pumped to the Bioreactors and the settled sludge is pumped to the Fermenters. The scum from the top of the basins is collected in hoppers and pumped to the Digesters.

DIFFERENT STAGES OF SECONDARY(BIOLOGICAL) TREATMENT OF WASTE WATER

Secondary treatment usually consists of two steps which remove the dissolved & colloidal organic material not removed by the primary treatment. There are various secondary treatment processes in use. This treatment use microorganisms to convert organic wastes into stabilized compounds - similar to self-purification process in streams.

It is used to reduce Solids (85-95%), BOD (80-95%), and Coliform (90-95%). The commonly used units in this treatment are Trickling filter, Activated sludge,and Water treatment ponds.

The biological treatment of water is based on the natural self-purification capacity of natural waters.This treatment is designed to substantially degrade the biological content of the sewage which are derived from human waste, food waste, soaps and detergent. The majority of the municipal & industrial water treatment plants treat the settled sewage liquor using aerobic biological processes. For this to be effective, the biota require nutrients and a substrate on which to live. There are number of ways in which this is done. In all these methods, the bacteria and protozoa consume biodegradable soluble organic contaminants (e.g. sugars, fats, organic short-chain carbon molecules, etc.) and bind much of the less soluble fractions into floc.

Secondary treatment systems are classified as **fixed film** or **suspended growth**, such as rock filters where the biomass grows on media and the sewage passes over its surface. In **suspended growth systems** - such as activated sludge - the biomass is well mixed with the sewage & can be operated in a smaller space than fixed film systems that treat same amount of water. However, fixed film systems cope better with drastic changes in the amount of biological material and can provide higher removal rates for organic material & suspended solids than suspended growth systems. This is reflected in two different possibilities for water treatment. The first type is a so-called fixed bed bio-reactor, usually referred to as a trickling filter.

- **Roughing filters** are intended to treat particularly strong or variable organic loads, typically industrial, to allow them to then be treated by conventional secondary treatment processes. Characteristics include typically tall, circular filters filled with open synthetic filter media to which sewage is applied at a relatively high rate. Designed to allow high hydraulic loading and a high flow-through of air. On larger installations, air is forced through the media using blowers. The resultant liquor is usually within the normal range for conventional treatment processes.

BIOLOGICAL TREATMENT OF WASTE WATER

Treatment of industrial effluents using microorganisms is is called biological treatment of waste water. Secondary treatment typically utilizes biological treatment processes, in which microorganisms convert non settleable solids to settleable solids. Biological treatment processes followed by settling tanks removes approximately 85% of the BOD & TSS in wastewater. Secondary treatment for municipal wastewater is the minimum level of treatment required. Its objective is to produce a stream (or treated effluent) and a solid waste or sludge also suitable for discharge or reuse back into the environment.

Secondary treatment usually consists of two steps which remove the dissolved and colloidal organic material not removed by the primary treatment. There are various secondary treatment processes in use today. This treatment uses microorganisms to convert organic wastes into stabilized compounds - similar to self-purification process in streams.

There are a number of compounds in the environment which are transformed by microorganisms, yet it has been difficult or impossible to find organisms that can use them as a source of carbon and/or energy. The compounds may be transformed sequentially by a series of bacteria or other microorganisms such that no organisms gained energy sufficient to allow growth or cell division, from the reactions it is necessary to have an alternate or primary substrate for growth under these conditions.

KINETICS OF BIOLOGICAL GROWTH

Biofilms are composed of populations or communities of microorganisms adhering to environmental surfaces. These microorganisms are usually encased in an extracellular polysaccharide that they themselves synthesize. Biofilms may be found on essentially any environmental surface in which sufficient moisture is present. Their development is most rapid in flowing systems where adequate nutrients are available.

Biofilms may form:

- On solid substrates in contact with moisture.
- On soft tissue surfaces in living organisms.
- At liquid air interfaces.

Typical locations for biofilm production include rock and other substrate surfaces in marine or freshwater environments. Biofilms are also commonly associated with living organisms, both plant and animal. Tissue surfaces such as teeth and intestinal mucosa which are constantly bathed in a rich aqueous medium rapidly develop a complex aggregation of microorganisms enveloped in an extracellular polysaccharide they themselves produce.

Typically, within minutes, an organic monolayer adsorbs to the surface of the slide substrate. This changes the chemical and physical properties of the glass slide or other substrate. These organic compounds are found to be polysaccharides or glycoproteins. These adsorbed materials condition the surface of the slide and appear to increase the probability of the attachment of planktonic bacteria. Free floating or planktonic bacteria encounter the conditioned surface and form a reversible, sometimes transient attachment often within minutes.

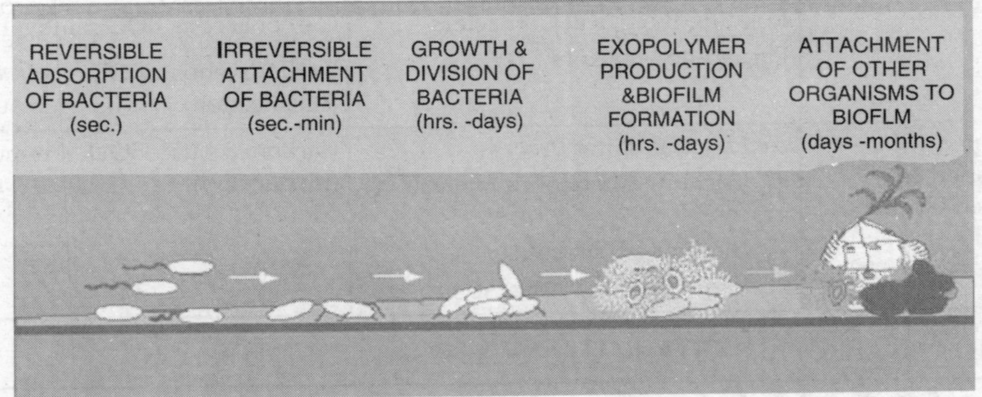

Fig.5.4: Formation of biofilms

This attachment called **adsorption** is influenced by electrical charges carried on the bacteria, by Van der Waals forces and by electrostatic attraction although the precise nature of the interaction is still a matter of intense debate. In some instances, as for example, in the association between a pathogen and the receptor sites of cells of its host there may be a stereospecificity which though still reversible is stronger than that achieved strictly by ionic or electrostatic forces.

If the association between the bacterium and its substrate persists long enough, other types of chemical and physical structures may form which transform the reversible adsorption to a permanent and essentially **irreversible attachment**.

The final stage in the irreversible adhesion of a cell to an environmental surface is associated with the production of **extracellular polymer substances** or EPS. Most of the EPS of biofilms are polymers containing sugars such as glucose, galactose, mannose, fructose, rhamnose, N-acetylglucosamine and others.

This layer of EPS and bacteria can now entrap particulate materials such as clay, organic materials, dead cells and precipitated minerals adding to the bulk and diversity of the biofilm habitat. This growing biofilm can now serve as the focus for the attachment and growth of other organisms increasing the biological diversity of the community.

The major aerobic biological processes are listed in the following Table 5.1.

Type of Growth	Common Name	Use
Suspended Growth Attached Growth	Activated Sludge (AS)	Carbonaceous BOD removal (nitrification)
	Aerated Lagoons	Carbonaceous BOD removal (nitrification)
	Trickling Filters	
	Roughing Filters (trickling filters with high hydraulic loading rates	Carbonaceous BOD removal (nitrification)

	Rotating Biological Contactors Packed-bed reactors	Carbonaceous BOD removal Carbonaceous BOD removal (nitrification)
Combined Suspended & Attached Growth	Activated Biofilter Process ✓ Trickling filter-solids contact process ✓ Biofilter-AS process ✓ Series trickling filter-AS process	Carbonaceous BOD removal (nitrification)

MAJOR ANAEROBIC BIOLOGICAL PROCESSES ARE LISTED BELOW.

Type of Growth	Common Name	Use
Suspended Growth	Anaerobic Contact Process Upflow Anaerobic Sludge-Blanket	Carbonaceous BOD removal Carbonaceous BOD removal
Attached Growth	Anaerobic Filter Process	Carbonaceous BOD removal, waste stabilization (denitrification)
	Expanded Bed	Carbonaceous BOD removal, waste stabilization

FLOCCULATION TECHNIQUE OF WASTE WATER TREATMENT

Generally aluminium compounds are used as flocculants. The aluminium ion, when released into the water, will form aluminium hydroxide, $Al(OH)_3$, flocs, which will partly block the pores of a sand filter and thus enhance filtering action. A flocculant should in principle also be able to destabilize colloid solutions. A colloid in water containing a low concentration on some ions, will be surrounded by charged particles, that create an electrical double layer around it. The electrical potential of that layer is called the zeta-potential. The higher this zeta potential is, the stronger the repulsion between the particles and thus the smaller the chance that two particles will collide and then coagulate. If to such a solution aluminium ions are added, the zeta-potential of the colloid will be strongly reduced, the colloid solution becomes unstable and flocculation will occur. This process is called salt flocculation.

ACTIVATED SLUDGE PROCESS OF WASTE WATER TREATMENT

The activated sludge process is a biological process in which air or oxygen is forced into sewage liquor to develop a biological floc utilizing an active biomass. The activated sludge water treatment plants use a variety of mechanisms and processes to use dissolved oxygen to promote the growth of biological floc that substantially removes organic material. It also traps particulate material and can, under ideal conditions, convert ammonia to nitrite & nitrate and ultimately to nitrogen gas. The biomass is largely composed off saprophytic bacteria but also has an important protozoan flora mainly composed of amoebae, Spirotrichs, Peritrichs including Vorticellids and a range of other filter feeding species. In poorly managed activated sludge,a range of mucilaginous filamentous bacteria can develop including *Sphaerotilus natans* which produces a sludge that is difficult to settle and can result in the sludge blanket decanting over the weirs in the settlement tank to severely contaminate the final effluent quality. This material is often described as sewage fungus but true fungal communities are relatively uncommon to neutralize the organic matter in the wastewater.

The active biomass (called return activated sludge) is mixed with the wastewater leaving the primary clarifiers and with air in the aeration basins. Part of the settled material, the sludge, is returned to the head of the aeration system to re-seed the new sewage entering the tank. This fraction of the floc is called R.A.S - Return Activated Sludge. The remaining sludge, also called W.A.S - Waste Activated Sludge, is further treated prior to disposal. (W.A.S is also sometimes called S.A.S - Surplus Activated Sludge) The microorganisms, in the presence of oxygen, use organic matter & nutrients in wastewater to sustain their life processes, including reproduction. Effluent from the aeration basins contains large quantities of suspended biological solids (called activated sludge) that must be removed prior to discharge. Therefore, the second step of the secondary treatment process is sedimentation in the secondary clarifiers. Most of the solids removed in the secondary clarifiers are returned to the aeration basins as returned activated sludge to maintain the appropriate population of microorganisms needed to assimilate the organic matter entering the aeration basins.

Activated sludge process is shown below in figure.

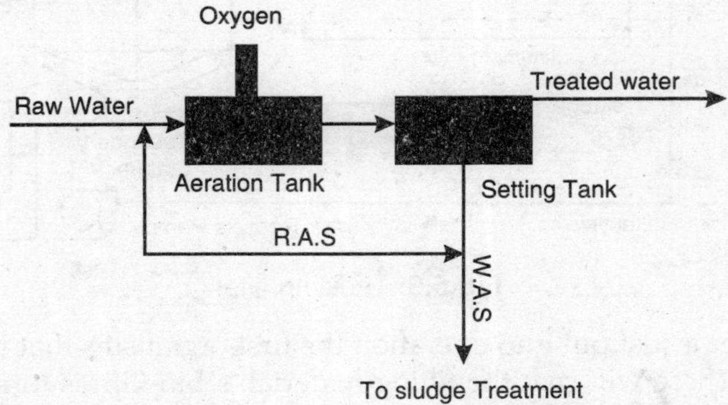

Fig.5.5: The activated sludge process

TRICKLING FILTER METHOD OF WASTE WATER TREATMENT

In the trickling filter method of waste disposal, a bed of crushed rock or synthetic media supports a film of aerobic microorganisms. The liquid waste water is sprayed over the bed and the microorganisms act on the contaminants as they pass through the filter. This is a common method used in sewage treatment plants to treat acetaldehyde, acetic acid, benzene, chlorinated hydrocarbons, cyanides, formaldehyde, ketones, and resins. Trickling filterization is generally effective for aqueous wastes where concentrations are $< 1\%$. Like all biological treatment, it requires stable, consistent operating conditions. It is not suitable for removal of highly chlorinated organics, aliphatics, amines, and aromatic compounds. Heavy metals and organic chemicals may kill the microorganisms. Heavy metals and non-biodegradable organics may also concentrate in the sludge. Hydrogen sulfide gas may also be released.

In a trickling filter, water is divided over the top and is allowed to flow slowly over the filter bed. In the course of that flow organic matter, both dissolved & suspended, is attacked by micro-organisms. In the first part of the filter (the top) mainly carbon decomposition takes place. The depth of that carbon decomposition zone depends on the load: if the water contains much organic carbon the zone will be very deep. In heavily loaded filters carbon decomposition may be the only process taking place. In the lower part of the trickling filter nitrification can occur. It will take rather long before a population of nitrifying bacteria has been established, since they grow slowly. It will take several weeks before a filter is properly conditioned.

Trickling filter is shown below in figure.

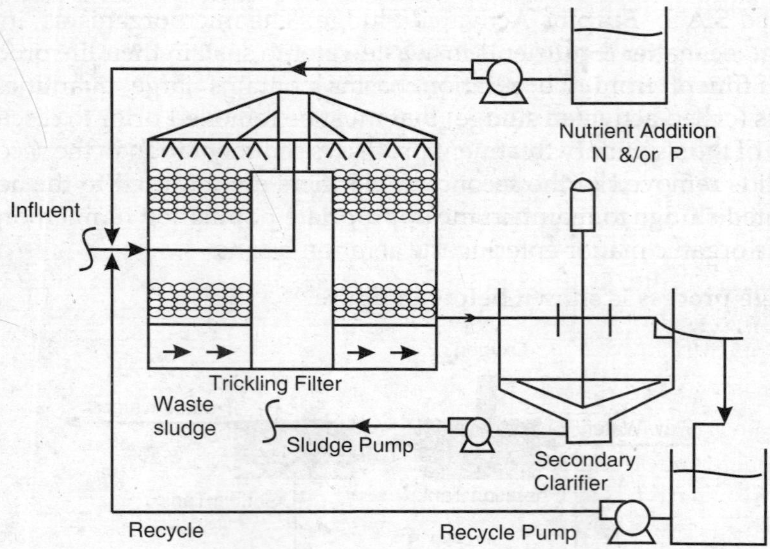

Fig.5.6: Trickling filter

If a trickling filter is just put into operation the first organisms that will grow in it will be bacteria and fungi. These will coat the filling material: a bio-film is formed. On this film the secondary colonizers will start growing. To this group belong among others protozoa and nematodes. The growth of organisms in and on the bio-film will increase its thickness. If a film becomes too thick it will start sloughing. This sloughed film will appear in the water as flocs. These flocs settle easily, so if water from the trickling filter is led through a settling area first, nearly all organic matter can be removed.

Processes happening in a trickling filter are the following:

- Dissolved organic matter is mineralized.
- Particulate organic matter & colloids are either converted into dissolved organic matter and then mineralized, or are adsorbed onto the bio-film and removed from the water as settleable flocs.
- Inorganic matter, particulate, colloidal or dissolved can react with or be adsorbed onto the bio-film and can thus be removed with the flocs.

The organisms that can be found in a trickling filter do not differ much from the ones found in activated sludge. No differences have been found between the bacterial flora of activated sludge and that of trickling filters. In activated sludge most are attached to suspended flocs, whereas in a trickling filter they are attached to the filter bed.

PROCESS OF ANAEROBIC DIGESTION OF SEWAGE SLUDGE

Waste treatment in this fashion uses the same process as that which naturally occurs in landfills. In an oxygen free environment it reduces organic waste to a relatively stable solid residue (digestate) similar to compost. Anaerobic digestion is particularly suited to wet, organic material and as such has been used for the treatment of sewage sludge.

The organic portion is separated from the waste to remove plastic, glass & metals and then placed in a sealed reactor. The conditions necessary for biological degradation are created which produces biogas, a mixture of methane and carbon dioxide. This can then be used as a fuel. The methane is generally burnt on site for heating or to produce electricity on a small scale. The digester will require an energy input to retain the material at elevated temperatures (between 20 and 40°C - the mesophilic range, or up to 60°C - the thermophilic range) in order to speed digestion. Some of the energy available from the methane generated can be recycled for this purpose, typically 20-40% of the total energy produced. Alternatively the process can produce solid fuel, a fibre product for board manufacture, or more commonly the residue is used as a soil conditioner.

The wastewater exiting the anaerobic digestion facility have elevated levels of BOD and COD. Some of this material is termed as 'hard COD' meaning it cannot be accessed by the anaerobic bacteria for conversion into biogas. If this effluent was put directly into water courses it would negatively affect them by causing eutrophication. As such further treatment of the wastewater is often required. This treatment will typically be an oxidation stage where air is passed through the water in a sequencing batch reactors or reverse osmosis unit.

Various Pond Treatment Processes for Waste Water Treatment

The treatment of wastewater could also be accomplished by various biological pond treatment processes which are listed below:

Aerobic Stabilization Ponds. Treatment is done with aerobic bacteria; oxygen is supplied by algal photosynthesis and natural surface reaeration; depth of 0.15 to 1.5 m is maintained ,

Carbonaceous BOD removal is achieved.

Maturation (tertiary) Ponds. Use aerobic treatment; applied loadings are low to preserve aerobic conditions , Secondary effluent polishing and seasonal nitrification.

Facultative Ponds. Treatment with aerobic, anaerobic and facultative bacteria; the pond has 3 zones: a surface aerobic zone, a bottom anaerobic zone, and an intermediate zone partly aerobic-anaerobic , Carbonaceous BOD removal is achieved.

Anaerobic Ponds. Treatment done with anaerobic bacteria; depths of up to 9.1 m to conserve anaerobic conditions , Carbonaceous BOD removal (waste stabilization) is achieved

The features of various pond treatment processes are tabulated below.

Common Name	Description	Use
Aerobic Stabilization Ponds	Treatment with aerobic bacteria; oxygen is supplied by algal photosynthesis and natural surface reaeration; depth of 0.15 to 1.5 m	Carbonaceous BOD removal
Maturation (tertiary) Ponds	Use aerobic treatment; applied loadings are low to preserve aerobic conditions	Secondary effluent polishing and seasonal nitrification
Facultative Ponds	Treatment with aerobic, anaerobic and facultative bacteria; the pond has 3 zones: a surface aerobic zone, a bottom anaerobic zone, and an intermediate zone partly aerobic-anaerobic	Carbonaceous BOD removal
Anaerobic Ponds	Treatment with anaerobic bacteria; depths of up to 9.1 m to conserve anaerobic conditions	Carbonaceous BOD removal(waste stabilization)

USE OF FERMENTERS , BIOREACTORS AND SECONDARY CLARIFIERS IN BIOLOGICAL TREATMENT OF WASTE WATER

Fermenters are where the primary sludge settling out from the Primary Settling Basins is broken down. The fermentation process converts the organic material into volatile acids called Volatile Fatty Acids (VFA's). An example of a VFA is acetic acid or vinegar. The VFA's are sent to the Bioreactors and utilized by the nutrient consuming microorganisms.

Bioreactors are where the Biological Nutrient Removal (BNR) process takes place. This natural process greatly reduces the concentration of dissolved organic compounds in the effluent and removes unwanted carbon, phosphorus, & nitrogen from the wastewater without the addition of chemicals. In the Bioreactor, the effluent from the Primary Settling Basins is mixed with micro-organisms and Volatile Fatty Acids (VFA's). The microorganisms come from the secondary clarifiers , where they have settled out and are then returned to the Bioreactors. The VFA's are created in the Fermenter. The micro-organisms naturally break down excess carbon and nutrients present in the wastewater. The effluent is moved through carefully controlled anaerobic (absence of available oxygen), anoxic (chemically available oxygen only) & aerobic (abundance of free oxygen) zones of the Bioreactors.Each of these areas remove specific organic compounds as the micro-organisms continue to grow and flourish, consuming impurities in the wastewater. After approximately 9 hours in the Bioreactors, the effluent flows to the Secondary Clarifiers.

Secondary Clarifiers are where any remaining solids along with the microorganisms from the Bioreactors settle to the bottom and the clear effluent flows out the top of the basins. Of the settled micro-organisms, about 97% are returned to the Bioreactors to be used again in the BNR process and 3% are sent to the DAF Thickener and then to the digesters for further treatment. The clarified final effluent flows from the Secondary Clarifiers to the Chlorine Contact Chamber before being released to the river.

WORKING OF ROTATING BIOLOGICAL CONTACTORS (RBCS) USED IN SECONDARY WASTE WATER TREATMENT SYSTEMS

Rotating Biological Contactors (RBC's) , used in waste water treatment plants are mechanical secondary treatment systems, which are robust and capable of withstanding surges in organic load. The rotating disks support the growth of bacteria and micro-organisms present in the sewage, which breakdown and stabilise organic pollutants. To be successful, micro-organisms need both oxygen to live and food to grow. Oxygen is obtained from the atmosphere as the disks rotate. As the micro-organisms grow, they build up on the media until they are sloughed off due to shear forces provided by the rotating discs in the sewage.

The primary function of these bio-reactors is the reduction of organic matter. It consists of series of closely spaced discs mounted on a horizontal shaft and rotated while 40% of each disc is submerged in wastewater. The discs are made up of light-weight plastic and are covered with a biofilm. The slime is 1-3 mm in thickness on disc. As the disk rotates the biofilm is exposed to the wastewater only part of the time then with atmosphere for adsorption of oxygen. The rotation in and out of the wastewater serves to vary the feeding cycle of the microorganisms that make up the biofilm. The shaft rotates at 1-2 rpm (slowly). Excess solids are removed by shearing forces created by the rotation mechanism. Effluent from the RBC is then passed through final clarifiers where the micro-organisms in suspension settle as a sludge. The sludge is withdrawn from the clarifier for further treatment.

The primary and secondary treatment processes generally remove at least 85% of the total suspended solids and biochemical oxygen demand. The secondary process is a very sensitive biological process and can be adversely impacted by the discharge of incompatible or toxic wastes into the sewer system. Precise control of this process is necessary to effectively treat the wastewater.

A rotating biological contactor is shown below in figure.

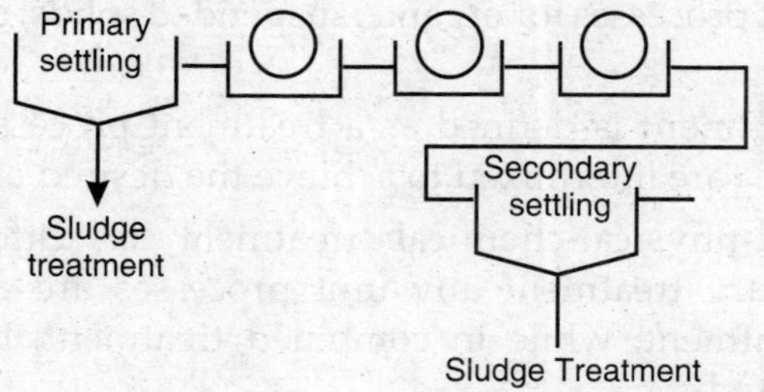

Fig.5.6: Rotating Biological Contactor (RBC)

Advanced waste water treatment

Advanced Wastewater Treatment is defined as any process designed to produce an effluent of higher quality than normally achieved by secondary treatment processes or containing unit operations not normally found in Secondary Treatment units. Primary and secondary waste water treatment processes removes the majority of BOD & Suspended Solids found in waste waters. However, in many cases this level of treatment is insufficient to protect the receiving waters or to provide reusable water for industrial and/or domestic recycle. Thus, additional treatment steps have been added to wastewater treatment plants to provide for further organic & solids removals or to provide for removal of nutrients and toxic materials.

Once freshwater has been used for an economic / beneficial purpose, it is generally discarded as waste. These wastewaters are discharged, either as untreated waste or as treated effluent, into natural watercourses, from which they are abstracted for further use after undergoing "self purification" within the stream. Through this system of indirect reuse, wastewater may be reused up to a dozen times or more before being discharged to the sea. Such indirect reuse is common in the larger river systems. However, more direct reuse is also possible: the technology to reclaim wastewaters as potable or process waters is a technically feasible option for agricultural & some industrial purposes (such as for cooling water or sanitary flushing), and a largely experimental option for the supply of domestic water. Waste water reuse for drinking raises public health, and religious, concerns among consumers. Due to stringent requirement of the quality of potable and process water, tertiary treatment of water is not a luxury any more rather it has emerged out as the only way to provide adequate safe guard for community health.

DIFFERENT TYPES , PURPOSE AND STAGES OF ADVANCED WASTEWATER TREATMENT

Advanced Wastewater Treatment is categorised by the type of process flow scheme as:

- Tertiary Treatment
- Physical-Chemical Treatment
- Combined Biological-Physical Treatment

Tertiary treatment is defined as any treatment process in which unit operations are added to the flow scheme following conventional secondary treatment. Additions to conventional secondary treatment could be the addition of a filter for suspended solids removal or the addition of many other unit processes for organic, suspended solids, nitrogen and phosphorous removal.

Physical-chemical treatment is defined as a treatment process in which biological and physical-chemical processes are intermixed to achieve the desired effluent characteristics.

Combined biological-physical-chemical treatment is differentiated from tertiary treatment in that in tertiary treatment any unit processes are added after conventional biological secondary treatment, while in combined treatment, biological and physical – chemical treatment are mixed.

Another way to classify advanced wastewater treatment is to differentiate on the basis of desired treatment goals. Advanced wastewater treatment is used for:

- Additional organic and suspended solids removal.
- Removal of nitrogenous oxygen demand (NOD).
- Nutrient removal.
- Removal of toxic materials.

The performance of secondary treatment plants is measured in terms of BOD and SS removals. A well designed & operated secondary water treatment plant will remove from 85 to 95% of the influent BOD & SS. However, the BOD test does not measure all of the organic material present in the wastewater. An average secondary effluent may have a BOD of 20 mg/l and a COD of 60 to 100 mg/l. The average secondary plant removes approximately 65% of the influent COD. Thus, when high quality effluents are required, additional organic removals must be accomplished. In addition to the organic materials remaining in most secondary effluents, there is an additional oxygen demand resulting from the nitrogen present in the wastewater.

In wastewaters, much of the nitrogen is found in the form of ammonia. When secondary treatment is used, a great deal of this ammonia is discharged in the effluent. Bacteria can utilize this ammonia as an energy source and convert ammonia to nitrite & nitrate. Another reason for advanced wastewater treatment may be to remove nutrients contained in discharges from secondary treatment plants. The effluents from secondary treatment plants contain both nitrogen (N) & phosphorous (P). which are ingredients of fertilizers. When excess amounts of N and P are discharged, plant growth in the receiving waters may be accelerated. Algae growth may be stimulated causing blooms which are toxic to fish life as well as aesthetically unpleasing. Fixed plant growth may also be accelerated causing the eventual process of a lake becoming a swamp to be speeded up. Therefore, is necessary to remove nitrogen and phosphorous prior to discharge in some cases.

Toxic materials, both organic & inorganic are discharged into many sewage collections systems. When these materials are present in sufficient quantities to be toxic to bacteria, it is necessary to remove them prior to biological treatment. In other cases, it is necessary to remove even small amounts of these materials prior to discharge to protect receiving waters or drinking water supplies. Thus, advanced wastewater treatment processes have been used in cases where secondary treatment was not possible due to materials toxic to bacteria entering the plant as well as in cases where even trace amounts of toxic materials are unacceptable in plant effluents.

ION EXCHANGE

Ion exchange is a process for water purification in which ions are exchanged between a solution and an **ion exchanger**, a non-aqueous solid or gel. Typical ion exchangers are ion exchange resins, zeolite, montmorillonite, clay, and humus. Ion exchangers are either **cation exchangers** for positively charged cations or **anion exchangers** for negatively charged anions. There are also exchangers called mixed beds which have anion and cation exchanging resins within them. Ion exchange is a reversible process and the ion exchanger can be *regenerated* or *loaded* by washing with an excess of the ions to be exchanged.

Ion exchange is a method widely used in household and industrial water purifications to produce soft water. This is accomplished by exchanging calcium Ca^{2+} and magnesium Mg^{2+} cations against sodium Na^{+} or hydrogen H^{+} cations.

Ion exchange chromatography is a chromatographical method that is widely used in biochemistry to separate charged molecules such as proteins. Ion exchange is also applied in water softeners.

Ion Exchange can be used in the food & beverage, hydrometallurgical, metals finishing, chemical & petrochemical, pharmaceutical, sugar & sweeteners, ground & potable water, nuclear, softening & industrial water, semiconductor, power, and a host of other industries.

Fig.5.7: Ion exchange resin beads

An **ion exchange resin** is an insoluble matrix (or support structure) normally in the form of small (1-2 mm diameter) beads, usually white or yellowish, fabricated from an organic polymer substrate on the surface of which are sites with easily trapped and released ions in a process called ion exchange. There are multiple different types of ion exchange resin which are fabricated to selectively prefer one or several different types of ions.

This resin is used extensively for water softening during water purification. They were developed as a more flexible alternative to the use of natural or artificial zeolites.

Most ion exchange resins are based on crosslinked polystyrene. The required active groups can be introduced after polymerization, or substituted monomers can be used. The crosslinking is usually achieved by adding a small proportion of divinyl benzene to styrene. Non-crosslinked polymers are used only rarely because of their tendency to change dimensions in dependence on the ions bonded. However, crosslinking somewhat decreases the capacity of the resin and prolong the time for reaching the equilibrium for the ions in solution and in the resin. Particle size also influences the resin parameters; smaller particles have larger surface, but cause larger head loss in the column.

Membranes made of highly cross linked ion exchange resins that allow passage of ions but not of water are used for electrodialysis.

There are four main types differing in their functional groups:

- Strongly acidic (sulfonic acid groups, eg. sodium polystyrene sulfonate or polyAMPS)
- Strongly basic, (trimethylammonium groups, eg. polyAPTAC)
- Weakly acidic (carboxylic acid groups)
- Weakly basic (amino groups, eg. polyethylene amine)

There are also specialised types:

- chelating resin

REVERSE OSMOSIS

To desalinate water, we must create a flow through a membrane, causing the water to leave the salty side of the membrane, flowing into the unsalted side. To achieve this, pressure must be created upon the water column on the salt side of the membrane; firstly, to remove the natural osmotic pressure and secondly, to create extra pressure on the water column, in order to push the water through the membrane. For the desalination of seawater, the pressure must be about 50-60 bars.

There are several different techniques that can be applied for water desalination.

Examples are reverse osmosis, electro dialysis, distillation and ion exchange:

Reverse osmosis is the most economic process for the desalination of brackish water and seawater. When we compare this process to the traditional thermic process of distillation, the capital investments and the energy use are much lower. Membranes occupy through a selective separation wall. Certain substances can pass through the membrane, while other substances are caught.

Membrane filtration

This can be used as an alternative for flocculation, sediment purification techniques, adsorption (sand filters and active carbon filters, ion exchangers), extraction and **distillation.**

There are two factors that determine the affectivity of a membrane filtration process that are membrane-dependent.:

- Selectivity
- Productivity.

Selectivity is expressed as a parameter called retention or separation factor (expressed by the unit l/m^2h).

Productivity is expressed as a parameter called flux (expressed by the unit l/m^2h). Membrane filtration can be divided up between micro and ultra filtration on the one hand and nano filtration and Reverse Osmosis (RO or hyper filtration) on the other hand. When membrane filtration is used for the removal of larger particles, micro filtration and ultra filtration are applied. Because of the open character of the membranes the productivity is high while the pressure differences are low.

When salts need to be removed from water, nano filtration and Reverse Osmosis are applied. Nano filtration and RO membranes do not work according to the principle of pores; separation takes place by diffusion through the membrane. The pressure that is required to perform nano filtration and Reverse Osmosis is much higher than the pressure required for micro and ultra filtration, while productivity is much lower.

Electrodialysis

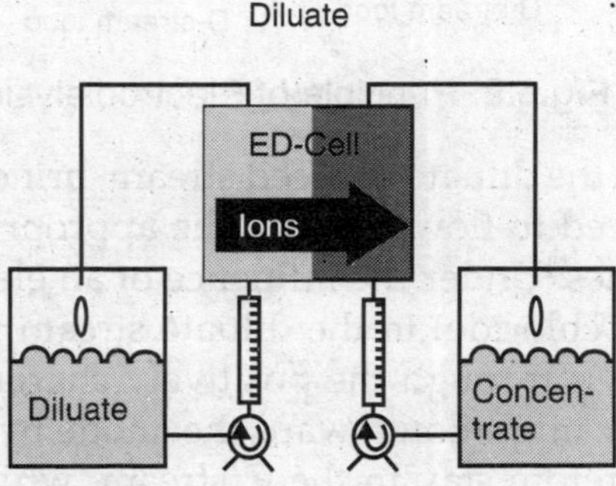

Fig.5.8: Electrodialysis

Electrodialysis (ED) is used to transport salt ions from one solution through ion- exchange membranes to another solution under the influence of an applied electric potential difference.

This is done in a configuration called an electrodialysis cell. The cell consists of a feed (diluate) compartment and a concentrate (brine) compartment formed by an anion exchange membrane and a cation exchange membrane placed between two electrodes. In almost all practical electrodialysis processes, multiple electrodialysis cells are arranged into a configuration called an electrodialysis stack, with alternating anion and cation exchange membranes forming the multiple electrodialysis cells. Electrodialysis processes are unique compared to distillation techniques and other membrane based processes (such as reverse osmosis) in that dissolved species are moved away from the feed stream rather than the reverse. Because the quantity of dissolved species in the feed stream is far less than that of the fluid, electrodialysis offers the practical advantage of much higher feed recovery in many applications.

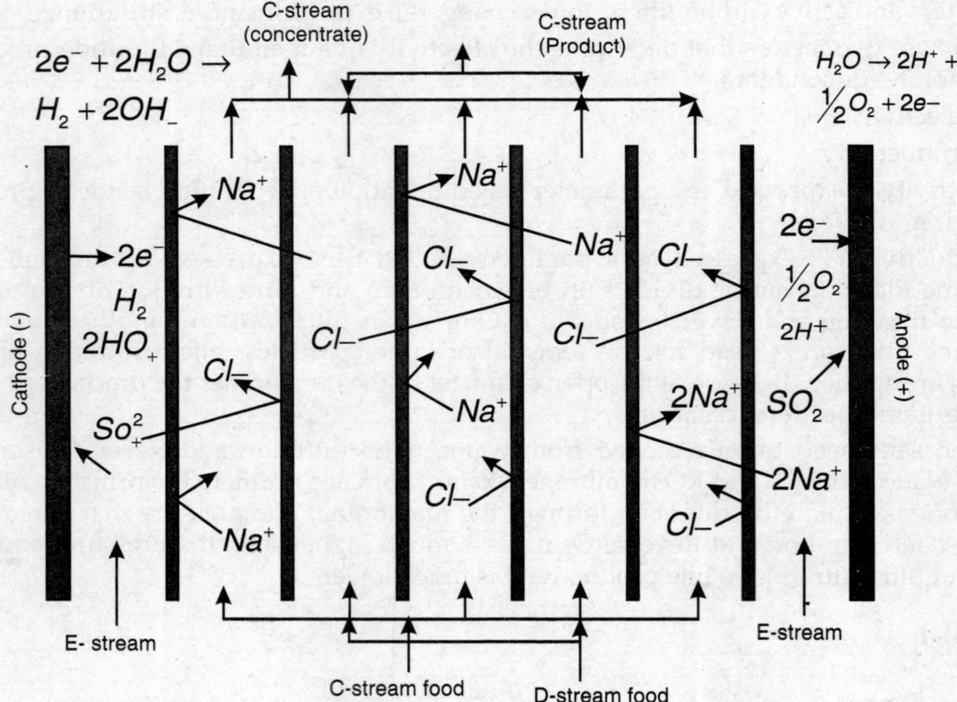

Fig.5.9: Principle of Electrodialysis

In an electrodialysis stack, the diluate (D) feed stream, brine or concentrate (C) stream, and electrode (E) stream are allowed to flow through the appropriate cell compartments formed by the ion exchange membranes. Under the influence of an electrical potential difference, the negatively charged ions (e.g., chloride) in the diluate stream migrate toward the positively charged anode. These ions pass through the positively charged anion exchange membrane, but are prevented from further migration toward the anode by the negatively charged cation exchange membrane and therefore stay in the C stream, which becomes concentrated with the anions. The positively charged species (e.g., sodium) in the D stream migrate toward the negatively charged cathode and pass through the negatively charged cation exchange membrane. These cations also stay in the C stream, prevented from further migration toward the cathode by the positively charged anion exchange membrane. As a result of the anion and cation migration, electric current flows between the cathode and anode. Only an equal number of anion and cation charge equivalents are transferred from the D stream into the C stream and so the charge balance is maintained in each stream. The overall result of the electrodialysis process is an ion concentration increase in the concentrate stream with a depletion of ions in the diluate solution feed stream.

The E stream is the electrode stream that flows past each electrode in the stack. This stream may consist of the same composition as the feed stream (e.g., sodium chloride) or may be a separate solution containing a different species (e.g., sodium sulfate). Depending on the stack configuration, anions and cations from the electrode stream may be transported into the C stream, or anions and cations from the D stream may be transported into the E stream. In each case, this transport is necessary to carry current across the stack and maintain electrically neutral stack solutions.

Anode and Cathode Reactions in electrodialysis:

Reactions take place at each electrode. At the cathode,

$$2e^- + 2H2O \rightarrow H2\,(g) + 2\,OH^-$$

While at the anode,

$$H2O \rightarrow 2\,H^+ + ½\,O2\,(g) + 2e^- \text{ or } 2Cl^- \rightarrow Cl2\,(g) + 2e^-$$

Small amounts of hydrogen gas are generated at the cathode and small amounts of either oxygen or chlorine gas (depending on composition of the E stream and end ion exchange membrane arrangement) at the anode. These gases are typically subsequently dissipated as the E stream effluent from each electrode compartment is combined to maintain a neutral pH and discharged or re-circulated to a separate E tank. However, some have proposed collection of hydrogen gas for use in energy production.

Applications of electrodialysis

Electrodialysis systems can be operated as continuous production or batch production processes. In a continuous process, feed is passed through a sufficient number of stacks placed in series to produce the final desired product quality. In batch processes, the diluate or concentrate streams are re-circulated through the electrodialysis systems until the final product or concentrate quality is achieved.

Electrodialysis is usually applied to deionization of aqueous solutions. However, desalting of sparingly conductive aqueous organic and organic solutions is also possible. Some applications of electrodialysis include:

- Large scale brackish and seawater desalination and salt production.
- Small and medium scale drinking water production (e.g., towns & villages, construction and military camps, nitrate reduction, hotels and hospitals)
- Water reuse (e.g., industrial laundry wastewater, produced water from oil/gas production, cooling tower makeup & blowdown, metals industry fluids, wash-rack water)
- Pre-demineralization (e.g., boiler makeup and pretreatment, ultrapure water pretreatment, process water desalination, power generation, semiconductor, chemical manufacturing, food and beverage)

- Food processing
- Agricultural water (e.g., water for greenhouses, hydroponics, irrigation, livestock)
- Glycol desalting (e.g., antifreeze or engine-coolants, capacitor electrolyte fluids, oil and gas dehydration, conditioning and processing solutions, industrial heat transfer fluids, secondary coolants from heating, venting, and air conditioning (HVAC))
- Glycerin Purification

The major application of electrodialysis has historically been the desalination of brackish water or seawater as an alternative to RO for potable water production and seawater concentration for salt production. In normal potable water production without the requirement of high recoveries, RO is generally believed to be more cost- effective when total dissolved solids (TDS) are 3,000 parts per million (ppm) or greater, while electrodialysis is more cost-effective for TDS feed concentrations less that 3,000 ppm or when high recoveries of the feed are required.

Another important application for electrodialysis is the production of pure water and ultrapure water by electrodeionization (EDI). In EDI, the compartments (diluate, concentrate, or both) of the electrodialysis stack are filled with ion exchange resin. When fed with low TDS feed (e.g., feed purified by RO), the diluate can reach very low levels (e.g., 18 Megaohms). The ion exchange resins act to retain the ions, allowing these to be transported across the ion exchange membranes. The main usage of EDI systems are in electronics, pharmaceutical, and cooling tower applications.

LIMITATIONS OF ELECTRODIALYSIS

Electrodialysis has inherent limitations, working best at removing low molecular weight ionic components from a feed stream. Non-charged, higher molecular weight, and less mobile ionic species will not typically be significantly removed. Also, in contrast to RO, electrodialysis becomes less economical when extremely low salt concentrations in the product are required and with sparingly conductive feeds: current density becomes limited and current utilization efficiency typically decreases as the feed salt concentration becomes lower, and with fewer ions in solution to carry current, both ion transport and energy efficiently greatly declines. Consequently, comparatively large membrane areas are required to satisfy capacity requirements for low concentration (and sparingly conductive) feed solutions. Innovative systems overcoming the inherent limitations of electrodialysis (and RO) are available; these integrated systems work synergistically, with each sub-system operating in its optimal range, providing the least overall operating and capital costs for a particular application.

As with RO, electrodialysis systems require feed pretreatment to remove species that coat, precipitate onto, or otherwise "foul" the surface of the ion exchange membranes. This fouling decreases the efficiency of the electrodialysis system. Species of concern include calcium and magnesium hardness, suspended solids, silica, and organic compounds. Water softening can be used to remove hardness, and micron or multimedia filtration can be used to remove suspended solids. Hardness in particular is a concern since scaling can build up on the membranes. Various chemicals are also available to help prevent scaling. Also, electrodialysis reversal systems seek to minimize scaling by periodically reversing the flows of diluate and concentrate and polarity of the electrodes.

Distillation

This involves boiling the water to produce water vapour. The water vapour then rises to a cooled surface where it can condense back into a liquid and be collected. Because the solutes are not normally vaporized, they remain in the boiling solution. Even distillation does not completely purify water, because of contaminants with similar boiling points and droplets of unvaporized liquid carried with the steam. However, 99.9% pure water can be obtained by distillation.

Adsorption

Wastewater treatment, using waste materials as adsorbents, represents a path for developing sustainable processes and for reducing the environmental burden. A number of technologies have been developed over the years to remove organic matter (expressed as chemical oxygen demand, COD) from industrial wastewater. The most important technologies include coagulation or flocculation process, membrane filtration, and oxidation process. These methods are generally expensive, complicated, time consuming and requires skilled personnel. The high cost of coal-based activated carbons has stimulated the search for cheaper alternatives. Low cost and non-conventional adsorbents including agricultural by products such as nut shells, wood, bone, peat processed into activated carbons have been reported to be the important adsorbents for the removal of metals and organics from municipal and industrial wastewater. Activated carbon is a commonly used adsorbent in sugar refining, chemical and pharmaceutical industries, water and wastewater treatment, and as an adsorbent in point-of-use (POU) and point-of-entry (POE) of home water filtration systems. Increasing requirements for clearer and more polished effluent from many processes suggest that, barring the development of new technologies, industrial need for activated carbon will only increase in future.

Commercial activated carbon is regarded as the most effective material for controlling the organic load. However due to its high cost and about 10-15 % loss during regeneration, unconventional adsorbents like fly ash, peat, lignite, bagasse pith, wood, saw dust etc. have attracted the attention of several investigations and adsorption characteristics have been widely investigated for the removal of refractory materials for varying degree of success.

Thus the removal of organic material by adsorption onto low cost waste material has recently become the subject of considerable interest. This approach offers a potentially simple and economic "End of Pipe" solution to the challenges set by new legislation covering effluent discharges.

Fly ash has a significant capacity for adsorption of organic compounds from aqueous solutions. The carbon content of fly ash plays a significant role during the adsorption of organic compounds by fly ash. The adsorption capacity increases with the increasing carbon content of fly ash. Fly ash is also suited for the removal of COD from the domestic wastewater.

Fly ash is a residue that results from the combustion of coal in power plants. One of the main advantages of COD removal by using fly ash over the other chemical treatment methods is that it is in abundance and easy availability makes it a strong choice in the investigation of an economic way of COD removal. Other advantage is that it could easily be solidified after the pollutants are adsorbed because it contains pozzolanic particles that react with lime in the presence of water, forming cementation calcium-silicate hydrates.

Biosorption

Biosorption, biopreciptation, and uptake by purified biopolymers derived from microbial cells provide alternative and/or additive processes for conventional physical and chemical methods. Intact microbial cells, live or dead, and their products can be highly efficient bioaccumulators of both soluble and particulate forms of metals. The cell surfaces of all microorganisms are negatively charged owing to the presence of various anionic structures. This gives the bacteria the ability to bind metal cations. Various microbial species, mainly *Pseudomonas,* have been shown to be relatively efficient in metal uptake from polluted effluents.

Biosorption is a promising method for removal of toxic ions from wastewater. Its advantage is especially in the treatment of large volumes of effluents with low concentration of pollutants. In the past few years a lot of effort has been made on screening of efficient biomass types, its preparation and the biosorption mechanism determination. Biosorption is a fast and reversible process which resembles adsorption and in some cases ion exchange.

Apparently there are many modes of non-active metal uptake by (microbial) biomass. Any one or a combination of them can be functional in immobilizing metallic species on biosorbents. A number of anionic ligands participate: phosphoryl, carbonyl, sulfhydryl and hydroxyl groups can all be active to various degrees in binding the metal.

Oxidation

Secondary treatment begins in oxidation pond one and then into oxidation pond two and sometimes into oxidation pond three. Treating wastewater in oxidation pond is of two reasons: the settling of solids and to breakdown the organic matter in the sewage. There are virtually no plants in the oxidation ponds. The oxidation ponds are virtually void of macrophytic plant life to promote the growth of algae.

The large size of the oxidation ponds and the 1.5 metre depth are perfect for settling out solids leftover from primary treatment. As the wastewater flows slowly from oxidation pond 1 to pond 2 to the treatment marshes (or oxidation pond 3 depending on the season) more and more solids settle out. Approximately half of the solid waste left after primary treatment settles out in the oxidation ponds.

Breakdown of the floating and settling organic matter in the sewage is aided by algae growth and microbial/bacteria activities in the ponds. The microbes and bacteria consume the unwanted nutrients present in the wastewater. To do this, they need oxygen. That's where the algae comes in. Algae, through the process of photosynthesis, produces oxygen. Algae also consumes Nitrogen and Phosphorous, two common nutrients found in the wastewater. The microbes and bacteria use the oxygen to breakdown and use the nutrients in the waste through their metabolism. Much like the way the humans consume food to create energy to be burned by your own metabolism. The microbes do the same thing. Again like the humans, they produce carbon dioxide through this process. The carbon dioxide is then used by the algae to produce more oxygen for the microbes.

Since the wastewater is still very high in nutrients or "waste", the city also uses mechanical aerators in oxidation pond one to ensure that there is enough oxygen in the water for the microbes to consume the waste. The amount of oxygen being consumed by the microbes is called BOD. BOD stands for Biological Oxygen Demand, and is used in determining how much "waste" or nutrients are still in the wastewater. If there is a lot of waste, then there will be many, many, many microbes who will consume lots of the oxygen in the wastewater. Humans at the treatment plant monitor the wastewater coming into the plant, the clarifier, the oxidation ponds, the marshes and the chlorine contact basin.

The solids that have settled out sink to the bottom where microbes continue to break them down.

Treatment Marshes

From oxidation pond the treated wastewater flow into the treatment marshes. The treatment marshes cover about 7.5 acres and are 0.5 metre deep on average. The treatment marshes appearance are the absolute opposite of the oxidation ponds. The treatment marshes are packed with plants, for good reasons too.

After the oxidation ponds, it would appear that the wastewater had been more polluted than it was before! But it's not actual "waste" that is doing that, but the treatment process in the oxidation ponds. In the oxidation ponds the wastewater tend to promote algal growth. The algae eventually dies. Algae, though a necessary part of the process, becomes a pollutant when it dies. One of the functions of the treatment marshes is to filter out this dead algae and to shade the water so the algae will not grow. The plants in the treatment marshes also slow down the flow of wastewater so even more solids can settle to the bottom.

The plants are very important in the wastewater treatment process. The treatment marshes are the opposite of the oxidation ponds, algae is an unwanted element in the treatment marshes. Some of the plants that we have for this are the freshwater *Scirpus Acutus, which* is good for pulling the unwanted nutrients out of the wastewater. They will incorporate the nutrients into their bio mass for their own purposes.

The Oxidation Ditch Process

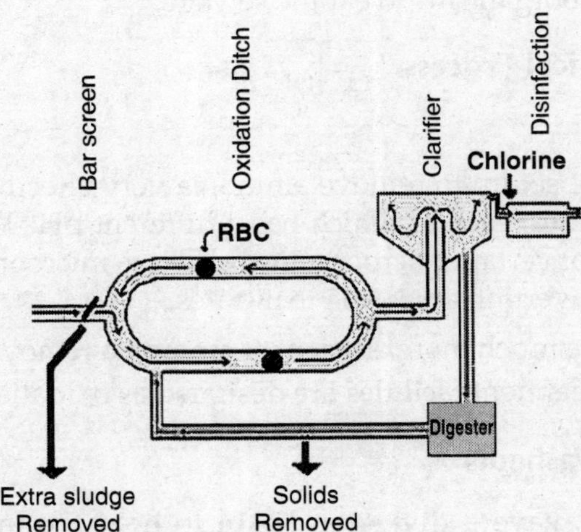

Fig.5.10: Oxidation ditch

Treatment of wastewater using an oxidation ditch is relatively similar to wastewater treatment in a packaged plant. But the oxidation ditch replaces the aeration basin and provides better sludge treatment.The only pretreatment typically used in an oxidation ditch system is the bar screen. After passing through the bar screen, wastewater flows directly into the oxidation ditch.

The **oxidation ditch** is a circular basin through which the wastewater flows. Activated sludge is added to the oxidation ditch so that the microorganisms will digest the B.O.D. in the water. This mixture of raw wastewater and returned sludge is known as **mixed liquor**. Oxygen is added to the mixed liquor in the oxidation ditch using rotating biological contactors (RBC's.) RBC's are more efficient than the aerators used in packaged plants. In addition to increasing the water's dissolved oxygen, RBC's also increase surface area and create waves and movement within the ditches.

Once the B.O.D. has been removed from the wastewater, the mixed liquor flows out of the oxidation ditch. Sludge is removed in the clarifier. This sludge is pumped to an aerobic digester where the sludge is thickened with the help of aerator pumps. This method greatly reduces the amount of sludge produced. Some of the sludge is returned to the oxidation ditch while the rest of the sludge is sent to waste.

Comparison to a Packaged Plant

The treatment of wastewater in an oxidation ditch is similar to treatment in a packaged plant. The two main differences between the processes are the retention time and the type of organisms which digest the wastewater.

Retention time is much longer in an oxidation ditch. A packaged plant usually has a retention time of two to four hours while an oxidation ditch retains the wastewater for two days.

Since the D.O. is higher in the oxidation ditch than in a packaged plant, a greater variety of microorganisms live in the oxidation ditch. In contrast, packaged plants usually depend upon only a few types of microorganisms to eat the sewage.

Variations from the Typical Process

Ammonia Removal

Oxidation ditches can be set up to remove ammonia very effectively. Wastewater can be sent through two sets of ditches, each of which has a different pH. The different pH in the two ditches creates a niche for certain microorganisms. These microorganisms are very efficient at removing B.O.D and converting ammonia to nitrates.

Oxidation ditches are much more efficient at ammonia removal than packaged plants are. As a result, most new treatment facilities are designed as oxidation ditches.

Diversion Basins and Washouts

Some oxidation ditches, have a **diversion basin** to hold the influent when flows increase because of excessive rainfall. The diversion basin holds the excess influent and allows more time for treatment.

Without a diversion basin, heavy rains could cause a **washout** to occur. A washout occurs when a large influx of influent rushes into the oxidation ditches. The ditches are unable to contain the extra water, so microorganisms, sludge, and wastewater are pushed through the plant and out into a river or stream before being properly treated.

Without diversion basins, total washouts can be prevented by shutting off the inner ditches and allowing the outside ditch to circulate the influent, providing primary treatment to the water before it is released. In periods of excessive rainfall, oxidation plants can be operated on high flow settings for a month at a time.

Advantages and Disadvantages of oxidation ditch

The greatest advantage of an oxidation ditch is the efficiency of sludge removal. In an oxidation ditch, only about 15% of the original B.O.D. ends up as sludge, compared to a packaged plant where about 60% of the B.O.D. becomes sludge.

However, oxidation ditches are expensive to maintain. The monetary cost is very high per ton of B.O.D. removed. In some cases, the cost may reach nearly 350 dollars per ton.

Oxidation ditches have an additional environmental drawback. The water is moved through the ditches using rotors, and these rotors in turn use electricity. The electricity used to operate the plant causes sulphur dioxide and other contaminants to be released into the atmosphere from coal-burning electrical plants.

Oxidation ditches provide the most thorough process for treating sewage, but oxidation ditches are also one of the most costly forms of treatment.

REMOVAL OF NUTRIENTS FROM WASTEWATER

Wastewater may contain high levels of the nutrients like nitrogen and phosphorus. Excessive release to the environment can lead to a build up of nutrients, called eutrophication, which can in turn encourage the overgrowth of weeds, algae, and cyanobacteria (blue- green algae). This may cause an algal bloom, a rapid growth in the population of algae. The algae numbers are unsustainable and eventually most of them die. The decomposition of the algae by bacteria uses up so much of oxygen in the water that most or all of the animals die, which creates more organic matter for the bacteria to decompose. In addition to causing deoxygenation, some algal species produce toxins that contaminate drinking water supplies.

Different treatment processes are required to remove nitrogen and phosphorus.

Nitrification and Denitrification

Nitrification, the conversion of ammonia to nitrate, is a process that takes place under aerobic conditions. This is done by bacteria. The first step in the process is the conversion of ammonia, NH_4^+ to nitrite, NO_2^-. The next step in the nitrification process is the conversion of nitrite, NO_2^-, to nitrate, NO_3^-. There are 4 species of nitrite oxidizers: *Nitrobacter* (2 species), *Nitrococcus* and *Nitrospina*. It was found that Nitrobacter sp. adapted fast to different salinities and grew equally well over the whole range from 0 to 3.5%. Under aerobic conditions, nitrate is the end stage of nitrogen metabolism.

For the removal of nitrate there are 2 possible pathways. There are several large algae, among others *Ulva* sp. that can use nitrate nitrogen for the formation of organic nitrogen compounds, provided there is plenty of light available. If there is not enough light, they will start metabolizing organic nitrogen and in that way increase the nitrate concentration.

Under anaerobic conditions, several bacteria species can use nitrate in respiration instead of oxygen. Among these, bacteria from the *Pseudomonas* group are predominant. In the process elementary nitrogen, N_2, is formed, which will escape as a gas. Since the energy gain from nitrate- induced respiration is about 10% less than that of oxygen- induced respiration, the latter will have preference if oxygen is present.

Biological nitrification may be used to prevent oxygen depletion from nitrogenous demand (NOD) in the receiving waters. Nitrification is simply the conversion of ammonia to nitrate in the treatment plant rather than in the receiving water. Nitrification may be carried out in the same tank as BOD removal or in a separate stage. Nitrification may be carried out either in activated sludge flocs or in fixed films.

Regardless of the particular scheme chosen, the same basic requirements for nitrification must be maintained:

- Oxygen level
- Loading rates
- Solids retention time
- Alkalinity
- pH
- Freedom from toxic materials
- Temperature

Sufficient oxygen must be available for nitrification to occur. Approximately 4.5 kg of dissolved oxygen are required for the conversion of 1 kg of ammonia to nitrate. Dissolved oxygen sufficient to satisfy the remaining BOD is also required. In activated sludge plants the mixing requirement of the basins must also be considered. Generally, dissolved oxygen levels of approximately 2 - 3 mg/l are recommended for nitrification. The bacteria responsible for nitrification reproduce at a much slower rate than those responsible for BOD removal. Thus, the danger always exists for the "wash out" of the nitrifying organisms. That is, unless the nitrifying bacteria reproduce at the same or greater rate than they are removed from the system (by waste sludge) then the population of bacteria will be insufficient to carry out nitrification. For this reason, nitrification systems are operated at higher return sludge rates than conventional secondary treatment. The amount of sludge to be wasted is significantly less than from a conventional activated sludge system.

Nitrification systems are sensitive to pH variation. Optimum pH has been found to be approximately 7.8 to 9.0. Reductions in nitrification have been found outside this range. Alkalinity is also destroyed during nitrification. Theoretically, 7.2 kg of alkalinity are destroyed in converting 1 kg of ammonia to nitrate. In low alkalinity wastewaters, Quick lime (CaO) or $Ca(OH)_2$ is often used to provide alkalinity and pH control.

Generally, the influent BOD to nitrification systems has not been found to effect performance. However, sufficient oxygen must be provided for the organic demand and organic shock loads must be avoided.

Biological Denitrification

Biological nitrification satisfies the nitrogenous oxygen demand by converting NH_3 to NO_3. In some applications, such as discharge into enclosed bodies of water or recycle to water supplies, nitrification may not be sufficient. When nitrogen removal is required, one of the available methods is to follow biological nitrification with biological denitrification.

Denitrification is accomplished under anaerobic or near anaerobic conditions by bacteria commonly found in wastewater. Nitrates are removed by two mechanisms: (1) Conversion of NO_3 to N_2 gas by bacterial metabolism and (2) conversion of NO_3 to nitrogen contained in cell mass which may be removed by settling.

In order for denitrification to occur, a carbon source must be available. Most commonly, methanol is used. The methanol must be added in sufficient quantity to provide for cell growth and to consume any dissolved oxygen which may be carried into the denitrification reactor.

Usually 3 to 4 kg of methanol per kg of nitrate are required. Careful control of methanol feed is necessary to prevent waste of chemicals. In addition, if excess methanol is fed to the system, unused methanol will be carried out in the effluent causing excessive BOD.

Denitrification may be carried out in either a mixed slurry reactor or in fixed bed reactors. Denitrification filters carry out both denitrification and filtration in the same unit. Mixed slurry systems consist of a denitrification reactor, reaeration basin and clarifiers. Reaeration prior to clarification is required to free the sludge from trapped bubbles of nitrogen gas. Denitrifying bacteria grow very slowly and are extremely sensitive to temperature.

Denitrification rates have been shown to increase five fold when the temperature is increased from 10 to 20°C. Thus, operating parameters such as sludge age and retention time must be varied with temperature.

The pH in denitrification systems must be carefully controlled. The optimum pH is from 6.0 to 8.0.

Denitrification is a very sensitive and difficult process to operate. Little full scale operational experience is available. Constant monitoring of pH, methanol feed and temperature is essential to successful operation.

Nitrogen Removal

The removal of nitrogen is effected through the biological oxidation of nitrogen from ammonia (nitrification) to nitrate, followed by denitrification, the reduction of nitrate to nitrogen gas. Nitrogen gas is released to the atmosphere and thus removed from the water.

Nitrification itself is a two-step aerobic process, each step facilitated by a different type of bacteria. The oxidation of ammonia (NH_3) to nitrite (NO_2^-) is most often facillitated by *Nitrosomonas* spp. (nitroso=ammonium). Nitrite oxidation to nitrate (NO_3^-), though traditionally believed to be facilitated by *Nitrobacter* spp. (nitro=nitrite), is now known to be facilitated in the environment almost exclusively by *Nitrospira* spp..

Denitrification requires anoxic conditions to encourage the appropriate biological communities to form. It is facillitated by a wide diversity of bacteria. Sand filters, lagooning and reed beds can all be used to reduce nitrogen, but the activated sludge process (if designed well) can do the job the most easily. Since denitrification is the reduction of nitrate to dinitrogen gas, an electron donor is needed. This can be, depending on the wastewater, organic matter (from faeces), sulfide, or an added donor like methanol.

Sometimes the conversion of toxic ammonia to nitrate alone is referred to as tertiary treatment.

Phosphorus Removal

Phosphorus can be removed biologically in a process called enhanced biological phosphorus removal. In this process, specific bacteria, called polyphosphate accumulating organisms, are selectively enriched and accumulate large quantities of phosphorus within their cells (up to 20% of their mass). When the biomass enriched in these bacteria is separated from the treated water, these biosolids have a high fertilizer value.

Phosphorus removal can also be achieved by chemical precipitation, usually with salts of iron (e.g. ferric chloride) or aluminum (e.g. alum). The resulting chemical sludge is difficult to handle and the added chemicals can be expensive. Despite this, chemical phosphorus removal requires significantly smaller equipment footprint than biological removal, is easier to operate and can be more reliable in areas that have wastewater compositions that make biological phosphorus removal difficult.

REMOVAL OF HEAVY METALS FROM WASTEWATER

Metals play an integral role in the life processes of living organisms. Some metals (e.g. Ca, Co, Cr, Cu, Fe, K, Mg, Mn, Na, Ni and Zn) are essential, serve as micronutrients and are used for redox-processes, to stabilize molecules through electrostatic interactions; as components of various enzymes; and regulation of osmotic pressure. While many other metals have no biological role (*i.e.* Ag, Al, Cd, Au, Pb, and Hg), and are nonessential and potentially toxic to living organism specially microorganisms. Toxicity of nonessential metals occurs through the displacement of essential metals from their native binding sites or through ligand interactions. In addition, at high levels both essential and nonessential metals can damage cell membranes; alter enzyme specificity; disrupt cellular functions, and damage the structure of DNA.

Heavy Metals

Heavy metals are metals with densities higher than 5 g/cm^3. Heavy metals in wastewater come from industries and municipal sewage, and they are one of the main causes of water and soil pollution. Accumulation of these metals in wastewater depends on many local factors such as type of industries in region, people's way of life and awareness of the impacts done to the environment by careless disposal of wastes. Therefore the presence of heavy metals in wastewater is not only of great environmental concern but also strongly reduces microbial activity, as a result adversely affecting biological wastewater treatment processes. Moreover the toxicity of heavy metals in wastewater was shown to be dependent on factors like metal species and concentration, pH, wastewater pollution load and solubility of the metal ions.

Metal Removals

Biological removal processes has been attracting considerable attention for removing heavy metals from aqueous wastes and screening for microorganisms having higher potential for removing heavy metals from wastes. Microbial removal of heavy metals offers the advantages of low operating cost, minimizing secondary problems with metal-bearing sludge and high efficiency in detoxifying very dilute effluent.

Cyanide and Heavy Metal Removal

Cyanide and heavy metals removal from wastewater streams are only a few of the extremely important fields required to insure the protection of global bio diversity through responsible environmental management. With the enactment of the "Clean Water Act", and the need for a more sustainable level of industrial development, ever more stringent levels of water quality in industrial wastewater are required.

Waste Segregation

Prior to treatment, it is necessary to segregate the different wastewater streams according to the type of treatment that will be required. As a minimum, it is imperative to segregate all cyanide bearing wastes, and all chrome bearing wastes, from any other waste streams. If possible, it is also desirable to segregate acidic streams, from basic streams, and highly chelated streams.

Cyanide Removal

The first step in any waste treatment process is to remove cyanide. Cyanide is used to complex metals such as cadmium, gold, platinum, etc. so that they will remain in solution at high pH ranges.

There are three different methods for removing cyanide. The first, used in Europe but not in the US, is acid hydrolysis. In this process, the pH of the solution is lowered with sulfuric acid and the resulting hydrogen cyanide is captured and recovered. This process is extremely dangerous due to the production of cyanide gas.

The second method is to oxidize the cyanide to cyanate with ozone. This method has been demonstrated to work in the laboratory but has not been as successful in the field. The reason for this is most likely due to the short half life of ozone.

The most common method of cyanide destruction is known as alkaline chlorination. This is usually accomplished in two stages. In the first stage, the solution pH is raised to approximately 10 – 11 and bleach is added. The solution is then allowed to react for 30 - 60 minutes while the cyanide is oxidized to cyanate. In the second stage, solution pH is lowered to 8.5 - 9.0. Additional bleach is added if required. This reaction takes 45 - 90 minutes and will oxidize the cyanate to CO_2 and nitrogen. At this point the cyanide bearing waste streams can be mixed with the normal non-cyanide bearing waste streams for additional treatment.

It is important to note that cyanide can form complexes that are not amenable to alkaline chlorination. It is therefore important to prevent these complex cyanides from forming in the first place. The metals that usually cause these complex cyanides to form are, iron, chrome, and nickel. Once these complexes have formed it may be possible to break them by adding excess bleach and increasing the reaction time.

Chrome Destruction

Chrome exists in two valence states; +6 and +3. The +6 or hexavalent state is very soluble over a broad pH range. This is also the most toxic form of chrome. The +3, or trivalent state is insoluble at higher pH levels and is fairly easy to remove from water.

The first step in removing chrome is to reduce the hexavalent chrome to the trivalent form. This is usually accomplished by reducing the pH to 2.5 - 3.0 with sulfuric acid and then adding sodium metabisulfite. This reaction is fairly quick and will normally be accomplished in approximately 15 - 20 minutes.

An alternative method of reducing chrome is done at a high pH with the addition of hyrosulfide. Eliminating the use of sulfuric acid is an advantage; however this method is relatively expensive.

After all of the hexavalent chrome has been reduced to trivalent chrome, the solution can be added to the normal metal bearing waste streams for additional treatment.

Mercury Removal

Many types of mercury treatment technology have been reported and the majority are based upon laboratory or piolet scale studies. Hydroxide precipitation is ineffective, with soluble levels exceeding 75 mg/l mercury over a pH range of 3.5 to 11.5.

Sulfide precipitation is the most common method of removing mercury. This method has all of the problems associated with sulfide precipitation.

Precipitation with sodium borohydride has also been reported. This method has many problems associated with it that include the formation of explosive hydrogen gas and the tendency for the mercury to go back into solution.

The most successful method is to use organometallic precipitation. This method can be used to reduce the level of mercury to 10 µg/l or less. Frequently we are able to achieve results of non-detectable levels.

Chelated Wastes

The most common chelating agents found in metal bearing wastes are cyanide, EDTA, NTA, ammonia, and citric acid. Chelates function by complexing with metal ions thereby keeping them in solution at elevated pH levels. In order to remove the metals it is necessary to break the chelated complexes. In the past, this has typically been accomplished by either adding large quantities of coagulants such as alum or ferric chloride or by raising or lowering the pH to extreme levels. The use of large amounts of coagulants is called "salting out", and although it will work, it creates a large volume of "RECRA" sludge that is typically hauled to landfills.

TRADITIONAL METAL REMOVAL PROCESSES

Hydroxide Precipitation

Traditionally, dissolved metals have been removed from water by the process of hydroxide precipitation. Since most metal hydroxides are insoluble, it would appear easy to remove metals by this process. The main problems associated with this process are that hydroxides of different metals have different pH levels for minimum solubility and the reactions are of an equilibrium type, i.e., some of the metal hydroxide will disassociate with the resulting metal ions going back into solution.

For example, nickel has its lowest solubility at a pH of 10.5 - 11.0, but chrome is at a minimum solubility at a pH of 7.5 - 8.0. Therefore, this method of metal removal can leave high levels of some metals still in solution or require an additional neutralization step. Additionally, mercury cannot be removed by this method at all.

The most common reagents used for hydroxide precipitation are caustic soda, lime, and magnesium hydroxide.

Sulfide precipitation

Soluble metals can also be removed by precipitating them as a sulfide by the addition of sodium sulfide to the solution. This method yields more complete metal removal than hydroxide precipitation but can easily leave toxic sulfides in solution. This method is much more expensive than hydroxide precipitation since the excess sulfides are usually regulated and the resulting sludge may be difficult to landfill. Therefore, it is not as widely used as hydroxide precipitation.

MODERN METAL REMOVAL PROCESSES

Carbamates

Carbamates are chemical reducing agents that can be obtained as either sodium dimethyldithiocarbamate or sodium diethyldithiocarbamate. Carbamate precipitation is again an equilibrium reaction that does not go to completion. Metal residuals of 1.0 - 1.5 mg/l can usually be obtained. Carbamates are not effective at acidic pH levels and are not always effective at treating chelated wastes.

Sodium Borohydride

Sodium borohydride is an extremely strong reducing agent and can reduce both chelated and non chelated metals. This process has the advantage of producing the least amount of sludge of any process but it has a number of disadvantages that almost always preclude its use in an efficient and cost effective system.

A major disadvantage, is that unless the liquid is removed from the sludge immediately, metals tend to go back into solution with the water. Another problem is that pH control is critical. Explosive hydrogen gas is evolved at acidic pH values.

High cost of this reagent has also been a problem, and as a result, it has been very difficult to justify its use.

Organometallic Precipitation

The synthesis of organometallic and coordination compounds has attracted interest among chemists during the past few decades. Organometallic compounds are organic molecules containing at least one atom of a metal bonded to a carbon atom. A familiar example of this class of substance is tetraethyl lead, which was often added to the gasoline that fuels internal-combustion engines. Other organometallic compounds include catalysts used in plastic manufacture and in organic synthesis.

In 1991 Steve Holtzman, pioneered a new more environmentally responsible method of removing heavy metals. This process revolves around the formation of insoluble organometallic compounds formed by reacting metal bearing wastes with a proprietary organic agent.

By forming specific types of insoluble organometallic compounds, all regulated metals can be reduced to non-detectable levels. The process is easily controlled with an inexpensive ORP controller and can adapt to changing levels of contaminants in the waste stream influent.

This process works over an extremely broad pH spectrum (1.5 - 12) and has the ability to break most chelates in extremely high concentrations. Since the metals are precipitated as an organometallic complex at all pH values, there is no problem with different levels of solubility based on pH. The only reasons for pH control are to make certain that the waste effluent is in a range that is allowed by the discharge permit and to allow the polymer flocculents to be in a pH range where they are effective.

The volume of sludge produced is comparable to that produced by borohydride. The sludge volume produced by this method and caustic soda is approximately 50% of the sludge that is produced by using caustic soda alone and precipitating the metals as hydroxides. If magnesium hydroxide is used as the caustic agent, sludge volumes can be as low as 25% of the sludge volume produced by hydroxide precipitation.

The best treatment approach is to first do any cyanide or chrome destruction that may be required. Adjust pH (if found necessary during bench testing) and add the organic reducing agent on a 2.5 to 1 stoichiometric basis. The pH is next adjusted to 7.5 - 8 with magnesium hydroxide and any coagulants required are added. Anionic polymer is then added to flocculate the metals. Metal levels in the effluent can be "dialed in" according to customer requirements. Typical residuals of all regulated metals are 0.01 mg/l or less.

A typical waste treatment facility uses alkaline chlorination for cyanide destruction, sodium metabisulfite for chrome destruction and, organometallic precipitation for the removal of heavy metals.

A TYPICAL OVERALL METHOD OF SEWAGE WATER TREATMENT

Sewage can be treated close to where the sewage is created, in a decentralised system (in septic tanks, biofilters or aerobic treatment systems), or be collected & transported by a network of pipes & pump stations to a municipal water treatment plant, a centralised system. Sewage treatment generally involves three stages, called primary, secondary and tertiary treatment.

Primary treatment consists of temporarily holding the sewage in a quiescent basin where heavy solids can settle to the bottom while oil, grease and lighter solids float to the surface. The settled and floating materials are removed.

Secondary treatment removes dissolved and suspended biological matter. Secondary treatment is typically performed by indigenous, water-borne micro-organisms in a managed habitat.

Tertiary treatment ,Treated water is sometimes disinfected chemically or physically (for example, by lagoons and microfiltration) prior to discharge into a stream, river, bay, lagoon or wetland, or it can be used for the irrigation of a golf course, green way or park.

Primary treatment includes Screening , Flow Equalization, Neutralization, Coagulation , Primary sedimentation.

Secondary treatment includes Biological treatment like Activated sludge process, Trickling filter, Membrane Bio Reactor,Secondary clarifier

Tertiary treatment includes Heavy metal removal ,Disinfection (UV treatment).

Pretreatment

Pretreatment of sewage water removes materials that can be easily collected from the raw sewage before they damage or clog the pumps and sewage lines of primary treatment clarifiers.

Screening. The influent sewage water passes through a bar screen to remove all large objects like cans, rags, sticks, plastic packets etc. carried in the sewage stream. This is done with an automated mechanically raked bar screen in modern plants serving large populations, whilst in smaller or less modern plants, a manually cleaned screen may be used.

Grit removal. Pretreatment may include a sand or grit channel or chamber, where the velocity of the incoming sewage is adjusted to allow the settlement of sand, grit, stones, and broken glass.

Flow equalization. Clarifiers and mechanized secondary treatment are more efficient under uniform flow conditions. Equalization basins may be used for temporary storage of diurnal or wet weather flow peaks. Basins provide a place to temporarily hold incoming sewage during plant maintenance and a means of diluting.

Fat & grease removal. In some larger plants, fat & grease are removed by passing the sewage through a small tank where skimmers collect the fat floating on the surface. Air blowers in the base of the tank may also be used to help recover the fat as a froth.

Primary clarifier

In the primary sedimentation stage, sewage flows through large tanks, the tanks are used to settle sludge while grease & oils rise to the surface and are skimmed off.

Secondary treatment

Secondary treatment systems are classified as *fixed-film* or *suspended-growth* systems.

Surface-aerated basins (lagoons). Most biological oxidation processes for treating industrial wastewaters have in common the use of oxygen (or air) and microbial action. Surface-aerated basins achieve 80 to 90 percent removal of BOD with retention times of 1 to 10 days.

Filter beds (oxidizing beds). In older plants and those receiving variable sewage loadings, trickling filter beds are used where settled sewage liquor is spread onto the surface of a bed made up of coke limestone chips or specially fabricated plastic media. The liquor is typically distributed through perforated spray arms. The distributed liquor trickles through the bed and is collected in drains at the base. Biological films of bacteria, protozoa and fungi form on the media's surfaces and eat or otherwise reduce the organic content.

Membrane bioreactors combine activated sludge treatment with a membrane liquid-solid separation process. The membrane component uses low pressure microfiltration or ultrafiltration membranes and eliminates the need for clarification and tertiary filtration. The membranes are immersed in the aeration tank.

Secondary sedimentation. The final step in the secondary treatment stage is to settle out the biological floc or filter material through a secondary clarifier and to produce sewage water containing low levels of organic material and suspended matter.

Tertiary treatment

The purpose of tertiary treatment is to provide a final treatment stage to further improve the effluent quality before it is discharged to the receiving environment (sea, river, lake, ground, etc.).

Filtration. Sand filtration removes much of the residual suspended matter. Filtration over activated carbon, also called *carbon adsorption*, removes residual toxins.

Disinfection

The purpose of disinfection in the treatment of waste water is to substantially reduce the number of microorganisms in the water to be discharged back into the environment for the later use of drinking, bathing, irrigation, etc. The type of disinfectant used, the disinfectant dosage (concentration & time), and other environmental variables affect the disinfection. Cloudy water will be treated less successfully, since solid matter can shield organisms, especially from ultraviolet light or if contact times are low. Generally, short contact times, low doses and high flows all act against effective disinfection. Common disinfectant include ozone, chlorine, ultraviolet light, or sodium hypochlorite. Chloramine, which is used for disinfecting drinking water, is not used in the treatment of waste water because of its persistence.

Ovarall process of sewage water treatment is shown below in the figure.

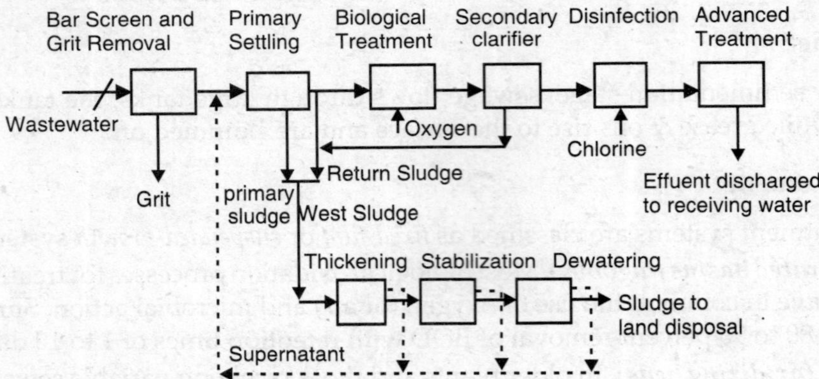

Fig.5.11: Overall process of sewage water treatment

SEWAGE SLUDGE TREATMENT & DISPOSAL METHOD

The sludges accumulated in a wastewater treatment process must be treated and disposed of in a safe & effective manner. The purpose of digestion is to reduce the amount of organic matter and the number of disease causing microorganisms present in the solids. The most common treatment options include anaerobic digestion, aerobic digestion, and composting. The choice of a wastewater solid treatment method depends on the amount of solids generated and other site-specific conditions. However, in general, composting is most often applied to smaller-scale applications followed by aerobic digestion and then lastly anaerobic digestion for the larger-scale municipal applications.

Anaerobic digestion

Anaerobic digestion is a bacterial process carried out in the absence of oxygen. The process can either be *thermophilic* digestion, in which sludge is fermented in tanks at a temperature of 55°C, or *mesophilic*, at a temperature of around 36°C. Though allowing shorter retention time (and thus smaller tanks), thermophilic digestion is more expensive in terms of energy consumption for heating the sludge.

One major feature of anaerobic digestion is the production of biogas, which can be used in generators for electricity production and/or in boilers for heating purposes.

Aerobic digestion

Aerobic digestion is a bacterial process occurring in the presence of oxygen. Under aerobic conditions, bacteria rapidly consume organic matter and convert it into carbon dioxide. The operating costs are much greater than for anaerobic digestion because of the energy costs needed to add oxygen to the process.

Composting

Composting is also an aerobic process that involves mixing the wastewater solids with sources of carbon such as sawdust, straw or wood chips. In the presence of oxygen, bacteria digest both the wastewater solids and the added carbon source and, in doing so, produce a large amount of heat.

Sludge disposal

When a liquid sludge is produced, further treatment may be required to make it suitable for final disposal. Typically, sludges are thickened (dewatered) to reduce the volumes and transported off site for disposal. No process completely eliminates the need to dispose of biosolids. However, an additional step some cities are taking to superheat the wastewater sludge and convert it into small pelletized granules that are high in nitrogen and other organic materials. This product is then sold to local farmers and turf farms as a soil amendment or fertilizer, reducing the amount of space required to dispose of sludge in landfills.

APPENDIX 5.1

Composition of Sewage in Some Indian Cities

S.N	Constituents	Unit	Delhi	Kanpur	Madras	Hyderabad	Ahmedabad
1	SS	Mg/litre	354	560	550	985	148
2	TDS	Mg/litre	803	1000	1170	723	1600
3	Nitrogen (total as N)	Mg/litre	28.5	73.9	59.2	50.9	27.3
4	POA	Mg/litre	13.7	2.5	25.3	-	-
5	Chlorides	Mg/litre	147	114	259	104	300
6	BOD, 5 days, 20°C	Mg/litre	203	255	352	339	133
7	COD	Mg/litre	377	532	-	-	-
8	pH value	-	7.5	7	7.4	7.4	7.5
SS- Suspended solids, TDS- Total dissolved solids							
BOD- Biochemical oxygen demand, COD-Chemical oxygen demand							

APPENDIX 5.2

Waste water characteristics (Unit-mg/litre)

S.N	Parameters	Typical waste water	Primary effluent	Secondary effluent
1	Suspended solids	230	85	20
2	BOD	200	135	20
3	Ultimate oxygen demand	383	276	99
4	Total nitrogen	30	25	20
5	Ammonia nitrogen	18	16	15
6	Phosphorous	13	11	10

Note: Wastewater collected from urban areas (cities, towns, villages) and from different industries, must ultimately be returned to receiving waters or to the land.

The presence of different pollutants in the wastewater makes it almost impossible to treat all the wastewater in the same manner. In order to treat the wastewater it is generally necessary to combine a number of basic processes, which may be **physical, chemical or biological** in character and which have the effect of removing first of all the *suspended solids*, followed by the *colloids & any dissolved inorganic or organic pollutants, and the elimination of pathogenic organisms.*

Finally, some characteristics of water and wastewater need adjustment.

In addition to these, basic processes, some new directions are also evident in various specific areas of wastewater treatment, including:

- modification in treatment operations, processes and concepts;
- the changing nature of the wastewater to be treated
- the problem of industrial wastes;
- wastewater treatability studies;
- environmental and energy concerns;
- land treatment;
- small and individual onsite systems.

The wastewater originating from various sources can be broadly divided into two categories:

- biodegradable wastewater,
- non-biodegradable wastewater.

Biodegradable wastewater

The biological wastes in general have a predominance of *biodegradable organic matter*, and are generally treated in a similar manner.

The *stabilization of organic matter* is accomplished biologically using a variety of microorganisms. The microorganisms are used to convert the colloidal and dissolved carbonaceous organic matter into various gases and into cell tissue. Because cell tissue has a specific gravity slightly greater than that of water, the resulting tissue can be removed from the treated liquid as a sludge by gravity settling. Based on bacterial relationship to oxygen (ability or inability to utilize oxygen as a *terminal electron acceptor* in oxidation/reduction reactions), the microorganisms can be:

- *obligate aerobes,*
- *obligate anaerobes,*
- *facultative anaerobes*
- *denitrifiers.*

The general term that describes all of the chemical activities performed by a bacterial cell is metabolism which is divided into *catabolism* and *anabolism.* Catabolism includes all the biochemical processes by which a substrate (food) is degraded to end products with the release of energy. Anabolism includes all the biochemical processes by which the bacterium synthesizes new cells.

The type of electron acceptor available for catabolism determines the type of decomposition used by a mixed culture of microorganisms.

Decomposition of wastes and particularly of wastewater can be:

- *aerobic decomposition,*
- *anaerobic decomposition,*
- *anoxic decomposition.*

For ***aerobic decomposition*** the molecular oxygen (O_2) must be present as the terminal electron acceptor to proceed by aerobic oxidation. The chemical end-products of aerobic decomposition are primarily carbon dioxide ($C0_2$), water, and new cell material.

Anoxic decomposition occurs when some microorganisms will use nitrate (NO_3) as the terminal electron acceptor in the absence of molecular oxygen. Oxidation by this route is called *denitrification.*

In order to achieve *anaerobic decomposition,* molecular oxygen and nitrate must not be present as terminal electron acceptors. Sulfate (SO_4^{2-}), carbon dioxide, and organic compounds that can be reduced serve as terminal electron acceptors. The end-products of anaerobic decomposition are hydrogen sulfide (H_2S), mercaptans (RSH) , methane (CH_4), carbon dioxide, ammonia and water.

Non- biodegradable wastewater

The **non-biological wastes** in general and the wastewater in particular are rich in *non-biodegradable matter* consisting of solids and liquids in suspended or dissolved form, including various inorganics and organics, many of which may be highly toxic.

Examples are domestic or industrial wastewater containing excessive dissolved solids (minerals), inorganic or organic compounds or naturally occuring organics such as humic and fulvic acids.Treatment processes are available for removing these contaminants. The *physical processes* frequently used include sedimentation, filtration and gas-transfer.

Chemical processes includes the usage of different chemicals for wastewater treatment. Chemicals may be added to alter equilibrium conditions and cause precipitation of undesirable species. It should be kept in mind that chemical processes are conversion processes and that actual removal is accomplished by physically separating the solid, liquid, or gaseous products of the chemical reactions. The chemical processes frequently used in engineered systems include neutralization, coagulation, flocculation, chemical precipitation and oxidation & reduction.

Some wastewater must be treated by means of highly sophisticated processes and equipment, requiring highly skilled operators, and therefore quite expensive. Such processes are *physico-chemical processes* and include: demineralization, desalinization, ion-exchange, reverse osmosis, electro-dialysis, adsorption, evaporation, incineration, etc.

APPENDIX 5.3

Typical composition of domestic waste water in different Indian cities

S. N.	Parameters	Unit	Kolkata	Delhi	Chennai	Kanpur	Nagpur	Ahemdabad	USA average
1	Total solids	Mg/ litre	785	-	-	-	-	-	-
2	Suspended solids	Mg/ litre	287	354	530	560	292	206	295
3	Chlorides	Mg/ litre	-	147	259	114	40	352	-
4.	BOD, 5 days, 20°C	Mg/ litre	125	203	352	255	334	186	180
5	COD	Mg/ litre	290	377	-	532	-	-	-
6	pH value	Unit	7.0	7.4	7.4	7.0	7.0	-	-
7	Temperature	°C	24	-	-	27	30	30	-

APPENDIX 5.4

Features of chemical waste water Treatment methods

S.N	Treatment method	Type of waste	Mode of operation	Degree of treatment	Remarks
1	Ion exchange	Plating, nuclear	Continuous filtration with resin regeneration	Demineralised water recovery, product recovery	May require neutralization and solids removed from spent regenerate
2	Reduction precipitation	Plating, heavy metals	Batch or continuous treatment	Complete removal of chromium and heavy metals	One day's capacity for batch treatment, 3 hours retention for continuous treatment, sludge disposal or dewatering required
3	Coagulation	Paperboard, refinery, rubber, paint, textile	Batch or continuous treatment	Complete removal of suspended and colloids matter	Flocculation and settling tank or sludge blanket unit, PH control required
4	Adsorption	Toxic or organics, refractory	Granular coloumn of powdered carbon	Complete removal of most organics	Powdered oxidation to render organic more biodegradable
5	Chemical oxidation	Toxic and refractory organics	Batch or continuous ,ozone or catalyzed hydrogen peroxide , advance oxidation processes	Partial or complete oxidation	Partial oxidation to render organics more biodegradable

APPENDIX 5.5

Features of biological waste water treatment methods

S.N	Treatment method	Mode of operation	Degree of treatment	Land requirement	Equipment	Remarks
1	Lagoons	Intermittent or continuous discharge, facultative or anaerobic	Intermediate	Earth dug, 10-60 days retention	-	Odour control frequently required
2	Aerated lagoons	Completely mixed or facultative continuous basins	High in summer, less in winter	Earth basin 2.44-4.88 M deep, 8.55-17.1m^2 (m^3/d)	Diffused or mechanical aerator, clarifier for sludge separation & recycle	Excess sludge dewatered and disposed off
3	Activated sludge	Completely mixed or plug flow, sludge recycle	>90% removal of organics	Earth or concrete basin, 3.66-6.10m, 0.561-2.262m^2(m^3/d)	Diffused or mechanical aerator, clarifier for sludge separation & recycle	Excess sludge dewatered and disposed off
4	Trickling filler	Continuous application may employ effluent recycle	Intermediate or high depending on loading	5.52-34.4 m^2(m^3/d)	Plastic packing 6.10-12.19 m deep	Pretreatment before activated sludge plant
5	Rotating biological contactor (RBC)	Multistage continuous	Intermediate or high	-	Plastic disks	Solids separation required
6	Anaerobic	Complete mix recycle up flow or down flow filter, fluidised bed, up flow sludge blanket	Intermediate or high	-	Gas collection required pretreatment before POTN or activated sludge plant	-
7	Spray irrigation	Intermittent application of waste	Complete water percolation into ground water and runoff to stream	6.24×10^{-7}-4.68×10^{-6} m^3(s.m^2)	Aluminum irrigation and spray nozzles movable for relocation	Solids separation required , salt content in waste limited

APPENDIX 5.6

Permissible limiting parameters of the effluents discharged into surface waters

S.N	Parameters	Upper permissible limit
1	Suspended solids	30 mg/litre
2	Particulates	15 mg/litre
3	Oil and grease	0.1 mg/litre
4	BOD (5 days, 20°C)	20 mg/litre
5	Phenols	0.5 ppm
6	Toxic metals	0.5 ppm
7	Sulphate	1 ppm
8	Cyanimide	0.1 ppm

APPENDIX 5.7

Various waste water treatment process methods for removal of toxic compounds from the effluent

S.N	Waste water treatment processes	Toxic compound removal applications
1	Chemical coagulation, sedimentation and filtration	Heavy metals
2	Air stripping	Volatile organic compounds (VOC's) and ammonia
3	Activated carbon adsorption	Nature & synthetic organic compounds including VOC's pesticides, & heavy metals
4	Activated sludge , powdered activated carbon	Ammonia, Refractory and toxic halogenated aliphatic & aromatic compounds
5	Chemical Oxidation	Ammonia, Refractory and toxic halogenated aliphatic & aromatic compounds
6	Conventional biological treatment (e.g activated sludge, Trickling filter	Phenol and selected hydrogenated hydrocarbons

Note: Classification of common wastewater treatment processes according to their level of advancement is tabulated below.

Primary treatment	Secondary treatment	Tertiary treatment	
Bar or bow screen	Activated sludge	Nitrification (Chemical treatment)	
Grit removal	Extended aeration	Denitrification (Reverse osmosis)	
Primary sedimentation	Aerated lagoon	Chemical , Electrodialysis , precipitation	
Comminution	Trickling filter	Disinfection (Carbon adsorption)	
Oil/fat removal	Rotating bio-discs	(Direct) filtration (Selective ion exchange)	
Flow equalisation	Anaerobic treatment/UASB	Chemical oxidation	Hyperfiltration
pH neutralisation	Anaerobic filter	Biological P removal	Oxidation
Imhoff tank	Stabilisation ponds	Constructed wetlands	Detoxification
—	Constructed wetlands	Aquaculture	—
—	Aquaculture	—	—

In primary and secondary treatment sludges are produced with a volume of less than 0.5 per cent of the wastewater flow. Heavy metals and other micro-pollutants tend to accumulate in the sludge because they often adsorb onto suspended particles. Nowadays, the problems associated with wastewater treatment have shifted gradually from the wastewater treatment itself towards treatment and disposal of the generated sludge.

Non-mechanised wastewater treatment by stabilisation ponds, constructed wetlands or aquaculture using macrophytes can, to a large extent, provide adequate secondary and tertiary treatment. As the biological processes are not intensified by mechanical equipment, large land areas are required to provide sufficient retention time to allow for a high degree of contaminant removal.

Tertiary treatment is designed to remove the nutrients, total N (comprising Kjeldahl-N, nitrate & nitrite) and total P (comprising particulate and soluble phosphorus) from the secondary effluents. Additional suspended solids removal and BOD reduction is achieved by these processes. The objective of tertiary treatment is mainly to reduce the potential occurrence of eutrophication in sensitive, surface water bodies.

Advanced treatment processes are normally applied to industrial wastewater only, for removal of specific contaminants. Advanced treatment is commonly preceded by physico-chemical coagulation and flocculation. Where a high quality effluent may be required for reclamation of groundwater by recharge or for discharge to recreational waters, advanced treatment steps may also be added to the conventional treatment plant.

Removal of Contaminants by Sewage Treatment methods is shown below in figures

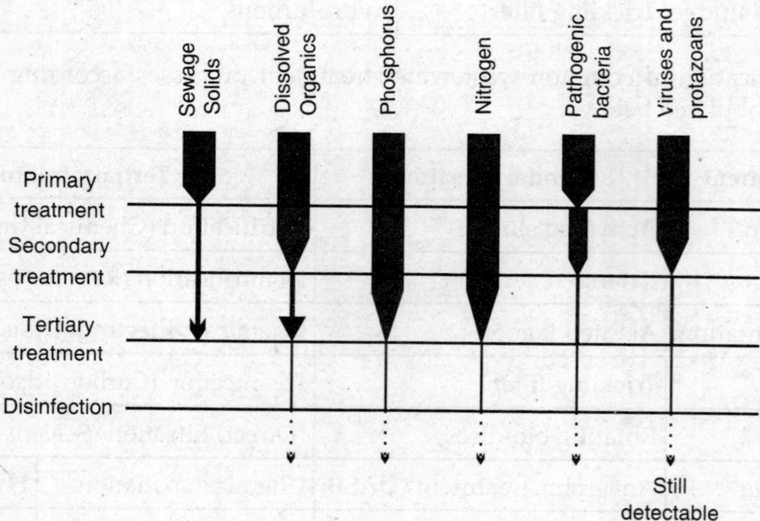

Fig.5.7F$_1$: Removal of Contaminants by various Sewage Treatment methods

Simplified flowchart of Wastewater Treatment is given below in figure.

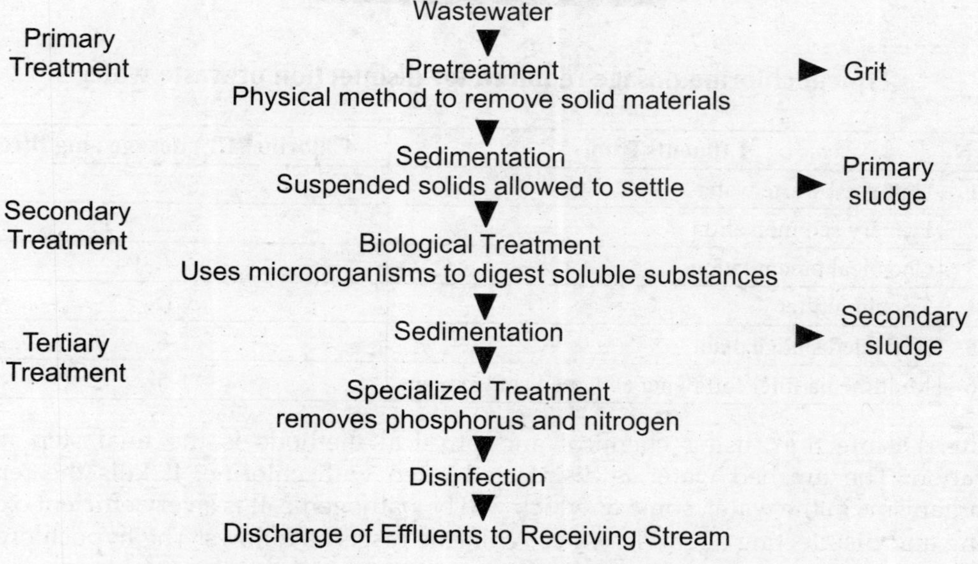

Fig.5.7F$_2$: Flowchart of Wastewater Treatment

A typical flowsheet of sewage treatment plant employing primary & secondary treatment is shown below in figure.

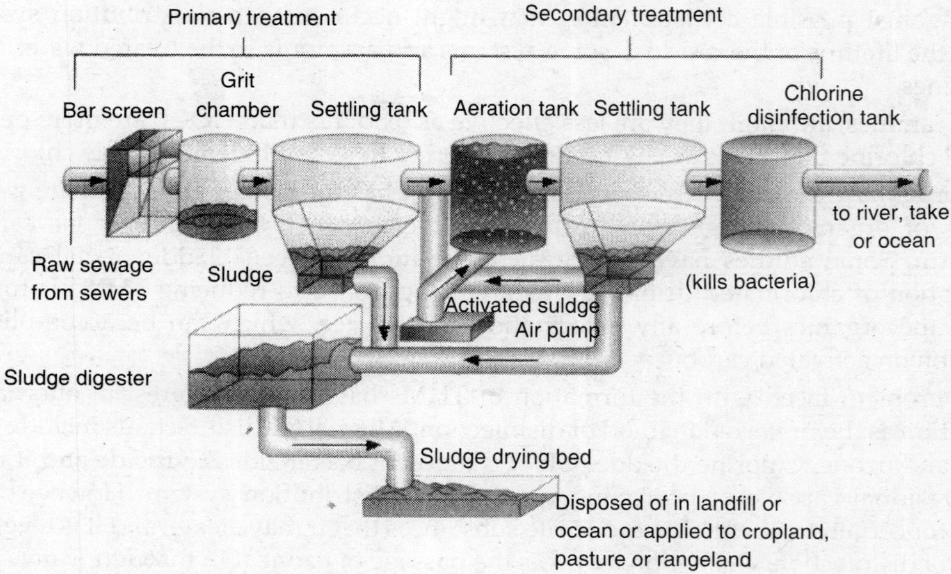

Fig.5.7F$_3$: Flowsheet of sewage treatment plant employing primary & secondary treatment

APPENDIX 5.8

Typical chlorine dosage required for disinfection of waste water

S.N	Effluents from	Chlorine (Cl_2) dosage , mg/litre
1	Untreated waste water	6-25
2	Primary sedimentation	5-20
3	Chemical precipitation	2-6
4	Trickling filter	3-15
5	Activated sludge plant	2-8
6	Multimedia filter following activated sludge plant	1-5

Note: Disinfection, using chemical and physical methods is the final step in water purification. The finished water is disinfected often with chlorine. It kills the remaining microorganisms in the water, some of which will be pathogenic. It is a very efficient oxidizing, beaching and disinfecting agent. In water , chlorine reacts as follows: The hypochlorous acid (HOCI) is the prime disinfecting agent. Its dissociation is pH dependent yielding less effective hypochlorite ions (OCI) at higher pH Values: Together, HOCI and OCI are called the free available chlorine.

A principal advantage of chlorination over other forms of disinfection is that a chlorine residual is created that can protect the treated water after leaving the treatment plant. This guards against possible contamination that might occur in water distribution system. To increase the lifetime of the residual, some systems add ammonia to the treated water, forming chloramines.

Chloramines, although they are less effective as oxidants than HOCI, are more persistent. Residual chlorine that exists as chloramine is referred to as combined available chlorine.

Chlorine may have adverse secondary effects. It has the potential to combine with trace amounts of organic substances to form trihalomethanes (THMs) such as *the* carcinogen chloroform. Some studies have shown an association between bladder rectal cancer and consumption of chlorinated drinking water. One approach to reducing THMs is to remove more of the organics before any chlorination takes place, which can be accomplished by adsorption on activated carbon.

The problem faced with the formation of THMs has spurred interest in alternatives to chlorination as the preferred method of disinfection. Alternative disinfectants include chlorine dioxide and ozone. Chlorine dioxide (ClO_2) is a potent bactericide & viricide and it does not form a residual capable of protecting water in the distribution system. However, there is concern for certain toxic chlorate & chlorite substances that it may create, and it is a very costly method of disinfection. Ozonation involves the passage of ozone (O_3) through water.

Ozone is a very powerful disinfectant that is even more effective against cysts & viruses than chlorine, and it has the added advantage of having no taste or odour problems. Unfortunately, the disinfective power of ozone is limited by its relatively low solubility in water.

Chlorination became the accepted means of disinfection, and it is the single most important discovery in potable water treatment. However, the concern over disinfection by-products (DBPs) produced by chlorine has given new impetus to investigating alternative disinfectants. Disinfection of potable water is the specialized treatment for destruction or removal of organisms capable of causing disease; it should not be confused with sterilization, which is the destruction or removal of all life.

Pathogens (disease producing organisms) are present in both groundwater and surface water supplies. These organisms, under certain conditions, are capable of surviving in water supplies for weeks at temperatures near 21° C , and for months at colder temperatures. Destruction or removal of these organisms is essential in providing a safe potable water supply. While the exact effect of disinfection agents on microorganisms is not clearly understood, some factors that affect the efficiency of disinfection are as follows:

- Type and concentration of microorganisms to be destroyed
- Type and concentration of disinfectant.
- Contact time provided
- Chemical characteristics
- Temperature of the water being treated.

Chlorine is the chemical predominantly used in the disinfection of potable water supplies. The first application of chlorine in potable water treatment was for taste and odour control in the 1830s. At that time, diseases were thought to be transmitted by odour. Currently, chlorine is used as a primary disinfectant in potable water treatment. Other use include taste & odor control, algae control, filter-media conditioning, iron & manganese removal, hydrogen sulfide removal, and color removal.

Chlorine is available in a variety of forms, including elemental chlorine (liquid or gas), solid hypo chlorine compounds of calcium or sodium, and gaseous chlorine dioxide. A chlorination system for disinfection of water supply consists of six separate subsystems: (i) chlorine supply; (ii) storage and handling; (iii) safety provisions; (iv) chlorine feed and application; (v) diffusion, mixing contact; and (vi) the control system.

Chlorine is usually supplied as a liquefied compressed gas under pressure. Chlorine can be supplied in containers or in bulk shipment. Selection of the size of chlorine containers or method of bulk shipment mainly depends on (a) the quantity of chlorine used, (b) the technology used in the chlorination system, (c) the space available for storage, (d) transportation handling costs, and (e) the preference of the plant operator. The cylinders are most likely applied to small water supply systems. The use of 900 kg containers is generally desirable for moderate size users. Bulk shipment may be the cost-effective for large scale water utilities.

The Chlorine storage & handling systems must be designed with full safety consideration as chlorine gas is very poisonous and corrosive. The cylinders and containers storage are usually housed in an enclosure or building. A designer's checklist for ton container storage handling facilities should include, the following: (a) appropriate auxiliary ton-container valves (captive Yoke type), flexible copper tubing, & a rigid black seamless steel manifold header with valves, fitting, & shut-off valves; (b) container weighing scales or load cells; (c) trunnions for ton containers; (d) ton container lifting bar; (e) overhead crane or mono rail with capacity; (f) chlorine-gas filter; (g) external chlorine pressure reducing valve as necessary; (h) pressure gauges; (i) drip legs ; (i.e, condensate traps) at inlet to chlorinators; (j) continuous chlorine-leak detector with sensors alarms; and (k) emergency-repair kit for ton-container.

USEPA has set forth the regulations that intended to minimize the risk of injury, death, or damage to the operation personnel and potential off site impact on public & Environmental receptors during an accidental release of chlorine.

A simple orifice controlled constant head arrangement or low capacity proportioning pumps are used to feed the chlorine solution. Often, constant speed feed pumps are programmed by time clock arrangement to operate the pump at the desired intervals.

At large facilities, complex automatic proportional control systems with recorders are used. Signals from a flow meter transmitter and chlorine residual analyzer are transmitted to the chlorinator to adjust the chlorine feed rate and to maintain a constant chlorine residual that is preset in accordance with the design criteria and standard operating procedures (SOPs). The chlorine analyzers and automatic control loop chlorinator systems are supplied by many manufactures. Several alarms are also considered as essential part of the control system. These include high & low pressures in storage vessels, liquid or gas chlorine lines, high & low injector vacuum lines, high & low temperatures for evaporator water bath, high & low chlorine residual and chlorine leaks.

Ozonation: Ozone has been used extensively in Europe for disinfection and for taste & odor control in water supplies. Interest in the United States and Canada has increased in recent years because of a growing concern about Trihalomethane (THM) formation during chlorination of drinking water. In addition to its use as a disinfectant, pre ozonation is also used for (a) removal of taste & odor, (b) removal of colour, (c) removal of iron & manganese, (d) enhanced removal of organic matters and (e) oxidation volatilization of organics.

Ozone is an unstable gas; therefore, it has to be generated on site. In addition, ozone cannot be used as a secondary disinfectant, because an adequate residual in water can be maintained for only a short period of time. Because of its high oxidation potential, ozone requires certain contact time between the dissolved ozone and water. The challenge is to reduce the spreading in contact time (CT: concentration times hydraulic residence time). This spreading is mainly caused by turbulent flow and mixing properties. As a micro flocculation aid, ozone is added during or before rapid mix followed by coagulation.

Many studies have shown that pre ozonation enhances coagulation & flocculation and improves performance of sedimentation & filtration processes. The advantages and disadvantages of ozonation in water treatment are given in Table below.

Table 5.8T_1: Advantages Vs Disadvantages of Ozonation in water treatment

S.No	Advantages	Disadvantages
1.	Complex taste, odor, and color problems are effectively reduced or eliminated.	The residual does not last long.
2.	Organic impurities are rapidly oxidized.	High electric energy input and / or eliminated high capital and about 10 to 15 times higher than chlorine is required.
3.	Effective disinfection is achieved over a wide temperature and pH range.	High temperature and humidity may complicate ozone generation.

4.	Bactericidal and sporicidal action is rapid (300 to 3000 times faster than chlorine); only short contact periods are required.	The process is less flexible than those for chlorine in adjusting for flow rate and water quality variations.
5.	Odors are not created or intensified by formation of complexes.	Analytic techniques are not sufficiently specific or sensitive for efficient process control.
6.	It reduces chlorine demand and in turn lowers chlorine dosage and so Tri Halo Methane formation potential.	Waters of high organic and algae content may require pretreatment to reduce to ozone demand.
7.	It improves overall treatment efficiency.	The overall cost of treatment is high.

APPENDIX 5.9

Typical Concentration of heavy metals in sewage sludge

S.N	Elements	Concentration, ppm
1	Selenium	1-3.9
2	Cobalt	5.4-10.8
3	Antimony	2.9-9.3
4	Cadmium	18
5	Mercury	4-320
6	Chromium	267-3590
7	+ Lead	2000-8000
8	Iron	5200-23100

Note: The main objective of any type of wastewater treatment is to get clear effluent and good sludge. The sludge which is in the form of slurry is produced in bulk quantities.

It is essential to classify a sludge in order to select the treatment method applicable to it, and to forecast the performance of the equipment' to be used.

The main objectives of sludge treatment are:

- **reduction** in the volume of sludge for disposal by removing some of the water,
- **stabilization** of the organic matter contained in the sludge,
- **destruction** of pathogenic organisms,
- **collection** of by-products which may be used (or) sold to get some revenue,
- **disposal** of the sludge in a safe and aesthetically acceptable manner.

Sludge treatment and disposal comprise any of the following method or combination of methods:

(a) *Concentration* - it is the reduction in the volume of sludge to be treated by pressing the sludge to a higher solids content.

(b) *Treatment*- it is intended to stabilise organic matter, destroy pathogens or to get by-products of the process.

(c) *Dewatering and drying* - it is the removal of water from sludge which decreases the sludge volume.

(d) *Disposal*- if the receiving environment is legally, aesthetically and ecologically acceptable, the sludge can be disposed on the land or into the water.

The principal methods used for sludge processing and disposal are tabulated below.

The most common sewage sludge treatment process chain is shown below in figure.

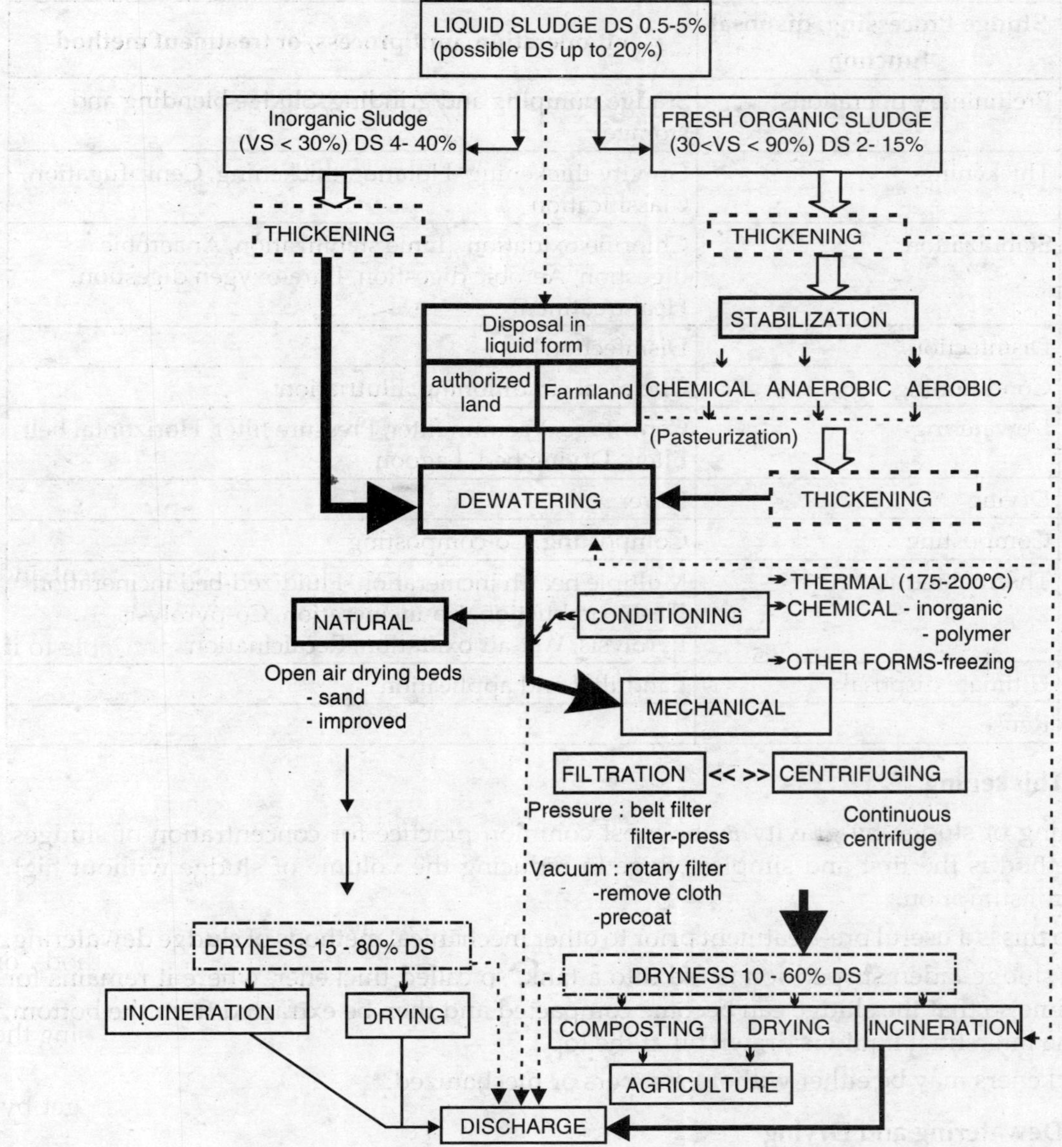

Fig.5.9F$_1$: Sewage sludge treatment process chain

Table 5.9T_1: Sludge processing and disposal methods

S. No.	Sludge Processing, disposal function	Unit operation, unit process, or treatment method
1	Preliminary operations	Sludge pumping and grinding, Sludge blending and storage
2	Thickening	Gravity thickening, Flotation thickening, Centrifugation, Classification
3	Stabilization	Chlorine oxidation , Lime stabilization, Anaerobic digestion, Aerobic digestion, Pure-oxygen digestion, Heat treatment
4	Disinfection	Disinfection
5	Conditioning	Chemical conditioning, Elutriation
6	Dewatering	Centrifuge, Vacuum filter, Pressure filter, Horizontal belt filter , Drying bed, Lagoon
7	Drying	Dryer
8	Composting	Composting, Co-composting
9	Thermal reduction	Multiple hearth incineration, Fluidized-bed incineration Flash combustion, Co-incineration, Co-pyrolysis, Pyrolysis, Wet-air oxidation, Recalcination
10	Ultimate disposal	Landfill, Land application
11	Reuse	-

Sludge Thickening

Thickening of sludge by gravity is the most common practice for concentration of sludges. This method is the first and simplest stage in reducing the volume of sludge without high energy consumption.

Also this is a useful pre-treatment prior to other mechanical methods of sludge dewatering.

The sludge-laden suspension is fed into a tank, so called thickener, where it remains for a long time so that the sludge can become compacted and then be extracted from the bottom, while the interstitial liquid is drawn off at the top.

Thickeners may be either without scrapers or mechanized.

Sludge Dewatering and Drying

Filtration is the commonest method of dewatering sludge produced by water or wastewater treatment. Filtration may consist solely of draining through beds of sand or it may be "mechanical", under vacuum or under pressure, which requires more complicated equipment.

Second stage: Evaporation of part of the bound water. Atmospheric drying can then produce dryness values of up to 65% dry solids depending on retention time, weather conditions, and sludge characteristics.

- **Drying beds-** This is one of the most common methods of sludge dewatering and drying. Operation of drying beds is based on the following principle:

- **First stage:** dewatering by drainage or filtration at very low pressure. The free water is eliminated from the sludge suspension by flowing down through the sand/gravel layer.

Vacuum filtration– This is the most widely used type of mechanical dewatering techniques. The vacuum filters most commonly used for wastewater treatment sludge are of the rotary drum type. Other types of filters, used in some industries, art disc filters, vacuum belt filters, horizontal table filters and paper filters.

Pressure filtration– Pressure filters can apply a large pressure differential across the septum to give economically rapid filtration with viscous liquids or fine solids. The most common type of pressure filtration equipment are filter presses.

Centrifuges are used both to thicken and to dewater sludges. Thickening or dewatering by centrifugation involves the settling of sludge particles under the influence of centrifugal forces. The three basic types of centrifuges currently available for sludge thickening and dewatering are nozzle-disk, solid-bowl, & basket centrifuges.

Sludge stabilization or conditioning includes a variety of processes such as anaerobic & aerobic digestion, chemical coagulation, and heat treatment.

Aerobic digestion may be employed to treat the sludge solids particularly in small installations. The end product is very stable, and organic content of the supernatant liquor is very low. Moreover, the digested solids are odourless and can be disposed off easily. Cost of installation of such units is low, but due to the requirement of power for the supply of air, the running cost may be more compared to that in anaerobic digester. **Anaerobic digestion** is usually accomplished in open tanks equiped with diffused-air or mechanical aerators.

Chemical conditioning may be applied to sludges which have been digested or to raw sludges. The chemicals used include the metallic & polymeric coagulants, and lime.

Heat treatment actually destroys *the* cell structure releasing even the bound water.

Sludge Disposal

Whatever processes are used in handling the solids from wastewater treatment , a residue will still remain which must be disposed off. This material may range from raw solids at a moisture content of over 95 percent to incinerator ash, and its handling will depend, in part, upon its nature.

Sanitary landfills are the preferred disposal technique for solid wastes in general. This method of disposal is most suitable , if it is also used for disposal of the other solid wastes of the community.

Land disposal of sludges is subject to biological,chemical and physical constraints. Disease transmission, odours, heavy metal accumulation, and social & aesthetic problems must be considered before land application is selected.

Ocean disposal has been commonly practiced by coastal cities, but now a days is being phased out because of changes in water pollution control regulations.

Dumping, such as in an abandoned mine quarry, is a suitable disposal method only for sludges and solids that have been stabilized so that no decomposition or nuisance conditions will result.

APPENDIX 5.10

Process efficiency (%) of various tertiary process for domestic effluents waste water

S.N	Treatment Process	Removal of components	Process efficiency , %
1	Electro dialysis	Organic matter	10-40
2	Algae treatment	Ammoniacal nitrogen	50-90
3	Reverse osmosis	Dissolved solids	65-95
4	Anaerobic treatment	Nitrate nitrogen	80-90
5	Ion exchange	Dissolved solids	80-92
6	Denitrification	Nitrate nitrogen	80-95
7	Ammonia stripping	Nitrogen-phosphorous	80-95
8	Carbon adsorption	Dissolved solids	90-98
9	Distillation	Surfactants	90-98

APPENDIX 5.11

Typical properties of domestic effluent after tertiary treatment (mg/litre)

S.N.	Properties	Solid effluent	Liquid effluent
1	Suspended solids	70-100	-
2	Total dissolved solids (TDS)	400-600	400-600
3	BOD (5 days, 20°C)	150-300	1-2
4	COD	200-500	3-6
5	Hardness (as $CaCO_3$)	150-200	100-150
6	Alkalinity	150-200	100-150
7	Chlorides	60-80	60-80
8	pH value	6.9-7.2	7.2-7.9

APPENDIX 5.12

Features of Liquid waste water disposal methods

S.N	Liquid waste disposal method	Features/ Remarks
1	Filtration	Limited in application, may require flocculation and clarification
2	Concentration	Done using evaporators, decanters, centrifuges etc.
3	Solvent extraction	Liquid- liquid extraction systems employed
4	Biological oxidation	Anaerobic process is slow and causes odours; aerobic process requires good control and creates a solid waste disposal problem (sludge disposal)
5	Adsorption	Limited to small dilute systems, large system not yet commercially feasible
6	Ion exchange	It is restricted to special applications only
7	Dumping	Involves possibility of fire, odour and ground water contamination
8	Storage	Restricted to small volumes and subjected to container leakage
9	Deep well injection	Dependent on proper geological factor
10	Incineration	Most feasible if combustion is self supporting
11	Reclaiming	Specialized but often feasible

Note: The materials in waste waters may be classified as follows:

- **Soluble organics**: These cause depletion of oxygen, and hence their quantity must be restricted to the capacity of the receiving waters for assimilation.
- **Suspended solids:** Deposition of solids in quiescent stretches of a stream impair the normal aquatic life of the stream. Sludge blankets containing organic solids decompose, leading to oxygen depletion and production of noxious gases.
- **Trace organics**: When a receiving water is to be used as a drinking water supply, phenol and other organics discharged in industrial wastes cause unpleasant tastes & odours.
- **Heavy metals, cyanides and toxic organics:** These are deleterious to aquatic life and usually have prescribed limits for discharge.
- **Colour and turbidity:** Though colour and turbidity may not be deleterious, these may have to be removed for aesthetic considerations. In some industries, such as pulp and paper, economical methods are not presently available for colour removal.
- Nitrogen & phosphorus: When waste waters are discharged to lakes and ponds, the presence of nitrogen & phosphorus leads to undesirable plant growths.
- **Substances resistant to biodegradation:** Nondegradable substances such as ABS (alkyl benzene sulphonate) from detergents frequently lead to persistent foam.

- **Oil and floating material:** These lead to unsightly and hazardous conditions.
- **Volatile compounds:** Substances such as gasoline, alcohols, aldehydes and ethers may cause explosion in sewers; toxic substances and sulphides may cause odour nuisances.

The first step in control of water pollution is to study each process within the industry in order to stop careless operation which causes losses of water, processed material, and end products. Large reduction in unnecessary losses can be achieved by ' good housekeeping'. Another preliminary step is the separation of relatively pure cooling & condenser waters from those waste waters which are concentrated and which in turn must be treated.

In some cases, change-over of process from wet to dry, modification of process or product formulation or manufacture of byproducts from wastes may be necessary. Some examples of these are the change over of the pulp and paper industry to sulphate process, development of biodegradable products and substitution of reclaimable hydrochloric acid instead of sulphuric acid in metal pickling liquors.

Characterisation Of Waste Waters

The process to be followed for the treatment of a given waste water depends on its characteristics which is tabulated below.

Table 5.12T_1: Characteristics Of Waste Water

• **Flow data:** Rate of discharge during production run, Duration and frequency of production run, Possibility of abnormal discharge, Average daily Duration and level of maximum flow rate
• **Physical Properties:** Temperature range, Insoluble components (colloidal, settleable, floatable), Colour, Odour, Radioactivity, dissolved oxygen, Corrosiveness
• **Chemical composition:** Known organic & inorganic components, Chemical oxygen demand. Total carbon and extractables, pH , Acidity & alkalinity, Oxidising or reducing agents (sulphides), Chloride ion, Hardness (calcium & magnesium), Nitrogen & phosphorus, Surfactants, Chlorine demand, Total dissolved salts, Specific ions (As, Ba, Cd, Cr, CN, Pb, Hg),
• **Biological data:** Biological oxygen demand (BOD), Pathogenic bacteria, Chemical toxicity (acquatic life, bacteria, plants, man)

The effect of waste waters on receiving system are summarized in Table below.

Table 5.12T_2: Effects Of Waste Water On Receiving Systems

• **Effects on Natural bodies of water:** Oxygen depletion, Interference with aquatic life, Danger to public health and safety, Economic damage as recreational losses.
• **Effects on Municipal sewage system:** Hazards to maintenance personnel by toxic gases and explosions, Damage to sewers by corrosion, deposition and overload, Decrease in efficiency of sewage treatment processes by overload of unit capacities and biological toxicity
• **Effects on Underground strata:** Contamination of potable water supplies, brines, and minerals.
• **Effect on Land:** Seepage to ground or surface water

Waste Water Treatment Methods & Equipments

Some of the method and requirement used for treatment of waste waters are similar to those used for treatment of municipal (or domestic) waste. The oldest method of sewage treatment is irrigation or spraying of agricultural lands with sewage. The oxygen present in the soil oxidises the organic matter in the sewage and purifies it. The mineral constituents or the use of sewage in agriculture may reduce the requirement of chemical fertilisers. But the difficulties involved are: (1) the location of sufficiently large agricultural land near the source of sewage & (2) the impracticability of utilizing diseases. Hence use of raw sewage for agriculture is not favoured.

The different methods used in treatment of sewage and waste water are:

- **Preliminary treatment**: Removal of gross solids, grit, oil and grease
- **Primary treatment:** Sedimentation, Aids to sedimentation
- **Secondary treatment:** Biological oxidation
- **Tertiary treatment:** Chlorination
- **Sludge treatment and disposal**

The actual method used for handling a particular waste depends on the characteristics of the waste.

Preliminary Treatment

The waste water may need a preliminary treatment to remove the larger floating and suspended solid matter, grit and removal of oil & grease , if present in an appreciable amount.

Removal Of Gross Solids: The most common method of removing the larger floating and suspended solids is by passing through fixed or moving screens. An alternative method is to pass the wastewater through a disintegrator or comminutor in which the solids are cut into small pieces and then settled out in a sedimentation tank.

The fixed screen consist of metal bras with spaces in between. Fine screens have spaces of 10-25 mm while coarse screens have spaces of 50-150 mm. The latter type are usually installed as a protection for fine screens. The velocity of flow through the screening chamber should be more than 0.5m/s to prevent sedimentation.

The simplest type of fixed screen is the hand-raked screen in which the bars are usually set at an angle of 60° to the horizontal. Mechanical raked screens are often preferred to overcome the unpleasant task of manual task of manual raking and are often placed vertically.

The moving screen may be of the wire rope or drum type. The wire rope screen consists of endless wire ropes moving over rollers and passing through a screening chamber.

The waste water passes through the spaces between the ropes; the gross solids which are arrested are carried from the screening chambers on the ropes and cleaned off by revolving brushes before the ropes pass back into the screening chamber.

Drum screen consists of a revolving drum covered by a wire mesh partially submerged in the settling chamber. The waste water enters axially through one end of the drum (the other end being closed) and passes out through the wire mesh. As the drum revolves, the solid matter on the inside of the mesh is raised above water level and swept into receiving traps by jets of water from sprayer pipes. The screenings are removed by burrows, hopper trucks or conveyor belts. They can be disposed off by burying underground or by burning in an incinerator. Another preferred practice is to chop the screenings to small sizes in a macerator and then to return them to the waste water.

A method adopted in place of screening is to cut the solids into small pieces without first removing them from the sewage by a comminutor or macerator which consists of a metal drum installed in a chamber within the waste water channel. The slots in the drum allow small pieces of solids to pass through. The larger pieces are cut by the cutting teeth fixed to the drum.

Grit Removal: Inorganic gritty matter for detritus is removed in the early stages of treatment (to avoid damage to pumps & other equipment and settling in channels & pipe bends) in grit channels or tanks. Grit being heavier than the organic solids , careful regulation of the velocity of flow through the grit tanks allows the prevention of most of the organic solids from settling out with it. Velocity through grit tanks may be 0.3 m/s. There are various types of constant velocity grit tanks. Many take the form of channels with a cross section to ensure constant velocity. Most of the grit falls 0.3 m in 10 s. The length of the grit channels is usually 20 times the maximum depth to compensate for turbulence and allow for some of the finer particles to settle.

Grit can be pumped with water into a pond if land is available. If clean, the grit can be used for filling in holes, for road making and on sludge drying beds.

Removal of Oil & Grease: Some waste waters contain appreciable amounts of oil and grease, especially from areas with a large number of engineering works. It is advisable to remove as much of these as possible in the early stages of treatment, in order to prevent any adverse effects on the rest of the plant. The waste water is usually passed through small skimming tanks where the oil and grease are skimmed off. This process is aided by aeration, chlorination or vacuum flotation.

Oil and grease can also be removed from waste water by causing them to settle out in primary sedimentation tanks by the use of chemical reagents. This method is useful for waste waters containing wool scouring waste liquors, in which the grease is in an emulsified state and so cannot be removed by ordinary skimming methods.

PRIMARY TREATMENT

After the gross solid matter, grit and excessive amounts of oil & grease have been removed, as much as possible of the remaining suspended solid matter must be removed in order to reduce the strength of the waste water and to facilitate secondary treatment. Sedimentation is the most practical and economic method of removing suspended solid matter.

Sedimentation: The size of the sedimentation tanks or clarifiers for the treatment of waste waters (particularly sewage) are based on the following detention times:

	Detention hours
Quiescent flow sedimentation tanks	15
Quiescent flow sedimentation tanks with chemical coagulation	8
Continuous sedimentation tanks	2
Continuous flow sedimentation tanks with chemicals coagulation	2

Figure below shows a horizontal flow sedimentation tank a center feed circular clarifier.

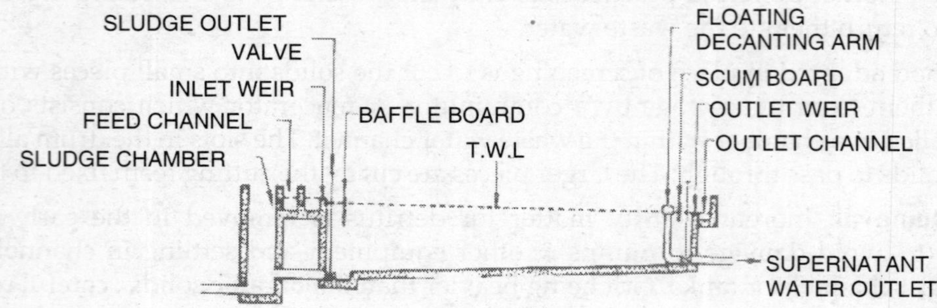

Fig.5.12F$_1$: Horizontal flow sedimentation tank

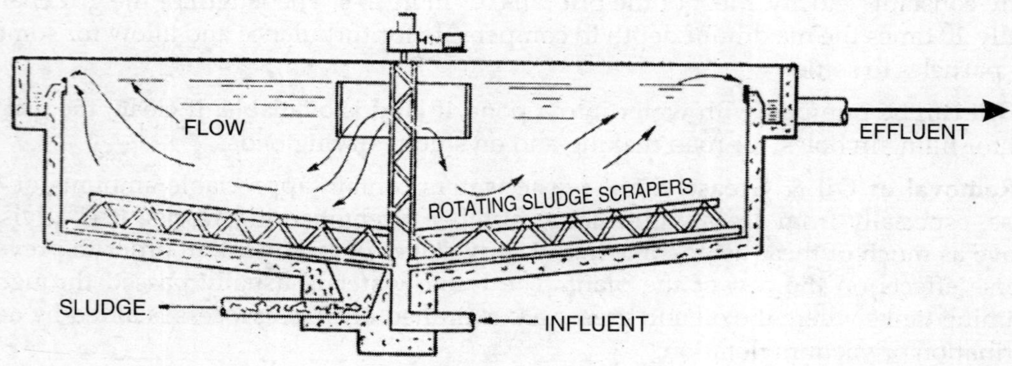

Fig.5.12F$_2$: Centre-feed circular clarifier

In a well designed continuous flow sedimentation tank, about half the suspended solid matter is settled out in the first two hours of sedimentation. But in waste waters containing a large proportion of industrial waste liquors, a long detention time assists in the mixing & balancing of the various wastes and helps in preventing excessively heavy loads from being passed on to the biological purification plant. An efficient sedimentation system removes about 90% of the suspended solids and 40% of the organic matter.

The settled sludge is removed from the sedimentation tanks by scraping it into hoppers by mechanical means and then removing it by pumps or by hydrostatic head. In the case of hopper bottomed sedimentation tanks, the sides are built at an angle steep enough to allow the sludge to fall to the bottom from where it can be removed.

AIDS TO SEDIMENTATION: Some waste waters contain very finely divided suspended solids and colloidal matter of which very little is removed by plain sedimentation. To remove some of these during sedimentation, it is necessary, by mechanical flocculation or chemical coagulation, to cause the finely suspended matter to form large particles and entrain.

Mechanical flocculation: If sewage is stirred gently, the finely divided matter coalesces into larger particles which settle more easily. To carry this out, the waste water is passed through a tank with a 30 minute detention time and fitted with paddles rotating at an optimum peripheral speed of 0.40 m/s.

The flocculating chamber is sometimes constructed as part of a sedimentation tank as in the case of Clariflocculator. This is a cylindrical radial flow tank with mechanical sludge scrapers and a circular chamber in the centre fitted with paddles mounted on a power driven vertical shaft. A clariflocculator is shown below in figure.

Chemical coagulation: The addition of certain chemicals to sewage causes the formation of an insoluble flocculent precipitate which absorbs and carries down some of the suspended & colloidal matter. The chemicals are added as a solution or suspension in water which requires mixing and dosing gear. The chemicals must be well mixed with the waste water and this can be carried out in baffled channels. Addition of chemicals is often followed by mechanical flocculation before sedimentation, which improves the results.

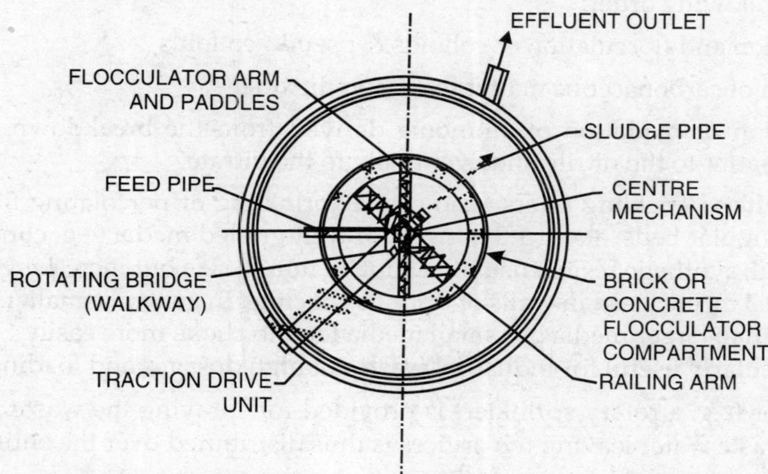

Fig.5.12F$_3$: Clariflocculator

The chemicals used for coagulation include hydrated lime, aluminium sulphate, and iron salts - ferrous sulphate, ferric chloride & chlorinated copperas $FeClS0_4$. Acids are used in special cases, e.g. where industrial wastes have a high pH or large amount of grease.

In recent years, polyelectrolytes are widely used for assisting flocculation. These svnthetic substances fall into two main groups, namely (1) those which have a long chain molecular structure and have the property of being absorbed on two or more particles, thus drawing the particles together and (2) those which reduce the charge on the particles and consequently the repulsive power of the like charges on the particles. Polyelectrolytes are selective and therefore the most suitable polyelectrolyte must be chosen. For example, polymers of cyanimide, acrylic & methacrylic acids or their derivatives, and hydrolysed polymers of acrylonitrile or acrylamide with molecular weight of 10^4 - 10^6 are used in dilute solutions. Very small quantities (0.25mg/1) are required but they are often expensive compared with ordinary chemical coagulants.

Preparation: Preparation for about 10 minutes prior to sedimentation is sometimes used as a means of assisting treatment. Aeration assists in the removal of entrained gases such as H_2S and CO_2 and aids flocculation and separation of oils & greases. Plants employing pre aeration claim about 8 % improvement in the reduction of BOD.

Secondary Treatment

The organic matter in waste waters can be removed biologically by utilizing the activities of bacteria and other micro-organisms. These processes can be classified into two groups aerobic and anaerobic.

Aerobic processes require oxygen for their metabolism while anaerobic processes operate in the absence of oxygen (or air). The effluent from primary sedimentation tanks is subjected to aerobic oxidation in *trickling filters, activated sludge units* or *oxidation ponds;* the sludge from the sedimentation tanks and aerobic processes is subjected to anaerobic process in *sludge digesters.*

In the aerobic processes, bacteria and other micro-organisms bring about a number of changes in the following order:

- Coagulation and flocculation of colloids & pseudo-colloids
- Oxidation of carbonaceous matter to carbon dioxide
- Nitrification or oxidation of ammonia derived from the breakdown of nitrogenous organic matter to the nitrite and eventually to the nitrate.

Trickling Filters: Trickling filters (also called sprinkling or percolating filters) consist of circular or rectangular beds, about 1.8 m deep, of well-graded media (e.g. clinker, slag, stone, gravel, coke, coal, synthetic resins) usually about 40 mm in size but increasing to 100-150 mm for the bottom 0.3 m, enclosed in walls of brick or concrete. In general, smaller media produce better effluents than larger media but small media tend to choke more easily. Synthetic plastic media are particularly useful for industrial wastes of high dosages and loadings.

A distributor (e.g. a rotary sprinkler) is provided for spraying the waste water on to the medium. The waste water leaving the orifices is thus distributed over the entire bed.

In the process, a gelatinous film consisting of bacteria and other micro-organisms forms on the surface of the medium. The impurities in the wastewater are adsorbed on to the film and then oxidised by the bacteria. As oxygen is necessary, the filter is arranged in such a way that air can enter at the bottom and a natural draft is produced; oxygen can thus be absorbed by the thin film of liquid at all points in the bed.

When the film on the medium reaches a certain thickness, it gets detached and is carried away by the etfluent which is therefore passed through a settling tank to retain the sludge particles. The sludge is pumped to the sludge digestion process.

The trickling filters are simple to use. But they are costly to construct and their loading per unit volume is low. If the loading is increased, the efficiency decreases. To overcome this, the waste water may be diluted with previously treated effluent. Another method is to use two filters in series; the film grows rapidly in the first filter and after one or two weeks, the order is reversed.

ACTIVATED SLUDGE PROCESS. (shown below):

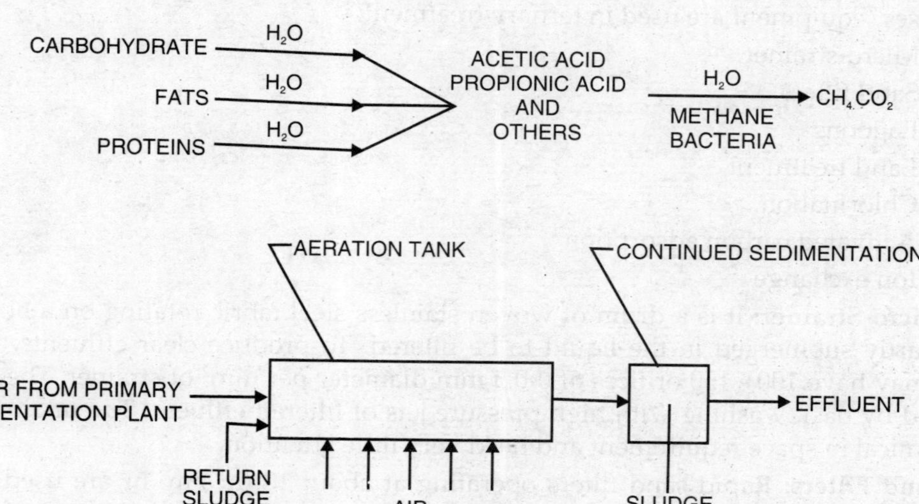

Fig.5.12F_4: Activated sludge process

In this process waste water is aerated in a tank in which the aerobic bacterial flora present oxidise a part of the impurities in the waste water to carbon dioxide and water while the other organic substances are synthesised into bacteria. The bacterial flora grows and remains suspended in the form of a floc which is called the activated sludge. The effluent is separated from the sludge by settling and is discharged. A part of the sludge is returned to the process, the rest of the sludge is removed and digested in a sludge digester along with the primary sludge obtained during primary sedimentation. This process removes about 90- 95% of the BOD of a waste water which has undergone primary sedimentation. An aeration period of 3 to 6 hours is adequate, if sufficient oxygen is supplied.

The activated sludge process is very widely used. Several modifications of the process (such as compressed air process, Swedish INKA process, simplex process, aeration rotor process) have been developed; these differ in the method of air supply.

The presence of synthetic detergents such as alkyl benzene sulphonates and polyethylene glycols which are not easily broken down by bacteria, lead to considerable foaming and make the process difficult & dangerous. In such cases, antifoams are to be used during the process.

OXIDATION DITCH: Oxidation ditch is an oval shaped closed circuit channel of depth 1-1.5 m and simple construction. Mechanical rotors in the channel intensely aerate the MLSS (Mixed liquor containing sludge solids) and keep the liquor in circulation. Oxidation ditches are used for wastes having low BOD.

Tertiary Treatment

It is the final treatment or 'polishing' the effluents of the secondary treatment for further improvement of the waste water quality.

Depending upon the quality of the final effluent required, one or more of the following processes/equipment are used in tertiary treatment:

- Micro-strainer
- Sand filters
- Lagoons
- Land treatment
- Chlorination
- Activated carbon adsorption
- Ion exchange

Micro-Strainer: It is a drum of woven stainless steel fabric rotating on a horizontal axis and partly submerged in the liquid to be filtered. To produce clear effluents, the strainers used may have 100 - 150 orifices of 0-0.5 mm diameter per mm^2 of strainer. The strainers are cleaned by back washing with high pressure jets of filtered effluent. The micro-strainers are economical in space requirement and need very little attention.

Sand Filters: Rapid sand filters operating at about 10 m^3 /m^2 hr are used for removal of fine suspended solids. If operated at slow rates, they effect good removal of bacteria and reduce BOD.

Lagoons: Lagoons (also called clarification lakes, maturation ponds or oxidation ponds) are large shallow artificial lakes used for purification of water by micro-organisms in the presence of dissolved oxygen. They are 1 -1.5 m deep, and may have a total capacity for 3 days retention. Three or four lagoons arranged in series give an excellent final effluent (BOD: 3.8 mg/1; suspended solids: less than 1 mg/1).

Aeration lagoon is a single unit treatment system without recirculation of solids or sludge return, consisting of simple earthen basins 3 - 4 m deep in which a mechanical surface aerator is installed either on floats or on a permanent base to aerate the liquid content.

Land Treatment: If land of a suitable nature and slope (less than 1 in 60) is available, land treatment gives good reduction of BOD and suspended solids with dosages upto 0.85 m^3 /m^2 day (8.5 X 10^6 1/hectare day).

Chlorination: A well operated aerobic biological plant destroys or removes a high percentage of pathogens from the sewage. But it is sometimes desirable to sterilise the effluent of an aerobic plant with chlorine. In some cases, the effluent from primary sedimentation is chlorinated to destroy the pathogens but this is more expensive as larger amounts of chlorine are required.

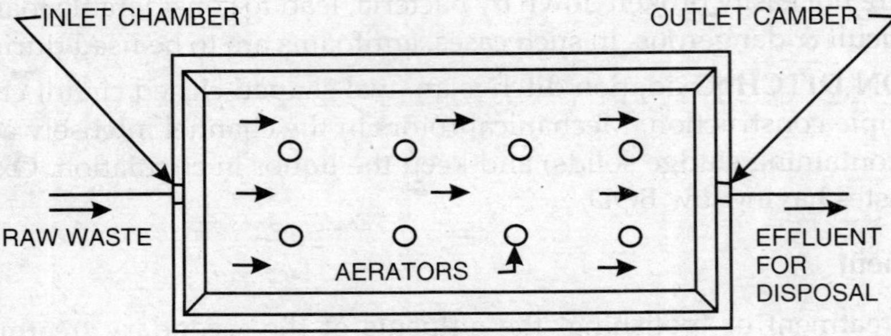

Fig.5.12F$_5$: Aerated lagoon

Activated Carbon Adsorption: This method can be considered when small quantities (less than 200 mg/1) of organic contaminants are to be removed. Activated carbon can be used in pulverised or granular form. In the first case, the finely powdered carbon is slurried with the liquid to be treated and is subsequently filtered out. In the latter case, the liquid is passed through a bed of carbon granules. A flow sheet of this process is given in Figure below.

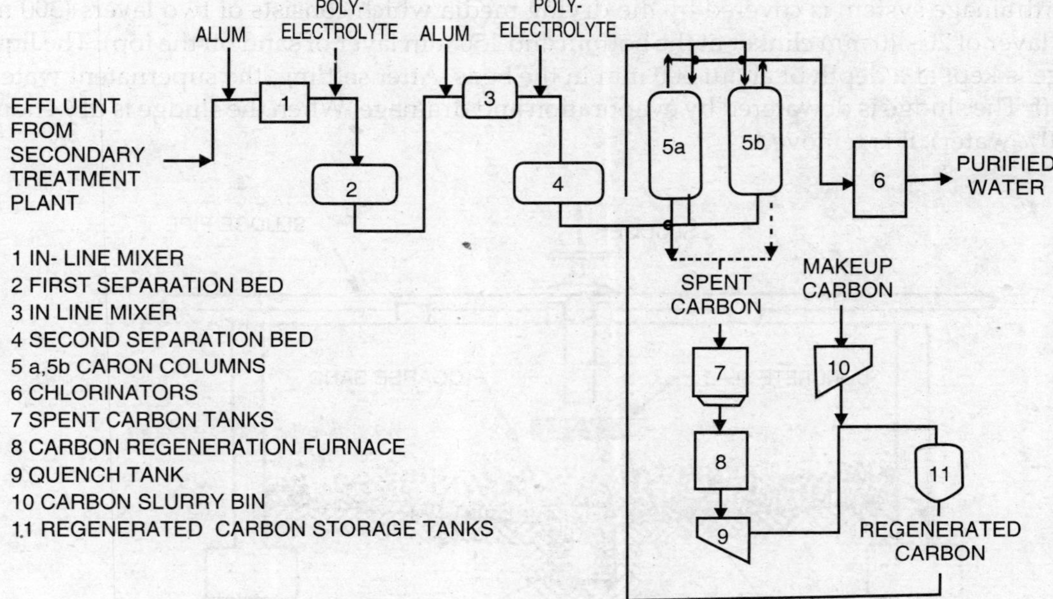

Fig.5.12F$_6$: Activated Carbon Adsorption

Ion Exchange: The use of ion exchange resins for demineralising water for boiler feed and other purposes is well known. Their use is now extended in the field of waste treatment for the recovery of costly materials and removal of toxic compounds.

Sludge Treatment/Disposal

The three main types of sludge produced in a sewage work are:

- Primary sludge settled out in primary sedimentation tanks (moisture content about 97.5%)
- Sludge from trickling filters (moisture content 94-97%)
- Surplus activated sludge from activated sludge process (moisture content 98-99.5%).

The volume of a sludge varies inversely as the percentage of dry solids; thus a sludge with 94% moisture content occupies only one-third of the-volume of a sludge having 98% moisture content. It is therefore advantageous to reduce the moisture content as much as possible. A definite thickening of the sludge is noticeable at a moisture content of 85%. Sludge with a moisture content of 70-80% does not flow and is called sludge cake. The moisture content of sludge cake is about 10 -15%.

Disposal of Liquid Sludge: Liquid sludge is disposed into sea or on land. Land disposal methods include (1) trenching (filling trenches partially with sludge and then with soil), (2) land plots (drying the sludge in furrows on land plots and reploughing so that the plots can be used for growing crops) and (3) lagoons or ponds constructed with high banks around an area of land into which the sludge is let in, and allowed to dry.

Dewatering of Sludges: Sludges are dewatered in drying beds, filter presses or vacuum. Drying beds (shown below) are large tanks with an under drainage system of tlles. The underdrainage system is covered by the drying media which consists of two layers (300 mm thick layer of 20 - 40 mm clinker at the bottom and 150 mm layer of sand on the top). The liquid sludge is kept at a depth of about 300 mm in the beds. After settling, the supernatent water is run off. The sludge is dewatered by evaporation and drainage. When the sludge is dry enough (45-60% water), it is removed.

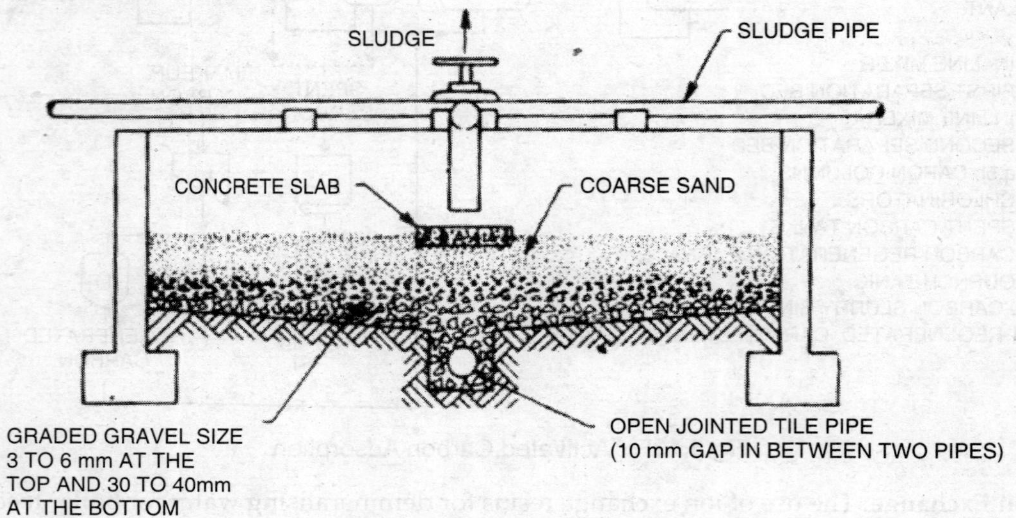

Fig.5.12F_7: Sludge Drying Bed

Filtration: It is carried out in filter presses at a pressure of 0.4 - 0.8 MN/m^2 (ca. 4 to 8 kg/cm^2). The moisture content of filter cake is 50-65%.

In *vacuum filtration*, sludge cakes of 3 - 20 mm thickness and 70 - 80% moisture are built on a suitable filter medium (wool, nylon or glass fibre) in rotary drum vacuum filters. It is essential to add chemical conditioners (such as aluminium chlorohydrate, aluminium chloride or lime and ferric chloride) before vacuum filtration.

Sludge Digestion by Anaerobic Process: Sludge undergoes slow fermentation or digestion by anaerobic bacteria when stored in tanks for a long time (3 to 6 months). The process is accelerated at higher temperatures. In medium & large works, the digestion is carried out in two stages: (1) primary digestion for 7-30 days at 27 - 35°C in closed tanks with fixed or floating roofs and (2) secondary digestion for 10 - 60 days at ordinary temperature in open tanks. Most of the digestion and gas evolution occurs in the primary stage. In the secondary stage, further digestion occurs, the sludge gets consolidated and the liquor separates.

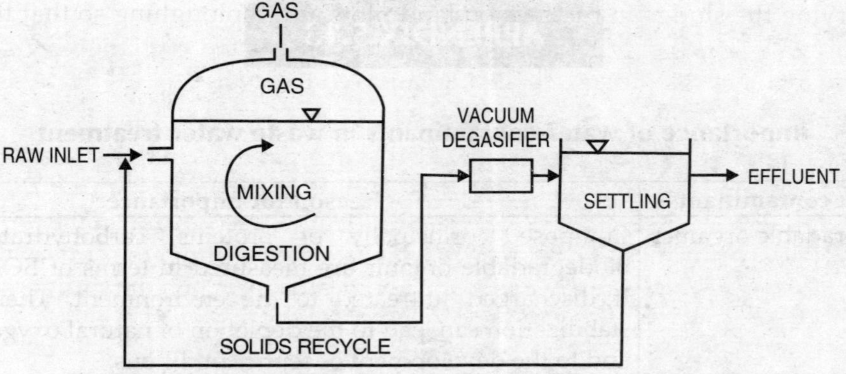

Fig.5.12F_8: Anaerobic Sludge Digestion

The advantage of sludge digestion are:

- Reduction in volume by about 65%.
- Improvement in odour and drying characteristics.
- Digested sludge can be used as manure more safely than undigested sludge.
- The gas produced (0.6-1.25m^3/kg of organic matter destroyed) has a heat of combustion of about 26 MJ/m^3

Drying Of Sludge: The method used for sludge drying are:

- Drying in rotary kiln drier
- Air drying in bags and grinding in ball mills.
- Drying in multiple hearth furnaces to low moisture content or burnt to ash.
- Mixing dewatered sludge with thrice its volume of dried sludge of 10% moisture content and processing in a chamber through which hot furnace gases pass to reduce its moisture content to 5-10%
- Feeding dewatered sludge containing 40-60% moisture to an attritor in which it is dried by a current of hot gases to reduce moisture content to 10% and ground.

APPENDIX 5.13

Importance of water contaminants in waste water treatment

S.N	Water contaminants	Reason for importance
1	Biodegradable organics	Composed principally of proteins, carbohydrates & fats, biodegradable organic are measured in terms of BOD and COD If discharged untreated to the environment. Their biological stabilization can lead to the depletion of natural oxygen resources and to the development of septic conditions.
2	Dissolved inorganic	Inorganic constituents like calcium, sodium and sulfate are added to the original domestic water supply as a result of water use and may have to be removed , if the waste is to be reused
3	Heavy metals	Commercial & industrial activities add heavy metals to water and they must be removed before waste is reused.
4	Nutrients	Both nitrogen & phosphorus along with carbon are essential nutrients for growth. When discharged to the aquatic environments, These nutrients can lead to the growth of undesirable aquatic life. When discharged in excessive amount to land, They can also lead to the pollution of ground water.
5	Pathogens (Disease causing bacteria)	Communicable diseases can be transmitted by the pathogenic organism in waste water.
6	Priority pollutants	Organic and inorganic compounds are selected on the basic of their known or suspected carcinogenicity, mutagenicity, teratogenicity or high acute toxicity.Many of these compounds are found in waste water
7	Refractory organics	These organic tend to resist conventional method of waste water treatment. Typical example include surfactants, phenols and agricultural pesticides.
8	Suspended solids	Suspended solids can lead to the development of sludge deposits and aerobic & anaerobic condition when untreated waste water is discharged in the aquatic environment.

Note: The contaminants (pollutants) in waste water are removed by physical chemical and/or biological means, and the individual method usually are classified as physical, chemical and biological unit processes or operations.

Treatment methods in which the application of physical forces predominate are known as physical unit operation are: screening, mixing, flocculation, sedimentation, flotation, filtration and membrane filter operations.

Treatment methods in which the removal or conversion of contaminants is brought about by the addition of chemicals or by other chemical reactions are known as chemical unit processes. Neutralization, oxidation, reduction, precipitation, gas transfer, adsorption, ion-exchange, electro-dialysis etc. are the most common examples of these unit processes used in water and wastewater treatment.

Treatment methods in which the removal of contaminants is brought about by biological activity are known as biological unit processes. Biological treatment is used primarily to remove the biodegradable organic substance (colloidal or dissolved) in wastewater. Basically, These substance are converted into gases that can escape to the atmosphere and into biological cell tissue that can be removed by settling. The most common approaches in the biological wastewater treatment are: aerobic processes such as trickling filter, activated sludge, oxidation ponds (or lagoons), and anaerobic processes such as anaerobic lagoons, sludge digestion, etc.

Usually in the municipal wastewater treatment, but also in other wastewater processing all the above mentioned unit operation & processes are grouped together to provide what is known as primary, secondary and tertiary (or advanced) treatment.

The term primary refers to physical unit operations and in some cases to chemicals unit processes; secondary refers to biological unit processes and tertiary refers to combinations of all three.

Main combination of all alternatives for municipal wastewater treatment is shown in figure.

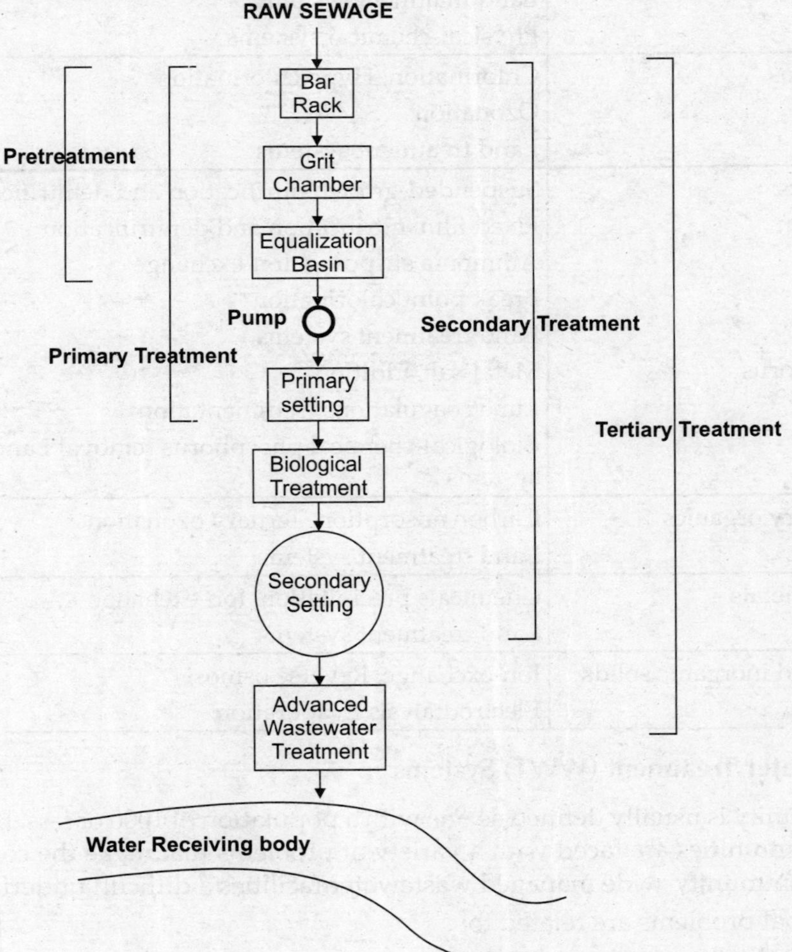

Fig.5.13F$_1$: Combination of all alternatives for municipal wastewater treatment

Unit operations & unit processes and treatment systems used to remove the major contaminants in wastewater is tabulated below.

S.N	Waste water contaminants	Unit operation, unit processes or treatment system used
1	Suspended solids	Sedimentation Screening and comminution Filtration variations, Flotation Chemical addition, Coagulation/Sedimentation Land treatment systems
2	Biodegradable organics	Activated-sludge variations Fixed film: trickling filters, rotating biological contactor Lagoon variations Intermittent sand filtration Land treatment systems Physical-chemical systems
3	Pathogens	Chlorination, Hypo chlorination Ozonation Land treatment systems
4	Nutrients: -Nitrogen -Phosphorus	Suspended-growth nitrification and denitrification Fixed film: nitrification and denitrification Ammonia stripping, Ion-exchange Break point chlorination Land treatment systems Metal-salt addition Lime coagulations/sedimentation Biological-chemical phosphorus removal Land treatment systems
5	Refractory organics	Carbon adsorption, Tertiary ozonation Land treatment systems
6	Heavy metals	Chemicals precipitation, Ion-exchange Land treatment systems
7	Dissolved inorganic solids	Ion-exchange, Reverse osmosis Electrodialysis Evaporation

Small Wastewater Treatment (WWT) Systems

A small community is usually defined as one with a population of 1000 or less. Because of their size, small communities are faced with a variety of problems that make the construction and operation of community-wide managed wastewater facilities a difficult undertaking.

The principal problems are related to:

- Stringent discharges requirements;
- High per capita costs;

- Limited finances;
- Limited operation and maintenance budgets.

Characteristics Of Small Wwt (Waste Water Treatment) Systems

Small system flowrate and wastewater characteristic differ significantly from those of large systems.

Per capita flowrates and variations must be considered for the effective design of waste water facilities for individual residence as well as for cluster of home and small communities.

Small wastewater system vary in size from systems designed to serve individual residences with a flow of 190 to 1900 1/d, to system designed for wastewater flow up to 380 m^3/day.

Two general types of small WWT system are usually implemented:

(a) system for individual residence and other community facilities in unsewered areas.

Alternative wastewater management options for unsewered areas are given in table below.

Table 5.13T_1: Wastewater management options for unsewered areas

Source of wastewater	Wastewater treatment and/or containment	Wastewater disposal
Individual residences combined waste water, black water, Gray water ***Public facilities commercial establishment***	***Primary treatment*** Septic tank , Imhoff tank ***Secondary treatment*** Aerobic/ anaerobic unit Aerobic unit Intermittent sand filter Recirculating granular filter Constructed wetlands ***Recycle treatment system*** ***Onsite containment*** Holding tank privy	***Subsurface disposal*** Disposal fields Seepage beds Shallow sand-filled disposal trenches Mound systems Evapotranspiration/percolation beds , Drip application ***Evaporation system*** Evapotranspiration beds Evaporation pond ***Wetland*** (marsh) ***Discharge to water bodies*** ***Combinations of the above***

(b) systems for clusters of homes and small communities that are to be sewered or already sewered.

Commercially available prefabricated treatment plants, known as package plants, are often used for the treatment of wastewater for individual properties and small communities.

The most common types of package plants are: extended aeration, contact stabilization, sequence batch reactors, rotating biological contactor and physical/chemical plants.

Depending on the pollutants that must be removed from wastewater a great number of different treatment processes flowsheets can be developed, using preliminary and/or primary, secondary and tertiary treatments.

To develop the proper treatment flowsheet , it is necessary that detailed process analysis of the suitability of ever possible individual treatment method be prepared.

The detailed process analysis of the suitability of an individual treatment method (unit operations and processes) is given in the Table below.

Important factors that must be considered when selecting and evaluating unit operations and processes for waste water treatment

S.N	Factors/Parameters	Remarks
1	Process applicability	The applicability of a process is evaluated on the basis of past experience data from full-scale plants, and pilot data from plant studies. If new unusual conditions are encountered, pilot-plant studies are necessary.
2	Applicable flow range	The process should be matched to the expected flow range. For example, Stabilization ponds are not suitable for extremely large flows.
3	Applicable flow variation	Most unit operation and processes work best with a constant flow rate, although some variation can be tolerated. If the flow variation is too great, flow equalization may be necessary.
4	Influent-wastewater characteristics	The characteristic of the influent affect the types of processes to be used (e.g chemical or biological) and the requirement for their proper operation.
5	Inhibiting and unaffected constituents	What constituents are present that may be inhibitory, and under what conditions? What constituents are not affected during treatment?
6	Climatic constraints	Temperature effects the rate of reaction of most chemicals and biological processes. Freezing conditions may affect the physical operation of the facilities.
7	Reaction kinetics and reactor selection	Reactor sizing is based on the governing reaction kinetics. Data for kinetic expression usually are derived from experience, the literature, and the results of pilot-plant studies (e.g the complete mixing reactor with continuous flow, or plug-flow reactor).
8	Performance	Performance is most often measured in terms of effluent quality which must be consistent with the given effluent-discharge requirements.
9	Treatment residuals	The types and amounts of solid, liquid and gaseous residuals produced must be known or estimated. Often, pilot-plants studies are used to identify residuals properly.
10	Sludge handling constraints	Are there any constraints that would make sludge handling expensive or infeasible? In many cases, a treatment method should be selected only after the sludge processing and handling options have been explored.
11	Environmental constraints	Nutrients requirements must be considered for biological treatment processes. Environmental factors. Such as the prevailing winds and wind direction, may restrict the use of certain processes, especially where odors may be produced.
12	Chemicals requirements	What resources and amount must be committed for a period of time for the successful operation of the unit operation or process?

S.N	Factors/Parameters	Remarks
13	Energy requirement	The energy requirements, and probable future energy costs, must be known, if cost-effective treatment systems are to be designed.
14	Other resources requirements	What, if any, additional resources must be committed to the successful implementation of the proposed treatment system using the unit operation or process question?
15	Reliability	What is the long-term record of the reliability of the unit operation or process under emergency condition? Is the operation process easily upset? Can it stand periodic shock loadings? If so, how do such occurrences affect the quality of the effluent?
16	Complexity	How complex is the process to operate under routine conditions and under emergency conditions such as shock loadings? What level of training must be given to operators who have to operate the process?
17	Ancillary processes	What support processes are required? How do they affect the effluent quality, especially when they become inoperative?
18	Compatibility	Can the unit operation or process be used successfully with existing facilities? Can plant expansion be accomplished easily? Can the type of reactor be modified?

General flowsheet for municipal WWT is shown below in figure.

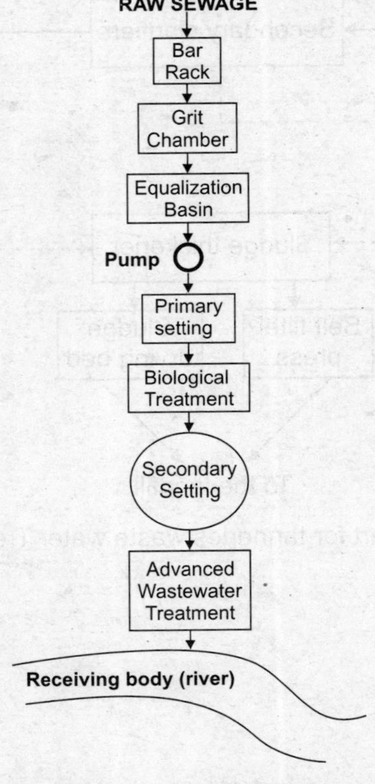

Fig.5.13F$_2$: Flowsheet of Municipal Waste Water Treatment

Typical waste water Treatment (WWT) plant flow chart for tanneries in India is shown below.

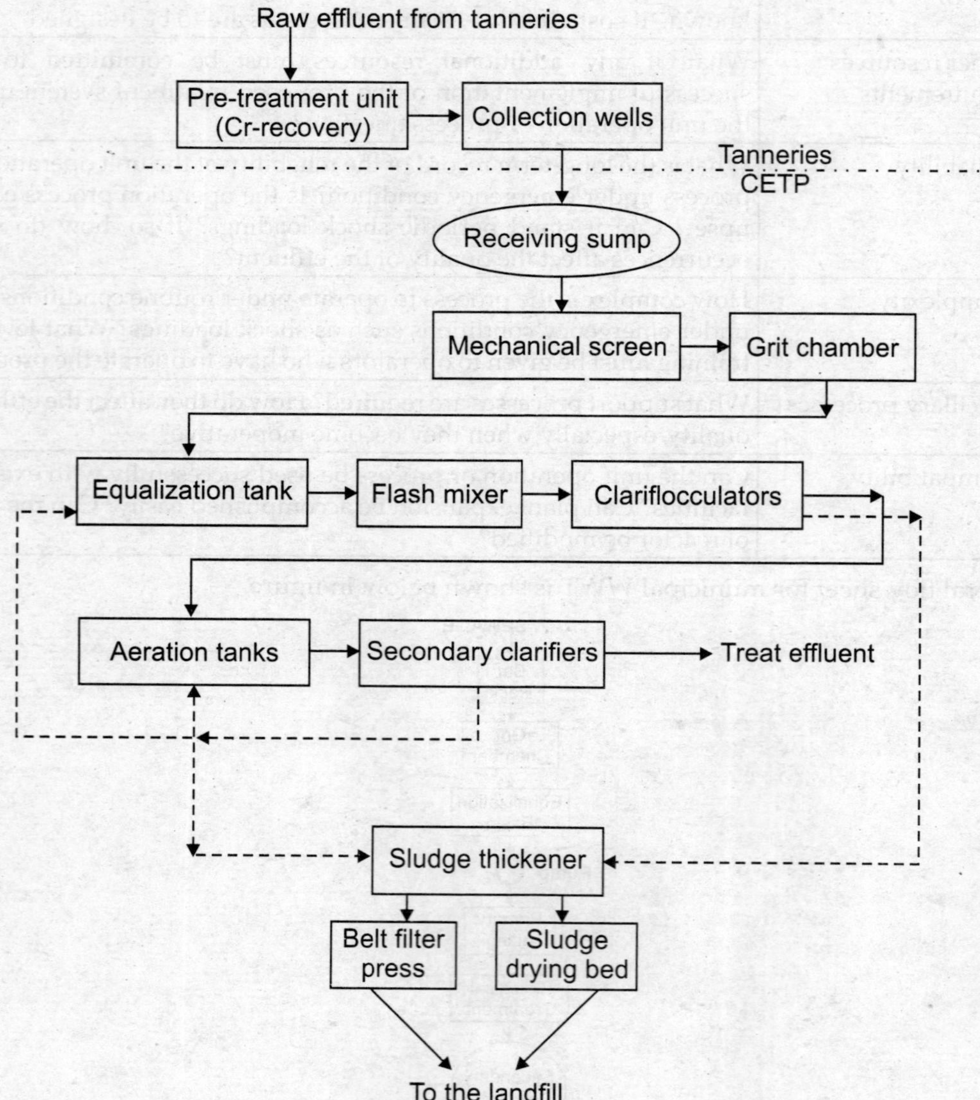

Fig.5.13F_3: Flow chart for tanneries waste water Treatment plant in India

Typical Process flowsheet for an WWT plant in USA is shown below

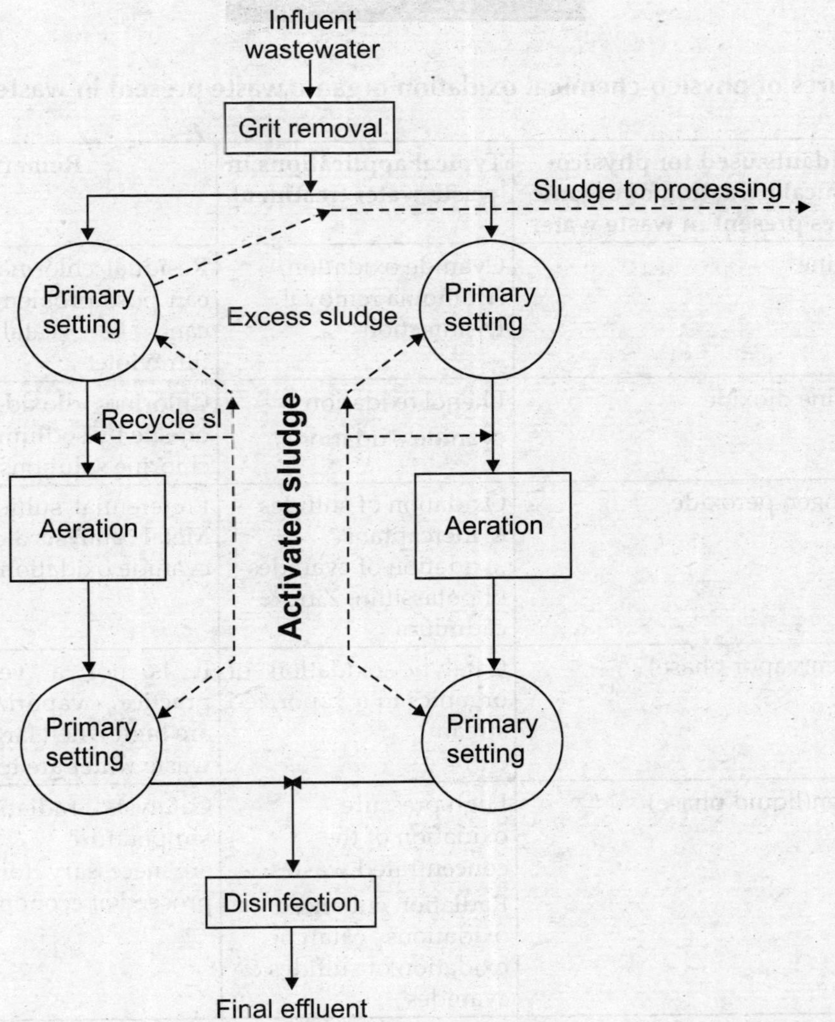

Fig.5.13F$_4$: Typical Process flowsheet for a wastewater treatment plant in USA

APPENDIX 5.14

Features of physico-chemical oxidation organic waste present in waste water

S.N	Oxidants used for physico-chemical oxidation of organic wastes present in waste water	Typical applications in waste water treatment	Remarks
1	Chlorine	Cyanide oxidation, Ammonia removal, Disinfection	Residual chlorinated organics can pose problems. Oxidation can be catalyzed with ultraviolet
2	Chlorine dioxide	Phenol oxidation cyanide oxidation	Chlorine dioxide generated on-site by sodium chlorite and chlorine solutions.
3	Hydrogen peroxide	Oxidation of sulfides & mercaptans ,oxidation of cyanides of potassium, Zinc & cadmium	Preferential sulfide oxidation, Metal catalysts aid phenol and cyanide oxidation.
4	Oxygen(vapor phase)	Catalytic oxidation of organics in a vaporized stream	It is not a very frequent practice, vaporization costs are high when large volume of waste water are to be treated.
5	Oxygen(liquid phase)	High pressure oxidation of the concentrated wastes. Radiation enhanced oxidations , catalytic oxidation of sulfides & cyanides	Catalysts, radiation or high temperature & pressures are necessary for reaction to proceed at economical rates.
6	Ozone	Decolorization Disinfection Trace organic removal Cyanide removal Phenol oxidation	Ozone is generated on site. Oxidations can be catalyzed with ultraviolet light or activated carbon Normally "no problems" by products are produced
7	Potassium permanganate	Specialized partial oxidations	Aldehydes, amines, aromatic , alcohols, keto acids , alkenes & phenol are amenable to oxidation.

APPENDIX 5.15

Alternative treatment for liquid wastes (dilute /concentrated) generated in manufacturing operations

S.N	Type of treatment of liquid wastes	Specific alternative treatments for liquid wastes
1	Physical Treatment	• Equalization/ holding • Adsorption • Sizing- Screening/straining, depth filtration, grinding/ commuting • Phase change- stripping, distillation/evaporation, freezing sublimation • Ion exchange • Membrane processes-ultrafiltration, reverse osmosis, dialysis, liquid membranes pre vaporization • Force filed separation- centrifugation, gravity- sedimentation, dissolved air floatation, dispersed-air floatation • Surface methods- foaming, non-foaming bubble separation, floatation, skimming • Extraction/ absorption
2	Chemical Treatment	• Acid base treatment • Chemical precipitation, Coagulation, flocculation, polymerization • Oxidation- Chlorination, ozonation, peroxides, permanganates, Catalytic, Oxidation, wet combustion, incineration • Reduction reactions • Complexation • photochemical reaction • Hydration and clathrates • Electromotive displacement • Thermal decomposition
3	Biological Treatment	• Activated sludge- completely mixed, multistage/multistep, contact stabilization, oxygen-based • Fixed film processes- Trickling filter, rotating disc • Aerated stabilization • Anaerobic –lagoons, contact stabilization, submerged filter • Algae stabilization ponds • Balanced ecosystem- aquatic plants, animals • Enzyme conversions- immobilized enzymes/cells

APPENDIX 5.16

Concentration of water pollutants that make Pre biological or primary treatment of wastewater desirable

S.N	Water pollutants or system condition	Limiting concentration of pollutants	Type of pretreatment required by waste water
1	Suspended solids	>50-125 mg/litre	Sedimentation, floatation, lagooning
2	Dissolved salts	>10-16 gm/litre	Dilution, ion exchange
3	Oil or grease	>35-50 mg/litre	Skimming tank or separator
4	Alkalinity	0.5kg alkalinity as $CaCO_3$/kg BOD removed	Neutralization for excessive alkalinity
5	Acidity	Free mineral acidity	Neutralization
6	Sulfides	>100 mg/litre	Precipitation or stripping with recovery
7	Phenols	>70-300 mg/litre (upto at least 1500 mg/litre, if bacterial acclimation is maintained)	Extraction, adsorption, internal dilution
8	Ammonia	>1.6 gm/litre	Dilution, ion-exchange, PH adjustment and stripping
9	Organic load variation	>2:1-4:1 (higher levels , if constituents remain in the same ratio)	Equalization
10	PH value	6-9	Neutralization
11	Temperature	13-38°c in the reactor	Cooling, steam addition
12	Toxic ions:- Pb Cu+Ni+CN Cr^{+6}+ZN $Cr+^3$	<0.1 mg/litre <1 mg/litre <3 mg/litre <10 mg/litre	Precipitation or ion exchange

APPENDIX 5.17

Function of various sequential unit processes used in sludge treatment

S.N	Unit Process	Functions
1	Thickening (blending)	-water removal -Volume reduction -Post process efficiencies -Blending
2	Stabilization (reduction)	-pathogen (disease causing bacteria) destruction -Volume & weight reduction -Odor control -Putrescibility (decomposition) control -Gas production
3	Conditioning (stabilization)	-Improve dewatering or Thickening rate -Improve solids capture -Improve compact ability -Stabilization
4	Dewatering	-Water removal -Volume & weight reduction -Change to damp cake -Reduces fuel requirement for incineration/ drying
5	Heat drying	-Water removal -Sterilization -Utilization
6	Reduction (stabilization)	-Destruction of solids -Water removal -Conversion -Sterilization
7	Final disposal	-Utilization (crop land) -Utilization (energy) -Utilization (land reclamation) -Disposed (land fill) -Disposal (ocean)

APPENDIX 5.18

Various oxygen demand & organic carbon parameters of waste water and their significance

S.N	Oxygen demand and organic carbon parameters of wastewater	Abbreviated form	Significance
1	Biochemical/ biological oxygen demands	BOD_5	Biological or biological oxygen demand excreted in 5 days; the oxygen consumed by a waste through bacterial action; generally about 45-55% of the THOD (Theoretical Oxygen demand of waste water)
2	Chemicals oxygen demand	COD	It is the amount of a strong chemical oxidant (chromic acid) that is reduced by a waste; the result are expressed in terms of an equivalent amount of oxygen; generally about 80% of the THOD.
3	Theoretical oxygen demand	THOD	It is the amount of oxygen theoretically required to completely oxidize a compound to CO_2, H_2O, PO_4^{-3}, SO_4^{-2}, and NO_3.
4	Total organic carbon	TOC	It is generally about 30% of the THOD.
5	Biological oxygen demand (ultimate)	BOD_{ult}	The ultimate BOD exerted by a waste in an infinite time.
6	Immediate Oxygen demand	TOD	Immediate oxygen demand. The amount of oxygen consumed by a waste within 15 minute (Chemical oxidizers and bacteria not used)
7	Dissolved oxygen	DO	A negative "DO" is a positive TOD.

APPENDIX 5.19

Alternative unit operation/ process for sludge/ brine processing and disposal

Unit operations/ process	Thickening	Stabilization	Conditioning	Dewatering	Disposal
Purpose	Concentrates solids	Reduces pathogens, and eliminates offensive odors	Enhances water removal	Removes moisture and produces sludge cake	Used for ultimate disposal of residues
Treatment processes	1. Gravity 2. Flotation 3. Centrifug-ation	1. Chlorine oxidation 2. Lime stabilization 3. Heat treatment 4. Aerobic digestion 5. Anaerobic digestion	1. Chemical 2. Elutriation 3. Heat treatment	1. Vacuum filter 2. Filter press 3. Horizontal bed filter 4. Centrifugation 5. Drying beds	1. Evaporation of brine 2. Incineration 3. Wet Oxidation 4. Pyrolysis 5. Composting 6. Land filling 7. Deep well injection

APPENDIX 5.20

Chlorine dosage level for disinfection of various categories of wastewater

S.N	Wastewater to be disinfected	Chlorine dosage, mg/l
1	Raw sewage	6-12
2	Raw sewage (septic)	12-25
3	Settled sewage	5-10
4	Settled sewage (septic)	12-40
5	Chemical precipitation effluent	3-10
6	Tricking filter effluent	3-10
7	Activated sludge effluent	2-8
8	Sand filter effluent	1-5

APPENDIX 5.21

Typical feed rates of dechlorination Chemicals for dechlorination of water supplies & wastewater effluents

S.N	Dechlorination chemicals	Typical rates, kg/kg
1	Sulfur dioxide	1.1
2	Sodium sulfite	0.57
3	Sodium bisulfite	0.68
4	Sodium thiosulfate	1.43

Note: Contact time: 1-5 minutes

APPENDIX 5.22

TSS & BOD reduction efficiency of various sewerage treatment plant units combination

S.N	Combination of sewage treatment plants units	Total % reduction (approximate) in	
		Total Suspended solids (TSS)	Biochemical oxygen demand (BOD)
1	Sedimentation plus sand filter	90-98	85-95
2	Sedimentation plus standard Trickling filter, 600 Ib BOD per acre-foot maximum loading	75-90	80-95
3	Sedimentation plus single stage high rate tricking filter	50-80	35-65
4	Sedimentation plus two stage high rate tricking filter	70-90	80-95
5	Activated sludge	85-95	85-95
6	Chemical treatment	65-90	45-80
7	Pre-aeration (1 hr) plus sedimentation	60-80	40-60
8	Plain sedimentation	40-70	25-40
9	Fine screening	2-20	5-10
10	Stabilization pond , aerobic	-	70-90
11	Anaerobic lagoon	60-70	40-70

[a] No recirculation.

Efficiencies can be increased within limits by controlling organic loading, efficiencies of settling tanks, volume of recirculation, and the number of stages: however, effluent will be less nitrified than from standard rate filter, but will usually contain dissolved oxygen. Filter files and odors are reduced. Study first cost plus operation and maintenance cost.

APPENDIX 5.23

Characteristic of various processes used for oil & grease removal from waste water

S.N	Process	Description	Advantages	Disadvantages
1	Gravity separation	API, CPI, TPS PPI	Potential for treatment of suspended solids; effective removal of dispersed oil; simple and economical treatment operation	Limited efficiency for removal of emulsified oil; will not remove soluble oil; restricted to the removal of oil droplets greater than 20 micrometre
2	Air flotation	DAF, IAF	Potential for treatment of suspended solids; effective removal of dispread and emulsified oil with chemical addition; reliable process which effectively treats shock loads	Chemical sludge produced
3	Chemical flocculation	Used in conjunction with gravity separation and air flotation	Potential for treatment of high suspended solids	Chemical sludge produced
4	Filtration	Sand, anthracite, multimedia, crushed graphite	Effective suspended solids removal; has application to the separation of free, dispersed and emulsified oil	Requires backwashing and the backwash creates a subsequent treatment problem
5	Coalescence	Fibrous media	Effective removal of all oil components except should oils	High levels of suspended solids will induce fouling potential for biological fouling; Needs extensive pretreatment; not demonstrated as a practical process in full-scale operation

S.N	Process	Description	Advantages	Disadvantages
6	Membrane process	Reverse osmosis, ultrafiltration, hyperfiltration	Soluble oil removal demonstrated in laboratory tests	Low flux rates; membrane fouling and limited membrane life; not demonstrated as a practical processes in full-scale operation; needs extensive pretreatment
7	Biological processes	Activated sludge, RBS	Effective soluble oil removal	Requires extensive pretreatment to reduce influent oil levels to less than 40 mg/l
8	Carbon adsorption	GAC will act both as effective filtration medium and coalescing separator; PAC for removal of soluble oils only	Effective removal of all oil components including soluble oils	Requires extensive pretreatment; process is expensive; carbon must be regenerated or replaced; not demonstrated as a practical process for full-scale operation

API- American petroleum institute Gravity separation method

DAF –Dissolved air flotation technique

IAF –Including air floatation technique

APPENDIX 5.24

Characteristics of various neutralization reagents used for water neutralization

S.N	Chemical Neutralization reagent	Chemical Formula	Neutralization requirements mg/l*	Neutralization factor
Basic				
1	Calcium carbonate	$CaCO_3$	1.0	1.0/0.56=1.786
2	Calcium oxide	CaO	0.560	0.56/0.56=1000
3	Calcium hydroxide	$Ca(OH)_2$	0.740	0.74/0.56=1.321
4	Magnesium oxide	MgO	0.403	0.403/0.56=0.720
5	Magnesium hydroxide	$Mg(OH)_2$	0.583	0.583/0.56=1.041
6	Dolomitic quicklime	$[(CaO)_{0.6}(MgO)_{0.4}]$	0.497	0.497/0.56=0.888
7	Dolomitic hydrated	$\{[Ca(OH)_2]_{0.4}$	0.677	0.677/0.56=1.209
8	Lime	$[Mg(OH)_2]_{0.6}$		
9	Sodium hydroxide	NaOH	0.799	0.799/0.56=1.427
10	Sodium carbonate	Na_2CO_3	1.059	0.059/0.56=1.891
Acid				
11	Sulfuric acid	H_2SO_4	0.98	0.98/0.56=1.750
12	Hydrochloric acid	HCI	0.72	0.72/0.56=1.285
13	Nitric acid	HNO_3	0.63	0.63/0.56=1.125

*The quantity required to neutralize 1 mg/l of acidity or alkalinity, expressed as a calcium carbonate. Assumes 100% purity of all compounds.

APPENDIX 5.25

Alternative Technologies for waste water treatments

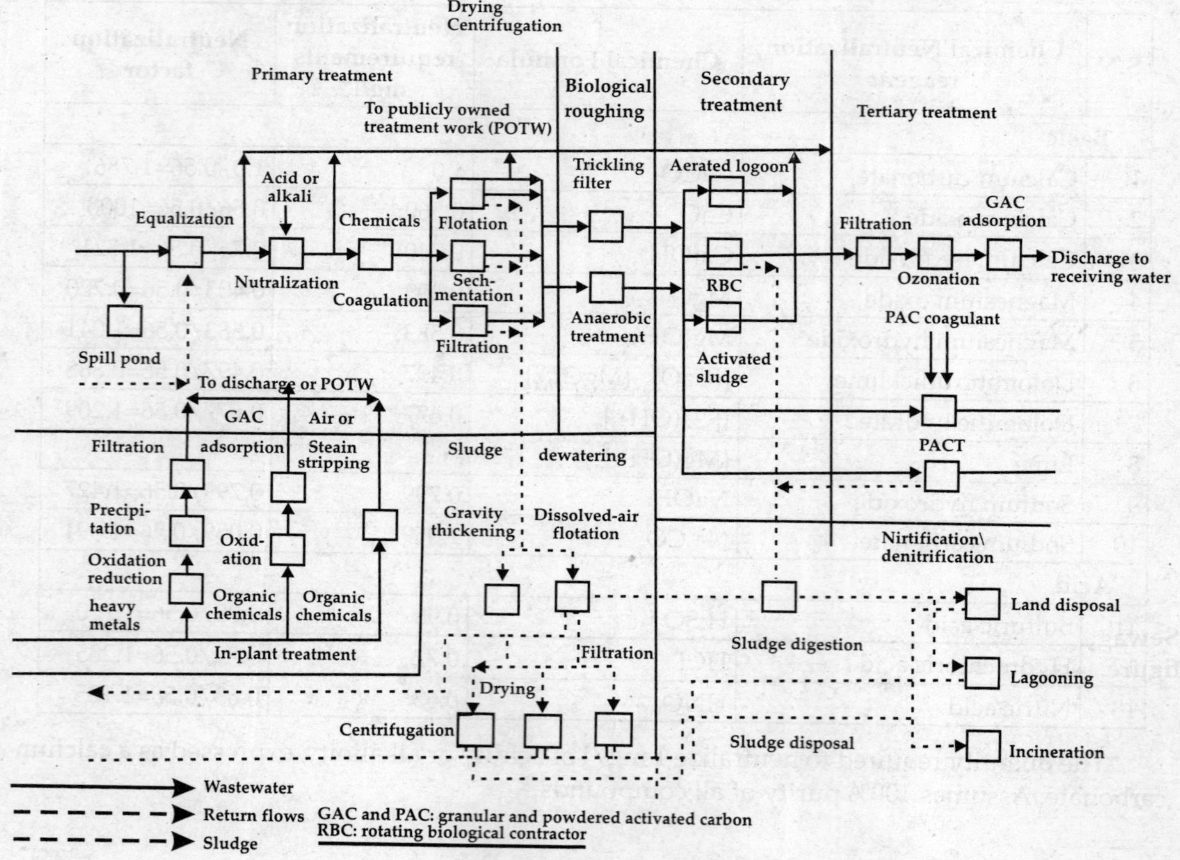

Three types of sewage treatment scheme is shown below in figure.

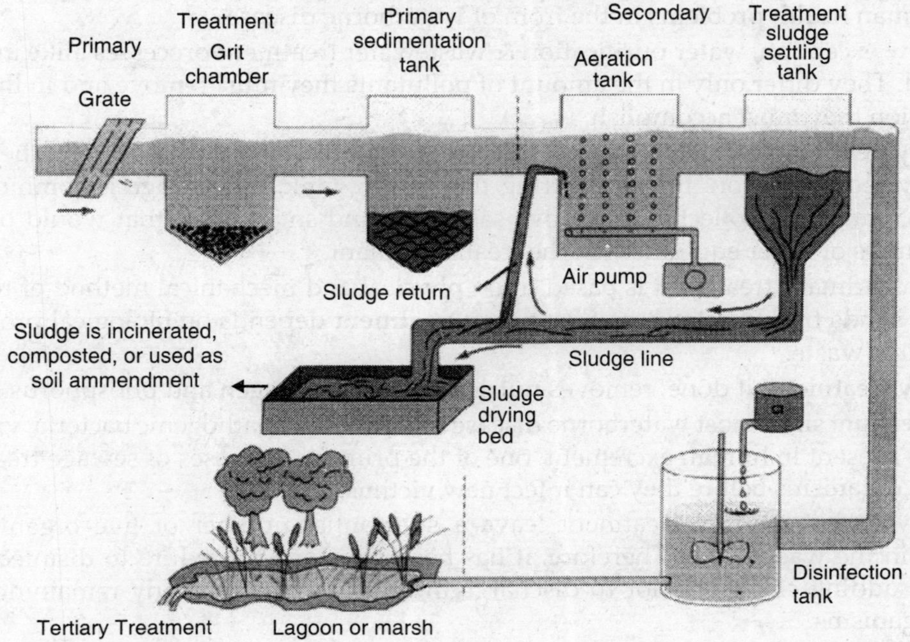

Fig.5.25F_1: Three types of sewage treatment scheme

Sewage Treatment option for removal of various water contaminants is shown below is figure.

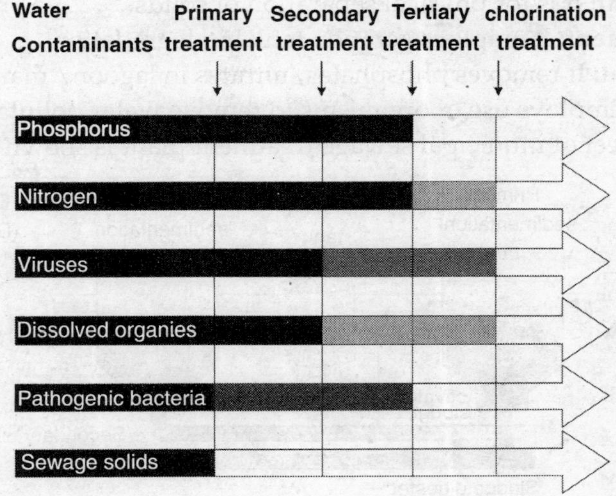

Fig.5.25F_2: Sewage Treatment option for removal of various water contaminants

The aim of sewage treatment is to improve the quality of wastewater to the point that can be discharged into a waterway without seriously disrupting the aquatic environment or causing human health problems in the from of waterborne disease.

With few exception, water purification & wastewater treatment processes alike in concept and in kind. They differ only in the amount of pollutants they must remove and in the degree of purification they must accomplish.

Primary treatment consists of simply holding sewage in a large bank to permit the removal of solids by sedimentation. Before entering the settling tank, the sewage is commonly sent through a chamber or collector to remove sand, grit and small rocks that would otherwise damage pumps or other equipment in the treatment plant.

Whereas primary treatment is based upon physical and mechanical method of removing suspended solids from wastewater, **Secondary treatment** depends on biological processes to digest organic waste.

Tertiary treatment, if done, removes nutrients such as nitrogen and phosphorus.

Disinfection: since most waterborne disease are caused by pathogenic bacteria, viruses, or protozoans present in human excrement, one of the primary purposes of sewage treatment is to kill such organisms before they can infect new victims.

Primary and secondary treatment leave a substantial number of live organisms still remaining in the wastewater. Therefore, it has been standard procedure to disinfect treated effluent by adding chlorine prior to discharge in order to eliminate any remaining disease causing organisms.

The policy of chlorinating all sewage treatment plant discharged has met with increasing resistance and today they no longer require chlorination of wastewater.

There are several reason for this changes in accepted practice:

- Chlorine is effective in killing bacteria but less so in relation to protozoans and viruses.
- Chloramines are formed which may be toxic to aquatic life.
- Chlorine treatment is expensive.

Primary treatment: It is for physical separation of solids.

Secondary treatment: It employs aeration tank biodegradation.

Tertiary treatment: It removes phosphates/nitrates in lagoon/ marsh or trickling filter.

Bioremediation employs use of organisms to remove water pollutants.

A typical flow sheet of municipal sewage treatment plant is shown below in figure.

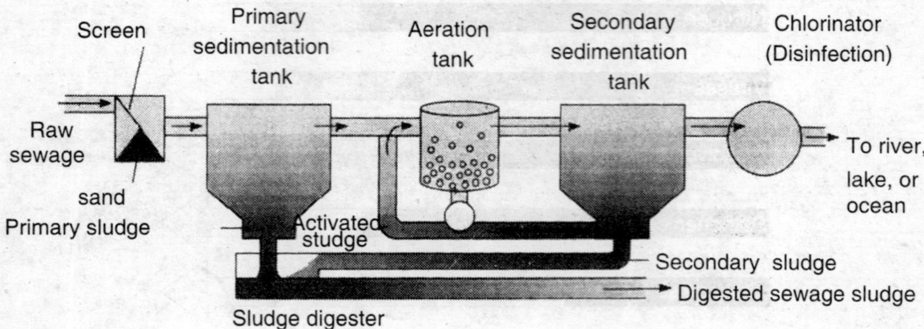

Fig.5.25F$_3$: Typical flow sheet of municipal sewage treatment plants

A Domestic sewage treatment scheme employing Septic tank and drain fields is shown below in figure.

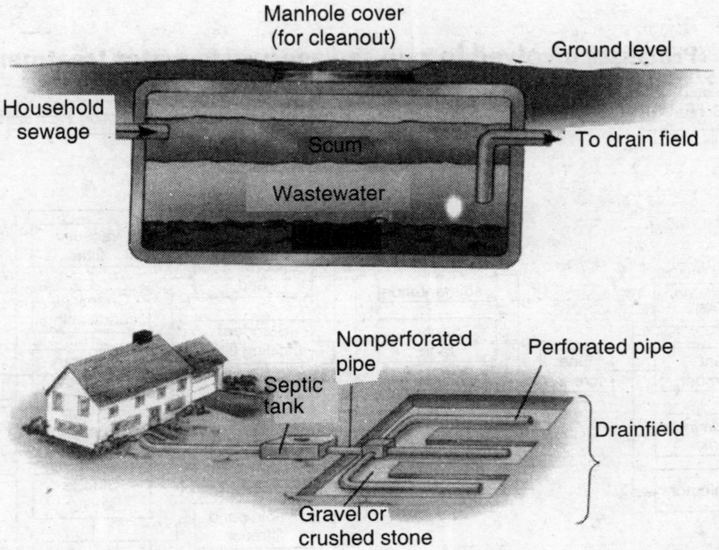

Fig.5.25F$_4$: Domestic sewage treatment scheme employing Septic tank and drain fields

APPENDIX 5.26

Processes involved in raw sewage/waste water treatment

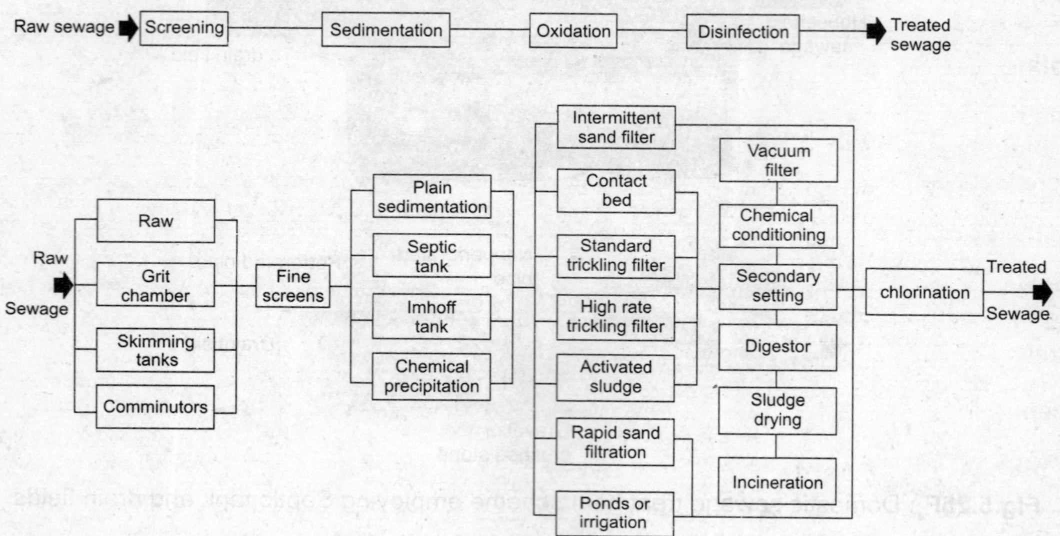

Fig.5.26F_1: Processes involved in raw sewage/waste water treatment

Note:- Wastewater Treatment Methods includes physical, chemical & biological treatment methods , which are described below.

Physical Treatment of Wastewater

Physical treatment usually includes cooling, equalization, screening, comminuting, grit removal, oil greases removal, mixing, sedimentation and filtration.

The principal applications of these unit operations are tabulated below.

Table 5.26T_1: Application of physical unit operations in wastewater treatment

S.N	Unit operation	Application
1	Screening	Removal of coarse and settable solids by interception (surface straining)
2	Comminution	Grinding of coarse solids to a more or less uniform size
3	Flow equalization	Equalization of flow and mass loading of BOD & suspended solids.
4	Mixing	Mixing of chemicals & gases with wastewater and maintaining solids in suspension
5	Flocculation	Promotes the aggregation of small particles into large particles to enhance their removal by gravity sedimentation
6	Sedimentation	Removal of settleable solids and thickening of sludges.

7	Floatation	Removal of finely divided suspended solids and particles with densities close to that of water. Also thickens biological sludges.
8	Filtration , Micro screening	Removal of fine residual suspended solids remaining after biological or chemical treatment. Same as filtration. Also removal of large from stabilization pond effluents.

Cooling

Some hot process effluents will require cooling (temperature reduction) before being fed to any biological or some other treatment or disposed into the receiving body.The devices like different types of heat-exchangers, cooling towers, etc. are employed in cooling processes.

Equalization

Wastewater from some industrial activities generally show variations in flow and some other characteristic like pH, BOD, COD, TSS, etc. These could be regulated by retaining the wastewater in a basin so that the resulting effluent is fairly uniform in flow and characteristics.

Screening

This unit operation is envisaged to protect equipments like pumps, valves, pipelines, etc. from damage or clogging by large objects. The screening element may consist of parallel bars, rod or wires, grating wire mesh or perforated plates.

Comminuting

This is a process where devices like grinders, cutter or shredders are employed to break up solid material into smaller sizes.

Girt removal

This process is employed to remove sand, dust, gravel, stones, cinders and other heavy inorganic settleable material. Grit chambers are provided to protect mechanical equipment from abrasion & abnormal wear, to reduce formation of heavy deposits in pipes & conduits, to reduce the size or frequency of cleaning of reactor (e.g digestors).

Oil and grease removal

The wastewater contains floating substances which seriously affect the working of various treatment units. The devices like a skimming tank or different types of oil separators are used to remove such floating substance which include oil, grease, soap, etc. from the effluent. Another method for the oil and grease (also removal of some suspended solids) from wastewater is the *flotation*. The waste flow or a portion of clarified effluent is pressurized in the presence of sufficient air to approach saturation. When this pressurized air-liquid mixture is released to atmospheric pressure in the flotation unit, minute air bubbles are released from solution. The sludge flocs, suspended solids, or oil globules are floated by these minute air bubbles, which attach themselves to and become enmeshed in the floc particles. The air-solids mixture rises to the surface, where it is skimmed off. The clarified liquid is removed from the bottom of the flotation unit; at this time a portion of the effluent may be recycled back to the pressure chamber. Emulsified oil and grease usually require some special treatment to break the emulsions so that the oily materials will be free and can be separated by gravity, coagulation or air-flotation.

Mixing

Mixing is an important unit operation in may phases of wastewater treatment where one substance must be completely intermingled with another. An example is the mixing of chemicals with wastewater (chlorine, hypochlorite, lime, etc.). Chemicals are also mixed with sludge to improve its dewatering characteristics before vacuum filtration or centrifugation.

In the biological process tank, air must be mixed with the activated sludge to provide the organisms with the oxygen required. In this case diffused air is introduced in such a way as to fulfill the mixing requirements, or mechanical turbine aerator-mixers may be used.

Sedimentation

This is one of the most widely used unit operation in wastewater treatment. In the sedimentation process, comparatively bigger size particles are allowed to settle down due to the force of gravity. Very fine particles takes very long time to settle and sometimes are not settleable at all. The process of settlement of a particle is obstructed or opposed by the following three forces: *velocity of flow size & shape of particle and viscosity of water.* The wastewater having high settleable/suspended solids are allowed to pass through a settling tanks. The tanks could be either rectangular or circular.

Filtration

This is a unit operation where the liquid/solid mixture under gravitational force or under pressure is made to pass through a natural or synthetic medium, leaving behind solids or a colloids of a very finely divided nature. Some of the common filter media are sand, coal, activated carbon, fiber glass screen, wire mesh, cloth, etc.

Another type of filtration is micro-straining and is usually used for removal of micro organisms and some other suspended solids from a wastewater.

Chemical Treatment Of Waste Water

Those processes used for treatment of wastewater in which change is brought about by means of or through chemical reaction are known as chemical unit processes.

Chemicals treatment processes usually includes neutralization, coagulation flocculation, chemical precipitation, oxidation, reduction, ion-exchange, chemical adsorption, reverse osmosis and electro dialysis.

The chemical unit process and their principal application are tabulated below.

Table 5.26T_1: Application of chemical unit processes in wastewater treatment (WWT)

S.N	Wastewater treatment process	Application
1	Chemicals precipitation	Removal of phosphorus and enhancement of suspended solids removal in primary sedimentation facilities used for physical- chemical treatment
2	Gas transfer	Addition and removal of gases

3	Adsorption	Removal of organics not removed by conventional chemical and biological treatment methods. Also used for dechlorination of wastewater before final discharge of treated effluent
4	Disinfection	Selective destruction of disease-causing organisms. Chlorine is the most commonly used chemical. Disinfection with ozone.
5	Dechlorination	Removal of total combined chlorine residual that exist after chlorination.
6	Other	Various other chemicals for specific objectives in WWT.

Neutralization

This is a unit process where the acid or alkaline waste should be neutralized before being discharged to transported to some additional treatment. The regents usually used in the neutralization are caustic soda, lime and sodium carbonate. Some other regents are also employed for specific applications (eg, ammonia, carbon, dioxide , calcium carbonate, sodium carbonate, sulphuric acid, etc.)

Coagulation and flocculation

In these unit processes , the wastewater is treated with certain chemicals which form a floc that absorbs and entrains the suspended & colloidal particles present. In coagulation individual particles/agglomerate or combine together. When a coagulant is used in water, it forms a spongy gelatinous precipitate which absorbs fine size particles in water and bind them together. The whole process result into bigger particles which are easily settleable.

Some important factors have to be followed: proper coagulant & flocculent selection, adequate mixing and efficient detention time. In order to determine approximately the dosage of coagulant, the usual test which is performed in the laboratory, is the jar test.

Chemical precipitation

Chemical precipitation in the wastewater treatment involves the addition of chemicals to alter the physical state of dissolved & suspended solids and facilitate their removal by sedimentation.They are mainly used for removal of heavy metals, phosphates and certain soluble anionic colour components.

The toxicity of some contaminants can be reduced or destroyed by oxidation. The oxidizing agents used in the process are sodium hypochlorite peroxide, ozone, etc.The cyanides are generally treated in this manner.

Reduction

Usually chromium is treated by reduction where the hexavalent from of chromium is converted into trivalent chromium by means of liquid sodium bisulphite or di sodium meta bisulphite, subsequently precipitated as chromium hydroxide, and removed by sedimentation.

Ion-exchange

This method is generally employed for the purification of raw water or for upgrading the treated water on an acceptable level or for removal of undesirable anions and from a wastewater.The process involves the exchange of cations by hydrogen ions (H^+) and anions by hydroxyl ions (OH^-). To facilitate these exchange the use of certain resins is made.To regenerate the used resins some acid or alkali are necessary.

Carbon adsorption

This method is chiefly employed for removal of dissolved organics such as saturated oils, alkenes & alkenes, dyes, phenols, etc. Certain colours and odours are also removed by this method. The method is based on the principle that charcoal of the charcoal filter where the surface tension on the activated carbon particles causes molecules to adhere on it.

Reverse Osmosis

This is the physico-chemical process that separates pure water from its pollutants. This membrane filter operation is also called as hyper filtration or super filtration. The method requires a high pressure to be applied to the waste to force out the liquid (water Only) through a semi-permeable membrane, leaving behind the dissolved solids.

Electro dialysis

It is a method of separation of dissolved solids from the solution based on the difference in the rates of diffusion. The permeation through the membranes is enhanced by the application of electrical energy. The membranes commonly employed are cellulose nitrate, cellophane and parchment.

Other methods

Some other methods of physico-chemical treatment includes: gas stripping, evaporation, distillation, etc.

Biological Treatment of Waste Water:

Biological wastes are rich in decomposable organic matter and can be treated by biological processes involving microbial organic matter, *aerobically or anaerobically.*The main difference between aerobic and anaerobic processes is in the presence or absence of oxygen.

The common methods of biological wastewater treatment are:

- Aerobic processes such as trickling filters, rotating biological contactors, activated sludge process oxidation ponds and lagoons, oxidation ditches,
- Anaerobic processes such as anaerobic digestion, and
- Anoxic processes such as denitrification.

Aerobic Processes

The basic principle of aerobic processes is the usage of aerobic microorganisms which need oxygen for their metabolism in the process of decomposition of organic matter from the wastewater.

While the basic principles remain the same in all aerobic processes , the techniques used in their application may vary widely, and may be classified as either attached (film) growth or suspended growth processes.

Attached growth processes utilize a solid medium upon which bacterial solids are accumulated in order to maintain a high population. Suspended growth processes include intermittent sand filter, trickling filter, rotating biological contactors, and a variety of similar devices.

S.N	Type of waste water	Common name of the WWT process	Uses
1	**AEROBIC PROCESS:** Suspended growth	**ACTIVATED SLUDGE PROCESS** Conventional (Plug flow) Step aeration, Modified aeration Contact Stabilization Extended Aeration, Oxidation Ditch **AERATED LAGOONS** **AEROBIC DEGESTION HIGH RATE ALAGAL PONDS**	Carbonaceous BOD " " " +Nitrification Carbonaceous BOD Carbonaceous BOD
	Attached growth	TRICKLING FILTERS Low rate High rate ROATING BIOLOGICAL CONTACTORS	Carbonaceous BOD " "
2	**ANAEROBIC PROCESSES:** **Suspended growth**	**ANAEROBIC DIGESTION** Standard rate High rate Single	Stabilization "
	Attached growth	**ANAEROBIC CONTACT PROCESS.** **ANAEROBIC FILTER PROCESS ,** **ANAROBIC LAGOONS**	Carbonaceous BOD " "
3	**ANOXIC PROCESSES**	- suspended growth - Fixed film	Denitrification "

This equipment consists of a bed of highly permeable media to which microorganisms are attached and through which wastewater is percolated. The surface of the media that comes in to contact with wastewater develops a l film rich in microorganism. The microorganisms in this film degrade the organic material present in the wastewater.

As the microorganisms grow, the simple layer increases. The adsorbed organic matter is metabolized, before it can reach the microorganisms near the media face. As a result of having no external organic sources available for cell carbon, the microorganisms near the media face enter into an endogenous phase of growth and lose their ability to slime layer stalks to grow. This phenomenon of losing the slime layer is called sloughing and primarily a function of the organic and hydraulic loading on the filter.

The tricking filter should also have an underdrain arrangement for collecting the filter effluent and a proper ventilation system to maintain anaerobic conditions.

Rotating biological contactors (RBCs)

A rotating biological contractor consists of a series of closely spaced circular disks of polystyrene or PVC, or other materials. The disks are partially submerged in wastewater and rotated slowly through it.

In operation, biological growths become attached to the surface of the disks and from a similar layer over the entire wetted surface of the disks. The rotation of the disks alternately contacts the biomass with the organic material in the wastewater and then with the atmosphere for adsorption of oxygen.

Suspended growth processes maintain an adequate biological mass in suspension within the reactor by employing either natural or mechanical mixing. In most processes , the required volume is reduced by returning bacteria from the secondary clarifier in order to maintain a high solids concentration. Suspended growth processes include activated sludge & its various modifications, oxidation ponds, and sludge digestion systems.

Activated sludge process

It is a treatment process in which biologically active mass, called activated sludge is continuously mixed with the sewage flow in an aeration basin in the presence of abundant oxygen. The liquid in the aeration basin is called mixed liquor. The oxygen is supplied to the mixed liquor either by diffused compressed air or by mechanical agitation. The activated sludge is subsequently separated from the mixed liquor by sedimentation process in a clarifier and a part of this sludge is re-circulated to the aeration basin. The rest of the sludge, which is the excess production of biological cell material, is disposed-off.

Many modifications in basin configuration and aeration techniques have been made in past. The ~~more~~ important types of activated sludge process are described below:

- **The *conventional process*** (known also as the ***plug flow*** type) consists of a aeration basin, a clarifier and a solid-return line from the clarifier bottom. The return solids are mixed with the incoming wastes and pass through the reactor in a *plug flow* fashion. Air is provided uniformly along the aeration basin.
- ***Tapered aeration* processes** attempt to match the oxygen supply to demand by introducing more air at the head end. The process is otherwise the same as described for conventional aeration process. The main advantage of tapered aeration is the optimum use of air.
- ***Step aeration* processes** distribute the incoming flow to a number of points along the aeration basin, thus avoiding the locally high oxygen demand encountered in conventional and tapered aeration process.
- ***Contact stabilization* or *biosorption*** takes advantage of the observed adsorptive properties of activated sludge. The sewage is mixed & aerated with activated sludge for a short period (1/2 -1 hours) and the mixed liquor then passes into a clarifier, where final effluent and sludge are separated by gravity. The settled sludge in the clarifier is re-aerated in a separate sludge re-aeration (stabilization) tank for a period of some

3-6 hours. After the aerated sludge is fed into contact tank for intimate mixing with the inflow. This process is suitable for high organic loadings.

- ***Completely mixed* or a *high rate* aeration processes** disperse the incoming waste and return sludge uniformly throughout the aeration basin. The main objective of high rate aeration is to reduce the cost of construction.
- ***Extended aeration*** is a completely mixed process operated at a long hydraulic retention time and a high sludge age. The process is limited in application to small plants where its inefficiency is outweighed by its stability and simplicity of operation. The advantage of this process is that usually there is no provision for excess activated sludge necessary.
- ***Short-term aeration* or *high-rate* activated sludge** is a pretreatment process where retention times and sludge age are low, which leads to a poor effluent and relatively high solids production. A possible application of this process is as the first stage of a two-stage nitrification process.
- ***High purity oxygen activated sludge systems*** have been developed to permit easier matching of oxygen supply to oxygen demand and higher rate processes through maintenance of higher concentrations of biological solids.

Waste Stabilization Ponds

Waste stabilization ponds are low cost wastewater treatment units which usually depend on natural processes. These systems are sometimes also called oxidation ponds or lagoons. In these ponds the organic matter from the wastewater is degraded by natural oxidation processes by the action of microorganisms, both algae and bacteria. If these processes take place in a stream water then it will be known as a *self-purification* process.

Waste stabilization ponds or lagoons are classified according to the biological process taking place as *aerobic, anaerobic, facultative and maturation ponds.*

Aerobic ponds are shallow ponds, less than 1 m in depth, where dissolved oxygen is maintained throughout the entire depth, mainly by the action of photosynthesis.

Facultative ponds are ponds between 1 to 2.5 m depth, which have an anaerobic lower zone, a facultative middle zone, and an aerobic upper zone maintained by photosynthesis & surface reaeration.

Anaerobic ponds are deep ponds that receive high organic loadings such that anaerobic conditions prevail throughout the entire pond depth.

Maturation or **tertiary ponds** are ponds used for polishing effluents from other biological processes. Dissolved oxygen is furnished through photosynthesis and surface reaeration. This type of pond is also known as a *polishing pond.*

Aerated lagoons are these ponds which are oxygenated through the additional action of surface or diffused air aeration.

Oxidation ditches are simple artificial ditches, dug in the ground for the purification of wastewater using a very simplified technique. The basic form of the plant is a ring shaped circuit or ditch. Oxygen as well as circulation is provided by mechanical rotors. In fact, oxidation ditch is a modified form of "extended aeration" of activated sludge process.

Natural systems

Natural treatment systems are designed to take advantage of physical, chemical and biological processes which occur in the nature when water, soil, plants, microorganisms& the atmosphere interact.

The natural treatment systems are:

- the soil-based or land-treatment systems, (infiltration, irrigation, etc.),
- the aquatic-based systems (constructed & natural wetlands and aquatic plant treatment systems).

Anaerobic Biological Processes

The anaerobic waste treatment is an effective method for the treatment of highly concentrated organic wastes. In the absence of oxygen, anaerobic bacteria convert organic material into gaseous end-products such as $C0_2$, CH_4 and H_2S.

This biological process is called as ***anaerobic digestion*** and is one of the oldest processes used for the stabilization of sludges.

Anaerobic digestion* is a *fermentation process which involves the decomposition of organic and inorganic matter in the absence of molecular oxygen. The process stabilizes the organic matter by converting it as completely as possible into methane and carbon-dioxide gas through the two stages:

- ***acid fermentation or liquefaction,***
- ***methane fermentation or gasification.***

In the first stage, *liquefaction or acid fermentation,* acid forming bacterial group consisting of acid-producing bacteria is responsible for converting complex organic compounds (like fats, proteins, carbohydrates) into more simple organic compounds (like acetic, propionic and butyric acids) which become a source of food for a second group, the **methane-forming bacteria.**

In the second stage, so called *gasification or methane fermentation,* the organic acids will be converted into gases, mainly methane and carbon dioxide

Two types of digesters are at present in use, standard-rate and high-rate.

In the *standard-rate digestion* process, the content of the digester are usually unheated and unmixed. Detention times for this process vary from 30 to 60 days.

In a *high-rate digestion* process the contents of the digester are heated and completely mixed. The required detention time is 15 days or less. A combination of these two basic processes is known as *the two-stage process digestion.*

Anoxic Processes

Under anoxic conditions, a group of facultative anaerobes called denitrifiers utilizes nitrites ($N0_2^-$) and nitrates ($N0_3$-) as the terminal electron acceptor.

The removal of nitrogen in the form of nitrate from the wastewater by conversion to nitrogen gas in the absence of oxygen is a biological process called anoxic denitrification.

APPENDIX 5.27

A secondary sewage treatment plant flowsheet

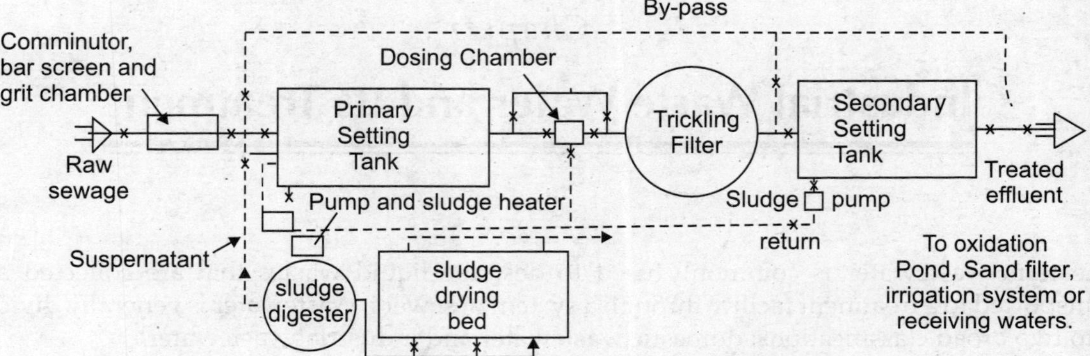

Fig.5.27F$_1$: Flowsheet of a secondary sewage treatment plant

6

Chapter

Industrial Waste Water and Its Treatment

The term wastewater is commonly used to describe liquid wastes that are collected and transported to a treatment facility through a system of sewers. Wastewater is generally divided into two broad classifications: domestic wastewater and industrial wastewater.

DOMESTIC WASTEWATER VS. INDUSTRIAL WASTEWATER

Domestic wastewater (sewage) is wastewater discharged from sanitary conveniences in residential, office, commercial, factories and various institutional properties. It is a complex mixture containing primarily water (approximately 99%) together with organic and inorganic constituents. These constituents or contaminants comprise of suspended, colloidal and dissolved materials. Domestic sewage, since it contains human wastes, also contains large numbers of micro-organisms and some of these can be pathogenic(disease causing). Waterborne bacterial diseases that can be present in sewage include cholera, typhoid and tuberculosis. Viral diseases can include infectious hepatitis. Inorganic constituents include chlorides & sulphates, various forms of nitrogen & phosphorous, as well as carbonates &bicarbonates. Proteins and carbohydrates constitute about 90% of the organic matter in domestic sewage. These arise from the excreta, urine, food wastes, and wastewater from bathing, washing, & laundering, and because of the latter, soaps, detergents, & other cleaning products can be found as well.

Domestic sewage has a flow pattern which typically shows two peaks — in the morning before the start of working hours and in the evening after the population has returned from work. Typically these hydraulic peaks would also become more distinct as the sewage flows considered come from smaller populations and consequently smaller sewer networks. Variations in sewage characteristics across a given community tend to be relatively small although variation across communities can be more readily detected. Notwithstanding these variations, the composition of domestic sewage is such that it lends itself well to biological treatment in terms of the availability of and balance between carbonaceous components and nutrients.

The biodegradability of sewage can be estimated by considering its Chemical Oxygen Demand (COD) and the corresponding BOD_5 (5 day BOD), and is indicated by its COD:BOD_5 and BOD_5:N:P ratios. This would typically be about 1.5:1 and 25:4:1 respectively. The nitrogen, N, would typically be in the form of organic nitrogen and ammonia nitrogen (Amm-N). Nitrates (NO_3-N) would not be expected to be present as conditions in the sewers would be such that nitrate formation is unlikely while nitrate degradation because of anoxic reactions is likely. The phosphorous (P) would be a combination of organic and phosphate (PO_4) forms. The pH of sewage would be within the range of 6–9 and this is generally considered suitable for biological processes. Examples of values of BOD_5, TSS (Total Suspended Solids), and TKN (Total Kjeldhal Nitrogen) which have been used for purposes of plant design are 250, 300 and 40 mg/ l respectively. As indicated earlier in this paragraph, sewage characteristics can vary across communities and a raw sewage BOD_5 of 500mg/ l has been encountered.

Industrial wastewater characteristics vary with each type of discharge. Food processing, dairy operations and meat packing all have wastes that are very high in BOD, sometimes over 1000 mg/l. This BOD is high in dissolved sugars, fats, & proteins and low in suspended material. It will dramatically increase the loading on the secondary treatment processes. Some food process wastes can also have a high pH, like potato processing and some vegetable canning, or a low pH, like green chili processing or fruit canning.

Toxins, like heavy metals coming from metal plating, battery shops,& heavy equipment manufacturing, and solvents from industries like body shops, dry cleaners, & furniture manufacturing can kill the microorganisms that are needed to treat the waste. Copper, chrome, lead, and cyanide are all chemicals that are commonly found in heavy industrial manufacturing discharges.

Hospitals and other medical facilities have the potential to discharge wastes that are radioactive or represent a biohazard.

In some cases, pretreatment of these industrial wastewater discharges may be required. Systems that deal with industrial discharges should adopt a pretreatment ordinance to cover conditions that the system may need to impose on the industrial dischargers in order to maintain their NPDES requirements on the treatment plant effluent. If local ordinances are different from specific categorical limits imposed in the Federal Regulations, the most stringent limits of the two must be followed. Pretreatment ordinances can require actual treatment of the industrial waste prior to discharge to the system, or it may levy a monetary surcharge based on the addition loading that results from the discharge.

DIFFERENT ZONES OF POLLUTION IN WATER STREAM

Zones of Purification

When sewage is discharged into water, a succession of changes in water quality takes place. If the sewage is emptied into a lake in which currents about the outfall are sluggish and shift their direction with the wind, the changes occur in close proximity to each other and, as a result, the pattern of changes is not crisply distinguished. If, on the other hand, the water moves steadily away from the outfall, as in a stream, the successive changes occur in different river reaches and establish a profile of pollution which is well defined. However, in most streams, this pattern is by no means static. It shifts longitudinally along the stream and is modified in intensity with changes in season and hydrography.

When a single large charge of sewage is poured into a clean stream, the water becomes turbid, sunlight is shut out of the depths, and green plants, which by photosynthesis remove carbon dioxide from the water and release oxygen to it, die off. Depending on the stream velocity, the water soon turns nearly black. Odorous sulfur compounds are formed and solids settle to the bottom, forming a sludge. The settled solids soon decompose, forming gases such as ammonia, carbon dioxide, and methane or marsh gas. Scavenging organisms increase in number until they match the food supply.

The oxygen resources are drawn upon heavily and, when overloaded, become exhausted. Life in such waters is confined to anaerobic bacteria (which exist when no oxygen is available), larvae of certain insects such as mosquitoes, and a few worms. There are no fish; turtles are generally the only forms of higher life present. This condition is known as the **zone of degradation**.

In a second zone, or **zone of decomposition**, more solids settle out, the water becomes somewhat clearer, and sunlight penetrates the surface. Oxygen is absorbed from the atmosphere at the air-water interface permitting the establishment of aerobic (oxygen available) conditions. The aerobic bacteria continue the conversion of organic matter into nitrates, sulfates. and carbonates. These, together with the carbon dioxide produced by decomposition as well as by bacteria and plant life, are food sources. With sunlight now penetrating the water, and with abundant food, algae begin to flourish and form a green scum over the surface.

In the third zone, or **zone of recovery**, algae become more numerous and self-purification proceeds more rapidly. Green plants utilizing carbon dioxide and oxygen will liberate in the say time more oxygen than is consumed, thus hastening the recovery of the stream. Simultaneously, the fish that require little oxygen such as catfish and carp, are also found. As the dissolved oxygen increases, more types of fish appear.

After recovery, in the **zone of cleaner water**, fish find the stream highly favorable, as the algae support various aquatic insects and other organisms on which fish feed. The water is clear or turbid according to concentration of algae, and may have odor for the same reason.

Throughout the stages of recovery of self-purification, disease organisms are greatly reduced in number because they lack proper food, and experience unfavorable temperatures and pH values of water. However, the water is still dangerous since all disease organisms have not perished.

SELF PURIFICATION OF WATER STREAM

The oxygen deficit (D) at the time in a polluted river stream is the difference between the actual DO content of water at that time and the saturation DO content. (Saturation DO means the amount of DO which will be present in water if it is fully saturated with oxygen. The normal saturation DO value for fresh water varies between 7.6mg/l to 14.6mg/l for temperature varying between 0 to 30^{0}C) at the water temperature.

Oxygen deficit (D) = Saturation DO - Actual DO

In order to maintain clean condition in a river stream, the oxygen deficit can be found out by knowing the rates of deoxygenation & reoxygenation.

Curve I - Deoxygenation curve or curve showing depletion of DO in the absence of aeration.

Curve II - Reoxygenation curve (DO content supplied)

Curve III - Oxygen sag curve or oxygen deficit curve (Actual DO Content or oxygen is balance)

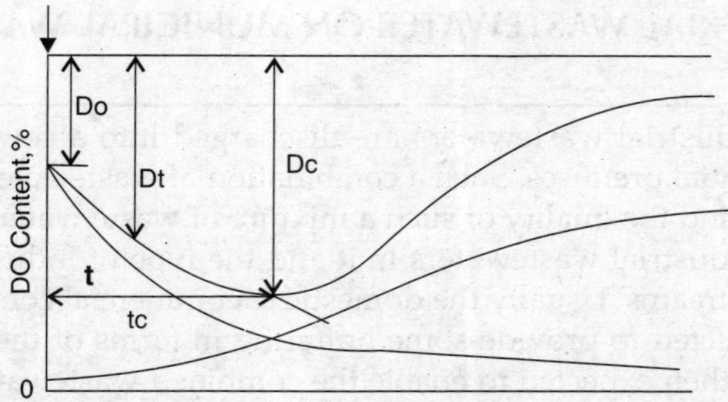

Fig.6.1: Deoxygenation, Reoxygenation & Oxygen sag(deficit) curve

Deoxygenation curve

In a polluted stream DO content goes on reducing due to decomposition of organic matter. The rate of deoxygenation depends upon the amount of the organic matter remaining to be oxidized at a given time and as well as on the temperature of the reaction. Hence at a given temperature , the curve showing depletion of DO with time *i.e.* deoxygenation curve is similar to the 1_{st} stage BOD curve.

Reoxygenation curve

In order to counter balance the consumption of DO due to deoxygenation atmosphere supplies oxygen to water and the process is called reoxygenation. The rate at which the oxygen is supplied by the atmosphere to the polluted water depends upon:

(a) The depth of the receiving water (rate is more in a shallow depth).

(b) The condition of the body of the water (rate is more in a running stream than in a quiescent pond).

(c) The saturation deficit or the oxygen deficit (ie. the deficit of DO below the saturation value).

(d) The temperature of water.

Depending upon these factors, the rate of reoxygenation can also be expressed mathematically and plotted in the form of the curve called reoxygenation curve.

Oxygen Deficit curve (oxygen sag curve)

In a running polluted stream exposed to the atmosphere, the deoxygenation as well as reoxygenation go hand in hand. If the deoxygenation is more rapid than the reoxygenation, as oxygen deficit results.

The amount of resultant oxygen deficit can be obtained by adding the deoxygenation & reoxygenation curves. The resultant curve so obtained is called the oxygen sag curve or the oxygen deficit curve. From this curve, the oxygen deficit and oxygen balance in the stream after a certain elapse of time can be find out.

It can also be seen that when the deoxygenation rate exceeds the reoxygenation rate, the oxygen sag curve shows increase in deficit of oxygen , but when both the rates become equal , the critical point is reached and then finally when the rate of deoxygenation fall below of reoxygenation, the oxygen deficit goes on decreasing till it becomes zero.

IMPACT OF INDUSTRIAL WASTEWATER ON MUNICIPAL WASTE WATER TREATMENT PLANTS

On some occasions, industrial wastewaters are discharged into a sewerage system serving commercial and residential premises. Such a combination of wastewater streams is known as municipal wastewater and the quality of such a mixture of waste waters can vary depending on the proportion of industrial wastewaters in it and the type of industries contributing the industrial wastewater streams. Usually the domestic & commercial components in municipal wastewater can be expected to provide some buffering in terms of the characteristics of the combined flow. This is then expected to enable the combined wastewater to be treated easily compared to the treatment of the industrial wastewater on its own. However, even where the option of discharging into a sewerage system is available, some degree of pretreatment is frequently required at the factory before such discharge is permitted. Such pretreatment may include pH adjustment to 6–9 and BOD_5 reduction to 400 mg/l. This is to protect the receiving sewers from corrosion and also protect the performance of the receiving treatment plant from an organic substrate overload.

All treatment plants should be planned giving serious consideration for the reduction of waste volume & strength, through process of chemical substitution, chemical recovery & recycling of water. The pollutional load is dealt with operations like segregation, neutralization, equalization, chemical precipitation, chemical oxidation & biological oxidation. Several chemicals are used to reduce the BOD by chemical coagulation such as alum, ferric sulphate, ferrous sulphate & ferric chloride. Lime or H_2SO_4 is used to adjust pH in this process. The dye waste may be economically treated by biological methods prior to equalization, neutralization & chemical oxidation.

The industrial waste when discharged into sewer not only chokes the sewer due to the deposition of solids but also reduces the cross section of the sewer arising out of the lime encrustature.

Chromium compounds in excess of 10-20 mg/l disrupt the operation of the trickling filter. Sulphides are toxic to the micro organisms and are removed along with the sludge. The sludge is dried over sand drying beds and can be used as good manure. Chemical coagulation (Alum, ferric chloride, and ferrous sulphate) with or without prior neutralization followed by biological treatment is necessary for better quality of effluent.

Deleterious effects of industrial effluents

Industries need a wide variety of raw materials and chemicals which are later discharged as effluents. Acids, alkalis, toxic metals, pesticides and other poisonous substances such as cyanide, dyes, oils, detergents, resins, rubbers are a few to mention. Heated effluents that impart thermal loading on receiving waters and effluents containing radioactive materials are also of prime concern. Some of the effluents such as from tanning and meat packing may also contain pathogenic bacteria. The nature and extent of pollution depends on the materials present in the effluent and on the quantity discharged.

Effects of industrial wastewater on water courses

Color: The effluents contain dyes in higher concentrations which impart color to the receiving streams and they persist for longer distances. Photosynthesis of phytoplankton is affected by these colors.

pH value: The extreme alkalinity makes the receiving water unfit for any purpose. Further, it is deleterious to most of the aquatic life.

Suspended impurities: The colloidal and suspended impurities produce turbidity in the receiving waters. The turbidity and color along with the oil & scum create an unsightly appearance.

Depletion of oxygen: Natural substances such as starch and dextrin and inorganic substances such as sulfide and nitrite present in the effluent exert an immediate oxygen demand. The stream will then be devoid of oxygen and the aquatic life are affected adversely.

Toxic substances: Chromium, sulfide, chlorine and aniline dyes present in these wastes are directly toxic to fish and microbial organisms which carry out purification. Thus the self purification of the water body is affected.

Oils: Various oils (mineral) in the effluent interfere with the oxygenation of stream as they form a blanket on the surface and prevents the entry of oxygen at air/water interface.

Dissolved minerals: The mineral materials, mostly sodium salts increase the salinity of the water and consequently it becomes unfit for irrigation.

Effects of industrial wastewater on land

- The excess content of sodium (60%) and boron (2 mg/l) are deleterious to crops.
- The high sodium alkalinity combined with salinity impairs the growth of plants.
- Texture of the soil is affected by sodium and penetration of roots is prevented.
- Soil permeability is also affected by sodium and ultimately the soil will lose its productivity.
- Suspended and colloidal impurities clog the pores and form a mat on the surface of soil preventing the passage of air, water etc.

Impact of industrial wastewater on water bodies

The impacts of of industrial waste water on water bodies is as follows:

(a) **Physical effects**– These include impact on clarity of the water and interference to oxygen dissolution in it. Water clarity is affected by turbidity which may be caused by inorganic suspended Solids and/or organic particulates suspended in the water (Volatile Suspended Solids). The latter may undergo biodegradation and thereby also have oxidation effects. Turbidity reduces light penetration and this reduces photosynthesis while the resultant loss in clarity would adversely affect the food gathering capacity of aquatic animals because these may not be able to see their prey. Very fine particulates may also clog the gill surfaces of fishes and thereby affecting respiration & eventually killing them. Settleable particulates may accumulate on plant foliage and bed of the water body forming sludge layers which would eventually smother benthic organisms. As the sludge layers accumulate, they may eventually become sludge banks and if the material in these is organic then its decomposition would give rise to malodours.

In contrast to the settleable material, particulates lighter than water eventually float to the surface and form a scum layer. The latter also interferes with the passage of light and oxygen dissolution. Because of the former, these scum layers affect photosynthesis. Discharge limits on wastewater or treated wastewater discharges typically have a value for TSS such as 30- 50 mg/l. Many industrial wastewaters contain oil and grease (O&G).

While some of the latter may be organic in nature, there are many which are mineral oils. Notwithstanding their organic or mineral nature, both types cause interference at the air-water interface and inhibit the transfer of oxygen. Apart from their interference to the transfer of oxygen from atmosphere to water, the O&G (particularly the mineral oils) may also be inhibitory. Unlike domestic sewage, industrial discharges can have temperatures substantially above ambient temperatures. These raise the temperatures of the receiving water and reduce the solubility of oxygen. Apart from this, rapid changes in temperature may result in thermal shock and this may be lethal to the more sensitive species. Heat, however, does not always have a negative impact on organisms as it may positively affect growth rates although there are limits here too , since the condition may favor certain species within the population more than others and over time biodiversity may be negatively affected.

(b) Oxidation and residual dissolved oxygen — Water bodies have the capacity to oxygenate themselves through dissolution of oxygen from the atmosphere and photosynthetic activity by aquatic plants. Of the latter, algae often plays an important role. However, there is a finite capacity to this re-oxygenation and if oxygen depletion, as a result of biological or chemical processes induced by the presence of organic or inorganic substances which exert an oxygen demand (i.e. as indicated by the BOD or COD), exceeded this capacity then the dissolved oxygen (DO) levels would decline. The latter may eventually decline to such an extent that septic conditions occur. A manifestation of such conditions would be the presence of malodours released by facultative and anaerobic organisms. An example of this is the reduction of substances with combined oxygen such as sulphates by facultative bacteria and resulting in the release of hydrogen sulphide. The depletion of free oxygen would affect the survival of aerobic organisms. DO levels do not, however, need to drop to zero before adverse impacts are felt. A decline to 3–4 mg/l, which still means the water contains substantial quantities of oxygen, may already adversely affect higher organisms like some species of fish. If inhibitory substances are also present, then the DO level at which adverse effects may be felt can be even higher than before. The case of elevated water temperatures due to warm discharges is somewhat different. The elevated temperatures can affect metabolic rates positively (possibly two fold for each 10^0 C rise in temperature) but elevated temperatures also reduce the solubility of oxygen in water. This would mean increasing demand for oxygen while its availability declines. Because of the impact of DO levels on aquatic life, much importance has been placed on determining the BOD value of a discharge. Typical BOD_5 limits set values are 20 and 50 mg/l.

(c) Inhibition or toxicity and persistence — These effects may be caused by organic or inorganic substances and can be acute or chronic. Examples of these include the pesticides and heavy metals. Many industrial wastewaters do contain such potentially inhibitory or toxic substances. The presence of such substances in an ecosystem may bias a population towards members of the community which are more tolerant to the substances while eliminating those which are less tolerant and resulting in a loss of biodiversity. For similar reasons, an awareness of the impact such substances have on biological systems is not only relevant in terms of protection of the environment but is of no less importance in terms of their impact on the biological systems used to treat industrial wastewaters. Even successful treatment of such a wastewater may not necessarily mean that the potability of water in a receiving waterbody would not be affected. For example small quantities of residual phenol in water can react with chlorine during the potable water treatment process giving rise to chlorophenols which can cause objectionable tastes & odors in the treated water. Apart from the organic pollutants which are potentially inhibitory or toxic, there are those which are resistant to biological degradation. Such persistent compounds can be bioaccumulated in organisms resulting in concentrations in tissues being significantly higher than concentrations in the environment and thereby making these organisms unsuitable as prey/food for organisms (including Man) higher up the food chain. While some organic compounds may be persistent, metals are practically non-degradable in the environment.

(d) Eutrophication– The discharge of nitrogenous and phosphorous compounds into receiving waterbodies may alter their fertility. Enhanced fertility can lead to excessive plant growth. The latter may include algal growth. The subsequent impact of such growth on a waterbody can include increased turbidity, oxygen depletion, and toxicity issues. Algal growth in unpolluted water bodies is usually limited because the water is nutrient limiting. While nutrients would include marco-nutrients like nitrogen, phosphorous, and carbon, and micro-nutrients like cobalt, manganese, calcium, potassium, magnesium, copper, & iron which are required only in very small quantities, the focus in concerns over eutrophication would be on phosphorous and nitrogen as quantities of the other nutrients in the natural environment are often inherently adequate. In freshwaters ,the limiting nutrient is usually phosphorous while in estuarine and marine waters it would be nitrogen. Treatment of industrial wastewater (or domestic sewage for that matter) can then target the removal of either phosphorous or nitrogen, depending on the receiving waterbody, to ensure that the nutrient limiting condition is maintained. When the nutrient limiting condition is no longer present in the water body, and when other conditions such as ambient temperature are appropriate, excessive algal growth or algal blooms (e.g. the red tide) may occur. Apart from aesthetic issues, such algal blooms may affect the productivity of the fisheries in the locale. It should be noted that not all industrial wastewaters contain excessive quantities of nutrients, macro and micro. This deficiency, if there is, results in process instability and/or the proliferation of inappropriate microbial species during biological treatment of the wastewaters.

Bulking sludge is a manifestation of such an occurrence. To address this deficiency, nutrients supplementation is required. The quantities used should be carefully regulated so that an excessive nutrients condition is not inadvertently created and these excess nutrients subsequently discharged with the treated effluent. In terms of BOD:N:P, the optimal ratio for biotreatment is often taken as 100:5:1 while the minimum acceptable condition can be 150:5:1.

(e) Pathogenic effects– Pathogens are disease-causing organisms and an infection occurs when these organisms gain entry into a host (e.g. man or an animal) and multiply therein. These pathogens include bacteria, viruses, protozoa, and helminthes. While domestic and medical related wastewaters may typically be linked to such micro-organisms (and especially the bacteria and viruses), industrial wastewaters are not typically associated with this category of effects. The exception to this is wastewaters associated with the sectors in the agro-industry dealing with animals. The concern here would be the presence of such organisms in the wastewater which is discharged into a receiving water body and diseases, if any, are then transmitted through the water. While many of these organisms can be satisfactorily addressed with adequate disinfection of the treated effluent and raw potable water supplies during the water treatment process, there are those which cannot be dealt with so easily. Two examples of such organisms, Cryptosporidum and Giardia, belong to the protozoa family. The difficulty is that the infected host does not necessarily shed the organism but is likely also to shed its eggs oroocysts. The latter can unfortunately be resistant to the usual disinfection processes.

Pre treatment of industrial waste water

Following pre treatment are applied to industrial wastewater prior to processing in sewage treatment plant:

(a) Volume reduction
(b) Strength reduction
(c) Neutralization
(d) Equalization

a. Volume reduction

In general the first step in minimizing the effects of industrial wastes on relieving streams & treatment plants is to reduce the volume of such wastes. This may be accomplished by:

(i) Classification of wastes.
(ii) Conservation of wastewater.
(iii) Change in production to decrease wastes.
(iv) Re using both industrial & municipal effluent as raw water supplies.

Classification of wastes

If wastes are classified so that manufacturing process waters are separated from cooling waters, the volume of water requiring intensive treatment may be reduced considerably. Sometimes it is possible to classify & separate the process waters themselves, so that only the most polluted once are treated & the relatively uncontaminated are discharged without treatment.

The 3 main classes of wastes are:

(a) Wastes from manufacturing process—These include waters used in forming paper, discharge from plating solutions in metal fabrication, those discharge from washing of milk cans in dairy plants etc.

(b) Waters used as cooling agents in industrial process—The volume of these wastes varies from one industry to another. Although cooling water can be become contaminated by small leaks, corrosion products or the effect of heat, these wastes contain little , if any organic matter & are classed as non pollution.

(c) Wastes from sanitary uses—The volume depends on many factors including size of the plant, amount of waste producing materials washed from the floors & the degree of cleanliness required & number of workers in the unit.

Unfortunately, in most old plants process, cooling & sanitary waste water are mixed in one pipeline & many industries are paying little attention to segregating wastes to avoid stream pollution.

b. Strength Reduction

Waste Strength reduction is the second major objective for an industrial plant concerned with waste treatment. The strength of wastes may be reduced by:

- Process Changes
- Equipment Modifications
- Segregation of Wastes
- Equalization of Wastes
- By-Product Recovery
- Proportioning of Wastes
- Monitoring Waste Streams

Process Changes

In reducing the strength of wastes through process changes, the sanitary engineer is concerned with wastes that are most troublesome from a pollution standpoint.

Equipment Modification

Changes in equipment can effect a reduction in the strength of the waste, usually by reducing the amounts of contaminants entering the waste stream. An outstanding example of waste strength reduction occurred in the dairy industry. The new cans were constructed with smooth necks so that they could be drained faster and more completely. This prevented a large amount of milk waste from entering streams and sewage plants.

Segregation of Wastes

Segregation of Wastes reduces the strength and/or the difficulty of treating the final waste from industrial plant. It usually results in two wastes: one strong and small in volume and the other weaker with almost the same volume as the original unsegregated waste. The small volume strong waste can then be handled with methods specific to the problem it presents. In terms of volume reduction alone, segregation of cooling waters and storm waters from process waste will mean a saving in the size of the final treatment plant.

Equalization of Wastes

Plants, which have many products, from a diversity of processes, prefer to equalize their wastes. This requires holding wastes for a certain period of time, depending on the time taken for the repetitive process in the plant. For example, if a manufactured item requires a series of operations that take eight hours, the plant needs an equalization basin designed to hold the wastes for that eight hours period. The effluent from an equalization basin is much more consistent in its characteristics than each separate influent to that same basin.

Stabilization of pH & B.O.D and settling of Solids & Heavy Metals are among the objectives of equalization. Stable effluents are treated more easily and efficiently, than unstable ones by industrial & municipal treatment plants.

By-Product Recovery

All wastes contain by products, the exhausted materials used in the process. Since some wastes are very difficult to treat at low cost, it is advisable to consider the possibility of building a recovery plant which will produce a marketable By-Product and at the same time solve a troublesome wastes problem.

Proportioning Wastes

By Proportioning its discharge of concentrated wastes into the main sewer , a plant can reduce the strength of its total waste to the point, where it will need a minimum of final treatment or will cause the least damage to the stream or treatment plant.

It may prove less costly to proportion one small but concentrated waste into the main flow. According to the rate of the main flow, than to equalize the entire waste of the plant in order to reduce the strength.

Monitoring Waste Streams

Accidental spills are often the sole cause of stream pollution or malfunctioning of treatment plants and these can be controlled & often eliminated completely, if all significant sources of wastes are monitored.

c. Neutralization

Excessively acidic or alkaline wastes should not be discharged without treatment into a receiving stream. A stream is adversely affected by low or high pH values. This adverse condition is even more critical when sudden sludge of acids or alkalis are imposed upon the stream.

Acceptable Methods of Neutralization are:

- Mixing wastes so that the net effect is a neutral pH.
- Passing acid wastes through beds of limestone.
- Mixing acid wastes with lime slurries.
- Adding the proper proportions of concentrated solutions of caustic soda (NaOH) or soda ash (Na_2CO_3) to acid wastes.
- Adding compressed CO_2 to alkaline wastes.
- Adding sulfuric acid to alkaline wastes.

The material and method used should be selected on the basis of the overall cost, since material costs vary widely and equipment for utilizing various agents will differ with the method selected. The volume, kind and quality of acid or alkali to be neutralized are also factors in deciding which neutralizing agent to use.

d. Equalization

Equalization is a method of retaining wastes in a basin so that the effluent discharged is fairly uniform in its characteristics (pH, color, turbidity, alkalinity, B.O.D etc). A secondary but significant effect is that of lowering the concentration of effluent contaminants. A retention pond serves to level out the effects of peak loadings on the plant while substantially lowering the B.O.D and suspended solids load to the aeration unit.

Air is sometimes injected into these basins to provide:

- Better mixing.
- Chemical oxidation of reduced compounds.
- Some degree of biological oxidation.
- Agitation to prevent suspended solids from settling.

The size and shape of the basins vary with the quantity of waste and the pattern of its discharge from the industry. The capacity should be adequate to hold and render homogeneous, all the wastes from the plant. Almost all industrial plants operate on a cycle basis; thus if the cycle of operations is repeated for every two hours, an equalization tank which can hold a two hour flow will usually be sufficient.

The mere holding of waste, however is not sufficient to equalizing it. Each unit volume of waste discharged must be adequately mixed with other unit volumes of waste discharged many hours previously.

This mixing may be brought about in the following ways:

- Proper distribution and baffling
- Mechanical agitation
- Aeration and
- Combination of all three above.

METHODS OF NEUTRALIZATION OF INDUSTRIAL WASTE WATER

Excessively acidic or alkaline wastes should not be discharged without treatment into a receiving stream. A stream is adversely affected by low or high pH values. This adverse condition is even more critical when sudden sludge of acids or alkalis are imposed upon the stream.

Acceptable Methods of Neutralization are:

- Mixing wastes so that the net effect is a neutral pH.
- Passing acid wastes through beds of limestone.
- Mixing acid wastes with lime slurries.
- Adding the proper proportions of concentrated solutions of caustic soda (NaOH) or soda ash (Na_2CO_3) to acid wastes.
- Adding compressed CO_2 to alkaline wastes.
- Adding sulfuric acid to alkaline wastes.

The material and method used should be selected on the basis of the overall cost, since material costs vary widely and equipment for utilizing various agents will differ with the method selected. The volume, kind and quality of acid or alkali to be neutralized are also factors in deciding which neutralizing agent to use.

Importance of equalization and proportioning in industrial wastewater treatment

Equalization

Equalization is a method of retaining wastes in a basin so that the effluent discharged is fairly uniform in its characteristics (pH, color, turbidity, alkalinity, B.O.D etc). A secondary but significant effect is that of lowering the concentration of effluent contaminants. A retention pond serves to level out the effects of peak loadings on the plant while substantially lowering the B.O.D and suspended solids load to the aeration unit.

The size and shape of the basins vary with the quantity of waste and the pattern of its discharge from the industry. The capacity should be adequate to hold and render homogeneous, all the wastes from the plant. Almost all industrial plants operate on a cycle basis; thus if the cycle of operations is repeated for every two hours, an equalization tank which can hold a two hour flow will usually be sufficient.

The mere holding of waste, however is not sufficient to equalizing it. Each unit volume of waste discharged must be adequately mixed with other unit volumes of waste discharged many hours previously.

Proportioning

Proportioning means the discharge of industrial wastes in proportion to the flow of municipal sewage in the sewers or to the stream flow in the receiving river. In most cases it is possible to combine equalization and proportioning in the same basin. The effluent from the equalization basin is metered into the sewer or stream according to a predetermined schedule. The objective of proportioning in sewers is to keep constant the percentage of industrial wastes to domestic sewage flow entering the municipal sewage plant. This procedure has several purposes:

- To protect municipal sewage treatment using chemicals from being impaired by a sudden overdose of chemicals contained in the industrial waste.
- To protect biological treatment devices from strong loads of industrial wastes which may inactivate the bacteria.
- To minimize fluctuations of sanitary standards in the treated effluent
- The rate of flow of industrial waste varies from instant to instant, as does the flow of domestic sewage system. Therefore the industrial waste must be equalized and retained, then proportioned to the sewer or stream according to the volume of domestic sewage or stream.

The size and shape of the basins vary with the quantity of waste and the pattern of its discharge from the industry. The capacity should be adequate to hold and render homogeneous, all the wastes from the plant. Almost all industrial plants operate on a cycle basis; thus if the cycle of operations is repeated for every two hours, an equalization tank which can hold a two hour flow will usually be sufficient.

The mere holding of waste, however is not sufficient to equalizing it. Each unit volume of waste discharged must be adequately mixed with other unit volumes of waste discharged many hours previously.

Techniques applied in removal of suspended solids from industrial wastewater

There are three methods for the removal of suspended solids, which are as follows:

- Screening.
- Flotation
- Sedimentation

Screening

Screening of industrial wastes is generally practiced on waste containing larger suspended solids of variable sizes. For e.g from canneries, pulp and paper mills, or poultry processing plants. It is an economical and effective means of rapid separation of these larger suspended solids from the remaining waste material. In many cases screening alone will reduce the suspended solids to a low enough concentration to be acceptable for discharge into a municipal sewer or a nearby stream.

Screens are available in sizes ranging from coarse (10 or 20 mesh) to fine (120 to 320mesh). In the industry, generally two types of screens are used:

(i) North Screen
(ii) Sweco Screen

North screen

They are generally rotary, self-cleaning gravity-type, units. It has been used successfully in treating wastes from meat-packing, grain-washing, tanning, making, woolen and sea food plants.

Sweco screen

They are circular, vibratory screens. They have been quite effectively used for screening wastes from food-packing processes such as meat and poultry packing or fruit and vegetable canning.

Floatation

The floatation process consists of creation of air bubbles in the waste body or tank by introducing air to the system. These air bubbles increase the buoyancy of the finer particles which could be easily lifted by skimming (i.e. the process of removing floating grease or scum from the surface of the effluent in a tank).

Advantages of floatation as wastewater process are:

- Grease, and light solids rising to the tops and grit and heavy solids settling to the bottom are all removed in one unit.
- Odor nuisance are minimized, because of short detention period.
- High overflow rates and short detention periods means smaller tank sizes, resulting in decreased space requirements and possible saving in construction costs

Disadvantages of floatation are:

- The additional equipment required results in higher operating costs.
- Floatation units generally do not give as effective treatment as gravity settling units, although the efficiency varies with the waste.
- More skilled maintenance is required for a flotation unit than for a gravity settling unit.

Sedimentation

Sedimentation processes are very effective in removing suspended solids in industrial wastewater. The efficiency of sedimentation tank depends, in general on the following factors:

- Detention periods.
- Wastewater characteristics
- Tank design.
- Particle size.
- Velocity & density of particles.

The clarifiers/tanks either rectangular or circular in shape are most commonly used in the application of sedimentation in wastewater treatment. The design of the clarifiers are based on several factors, which are as follows:

- Influent TSS concentration.
- Effluent TSS concentration.
- Surface loading.
- Detention time.
- Sludge generation.

Removal of dissolved inorganic solids from industrial wastewater by:

(a) **Dialysis** (b) **Reverse osmosis** (c) **Ion exchange**

a. **Dialysis** -It is the separation of solute by means of their unequal diffusion through membranes. It is most useful in recovering pure solutions for reuse in manufacturing processes for e.g. caustic soda in textile industry. Dialysis work on the simple principle of passing a concentrated, impure caustic solution upward, counter current to a downstream water supply, from which it is separated by a semi-permeable membrane. The caustic soda permeates the membranes and goes into water more rapidly than the other impurities contained in the water The quantity of NaOH diffusing through the membrane diaphragm depends upon the time, the area of the dialyzing surface, the mean concentration difference and the temperature. Dialysis is an operation requiring very little operator attention and although its main role is to conserve raw materials and to reduce plant waste, at the same time it aids in waste treatment. With the introduction of acid resistant membranes, dialysis has been used successfully in the recovery of sulphuric acid in the copper, stainless steel and other industries. Some operation can recover as much as the 70 to 75% of the acid, but recovery as little as 20% may be justifying the process. In dialysis, the driving force of separation is natural diffusion because of concentration gradient. Electro dialysis is another form of dialysis in which the natural driving force is enhanced by the application of electrical energy. Electro-dialysis can achieve as much as 44 % reduction in concentration of dissolved solids in industrial effluents.

b. **Reverse osmosis** -It is process for separating relatively pure water or some other solvent from a less pure solution. The solution is passed over the surface of a specific semi-permeable membrane at a pressure in excess of the effective osmotic pressure of the feed solution. The permeating liquid is collected as the product and the concentrated feed solution is generally discarded. The membrane must be (1) highly permeable to water (2) highly impermeable to solutes (3) capable of withstanding the applied pressure without failure, (4) chemically inert, mechanically strong. Although several types of membranes have been developed, two types of membranes are generally used in commercial equipment.(1). The first is a symmetric of skinned cellulose acetate membranes made in flat or Tubular forms.(2). The second type of membrane is an aromatic polyamide or polyamide hydrazide. Reverse osmosis is an advanced unit operation in water treatment. Reverse osmosis membranes are capable of removing at least 90 percent of the dissolved solids in water as well as organics, bacteria and others impurities. Osmosis can be defined as the spontaneous passage of a liquid from dilute to a more concentrated solution across an ideal semipermeable membranes which allows the passage of the solvent (water) but not the dissolved solids (solutes). The transport of the water from one side of the membrane to the other continues until the pressure (p) is large enough to prevent any net transfer of the solvent (water) to the more concentrated solution. At equilibrium, the quantity of water passing in the either direction is equal, and pressure (p) is then defined as the osmotic pressure of the solution having that particular concentration and make up of dissolved solids. If a piston is placed on the more concentrated solution side of a semipermeable membrane and a pressure (P) is applied to the solution, the following conditions can be realized:(i) P is less them the osmotic pressure of the solution and solvent flows (in the same direction) spontaneously toward the more concentrated solution. (ii) P equals the osmotic pressure of the solution and solvent flows at the same rate in both directions that is no net change in water levels. (iii) P is greater than the osmotic pressure of the solution and solvent flows from the more concentrated solution to the pure solvent side of the membrane leaving a higher concentration behind. This condition represents the phenomenon of reverse osmosis. Substantial energy is needed to constantly pressurized the incoming salty water is one of the drawback of the reverse osmosis.

c. **Ion exchange-** Ion exchange is basically a process of exchanging certain undesirable cations and anions of the wastewater for sodium, hydrogen or other ions in resinous materials. The resins, both natural and artificial are commonly referred to as Zeolites. When the ability of the zeolite bed to produce soft water is exhausted, the softener is temporarily cut out of service. It is then backwash to clean and hydraulically regrade the bed, regenerated with a solution of common salt; which remove the calcium and magnesium in the form of their soluble chlorides and restores the zeolite to its original conditions, rinsed free of these and the excess salt, and finally returned to service.

Methods of sludge disposal -Both primary & secondary sewage treatments involve settling of particulate matter, and thus produce sludge. The concentration of solids in the primary sewage sludge is about 5%; the activated sludge contains about 1%; and the sludge from trickling filters has about 2% solids.

Thus the sludge is composed of almost entirely of water and volume reduction is the key to economic disposal. In addition to reducing its high water content, the sludge must be stabilized so that its biological activity and tendency towards putrefaction are reduced drastically.

The sludge is concentrated by gravity settling and floatation. After concentration , the sludge is subjected to anaerobic digestion in a digester in which the organic content of the sludge decomposes to give mainly methane and carbon dioxide and at the same time the bound water is released from the sludge.

The sludge is then conditioned to improve its dewatering characteristics by adding chemicals like iron salts and polyelectrolytes. These chemicals bind the sludge particles together and encourage the release of water. The sludge is then heated under pressure and after a period of time the gel structure of the sludge breaks down so that the water is released.

The thickened sludge is then dewatered for efficient handling and disposal. The dewatering is accomplished by mechanical methods, the most common being centrifugation and filtration.

The dewatered sludge is then subjected to oxidation to reduce the organic content, with the consequent destruction of bacteria and significant reduction in their volume.

Incineration and oxidation are the two common methods employed for oxidation. Several methods are employed for the ultimate disposal of sludge. The wet digested sludge may be sprayed on to a cropland where it functions as fertilized Dried sludge may be used a land fill or soil conditioner. Wet or partially dewatered sludge or ash from incineration may be transported from the shore to dumping grounds at sea. The potential drawback to the use of sewage sludge as fertilizer in agricultural fields is the presence of both organic and inorganic toxic compounds. The former compounds are oxidation-resistant organic substances, such as organochlorine species which become bound in the organic matrix of the sludge. The inorganic toxicants are represented by heavy metals, mainly arsenic, cadmium, lead, mercury and zinc. These metals can be taken up by crops and introduced into the food chains or leached to the ground water.

Drying beds / sand bed drying

For small industrial waste treatment plants, sludge can be dewatered on open or covered sand beds. Drying of the sludge occurs by percolation and evaporation. The proportion of the water removed by percolation may vary from 20- 55%. Depending on the initial solids contents of the sludge & on the characteristics of the solids. The design and use of drying beds are affected by climatic conditions (rainfall & evaporation)

Sludge drying beds usually consists of 10-23 cms of sand, 20-46 cms of gravel or stone. The sand has an effective size of 0.3 to 1.2 mm & a uniformity coefficient < 5.0. Gravel is graded from 0.32 to 2.54 cm. The beds are provided with under drains spaced from 2.7 to 6.1 m apart. The under drain piping having a mini dia of 10 cm & minimum slope of 1% The filtration is returned to the treatment plants. Wet sludge is usually applied to the drying beds at depths of 20 -30 cms. Removal; of the dried sludge varies with both individual judgment and final disposal means many cases, the bed turn over can be substantially increased by the use of chemicals. Sludge treatment can be reduce the sludge drying time by 50%.The use of polymers can increase the rate of bed dewatering and also increase the depth of application.

Drying and incineration

A large volume of sludge can be reduced to a small volume of ash, which is free from organic matter & therefore easily disposable, by combination of heat drying & incineration. Incineration involves drying & combustion. Various types of incineration units are available to accomplish these reactions in single or combined units. In the incineration process, a sludge temperature is raised to 100 °C, at which point moisture is evaporated from the sludge. The water vapour & air temperature are increased to ignition point. Some excess air is required for complete combustion & of the sludge. Self sustaining combustion is often possible with dewatered waste sludges ones the burning auxiliary fuel raises the incineration temperature to ignition point. The primary end products of combustion are CO_2, SO_2 & ash.

Incineration can be accomplished in multiple hearth furnaces in which the sludge passes vertically through a series of hearths. In fluidized bed sludge particle are fed into a bed of sand fluidized by upward moving air. A temperature of 760-815 °C is maintained in the beds, resulting in rapid drying and burning of sludge. Ash is removed from the bed by the upward flowing combustion gases.

Sanitary landfill

Sanitary landfill is used to burry garbage, refuse & sludge in a planned and methodical manner. It is a simple, effective & inexpensive method for disposing of dry matter such as refuse, but sludge is usually too liquid for this procedure. However, vacuum filter or sand bed dried sludge can be disposed off in this manner. The area proposed for the sanitary land fill should be easily accessible, yet removed from sources of water supply & recreational areas at the same time on the land which is not too costly. The suitability of soil & possible future use of the property are also imp. Consideration sanitary landfills should be located above the ground water level and 152 m away from water sources, particularly when the soil is sandy, gravely or of limestone. At the end of each days dumping, the sludge should be covered and compacted by a bulldozer or tractor.

Sludge barging

Sludge barging is a means of final disposal of sludge to a deep body of water usually an ocean. There is little theory involved in this method of treatment. Raw precipitated, digested or filtered sludge solids are pumped into a waiting ship barge. When the barge is fully loaded it transports the sludge to a suitability site far from the shore, where it is discharged usually by being pumped out deep under the surface of the water. The sludge should be somewhat concentrated before being loaded on the barge (to conserve space) but not too much that pumping difficulties arise.

Pumping raw or digested sludge directed into the ocean & disposing it by dilution technique is usually practiced in coastal cities. These sludges however, should not be contaminated by oil & grease, odor high coliform counts, toxic materials discolouration or be likely to create sludge banks or deposits.

Sludge thickening methods

Thickening is a procedure used to increase the solids content of sludge by removing a portion of the liquid fraction. To illustrate, if waste activated sludge, which is typically pumped from secondary tanks with a content of 0.8% solids, can be thickened to a content of 4% solids, then a five fold decrease in sludge volume is achieved. Thickening is generally accomplished by physical means including co-settling, gravity settling, flotation, centrifugation, gravity belt, and rotary drum. The volume reduction obtained by sludge concentration is beneficial to subsequent treatment processes, such as digestion, dewatering, drying and combustion from the following standpoints:

- Capacity of tanks and equipment required.
- Quantity of chemicals required for sludge conditioning.
- Amount of heat required by digesters and amount of auxiliary fuel required for heat drying or incineration, or both.

Most common techniques of thickening in sludge processing are:

- Gravitational settling
- Dissolved Air Flotation(DAF)
- Basket centrifuge

- Solid-bowl centrifuge
- Gravity belt filter
- Rotary drum

Thickening is the process by which bio solids are condensed to produce a concentrated solids product and a relatively solids free supernatant. Thickening wastewater solids reduces the volume of residuals, improves operation, and reduces costs for subsequent storage, processing, transfer, end use, or disposal. For example, thickening liquid-solids (slurry) from 3 to 6 % will reduce the volume by 50 %.

There are several different methods for thickening bio solids, including dissolved air floatation (DAF), centrifugal thickening, gravity belt thickening, and gravity thickening. Gravity thickening uses the natural tendency of higher density solids to settle out of liquid to concentrate the solids.

Gravity thickeners consist of a circular tank (usually with a conical bottom) that is fitted with collectors or scrapers at the bottom. Primary and/or secondary solids are fed into the tank through a center well, which releases the solids at a low velocity near the surface of the tank. The solids settle to the bottom of the tank by gravity, and the scrapers slowly move the settled, thickened solids to a discharge pipe at the bottom of the tank. A v notch weir located at the top of the tank allows the supernatant to return to a clarifier. In addition, many systems also use a skimmer to collect and remove any floatables and grease that have accumulated at the top of the tank.

The biosolids concentration and thickening that occurs in the tank is achieved through three different settling processes, which include gravity settling, hindered settling, and compaction settling. Gravity settling occurs when solid particles travel downward due to their weight. Settlement continues as solids begin to concentrate near the bottom of the tank, but the settlement rate decreases as the solids concentrations increase. This is known as hindered settling.

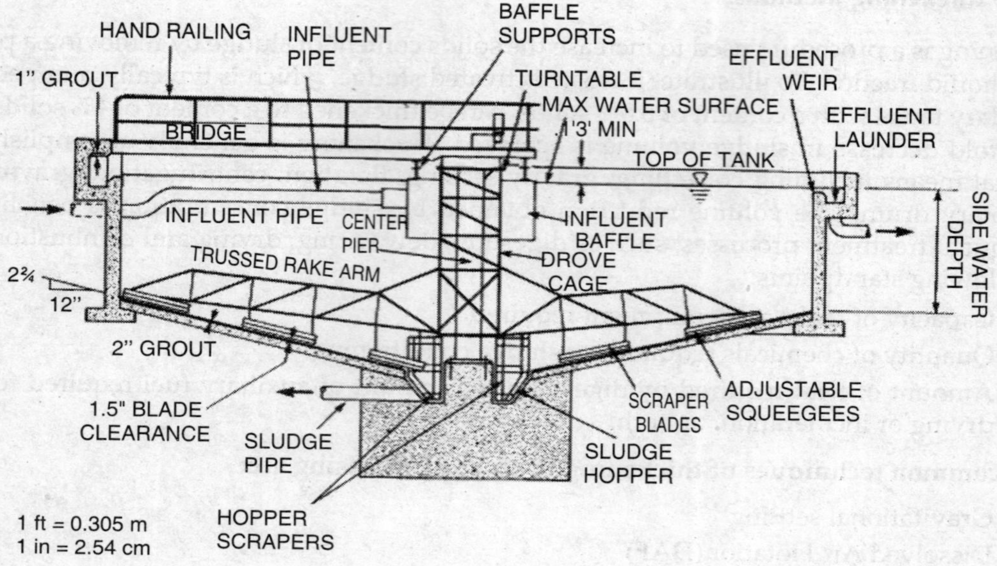

Fig.6.2: Gravity Thickener

Compaction settling occurs when bottom solids are further concentrated due to the pressure of solids on top of them.

Solids at the bottom of the tank can reach as high as 15 percent total solids (TS). A more typical result is 4 to 6 % TS. Liquid at the surface of the tank is nearly clear, with suspended solids concentrations as low as 200 mg/l.

Advantages and disadvantages of combined treatment of Industrial raw waste with the domestic waste

It is often possible & advisable for an industry to discharge its waste water directly into a municipal sewage treatment plant, where a certain portion of the pollution can be removed. A municipal sewage plant, if designed & operated properly can be handle almost any type & quantity of industrial waste. Hence one possibility that should be seriously considered is the cooperation of industry & municipality in the joint construction and operation of a municipal waste water treatment plant.

There are many advantages to be gained from such a joint venture:

- Here the responsibility is placed with one owner, while at the same time, the cooperative spirit between industry & municipality increases, particularly if the division of costs is mutually satisfactory.
- Only one chief operator is required, whose sole obligation is the management of the treatment plant i.e. he is not burden by the miscellaneous duties often given to the industrial employee in charge of waste disposal & the chances of mismanagement and neglect which may result if industrial production men operate waste treatment plants, are eliminated.
- Since the operator of such a large treatment plant usually receives higher pay than separate domestic plant operators, better trained people are available.
- Even if identical equipment is required construction costs are less for a single plant than for two or more. Furthermore, municipalities can apply for state & or federal aid for plant construction, which private industry is not eligible to receive.
- The land required for plant construction & for disposal of waste products is obtained more easily by the municipality.
- Operating costs are lower, since more waste is treated at a lower rate per unit of volume.
- Possible cost advantages resulting from lower municipal financing cost & federal grants.
- Some wastes may add valuable nutrient for biological activity to counter act other industrial wastes that are nutrient deficient. Thus bacteria in the sewage are added to organic industrial wastes as seeding material. These micro organisms are vital to biological treatment. Also, acids from one industry may help to neutralization alkaline wastes from another industry.
- The treatment of all waste water generated in the community in a municipal plant, enables the municipality to assure a uniform level of treatment to all the users of the river & even to increase the degree of treatment given to all waste water to the maximum level obtainable with technological advance.

Effluent standards and Stream standards

EFFLUENT STANDARDS

- Effluent standards pertain to the quality of the discharge water itself.
- They are based on economics than on absolute protection of the stream.
- Easy to control.
- Detailed stream analysis are not required.
- They do not establish an overall level of pollutant loading for a given water body.
- Ratio of wastewater to stream flow are not considered.
- Treatment is obligatory irrespective of the size of industry.
- For effective protection of an overloaded stream, the effluent standards are required to be upgraded.
- Large industries have an edge over small industry.

STREAM STANDARDS

Stream standards refer to the quality of the receiving water downstream from the origin of the wastewater discharge.

- They are based on establishing classification of quality for a stream.
- The quality of the receiving water is regulated to maintain established stream classification.
- Prevention of excessive pollution/ Loading is limited to what the stream can assimilate.
- No consideration of type and location of industry.
- Allows public to establish goals for present and future water quality.
- Confusion of zone of different classification.
- Controversy over proportion of stream to be reserved for future usage (municipal, industrial, agriculture etc.).
- Opposition from industry/ public to change the established classification.
- A detailed stream analysis is required to determine the level of wastewater treatment required to maintain the health of the ecosystem.
- Cost of treatment may affect the survival of industry.

Factors to be considered for stream sampling

Before embarking on designing of any industrial wastewater treatment facility a comprehensive stream sampling program should be conducted. This will in turn help considerably for efficient operation of facility.

The factors to be considered before starting a sampling program are:

- Overall objectives of the program
- Total number of samples
- Points of collection
- Method of collection
- Data to be obtained

- Frequency of sample collection
- Time of the year for sampling
- Statistical handling of data
- Care of samples, prior to analysis

Overall objectives of the program

There are various objectives of stream sampling. In one instance the engineer may be concerned with the effect of an upstream industry on the water quality downstream, of special interest in this case might be the color of the water in the receiving stream. In another instance, he may be attempting to ascertain the dissolved oxygen sag characteristics of the stream during the summer season only. In another case, he could concern that the stream characteristics comply with the standards stipulated by the state pollution control board. The importance of the 8 factors will depend to some degree on the overall objectives of the sampling program.

Total number of samples

The no. of samples required depends on the objectives of the programme & the amount of time alloted for completion of the same. The use of the few locations & more samples to define the results is usually much more reliable than using many stations with only a few samples from each. Also, samples are frequently taken over a long time interval, during which the condition of the water course is subject to variation. In many instances, an attempt to test all condition by infrequent, random sampling produces no definite pattern & in fact may be misleading. It may be better to concentrate on well defined frequent & intensive sampling

A well planned survey with a specific objective will require a minimum samples. If for ex, someone wishes to determine the river characteristics, especially the DO profile during low flow periods, 2 or 3 samples collected at the proper time will suffice. These samples must be collected, however during extreme drought periods, if factors influencing the character of the stream to be clearly established

However, since the industrial processing is so varied & usually unpredictable, that many samples of the receiving water under all conditions must be collected in order to evaluate the effect of waste upon the stream.

Points of sample collection

Sampling points should be selected with great care & due consideration should be given to the sources of pollution, dilution by branch streams, changes in surrounding topography & slope of the river significant river side features also influent the choice of sampling points such as a municipal water intake, a park, an industrial area, a good fishing spot, a hotel or a camping site would each have a definite bearing on the usage of the stream. Since the acceptable pollution limits for water vary according to usage, samples should to collected & the record made of the condition of the stream just above & just below all such points of stream use are changed.

A minimum of four stream stations is recommended:

(a) An upstream site where the water is uncontaminated.

(b) Just below the source of pollution or dilution.

(c) Where the stream is in worst condition due to the source of pollution.

(d) A point midway between bottom of oxygen sag & the recovery of oxygen level.

Method of sample collection

Samples should be taken from 0.6 ft depth in streams less than 2 ft. The volume of samples depends on the number and type of analysis to be carried out. A standard type dissolved oxygen sampler is recommended for collecting most samples. Glass bottles with glass caps or polyethylene containers are most widely used. Any doubt about the cleanliness of the sample bottle can usually be dispelled by rinsing it first with some of the actual stream water, but special bottles are required for bacteriological samples.

Data to be collected

The scientific collection of data for stream analysis may be divided into three major categories i.e., hydrologic factors, sources of pollution and water course sampling. The type of data to be obtained depends on the objectives of sampling programme and the amount of time and money available for the investigation. For example, if the oxygen resources of a section of a stream is required then, DO, water temperature and stream flow should be measured. If the sampling programme is of general nature, the stream analyst should undertake as many chemical, physical and biological tests as possible. For design of wastewater treatment plants, the rate of flow, temperature, BOD and DO can supply sufficient information. In addition, data on pH, color and turbidity may indicate the general physical condition of the stream. Biological analysis is required when the stream water is used for drinking, bathing or flushing. In this case, the coliform count is usually determined.

Frequency of sample collection

Samples should be collected as frequently as possible to provide a representative of total sample. The master sample should contain individual constituents of variation expected. For ex, if the pH is known to vary from 4 to 10, individual samples with pH values of 4,5,6,7,8,9 and 10 should appear in the composite sample at least once during each sampling period. If the situation requires instantaneous analysis, more individual samples with little or no composting are collected.

Time of the year for sampling

Time of the year is of at most importance when there is a deadline for producing results. In stream studies, dealing with industrial waste treatment is concerned primarily with the critical condition of pollution which generally exists when the environment is at its warmest & its stream flow slowest, & the man made pollution greatest. But many studies must be undertaken in the spring & autumn seasons due to stream conditions and man power problems. However, many times due to the urgency of the problem or unusual conditions of pollution, the investigations may have to be carried in any of the seasons. The objective of every stream analysis should however to be collect the data during critical stream conditions of temperature, flow and pollution load. It may sometime be necessary to project stream analysis to conditions which might obtain during future critical points. This is sometimes requiring even when a survey is made during the summer. Because the intensity of the problem varies to some degree from year to year & even from one period of the year to another.

Statistical handling of data

It is a well known fact that data can be manipulated to emphasize that aspect of the survey which the analyst fill the most important. For this the engineer must have a working knowledge of statistic and mathematics in order to convey this information in the best form to the layman.

The rate of flow during critical conditions is of importance to the industrial waste engineering. If he incorporates the minimum flow ever recorded in the stream, he may end up with unrealistic evaluation. On the other hand the use of the mean summer or low flow value can be dangerous, since lower than this often occurs. Therefore, some regulatory agencies make use of minimum of 7 days to review once in 10 years as the criteria for designing waste water treatment facilities.

SOURCES , CHARACTERISTICS & TREATMENT OF SUGAR MILL EFFLUENT

Sources and characteristics of wastes. Waste from the sugar mill house includes the water used as splashes to extract maximum amount of juice & those used to cool the roller bearings. As such the mill house waste contains high BOD due to the presence of sugar & oil from the machineries. The filter cloths used for filtering the juice needs occasional cleaning. The wash waster thus produced though small in volume, contains high BOD & suspended solids.

IGN SCHEMATIC BLOCK DIAGRAM OF SUGAR PROCESSING

CANE
CANE WEIGHING
BOILER
bagasse
MILLING STATION
JUICE SCALE
nitrate
raw juice
JUICE HEATER I (T- 70°C)
Lime milk
DEFEKATOR
JUICE HEATER II (T-102 - 105°c)
RVF
Sludge
JUICE CLARIFIER
Cake mud
CLARIFIER
B-sugar
EVAPORATOR Brix - 60 - 65
A-VACUUM PAN
CENTRIFUGAL
Molasse
SUGAR FROM CANE
Final Molasse
C-PAN
B-Molasse
B-LGF
C-sugar
B-PAN
RAW SUGAR
RS WEIGHING
MELTER (T-65º brix : 60)
A-sugar
sweet water
DEFEKATOR (pH-11)
CARBONATOR I (pH-10.5 : T-85ºC)
Cake
FILTER PRESS
CARBONATOR II (pH-8-8.5)
FILTERING
Clear liquor
WHITE VACUUM PAN
white molasses
HIGH GRADE FUGAL
WHITE SUGAR (FINAL PROOUCT)
Final molasse
C-VACUUM PAN
B-CENTRIFUGAL
C-sugar
B-MOLOSSE
C-VACUUM PAN
A-MOLOSSE
A-CENTRIFUGAL
A-VACUUM PAN
B-sugar
note:
main flow
by product flow

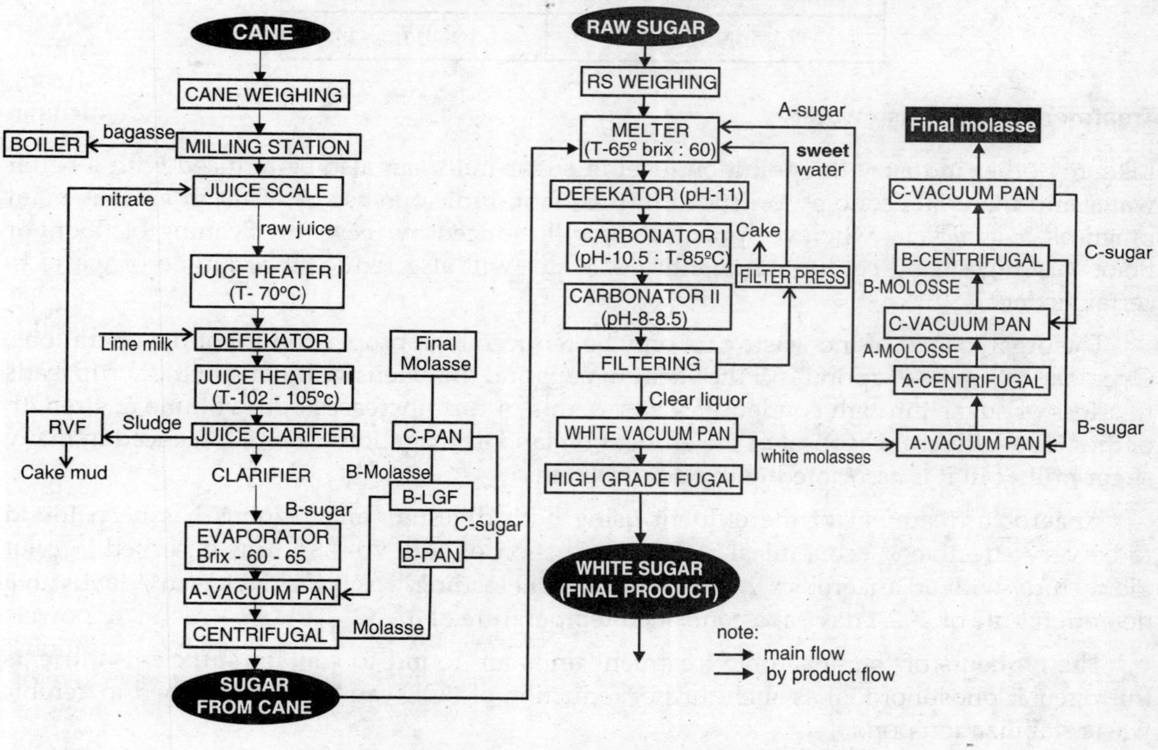

Fig.6.3: Schematic diagram of sugar processing

A large volume of water is required in the barometric condensers of the evaporators & vacuum pan. The water is usually partially or fully circulated after cooling through a spray pond. This cooling water gets polluted as it picks up some organic substances from the vapour of boiling syrup in evaporation & vacuum pans. This polluted water, instead of recirculated is discarded as excess condenser water. These discharges contribute substantially to the waste volume & modulated to BOD in many sugar mills.

Additional waste originates due to the leakages & spillages of juice, syrup & molasses in different sections. The periodical washings of the floor through small in volume have got very high BOD. The periodic blow off of the boilers produces another intermittence waste discharge. This waste is high in suspended solids ,low in BOD & usually alkaline.

The characteristic of waste water is tabulated below.

Characteristics	Value
pH	4.6-7.1
Total solids	870-3500 mg/l
BOD	300-200 mg/l
COD	600-4380 mg/l
Total suspended solids	220-800 mg/l
Total Nitrogen	10-40 mg/l

Treatment of the wastewater

Like any other industry the pollution level in sugar mills can also be reduced with a better water and material economy practiced in the plant. Judicious use of water in various plant practices & it recycle, wherever practicable, will reduced by recycling cleaning of floors or floor washings using controlled quantity of water will also reduce the volume of waste to certain extent.

The organic load of the waste cam only be reduced by a proper control of the operations. Over loading of the evaporates & the vacuum pans and the extensive boiling of the syrup leads to a loss of sugar through condenser water & this in turn increases both volume & strength of the waste effluent. Disposal of the effluent on land as irrigation water is practiced in many sugar mills, but it is associated with odor problems.

Anaerobic treatment of the effluent using both digesters and lagoons has been found to be very effective & economical. A BOD reduction of both 70- 88% was observed in pilot plant study with an anaerobic digester, where BOD loading was 0.65-1.79 kg/m^3/day with a detention time of 2- 2.4 days at a controlled temperature of 37 °C.

The effluents of the anaerobic treatment units are found to contain sufficient nutrients (nitrogen & phosphorous) as such further reduction of BOD can be accomplished in aerobic waste stabilization ponds.

Where sufficient land is available a two stage biological treatment with anaerobic lagoons followed by aerobic waste stabilization ponds is recommended for Indian conditions. The mill effluent however is to be pretreated primarily in bar screens & grease trap. The flow sheet for the complete treatment for the sugar mill waste is shown in the fig, it is expected that the BOD reduction in the anaerobic process will be in order of 60% while overall BOD reduction may be in the order of 90%.

SOURCES AND CHARACTERISTICS OF TANNERY WASTEWATER

The waste water originates from the all operations in the tanning process. The waste may be classified as continues flow water, and intermittent flow waste. Continues flow waste consists of wash water after various process and comprises of large portions of the total waste and are relatively and less polluted then the other one. Spent liquors belonging to soaking, liming and batting, pickling, tanning and finishing operation are discharged intermittently. Although small in volume, they are highly polluted and contain varieties of solute and organic and inorganic substances.

The spent and soaked liquor contains soluble proteins of the hides, dirt and large amount of common salt where salted hides are process. This spent liquor under goes putrefaction very rapidly as it offers a good amount nutrient and favorable environment of bacterial growth. The growth of pathogenic - anthrax bacteria in this waste is also reported.

Flowsheet of tanning industry is shown below.

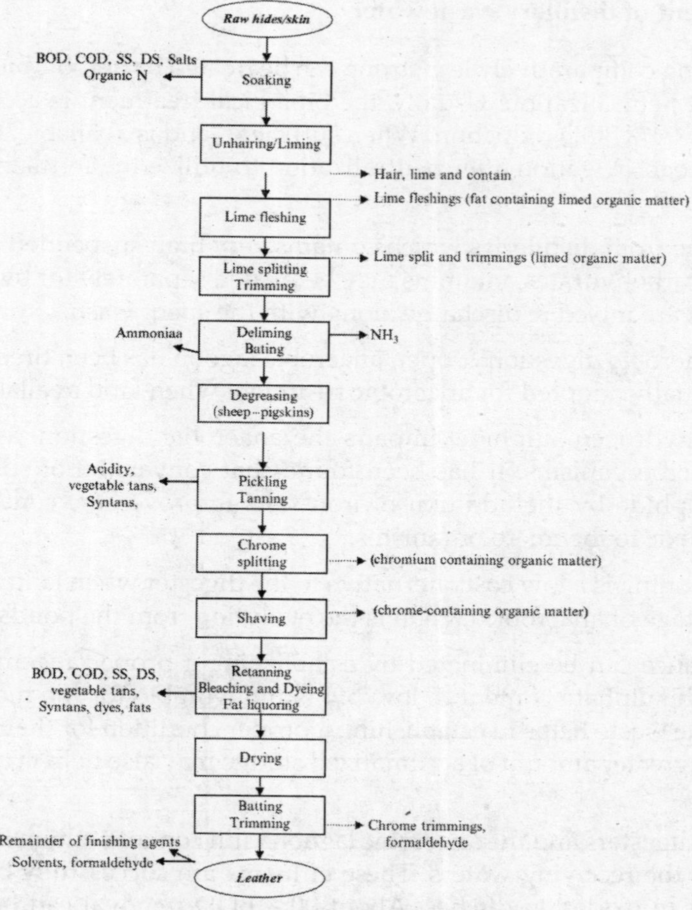

Fig.6.4: Flowsheet of tanning industry

The spent lime liquor contains dissolved and suspended lime, colloidal proteins and their degradation products, sulphides emulsified fatty matters and also carrying a sludge composed of unreacted lime, Calcium sulphide and calcium carbonate. As such, the spent lime liquor as a high alkalinity and moderate BOD and high ammonia nitrogen content.

The spent bate liquor contains high amount of organic and ammonium nitrogen due to the presence of soluble skin proteins and ammonia salts used in bating.

The vegetable ton exact contains tannins and also non tannins. Tannins are of high COD but relatively low BOD value. While non tannins including inorganic salts, organic acids and salts and sugar are of high BOD and COD. The spent vegetable tanning liquor is the strongest individual waste in the vegetable tannin having the highest BOD and very strong dirty brown color.

The spent pickling and chrome tanning waste comprises of as small volume, having a low BOD and contains traces of proteins impurities, sodium chloride and mineral acids and chromium salts.Chromium is highly toxic to the living aquatic creatures.

Origin and treatment of distillery wastewater

Brewery wastes being comparatively less strong can be treated by aerobic biological treatment, after screening and neutralization. Usually, the biological treatment is accomplishes by two stage process for 90-94% BOD reduction. When sufficient land is available, the brewery waste may be used for broad irrigation after neutralization to utilize the fertilizing components of the waste.

The yeast sludge from distilleries which contains very high suspended solids & BOD and is rich in proteins, carbohydrates, vitamins may be treated separately for by product recovery. But in practice they are mixed & discharge along with the spent wash.

Both closed anaerobic digestion & open anaerobic lagoon has been tired in India. A single stage digester is usually adopted for anaerobic treatment when land available is limited.

Production of hydrogen sulphide, impairs the anaerobic digestion, as soluble sulphide are toxic to the microorganisms. It has been found that conversion of soluble sulphides to insoluble ferric sulphide by the addition of iron salts improves the condition as the ferric sulphides are not toxic to the micro organisms.

Anaerobic lagooning is a low cost alternative to the digester when land is available plenty. The only disadvantage of anaerobic lagoon is the evolution from the ponds.

The odor nuisance can be eliminated by establishing a proper anaerobic activity in the lagoon. As the high sulphate content & low pH is unfavorable for the methane fermentors, neturalization of the waste helps in establishing a proper condition for their activity. A greater initial dilution and greater amount of acclimatized sludge may also help in establishing proper anaerobic activities.

Effluent of the digesters and the anaerobic lagoons still contains a high BOD, which cannot be discharged into the receiving waters. These effluents are successfully be treated either in aerated lagoons or in oxidation ditches. About 90% BOD removal can be accomplished in aerated lagoons. Aerated lagoon effluent requires further treatment in a polishing lagoon of about 24hrs detention time. Single stage anaerobic treatment either in lagoon or in digester is followed by 2 stage aerobic treatment is also proposed for distilleries wastes.

Sources and characteristics of waste water from pulp & paper mill

The volume & characteristics of the waste depends on the type & manufacturing process adopted & the extent of reuse of water employed in the plant. In the pulp making phase, the chipped cellulose raw materials are digested with different chemicals in one tank under high temperature & pressure. The process thus loosens the cellulose fibers & dissolves the lignin & resin and other non cellulosic materials in the raw material. The kraft process or sulphate process of pulp making uses sodium, sulphate, sodium hydroxide & sodium sulphide as the above mentioned digestive chemicals.

Another process of pulp making known as sulphide process uses magnesium or calcium bisulphate & sulphurous acid as the digestive chemicals. The alkali process uses sodium hydroxide or lime for pulp making.

The spent liquor produced by the above process of digestion is known as Black Liquor. This is not only very rich in lignin content, but also contains a large amount of utilized chemicals. Therefore this black liquor is treated separately for the recovery. So while the entire quantity of the liquor is up hide process makes the colored waste from the section, in the kraft process same is produced due to the leakages, spillages or overflow only from the digester. The cellulosic fiber after being separated from the black liquor is washed & then partially dewatered.

A concentrated wash water may be sent for chemical recovery) in the kraft process. The dilute wash water form the waste water. This volume of waste water is known as Brown stock wash or Unbleached Decter waste.

The washed cellulosic fibers are then sent for the bleaching in three stages, where chlorine, caustic & hypochloride are used in successive stages. Waste waters from the first & the last stages are light yellow in color while that from the caustic extraction stage is highly colored. The bleached pulp is then sent for the paper mill. In the paper mill, the pulp is disintegrated & mixed with various filter materials like alum talc, etc and dyes on an along shaped especially pulp is refined in a machine. The refined pulp is then diluted to proper consistency for paper making and passed through a screen to remove lumps or knots. Now this pulp is carried by a travelling belt of fine screen to a series of rolls, where the final product, the paper is produced. The drained water often called as White water forms the waste water from the paper mill section. This waste water contains fine fibers, alum, talc, etc. Usually the fibers are recovered from this waste; the treated liquid is reused for the wet chipping process.

The black liquor of the kraft process concentrated by evaporation and then incineration with the addititon of sodium sulphate. The organic like lignin, resin., etc are burnt off & the smell is dissolved in water. The resulting liquid is known as green liquor ; lime is now added to this liquor resulting in the formation of white liquor & lime mud containing chiefly digesting chemicals & is sent for use in digester. The lime mud is calcined (by burning) to form calcium oxide, which is reused to recausticised other green liquors in to the white liquors. Beside those mentioned about, the small volume of waste water is also produced when the bark is removed from the raw wood & the latter is reduced to chips by wet process.

A schematic diagram of pulp & paper industry is shown below.

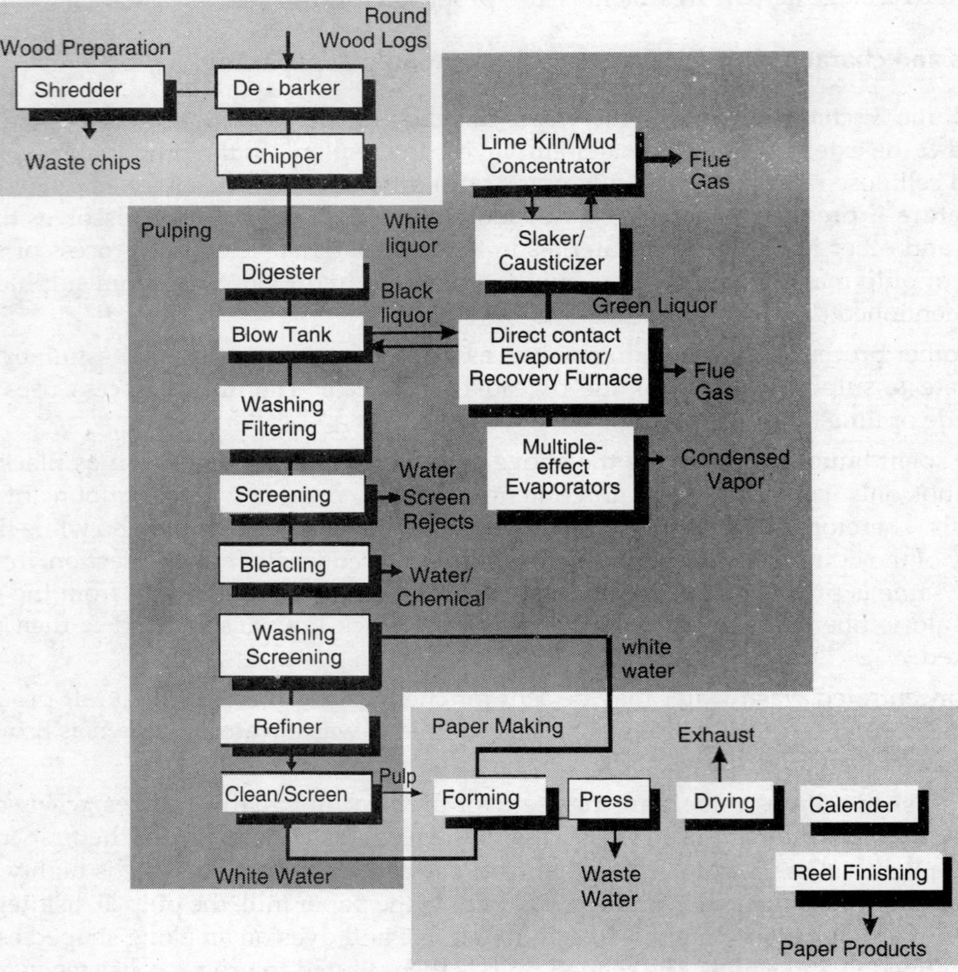

Fig.6.5: Schematic diagram of pulp & paper industry

Some very toxic waste materials are also generated during the process of chemical recovery from black liquor. Toxic materials like dimethyl sulphide, methyl mercaptan etc also comes out with digester gases & forms a colorless waste water after condensation.

Characteristics of pulp & paper mill wastes

The chemical composition of the wastes depends on the size of the plant, manufacturing process and to a great extent on the material economy practiced (by way of recovery of chemicals and fibers) in the plant. In most of the small paper mills in India, the chemical recovery is not practiced due to economical reasons. As such, the pollution potential of the waste of smaller mills is higher than that of the larger mills. Generally, the pulp and paper mills wastes are characterized by very strong color, high BOD, high suspended solids, and a high COD/BOD ratio.

It may be noted that too much difference in percent suspended solids contribution from different sections b/w (bleaching/washing) of large & small mills arises due to following reasons.

Large mills produce large amount of lime mud and without being calcined, it is discharged as a waste. This lime mud in the large mills contributes 86.5% of the total suspended solids in the effluent. As such, the percent suspended solids from the digester section; bleaching paper mill section etc. assumes a very low value. On the other hand, no question of lime mud rises in small mill, as it does not have the chemical recovery plant. The waste volume percentage from the digester section is also higher in the case of small mill as the entire quantity of black liquor is wasted in such mills.

Characteristics of wastes of pharmaceutical industry

Pharmaceutical industry produces varied type of products. They range from vitamins, synthetic drugs to antibiotics. The raw materials used are includes both organic and inorganic compounds. Some of the pharmaceutical plants do not generate any liquid effluents, while some others discharge little quantities of strong waste & others let out larger volumes. Due to these wide variations a generalization cannot be drawn on the effluents of pharmaceutical industry.

Most of the antibiotics such as penicillin. Streptomycin, lysine, sulfaquinazoline, nicarbazine & vitamins such as B_1, B_2, B_{12} and many steroids are prepared in the fermentation. The most waste produced in the fermentation process is the spent beer liquor. The spent beer liquor is the fermented broth remaining after the recovery of antibiotics and other valuables. It contains large amounts of organic materials, proteins and other nutrients and consequently the BOD of these effluents is abnormally high.

Five main pharmaceutical wastes and their characteristics are as follows:

1. Strong fermentation beers (small in volume but having 4000 to 8000 mg/l BOD)
2. Inorganic solids (waste slurry with little BOD)
3. Washings of floor and equipment (large percentage of total volume and BOD from 600 to 2500 ppm)
4. Chemical waste– solution or solvents which exert a substantial BOD when diluted with other wastes
5. Barometric condenser wastewater - resulting from solids and volatile gases being mixed with condenser wastewater causing 60 to 120 ppm BOD.

The antibiotic wastes impart objectionable odors to stream and inhibit biological population and action. If they are discharged into sewer, they must be properly diluted, otherwise they affect the sewage treatment.

The volume and composition vary from unit to unit. Approximately 1000 to 3000 litres of waste will be discharged per 100 kg of products manufactured. No specific conclusion on the characteristics of the effluents can be drawn. In general, they are either highly acidic or highly alkaline and possess a high BOD and COD. Some of the effluents contain toxic substances like cyanides.

If the wastes are discharged into stream, they deplete the dissolved oxygen immediately. These are corrosive due to their high acidity/alkalinity. Further, some of the substances present in them are toxic to aquatic life.

Wastes produced by food processing industry (Breweries, wineries etc.)

While breweries and wineries produce beer and wine respectively, a large no of products are obtained in distilleries. The range of products from distilleries includes industrial alcohols, rectified spirit absolute alcohol, spirit, beverage alcohol etc. But two things are common in all the products mentioned above.

1. All the above products are obtained through the biochemical process of fermentation by yeast using carbohydrates as raw materials and
2. All the products contain ethyl alcohol in different proportions.

In all the industries mentioned above are all characteristically of high BOD and they present a threat to the environment when discharged into the water sources or to the land without treatment.

Making of beer essentially consists of two stages:

1. Preparation of molt from grains like barley.
2. Brewing the barley.

In the malt making the barley grain are steeped (soaked) to bleach out color and then made to sprout under aerobic conditions. The grain malt is then dried and stored after screening the sprouts out.

The malt from the malt house is then transported to the brewing section, where the wart the medium for fermentation is prepared by mixing the course grain malt with hot water and by transforming the starch to. Sugars boiling in hops. The wort is then inoculated with a prepared suspension of yeast which common the sugar to alcohol.

When the fermentation is complete, the yeast and mall residue is filtered out and finally the beer is carbonated before packing for sale.

As the flavor of the product is of prime importance selection of raw materials & control of process is done accordingly.

Origin and treatment of dairy waste

The liquid waste from a large dairy originates from the following sections or plants. Receiving station bottling plant, cheese plant, butter plant, casein plant, condensing plant, dried milk plant & ice cream plants. Wastes also come from water softening plant & from bottle and washing plants. At the receiving station, the milk is received from the farmers and is emptied into large containers for transport to bottling of other processing plants. The empty is rinsed, washed and sterilized before returning to the farmers. At the bottling plants the raw milk delivered by receiving station is stored. The processing includes cooling, filtration, clarification, pasteurization and bottling.

In the above sections, the liquid wastes originate out of rinse & washings of bottles, cans & equipments & thus contain milk drippings and chemicals used for cleaning containers & equipments.

In a cheese plant, the milk (whole milk or skimmed milk) is pasteurized and cooled and placed in a vat, where a starter (lactic acid producing bacterial culture) and rennet are added. This separates the casein of the milk in the form of curd. The whey is then withdrawn and the curd compressed to allow excess to drain out. Other ingredients are now added and the cheese blocks are cut and packaged for sale. Waste water from this plant includes mainly the discarded whey and wash water used for cleaning vats, equipments floor.

In the creamery process the whole milk is pre heated to about 30^{0}C to separate the cream from the milk. In a butter plant, the cream is pasteurized and may be ripened with a selected acid and a bacterial culture. This is then churned at a temperature of about 7- 10^{0}C to produce butter granules. At the proper time the butter milk is drained out of the churned & the butter is washed & after standardization packaged for sale. Butter milk and wash waters used to clean the churns and small quantity of butter comes out as waste from the butter plants.

The skimmed milk may now be sent for bottling for human consumption or for further processing in the dairy for other products like non fat milk powders. Milk powders are produced by evaporation followed by drying by either roller process or spray process. The dried milk plant wastes consist of chiefly of wash waters used to clean containers & equipments.

The scoured or spoiled milk and some time the skimmed milk are processed to produce caseins used for preparation of some plastics. The process involves the coagulation & precipitation of the casein by the addition of some mineral acids. The wastes from this section include whey, washings and the chemicals used for precipitation. Very large dairies also produce condensed milk & ice creams.

In addition to the wastes from all the above milk processing units, some amount of uncontaminated cooling water comes as waste. These are usually recirculated. The dairy wastes are very often discharged intermittently. The nature & composition of the waste also depends on the types of products produce & the size of the plants.

Due to low COD, BOD ratio the dairy wastes can be treated efficiently by biological processes. Moreover, these wastes contain sufficient nutrients for bacterial growth. But for economical reasons, attempt should be made to reduce the volume & strength of the waste. This can be accomplished by:

- Prevention of spills, leakages & dropping of milk from the cans.
- By reducing the amount of water for washes.
- By segregating the uncontaminated cooling water and recycling the same.
- By utilizing the buttermilk and whey for the production of dairy bi products of good market value.

Due to the intermittent nature of the waste, it is desirable to provide equalization tank, with or without aeration, before the same is sent for biological treatment. A provision of grease trap is also necessary as a pretreatment to remove fat & other greasy substances from the wastes. A aeration for a day not only prevents the formation of lactic acid, but also reduces the BOD by about 50%.

Both high rate trickling filters & activated sludge plants can be employed very effectively for a complete treatment of the dairy waste. But these convention methods involve much maintenance, skilled methods like oxidation ditch, aerated lagoon, waste stabilization pond etc., can be employed with simpler type of equipments & less maintenance.

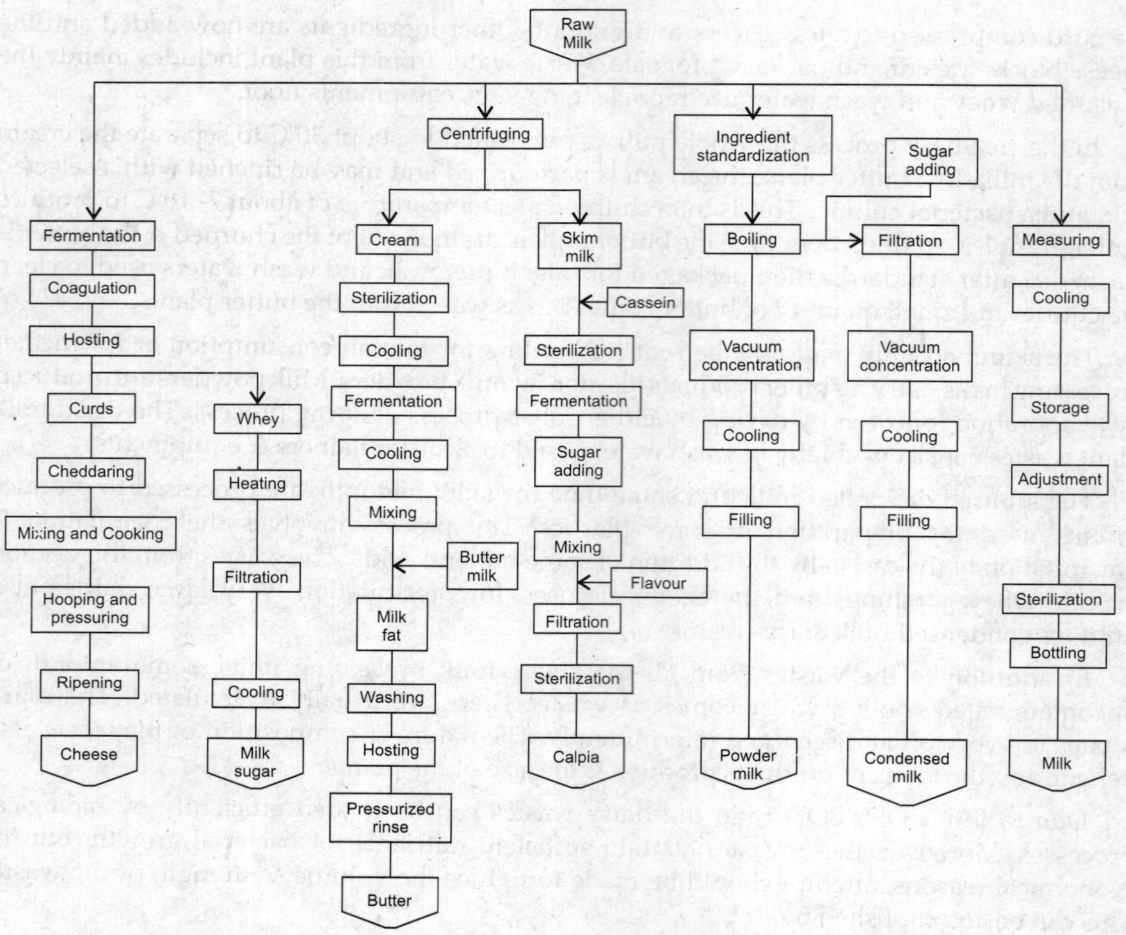

Fig.6.5: Treatment of dairy waste

Use of the dairy waste for irrigation after primary treatment in an aerated lagoon may be a good answer for the disposal of dairy wastes.

Effect of dairy waste in receiving streams

As the dairies are usually situated in rural areas or in small towns, the question of discharging the dairy waste in to the sewers does not arise.

As observed from waste is basically organic in nature. This is also slightly alkaline when fresh. When these wastes are allowed to go into the stream without any treatment, a rapid depletion of DO of the stream occurs along with growth of sewage fungi covering the entire bottom of the stream. The waste is said to carry occasionally, the bacteria responsible for tuberculosis (TB). Though alkaline in fresh condition the milk waste becomes acidic due to the decomposition of lactose into lactic acid under anaerobic condition.

The resulting condition precipitates casein from the waste, which decomposes further into a highly odors black sludge. At certain dilution the dairy waste is found to be toxic to fishes also.

EFFECT OF UNTREATED/PARTIALLY TREATED WASTES FROM COTTON TEXTILE MILL ON THE RECEIVING STREAM

Effect of textile mill waste on receiving streams/sewers

If the mill waste water is discharged into streams, it causes depletion of DO of the stream. This is due to the settlement of the suspended substances and subsequent decomposition of the same in anaerobic condition. The alkalinity and toxic substances like sulphides & chromium affects the aquatic life and also interferes with the biological treatment processes. Some of the dyes are also found to be toxic. The color often renders the water unfit for use for side. The presence of sulphides makes the waste corrosive particularly to concrete structures. All treatment plants should be planned giving serious consideration for the reduction of waste volume & strength, through process of chemical substitution, chemical recovery & recycling of water.

The pollution load from a textile, mill is dealt with operations like segregation, neutralization, equalization, chemical precipitation, chemical oxidation & biological oxidation. Several chemicals are used to reduce the BOD by chemical coagulation such as alum, ferric sulphate, ferrous sulphate & ferric chloride, lime or H_2SO_4 is used to adjust pH in this process. The dye waste may be economically treated by biological methods prior equalization, neutralization & chemical oxidation.

All treatment plants should be planned giving serious consideration for the reduction of waste volume & strength, through process of chemical substitution, chemical recovery & recycling of water. The pollution load from a textile, mill is dealt with operations like segregation, neutralization, equalization, chemical precipitation, chemical oxidation & biological oxidation. Several chemicals are used to reduce the BOD by chemical coagulation such as alum, ferric sulphate, ferrous sulphate & ferric chloride, lime or H_2SO_4 is used to adjust pH in this process. The dye waste may be economically treated by biological methods prior equalization, neutralization & chemical oxidation.

A Composite waste, when free from toxic substances may be treated as efficiently as domestic sewage, as most of the textile mill wastes contain sufficient nutrients like nitrogen & phosphorous. Trickling filters, activated sludge process & stabilization ponds have been effective in treating textile mill wastes. Extended aeration is found to be very effective in treating strong wastes even without equalization & pretreatment.

MANUFACTURING PROCESS OF COTTON TEXTILE MILL AND WASTE GENERATION

Manufacturing process

An integrated cotton textile mill produces its own yarn from the raw cotton. Production of yarn from raw cotton includes steps like opening & cleaning picking, carding, and drawing spinning, winding & warping. All these sequences are dry operations and as such do not contribute to the liquid waste of the mill. The entire liquid waste from the textile mills comes from the following operation of slashing (sizing), scouring, desizing, bleaching, mercerizing, dyeing & finishing.

In slashing the yarn is strengthen by loading it with starch or other substances wastes originates from the sections due to spills & floor washings. The substitution of low BOD sizes (such as carboxy methyl cellulose) for the high BOD of the mill effluent by 40 to 90%.

After slashing, the yarn goes for weaving. The prepared cloth now requires scouring & desizing to remove natural impurities and the slashing compounds. Enzymes are usually used in India to hydrolyze

the starch, acids may also be used for this purpose. Caustic soda, soda ash, detergents etc. are also used in this section. Bleaching operations use oxidizing chemicals like peroxides & hyper chloride to remove natural coloring material. The section contributes about 10% of the total pollution load.

Mercerizing, consists of passing the sloth through 20% caustic soda solution. This process includes the strength elasticity luster & dye affinity. Waste from this section is recycled after sodium hydroxide recovery. Negligible waste which may come out of this section contributes little BOD but a high degree of alkalinity.

Dyeing may be done in various ways, using different types of dyes and chemical. Classes of dyes include Vat dyes, developing dyes etc. Color from the dyes vary widely and although these are not usually toxic, they are treated separately. Thickened dyes are used for probing and subsequent fixation. After fixation of the prints, the fabric is given a thorough wash to remove the unfixed dyes.

Flow diagram of a cotton textile mill is shown below.

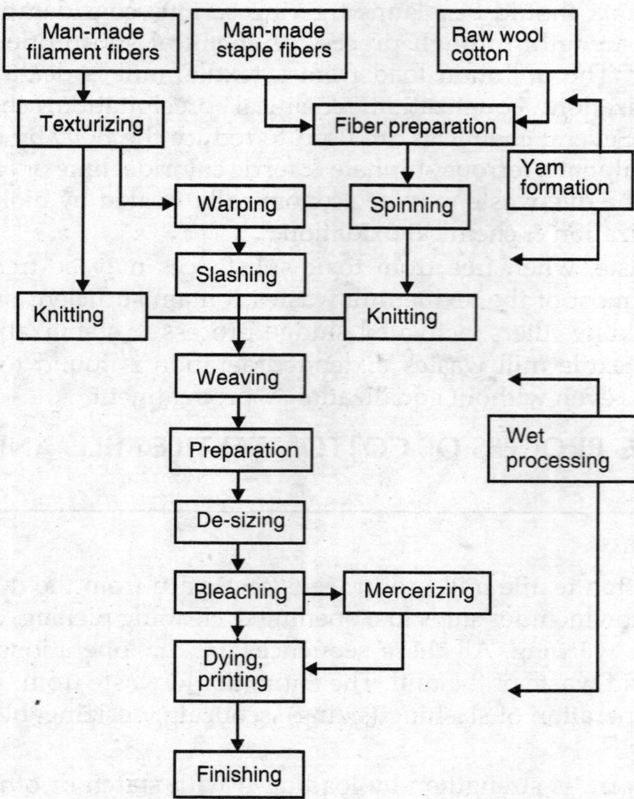

Fig.6.7: Flow sheet of a cotton textile mill

The finishing section of the mill imparts various types of chemicals are used for various objectives. These include starches, dextrines, natural & synthetic waxes, synthetics etc. Therefore a composite waste from an integrates cotton textile mill may include the following organic & inorganic substances starch, carboxyl methyl cellulose, sodium hydroxide, detergents, peroxides , hyperchloride dyes & pigments, sodium gums, dextrines, waxes, sulphides, soap etc. Depending on the process & predominant dye used, the characteristics of the mill waste varies widely.

APPENDIX 6.1

Water requirement of selected industries per unit of product

S.N.	Plant/Industry	Unit of product	Water requirement, kiloliters (KL) or m^3
1	Bread	Ton	2.1- 4.1
2	Bear	KL	15
3	Milk Products	Ton	20
4	Sulphuric acid (H_2SO_4)	Ton of 100% H_2SO_4	10.4
5	Petrol (gasoline)	KL	7.10
6	Wood pulp	Ton of pulp and paper	236
7	Finished steel	Ton	300
8	Cotton bleaching	Ton	300
9	Acetic acid	Ton	450-1100
10	Alcohols	Kilo litres	50-150
11	Ammonia	Ton of NH_3	150
12	Ammonium sulphate	Ton	900
13	Calcium carbide	Ton	150
14	Calcium metaphosphate	Ton of Ca $(PO_3)_2$	10
15	Carbon dioxide	Ton	100
16	Caustic nitrate	Ton of NaOH	100
17	Cellulose nitrate	Ton	50
18	Charcoal and wood chemicals	Tons of CaC_2	300
19	Corn refining	Ton of starch	1.5
20	Gun powder	Ton	750
21	Hydrochloric acid	Ton of 20% be HCl	130
22	Hydrogen	Ton of H_2	3000
23	Lactose	Ton	900
24	Oxygen	$100m^3$	10
25	Soap	Ton	2

S.N.	Plant/Industry	Unit of product	Water requirement, kiloliters (KL) or m^3
26	Soda ash	Ton	75
27	Sulphur	Ton	10
28	Sugar	Ton of sugar cane	1-5
29	Oil refinery	Ton of dry pulp	10-30
30	Paper	Ton of dry pulp	175
31	Artificial silk	Ton	2000
32	Cotton	Ton	10-300
33	Cotton dyeing	Ton	35-70
34	Rayon	Ton of yarn	400-900
35	Wool	Ton of yarn	150-900
36	Aluminium	Ton	1400
37	Bauxite mining	Ton of ore	10
38	Gold. Ore mining	Ton of ore	1
39	Iron ore mining	Ton of ore	5
40	Automobiles	Per vehicle	40
41	Coal	Ton	5-15
42	Electric power	Kwh	0.3-0.7
43	Explosive	Ton	850
44	Fertilisers	Ton	300
45	Glass	Ton	75
46	Leather Tanning	Ton of raw hide	75
47	Synthetic rubber	Ton	900-3000
48	Laundering	Ton	30

Note: Ordinarily an industry uses the lowest cost water supply which will satisfy its requirements. Major users such as the chemical and metallurgical industries withdraw 20 to 30 times as much as the minor users such as the rubber industry, one plant may use 2 to 4 times as much water per unit output as another making identical products. Numerous factors contribute in intra-industry variation. Factories in arid climates have higher water-use-efficiencies. Large plants tend to use less water per unit production than small plants. Because of the economic pressures and legislation, modern plants are generally more water conscious than older plants. Water is one substance which industry uses in such vast quantities that in sheer tonnage and bulk, it far overshadows all other materials.

The variety of consumers of water for industrial purposes is so great and the specific physical, chemical and bacteriological requirements cover such broad limits that no standard specifications for quality are possible. Water suitable for one industry may be totally unsuitable for a dissimilar Industry. Bureau of Indian Standards has prescribed standards for water for various industries.

Quantities Of Water Requied For Some More Specific Sections Of Industries Are Tabulated Below.

S.N.	Industry	Water required in m^3 per unit of product
1	Pulp and paper mills: • Paper board • Straw board • Sulphate pulp	60 – 70/t 140 – 150/t 270 – 360/t
2	Steel plant: • Fabricated steel • Ingot steel • Pig iron	180 – 200/t 80 – 90/t 18 – 24/t
3	Oil refinery	21 – 25/m^3 crude
4	Oil, Fisher–Tropsch synthesis	14 – 15/ m^3 crude
5	Gasoline, natural	10 – 15/ m^3
6	Coal carbonization	15 - 20/t
7	Cement	3 – 4/t
8	Tannery	30 – 40/t
9	Cotton textile: • Bleaching • Dyeing • Finishing • Processing	 130 – 190/km 100 – 120/t 25 – 70/km 30 – 450/t
10	Rayon: • Cupro ammonium yarn • Dissolved pulp • Viscose rayon	 800 – 880/t 830 – 950/t 8 000 – 9 000/t

PULP AND PAPER INDUSTRY

- Water is used in pulp and paper industry in (1) the flume for carrying the bamboo to the chipper, (2) compaction of the bamboo chips, (3) preparation of the digester liquor, (4) countercurrent washing in the knotters, (5) thinning of the pulp, (6) barometric condensers of the soda recovery section and (7) turbine cooling feed pumps.
  The quality requirements of process water used in manufacturing pulp and paper vary with the desired quality of the product. Fine papers require much purer water than coarse or crude paper. Turbidity & colour affect the brightness & colour of the final paper. Constituents that cause hardness react with certain types of sizing materials producing insoluble deposits causing uneven sizing. Iron and manganese cause staining and loss of brightness. Algae and mould spores cause semi-transparent spots. Quality tolerances for water for pulp and paper industry are given in Table below.

Maximum Tolerances for water for pulp and paper industry (Units of items 3 to 8: mg/1)

S.N.	Parameters/Characteristics	Tolerances for				
		Ground wood bleached	Kraft paper unbleached and pulp bleached	Kraft paper unbleached and pulp unbleached	Soda and sulphite paper	High grade paper
1	Colour (Hazen units)*	20	15	100	10	5
2	Turbidity ** (silica scale units)	50	25	100	15	10
3	Total dissolved solids	500	300	500	300	300
4	Total hardness (as $CaCO_3$)	200	100	200	100	100
5	Iron (as Fe)	1.0	0.25	1.0	0.25	0.25
6	Manganese (as Mn)	0.5	0.1	0.5	0.05	0.05
7	Iron (as Fe) and Manganese (as Mn), added together	1.00	0.25	-	0.25	0.25
8	Total residual chlorine (as Cl_2)	2.0	2.0	-	2.0	2.0

* The colour produced by 1 mg/1 of platinum (as K_2 Pt CI_6) is taken as the standard unit of colour (Hazen unit). It is known that solutions of potassium chloro platinate tinted with small amounts of cobalt chloride yields colours that are very much like the natural colours.

** Because of the wide variety of materials that cause turbidity in natural waters, it has been necessary to use an arbitrary standard. The standard chosen is I mg SiO_2-I unit of turbidity and the silica used must meet certain specifications as to particle size.

Textile Industry

In textile industry water is used for washing, wool scouring, kiering, dyeing, finishing, cooling and steam raising. Quality tolerances for water for textile industry are given in table below.

Maxiumum General Tolerances For Water For TEXTILE Industry (Units of items 4 to 7 - mg/1)

S.N	Characteristics /Parameters	Tolerance for		
		General purposes	Bleaching, dyeing subsequent processing	Wool scouring
1	Colour (Hazen units)	20	5	70
2	Turbidity (silica scale units)	5	5	-
3	pH	6.0-8.5	6.0 to 8.5	6.0 to 8.5
4	Total hardness (as $CaCO_3$)	50	50	50

S.N	Characteristics /Parameters	Tolerance for		
		General purposes	Bleaching, dyeing subsequent processing	Wool scouring
5	Iron (as Fe)	0.25	0.25	0.25
6	Manganese (as Mn)	0.10	0.10	0.10
7	Iron (as Fe) and Manganese (as Mn) added together	0.25	0.25	0.25

The water used in the wet processing of textiles, should be free from turbidity, colour, iron , manganese, hardness & organic growths and it should not be corrosive. For dyeing purposes the quality is of great importance when delicate fines are to be produced. Freedom from iron is absolutely essential; the smaller the saline constituents, the better is the water for the industry. Water should also be quite free from colour and turbidity.

TANNNING INDUSTRY

Water is used in tanning in a number of operations , namely soaking the hides & skins, liming, deliming, bating , pickling, vegetable & chrome tanning, bleaching, dyeing and fat liquoring. The quality requirements of water vary from one operation to another. In the individual operations, the tanners take the quality of water into consideration while fixing the treatment period, quantity of chemicals being added etc. The quality tolerances for water for tanning are given in table below.

For tanning purposes, water must be free from iron and manganese in liming as adherent deposit of calcium carbonate ('kiss spots') are formed which are dye resistant. Bicarbonate hardness is also objectionable in rinsing after liming, bating and dyeing. Zero-hardness water is preferable for leaching tannin extract, vegetable tanning operations, dyeing with basic dyes and fat liquoring. Hardness has no apparent effect in chrome tanning and does not waste dyestuffs in acid baths.

Maxumum General Tolerances For Water For Tanning Industry
(Units Of Items 4 To 6 - Mg/1)

S.N	Characteristics /Parameters	Tolerance
1	Colour (Hazen units)	25
2	Turbidity (silica scale units)	20
3	pH	6.5-8.0
4	Total alkalinity (as $CaCO_3$)	150
5	Total hardness (as $CaCO_3$)	500
6	Iron (as Fe) and manganese (as Mn), together	1.0

FERMENTATION INDUSTRY

Water is used in the fermentation industry for brewing and distillation. In the brewery, water is used for brewing, boiler feed, bottle keg washing, pasteurizing and cooling. In the distillery,

Water is used for mash making, cooling of masher & condenser, boiler feed and deproofing. Quality tolerances for water for fermentation industries is given below in Table.

Maximum Physical and Chemical Tolerances of Water for Fermentation Industry (Units of items 5 to 21 - mg/1)

S.N.	Characteristics/Parameters	Tolerance
1	Turbidity (silica scale units)	10
2	Colour (Hazen units)	20
3	Odour	None
4	pH	6 to 9.2
5	Total solids	1,000
6	Total hardness (as $CaCO_3$)	500
7	Chlorides (as CI)	250
8	Sulphates (as SO_4)	200
9	Iron (as Fe)	0.3
10	Manganese (as Mn)	0.2
11	Copper (as Cu)	1.0
12	Zinc (as Zn)	15
13	Magnesium (As Mg)	125
14	Fluorides (as F)	1.5
15	Nitrates (as N)	20
16	Phenolic	0.001
17	Arsenic (as As)	0.2
18	Chromium (As Cr^{6+})	0.05
19	Cyanide (as CN)	0.01
20	Lead (as Pb)	0.1
21	Selenium (as Se)	0.05

A commonly recommended quality requirement of water used in brewing is that it be of purity at least equal to that equal to that of drinking water. Alkalinity caused by carbonates and bicarbonate is generally considered to be disadvantageous in brew water. The water should be free from nitrates, iron and manganese. The presence of calcium, especially calcium sulphate in high concentrations, is advantageous. Moderate amounts of magnesium sulphate and calcium, magnesium & sodium chlorides are beneficial.

Water used in distilleries should be clear, colourless, free from iron, manganese, odours & tastes and of approved bacteriological quality. Some hardness is usually desirable in the water used for mashing, and moderate amounts of bicarbonate hardness are frequently of value as a buffer in maintaining the proper pH value during fermentation, The water used for deproofing should be free from matter, tastes and odour.

FOOD INDUSTRY

In processed food industry, water is used for processing, washing, flushing, boiler feed , cooling and general usage. The quality of water required varies for the different processed food industries. The quality tolerances for water for processed food industry is given below in table.

Maximum Physical and Chemical Tolerances of Water for Food Industry (Units of items 5 to 21: mg/1)

S.N.	Characteristics/Parameters	Tolerance
1	Colour (Hazen units)	10
2	Turbidity (silica scale units)	20
3	Odour	None
4	pH	6 to 9.2
5	Total solids	1 000
6	Total hardness (as $CaCO_3$)	600
7	Sulphates (as SO_4)	200
8	Fluoride (F)	1.5
9	Chloride (as CI)	250
10	Cyanide (as CN)	0.01
11	Selenium (as Se)	0.05
12	Iron (as Fe)	0.3
13	Magnesium (as Mg)	75.0
14	Manganese (as Mn)	0.2
15	Copper (as Cu)	1.0
16	Lead (as Pb)	0.1
17	Chromium (as Cr^{6+})	0.05
18	Zinc (As Zn)	0.03
19	Arsenic (as As)	0.02
20	Nitrate (as N)	20
21	Phenolic substances (as C_6H_5OH)	0.001

For food process industries the water should be clear, colourless, free from iron, manganese, objectionable tastes & odours and of approved bacteriological quality, even when the food products are put through a sterilizing operation. Other requirements for process waters may differ according to their end uses. For instance, a water high in sulphate hardness may be excellent for certain fermentation process but unusable for canning peas, bens or lentils.

RAYON INDUSTRY

The rayon fibers are manufactured by viscose, cuprammonium or cellulose acetate processes. Considerable quantities of water are used in various steps such as steeping, washing and make up solutions. The quality tolerances for water for rayon industry is tabulated below.

Maximum Tolerances for Water for Rayon Manufacturing Industry
(Units of items 3 to 10: mg.1)

S.N.	Characteristics / Parameters	Tolerance
1	Colour (Hazen units)	5.0
2	Turbidity (silica scale units)	1.0
3	Total dissolved solids	200
4	Total alkalinity (As Ca CO_3)	75.0
5	Total hardness (as $CaCO_3$)	5.0
6	Silica (as SiO_2)	10.0
7	Iron (as Fe)	0.05
8	Manganese (as Mn)	0.03
9	Iron (as Fe) and manganese (as Mn) , added together	0.05
10	Heavy metals (as Pb)	0.01

Processing in rayon manufacturing industry is greatly affected by certain impurities present in water. The presence of suspended mineral matter is disadvantageous.

ICE-MAKING

Ice is manufactured in food industries and in cold storage plants. A good saleable ice should be crystal clear, colourless, free from bubbles, snowy & heavy cores, and must not shatter when handled. Water used for the preparation of ice must be free from colour & suspended matter and contain no metallic impurities, such as iron & manganese. The presence of these substances leads to discolouration and opacity in ice. The slime matter contained in solution should be low and the water should be of high degree of organic & bacterial purity expected of drinking water. The freezing of water does not kill bacteria and may even prolong their existence.

CHEMICAL INDUSTRY

Chemical plants, in general, are large users of water and most of them have their own water supplies. Most chemical plants use large quantities of steam for heating, boiling, digesting, evaporating, distilling, drying and other processing operations on a variety of products. The amounts of cooling water required in various chemical industries vary from a small fraction to over 90% of the total water used.

The requirements for process waters vary widely but in general these waters should be clear, colourless and free from iron, manganese, hydrogen sulphide & organic growths. For most of the chemical industries using water as the raw material, there are limiting concentrations of impurities that cannot be exceeded without damaging the product but many such concentrations are so high that they are rarely encountered in normal domestic supplies. In other instances, however, special water treatment such as distillation is required.

IRON AND STEEL INDUSTRY

In iron and steel mills, large quantities of water, are for a variety of purposes, as in boiler feed water, cooling condensers & furnace fronts, pickling, rinsing, plating operations etc. and for general use. The characteristics operations etc. and for general use which determine the suitability of water for steel manufacture are: (1) temperature preferable below 45 °C , (2) chlorides below 175 mg/1, (3) pH between 6.8 and 7.0 (4) hardness below 50 mg/1, (5) suspended matter below 25 mg/1, (6) organic content as low as possible and (7) corrosion potential at the lowest possible level.

Where surface waters used, treatment processes include sedimentation, chlorination, coagulation , settling, filtration, iron & manganese removal and softening. The water used for pickling is usually not softened: but the pickling water used for cleaning before hot plating operations and the water for cleaning with alkaline and subsequent rinsing, are best softened.

FEED WATER FOR BOILERS

The problem of making up feed water boilers is common to almost all industries.

In the operation of boilers, foaming, priming, scale formation, caustic embrittlement and corrosion increase with operating pressure. Foaming is caused by impurities in the water forming a surface film which stabilizes bubble formation. Forming and priming entrain moisture and solids in steam. Caustic embrittlement (the inter crystalline cracking of boiler metal), is associated with localized stresses that have strained the metal beyond the elastic limits and with a high concentration of caustic soda in the absence of adequate concentration of sulphate in the feed water. The suggested limits of tolerance for boiler feed waters are given below in Table.

Suggested Limits of Tolerances for Boiler Feed Waters

(Units for items 3 to 13: mg/1)

S.N.	Characteristics / parameters	Pressure, MN/m^2			
		0-1.0	**1.0-1.7**	**1.7-2.8**	**Over 2.8**
1	Turbidity (silica scale units)	20	10	5	1
2	Colour (Hazen units)	80	40	5	1
3	Oxygen consumed	15	10	4	3
4	Dissolved oxygen	2.0	0.2	0	0
5	Hydrogen sulphide	5	3	0	0
6	Total hardness ($CaCO_3$)	80	40	10	2
7	Sulphate-carbonate ratio	1:1	2:1	3:1	3:1
8	Aluminium oxide	5	0.5	0.05	0.01

9	Silica	40	20	5	1
10	Bicarbonate	50	30	5	0
11	Carbonate	200	100	40	20
12	Hydroxide	50	40	30	20
13	Total solid	3 000 – 1 500	2 000 – 1 500	1 500 – 100	50
14	pH value (min)	8.0	8.4	9.0	9.6

These tolerance limits take into account that (1) reductions is turbidity, colour, oxygen consumed value and total solid decrease foaming & priming, (2) reductions in hardness, silica and alumina alleviate scale formation, (3) maintenance of a high of sulphate to carbonate ratio controls caustic embrittlement and (4) elimination of oxygen, reduction of bicarbonate ions and increase in pH suppresses corrosion.

APPENDIX 6.2

Major organic & inorganic water pollutants in various industries

S.N.	Major water pollutants	Industry type
	INORGAING POLLUTANTS	
1	Acids	Textile, chemicals, metallurgical, Engineering industries
2	Alkalis	Textile, chemical, Petrochemical, coke ovens, Metallurgical, Laundry, Engineering industry
3	Metals & metallic salts	`Textile, Leather tanning, Paper making, chemical, metallurgical, Agriculture, Engineering industry
4	Other salts	Dairy, tanning, textiles, Paper making, chemical industries
5	Minerals (chin clay and soils)	Dairy, paper making, Agriculture, industry
6	Chromates	Tanning, Petrochemicals, Metallurgical, Engineering industries
7	Cynides, Cyanates	Coke ovens, Metallurgical, Engineering industries
8	Sulphides	Textile, Tanning, Petrochemical, Coke, ovens
9	Phosphates, Nitrates	Dairy, Textiles, Tanning, Paper, making, chemical industries
10	Bleaches	Textile, Paper making, Laundary
	ORGANIC POLLUTANTS	
11	Dyestuffs	Textile, Tanning, paper Making, chemical industries
12	Organic acids	Chemical, Agriculture industries
13	Organo-pesticides	Food processing, Agriculture Industries
14	Detergents	Dairy, Food Processing, Textile, Tanning, Agriculture industries
15	Proteins	Dairy, Food Processing, Textile, paper making. Chemical, Agriculture industries
16	Carbohydrates	Dairy, Food processing, Textile, paper making chemical, Agriculture industries
17	Phenols	Textile, chemical, Photochemical, coke ovens, metallurgy, Industrial oil production industries
18	Oil & fats	Dairy, Food processing, Textiles, Tanning, petrochemicals, Industrial oil production industries Metallurgical, Laundry, Agriculture, Engineering industries

Note: Treatment of Industrial Wastes

The most satisfactory and efficient means of disposing industrial wastes from manufacturer's premises is to discharge them into the sewers of the local authority, provided (1) the sewers and sewage woks are of adequate capacity to deal with them and (2) the waste has no damaging effect on the sewage workers, sewers, sewage works or treatment process. But this is not feasible if the effluents contain highly toxic substances or excessive amount of mineral acids or alkalis.

In large towns, industrial plants of all kinds discharge effluents into the sewage system, Though the character of the sewage may consequently change, it is usually possible to treat the combined wastes by the normal methods used for domestic sewage treatment.

Industrial wastes may be broadly classified into four groups:

- Wastes from industries based on natural organic products
- Wastes from organic chemical industry
- Wastes containing inorganic pollutants
- Wastes with radioactive substances

Effluents form industries based on natural organic products: Industries based on natural organic products produce large quantities of highly polluted effluents. Examples of these are the textile, viscose rayon, pulp & paper, distillery, dairy, tanning and meat packing industries.

Effluents of these types can sometimes be satisfactorily treated by the processes used for treatment of domestic sewage. Some of these industries treat the effluents partially or fully before discharging. In a few cases, they discharge the effluents into the municipal sewage system.

Effluents from organic chemical industries: Examples of these are from the petroleum & petrochemical industries, coke ovens and gas works. Some of these effluents can be treated satisfactorily by biological methods if the concentrations of the toxic substances are below certain limiting values. For instance, phenol in concentrations of 100 mg/1 can be decomposed by biological processes.

Effluents containing inorganic compounds: Examples of these effluents are form the electroplating, coal mining, blast furnace and inorganic chemical industries, As far as possible, the pollutants of these industries are removed by sedimentation and chemical processes resulting in precipitation or neutralization of the pollutants; if necessary, the resulting effluent is diluted so that the dissolved salts have no harmful effect on the receiving waters.

Effluents with radioactive substances: Several methods have been investigated or developed to handle these effluents. One method involves the adsorption of the radioactive substance in ion exchange materials which are then made into a solid mass with cement; the cement blocks are than buried or sunk in sea.

TEXTILE MILL WASTE WATER

The manufacture of finished cloth in cotton textile mills is carried out by dry or wet processes. The dry processes do not produce any liquid water. The liquid wastes of wet processes are to be treated before discharging them.The sources and characteristics of these liquid wastes are given below in Table. The pH of composite wastes resulting from various operations is 8.0-9.5 and other analytical values in mg/1 are: total solids 2 900 -3 200, suspended solids 100 – 720. Alkalinity 900 – 1 600 and BOD (20°C, 5 days) 300 – 620.

Table 6.2T_1: Waste Waters from Textile Mills

S.N.	Source	Process	% of total liquid wastes	% of total BOD	% of total solids	Remarks
1	Slashing	Strengthening yarn by loading with starch	10 – 20	50 – 55	30	Slightly
2	Scouring and desizing	Removal of certain natural ingradients and starch; Addition of chemicals and enzymes	40 – 60	35 – 40	40 – 45	Highly alkaline
3	Bleaching, mercerizing	Use of chlorine bleaching, caustic soda for mercerizing, chemicals and dyes for dyeing	50 – 60	5 – 15	17 - 20	Alkaline and coloured

Chemical treatment: The alkalinity of the waste water from the equalization or storage tank is neutralized by acid. The waste water is then treated with alum and sent to clarifloculators. The clarified effluent is sent for disposal or for biological treatment, depending upon local regulations. The chemical treatment gives a BOD reduction of 10-25%.

Biological treatment: To reduce the BOD further, the effluent is treated biologically in trickling filters or activated sludge plants. As textile mill wastes are deficient in nitrogen and phosphorus, one part of nitrogen for 20 parts of BOD and one part of phosphorus for 100 parts of BOD are added. These additional nutrients will not be required if wastes are led into municipal sewage.

About 50 – 60% reduction of BOD occurs in trickling filters with a high loading of 1.60 kg/m^3. Activated sludge process reduces BOD by 90 – 95 % and removes colour with aeration period 12-36 hours. The sludge from the biological treatment is dried in drying beds before disposal.

VISCOSE RAYON WASTES

Viscose rayon is made from cellulose pulp by allowing it to stand in a solution of caustic soda, After pressing to remove the alkaline liquor, the pulp is broken up, mixed with carbon disulphide and dissolved in a weak solution of caustic soda. It is then extended into a bath of dilute sulphuric acid with sodium sulphide to remove the precipitated sulphur, washed again and bleached. The effluents from these processes are (1) alkaline liquors from the initial treatment of cellulose, (2) alkaline liquors containing polysulphides, (3) acidic liquors containing sulphates & zinc and (4) bleach liquors and wash containing appreciable quantities of Zinc having a high pollutional effect. So the wastes are segregated and only the acidic waste is treated. The other waster are added to the treated acidic waste for dilution.

The pH of the acidic waste is 1.5 to 1.8. The range of other analytical values in mg/1 are: total solids: 18000 = 30 000, dissolved solids: 2700 – 30000, Acidity as $CaCO_3$: 380 – 4400, Zinc: 200, sulphates: 20 000 – 23 000, COD (dichromate): 400 – 680 and chlorides: 130-150.

Removal of zinc: Zinc precipitates as hydroxide when the acidic wastes are treated with lime or sodium hydroxide. The process of precipitation commences at pH 6.0 and is 99% complete at pH 9.5. In the first phase of the reaction, lime is used to increase the pH to 6 and in the second phase sodium hydroxide is employed to raise the pH to 9.5.

The reaction is carried out in three neutralization tanks, fitted with high speed agitators for efficient mixing of the chemicals with the wastes. The neutralized wastes are then sent to a clariflocculator for settling fluffy zinc sludge. The treated wastes are mixed with the other wastes before discharging. The economics of the zinc recovery depends on the concentration of the metal and on the acidity of the wastes.

PULP AND PAPER MILL WASTES

About 230 - 455 m^3 of water is used per tonne of paper produced; almost the entire quantity comes out as waste water which requires treatment before disposal. The sources of the waste water are: (1) highly coloured dilute wash water resulting from washing the pulp (concentrated washes are sent for chemical recovery), (2) acidic water containing chlorine from bleaching operations, (3) 'white water' collected under the paper making machine and containing fibres, sizers, dyes & loading materials used in paper manufacture and (4) water used for washing the machines & floors.

The pH of water from an integrated pulp & paper mill is 8.4 - 9.4 and the range of other analytical figures in mg/1 are – alkalinity: 70 - 800, suspended solids: 600-1400, total solids:2 500 - 3 000 and BOD (20 °C, 5 days): 170 - 500.

The methods of treatment include:

- **Primary treatment** in a clarifier or clariflocculator to remove 60 - 80% of suspended solids and 10-25% of BOD. Clarifiers have overflow rate of 20- 24 m^3 / m^2 per day with detention time of 3 - 4 hours
- **Secondary treatment** for removal of BOD using trickling filters or activated sludge process, oxidation pond, anaerobic digestion or lagooning.

Trickling filters reduce only 40-60% BOD even at low loadings of 0.5-1.0 kg BOD /m^3 per day. Activated sludge plants remove upto 95% BOD with a loading of 0.2 - 0.4 kg BOD / kg MLSS (mixed liquor with sludge solids) and operate at a MLSS concentration upto 4 000 mg/1.

Diammonium phosphate, phosphoric acid and ammonium hydroxide are used to have a ratio of BOD to nitrogen to phosphorus of 100: 5: 1 , to facilitate the biological reaction to take place.

In anaerobic lagoons, BOD reduction is 45-72% with a detention period of 15 days. To reduce the BOD to 30 mg/1, the effluent is treated in aerobic lagoons for 3-4 days.

Disposal of_sludge: The sludge is separated in a rotary vacuum precoat drum filter or centrifuge. The resulting coke is burnt in a rotary oil fired kiln. The kiln gases are scrubbed for recovery of sulphur dioxide.

A flowsheet for treatment of pulp and paper mill wastes is given below in figure.

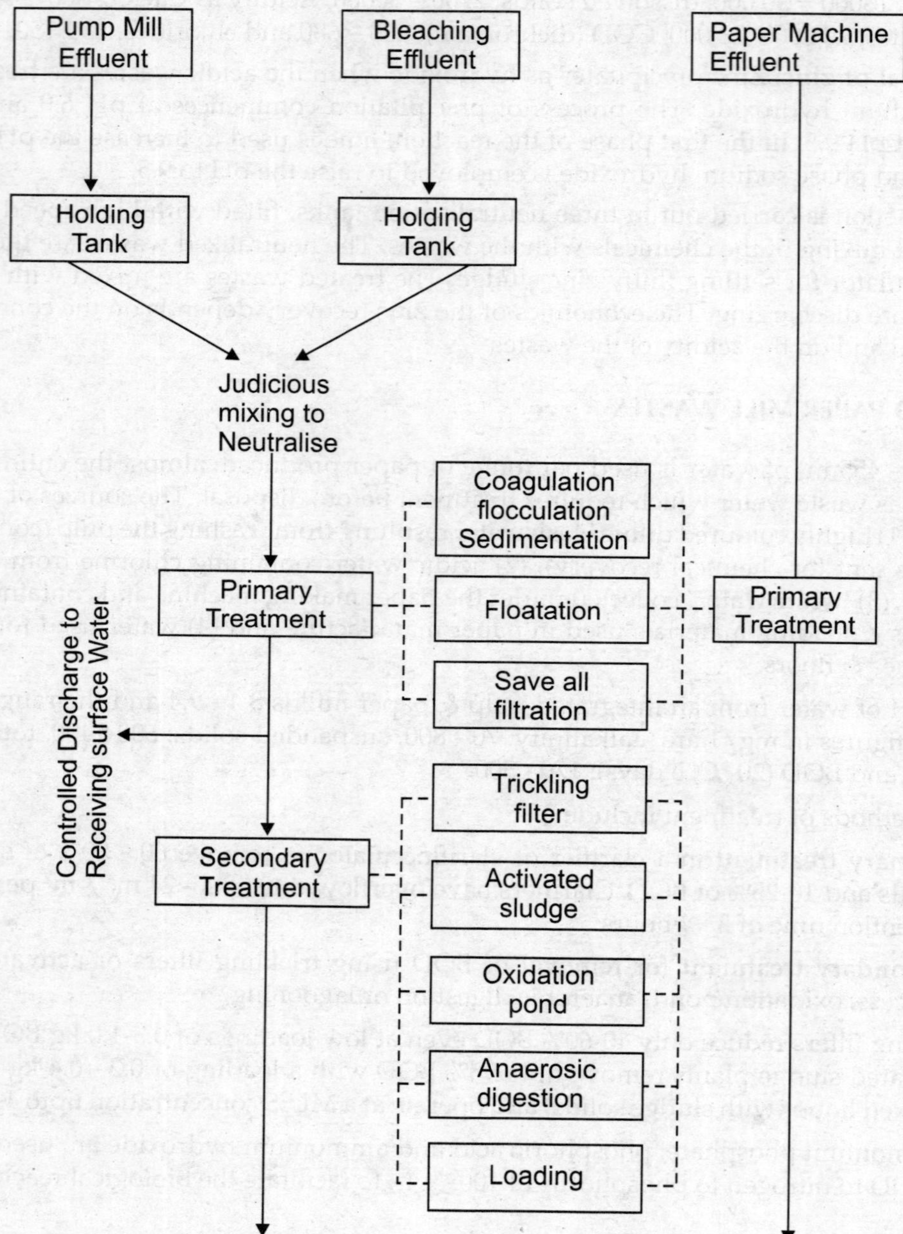

Fig.6.2F$_1$: Treatment Of Pulp And Paper Mill Liquid Wastes

DISTILLERY WASTES

Alcohol distilleries in India use sugar cane molasses as raw material. The molasses is diluted and after addition of acid & chemicals, it is fermented. The fermented liquor containing 6 to

8% alcohol is distilled to obtain alcohol and the rest of it is discharged as spent waste or 'slop' which is the major source of pollution from the distilleries. 60-80 litres of slop are obtained per litre of alcohol produced. The characteristics of the slop are: temperature:90- 100 °C, pH: 4.5, BOD: 32 000-45 000 mg/l, sulphates/chlorides:9 000- 12 000 mg/l. Other waste waters result from washings of fermentation vats, floors etc. The entire waste water flow is finally let out as distillery effluent.

One method of handling the waste water is to evaporate in multiple stages but this is expensive. Alternatively, the waste can be discharged into large settling tanks; the clear liquor is diluted and sent for disposal while the settled slurry containing 1-3% nitrogen, 2-4% phosphorus and 2-6% potash can be used for irrigating the surrounding areas.

The second method is anaerobic digestion for 10-12 days which reduces the BOD by 95%. 350-550 m^3 of gas (containing 50-60% methane) of heat of combustion 2.1 MJ/m^3 is obtained per kg BOD. The digested sludge is used as a manure and the effluent is sent for irrigation.

The third method is anaerobic digestion in lagoons for 65 days resulting in 90-95% BOD reduction followed by aerobic treatment of effluent before final disposal. But this method requires large areas and foul odours are emitted at times.

DAIRY WASTES

Dairies collect milk, pasteurise it and supply hygienic wholesome milk to the consumers. The other products from the dairies are powdered milk, baby food, cheese, casein etc.

The main sources of dairy wastes are:

- **Receiving station:** This waste is mainly wash water consisting of milk drippings, can washings etc.
- **Pasteurisation plant:** This waste mainly consists of water used for washing the equipment, spillages and leaks.
- **Butter manufacturing section:** The main wastes from this section consist of washing of chums and small quantities of butter.
- **Cheese manufacturing section:** Major wastes from this section consist of wash water from cleaning and rinsing the vats and discarded whey.
- **Casein plant:** This plant works intermittently on batch basis. The wastes consists of whey, mineral acids and washings.
- **Washing plant:** This plant also works intermittently depending on the frequency of operation. The wastes usually consist of bottles, cans, caustic soda solution and detergents.

The volume & characteristics of the wastes depend on the size and number of processes employed in a dairy. Typical analyses of the composite wastes are shown below in table.

Treatment: The dairy wastes are rich in organic matter. These wastes are easily degradable, tend to ferment and become anaerobic, producing foul odours. They are amenable to both aerobic and anaerobic treatment although aerobic processes are preferred.

Table 6.2T$_2$: Composite Wastes from Dairies
(All figures except pH and temperature are in mg/1)

S.N	Parameters	Dairy 1	Dairy 2	Dairy 3
1	pH	9.7	7.2	
2	Temperature, °C	35-40	—	—
3	Suspended solids	360	760	560
4	BOD 5 days at 20 °C	663	1 240	1 890
5	Oil and greases	10	290	
6	Sulphates	955	—	
7	Total dissolved solids	935		3 956

After the removal of oils and greases, the wastes are sent for biological treatment by activated sludge process with extended aeration, in trickling filters or by anaerobic treatment in lagoons or digesters. Of these activated sludge process has proved to be the most successful method.

The degree of treatment depends on the ultimate disposal of the treated effluents. Usually it is required to remove more than 90% of the BOD. However, when the treated effluents are to be used for irrigation purposes, only 50 - 60% BOD removal is adequate.

The extended aeration system achieves over 90% BOD removal when using an aeration period of about 16 hours In this process the organic matter is completely stabilised and very little excess sludge is produced. The sludge recirculated is equal to 100% of the average flow. About two kg of oxygen are required for one kg of BOD. The excess sludge is dried and used as a manure.

TANNING

Tanning is the process of making leather from animal hides. The process involves various stages which in turn produce waste waters of varying compositions. These waste waters are broadly classified as beam house wastes and tan house wastes. The beam house wastes are produced from the processes of soaking, liming & deliming and the tan house wastes are produced from tanning & finishing processes.

The volume of composite wastes is 3 000-3 400 kg of hides per day. The pH of composite waste is 7-11. The range of other characteristics in mg/l are: suspended solids: 1 200-6 000, total solids: 11000-32 000, BOD (5 days, 20 °C): 700-2 006, COD: I 400-4 200 and chrome: 0-4.

The pH of the composite waste is in the alkaline range, though some of the individual processes produce wastes of acidic nature, due to the chemical reactions involved. It is advisable tō mix all the waste waters and treat them together.

The treatment of tannery wastes consists of:

- Primary treatment for removal of the larger solids and equalisation.
- Primary sedimentation which removes 80 - 90% of the solids.

- Secondary treatment: Biological treatment using trickling filters or activated sludge process, resulting in 90% BOD reduction. Oxygen requirement is 1 to 1.2 kg/kg of BOD. The sludge from the biological treatment is dried on drying beds and used as manure.

MEAT PACKING

The wastes from the meat industry originate from three sources: the stockyard, the slaughter house and the processing plant. In India, the Ministry of Defence has installed a factory for the production of freeze dried mutton for supplying to the armed forces. In addition to this, number of bacon factories have come up in different states.

Sources and characteristics of the wastes: la a meat packing plant, the animals are slaughtered with arrangements for recovery of byproducts. The meat is processed, precooked and canned for despatch. In certain plants the precooked meat is freeze dried before canning.

The main sources of the waste waters are the slaughter house, the skinning section, the deboning and the drying & freezing sections. The pH of the composite waste water is 7.1 - 7.8. The range of other characteristics in mg/1 are: total solid: 1 400 - 4 000, suspended solids: 80 - 2 600, BOD: 70 - 600 and grease: 20 - 1 350. The pollutional effects of the waste waters are similar to domestic sewage but the danger due to pathogenic organisms is much less.

Treatment: The treatment consists of.'

- Preliminary treatment involving grit removal (detention time is 1 min), grease removal (detention 3 to 5 min) and equalisation or storage.
- Secondary treatment by chemical or biochemical methods. Chemical methods are uneconomic due to the cost of chemicals involved. Both aerobic and anaerobic processes may be used but anaerobic process is more economical. The gas from the anaerobic digestion process is similar to that from sewage plants and can be profitably used. The effluent from the digester after removal of the entrained gas is sent to a clarifier. The clarifier has a detention time of 2.5 hours and over flow rate of 20 m^3/m^2 d.

The excess sludge is sent to drying beds. With a digestion time of about four days, BOD reduction after the clarifier is about 90%.

PETROLEUM REFINERY WASTES

The main pollutant of waste water from refineries is oil due to leakages and spills during the refining operations such as distillation, cracking & chemical treatment. During emergencies, considerable amount of oil are released. The oil may be in free, emulsified or dissolved form. In addition to petroleum hydrocarbon the wastes contain acids, alkalis, sulphides, phenols, ammonia, suspended solids and other organic matter such as coke fines. The range of analytical values in mg/1 of composite wastes from a few Indian refineries is free oil: 2 000 - 3 000, emulsified oil: 80 - 120, phenolic compounds: 12-30, BOD (20 °C, 5 days):100- 300, coke fines:15 000 (maximum), suspended solids: 200 - 400.

The treatment of refinery wastes involves the removal of (1) free oil (2) coke fines (3) emulsified oil & dissolved matter and (4) phenols and other organic matter.

Removal of free oil: The free oil is removed by rectangular or circular gravity separators with oil skimming devices and sludge scrapers. The separators remove 90 - 95% of oil globules of diameters 1.5 mm with a detention time of 1 hr and settling velocity of 75 mm/min. The design criteria given by American Petroleum Institute (API) are: (1) maximum horizontal velocity of 1 m/min or 15 times the vertical velocity whichever is less, (2) depth between 0.9-2.5 m, (3) width between 1.9 - 6 m and (4) ratio of depth to width 0.3 - 0.5.

Removal of coke fines: Over 95% of the coke fines can be removed in settling tanks with mechanical scrapers. Addition of coagulants leads to effective removal of fines, but is expensive.

Removal of emulsified oil and dissolved matter: The emulsified oil and dissolved matter are removed by use of coagulants, weighting agents such as calcium carbonate & coagulant aids such as Fuller's earth in a clariflocculator provided with a skimmer to remove floating oil and mechanical scrapers for sludge removal. Flocculation occurs in 30- 40 minutes. Good results are obtained with an overflow rate of 30 m^3/m^2 d and detention time of 2 - 3 hr.

Removal of phenols and organic matter: The phenols and organic matter are removed by biological treatment in oxidation ponds, aerated lagoons, trickling filters or activation sludge plants. More than 90% removal is achieved in activated sludge plants with a loading of 0.3 kg/kg MLSS and air supply of 1.0 - 1.5 kg oxygen/kg BOD.

Sludge disposal: The organic sludge is disposed by drying in lagoons or ponds.

ELECTROPLATING WASTES

The electroplating wastes are treated to reduce and/or to remove the heavy toxic metals like chromium, zinc, copper, iron and the cyanides. The quantity and the quality of the wastes vary from one plating shop to the other and sometimes within the same plating shop itself. Prior to the plating operation, cleaning is done to remove rust, grease and scales.

The wastes from the electroplating operations can be broadly classified into (1) cyanide bearing (dilute or concentrated alkalis with cyanides) and (2) chromium bearing (dilute or concentrated acids in the hexavalent chromium).

Cyanide Wastes: Several methods for the treatment of cyanide wastes have been reported. The most accepted method is 'Alkaline chlorination process' in which the wastes are oxidised by chlorine gas in the presence of sodium hydroxide at a pH of 9. Although oxidation can be caused to the intermediate stage to produce cyanates, it is usually carried to the final stage to produce CO_2 and N_2.

$$NaCN + 2\,NaOH + Cl_2 \rightarrow NaCNO + 2\,NaCl + H_20$$

$$2\,NaCNO + 4\,NaOH + 3\,Cl_2 \rightarrow 6\,NaCI + 2\,C0_2 + N_2 + 2\,H_20$$

The first reaction is normally instantaneous, while the second reaction takes nearly one hour. For complete reaction, it is found that about 8.5 kg of chlorine and 8.0 kg of sodium hydroxide are required for one kg of cyanide.

Some small plating shops treat the wastes with the addition of ferrous sulphate under alkaline conditions to form less toxic compounds such as ferro & ferri-cyanides. Ferrous hydroxide formed reacts with cyanide to form ferrocyanide. This is further oxidised by atmospheric oxygen to ferric ferrocyanide or prussian blue. This method is not widely practised due to its questionable efficiency and formation of large volumes of coloured sludge with pronounced odours.

Chromium Wastes: The highly toxic hexavalent chromium is present as chromates, dichromates and chromic acid in the wastes produced by metal finishing operations, plating and anodising processes. Due *to* the toxicity, the chromium ions should be completely removed from the wastes before they are discharged.

$$H_2Cr_20_7 + 6\ FeS0_4 + 6\ H_2S0_4 \rightarrow Cr_2(S0_4)_3 + 3\ Fe_2(SO_4)_3 + 7\ H_20$$

$$Cr_2(S0_4)_3 + 3\ Ca(OH)_2 \rightarrow 2\ Cr(OH)_3 + 3\ CaS0_4$$

$$Fe_2(S0_4)_3 + 3\ Ca(OH)_2 \rightarrow 2\ Fe(OH)_3 + 3\ CaS0_4$$

Depending on the quantity and quality, the chromium wastes are given either a continuous. or a batch treatment.

Hexavalent chromium which exists in solution is reduced to trivalent form before it can be removed. The pH of the solution is decreased to the desired level by addition of sulphuric acid. It is then treated with reducing agents such as ferrous sulphate, sulphur dioxide or sodium bisulphite. The reduction of hexavalent to trivalent form takes place in 1 hour at pH 2-3. The trivalent chromium precipitates out as chromium hydroxide when treated under thorough agitation with sodium hydroxide or lime at pH 9-10.

The cyanide treatment produces no sludge except that due to lime. On the other hand, chromium wastes produce sludge in the form of metallic wastes. The usual practice is to dry the sludge on drying beds and use the product for filling the low lying areas. Burning of sludge is also practised.

COAL MINING WASTES

The coal mining wastes originate from two sources - coal washing and coal mine drainage.

Coal washing wastes obtained during separation of coal from mineral impurities such as clay, slate, stone etc. in gravity separators, have about 3% by volume of suspended solids. Sedimentation tanks with mechanical scrapers are used for separation of the solids from the waste water. A detention time of about 1 hour is usually sufficient.

Mine drainage wastes result from the earth's water percolating through the voids created in coal beds during mining. They drain out naturally into water courses polluting them or are removed before starting the mining. The waste water is highly acidic due to presence of sulphuric acid resulting from oxidation of sulphides present in coal and is therefore neutralised by alkali treatment.

BLAST FURNACE WASTES

About 3 tons of gas containing flue dust result from the blast furnace for every tonne of iron produced.

Much of the flue dust is removed in dust catchers. The remaining flue dust is removed by washing with water in wet cleaners. The waste water thus obtained contains 500-5000 mg/1 of solids whose composition is 70% ferric oxide, 10-15% silica and rest alumina, carbon, lime & magnesia.

Over 90% of the suspended solids in the waste water are removed in thickeners. The overflow of 1.2-2.4 m^3/m^2 hr from thickeners is reused in the steel mill or is suitably disposed. The underflow containing 40 - 50% solids is filtered in vacuum filters. The cake containing 30 - 40% moisture is sintered for reuse in blast furnaces.

RADIOACTIVE WASTES

Waste waters containing radioactive isotopes arise from the premises of atomic energy centres, nuclear power stations, hospitals and research organisations. Of the three kinds of radioactive isotopes (a -, B-, if -emitters), the B-emitters (e.g. P-32, Na-24, 1-131 and Au-198) are the most likely to occur. Some isotopes such as Br-32 emit both B and radiations. Sr-90 is the most dangerous among the B- emitters. Ra-226 emits all the three radiations and is the most hazardous of emitter.

The upper limits of allowable radioactivity are given below in table.

Table 6.2T_3: Upper Limits of Allowble Radioactivity

	µC*/ml
1. Momentary quantity of radioactivity in discharge	0.1
2. Sewage from hospitals and other establishments	
3. Direct discharge to river	10 – 4
	10 – 5
*µc (microcurie) is the amount of radio-isotope with a disintegration rate of 3.7×10^4 per second. For Sr-90, the most dangerous B-emitter, the upper limit is 10 – 6 µ c/ml	

Bacteria and other low forms of life are resistant to the effects of radiation. Hence biological processes can be used for radioactive wastes.

Wastes with radioactive substances of low half life can be stored to bring about decay of radioactivity. The methods of treatment of radioactive wastes include evaporation, adsorption in ion-exchange materials, chemical precipitation, biological methods and final disposal by fixing into a solid mass with cement and sinking deep in the sea. In recent years, the emphasis is on complete containment and on conversion of the nuclear byproducts into useful forms.

APPENDIX 6.3

Range of Biochemical oxygen demand (BOD) levels of some industrial effluents

S.N.	Sources of industrial effluents	Range of BOD, mg/litre
1	Paper pulping	1500 - 25000
2	Wool scouring	500 – 10000
3	Pharmaceutical	400 – 1000
4	Mixed chrome tanning	500 - 5000
5	Food processing wastes	500 - 4000
6	Dairy wastes	400 - 2000

APPENDIX 6.4

Industrial water quality limits (USA)

S.N.	Industry	Turbidity (units)	Color (units)	Hardness	pH	TDS	SS	SiO_2	Fe	M_aCL	Cl	SO_4	Alkalinity
1	Textiles	**0.35-5**	**0.5**	**0.50**	**100-200**	**0-5**	**25**	**0.03**	**0.01-0.05**	**1000**	**100**	**100**	**50-200**
2	**Pulp and paper:**												
	Fine paper	10[6]	5	100	-	200	-	20	0.1	0.03	-	-	75
	Karft paper -Bleached	40[6]	25	100	-	300	-	50	0.2	0.10	200	-	75
	-Unbleached	100[6]	100	200	-	500	-	100	1.0	0.05	200	-	150
	Ground wood papers	50[6]	30	200	-	500	-	100	1.0	0.50	200	-	150
	Soda and sulfite paper	25[6]	5	100	-	250	-	20	0.1	0.05	75	-	75
3	Chemicals	-	500	1000	5.5-9	2500	1000	C	10	2	500	850	500
4	Petroleum	-	25	900	6-9	3500	5000	85	15	-	1600	900	500
5	Iron and steel	-		-	38	5-9	-	100[d]	-	-	-	-	-
6	Food canning	-	5	250	6.5-8.5	500	10	50	0.2	0.20	250	250	250
7	**Tanning:**												
	Tanning processes	e	5	150	6-8	-	-	-	50	-	250	250	C
	Finishing processes	e	5	f	6.8	-	-	-	50	-	250	250	C
8	Coloring	e	5	e	6.8	-	-	-	0.3	0.20	250	250	C
9	Soft drinks	-	5	G	G	G	-	-	0.3	0.05	500	500	

Units are mg/litre, unless otherwise specified, TDS-Total dissolved solids, SS-suspended solids

b- units in mg/litre as SiO_2

c- Not considered a problem of concentrations encountered

d- Settleable solids

e- Not detecteble by tests

f- Lime softened

g- Controlled by treatment for other constituents

APPENDIX 6.5

Major industrial sources of pollution of some Indian rivers

S.N.	Name of the rivers	Major industrial sources of pollution
1	Damodar river between Bokaro and Panchet (Jharkhand)	Fly ash from steel plants, fertilisers, suspended coal particles from coal washeries, thermal power stations
2	Sone river at Dalmianagar (Bihar)	Cement, pulp & paper mill
3	Siwan (Bihar)	Paper, sulphur, sugar mills, cement
4	Kali at Merrut (U.P)	Sugar mills, distilleries, paint, soap, rayon, silk, yarn, tin and glycerene industries
5	Ganga at Kanpur (U.P)	Jute, chemical, metal and surgical industries, tanneries, textile mills and great bulk of domestic sewage of highly organic nature.
6	Gomati near Lucknow (U.P)	Pulp & paper mills, sewage
7	Dajora in Bareilly (U.P)	Synthetic rubber factories
8	Suwao in Balrampur	Sugar industries
9	Yamuna river, near Delhi	DPT factory sewage, Indraprasth thermal power station
10	Hooghly river near Kolkata	Power station, paper, pulp, jute, textiles, chemicals, paint, varnishes, metal, steel, hydrogenated vegetable oil, rayon, soap, match, shellac, polythene industries and sewage
11	Bhadra (in Karnataka)	Pulp, paper and steel industries
12	Cauvery (in Tamilnadu)	Sewage, tanneries, distilleries, paper and rayon mills
13	Godavari river	Paper mill
14	Kulu river between Bombay and kalyan	Chemical factories, Rayon mills, tanneries
15	Coolum, Adaro and Buckingham Canal (Tamilnadu)	Domestic sewage, automobiles workshop.

APPENDIX 6.6

Various physical treatments of industrial effluents, their purpose & applications

S.N.	Physical Treatment Processes	Their Purpose	Applications areas
1	Screening	Removal of coarse solids	Paper, mills, Textile mills, Vegetable tanneries, slaughter houses.
2	Filtration	Concentration, fine solid removal	Sludge dewatering, food , manufacture, Metal Processing, Chemical manufacture, final polishing treatments.
3	Sedimentation	Simple solids removal	Coal washing, oil separation, metal hydroxide removal, emulsified fats-with froth floatation,
4	Centrifuging	Concentration	Sludge dewatering, soluble oils treatment, chemical manufacture, Food, manufacture, Electrochemical machining.
5	Freezing	Concentration	Recovery of pickle liquors, Non-ferrous metals.
6	Evaporation	Concentration	Food manufacture, drag-out recovery.
7	Adsorption	Concentration and removal of trace impurities	Pretreatment for ion-exchange, Pesticide manufacture, dye stuffs removal, , dechlorination, final effluent polishing.
8	Solvent extraction	Separation	Coal carbonizing, plastics, manufacture.
9	Ion-exchange	Separation and concentration	Metal finishing, motor car manufacture, Aircraft, engine manufacture, semiconductor manufacture, precious metals, Production of high quality water.
10	Reverse osmosis	Separation and concentration	Desalination, Metal finishing, chemical manufacture.

APPENDIX 6.7

Physico-chemical characteristics of untreated waste water from various industries

S.N.	Industries	Physico-chemical characteristics of untreated waste water
1	Oil refineries	Acids, alkalis, phenols , tarry or resinous materials, petroleum oil
2	Iron and steel plants	Low pH, alkalis, acids, phenols, limestone, cyanogen, oils, fine suspended solids, cyanides , cyanates, iron salts, ores and coke
3	Fertiliser plants	High pH, high ammonia, high fluoride content, organic matter, nitrogen, phosphorous potassium, acidity or alkalinity
4	Thermal power plants	Dissolved solids, heat, heavy metals, inorganic compounds
5	Pulp and paper plants	High or low pH, color, suspended solids, fibres, high temperature, high BOD & COD, dissolved substances.
6	Metal electroplating industry	Metalics, toxic cynides, low pH, cadmium , chromium, zinc, copper, aluminium
7	Iron foundaries	Coal, clay, iron and suspended solids
8	Cotton textile industries	Sodium, organic matter, color, high pH and fibers
9	Acid manufacturing plants	Low pH and organic content
10	Rubber industries	Suspended & dissolved solids, chlorides, variable pH and high BOD
11	Organic chemical industry	High acidity, alkalinity, high pH and Zinc
12	Synthetic fibers mills	Toxic substances, sulphides, high pH and zinc
13	Photographic products	Organic & inorganic reducing agents, silver and alkalis
14	Tanneries	Calcium, chromium, high salt content, color, dissolved & suspended matter.
15	Antibiotics	High acidity or alkalinity, toxic organics
16	Synthetic drugs	High acidity or alkalinity, toxic organics
17	Distillers	Low pH , suspended & dissolved organic matter including vitamins, high acidity or alkalinity
18	Fruits & vegetable processing plants	Dissolved & colloidal organic matter
19	Soft drink industry	High BOD, pH, high suspended solids
20	Dairy products	High BOD, alkalis, greases, fats, lactose, colloidal solids mainly protein
21	Beer plants	High dissolved solids containing nitrogen & fermented starch and allied products
22	Beet sugar industry	High BOD, high suspended & dissolved organic matter.
23	Slaughter houses	High suspended solids, dissolved organic matter, protein and blood

APPENDIX 6.8

Waste water generation standards for various industries in India

S.N.	Industries	Quantity of waster water produced
1	Integrated iron & steel plant	16 m^3/ton of finished steel
2	Sugar industry	0.4 m^3/ton of finished sugar
3	Tanneries	28 m^3/ton of cane couched
4	Starch, glucose & related products	8 m^3/ton of Maize crushed
5	Dairy	3 m^3/KL of milk
6	Natural rubber processing industry	4 m^3/ton of rubber
7	**Textile industries (Man made fibre):** (i) Nylon & polyester (ii) Viscose rayon	 120 m^3/ton fiber produced 150 m^3/ton of product
8	**Fertilizer industry:** (i) Straight nitrogenous fertilizer (ii) Straight phosphatic fertilizer (SSD & TSP) (iii) Complex fertilizer	 5 m^3/ton or equivalent produced 0.5 m^3/ton of TSP Standards of nitrogenous and phosphoric fertilizers are applicable depending upon the primary product
9	**Fermentation industry:** (i) Malty (ii) Breweries (iii) Distillery	 3.5 m^3/ton of grain produced 0.25 m^3/ton of beer produced 12 m^3/LK of alcohol produced
10	**Caustic soda industry** (i) Membrane cell process (ii) Mercury cell process	1 m^3/ton of Caustic soda produced excluding cooling tower 4 m^3/ton of Caustic soda produced (mercury bearing), 10% blow down permitted for cooling tower
11	**Pulp & paper industries:** A-Large Pulp & paper plant (i) Pulp & paper (ii) Viscose staple fibre **B. small pulp & paper plant** (i) Agro residue based (ii) Waste paper based	 175 m^3/ton of Paper produced 150 m^3/ton of fibre Produced 500 m^3/ton of paper Produced 150 m^3/ton of paper Produced 50 m^3/ton of paper Produced

APPENDIX 6.9

Characteristics of waste water produced in some chemicals manufacturing industries

S.N.	Chemicals manufactured	Quantity of waste water produced, m^3/ton of product	BOD, mg/litre	COD, mg/litre
1	Ammonia	1.14-11.4	25-100	50-250
2	Butadiene	0.38-7.6	25.200	100-400
3	Urea	0.38-7.6	50-300	100-500
4	Polyvinyl chloride (PVC)	0.38-7.6	50-500	1000-2000
5	Ethylene	0.19-5.7	100-1000	500-3000
6	Propylene	0.38-7.6	100-1000	500-3000
7	Toluene	1.14-11.4	300-2500	1000-4000
8	Ethanol	1.14-11.4	300-1000	500-2000
9	Methanol	1.14-11.4	300-1000	500-2000
10	Cellulose acetate	0.038-0.76	500-2000	2500-5000
11	Acetone	1.9-5.7	1000-5000	2000-10000
12	Phenol	1.9-8.5	1200-10000	2000-15000

APPENDIX 6.10

Sugar industry waste water characteristics

S.N.	Characteristic	Mill house waste water	Filter cloth washings	Floor & boiler washings	Combined waste water
1	Oil & grease	40	-	-	12
2	pH value	5.8	6.2	7.2	6.9
3	Odour	Oily	Organic	Pungent organic	Organic
4	Color	White	Milky white	Brown	Brown
5	Total solids	1686	6406	4535	3476
6	Dissolved solids	1210	3890	2486	2368
7	Dissolved oxygen	5.2	0	0	0
8	BOD, 5 days, 37°C	247	6556	3045	1348
9	COD	287	7810	3466	1566
10	COD/BOD ratio	1.16	1.05	1.1	1.1
11	Alkalinity (as Ca CO_3)	260	280	2260	1890
12	Sulphate (as SO_4)	12.8	-	-	80
13	Total nitrogen (as N)	0.10	Traces	4.8	2.8
14	Total Phosphate (as P)	-	Traces	0.97	0.84

Except pH value, Color & odour, all values are in mg/litre

APPENDIX 6.11

Effluent characteristics of typical fertilizer factory producing ammonia, urea & phosphoric acid (in ppm)

S.N.	Plant/units	Ammonia N_2	Urea N_2	Chromium, Cr	Phosphate as P_2O_5	Fluorides as F	pH value
1	Ammonia Plant	250	—	11	3	—	8-10
2	Urea plant	4100	3300	—	—	—	9-10.5
3	Phosphoric acid plant	—	—	—	800	4000	1.5-2.0

APPENDIX 6.12

Waste water generation from a pharmaceutical formulation unit

S.N	Sources of waste water generation	Average waste water production, m^3/hr	Average effluent characteristics, mg/litre
1.	Sterile products preparation	1-10	BOD = 50-100 TSS = 20 – 260
2	Syrup preparation	10-25	COD = 150 – 2500 Cl = 20-150
3	Malt preparation	1-5	BOD = 300 – 2000 SO_4^+ = 20 - 200
4	Pastilles preparation	10-25	BOD = 2000 – 2500 TSS = 100 – 300 Heavy metals = 1 – 20 PH value = 4 - 8

APPENDIX 6.13

Characteristics & generation of waste water from an antibiotic drug plant

S.N.	Sources of waste water generation	Nature of waste water & solid wastes	Average characteristics of combined effluents, mg/litre
1	Fermentation block	Floor & equipment washings, leakages of valves/machines, contaminated batches, cooling waters, laboratories & utility wasters.	pH value = 4 – 8 TSS = 100 – 1000
2	Filtration/ Centrifugation	Mycellium cakes, filter washings,floor washings	BOD = upto10000
3	Recovery & purification block	Faecal wastes , acid & filter washings, floor washings, laboratory wastes	Oil & grease = 20 – 50 Heavy metals = 1 – 5 CN = 0.1 – 1.0
4	Style and finishing block	Floor & equipments washings, other utility wastes	Phenol (after mycellium cake separation) = 1-5

APPENDIX 6.14

Maximum tolerances for water used in pulp & paper industry

S.N.	Characteristics	Tolerances of (in mg/litre , except for pH & turbidity)				
		Ground wood paper	Soda and sulfite paper	High grade paper	Kraft paper bleached	Kraft paper unbleached and pulp unbleached
1	Color (Hazen units)*	20	10	5	15	100
2	Turbidity (silica scale units)	50	15	10	25	100
3	Total dissolved solids (TDS)	500	300	300	300	500
4	Total hardness (as $CaCO_3$)	200	100	100	100	200
5	Iron (as Fe)	1.0	0.25	0.25	0.25	1.0
6	Manganese (as Mn)	0.5	0.05	0.05	0.10	0.5
7	Iron (as Fe) & Manganese (as Mn) added together	1.0	0.25	0.25	0.25	-
8	Total residual chlorine (as Cl_2)	2.0	2.0	2.0	2.0	-

- The color produced by 1mg/liter of platinum (as potassium chloro platinate – K_2PtCl_6) is taken as the standard unit of color (Hazen unit). It is known that solutions of potassium chloro platinate tinted with small amounts of cobalt yields colors that are very much like the natural colors.
- Because of wide varieties of materials that cause turbidity in natural waters, it has been necessary to use an arbitrary standard. The standard choosen is 1 mg SiO_2 = I unit of turbidity and the silica used must meet certain specification as to particle size.

APPENDIX 6.15

Characteristics of waste from various sections of pulp & paper plant

S.N.	Characteristics	Unit	Chipper house	Digestor house	Pulp varius	Chemical recovery	Pulp bleaching	Paper machine	Combined waste water
1	Flow	M^3/ ton	20-60	5-10	20-40	-	140-180	40-90	240-300
2	Color/ appearance	-	Muddy	Dark brown	Dark brown	Light brown	Brown	Whitish	Brown
3	pH value	Unit	6.4-8.0	9.0-10.0	8.5-9.6	7-9	6-9	5.3-8.1	6.5
4	Suspended solids	Mg/ litre	240-520	140-190	350-400	400-760	140-220	490-500	350-400
5	Total solids	Mg/ litre	540-900	1000-2500	1400-2500	1270-2800	2100-2900	850-1250	1200-20000
6	BOD, 5 Days, 20°C	Mg/ litre	175-450	1850-220	900-1700	320-610	550-790	520-780	600-750
7	COD	Mg/ litre	175-450	1850-2200	900-1700	320-610	550-890	520-780	600-750

APPENDIX 6.16

Typical characteristics of waste water from various sections/processes of a tannery (mg/litre)

S.N.	Effluent characteristics	From process			From tanyord process			From Finishing process		Composite waste water from raw hide to chrome tanning fin ished leather
		Soaking	Liming	Deliming and bating	Pickling	Chrome tanning	Vegetable tanning	Neutral-isation	Dyeing & fat liquoring	
1	pH value	7.4-7.9	10-12.5	8.1-9.0	2.8-3.3	3.4-4.2	3.5-4.5	4-6.5	3.8-4.5	7.5-8.5
2	Alkalinity as Ca CO_3	600-1500	8000-15000	1000-1200	500-1200	4000-5500	2000-4000	300-1200	800-2300	1100-2000
3	BOD, 5 days, 20°C	1000-1600	5000-9000	900-1500	250-600	400-650	12000-18000	400-1000	2200-4500	1200-2500
4	COD	2300-3500	1500-3000	1400-3300	900-2500	1800-4000	25000-35000	2000-4000	5000-10000	3000-6000
5	Chlorides as Cl	15000-21000	4000-7000	700-1200	2000-35000	11000-16000	1500-3000	600-1800	1200-3000	4500-6500
6	Total solids	22000-35000	30000-45000	15000-25000	45000-65000	40000-65000	55000-80000	7000-12000	7000-12000	17000-25000
7	Dissolved solids	19500-29500	25000-33000	13500-22000	43000-60000	38000-60000	5000-15000	6000-10000	62000-10000	14000-20500
8	Suspended solids	2500-5500	5000-12000	1500-3000	2000-5000	5000-15000	1000-2000	800-2000	800-2000	3000-4500
9	Sulphides as S	-	350-700	25-40	-	-	-	-	-	20-40
10	Total chromium	-	-	-	-	1700-3500	-	25-50	Treaes	80-250

APPENDIX 6.17

Typical characteristics of liquid wastes from different sections / operations of a finished leather processing unit (mg/litre)

S.N.	Effluent characteristics	Stripping	Washing	Rechroming	Basification	Fat liquoring	Composite waste
1	Suspended solids	1500	900	900	1500	400	800
2	Total solids	10120	4400	15680	11080	6980	5240
3	BOD, 5 days, 20°C	11000	2400	2000	1500	700	900
4	COD	18810	4730	4356	4138	2562	2760
5	Free ammonia cal nitrogen	5	4.5	3	3.5	20	6
6	Chlorides	560	480	740	480	92	600
7	Sulphates	3955	1483	5361	4944	4218	2044
8	Boron (as B)	96	48	3.2	2.8	3.2	3.6
9	Sodium (as Na)	1800	600	3600	3600	1400	760
10	Chromium (Cr)	Nil	Nil	990	450	18	20
11	Sodium (as Na)	1800	600	2600	3600	1400	760
12	Oil & grease	-	-	-	-	400	70
13	pH value	6.2	6.5	3.6	4.8	3.5	5.8

APPENDIX 6.18

Typical characteristics of waste water from various sections/processes of a cotton textile mill (mg/litre)

S.N.	Effluent characteristics	Bleaching section	Dyeing section	Printing section	Combined waste water from various typical cotton textile mills			
					Cotton mill I	Cotton mill II	Cotton mill III	Cotton mill IV
1	pH value	8.4-12.1	8.2-11.8	9.2-11.8	6.8-8.5	7.2	-	11.9
2	Alkalinity	400-6280	1150-1300	800-5360	70-1000	180	1600	7300
3	Suspended solids	200-530	160-620	290-950	35	570	1600	1730
4	Dissolved solids	3184-5532	1450-2600	1900-6180	830-2360	2180	3620	7710
5	BOD, 5 days, 20°C	70-525	4-122	50-820	135-1380	160	315	760
6	COD	-	-	-	-	400	925	2400
7	Chlorides	600-2960	71-680	40-1700	250-520	90	645	1390
8	Sulphates	168-205	78-350	15-180	35-60	35	175	320
9	Oil & grease	5.8-120	-	5.6-79	-	4	18	27
10	Percent-sodium	75-92	-	71-98	-	83	86	90
11	Phenols	-	-	-	-	0.03	0.06	0.08

APPENDIX 6.19

Typical characteristics of waste water from various processes/sections of woolen textile mills (mg/litre)

S.N.	Effluent	Wool combing section	Dyeing & furnishing section	Combined waste water from three different woolen mills		
				Woolen mill I	Woolen mill II	Woolen mill III
1	pH value	3.1-9.8	5.3-9.6	7.0	7.0	7.0
2	Acidity	160	200	-	-	-
3	Alkalinity	80	240	-	-	-
4	Suspended solids	5335	715	366	1070	40
5	Dissolved solids	-	-	5216	6410	1236
6	COD	8300	2440	387	7000	400
7	BOD, 5 days, 20°C	3000	610	204	6000	210
8	Oil & grease	2455	560	12	506	35
9	Surfactants (alkyl benzene sulfonate)	240	225	-	-	-

APPENDIX 6.20

Typical characteristics of combined waste water /composite effluent of synthetic textile mills (mg/litre)

S.N.	Effluent characteristics	Textile Mill I		Textile Mill II	
		Range	Average	Range	Average
1	pH value	7.3-8.0	7.5	6.7-9.5	8.5
2	Total alkalinity	522-630	590	500-795	650
3	Total dissolved (TDS)	1060-1080	1070	2180-3600	2675
4	Suspended solids (SS)	80-130	100	80-720	500
5	BOD, 5 days, 20°C	180-200	190	60-580	325
6	COD	420-630	525	590-800	695
7	Chlorides (as Cl)	130-205	170	490-1390	1000
8	Sulphates (as SO_4)	50-90	65	47-500	322
9	Calcium (as Ca^{++})	14-17	15	8-76	32
10	Magnesium (as Mg^{++})	17-22	19	7-53	38
11	Sodium (as Na^+)	830-1050	900	610-940	830
12	Potassium (as K^+)	8-10	9	0-18	18
13	Percent sodium	92.5-94.5	93	90-98	92

APPENDIX 6.21

Typical characteristics of waste water in dye & dye intermediate industries (mg/litre)

S.N.	Effluent Characteristics	Indian Dyestuff industries, Vadodara		Atul Products Ltd, Balsar, Gujrat		Sondoz (India) LTD; Thane , Maharastra	
		Untreated waste water	Treated waste water	Untreated waste water	Treated waste water	Untreated waste water	Treated waste water
1	pH value	3.4	7.2	1.9	6-8.5	3	7.5
2	Temperature, °C	45	30	-	-	29	29
3	Color	Dark brown	Dark brown	-	-	Brown yellow	Pale yellow
4	Total dissolved solids (TDS)	5455	5658	-	-	29	29
5	Suspended Solids (SS)	96	65	370	200	200	50
6	BOD, 5 days 20°C	503	49	350	275	100	90
7	COD	1518	824	930	500	1500	400
8	Chloride	1952	1927	1125	800	-	-
9	Sulphate	1675	1028	1985	-	-	-
10	Cyanide	0.075	0.04	-	-	3	Nil
11	Oil & grease	19.3	7.2	20		-	-
12	Phosphate	-	-	23	-	-	-
13	Total nitrogen	-	-	17	-	-	-
14	Phenol	-	-	10.8	4.0	-	-

APPENDIX 6.22

MINAS for waste water from dye & dye intermediate industries

S.N.	Effluent characteristics /Parameters	Permissible limits (mg/litre) unless otherwise stated
1	pH value	6-9
2	Color, Hazen units	400
3	Suspended solids (SS)	100
4	BOD, 5 days, 20°C	100
5	Oil & grease	10
6	Phenolic compounds	1
7	Cadmium	0.2
8	Copper	2
9	Lead	0.1
10	Mercury	0.1
11	Manganese	2
12	Nickel	2
13	Zinc	5
14	Hexavalent chromium	0.1
15	Total chromium	2
16	Bioassay test	90% survival of test animals after 96 hours in treated waster water (1: 10 dilution)

APPENDIX 6.23

Characterisitics of waste water from various sections of paint industry (mg/litre, unless otherwise stated)

S.N.	Effluent characteristics	Caustic cleaning operation	Resin house	Stiff paints	Lab section	Liquid paint	Combined waste water	MINAS for paint industry
1	pH value	8.5-13.5	3.2-6.3	6.5-13.5	5.1-6.7	13.5	6-10.5	6-9
2	Suspended solids (SS)	200-600	240-400	400-700	20-200	40	220-1200	100
3	COD	1100-3800	240-78000	1215-6000	150-200	560	300-5700	-
4	BOD, 5 days, 20°C	475-2400	225-60000	380-980	18-45	60	1700-3100	50
5	Oil & grease	32-150	14-25	252	-	6	22-138	10
6	Phenolics	12.5	6-68	6.4-100	0.84-5.5	5.5	18-55	-
7	Phenolics (as C_6H_5OH)	-	-	-	-	-	-	Optional
8	Lead (as Pb)	-	-	-	-	-	-	0.1
9	Hexavalent chromium	-	-	-	-	-	-	0.1
10	Total chromium	-	-	-	-	-	-	2
11	Copper (as Cu)	-	-	-	-	-	-	2
12	Nickel (as Ni)	-	-	-	-	-	-	2
13	Zinc (as Zn)	-	-	-	-	-	-	5
14	Bio-assay test	-	-	-	-	-	-	90% survival in 96 hours

APPENDIX 6.24

Typical characteristics of composite water from dairy industries (mg/litre)

S.N.	Characteristics	Dairy I	Dairy II	Dairy III
1	pH	9.7	7.2	-
2	Temperature, °C	35-40	-	-
3	BOD, 5 days, 20°C	663	1240	1890
4	Suspended solids (SS)	360	760	560
5	Total dissolved solids (TDS)	935	-	3956
6	Oil & grease	10	290	-
7	Sulphates	955	-	-

APPENDIX 6.25

Effluent characteristics of some typical chemical plants (mg/litre)

S.N	Effluent characteristics	Petrochemical plant	Textile processing unit	Distillery	Sugar factory
1	pH	5.5-9.5	5.5-8.5	-	4.5-7
2	Total solids	2000-14000	2500-3500	-	850-3500
3	Total suspended solids	-	-	200-300	200-800
4	BOD, 5 days, 20°C	600-6000	550-2000	40000	300-2000
5	COD	-	1500-5000	80000-100000	600-4500
6	Oil & grease	10-90	-	-	-
7	Total ammoniacal nitrogen	150-550	-	-	-
8	Total nitrogen	-	25-40	-	10-40
9	Phosphate	15	-	-	-
10	Chlorides	-	250-500	8500	-
11	Sulphates	-	280-400	2000	-
12	Volatile acids	-	-	5000-5500	-
13	Residual chlorine	-	Traces	-	-
14	Temperature	-	45-65	90	-

APPENDIX 6.26

Charcteristics of alkaline and acidic mine water (mg/litre)

S.N	Characteristics	Categories of alkaline mine water			Categories of acidic mine water	
		A-1	A-2	A-3	B-1	B-2
	Water quality factors:					
1	pH	7.5-8.5	7-8.5	6-8	5.5-7	2-4.5
2	Alkalinity (as Ca Ca CO_3)	200-700	100-500	10-30	-	-
3	Acidity (as H_2 SO_4)	–	-	-	25-150	100-250
4	Hardness (as Ca CO_3)	50-150	200-1500	1000	200-500	500-3500
5	Total dissolved solids (TDS)	300-700	500-1500	1500-2500	400-800	1000-4000
6	Total suspended Solids (TSS)	5-30	12-50	10-50	-	-
	Bio-chemical Parameters:					
7	Dissolved oxygen (DO)	5.5-8	5-8	5-8	5.5-8.5	3.8-7.5
8	BOD, 5 days, 20°C	< 1	< 1-2	< 1-2	< 1-2	1-5
9	COD	8-30	15-50	15-60	10-60	20-100
	Element or ion content:					
10	Calcium	10-50	50-200	100-150	20-80	100-250
11	Magnesium	10-30	40-180	100-150	10-70	50-100
12	Ferric ion	-	-	-	0-20	10-250
13	Total iron	0-2	0-5	0-5	-	20-200
14	Ferrous ion	-	-	-	20-50	20-200
15	Sulphates	10-100	100-800	1000	200-600 & above	100
14	**Trace elements:**			-		
	Al, Cu, Zn, Mg, Ni Pb Cr, sb	-	-	-	0.5	0.5

APPENDIX 6.27

Typical characteristics of raw sewage, reclaimed water and water supplied by Bombay Municipal Corporation

S.N	Characteristics	Raw sewage	Reclaimed water	Water supplied by Bombay Municipal Corporation
1	pH value	6.9-7.2	7.2-7.9	7.3-7.6
2	Total dissolved Solids (TDS)	400-600	400-600	17-150
3	BOD, 5 days, 20°C*	150-300	1-2	2-5
4	COD (dichromate)*	90-120	3-6	5-10
5	Alkalinity (as Ca CO_3)	150-250	150-160	25-20
6	Hardness (as Ca CO_3)	150-200	35-45	40-50
7	Chloride (as Cl)	60-80	60-80	20-30

* Biological oxygen demand (BOD) is the amount of oxygen demanded in 5 days by living organisms in the water to oxidise organic matter present as food for the organisms. BOD is a characteristic and not a constituent of the water.
* Chemical oxygen demand (COD) is a measure of chemically oxidisable organic matter present in water. Potassium dichromate in presence of sulphuric acid is used as oxidizing agent. It is performed by digesting known quantity of water with acid and potassium dichromate for specified time.

APPENDIX 6.28

Important constituents of effluents discharged from various industries

S.N.	Industries/Processes/ Plants	Important constituents of the effluent
1	Acetate rayon plant	Acetic acid
2	Aluminium anodizing	Chromium
3	Atomic energy plants	Fluorides
4	Battery manufacture	Lead, mineral acids
5	Bleaching industries	Sulfate, free chlorine
6	Breweries	Sugars, mineral acids
7	Ceramic plants	Fluorides
8	Chemicals Manufacturing plants	Phenols, mineral acids, ammonia, organic acids such as tartaric acid, nitro compounds etc.
9	Chocolate & sweet industries	Sugars
10	Chrome tanning	Chromium
11	Copper Pickling	Copper, mineral acids
12	Copper Plating	Copper
13	Cotton mercerizing	Alkalis
14	Cotton & straw kiering	Alkalis
15	Cupro ammonium rayon manufacture	Copper
16	Dairies	Sugar
17	DDT manufacture	Mineral acids
18	Distilleries	Organic acids
19	Dye Manufacture	Phenolic compounds
20	Dyeing industries	Tartaric acid
21	Engineering works	Oil, fats and grease
22	Explosive factories	Nitro compounds
23	Fertilizers plants	Phosphates, fluorides
24	Food processing	Starch
25	Galvanising industry	Zinc
26	Gas and coke manufacture	Ammonia, cyanides, sulfides, Phenols
27	Glass etching	Fluorides
28	Iron pickling	Iron, mineral acids
29	Laundries	Alkalis, free chlorine, fats, oils and grease
30	Leather manufacture	Tartaric acid
31	Metal cleaning	Cyanides

S.N.	Industries/Processes/ Plants	Important constituents of the effluent
32	Metal refineries	Mercaptans
34	Paint mills	Lead
35	Paper mills	Free chlorine
36	Penicillin manufacture	Formaldehyde
37	Peroxide bleaching of textiles	Hydrogen peroxide
38	Petroleum refineries	Hydrocarbons
39	Petroleum refineries	Hydrocarbons, Phenols, fats, oil and grease
40	Photo engraving	Mineral acids
41	Photography	Silver
42	Pickle manufacture	Acetic acid
43	Plating industries	Copper, chromium, nickel , cadmium, Zinc & Silver
44	Preservative manufacture	Sugar
45	Pulp mills	Mercaptans
46	Rock - motor testing	Hydrogen peroxide
47	Rubber processing	Zinc
48	Saw mills	Tannic acid
49	Scrubbing of flue gases	Fluorides
50	Sheet dipping	Arsenic, phenols
51	Soft drinks manufacture	Citric acid
52	Sugar factories	Sugars
53	Sweets manufacture	Phenols,formaldehyde
54	Synthetic resin manufacture	Phenols, formaldehyde
55	Synthetic rubber factories	Hydrocarbons
56	Tanneries	Sulfides, chromium, phenols, tannic acid
57	Tar distilleries	Phenols
58	Textile bleaching	Free chlorine, sulfites
59	Textile industries	Mineral acids, fats, oil and grease
60	Transistor factories	Fluorides
61	Viscose film manufacture	Sulfites
62	Viscose rayon manufacture	Sulfides, zinc
63	Wall paper manufacture	Starch
64	Wood processing	Sugars
65	Wood pulp processing	Alkalies and sulfites
66	Wool scouring	Alkalis, fats, oil and greases

APPENDIX 6.29

Industrial Wastewater Treatment Approaches-Control options & technologies

S.N.	Water pollutants/ parameters	Control option/principle	Common control technology
1	pH	Chemical, Equalization	Acid, Base addition, Flow equalization
2	Oil and Grease-TPH	Phase separation	Dissolved Air Floatation, oil separator, grease trap
3	TSS-Settleable	Settling , Size Exclusion	Sedimentation, Basin, clarifier, centrifuge, screens
4	TSS-Non-Settleable	Floatation, filtration-traditional and tangential filtration	Dissolved air floatation, Multi media filter, sand filter, fabric filter
5	High -BOD (>2 kg/m^3)	Biological- Anaerobic	Suspended growth, attached growth, hybrid
6	Low -BOD (< 2 kg/m^3)	Biological-Anaerobic	Suspended growth, attached growth, hybrid
7	COD-Non-Biodegradable	Oxidation, Adsorption , Size Exclusion	Chemical oxidation, Thermal oxidation, traditional and tangential
8	Metals-Particulate and soluble	Coagulation, flocculation, precipitation, size exclusion	Flash mix settling, filtration-traditional and tangential filtration
9	Inorganics/Non-metals	Coagulation, flocculation, precipitation, size exclusion, Oxidation, Adsorption	Flash mix with filtration-traditional and tangential
10	Organics-VOCs and SVOCs	Biological-Aerobic, Anaerobic, Facultative; Adsorption, Oxidation	Biological: Attached growth; chemical oxidation, Thermal oxidation, Activated Carbon
11	Emissions-Odors and VOCs	Capture-Active, Anaerobic, Facultative, Adsorption, Oxidation	Biological-Aerobic, Chemical oxidation, Activated Carbon

S.N.	Water pollutants/ parameters	Control option/principle	Common control technology
12	Nutrients	Biological Nutrient Removal, Chemical & Physical, Adsorption	Aerobic/Attached Growth: Chemical oxidation, Thermal oxidation, Activated Carbon
13	Color	Biological-Aerobic, Anaerobic Facultative; Adsorption, Oxidation	Biological Aerobic, Chemical oxidation, Activated Carbon
14	Temperature	Evaporative Cooling	Surface Aerobic, Chemical oxidation, Activated Carbon
15	TDS	Evaporative Size Exclusion	Evaporation, Oxidation, Reverse Osmosis
16	Active Ingradients/ Emerging Contaminants	Adsorption, Oxidation, Size Exclusion, Concentration	Chemical oxidation, Thermal oxidation, Activated Carbon, Ion Exchange, Reverse Osmosis, Evaporation, Crystallization
17	Radionuclides	Adsorption, Size Exclusion, Concentration	Ion Exchange, Reverse Osmosis, Evaporation, Crystallization
18	Pathogens	Disinfection, Sterilization	Chlorine, Ozone, Peroxide, UV, Thermal
19	Toxicity	Adsorption, Oxidation, Size Exclusion, Concentration	Chemical oxidation, Thermal oxidation, Activated Carbon, Evaporation, crystallization, Reverse Osmosis

Note:

VOCs-Volatile Organic Chemicals

SVOCs-Semi Volatile Organic Chemicals

APPENDIX 6.30

Industrial Sources & effects of major water contaminants/Pollutants

S.N.	Type of contaminants	Some Industrial sources	Some major effects of water contaminants/pollutants
1	Inorganic salts	Oil refinery, desalination plants, ammunitions manufacturing and pickle curing	Interferes with industrial usage, municipal (drinking water), and agriculture (irrigation water) uses
2	Acids and/or alkalies	Chemical manufacturing, tanneries	Corrosion of pipelines and equipments, kills fish
3	Organic matter	Tanneries, textile mills etc.	Food for bacteria and thus depletes oxygen
4	Suspended matter	Paper mills, tanneries, etc.	Food for bacteria and thus depletes oxygen
5	Suspended matter	Paper mills, tanneries, etc.	Suffocates fish , degrades stream , appearance
6	Heated water	Cooling waters from most industries and power plants	Accelerates bacterial action, lowers total oxygen saturation lavel
7	Color	Textile, tanneries, metal finishing, and chemical plants	Objectionable appearance
8	Toxic chemicals	ammunitions manufacturing, metal, steel mills, petrochemicals, etc.	Alters stream biota and animal diversity
9	Microorganism	Pharmaceutical, combined municipal- industrial plants.	Unsafe for drinking and swimming
10	Radioactivity	Nuclear power plants, chemical laboratories	Concentrates in fish, harmful for drinking
11	Foam producing matter	Glue manufacturing slaughter house ,detergent manufacturing	Aesthetically objectionable

APPENDIX 6.31

Origin, characteristics, treatment & disposal methods of industrial liquid wastes produced by some major chemical industries

S.N.	Industries Producing Liquid Wastes	Origin	Major Characteristics	Major Treatment and Disposal Methods
1	Acids and Alkalies industries	Dilute wash water; many varied dilute acids and bases	Low or high pH, low organic content	Upflow or straight neutralization, coagulation and sedimentation
2	Detergents industries	Washing and purifying soaps and detergents	High in BOD and saponified soaps	Flotarion and skimming, precipitation with $CaCl_2$
3	Explosive industries	Wahsing TNT and gun cotton for purification, washing and pickling of cartridge	TNT, colored, acid, odorous, and contains organic acids and alcohol from powder and cotton, metals, acid, oils, and soaps	Floatation, chemical Precipitation, biological treatment, aeration, chlorination of TNT, neutralization, adsorption
4	Pesticides industries	Washing and purification of products	High organic matter, benzene ring structure, toxic to bacteria and fish acid	Dilution, storage, activated carbon adsorption, alkaline chlorination
5	Phosphate and phosphorus industries	Washing, screening, floating rock, condenser bleed-off from phosphate reduction plant	Clays, slimes and oils, low pH, high suspended solids, phosphorus, silica and fluoride	Lagooning, mechanical clarification, coagulation and settling of refined activated charcoal
6	Formaldehyde plants	Residues from manufacturing synthetic resins and from dyeing synthetic fibers	Normally high BOD and formaldehyde, toxic to bacteria in high concentrations	Trickling filtration, absorption on activated charcoal

7	Plastics and resins industries	Unit operations from polymer preparation and use: spills and equipment wash downs	Acids, caustic, dissolved organic matter such as phenols, formaldehyde, etc.	Discharge to municipal sewer, reuse, controlled-discharge
8	Fertilizer	Chemical reactions of basic elements, Spills, cooling water, washing of products, boiler blowdown	Sulfuric, phosphoric & nitric acids; minerals elements, P,S,N, K, Al,NH_3, NO_3	Neutralization detain for reuse, sedimentation, air stripping of NH_3, lime precipitation
9	Toxic chemicals industries	Leaks, accidental spills, and refining of chemicals	Various toxic dissolved elements and compounds such as Hg and PCBs	Retention and reuse, change in production, neutralization and precipitation, carbon adsorption

APPENDIX 6.32

Origin, characteristics, treatment & disposal methods of liquid wastes generated in various crude oil refinery operations

S.N.	Major Industrial Operations in crude oil refineries	Origin of Major Wastes from various units	Major Characteristics of the liquid wastes	Major Treatment and Disposal Methods
1	Crude oil and product storage	Primary fractionation of oil and water, spills and leakages	High concentrations of emulsified oil, COD, TSS	API separation, DAF, settling, aeration
2	Crude desalting	Chemical desalting and gravity separation of oil	Emulsified and free oil, ammonia, phenol, sulfide, TSS, high BOD & COD	API separation, DAF, activated sludge, carbon adsorption
3	Cracking	Thermal cracking or catalytic cracking, fractionation, steam stripping, and overhead accumulators of fractionators	BODs, COD, ammonia, phenol sulfide, cyanides, and alkalinity	Chemical oxidation, biological treatment, carbon adsorption
4	Polymerization	Catalytic reaction, and removal action and gas stabilizer	Alkaline waste stream, high in sulfide, mercaptans, and ammonia	Acid catalysts recycled, carbon adsorption
5	Alkylation	Catalytic reaction caustic and water wastes, neutralization of hydrocarbon streams	Oil, sulfides, TSS, fluoride	Neutralization, chemical oxidation, sedimentation

APPENDIX 6.33

Origin, characteristics, treatment & disposal methods of liquid wastes generated in some major metal industries

S.N.	Industries Producing liquid	Origin of Major Wastes	Major Characteristics of the liquid water	Major treatment and Disposal Methods
1	Steel	Coking of coal, washing of blast-furnace flue gases, and pickling of steel	Low pH, acids, phenol, coke, limestone, alkali, oils, mill scale, and fine suspended coal	Selective screening, drying of reclaimed sand
2	Iron-foundry products	Washing of used sand by hydraulic discharge	High suspended solid, mainly sand; some clay and coal	Selective screening, drying of reclaimed sand
3	Metal container	Cutting and lubricating metals, cleaning can surface	Metal fines, lubricating oils, variable pH, surfactants, dissolved metals	Oil separation, lagoon storage, Final carbon adsorption
4	Metal plated products	Stripping of oxides, cleaning and plating of metals	Acid, metals, toxic, low volume, mainly mineral matter	Alkaline chlorination of cyanide; reduction and precipitation of chromium: lime precipitation of other metals

APPENDIX 6.34

Origin, characteristics, treatment & disposal methods of liquid wastes generated in operations in a pulp & paper industry

S.N.	Major Industrial Operations	Origin of Major Wastes	Major Characteristics	Major Treatment and Disposal Methods
	Pulp mill:			
1	Wood preparation	Log transportation and storage, debarking and chipping, Mechanical pulping, kraft pulping, sulfide pulping	Solid wastes, hydraulic debanking uses water	Incineration, water is recycled through lagoons
2	Pulping	Mechanical pulping, kraft pulping, sulfide pulping	Suspended solid, BOD, liquors containing high BOD, sulfide, mercaptans, high pH	Sedimentation, recovery if sodium hydroxide and sodium sulfide, aeration
3	Deinking	Removal of ink	Contains dirt and clay fillers, chemicals and usually alkaline	Sedimentation and coagulation
4	Pulp washing, screening, and thickening	Washing action, centrifugal cleaning, gravity and vacuum thickening Bleaching of pulp thickening	Liquor contains fibers	Pulp fibers are recovered and recycled
5	Bleaching	Bleaching of pulp fibers by chlorine solution	Contains chlorine, low pH,	Dechlorination neutralization
	Paper mill:			
6	Stock preparation	Addition of filler, colors, chemicals, and screening	Color, clay fillers, chemicals, high TDS	Coagulation and reuse
7	Paper machine	Removal of free water	Contains fillers, known as white water., TSS. TDS and BOD	Fibers reclaimed, liquid wastes treated by coagulation, activated sludge
8	Finishing and converting	Surface improving, size cutting, Mostly dry process	Little or no liquid waste, mainly solid waste	Recycled in pulping operation, incineration

APPENDIX 6.35

Origin, characteristics, treatment & disposal methods of liquid wastes generated in major food industries

S.N.	Food, Industries liquid Wastes	Origin of Major Wastes	Major Characteristics of the liquid wastes	Major Treatment and Disposal Methods
1	Canned goods	Trimming, culling, juicing and balancing of fruits and vegetables	High in suspended solids, colloidal and dissolved organic matter	Screening, lagooning, soil absorption by spray irrigation
2	Dairy Products	Dilutions of whole milk, buttermilk and whey	High in dissolved matter, mainly protein, fat and lactose	Biological treatment aeration, RBC, trickling filtration, activated sludge, whey recovery
3	Brewed and distilled beverages	Steeping and pressing of grain; residue from distillation of alcohol; condensate from stillage evaporation	High in dissolved organic solids, containing nitrogen and fermented starches or their products	Recovery, concentration by centrifugation and evaporation, RBC; trickling filtration; use in feeds; anaerobic biological treatment followed by aerobic treatment
4	Meat and poultry products	Stockyards; slaughtering of animals; residues in condensates: grease and wash water; picking chickens	Organic matter, blood, other proteins, and fats & oils	Screening, settling and/or floatation, trickling filtration, RBC, activated sludge, recovery for animal food products
5	Animal feedlots	Excreta from animals	High in organic suspended solids and BOD	Land disposal and anaerobic lagoons
6	Beet sugar	Transfer, screening, and juicing waters: drainings from lime sludge; condensates after evaporator; juice and extracted sugar	Variable, pH, soluble organic matter with relatively high BOD of carbonaceous nature.	Neutralization, recirculation, chemical treatment, some selected aerobic oxidation
7.	Cane sugar	Spillage from extraction, clarification, evaporation cooling, and condenser waters	Variable pH, soluble organic matter with relatively high BOD of carbonaceous nature	Neutralization, recirculation, chemical treatment, some selected aerobic oxidation

APPENDIX 6.36

Glossary of terms related to Industrial Waste Water & Its Treatment

Absorption. The passage of one substance into or through another, e.g. an operation in which one or more soluble components of a gas mixture are dissolved in a liquid.

Activated Carbon. A highly adsorbent form of carbon used to remove odorous and toxic substances from liquid or gaseous emissions. In waste treatment it is used to remove dissolved organic matter from wastewater.

Activated Sludge. Sludge that results when primary effluent is mixed with bacteria-laden sludge and then agitated & aerated to promote biological treatment. This speeds breakdown of organic" matter in raw sewage undergoing secondary waste treatment.

Adsorption. An advanced method of treating waste in which activated carbon removes organic matter from wastewater.

Advanced Wastewater Treatment. Any treatment of sewage that goes beyond the secondary or biological water treatment stage and includes the removal of nutrients such as phosphorus & nitrogen and a high percentage of suspended solids.

Aerobic Treatment. Process by which microbes decompose organic compounds in the presence of oxygen and use liberated energy for reproduction & growth. Types of aerobic processes include extended aeration, trickling filtration, and rotating biological contactors.

Agglomeration. The process by which precipitation particles grow larger by collision or contract with cloud particles or other precipitation particles.

Anaerobic. A process that occurs in, or is not destroyed by, the absence of oxygen.

Bar Screen. In wastewater treatment, a device used to remove large solids.

BOD (Biochemical Oxygen Demand). A measure of the amount of oxygen consumed in the biological processes that break down organic matter in water.

Biodegradable. The ability to break down or decompose rapidly under natural conditions and. processes.

Biological Treatment. A treatment technology that uses bacteria to consume waste. This treatment breaks down organic materials.

COD (Chemical Oxygen Demand). A measure of the oxygen required to oxidize all compounds in water, both organic and inorganic.

Chemical Treatment. Any one of a variety of technologies that use chemicals or a variety of chemicals processes to treat waste.

Clarifier. A tank in which solid is settled to the bottom and is subsequently removed as sludge.

Coagulation. A clumping of particles in wastewater to settle out impurities. It is often induced by chemicals such as lime, alum, and iron salts.

Contaminant. Any physical, chemical, biological, or radiological substance or matter that has an,. adverse effect on air, water, or soil.

Dechlorination. Removal of chlorine for a substance by chemically replacing it with hydrogen or hydroxide ions in order to detoxify the substances involved.

Decomposition. The breakdown of matter by bacteria and fungi. It changes the chemical makeup and physical appearance of materials.

Denitrification. The anaerobic biological reduction of nitrate nitrogen gas.

Diffused Air. A type of aeration that forces oxygen into sewage by pumping air through perforated pipes inside a holding tank and bubbling it through the sewage.

Digester. In wastewater treatment, a unit process for degradation of sludge.

Disinfectant. A chemical or physical process that kills pathogenic organisms in water. Chlorine is used to disinfect sewage treatment effluent, water supplies, wells, and swimming pools.

DO (Dissolved Oxygen). The oxygen freely available in water. Dissolved oxygen is vital to fish & other aquatic life and for the prevention of odors. Secondary and advanced waste treatment are generally designed to protect DO in waste receiving waters.

Dissolved Solids. Disintegrated organic and inorganic material contained in water. Excessive amounts make water unfit to drink or to use in industrial processes.

Evaporation Ponds. Areas where sewage sludge is dumped and allowed to dry out.

Filtration. A treatment process for removing solid (particulate) matter from water by passing the water through porous media such as sand or a manmade filter. The process is often used to remove particles that contain pathogenic organisms.

Flocculation. The process by which clumps of solids in water or sewage are made to increase in size by biological or chemical action so that they can be separated from the water.

Heavy Metals. Metallic elements with high atomic weights, e.g., mercury, chromium, cadmium, arsenic, and lead. They can damage living things at low concentration and tend to accumulate in the food chain.

Ion Exchange Treatment. A common water softening method often found on a large scale at water purification plants that remove some organisms and radium by adding calcium oxide or calcium hydroxide to increase the pH to a level where the metals will precipitate out.

Mechanical Aeration. Use of mechanical energy to inject air into water to cause a waste stream to absorb oxygen.

Nutrient. Any substance assimilated by living things that promote growth. The term is generally,, applied to nitrogen & phosphorus in wastewater, but is also applied to other essential and trace elements.

Organic Matter. Carbonaceous waste contained in plant or animal matter and originating from industrial sources.

Pathogens. Microorganism that can cause disease in other organisms or in humans, animals and plants. They may be bacteria, viruses, or parasites and are found in sewage. Fish and shellfish contaminated by pathogens, or the contaminated water itself, can cause serious illnesses.

pH, A measure of the acidity or alkalinity of a liquid or solid material.

Physical & Chemical Treatment. Processes generally used in large-scale wastewater treatment facilities. Physical processes may involve air stripping or filtration. Chemical treatment includes coagulation, chlorination, or ozone addition.

Pretreatment. Processes used to reduce, eliminate, or alter the nature of wastewater pollutants from non-domestic sources before they are discharged into publicly owned treatment works.

Primary Waste Treatment. First steps in wastewater treatment; screens and sedimentation tanks are used to remove most materials that float or will settle. Primary treatment results in the removal of about 30 per cent of carbonaceous biochemical oxygen demand from domestic sewage.

Raw Sewage. Untreated wastewater.

Residual. Amount of pollutant remaining in the environment after a natural or technological process has taken place, e.g., the sludge remaining after initial wastewater treatment.

Reverse Osmosis. A water treatment process used in small water systems by adding pressure to force water through a semi-permeable membrane.

Sand Filters. Devices that remove some suspended solids from sewage. Air and bacteria decompose additional wastes filtering through the sand so that cleaner water drains from the bed.

Screening. Use of screens to remove coarse floating and suspended solids from sewage.

Secondary Treatment. The second step in most publicly owned waste treatment systems in which bacteria consume the organic parts of the waste. It is accomplished by bringing together waste, bacteria, and oxygen in trickling filters or in the activated sludge process. This treatment removes floating & settleable solids and about 90 per cent of the oxygen demanding substances & suspended solids. Disinfection is the final stage of secondary treatment.

Sedimentation. Letting solids settle out in wastewater by gravity during wastewater treatment.

Settleable Solids. Materials heavy enough to sink to the bottom of a wastewater treatment tank.

Skimming. Using a machine to remove oil or scum from the surface of the water.

Sludge. A semi-solid residue from many number of air or water treatment processes.

Stabilization. Conversion of the active organic matter in sludge into inert, harmless material.

Suspended Solids. Small particles of solid pollutants that float on the surface of, or are suspended in sewage or other liquids. They resist removal by conventional means.

Tertiary Treatment. Advanced cleaning of wastewater that goes beyond the secondary or biological stage. It removes nutrients such as phosphorus &nitrogen and most BOD & suspended solids.

TDS (Total Dissolved Solids). The total amount of dissolved solid materials present in an aqueous solution.

TOC (Total Organic Carbon). TOC is a measure of the amount of carbon in a sample originating from organic matter only. The test is run by burning the sample and measuring the CO_2 produced.

TOD (Total Oxygen Demand). A measure of the oxygen demand of wastewater by injecting a small volume of the sample into oxygen containing carrier gas and passed through a catalyst bed at 900 °C. The carbon, nitrogen, & many minerals are converted to their oxides by consuming oxygen from the carrier gas, this amount of gaseous oxygen consumed is measured and given as the TOD of the sample in mg oxygen/liter of wastewater.

TSS (Total Suspended Solids). A measure of the suspended solids in wastewater, effluent, or water bodies.

Trickling Filter. A coarse, biological treatment system in which wastewater is trickled over a bed of stoned or other material covered with bacterial growth, leading to bacterial break down of the waste.

Turbidity. A cloudy condition in water due to suspended silt or organic matter.

VSS (Volatile Suspended Solids). The suspended organic fraction which will oxidize and will driven off as gas at temperature 550 ± 50 °C.

Wastewater Treatment Plant. A facility containing a series of tanks, screens, filters, and other processes by which pollutants are removed from water.

APPENDIX 6.37

Origin, and characteristics of Industrial waste water produced by various industries

S.N.	Industries producing waste water	Origin of major wastes	Major characteristics & waste water
1	Textile	Cooking of fibres, desizing of fabrics	Highly alkaline, colour, high BOD
2	Leather goods	Dehairing soaking and deliming of hides	High total solids, hardness, salt, sulphides, chromium, pH, precipitated lime and BOD
3	Laundary Trade	Washing of fabrics	High turbidity, and organic solids.
4	Canned goods	Trimming, culling, juicing and balancing of fruits & vegetables	High in dissolved organic matters, mainly protein, fat and lactose.
5	Dairy product	Dilution of whole milk, separated milk, butter milk and whey	High in dissolved organic matter
6	Brewed distilled beverages	Steeping and pressing grain residue from distillation of alcohol	High in dissolved organic solids, containing nitrogen, chlorine and fermented or their products
7	Meat and poultry products	Stock yards, slaughtering of bones and fats, residues in concentrated grease, wash water and pickling of chickens.	High in dissolved suspended organic matter, blood, other proteins and fats
8	Animal	Excreta from animal	High in organic suspended solids and BOD
9	Beet sugar	Transfer, screening and juicing water, drainage from lime	High in dissolved & suspended organic matters containing sugar and protein
10	Pharmaceutical products	Mycelium, spent filtrate and wash waters	High in suspended & dissolved organic matter including vitamins
11	Yeast	Residue from yeast filtration	High in solids (mainly organic and (BOD)
12	Pickles	Lime water, brine, alum and turmeric, syrup deeds, etc	Variable pH, suspended solids, colour and organic matter
13	Coffee	Pulping and fermenting of coffee bean	High BOD and suspended solids

S.N.	Industries producing waste water	Origin of major wastes	Major characteristics & waste water
14	Fish	Rejects from centrifuge, evaporators and other wash water wastes	Very high BOD, total organic solids and odour
15	Rice	Soaking, cooking and washing of rice	High BOD, total & suspended solids and BOD
16	Soft drink	Bottle washing, floor and equipment cleaning, syrup storage, tank drains	High pH, suspended solids and BOD
17	Bakeries	Washing and greasing of pans, floor, alum sludge.	Minerals and suspended solids.
18	Water product	Filter backwash, lime soda, sludge, brine, alum, sludge	Minerals and suspended solids
19	Pulp and paper	Cooking, refining, washing of fibres, screening of paper, paper pulp	High or low pH, colour, high suspended colloidal and dissolved solids
20	Photographic	Spent solution of developer and fixer	Alkaline, containing various organic and inorganic reducing contents
21	Steel	Coking of coal, washing of blast furnace flue gases, pickling of steel	Low pH, acids, phenol, and ore, cokes, limestone, alkali oils and fine suspended solid
22	Metal plating	Stripping of oxides, cleaning of steel	Acid metals, toxic low volume, mainly, mineral matter
23	Iron Foundary	Washing of used sand by hydraulic discharge	High suspended solid, mainly sand, some clay and coal
24	Oil fields and refineries	Drilling muds, salt, oil and some natural gas, acid sludge and miscellaneous oils from refining	High dissolved salts , high BOD odour, phenol and sulphur compounds from refining
25	Fuel oil use	Spills from fuel tank filling waste, auto crank case oils	High in emulsified and dissolved oils.
26	Rubber	Washing of latex, coagulated rubber, exuded impurities from crude rubber	High BOD, odour, high suspended solids, variable pH, high chlorides
27	Glass	Polishing and cleaning of glass	Red colour, alkaline, non settleable suspended solids
28	Naval stores	Washing solvent recovery	Acid, high BOD and oil recovery water
29	Glue manufacturing	Line wash acid, extraction of non-specific proteins	High COD, BOD, pH chromium, periodic strong mineral acids.
30	Wood preservation	Steam condensates	High in COD, BOD, solids, phenols

S.N.	Industries producing waste water	Origin of major wastes	Major characteristics & waste water
31	Candle manufacturing.	Waste spills, stearic acid condensates, dilute	Organic (fatty) acid, low, pH, low organic content
32	Plywood manufacturing	Glue washing	High BOD, pH, phenols, potential toxicity
33	Acid	Dilute wash waters, dilute acids	Low pH, low organic contents
34	Detergents	Washing & purifying soaps and detergents	High BOD and saponified soaps.
35	Corn starch	Evaporator condensate or bottoms, when not re-used or recovered syrup final washes	High BOD and dissolved matter, mainly starch & related materials
36	Explosives	Washing TNT and gun cotton for purification, washing and pickling or cartridges	TNT colour, acid ,odorous and contains organics acids & alcohol from powder and cotton, metals acids, oils and soaps
37	Pesticides	Washing and purification of products, such as 2, 4, D & DDT	High organic matter, benzene, ring structure, toxic to bacteria and fish acid
38	Phosphate and Phosphorous	Washing and screening of floating rock	Clays, slimes , oils low pH, high suspended solids, phosphorous, silica and fluorides.
39	Formaldehyde	Residues from manufacturing synthetic resins and from dying synthetic fibres	Normally high BOD and HCHO, toxic to bacteria in high concentrations
40	Plastic and Resins	Unit operations for polymer preparations and use, spills and equipment wash downs	Acids, caustic dissolved organic matter such as phenols, formaldehyde, etc
41	Coal processing	Cleaning and classification of coal leaching and use, spills and equipment wash started with water	High suspended solids, mainly coal, low pH, high H_2SO_4
42	Steam power	Cooling water, boiler blow down, coal drainage	Hot, high volume and high inorganic and dissolved solids
43	Nuclear power and radioactive materials	Processing ores, laundering of contaminated clothes, research lab wastes, processing of fuel, power plant, cooling water	Radio active elements can be very acid "hot"

Chapter

Water Quality and Standards

Water quality is a term used to express the suitability of water for various uses or processes. Any particular use of water will have certain requirements for the physical,chemical or biological characteristics of water; for example limits on the concentrations of toxic substances for drinking water use, or restrictions on temperature & pH ranges for water supporting animals. Consequently, water quality can be defined by a range of variables which limit the water. Although many uses have some common requirements for certain variables, each use will have its own demands and influences on water quality. It should be free from colour, turbidity, taste & odour, of moderate temperature and aerated.

"The water quality is that which a body of water should be able to support its designated uses, which include drinking, industrial uses ,shell fishing, swimming, other body contact, recreational boating, or fishing".

Any physical, chemical, or biological condition that prevents the designated use(s) of a body of water represents poor water quality.

The quality of water may be described in terms of the concentration and state (dissolved or particulate) of some or all of the organic & inorganic material present in the water, together with certain physical characteristics of the water.

Water temperature, pH, specific conductance, turbidity, dissolved oxygen and suspended sediments are important parameters, if we discuss about water quality. Potable water is an important parameter for human beings, because it is useful for maintaining human health.

PHYSICAL QUALITY CHARACTERISTICS OF WATER

The main physical quality characteristics of water are colour, odour , taste, turbidity, and temperature. These physical quality characteristics and their related quality standards are described below.

Colour - Colour in water is derived from coloured substances, such as vegetable matter, dissolved from roots & leaves, from humus, or from inorganic compounds such as iron & manganese salts. The colour standard is designed to make drinking water more palatable.

Odour and Taste - There is no set standards for odour & taste as there are no specific tests for these. Odour and taste found in water are most commonly caused by algae, decomposed organic matter, dissolved gases, or industrial waste.

Turbidity - Turbidity refers to a muddy or unclear condition of water caused by suspended clay, silt, organic & inorganic matter, plankton and other microorganisms. The turbidity standard was established to improve the efficiency of disinfection by reducing particles to which micro organisms could attach.

Temperature - Warm water tastes flat. Cooling water suppresses odours & tastes and makes it more palatable. Temperature also effects the chlorination & purification of water. Disinfection takes longer when water is colder, and purification capacity is reduced with reverse osmosis treatment equipment. Water having physical characteristics exceeding the limits or making it less palatable should not be used for drinking. Otherwise, reduced consumption and increased risk of dehydration may result. High water temperature in rivers and stream reduces dissolved oxygen content posing threat to survival of aquatic creatures,fishes etc.

CHEMICAL QUALITY CHARACTERISTICS OF WATER

The chemical quality of water depends on the chemical substances it contains. These substances include TDS, chlorides, sulphates and other ions. The chemical quality of water involves its hardness, alkalinity, acidity, and corrosiveness. Chemical substances having an adverse health effect have established standards that should not be exceeded.

Potential Hydrogen (pH)

The pH is a measure of the acidic or alkaline nature of water. It is defined as the negative logarithm of the hydrogen ion concentration. It ranges from 0 to 14 and pH of 7 is neutral. The pH influences the corrosiveness of the water, the amount of chemicals needed for proper disinfection, and the ability of an analyst to detect contaminants. Water with a pH below 7 is regarded as acidic while that with a pH above 7 is regarded as alkaline. The pH standard was established to ensure effective purification and disinfection.

Arsenic

Arsenic can be present in natural water sources in a wide range of concentrations. It can come trom either natural or industrial sources. Ingestion of low concentrations of arsenic can cause nausea, vomiting, abdominal pain, or nerve damage.

Chloride

Chloride exists in most natural waters. It is the main anion found in seawater. Chloride comes from natural salt deposits, domestic & industrial waste, and agricultural runoff. Even in low concentrations, chloride can produce an objectionable taste in water.

Calcium, Magnesium & Hardness

Calcium & magnesium in drinking water, is an important parameter to be taken into account. Water hardness is a measure of the capacity of water to precipitate soap, which is in practice the sum of concentrations of all polyvalent cations present in water (Ca, Mg, Sr, Ba, Fe, Al, Mn, etc.); nevertheless, since the other ions (apart from Ca & Mg) play a minor role , generally hardness is defined as the sum of the Ca & Mg concentrations.

Water calcium & magnesium result from decomposition of calcium & magnesium aluminosilicates and, at higher concentrations, from dissolution of limestone, magnesium , magnesite, gypsum & other minerals. Drinking water may be intentionally supplemented with these elements while treated, as happens with deacidification of underground waters by means of calcium hydroxide or filtration through different compounds counteracting acidity such as $CaCO_3$, $MgCO_3$ & MgO, and also with stabilization of low-mineralized waters by addition of CaO & CO_2. In low & medium mineralized underground & surface waters (as drinking waters are), calcium & magnesium are mainly present as simple ions Ca^{2+} & Mg^{2+}, the Ca levels varying from tens to hundreds of mg/l and the Mg concentrations varying from units to tens of mg/l.

Magnesium is usually less abundant in waters than calcium, which is easy to understand since magnesium is found in the Earth's crust in much lower amounts as compared with calcium. In common underground and surface waters the weight concentration of Ca is usually several times higher compared to that of Mg, the Ca to Mg ratio reaching up to 10. However, a common Ca to Mg ratio is about 4.

Sulphates

Sulphates occur naturally in water as the result of dissolution of sulfur bearing minerals. High concentrations also result from industry sources, such as coal mine drainage, pulp paper mills, tanneries, textile mills, and domestic waste water. When ingested, sulphates have a laxative effect. They also can produce a bad taste in water. The sulphate standard was established to prevent chemically induced diarrhea.

Total Dissolved Solids(TDS)

The TDS of water is composed of mineral salts and small amounts of other inorganic & organic substances. The proportion of each constituent is the result of weathering of rocks found in the drainage basin and of any industrial contributions. Since TDS is composed of chloride, magnesium, sulfate, and other ions, its ingestion in water has the same effects. Therefore, the TDS standard was established to prevent chemically induced diarrhea.Chemical water quality standards are based on the effect the water will have on the health of living organisms. The effect of a particular chemical substance determines if a limit is established for that substance. Chemical substances having a negative physical effect will have a mandatory limit that should not be exceeded. Some substances, such as iron and manganese, have no significant negative physical effect, but may restrict the use of the water, such as for the laundering of clothes.

MICROBIOLOGICAL QUALITY OF WATER

The microbiological quality of potable water shows its potential for transmitting waterborne diseases which are caused by viruses, bacteria, protozoa, or higher organisms. The testing for microorganisms in water is extremely difficult. The number of these organisms is usually very low, even in a badly polluted water supply, and the test used to find them is difficult. For these reasons, indicator organisms are used to detect the presence of contamination. The bacterial organisms used as an indicator of possible contamination are total coliform. These organisms occur in large quantities in the intestines of warm blooded animals. The presence of coliform organism in potable water is an indication of either inadequate treatment or the introduction of undesirable materials to the water after treatment. While the detection of many disease-causing microbes is difficult, the test to detect a surrogate organism, E. coli, is simple and effective for field use. Because of its relative simplicity and field adaptability, the membrane filter technique has gained wide acceptance as the preferred technique for the presumptive determination of the presence of coliform organisms in potable water. The microbiological standard was established to ensure that infectious microorganisms would not cause diseases.

Some of the important water quality parameters and regulatory limits for them are as follows:

Dissolved oxygen

Dissolved oxygen (DO) is critical for aquatic life and also controls chemical processes. DO is important to aquatic organisms to perform their decomposition function ("aerobic" decomposers). DO enter the water from the air and from aquatic plant photosynthesis. Oxygen is consumed by respiration (Biological Oxygen Demand) and some chemical reactions. Values of DO less than a daily mean of 5.0 mg/liter (ppm) or instantaneous value of 4.0 mg/l is considered to be low and less than 2.0 mg/l to be dangerous to aquatic life.

Turbidity

Turbidity is caused by suspended material that blocks light. High levels of turbidity car slow the growth of aquatic plants and reduce visibility for aquatic animals that need to see their prey. Turbidity is a measure of how much light is scattered by materials suspended in the water. Scattered light is much less available for aquatic plants, so high levels of turbidity can limit the growth of aquatic plants as well as restrict the visibility of prey and enemies to aquatic animals. Consequently, there are standards for turbidity in fresh (50 NTU) & salt (25 NTU) waters. An NTU is a Nephelometric Turbidity Unit.

Fecal coliform bacteria

Fecal Coliform Bacteria are indicators of recent contamination by fecal matter from warm blooded creatures, including humans, other mammals and birds. The MFC membrane filtration technique/ method, enumerate fecal coliforms in water samples. Most of the bacteria detected by this method are E. coli, the most abundant bacterium in the guts of humans. While E. coli itself is usually not a pathogen, its presence is an indicator of the possible presence of real pathogens that can cause disease.

Coliform bacteria themselves are not necessarily pathogenic, but they indicate the likelihood that more dangerous organisms, such as the organisms that cause salmonellosis, cholera or hepatitis, are present. Although contamination of waters by human wastes is especially serious, animal wastes can also transmit disease. Consequently, high levels of fecal coliform bacteria, regardless of the source, imply a high risk of disease transmission. Two additional warnings apply: a) some pathogens live longer in surface waters than fecal coliform bacteria, and b) many bacteria are alive in the water but cannot be cultured and detected by the standard techniques. Owing to the risk of disease transmission by contact with waters contaminated by fecal wastes, standards for this water quality parameter are quite stringent.

As guidelines, waters designated for shell fishing should not have average fecal coliform levels above 14 colony forming units (CFU) per 100 ml. Waters designated for most other uses, e.g., swimming, should not have average above 200 CFU/100 ml.

Total Nitrogen and Phosphorus

Total Nitrogen is a measure of the quantity of the important plant nutrient, nitrogen, present in the water in dissolved and particulate forms. It is one measure of the degree of nitrogen loading to surface waters.

Total Phosphorus is a measure of the important plant nutrient, phosphorus, which not only stimulates aquatic weed growth, but also blooms of nitrogen fixing blue-green algae (cyanobacteria) and bacteria.

These plant nutrients occur at variable but typically low levels in natural waters. Elevated levels of nitrogen and/or phosphorus can stimulate excessive plant growth ("eutrophication"), which can be very harmful to aquatic ecosystems. There are no general standards for nitrogen or phosphorus. Non-point sources of nitrogen and phosphorus are not currently regulated. Some countries have banned high-phosphate detergents, in order to control algae blooms in rivers receiving sewage effluents.

Total suspended solids

Sediment loading to surface waters raises turbidity & suspended solids levels and causes siltation, other pollutant loadings, and problems for aquatic life. Sedimentation is a serious problem in areas downstream of improperly managed land disturbing activity. Suspended solids cause siltation, clog the gills of aquatic animals, and carry toxic chemicals and nutrients with them when they enter the water. High levels of suspended solids are considered to be one of the most widespread water quality problems, and are frequently caused by human activities.

Chlorophyll

Chlorophyll-a is a measure of phytoplankton biomass. All plants have the photosynthetic pigment chlorophyll, so its concentration in the water is a measure of plant abundance. It is a good measure for phytoplankton, which is otherwise hard to quantify. Standard for chlorophyll is 40 micrograms per liter (μg/l), above which water quality is considered to be impaired by excess phytoplankton. However, chlorophyll measures only the biomass of small phytoplankton. Larger aquatic plants that grow on surfaces or the big aquatic weeds that fill many ponds are not measured by this technique. There is no accepted standard method for the measurement of these larger plants, so measures of chlorophyll in a body of water may grossly underestimate the actual amount of plant growth there. High levels of plant growth indicate excessive nutrient loading and can cause problems with low dissolved oxygen at night.

Because shallow waters usually have enough light for plants to grow, unusually high levels of Chlorophyll-a usually indicate excessive nutrient loading.

WATER QUALITY CLASSIFICATION SYSTEM FOLLOWED IN INDIA

In India, five water quality classes have been designated (A-E) on the basis of the water quality requirements for a particular use:

***Class A* waters**: for use as drinking water source without conventional treatment but after disinfection.

Class B **waters**: for use for organised outdoor bathing.

Class C **waters:** for use as drinking water source with conventional treatment followed by disinfection.

Class D **waters:** to maintain aquatic life (i.e. propagation of wildlife and fisheries).

Class E **waters:** for use for irrigation, industrial cooling and controlled waste disposal.

The five classes have been used to set quality objectives for stretches of the Yamuna and Ganga rivers, and surveys have been carried out to compare the actual river-quality classification with that required to sustain the designated best use. Where a river has multiple uses, the quality objectives are set for the most stringent (best) use requirements. After comparing ambient water quality with the designated water quality objective, any deficiencies will require appropriate pollution control measures on the discharges, including discharges in upstream stretches. This system is also helpful for the planning and siting of industry. No industries are permitted to discharge any effluent in stretches of rivers classified in Class A.

MAIN WATER QUALITY PARAMETERS

Water quality is determined by various water parameters like temperature,pH , specific conductance,turbidity,dissolved oxygen,hardness ,suspended sediments etc.

Water temperature- It is important to fisherman , industries, and growth of fish & algae. A lot of water is used for cooling purposes in power plants that generate electricity. They need cool water to start with, and they generally release warmer water back to the environment. The temperature of the released water can affect downstream habitats. Temperature also can affect the ability of water to hold oxygen as well as the ability of organisms to resist certain pollutants.

pH of water- pH is a measure of how acidic/basic a water sample is. The range goes from 0 - 14, with 7 being neutral. The pH of less than 7 indicates acidity, whereas a pH of greater than 7 indicates a base. Since pH can be affected by chemicals in the water, pH is an important indicator of water that is changing chemically. The pH is reported in logarithmic units. Each number represents a 10-fold change in the acidity/basicity of the water.Water with a pH of 5 is ten times more acidic than water having a pH of 6. Pollution can change a water's pH, which in turn can harm animals and plants living in the water. For instance, water coming out of an abandoned coal mine can have a pH of 2, which is very acidic and would definitely affect any fish and other micro organisms.

Specific Conductance-Specific conductance is a measure of the ability of water to conduct an electrical current. It is highly dependent on the amount of dissolved solids (such as salt) in the water. Pure water, such as distilled water, will have a very low specific conductance, and sea water will have a high specific conductance. Rainwater often dissolves airborne gases & dust while it is in the air, and thus often has a higher specific conductance than distilled water. Specific conductance is an important water-quality measurement because it gives a good idea of the amount of dissolved material in the water.

Turbidity-Turbidity is the amount of particulate matter that is suspended in water. Turbidity measures the scattering effect that suspended solids have on light: the higher the intensity of scattered light, the higher the turbidity. Material that causes water to be turbid include:

- Clay
- Silt
- Finely divided organic & inorganic matter
- Soluble coloured organic compounds
- Plankton
- Microscopic organisms

Turbidity makes the water cloudy or opaque. Turbidity is measured by shining a light through the water and is reported in Nephelometric Turbidity Units (NTU). During periods of low flow (base flow), many rivers are a clear green colour, and turbidities are low, usually less than 10 NTU. During a rainstorm, particles from the surrounding land are washed into the river making the water a muddy brown colour, indicating water that has higher turbidity values. Also, during high flows, water velocities are faster and water volumes are higher, which can more easily stir up and suspend material from the stream bed, causing higher turbidities.

Dissolved Oxygen-Although water molecules contain oxygen atom, this oxygen is not what is needed by aquatic organisms living in natural waters. A small amount of oxygen, upto about ten molecules of oxygen per million of water, is actually dissolved in water. This dissolved oxygen is breathed by fish & zooplankton and is needed by them to survive. Rapidly moving water, such as in a mountain stream or large river, tends to contain a lot of dissolved oxygen, while stagnant water contains little. Bacteria in water can consume oxygen as organic matter decays. Thus, excess organic material in the tanks, lakes and rivers can cause an oxygen-deficient situation to occur. Aquatic life can have a hard time in stagnant water that has a lot of rotting, organic material in it, especially in summer, when dissolved oxygen levels are at a seasonal low.

Hardness -The amount of dissolved calcium & magnesium in water determines its 'hardness.Where the water is relatively hard, it is difficult to get lather up when washing hands or clothes. And, industries in these area have to spend money to soften their water, as hard water can damage equipment. Hard water can even shorten the life of fabrics and clothes.

Suspended Sediment- It is the amount of soil moving along in a stream. It is highly dependent on the speed of the water flow, as fast-flowing water can pick up and suspend more soil than calm water. During storms, soil is washed from the stream banks into the stream. The amount that washes into a stream depends on the type of land in the river's watershed and the vegetation surrounding the river. If land is disturbed along a stream and protection measures are not taken, then excess sediment can harm the water quality of a stream. These bunds and fences are supposed to trap sediment during a rainstorm and keep it from washing into a stream, as excess sediment can harm the creeks, rivers, lakes, and reservoirs. Sediment coming into a reservoir is always a concern; once it enters it cannot get out - most of it will settle to the bottom. Reservoirs can 'silt in' if too much sediment enters them. The volume of the reservoir is reduced, resulting in less area for water storage for agriculture as well as reducing the power-generation capability of the hydel power plant in the dam.

PRIMARY WATER QUALITY STANDARDS IN INDIA FOR VARIOUS WATER USES

Indian water quality standards are tabulated below.

A. **Drinking water source without conventional treatment but after disinfections:**
- Total coliform organism MPN/100ml shall be 50 or less
- pH between 6.5 and 8.5
- Dissolved Oxygen 6 mg/l or more
- Biochemical Oxygen Demand 5 days 20°C, 2 mg/l or less

B. **Outdoor bathing (Organised)**
- Total coliforms organism MPN/100ml shall be 500 or less
- pH between 6.5 and 8.5
- Dissolved Oxygen, 5 mg/l or more

C. **Drinking water source after conventional treatment and disinfection:**
- Total coliforms organism MPN/100ml shall be 5000 or less
- pH between 6 and 9
- Dissolved Oxygen 4 mg/l or more
- Biochemical Oxygen Demand 5 days 20°C, 3 mg/l or less

D. **Propagation of wild life and fisheries:**
- pH between 6.5 to 8.5
- Dissolved Oxygen, 4mg/l or more
- Free Ammonia (as N), 1.2 mg/l or less

E. **Irrigation, industrial cooling, controlled waste disposal:**
- Electrical Conductivity at 25°C micro mhos/cm, Max. 2250
- Sodium absorption Ratio, Max. 26
- Boron, Max., 2mg/l

ACCEPTABLE WATER QUALITY STANDARDS

Water quality management should ensure that water supplied is free from pathogenic organisms, clear, potable , free from undesirable taste & odour, of reasonable temperature, neither corrosive nor scale forming and free from minerals which could produce undesirable physiological effect.

Physical & chemical standards for Potable and Safe water is tabulated below.

S.N.	Characteristics	* Acceptable	** Cause for Rejection
1.	Turbidity (Units on NTU scale)	2.5	10
2.	Colour (units on platinum-cobalt scale)	5.0	25
3.	Taste and odour	Unobjectionable	Unobjectionable
4.	pH	7.0 to 8.5	6.5 to 9.2

5.	Total dissolved solids(mg/l)	500	1500
6.	Total hardness (mg/l as $CaCO_3$)	200	600
7	Chlorides (mg/l as Cl)	200	1000
8	Sulphates (as SO_4, mg/l)	200	1000
9	Fluorides (as F, mg/l)	1.0	1.5
10	Nitrates (as NO_3, mg/l)	45	100
11	Calcium (as Ca, mg/l)	75	200
12	Magnesium (as Mg, mg/l)	30	150
13	Iron (as Fe, mg/l)	0.1	1.0
14	Manganese (as Mn, mg/l)	0.05	0.5

Bacteriological Standards

(a) Coliform count in any water sample of 100 ml should be zero. A sample of water entering the distribution system that does not conform to this standard calls for an immediate investigation into both efficacy of the purification and the method of sampling.

(b) Water in the distribution system should satisfy all the three criteria indicated below:

- *E,Coli* count in 100 ml of any sample should be zero.
- Coliform organisms not more that 10 per 100 ml shall be present in any sample.
- Coliform organisms should not be detectable in 100 ml of any two consecutive samples or more than 50% of the samples collected for the year.

World Health Organisation (WHO) Guideline Values for chemicals (in **mg/l**) that are of health significance in drinking water is as follows:

Arsenic – 0.01
Fluoride – 1.5
Manganese – 0.4
Nitrate – 50
Nitrite – 3
Chlorine – 5
Copper – 2000
Lead – 10
Nickel – 20

POTABLE WATER QUALITY

Potable water must be free of anything that would degrade human performance. Further, it should not damage the materials used in its transportation and storage. Potable water must be suitable for maintaining human health (personal hygiene and medical treatment). Water quality standards give a basis for selecting or rejecting water intended for human use. These standards provide minimum accepted values for safeguarding human health.

Impurities in Water

As water goes through the hydrologic cycle, it gathers many impurities. Dust, smoke, and gases fill the air and can contaminate rain, snow, hail, and sleet. The rain water picks up silt, chemicals, and disease organisms. As it enters the earth through seepage and infiltration, some of the suspended impurities may be filtered out. However, other minerals and chemicals are dissolved and carried along. As groundwater it may contain disease organisms as well as harmful chemicals. In addition to the impurities in water resulting from infiltration, many are contributed by an industrialized society. The garbage, sewage, industrial waste, pesticides etc., are all possible contaminants of raw water. Impurities in raw water are either suspended or dissolved. Suspended impurities include diseases organisms, silt, bacteria, and algae. They must be removed or destroyed before the water is consumed. Dissolved impurities include salts, (calcium, magnesium, and sodium), iron, manganese, and gases (oxygen, carbon dioxide, hydrogen sulphide, pH and nitrogen). These impurities must be reduced to levels acceptable for human consumption.

Raw Water Classifications

Water is classified as fresh, brackish, or salt water (seawater) based on the concentration of TDS (Total Dissolved Solids). Fresh water has a TDS concentration of less that 1,500 ppm. Brackish water is high in minerals and has a TDS concentration between 1,500 ppm and 16,000 ppm. Salt water has a TDS concentration greater than 15,000 ppm.

Generally, groundwater (subsurface) has less chemical or biological contaminants than surface water, provided reasonable care is exercised in the selection of the well site. Harmful micro organisms are usually reduced to tolerable levels by passage through the soil.

Treated Water Classifications(PalatableWater)

Palatable water is water that is pleasing in appearance and taste. It is significantly free from colour, turbidity, taste, and odour. It should also be cool and aerated.

Potable Water Quality Standards

Quality standards for treated water reflect the values of substances allowed in potable water. Standards exist to measure the physical, chemical, microbiologic, and radiologic quality of water and to test for the presence of chemical agents.

Physical Quality of potable water

The principal physical characteristics of water are colour, odour and taste, turbidity, and temperature. These characteristics and their related quality standards are described below:

Colour - Colour in water is derived from coloured substances, such as vegetable matter, dissolved from roots and leaves, from humus, or from inorganic compounds such as iron and manganese salts. The colour standard is designed to make drinking water more palatable.

Odour and Taste - There is no set standards for odour and taste as there are no specific tests for these. Odour and taste found in water are most commonly caused by algae, decomposed organic matter, dissolved gases, or industrial waste.

Turbidity - Turbidity refers to a muddy or unclear condition of water caused by suspended clay, silt, organic and inorganic matter, and plankton and other microorganisms. The turbidity standard was established to improve the efficiency of disinfection by reducing particles to which micro organisms could attach.

Temperature - Warm water tastes flat. Cooling water suppresses odours and tastes and makes it more palatable. Temperature also effects the chlorination and purification of water. Disinfection takes longer when water is colder, and purification capacity is reduced with reverse osmosis treatment equipment. Water having physical characteristics exceeding the limits or making it less palatable should not be used for drinking. Otherwise, reduced consumption and increased risk of dehydration may result. When water of low physical quality has not been used, the appropriate command level will make that decision based on medical recommendations.

Chemical Quality of potable water

The chemical quality of water depends on the chemical substances it contains. These substances include TDS, chlorides, sulphates and other ions. The chemical quality of water involves its hardness, alkalinity, acidity, and corrosiveness. Chemical substances having an adverse health effect have established standards that will not be exceeded without medical approval.

Potential Hydrogen - The pH is a measure of the acidic or alkaline nature of water. It is technically defined as the negative logarithm of the hydrogen ion concentration. It ranges from 0 to 14 and pH of 7 is neutral. The pH influences the corrosiveness of the water, the amount of chemicals needed for proper disinfection, and the ability of an analyst to detect contaminants. Water with a pH below 7 is regarded as acidic while that with a pH above 7 is regarded as alkaline. The pH standard was established to ensure effective purification and disinfection.

Arsenic - Arsenic can be present in natural water sources in a wide range of concentrations. It can come from either natural or industrial sources. Ingestion of low concentrations of arsenic can cause nausea, vomiting, abdominal pain, or nerve damage.

Chloride - Chloride exists in most natural waters. It is the main anion found in seawater. Chloride comes from natural salt deposits, domestic and industrial waste, and agricultural runoff. Even in low concentrations, chloride can produce an objectionable taste in water. The chloride standard ensures that potable water is also palatable.

Calcium, Magnesium and Hardness

Calcium and magnesium in drinking water, is an important parameter is to be taken into account: water hardness, even if this term is incorrect and obsolete from a strictly chemical point of view. That is to say that both of these elements largely have not been analysed individually in drinking water in the past, but just non-specifically in summary as hardness. This approach was applied in many studies focused on health effects of this 'water factor'.

Since the definition of water hardness is approached either analytically or technologically, it was not and still has not been defined in a unified manner, and as with other parameters, multiple definitions have been available and multiple units have been used to express it (German, French, and English degrees; equivalent $CaCO_3$ or CaO in mg/l).

Initially, water hardness was understood to be a measure of the capacity of water to precipitate soap, which is in practice the sum of concentrations of all polyvalent cations present in water (Ca, Mg, Sr, Ba, Fe, Al, Mn, etc.); nevertheless, since the other ions (apart from Ca and Mg) play a minor role in this regard, later it has been generally accepted that hardness is defined as the sum of the Ca and Mg concentrations.

From the technical point of view, multiple different scales of water hardness were suggested (e.g. very soft – soft – medium hard – hard – very hard). Expectedly, both extreme degrees (i.e. very soft and very hard) are considered as undesirable concordantly from the technical and health points of view, but the optimum Ca and Mg water levels are not easy to determine since the health requirements may not coincide with the technical ones.

Calcium and magnesium presence in waters

Water calcium and magnesium result from decomposition of calcium & magnesium alumino silicates and, at higher concentrations, from dissolution of limestone, magnesium limestone, magnesite, gypsum & other minerals. Anthropogenenic contamination of drinking water sources with calcium and magnesium is not common but drinking water may be intentionally supplemented with these elements while treated, as happens with deacidification of underground waters by means of calcium hydroxide or filtration through different compounds counteracting acidity such as $CaCO_3$, $MgCO_3$ and MgO, and possibly also with stabilization of low mineralized waters by addition of CaO and CO_2.

In low and medium mineralized underground and surface waters (as drinking waters are), calcium and magnesium are mainly present as simple ions Ca^{2+} and Mg^{2+} , the Ca levels varying from tens to hundreds of mg/l and the Mg concentrations varying from units to tens of mg/l.

Magnesium is usually less abundant in waters than calcium, which is easy to understand since magnesium is found in the Earth's crust in much lower amounts as compared with calcium. In common underground and surface waters the weight concentration of Ca is usually several times higher compared to that of Mg, the Ca to Mg ratio reaching up to 10. Nevertheless, a common Ca to Mg ratio is about 4, which corresponds to a substance ratio of 2.4.

Sulphates - Sulphates occur naturally in water as the result of dissolution of sulfur-bearing minerals. Significant concentrations also result from industry sources, such as coal mine drainage, pulp paper mills, tanneries, textile mills, and domestic waste water. When ingested, sulphates have a laxative effect. They also can produce a bad taste in water. The sulphate standard was established to prevent chemically induced diarrhea.

Total Dissolved Solids - The TDS of water is composed of mineral salts and small amounts of other inorganic and organic substances. The proportion of each constituent is the result of weathering of rocks found in the drainage basin and of any industrial contributions. Since TDS is composed of chloride, magnesium, sulfate, and other ions, its ingestion in water has the same effects. Therefore, the TDS standard was established to prevent chemically induced diarrhea.

Chemical water quality standards are based on the effect the water will have on the health of living organisms. The effect of a particular chemical substance determines if a limit is established for that substance. Chemical substances having a negative physical effect will have a mandatory limit that should not be exceeded. Some substances, such as iron and manganese, have no significant negative physical effect, but may restrict the use of the water, such as for the laundering of clothes.

Microbiological Quality of potable (drinking) water

The microbiological quality of potable water shows its potential for transmitting waterborne diseases. These diseases may be caused by viruses, bacteria, protozoa, or higher organisms.

A microbiological test will reveal the quality of the raw water source and aid in determining any treatment required. The test is necessary to maintain the quality of the water. The testing for microorganisms in water is extremely difficult. The number of these organisms is usually very low, even in a badly polluted water supply, and the test used to find them is difficult. For these reasons, indicator organisms are used to detect the presence of contamination. The bacterial organisms used as an indicator of possible contamination are total coliform. These organisms occur in large quantities in the intestines of warm-blooded animals. The presence of any coliform organism in treated potable water is an indication of either inadequate treatment or the introduction of undesirable materials to the water after treatment. While the detection of many disease-causing microbes is difficult, the test to detect a surrogate organism, E. coli, is simple and effective for field use. Because of its relative simplicity and field adaptability, the membrane filter technique has gained wide acceptance as the preferred technique for the presumptive determination of the presence of coliform organisms in potable water. The microbiological standard was established to ensure infectious microorganisms would not cause diseases.

APPENDIX 7.1

Maximum permissible limit of various pesticides in drinking water

S.N	Pesticides	Maximum Permissible Limit, mg/litre
1	Endrin	0.001
2	Chlordane	0.003
3	Toxaphene	0.005
4	Aldrin	0.017
5	Dieldrin	0.017
6	Heptachlor	0.018
7	Heptachlor epoxide	0.018
8	Methoxychlor	0.035
9	DDT	0.042
10	Lindane	0.056
11	Organic phosphates and carbonates	0.100
12	Herbicides	0.100

Note:- Drinking Water Standards as per BIS-10500 (2004) is tabulated below.

S.N	Parameter	Unit	Desirable limits	Permissible limits
1	pH		6.5 to 8.5	6.5 to 8.5
2	TDS	mg/l	500	2000
3	Total Alkalinity	mg/l	200	600
4	Total Hardness	mg/l	300	600
5	Nitrate	mg/l	25	45
6	Chlorides	mg/l	250	1000
7	Fluoride	mg/l	1	1.5
8	Iron	mg/l	0.3	1
9	sulphate	mg/l	200	400
10	Calcium	mg/l	75	200
11	Magnesium	mg/l	30	100
12	E-Coli	Number/ 100 ml	0	No relaxation

Note: Desirable & permissible limit, causes, effect & treatment for controlling various drinking water parameters is given below.

PH:

Desirable Limit: 6.5-8.5; Permissible limit: No Relaxation

Cause:

- Dissolved minerals of medium, the water passes through

Effect:

- Acidic: pH less than 7: Corrosion, Excessive iron content makes reddishness in plumbing fixtures and clothes, affects mucous membrane
- Basic: pH Greater than 7: Scaling in plumbing fixture and deposits, bitter taste, affects aquatic life

Treatment

- Increase pH by soda ash.
- Decrease pH by white vinegar/ citric acid.

TDS (Total Dissolved Solids)

Desirable Limit: 500 mg/l; Permissible limit: 2000 mg/l

Cause

- Livestock waste, septic system, Landfills, nature of soil, Dissolved minerals.

Effect

- Hardness, scaly deposits, sediment, cloudy colored water, staining, corrosion of pipes and fittings.
- Gastro intestinal irritations and salty or bitter taste.

Treatment

- Reserve Osmosis, Distillation, deionization by ion exchange.

Total Alkalinit

Desirable Limit: 200 mg/l; Permissible limit: 600 mg/l.

Cause:

- Pipes, landfills, Hazardous waste landfills.

Effect:

- Causes deterioration of plumbing and increases the change for many heavy metals in water.
- Boiled rice turns yellowish.

Treatment:

- Neutralizing Agent.

Total hardness

Desirable Limit: 300 mg/l; Permissible limit: 600 mg/l

Cause:

- Dissolved calcium and magnesium from soil.
- Aquifer mineral containing limestone or dolomite.

Effect:

- Scale in utensils and hot water system, soap scum.
- Skin irritation; boiled meat and food become poor in quality.

Treatment:

- Water Softener, Ion Exchanger, Reverse Osmosis.

Nitrate

Desirable Limit: 25 mg/l.

Cause:

- Livestock facilities, septic systems, manure lagoons, fertilizers Domestic waste water, Natural Deposits.

Effect:

- Carcinogenic, algae growth in water bodies, Blue Baby syndrome.

Treatment:

- Ion Exchange, Distillation, Reverse Osmosis.

Chlorides

Desirable Limit: 250 mg/l; Permissible limit: 1000 mg/l.

Cause:

- Fertilizers, Industrial wastes, minerals, seawater.

Effect:

- Bad taste, High blood pressure, salty taste, corroded pipes fixtures and appliances, blackening and pitting of stainless steel.

Treatment:

- Reverse Osmosis, Distillation, Activated Carbon.

Fluoride

Desirable Limit: 1.0 mg/l; Permissible limit: 1.5 mg/l.

Cause:

- Dissolved fluorides in soil the water passes through in groundwater.

Effect

- Brownish discoloration of teeth and bone damage (Dental or skeletal).

Treatment:

- Activated Alumina, Distillation, Reverse Osmosis, Ion exchange.

Iron

Desirable Limit: 0.3 Mg/l; Permissible limit: 1.0 Mg/l.

Cause:

- Leaching of cast iron pipes in water distribution systems
- Iron content in path medium dissolved in water.

Effect:

- Turbidity in water, Brackish color, rusty sediment.
- Bitter or metallic taste, brown-green stains, iron bacteria causing slime, discolored foods.

Treatment:

- Oxidizing filter, Green-sand Mechanical Filter.

Sulphate

Desirable Limit: 200 Mg/l; permissible limit: 400 Mg/l

Cause:

- Animal sewage, septic system, sewage.
- By- product of coal mining, industrial waste.
- Natural deposits , salt.

Effect:

- Bitter, medicinal taste, scaly deposits, corrosion.
- Rotten-egg" odor from hydrogen sulfide gas formation.
- Laxative effect; gastro intestinal irritation.

Treatment:

- Ion Exchange, Distillation, Reverse Osmosis.

Calcium

Desirable Limit: Mg/l; Permissible limit: 200 Mg/l.

Cause:

- Water in contact with limestone, reaction with carbon dioxide.

Effect:

- Chief sources of scale in equipment, boilers, pipe lines, etc.
- Forms scum with soap, interferes with dyeing, etc.
- Deterioration of the quality of clothes.

Treatment:

- Softening; demineralization; internal boiler water treatment; surface active agents.

Magnesium

Desirable Limit: 30 Mg/l; Permissible limit: 100 Mg/l

Cause:

- Deposits in rock and soil.
- Industrial pollution.

Effect:

- Poor lathering and deterioration of clothes;
- With sulfate laxative effect.

Treatment:

- Ion Exchange, Oxidizing Filter, distillation.

E-coli

Desirable Limit: nil/100 ml.

Cause:

- Livestock facilities, septic systems, manure lagoons, domestic waste water, open defecation and poor sanitary condition around sources.

Effect:

- Gastrointestinal illness, diarrhea, cholera.

Treatment:

- Chlorination, Ultraviolet, Distillation.

APPENDIX 7.2

Standards for drinking water prescribed by different agencies

A = Recommended maximum concentration (mg/litre) except where shown otherwise.

B = Maximum permissible concentration (mg/litre) except where shown otherwise.

S.N	Parameters	ICMR		WHO		USPHS	
		A	B	A	B	A	B
	Physical:-						
1	Turbidity	5	25	5	25	5	-
2	Colour (hazen units)	5	25	5	50	15	-
3	Odour	Nothing disagreeable		Unobjectionable		To = 3	
	Chemical:-						
4	pH, Units	7-8.5	6.5 or 9.2	7-8.5	6.5 or 9.2	-	-
5	Total solids	-	-	500	1500	500	-
6	Calcium	75	200	75	200	-	-
7	Magnesium	50	150	50	150	-	-
8	Iron	0.3	1.0	0.3	1	0.3	-
9	Manganese	0.1	0.5	0.1	0.5	0.05	-
10	Copper	1.0	0.5	0.1	1.5	1.0	-
11	Sulphates	200	400	200	400	250	-
12	Phenols	0.001	0.002	0.001	0.002	0.001	-
13	Fluorides	1.0	2.0	0.5	1-1.5	0.6-1.7	-
14	Nitrates	20	50	-	50-100	45	-
	Toxic:-						
15	Arsenic	-	0.2	-	0.2	0.01	0.05
16	Barium	-	-	-	-	-	1.0
17	Cadmium	-	-	-	-	-	0.01
18	Chromium	-	0.05	-	0.05	-	0.05
19	Cyanide	-	0.01	-	0.01	0.01	0.20
20	Lead	-	0.10	-	0.10	-	0.05
21	Selenium	-	0.5	-	0.05	-	0.01
22	Silver	-	-	-	-	-	0.05
23	**Bacteriological:-**	-	1 coliform per 100 ml	-	1 Coliform per 100 ml	-	1 coliform per 100 ml

ICMR: Indian Council of Medical Research

WHO: World Health Organisation

USPHS: United States Pubic Health Services

Note: Water Quality Standards

- Water quality indicates the physical, chemical and biological characteristic of water.
- It is a measure of the condition of water relative to the requirements of one or more biotic species and / or to any human need or purpose.
- It is most frequently used by reference to a set of standard against which compliance can be assessed.

  The most common standards used to assess water quality related to health of ecosystems, safety of human contact and drinking water.

Water Quality Categories

– **Human consumption**

- Drinking water, including bottled water, may reasonably be expected to contain at least small amounts of some contaminants.
- Presence of these contaminants dose not necessarily indicate that water poses a health risk.

– **Industrial use**

- Dissolved minerals may affect suitability of water for a range of industrial and domestic purposes.
- Presence of ions of calcium and magnesium which interfere with the cleaning action of soap, and can form hard sulfate and soft carbonate deposits in water heaters or boilers. Hard water may be softened to remove these ions.

  Softened but may sacrifice nutrition for cleaning effectiveness.

Water Quality Categories

– In the environments.

- Toxic substances and high populations of certain microorganisms can present a health hazard for non-drinking purposes such as irrigation, swimming, fishing, rafting, boating, and industrial uses.
- These conditions may also affect wildlife, which use the water for drinking or as a habitat.

– Irrigation purpose: Crop production- irrigation water quality requirements

– Agriculture- single largest user of freshwater - major cause of degradation of surface water and groundwater resources through erosion & chemical runoff.

Water Quality – Parameters

Following is a list of indicators often measured:

- Alkalinity; Color of water; pH; Taste and odor
- Dissolved metals and salts (sodium, chloride, potassium, calcium, manganese, magnesium)
- Microorganisms such as faecal coliform bacteria, cryptosporidium, and Giardia lamblia

- Dissolved metals & metalloids (lead, mercury, arsenic, etc.)
- Dissolved organics: colored dissolved organic matter (CDOM), dissolved organic carbon (DOC), Heavy metals
- Pharmaceuticals byproducts
- Parameters depend on type of use eg, Drinking, Industrial use, Irrigation etc.

Important Water Quality Parameters

Important physico-chemical parameters to be tested for ascertaining water quality are:pH: Colour, Taste and odour, turbidity, TDS, total hardness, Chlorides, sulphates, fluorides, nitrates, calcium, Heavy metals, dissolved oxygen, pesticides, detergents & radio-nuclides

- **pH** – neutral (pH -7); acidic (pH<7); basic (pH>7); drinking water (pH: 6.5-8)
- **Electrical conductivity** – water ability to conduct electrical current – depends on concentration of dissolved, associated substances – to find TDS – unit micro siemens cm
- Odour, Colour & Taste
  - Odour – classified as: very weak, weak, cleaner, strong
  - Color – tested – using colorimeter tubes – expressed in Hazen standard unit
  - Taste – purest form is tasteless
- **Turbidity** – caused by presence of suspended matter – ranges in size from colloidal to coarse dispersions – measured by Nephole turbidity meter & expressed in NTU (Nephelometric turbidity unit) – indicator necessity of treatment
- **Dissolved oxygen** – indicates amount of oxygen gas dissolved in water – solubility of atmospheric oxygen in fresh water ranges from 14.6 mg/l ac at 0°C to about 7 mg/l at 35°C under 1 atmospheric pressure – measured using DO meter.
- **BOD (Biochemical Oxygen Demand)** – amount of dissolved O_2 demanded by bacteria during stabilization of the decomposable organic matter under aerobic conditions – BOD expressed in mg of 0_2 consumed / litre of water sample during 5 days of incubation at 20° C & is often used as a robust surrogate of degree of organic pollution of water- pristine river BOD: <1mg/l.
- **Nitrate**- water soluble molecule made up of nitrogen & oxygen – natural constituent of plants.
- **In natural from water contains less than 1mg of nitrate**- nitrogen per litre – higher levels – contamination; common sources: fertilizers, animal waste etc. – max. permissible limit: 10 ppm
- **Chlorides** – from dissolved salt deposits, discharge of effluents etc. – max. limit: 250 ppm.
- **Fluorides** – fluorine containing compounds – fluorides – found naturally in low concentration in drinking water and foods –Fresh water –between 0.01-0.3ppm, ocean contains between 1.2-1.5 ppm; max. permissible limit: 1ppm.- more fluorine – disease like skeletal fluorosis
- **Hardness** – represents total concentration of Ca & Mg ions – ppm (weight/ volume) –Hard water is generally not harmful to one's health, but can pose serious problems in industrial settings.
- **Iron** – naturally occurring – not hazardous – recommended limit: 0.3mg/l.

- **Heavy Metals:** Arsenic, Cadmium, Chromium, Copper, Iron, Manganese, Mercury, Nickel, Silver, Zinc etc.- present as minerals in soils; also artificially from man made things – major contaminations – determined by AAS (atomic absorption spectrophotometer), polarography or colorimetry. -Arsenic – WHO guideline < 0.05 mg/l
- **Pesticides:** - harmful health effects such as cancer; eg. DDT, BHC, parathion, endosulphan etc.
- **Detergents, phenol, radionuclides etc.** – WHO guideline: < 0.5 mg/l
- **Halogenated chloro-organic compounds** – due to higher chlorination of water for disinfection
- **Microbes** – indicator of potential water-borne diseases – bacteria, viruses & pathogenic protozoa; eg. Coliform bacteria.

  Water quality –environmental indicators
- **Chemical assessment:** Dissolved oxygen (DO); Nitrate-N; Ortho phosphates; Chemical oxygen demand (BOD); Pesticides; Metals
- **Physical assessment:** pH; Temperature; Total suspended solids (TSS); Turbidity; Total dissolved solids (TDS)
- **Biological assessment:** Biological monitoring metrics have been developed in many places, and one widely used measure is the presence and abundance of members of the insect orders Mayfly, Stonefly and Caddisfly.

APPENDIX 7.3

Maximum permissible concentration of Constituents in raw water supplied for various industries (mg/litre)

S.N	Constituents	Boiler water (0-100 kg/cm²)	Cooling water (once through)	Textile industry	Pulp and paper industry	Chemical Industry	Petroleum industry
1	Silica	150	50	-	50	-	85
2	Aluminum	3	3	-	-	-	-
3	Iron	80	14	0.3	2.6	5	15
4	Manganese	10	2.5	1	-	2	-
5	Calcium	-	500	-	-	25	220
6	Magnesium	-	-	-	-	100	85
7	Bicarbonate	600	600	-	-	600	480
8	Sulphate	1400	680	-	-	850	900
9	Chloride	19000	600	-	200	500	1600
10	Nitrate	-	30	-	-	-	8
11	Phosphate	-	4	-	-	-	-
12	Ammonia	-	-	-	-	-	40
13	Alkalinity	500	500	-	-	500	500
14	Suspended solids	15000	1500	1000	-	10000	5000
15	Dissolved solids	35000	1000	15	1080	2500	3500
16	Hardness	5000	850	120	475	1000	900
17	COD	100	-	-	-	-	-
18	pH (units)	-	5-8.9	6-8	4.6-9.4	5.5-8.0	6-9
19	Color (units)	1200	-	-	360	500	25
20	Temperature, °C	49	38	-	35	-	-

APPENDIX 7.4

ISI (BIS) tolerance limits for the sewage, industrial effluents and that of inland surface water (mg/litre)

S.N	Characteristics/ parameters	Tolerance limits for sewage effluent discharged into inland surface water (IS: 4764-1947)	Tolerance Limit for industrial effluents discharged into		Tolerance limits for island surface water, when used as raw water for public water supplied and bathing Ghats (IS 2296-1947)
			Inland surface water IS: 2490-1947	Public sewers (IS 3306-1947)	
1	Total suspended solids	30	100	600	-
2	Total residual chlorine	-	1	-	-
3	BOD (5 days, 20° C)	20	30	500	3
4	COD	-	250	-	-
5	Oil and grease	-	10	100	0.1
6	Phenolic compounds	-	1	5	0.005
7	Cyanides (as CN)	-	0.2	2	0.01
8	Sulphides (as S)	-	2	-	-
9	Fluorides (as F)	-	2	-	-
10	Insecticides	-	0	-	0
11	Arsenic (as As)	-	0.2	-	0.2
12	Cadmium (as Cd)	-	2	-	-
13	Chromium hexavalent	-	0.1	2	0.05 (Total chromium)
14	Copper	-	3	3	-
15	Lead	-	0.1	1	0.1
16	Mercury	-	0.01	-	-
17	Nickel	-	3	2	-
18	Selenium	-	0.05	-	0.05
19	Zinc	-	5	15	-
20	Chloride (as Cl)	-	-	600	600

21	Sulphate	-	-	-	1000
22	Ammoniacal nitrogen	-	50	50	-
23	Nitrogen(as NO_3)	-		-	50
24	% Sodium	-		60	-
25	pH, unit	-	5.5-9.0	5.5-9.0	6.9
26	Temperature, °C	-	40	45	-
27	Radioactive materials: α-emitters, μc/ml β-emitters, μc/ml	 10^{-7} 10^{-6}	 - -	 10^{-9} 10^{-8}	
28	Dissolved Oxygen	-	-	-	40% of saturation value or, 3mg/litre, whichever is higher
29	Coliform organism (monthly average) MPN/100ml	-	-	-	Should not exceed 5000 (should not exceed 20000 with less than 5% samples 25000 with less than 20% samples

APPENDIX 7.5

International drinking water standard (mg/litre)

S.N	Parameters	USEPA, maximum contamination	EEC, Maximum admissible concentration	Germany, Maximum admissible concentration	WHO, Guideline limit	IS 10500/ BIS, Desirable (INDIA)
1	Aluminium	0.05	0.2	0.2	0.2	0.3
2	Antimony	0.006	-	0.01	0.005	-
3	Arsenic	0.05	0.05	0.01	0.01	0.05
4	Barium	2	-	1	0.3	1
5	Beryllium	-	-	-	-	-
6	Boron	-	-	1	0.3	1
7	Cadmium	0.005	0.005	0.005	0.003	-
8	Calcium	-	-	400	-	75
9	Chromium	0.1	0.05	0.5	0.05	0.05
10	Cobalt	-	-	-	-	-
11	Copper	1	-	3	1-2	0.05
12	Iron	0.3	0.2	0.2	0.3	0.3
13	Lead	0.015	0.05	0.05	0.1-0.5	0.1
14	Magnesium	-	50	50	-	30
15	Manganese	0.05	0.05	0.05	0.1-0.5	0.1
16	Mercury	0.002	0.001	0.001	0.001	0.001
17	Molybdenum	-	-	-	0.7	-
18	Nickel	0.1	0.05	0.05	0.02	-
19	Potassium	-	12	12	-	-
20	Selenium	0.05	0.01	0.01	0.01	0.05
21	Silver	0.1	0.01	0.01	-	-
22	Sodium	20	175	150	200	-
23	Vanadium	-	-	-	-	-
24	Zinc	5	-	5	3	-
25	Chloride (Cl)	250	-	250	250	250
26	Chlorine (Cl_2)	-	-	0.3	-	0.2
27	Chlorine (Total)	-	-	0.3	-	0.2

S.N	Parameters	USEPA, maximum contamination	EEC, Maximum admissible concentration	Germany, Maximum admissible concentration	WHO, Guideline limit	IS 10500/ BIS, Desirable (INDIA)
28	Chlorine dioxide (Cl_2O_2)	-	-	0.2	-	-
29	Cyanides(cy)	0.2	0.05	0.05	0.07	-
30	Fluoride (F)	2-4	0.7-1.5	1.5	1.5	1
31	Nitrate/ nitrite(Total N)	10	1(as N)	1(as N)	-	-
32	Nitrates	10(as N)	50	50(as NO_2)	50(as NO_2)	-
33	Nitrites	1(as N)	0.1	0.1(as No_2)	3(as No_2)	-
34	Sulphate(50A)	250	250	240	250	200
35	Ammonium (NH_A)	-	0.5	0.5	1.5	-
36	Hydrogen peroxide	-	-	0.1	-	-
37	Hydrogen sulphide	-	-	-	0.05	-
38	Ozone (O_3)	-	-	0.05	-	-
39	Phenals (C_6H_5OH)	-	0.0005	0.0005	-	0.001
40	Phosphorous (P_2O_5)	-	5	5	-	-
41	Silica dioxide (Sio_2)	-	10	40	-	-
42	Sodium thiosulphate Surfactants	-	-	2.8	-	-
43	Surfactants	-	0.2	0.2	-	0.2
44	BOD	-	-	-	-	-
45	COD	-	-	-	-	-
46	Dissolved oxygen (DO)	-	-	-	-	-
47	Hardness	-	50	-	-	300
48	Total dissolved solid	500	-	-	1000	500
49	Colour	15Cu	20mg	$0.5m^{-1}$	15cu	5Hz

S.N	Parameters	USEPA, maximum contamination	EEC, Maximum admissible concentration	Germany, Maximum admissible concentration	WHO, Guideline limit	IS 10500/ BIS, Desirable (INDIA)
50	Oxidization (Mn VIII-Mn II)	-	5	5	-	-
51	pH value	6.5-8.5	6.2-8.5	6.5-8.5	6.6-8.5	-
52	Turbidity (Non-microbial)	0.55NTU	4NTU	1.5NTU	5NTU	5
53	Coliform (E coli) (orgnisms/ 100ml	0	0	-	0	0
54	Coliforms (total) (organisms/ 100ml	<5% positive samples/ month	0 or MDN<1	-	0	10

USEPA – United states environmental protection agency
EEC – European economic community
WHO – world health organisation
BIS – Bureau of Indian Standards

APPENDIX 7.6

Comparison of standards for drinking water in different countries (mg/litre)

S.N	Parameters	WHO	EEC	USA	Canada	France	Acceptable value
1	Dissolved solids	-	-	-	-	-	500
2	Dissolved oxygen	-	5	-	-	1-1.5	-
3	Hardness	-	35	-	18	-	100
4	Turbidity	-	5	1-5	5	5	5
5	pH value	-	6.5-8.5	-	6.5-8.3	7-8.5	7-8.5
6	Color	-	5	3-15	15	5	5
7	Conductivity	-	400	-	-	2000	-
8	Free chlorine	-	-	0.2-0.3	-	-	-
9	Chloride	-	5	250	250	200-600	200
10	Sulphate	200-400	5	250	500	250	200
11	Nitrate	45	-	45	10	50-100	-
12	Cyanide	0.05	0.05	0.01-0.2	0.01-0.2	0.05	0.05
13	Fluoride	1.5	0.7-1.5	1.4-2.4	-	1.5	0.7-1.0
14	Ammonia	-	0.05	-	0.5	0.05	-
15	Zinc	1.5-5.0	0.1-0.2	5	2	-	5
16	Selenium	0.01	0.01	0.001	0.01	0.01	0.01
17	Antimony	-	0.01	-	-	-	-
18	Lead	0.1	0.05	0.05	0.05	0.05	0.10
19	Nickel	-	0.5	-	-	-	-
20	Manganese	-	0.5	0.01-0.05	0.05	0.05	0.10
21	Boron	-	-	-	5	-	-
22	Phosphorous	-	0.3	-	0.002	-	-
23	Calcium	-	100	80-100	200	-	5
24	Magnesium	-	30	80-100	150	-	30
25	Silver	-	0.01	0.05	0.05	-	-
26	Arsenic	0.05	0.05	0.05	0.01-0.05	0.5	0.05
27	Barium	-	0.1	1.0	1.0	-	-
28	Cadmium	0.01	0.005	0.01	0.01	0.005	0.01
29	Chromium	0.05	0.05	-	-	-	-
30	Copper	-	0.5	0.2-1.0	1	0.05	0.05
31	Iron	-	0.3	0.05-0.3	0.3	0.1	0.1
32	Mercury	0.001	0.001	0.002	0.05	0.01	0.001

APPENDIX 7.7

Concentration and amounts of major elements present in sea water

S.N	Element present in sea water	Concentration of elements, mg/litre	Amount of elements in sea water, tons/mile^3	Total amount present in the oceans, tons
1	Chlorine	19,000	89.5×10^6	29.3×10^{15}
2	Sodium	10500	49.5×10^6	16.3×10^{15}
3	Magnesium	1350	6.4×10^6	16.3×10^{15}
4	Sulphur	885	4.2×10^6	2.1×10^{15}
5	Calcium	400	1.9×10^6	0.6×10^{15}
6	Potassium	400	1.9×10^6	0.6×10^{15}
7	Bromine	65	0.306×10^6	0.1×10^{15}
8	Carbon	28	0.132×10^6	0.04×10^{15}
9	Strontium	8.0	38,000	12000×10^9
10	Boron	4.6	23,000	7100×10^9
11	Silicon	3.0	14,000	4700×10^9
12	Fluorine	1.3	6100	2000×10^9
13	Argon	0.6	2800	930×10^9
14	Nitrogen	0.5	2400	780×10^9
15	Lithium	0.17	800	260×10^9
16	Rubidium	0.12	570	190×10^9
17	Phosphorous	0.07	330	110×10^9
18	Iodine	0.06	280	93×10^9
19	Barium	0.03	140	47×10^9
20	Indium	0.02	94	31×10^9
21	Zinc	0.01	47	16×10^9
22	Iron	0.01	47	16×10^9
23	Aluminium	0.01	47	16×10^9
24	Molybdenum	0.01	47	16×10^9
25	Selenium	0.004	19	6×10^9
26	Tin	0.003	14	5×10^9
27	Copper, arsenic & uranium	0.003 each	14 each	5×10^9 each

S.N	Element present in sea water	Concentration of elements, mg/litre	Amount of elements in sea water, tons/mile^3	Total amount present in the oceans, tons
28	Nickel, vanadium, manganese	0.002 each	9 each	3×10 each
29	Titanium	0.001	5	1.5×10^9
30	Antimony, Cobalt, Cesium Cerium	0.0005 each	2	0.8×10^9 each
31	Cerium	0.0004	2	0.6×10^9
32	Yttrium,silver, lanthanum, krypton	0.0003 each	1 each	5×10^8 each
33	Neon,cadmium, tungsten, xenon	0.0001 each	0.5 each	150×10^6 each
34	Germanium	0.00007	0.3	110×10^6
35	Chromium, thorium	0.00005 each	0.2 each	78×10^6 each
36	Scandium	0.000045	0.2	62×10^6
37	Lead, mercury, gallium	0.00003	0.1 each	46×10^6 each
38	Bismuth	0.00002	0.1	31×10^6 each
39	Niobium, thallium	0.00001 each	0.05 each	15×10^6 each
40	Helium	0.000005	0.03	8×10^6
41	Gold	0.000004	0.02	6×10^6
42	Protactinium	2×10^{-9}	1×10^{-5}	3000
43	Radium	1×10^{-10}	5×10^{-7}	150
44	Radium	0.6×10^{-15}	3×10^{-12}	1×10^{-3}

APPENDIX 7.8

Significance of water quality indicating parameters

<table>
<tr><th>S.N</th><th>Water quality indicating parameters</th><th>Significance of parameters</th><th>Tolerable level of various parameters</th></tr>
<tr><td>1</td><td>Total suspended solids (TSS)</td><td>It clogs fish gills, reduces light penetration ,increases heat absorption</td><td>Dependent on location</td></tr>
<tr><td>2</td><td>Total suspended solids (TDS)</td><td>It represents total mineral content which may or may not be toxic.</td><td>A maximum of 400 mg/litre for diverse fish population.</td></tr>
<tr><td>3</td><td>Dissolved oxygen (DO)</td><td>It is the source of oxygen for respiration by aquatic creatures; it is a general indicator of water quality</td><td>Minimum acceptable level is 5 mg/ litre. 10-15 mg/litre is desirable for reproduction of fish</td></tr>
<tr><td>4</td><td>Biological oxygen demand (BOD)</td><td>It is the amount oxygen removed during decomposition of organic matter in a given time; it is a general indicator of the contamination of water due to biodegradable organics.</td><td><table>
<tr><th>BOD, mg/litre</th><th>Water quality</th></tr>
<tr><td>1</td><td>Very clean water</td></tr>
<tr><td>2</td><td>Clean water</td></tr>
<tr><td>3</td><td>Fairly clean water</td></tr>
<tr><td>5</td><td>Doubtful quality</td></tr>
<tr><td>10</td><td>Contaminated water</td></tr>
</table></td></tr>
<tr><td>5</td><td>Chemical oxygen demand (COD)</td><td>It indicates the concentration of organic material oxidisable by chemical reaction.</td><td>0-5 mg/litre indicates very clean water streams.</td></tr>
<tr><td>6</td><td>pH value</td><td>Indicates the addition of acids or bases.</td><td>pH depends on actual systems.</td></tr>
<tr><td>7</td><td>Nitrate</td><td>It is a major plant nutrient; in high concentrations it can promote excessive plant growth; its major source are fertilizers, sludge and sewage</td><td>A maximum of 0.03 mg/litre to prevent excessive fertilization of systems.</td></tr>
<tr><td>8</td><td>Phosphate</td><td>A major plant nutrients; its major sources are detergents, fertilizers & sewage.</td><td>A maximum of 0.3-0.4 mg/litre , total inorganic phosphate is a common criterion.</td></tr>
</table>

S.N	Water quality indicating parameters	Significance of parameters	Tolerable level of various parameters
9	Iron	Excessive amount can clog fish gills, indicates drainage from iron bearing sediments, mines, industrial processes etc.	A maximum of 0.7 mg/litre.
10	Manganese	Its concentration is low in natural water systems due to low solubility; high concentration indicates water contamination.	A maximum of 1 mg/litre , is a common criterion for water stream quality
11	Copper	Indicates drainage from copper bearing sediments, mine, plating or other industrial sources	A maximum of 0.02 mg/litre is a common criterion for water stream quality
12	Zinc	Indicate mine drainage or industrial input	A maximum of 1 mg/litre is a common criterian for water stream quality
13	Hg, cd, Pb, Ni, Cr, Ag etc.	Indicates industrial input	A maximum of 1 mg/litre is a common criterian for water stream quality

APPENDIX 7.9

Typical composition (in ppm) of ground water , surface water & sea water

S.N	Constituents	Ground water (well)	Surface water (river)	Sea water
1	Total hardness (as $CaCO_3$)	369	123	-
2	Total dissolved solids (TDS)	434	1.65	-
3	Sodium	8.2	6.5	10.5×10^3
4	Potassium	1.4	1.2	380
5	Calcium	92	36	400
6	Magnesium	34	8.1	1.35×10^3
7	Iron (III)	0.09	0.02	0.01
8	Silica	10	1.2	6.0
9	Bicarbonate	339	119	140
10	Chloride	9.6	13	19×10^3
11	Nitrate	13	0.1	-
12	Sulphate	84	22	2.65×10^3

APPENDIX 7.10

Effluent discharge/emission standards for various industries [as per Environmental protection Act, 1986 (INDIA)]

S.N	Industries	Effluent discharge parameters	Discharge/emission standards
1	Caustic Soda Industry	(a) Total concentration of mercury in the final effluent* (b) Mercury bearing waste generation (flow) (c) pH value of effluent Final affluent is the combined effluent from (i) Cell house (ii) brine plant (iii) chlorine handling plant (iv) hydrogen handling plant and (v) hydrochloric acid plant	Not to exceed 0.01 mg/litre 10 kiloliters/ton of caustic soda produced 5.5-9.0
2	Man made synthetic fibre plant	(a) Suspended solids (b) BOD, 5 days, 20°C (c) pH	< 100 mg/litre < 30 mg/litre 5.5-9.0
3	Oil refinery	 (a) Oil and grease (b) Phenol (c) Sulfide (d) BOD, 5 day, 20°C (e) Suspended solids (f) pH value	Quantum,Kg/1000 tons crude processed <10 mg/litre , <7 <1 mg/litre , <0.7 <0.5 mg/litre , <0.35 <15 mg/litre , <10 <20 mg/litre , <14 6-8.5
4	Sugar industry	(a) BOD, 5 day, 20°C (b) Suspended Solids	(i) <100 mg/litre for disposal on land (ii) <30 mg/litre for disposal in surface water (i) <100 mg/litre for disposal on land (ii) <30 mg/litre for disposal in surface water
5	Thermal power plants	Condenser cooling water (once through cooling system):- (i) pH value (ii) Temperature	6.5-8.5 Not more than 5°C higher than the intake water temperature.

S.N	Industries	Effluent discharge parameters	Discharge/emission standards
		Boiler blow downs:-	
		(i) Free available chlorine	<0.5 mg/litre
		(ii) Suspended solids	<100 mg/litre
		(iii) Oil and grease	<20 mg/litre
		(iv) Copper (total)	<1 mg/litre
		(v) Iron (total)	<1 mg/litre
		Cooling tower blow down:-	
		(i) Free available chlorine	<0.5 mg/litre
		(ii) Zinc	<1 mg/litre
		(iii) Chromium (total)	<0.2 mg/litre
		(iv) Phosphate	<5 mg/litre
		(v) other corrosion inhibiting material	Limited; to be established on case by case basis by central board in case of the union territories and state boards in case of states.
		Ash Pond effluent:-	
		(i) pH value	6.5-8.5
		(ii) Suspended solids	<100 mg/litre
		(iii) Oil and grease	<20 mg/litre
6	Cotton textile industries (composite and processing)	**Common:-**	
		(i) pH value	5.5-9.0
		(ii) Suspended Solids	<100 mg/litre
		(iii) BOD, 5 day, 20°C	<150 mg/litre
		(iv) Oil and grease	<10 mg/litre
		(v) Bio-assay test	90% survival of fish after 96 hours
		Special:-	
		(i) Total chromium (as Cr)	<2 mg/litre
		(ii) Sulphide (as S)	<2 mg/litre
		(iii) Phenolic Compounds (as C_6H_5OH)	<5 mg/litre

The special parameters are to be stipulated by the central pollution control board in case of union territories and state pollution control board in case states depending upon the dye used in the industry. Where the industry uses chrome dye, Sulphur dye and/or phenolic compounds in the dying/printing process, the limits on chromium of 2 mg/litre, sulphides of 2 mg/litre and phenolic compounds of 5 mg/litre respectively shall be imposed.

Where the quality requirement of the recipient system so warrants, the limit of BOD should be lowered upto 30 according to the requirement by the state boards for states and the central boards for the union territories.

A limit on sodium absorption ratio of 26 should be imposed by the state boards for the states and the central board for union territories, if the disposal of effluent is to be made on land.

S.N	Industries	Effluent discharge parameters	Discharge/emission standards
7	Composite woolen mills	**Common:-**	
		(i) Suspended solids	<100 mg/litre
		(ii) pH Value	5.5-9.0
		(iii) BOD, 5 days, 20°C	<100 mg/litre
		(iv) Oil and grease	<10 mg/litre
		(v) Bio-assay	90% survival of fish after 96 hours
		Special:-	
		(i) Total chromium (as Cr)	<2 mg/litre
		(ii) Sulphide (as S)	<2 mg/litre
		(iii) Phenolic compounds (as C_6H_5OH)	<5 mg/litre
Special parameters stipulated for textile industry as detailed above is applicable to the composite woolen mills also.			
8	Dye and dye intermediate industries	(i) Suspended solids	<100 mg/litre
		(ii) pH value	6-8.5
		(iii) Temperature	Shall not exceed 5°C above the ambient temperature of the receiving body
		(iv) Mercury (as Hg)	<0.01 mg/litre
		(v) Hexavalent Chromium (as Cr)	<0.1 mg/litre
		(vi) Total chromium (as Cr)	<2 mg/litre
		(vii) Copper (as Cu)	<3 mg/litre
		(viii) Zinc (as Zn)	<5 mg/litre
		(ix) Nickel (as Ni)	<3 mg/litre
		(x) Cadmium (as Cd)	<2 mg/litre
		(xi) Chloride (as Cl)	<1000 mg/litre
		(xii) Phenolic compounds (as C_6H_5OH)	<1 mg/litre
		(xiv) Oil and grease	<10 mg/litre
		(xv) Bio-assay test	90% survival of test animals after 96 hours.
The standards for chloride and sulphates are applicable for discharge into inland and surface water courses. However, When discharged on land for irrigation, the limit for chloride shall not be more than 600 mg/litre and the sodium adsorption ratio shall not exceed 26.			

S.N	Industries	Effluent discharge parameters	Discharge/emission standards
9	Electronic industries	(i) pH value	6-9
		(ii) Temperature	Shall not exceed 5°C above the ambient temperature of the receiving body
		(iii) Oil and grease	<10 mg/litre
		(iv) Suspended solids	<100 mg/litre
		(v) Cyanides (as CN)	<0.2 mg/litre
		(vi) Ammoniacal nitrogen (as N)	<50 mg/litre
		(vii) Total residual chloride (as Cl)	<1 mg/litre
		(viii) Cadmium (as Cd)	<2 mg/litre
		(ix) Nickel (as Ni)	<3 mg/litre
		(x) Zinc (as Zn)	<5 mg/litre
		(xi) Hexavalant chromium (as Cr)	<0.1 mg/litre
		(xii) Total chromium (as Cr)	<2 mg/litre
		(xiii) Copper (as Cu)	<3 mg/litre
		(xiv) Lead (as Pb)	<0.1 mg/litre
		(xv) Iron (as Fe)	<3 mg/litre
		(xvi) Total metal	<10 mg/litre
10	Cement plants: Plant capacity = 200 tons/day Plant capacity = > 200 tons/day	Total dust (all sections) Total dust (all section)	<400 mg/Nm^3 <250 mg/Nm^3
11	Stone crushing plant	Suspended particulate matter (SPM)	SPM measured between 3 meter and 10 meters from any process equipment shall not exceed 600 mg/m^3
12	Coke ovens		**For effluent to be discharged into inland surface water:**
		(i) BOD, 5 days, 20°C	<5.5-9.0
		(ii) Suspended solids	<30 mg/litre
		(iii) Phenolic compounds (as C_6H_5OH)	<100 mg/litre
		(iv) Cyanides (as CN)	<5 mg/litre
		(v) Oil and grease	<0.2 mg/litre
		(vi) Ammoniacal nitrogen (as N)	<10 mg/litre
		(vii) pH value	6-8

S.N	Industries	Effluent discharge parameters	Discharge/emission standards
13	Synthetic rubber industry	(i) Color (ii) pH value (iii) BOD, 5 days, 20°C (iv) COD (v) Oil and grease	**For effluent to be discharged into inland surface water:** Absent 5.5-9.0 <50 mg/litre <250 mg/litre <10 mg/litre
14	Small pulp & paper industries:- For discharge into inland surface water- For disposal on land-	(i) pH value (ii) Suspended solids (iii) BOD, 5 days, 20°C (iv) Sodium absorption ratio	5.5-9.0 <100 mg/litre <30 mg/litre 26
15	Fermentation industry (Distilleries, Maltries & Breweries)	(i) pH value (ii) Color and color (iii) Suspended solids (iv) BOD, 3 days, 20°C - for disposal into inland surface water - for disposal on land Waste water generation shall not exceed 250 m^3/ ton of paper produced	5.5-9.0 Absent <100 mg/litre <30 mg/litre <100 mg/litre
16	Leather tanneries:-		**Concentration in the effluent not to exceed mg/litre (except for pH & percent sodium)**

Effluent discharge parameters	Inland surface water	Public sewer	Land for irrigation	Marine coastal areas
(i) Suspended Solids	100	100	200	100
(ii) BOD, 5 days, 20°C	30	350	100	100
(iii) pH value	6-9	6-9	6-9	6-9
(iv) Chlorides (as Cl)	1000	1000	600	-
(v) Hexavalent chromium(Cr^{6+})	0.1	0.2	0.1	1.0
(vi) Total chromium (as cr)	2	2	2	2
(vii) Sulphides (as S)	2	5	-	5
(viii) Sodium percent	-	60	60	-
(ix) Boron (as B)	2	2	2	-
(x) Oil and grease	10	20	10	20

S.N	Industries	Effluent discharge parameters	Discharge/emission standards	
17	Fertilizer industry:- **(a) Effluents-** Straight nitrogenous fertilizers excluding calcium ammonium nitrate fertilizers		**For plants commiss-ioned 1.1.1982 onwards , unit - mg/litre except for pH**	**For Plants commissio-ned prior to 1.1.1982, unit - mg /litre except for pH**
		(i) pH value	6.5-8.0	6.5-8.0
		(ii) Nitrogen (total)	<50	<75
		(iii) Kjeldahl nitrogen (Free ammoniacal nitrogen)	<4	<4
		(iv) Nitrate nitrogen	<10	<10
		(v) Cyanides (as CN)	<0.2	<0.2
		(vi) Vanadium (as V)	<0.2	<0.2
		(vii) Arsenic (as As)	<0.2	<0.2
		(Viii) Suspended solids	<100	<100
		(ix) Oil and grease	<10	<10
		(x) Hexavalent chromium (as Cr) (To be complied with at the outlet of chromate removal unit)	<0.1	<0.1
		Total Chromium (as Cr) (To be complied with at the outlet of chromate removal unit)	<2	<2
	(b) Effluents- Straight nitrogenous fertilizers including Calcium ammonium nitrate and ammonium nitrate fertilizers.		Unit – mg/litre **For plants commiss-ioned 1.1.1982 onwards**	**For plant commissio-ned prior to 1.1.1982**
		(i) pH value	6.5-8.0	6.5-8.0
		(ii) Ammoniacal nitrogen (total)	<50	<75
		(iii) Kjeldahl nitrogen (total)	<100	<150
		(iv) Free ammoniacal nitrogen	<4	<4

S.N	Industries	Effluent discharge parameters	Discharge/emission standards	
		(v) Nitrate nitrogen	<20	<20
		(vi) Cyanide (as CN)	<0.2	<0.2
		(vii) Vanadium (as V)	<0.2	<0.2
		(viii) Arsenic (as As)	<0.2	<0.2
		(ix) Suspended solids	<100	<100
		(x) Oil and grease	<10	<10
		(xi) Hexavalent chromium (as Cr)	<0.1	<0.1
		(xii) Total chromium (as Cr) (To be complied with at the outlet of chromate removal unit)	<2	<2
	(c) Effluents- complex fertilizers excluding calcium ammonium nitrate, ammonium nitrate and ammonium nitro phosphate fertilizers	(i) pH value	6.5-8.0	6.5-8.0
		(ii) Ammoniacal nitrogen	<50	<75
		(iii) Free ammoniacal nitrogen	<4	<4
		(iv) Total kjeldahl nitrogen	<100	<150
		(v) Nitrate nitrogen	<10	<10
		(vi) Cyanide (as CN)	<0.2	<0.2
		(vii) Vanadium (as V)	<0.2	<0.2
		(viii) Arsenic (as As)	<0.2	<0.2
		(ix) Phosphate (as P)	<5	<5
		(x) Oil and grease	<10	<10
		(xi) Suspended Solids	<100	<100
		(xii) Fluoride (as F) (To be complied with at the outlet of the fluoride removal unit. If the recipient system so demands fluoride as F shall be limited to 1.5 mg/ litre)	<10	<10
		(xiii) Hexavalant chromium (as Cr)	<0.1	<0.1

S.N	Industries	Effluent discharge parameters	Discharge/emission standards	
		(xiv) Total chromium (as Cr) (To be complied with at the outlet of Chromate removal unit)	<2	<2
	(d) Effluent- Complex fertilizers including calcium ammonium nitrate, ammonium nitrate and ammonium nitro phosphate fertilizers		Unit- mg/litre except for pH value	
			For plant commiss-ioned on 1.1.1982 onwards	**For plant commissioned prior to 1.1.1982**
		(i) pH value	6.5-8.0	6.5-8.0
		(ii) Ammoniacal nitrogen	<50	<75
		(iii) Free ammoniacal nitrogen	<100	<100
		(iv) Nitrate nitrogen	<20	<20
		(v) Cyanide (as CN)	<0.2	<0.2
		(vi) Vanadium (as V)	<0.2	<0.2
		(vii) Arsenic (as As)	<0.2	<0.2
		(viii) Phosphate (as P)	<5	<5
		(ix) Oil and grease	<10	<10
		(x) Suspended solids	<100	<100
		(xi) Fluoride (as F) *(To be complied with at the outlet of the fluoride removal unit. If the recipient system so demands, fluoride as F should be limited to 1.5 mg/litre	<10	<10
		(xii) Hexavalant chromium (as Cr)	<0.1	<0.1
		(xiii) Total chromium *(To be complied with at the outlet of the chromate removal unit)	<2	<2
	(e) Effluent- Straight phosphatic fertilizers	(i) pH value	7-9	
		(ii) Phosphate (as P)	<5	

S.N	Industries	Effluent discharge parameters	Discharge/emission standards
		(iii) Oil and grease	<10
		(iv) Suspended solids	<100
		(v) Fluoride (as F)*	<10
		(vi) Hexavalant Chromium (as Cr)[xx]	<0.1
		(vii) Total chromium (as Cr)[xx]	<2
	Emissions- Phosphatic fertilizers and particulate matter emissions	(i) Phosphoric acid manufacturing unit; Granulation, mixing & grinding of rock phosphate.	20 mg/Nm^3 as total fluoride; 150 mg/Nm3 of particulate matter
	Urea plant- Particulate matter emission	**Prilling Tower-** – Plants Commissioned prior to 1-1-1982 – Plants Commissioned after 1-1-1982	 -150 mg/Nm^3 or, 2 kg/ton of product -50 mg/Nm^3 or, 0.5 kg/ton of product

APPENDIX 7.11

Liquid effluent discharge standard for battery manufacturing industry

Liquid Effluent Discharge Standards:

Pollutant	Concentration
pH	6.5-8.5
Suspended solids	50 mg/l
Lead	0.1 mg/l

Dry cell manufaturing industry:

Effluent Standards

Pollutant	Concentration
pH	6.5-8.5
Total suspended solids	100mg/l
Manganese as Mn	2 mg/l
Mercury as Hg	0.02 mg/l
Zinc as Zn	5mg/l

APPENDIX 7.12

Effluent discharge standard for edible oil & Vanaspati industry

Parameter	Concentration not to exceed
Temperature	Not more than 5°C above ambient temperature of the recipient water body
pH	6.5-8.5
Suspended solids	150 mg/l
Oil & grease	20 mg/l
BOD at 27°C, 3 days	100 mg/l
COD	200 mg/l

Wastewater discharge

(i)	Solvent extraction	2.0 m^3/ tonne of product
(ii)	Refinery/vanaspati (refined oil/vanaspati)	2.0 m^3 / tonne of product
(iii)	Intergrade unit of solvent extraction refinery/ vanaspati	4.0 m^3/ tonne of refined oil
(iv)	Barometric cooling water/deodorizer Water	15.0 m^3/tonne of refined oil/vanaspati

Note:

1. The above standards will be applicable to wastewater from processes and cooling.
2. BOD shall be made stringent upto 30 mg/l, if the recipient fresh water body is a source of drinking water supply.

APPENDIX 7.13

Liquid waste water discharge standards for gas / naphtha based thermal power plants

Liquid waste discharge limit (mg/litre)

Parameters	Maximum limit of concentration (mg/l expect for pH and temperature)
pH	6.5-8.5
Temperature	As applicable for other thermal power plants
Free available chlorine	0.5
Suspended solids	100.0
Oil and grease	20.0
Copper (total)	1.0
Iron (total)	1.0
Zinc	1.0
Chromium (total)	0.2
Phosphate	5.0

APPENDIX 7.14

Standard for liquid waste discharge from hospitals

Standard for Liquid Waste

The effluent generated from the hospital should conform to the following limits:

PARAMETERS	PERMISSIBLE LIMITS
pH	6.5-8.5
Suspended solids	100 mg/l
Oil and grease	10 mg/l
BOD (3 days at 27°C)	30 mg/l
COD	250 mg/l
Bio-assay test	90% survival of fish after 96 hours in 100% effluent

These limits are applicable to those hospitals which are either connected with sewers without terminal sewers with terminal facilities, the general standards as notified under the Environment (Protection) Act, 1986 shall be applicable.

APPENDIX 7.15

Inorganic Chemical Industry: Wastewater Discharge Standard

(Part I –metal Compounds of Chromium, Manganese, Nickel, Copper, Zinc, Cadmium, Lead & Mercury)

Parameter	Concentration in mg/l except pH (Not to exceed)
pH	6.0 to 8.5
Chromium as Cr Hexavalent (total)	0.1
Manganese as Mn	2.0
Nickel as Ni	2.0
Copper as Cu	2.0
Zinc as Zn	5.0
Cadmium as Cd	0.2
Lead as Pb	0.1
Mercury as Hg	0.01
Cyanide as CN	0.2
Oil & grease	10.0
Suspended solids	30.0

APPENDIX 7.16

Petrochemicals (Basic & Intermediates): Standards For Liquid Effluent

Parameters	Concentration not to exceed limits (unit -in mg/l except pH)
pH	6.5-8.5
BOD (3 days at 27°C)	50
Phenol**	5
Sulphide as S	2
COD	250
Fluorides as F***	2
Total suspended solids	100
Chromium****	Hexavalent Cr: 0.1
Total Cr:	2.0

* The state boards may prescribe the BOD value of 30 mg/l , if the recipient system so demands.

** The limit of phenol shall be conformed at the outlet of effluent treatment of phenol cumeme plant. However, at the final disposal point, the limit shall be less than 1 mg/l.

*** The limit for fluoride shall be conformed at the outlet of fluoride removal unit. However, at the disposal point fluoride concentration shall be lower than 5 mg/l.

The limits for total and hexavalent chromium shall be conformed at the outlet of the chromate removal unit

**** This implies that in the final treated effluent, total and hexavalent chromium shall be lower than prescribed herein.

APPENDIX 7.17

Effluent Discharge Standards For Soda Ash Industry

Parameter	Concentration not to exceed		
	Marine	Brackish	Inland surface water
pH	6.5-8.5	6.5-9	6.5-9
Temperature	45°C or less	45°C or less	45°C or less
Oil & grease	20 mg/l	20 mg/l	10 mg/l
Suspended solids (SS)	500 mg/l	200 mg/l	100 mg/l
Ammonical nitrogen	50 mg/l	50 mg/l	30 mg/l
Bio- assay (96 hours)	90% survival	90% survival	90% survival

Note: MINAS for disposal in brackish and inland surface water are without any dilution.

B: Dual Process Soda Ash plants

Parameters	MINAS (Inland Surface Water)
pH	6.5-8.0
Ammonical nitrogen, as N (mg/l)	50
Nitrate nitrogen, as as N (mg/l)	10
Cyanides, as N (mg/l)	0.2
Hexavalent chromium (mg/l)	0.1
Total chromium (mg/l)	2.0
Suspended solids, (mg/l)	100
Oil and grease (mg/l)	10

APPENDIX 7.18

Small Scale Industry: Wastewater Discharge Standards

S. No.	Parameters	Concentration not to exceed, limits in mg/l (expect pH , temperature and bioassay test)
1.	Mercury (Hg)	0.01
2.	pH	5.5-9.0
3.	Suspended solids	250
4.	Biochemical oxygen demand (3 days at 27°C)	150
5.	Temperature	Shall not exceed 5°C above the ambient temperature of the receiving body
6.	Free available chlorine	0.5
7.	Oil & grease	1.0
8.	Cu (total)	3.0
9.	Iron (total)	3.0
10.	Zinc	5.0
11.	Cr (total)	3.0
12.	Phosphate (as P)	5.0
13.	Bioassay test	90% of test animals survival after 96 hrs with 1.8 dilution
14.	Sulphide (as S)	2
15.	Phenolic compounds (as C_6H_5OH)	5
16.	Hexavalent chromium (as Cr)	0.1
17.	Nickel (as Ni)	3.0
18.	Cadmium (Cd)	2.0
19.	Chloride (as Cl)	1000
20.	Sulphate (as SO_4)	1000
21.	Cyanide (as CN)	0.2
22.	Ammoniacal nitrogen (as N)	50
23.	Lead (as Pb)	0.1
24.	Total Metal	10.0

Note: The standards may be relaxed in cases where the wastewater from small scale industrial units are collected and treated in a terminal plant. These standards are laid down without prejudice to the Board varying or modifying them while issuing consents pursuant to Section 25 and 26 of the Water (Prevention and Control of Pollution) Act, 1947.

APPENDIX 7.19

Jute Processing Industry: Liquid Effluent Standards

Parameters	Concentration not to exceed, in mg/l expect pH
pH	5.5 to 9.0
BOD at 27°C for 3 days	30
Total suspended solids	100
Oil & grease	10

Note:

1. Water consumption from the jute processing industries will be 1.5 m^3/tonne of product from January.
2. At present no limit for color is given for liquid effluent however, as far as possible color should be removed.

APPENDIX 7.20

Effluent discharge standards for coal washeries

Effluent discharge standards

- The coal washeries shall maintain the close circuit operation with zero effluent discharge.
- If in case due to some genuine problems like periodic clearing of the system , heavy rainfall etc. it becomes necessary to discharge the effluent of sewer/land/stream then the effluent shall conform to the following standards at the final outlet of the coal washery.

S.N.	Parameter	Limits
1.	pH	5.5-9.0
2.	Total suspended solids	100 mg/l
3.	Oil & grease	10 mg/l
4.	B.O.D. (3 days, 27°C)	30 mg/l
5.	COD	250 mg/l
6.	Phenolic	1.0 mg/l

Code of practice for Coal Washery

- Water or Water mixed Chemical shall be sprayed at all strategic coal transfer points such as conveyors , loading & unloading points etc. As far as practically possible conveyors, transfer point etc. shall be provided with enclosures.
- Water sprinkling by using fine atomizer nozzles arrangement shall be provided on the coal heaps and on land around the crushers/pulverizers.
- Area, in land around the coal washery shall be pucca either asphalted or concreted.
- Water consumption in the coal washery shall not exceed 1.5 cubic meter per tonne of coal.
- The efficiency of the settling ponds of the wastewater treatment system of the coal washery shall be not less than 90%.
- Green belt shall be developed along the road side, coal handling plants, residential complex, office building and all around the boundary line of the coal washery.
- Storage bunkers, hoppers rubber decks in chutes and centrifugal chutes shall be provided with paper rubber linings.
- Vehicles movement in the coal washery area shall be regulated effectively to avoid traffic congestion. High pressure horn shall be prohibited. Smoke emission from heavy duty vehicle operating in the coal washery should conform the standards prescribed under Motor Vehicle Rules 1989.

APPENDIX 7.21

Indian standards related to disposal of treated effluents (mg/litre) in various water bodies

S.N.	Parameters	Inland surface water	Discharges into public sewers	On land irrigation	Marine disposal
1.	Total suspended solids (TSS)	100	Relaxable upto 750 by local unit	200	100
2.	Total dissolved solids (TDS) (inorganic)	2100	2100	2100	-
3.	BOD (5 days, 20 °C)	30	350	100	100
4.	COD	100	-	-	250
5.	Oil and grease	10	20	10	20
6.	Pesticide	Absent	Absent	Absent	Absent
7.	Phenolic compounds	1	5	-	5
8.	Total residual chlorine	1	-	-	1
9.	Sodium	-	-	60	-
10.	Arsenic	0.2	0.2	0.1	0.2
11.	Chloride (as Cl)	1000	1000	600	-
12.	Cyanides (as CN)	0.2	0.2	0.2	0.2
13.	Sulphides (as S)	2	-	-	5
14.	Sulphate (as SO_4)	1000	1000	1000	-
15.	Phosphate (dissolved as P)	5	-	-	5
16.	Fluoride (as F)	1.5	15	-	10
17.	Ammoniacal nitrogen (as N)	50	50	-	50
18.	Total kjeldahl nitrogen (As N)	100	-	-	100
19.	Free ammonia	5	-	-	5
20.	Boron (as B)	2	2	0.75	-
21.	Copper (as Cu)	3	3	0.2	3
22.	Lead (as Pb)	0.1	1	5	1
23.	Hexavalent chromium (as Cr^{+6})	0.1	2	0.1	1
24.	Total chromium (as Cr)	2	2	1	2
25.	Mercury (as Hg)	0.01	0.01	0.01	0.01
26.	Nickel (as Ni)	3	3	0.2	4

S.N.	Parameters	Inland surface water	Discharges into public sewers	On land irrigation	Marine disposal
27.	Selenium (as Se)	0.05	0.05	0.02	0.05
28.	Zinc	5	15	2	15
29.	PH value	5.5-9	5.5-9	5.5-9	5.5-9
30.	Temperature, °C	≤40	45	40	45
31.	Color (Platinum cobalt scale)	100 units	-	-	100 units
32.	**Radioactive materials:-**				
	(a) α-emitters, μc/ml	10^{-7}	10^{-7}	10^{-8}	10^{-7}
	(b) β-emitters, μc/ml	10^{-6}	10^{-6}	10^{-7}	10^{-6}

APPENDIX 7.22

Quality of effluent water discharged from petroleum refineries

S.N.	Characteristics/ Parameters	Unit	Tolerance limit for effluent water discharge into		
			Inland surface water (IS: 2490)	Marine coast area (IS:7968)	MINAS
1.	pH value	Unit	5.5-9.0	5.5-9.0	6-8.5
2.	BOD, 5 days, 20°C	mg/litre	30	100	15
3.	COD	mg/litre	250	250	-
4.	Oil & grease	ppm	10	-	1020
5.	TSS	ppm	1000	-	20
6.	Phenolic compounds	ppm	1	5	1
7.	Sulphides	ppm	2	5	0.5
8.	Cadmium (as Cd)	ppm	2	2	-
9.	Chromium (as Cr)	ppm	0.1	1	-
10.	Copper (as Cu)	ppm	3	3	-
11.	Lead (as Pb)	Ppm	0.1	1	-
12.	Mercury (As Hg)	ppm	0..01	0.01	-
13.	Nickel (Ni)	ppm	3	5	-
14.	Selenium (as Se)	ppm	0.05	0.05	-
15.	Zinc (as Zn)	ppm	5	-	5

APPENDIX 7.23

Effluent quality of a typical petroleum refinery

S.N.	Effluent Parameters	Typical Values, mg/litre
1.	pH value	6.5-9.0
2.	Oil & grease	15
3.	Phenol	2
4.	Sulfide	3
5.	Total residual chlorine	0.5
6.	Total ammoniacal nitrogen	25
7.	Suspended solids	Influent +50%
8.	Dissolved oxygen	40% of saturation value
9.	BOD, 5 days, 20°C	100
10.	COD	250

APPENDIX 7.24

Prescribed schedule of maximum water consumptions for specified industries (India)

S. N.	Name of Industry	Category of industries	Maximum quantity of water to be consumed
1.	Ferrous Metallurgical industry	Integrated iron & steel plant	20 m^3/ton of finished product
2.	Non- ferrous Metallurgical industry	(a) copper smelter (b) Zinc smelter	100m^3/ton of copper produced 50m^3/ton of zinc metal produced
3.	Chemical industries	(a) caustic soda:- (i) Mercury cell process (ii) Membrane cell process	5m^3/ton of caustic soda produced (excluding cooling water) 5m^3/ton of caustic soda produced (including cooling water)
4.	Textile industries	(a) Man-made fibre:- (i) Nylon and Polyester (ii) Viscose rayon	170 m^3/ ton of fibre produced 200 m^3/ton of fibre produced
5.	Paper industries	(a) Small pulp & paper plant:- (i) Agro- residue based (ii) Waste paper based (b) Large pulp & Paper plant:- (i) Pulp & Paper (ii) Rayon grade paper	200m^3/ton of paper 75m^3/ton of paper 250m^3/ton of paper 200 m^3/ ton of paper
6.	Fertilizer Industries	(a) Straight nitrogenous fertilizer (b) Straight phosphatic fertilizer (SSP and TSP) including manufacture of any acid (c) Complex fertilizer	15m^3/ton of urea of equivalent produced 2m^3/ton of SSP/TSP 15m^3/ton in case primary product is nitrogenous fertilizer and 2 m^3/ton in case primary product is a phosphatic fertilizer

S. N.	Name of Industry	Category of industries	Maximum quantity of water to be consumed
7.	Processing of animal or vegetable products industry including processing of milk, hides, meat, and skins, all agricultural products and their waste	(a) Tanneries	30 m^3/ton of raw hide
		(b) Natural rubber	6m^3/ton of rubber
		(c) Starch, glucose and related products	10 m^3/ton of maize crushed
		(d) Dairy	4m^3/ton of milk
		(e) Jute	1.5 m^3/ton of jute produced
		(f) Sugar	2 m^3/ ton of cane crushed
		(g) Malty	8.5 m^3/ton of grain processed
		(h) Brewery	1m^3/kiloliter of bear produced
		(i) Distillery	15 m^3/kiloliter of alcohol produced

APPENDIX 7.25

Water quality characteristics of lakes & wells in various parts of India (mg/litre)

S.N.	Water Characteristics	Lake water quality			Wells water quality		
		Upper lake Bhopal	Lakes in Hyderabad	Red Hill lake Chennai	Bhopal	Jaipur	Pondicherry
1.	pH value	7.2-9.5	8.2-9.0	7.7-8.9	6.8-8.0	7.8-8.0	6.9-7.4
2.	Total Solids	196-676	200-470	-	220-682	-	360-1432
3.	Dissolved solids	142-272	13-80	9-43.6	-	-	-
4.	Alkalinity	62-182	127-358	-	54-773	96-488	-
5.	Total hardness (as $CaCO_3$)	26-92	-	1.5-14	32-832	104-1200	74-156
6.	Calcium hardness (as $CaCO_3$)	14-64	-	-	22-486	84-708	25-72
7.	Magnesium hardness(As $CaCO_3$)	12-28	-	-	10-346	20-492	49-84
8.	Chlorides (as Cl)	10.7-86.5	14-77	1.7-8.0	12-850	16-640	35-196
9.	Nitrate (As N)	-	-	-	0-27	0-6	0.04-0.21
10.	Iron(as Fe)	-	-	-	0-1.5	-	0.3-0.9

APPENDIX 7.26

Classification of hardness of water for practical purpose

S.N.	Classification of water hardness	Hardness, $CaCO_3$/litre of water
1.	Soft water	0-40
2.	Moderately hard water	10-100
3.	Hard water	100-300
4.	Very hard water	300-500
5.	Extremely hard water	500-1000

APPENDIX 7.27

Average fluoride level in drinking water in various parts of India (mg/litre)

S.N.	States	Districts	Average fluoride level in drinking (potable) water
1.	Tamilnadu	Chinglepet, salem, Madurai, Trichurapalli, Coimbatore	4.5
2.	Gujrat	Mehsana, Banaskantha, Sabarkantha, Panchamahal, surat	4.0
3.	Haryana	Gurgoan, Hissar, Karnal	3.6
4.	Punjab	Bhatinda, Ferojpur, Amritsar	3.0
5.	Andhra Pradesh	West Godavari, Nalangonda, Vellore, Anantpur	0.8-1.2
6.	Karnatka	Belgaum	0.8-1.2

Note: Safe limit of fluoride in potable (drinking) water is 0.8-1.2 mg/litre. Its excess concentration causes dental crevices, acute poisoning of central nervous system, osteoporosis of bone etc.

APPENDIX 7.28

BOD (5 days, 20°C) level of various typical samples of water

S.N.	Type of water sample	BOD, mg/litre	Remarks
1.	Rain water	1-3	Reasonable
2.	River water	5-20	Tolerable
3.	Sewage	50-100	Very bad
4.	Industrial waste water	100-10000	Extremely bad

Note: Definition & determination of BOD & COD

Microorganisms (e.g.) bacteria) are responsible for decomposing organic waste. When organic matter such as dead plants, leaves, grass clippings, manure, sewage, or even food waste is present in a water supply, the bacteria will begin the process of breaking down this waste. When this happens, much of the available dissolved oxygen is consumed by aerobic bacteria, robbing other aquatic organisms of the oxygen they need to live. Biological Oxygen Demand (BOD) is a measure of the oxygen used by microorganisms to decompose this waste. If there is a large quantity of organic waste in the water supply, there will also be a lot of bacteria present working to decompose this waste. In this case, the demand for oxygen will be high (due to all the bacteria) so the BOD level will be high. As the waste is consumed or dispersed through the water, BOD levels will begin to decline.

The Biological Oxygen Demand (BOD) test is used in order to determine the oxygen requirements of wastewaters. The BOD, unless treated with an ammonia inhibiting chemical, will indicate the total of carbonaceous and nitrogenous oxygen demand.

A 250 to 300 ml airtight bottle is filled with water sample to overflowing. Dissolved oxygen is measured initially after dilution and after incubation at 20°C for the time period of the test.It is necessary to have a population of microorganisms present that is capable of degrading the organic material in the sample, Many unchlorinated domestic or industrial wastewaters will afford sufficient bacteria for this purpose. This population is called "the seed".

One BOD test must be run on a bottle containing seed and dilution water. The dilutions should be made so that the least dilution results in at least a 50% dissolved oxygen depletion. The seed is also added and diluted so that the residual dissolved oxygen is at least 1 mg/1 &the uptake at least 2mg/1 after 5 days. Several dilutions of the prepared sample should be made to ascertain that these results would be obtained. Initial dilutions are as shown in Table below.

Table 7.28T_1: BOD Dilution Recommendations

Water Sample Source	% Sample
Strong industrial wastes	0.1-1.0
Raw or settled wastewater	1.0-5.0
Biologically treated wastewater	5.0-25.0
Polluted river waters	25.0-100.0

Oxygen depletion as a percent of total depletion will typically occur as shown in Table below.

Table 7.28T$_2$: Oxygen Depletion Estimation

Days	% Oxygen depletion
0	0
5	60 to 70
20	95 to 99
Ultimate	100

Chemical Oxygen Demand (COD) test is a measure of the oxygen required from a strong chemical oxidant for the destruction of an organic material. The chemical oxidant selected for the COD test is potassium dichromate ($K_2Cr_2O_7$).

The sample is oxidized by a boiling mixture of $K_2Cr_2O_7$ and sulfuric acid using an excess of $K_2Cr_2O_7$. After digestion, the remaining unreduced $K_2Cr_2O_7$ is measured to determine the amount consumed by titration with ferrous ammonia sulfate (FAS) until the end point is indicated by sharp colour change. The COD is calculated as:

COD = 8000M (A-B)/ml sample

Where:

COD–Chemical Oxygen Demand

A– ml FAS used for blank

B – ml FAS used for sample

M– molarity of FAS.

The blank is a test with reagents without the sample. The COD takes about 2 to 3 hours and can be run with a relatively simple and inexpensive kit.

Total Organic Carbon (TOC) test is more direct indication of organic content than the BOD or COD test.The TOC test indicates the total organic matter present and is independent of the oxidation state of the pollutant. This test converts and removes inorganic carbon (CO + CO_2) and measures total organic carbon by converting it into CO_2 in a heated reaction chamber. The CO_2 is measured using infrared methods.

Theoretical Oxygen Demand (ThOD) is the stoichiometrically determined oxygen needed to convert all carbon molecules in pollutants to $C0_2$, and all NH_3 by balancing equations. This calculation is possible on certain rather pure industrial waste, but is impractical for most wastewaters, especially those containing domestic sewage or containing vegetable or animal wastes.

Relationship between BOD, COD, TOC and ThOD: After biological oxidation, approximately ten percent of the original organics (completely biodegradable wastewater) remain as non-biodegradable cellular residues. Hence, the cellular residues are not measured by the BOD test. Therefore, depending on the type of wastewater:

BODu = 0.9 ThOD

where:

BODu-ultimate BOD , ThOD-theoretical oxygen demand

For domestic sewage and some biodegradable industrial wastes, the relationship between BOD_5 and BODu is:

$$BOD_5 = 0.77\ BOD_U$$

where:

BOD_5 = 5 day BOD

BODu-ultimate BOD

For most wastewaters: ThOD = COD. Where ThOD is the theoretical oxygen demand; COD is the chemical oxygen demand ; since the COD test oxidizes all organics except those that are totally resistant to dichromate oxidation.Stoichiometrically, the COD/TOC ratio should be approximately the molecular ratio of oxygen to carbon:

$$COD/TOC = 32/12 = 2.66$$

The ratio will actually range from zero, when organic material is resistant to dichromate oxidation, to as much as 6.0 when inorganic reducing agents are present.

For raw domestic sewage and some biodegradable industrial wastes, the following ratio occurs:

$$BOD_5/TOC = 32/12(0.90)(0.77) = 1.85$$

where:

BOD_5 = 5 day BOD

TOC = Total Organic Carbon

0.90 = BODu/ThOD as 0.77 = BOD_5/BODμ

BOD Effects on Water Quality is shown below

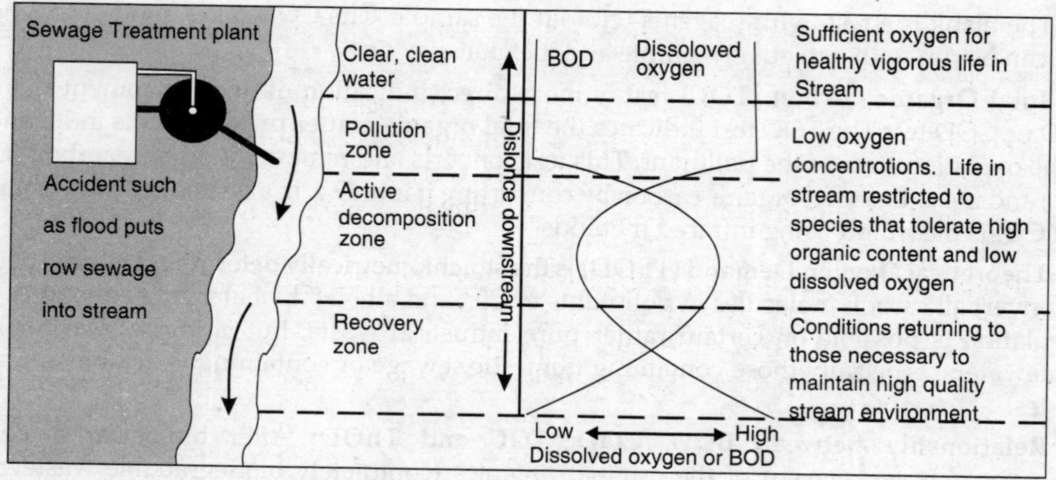

Fig.7.28F$_1$: Effects of BOD on Water Quality

All streams have some capability to degrade organic waste. Problem occur when stream is overloaded with biochemical oxygen-demanding waste

Water Quality Vs. Dissolved Oxygen (Do) in water at 20 °C is shown below

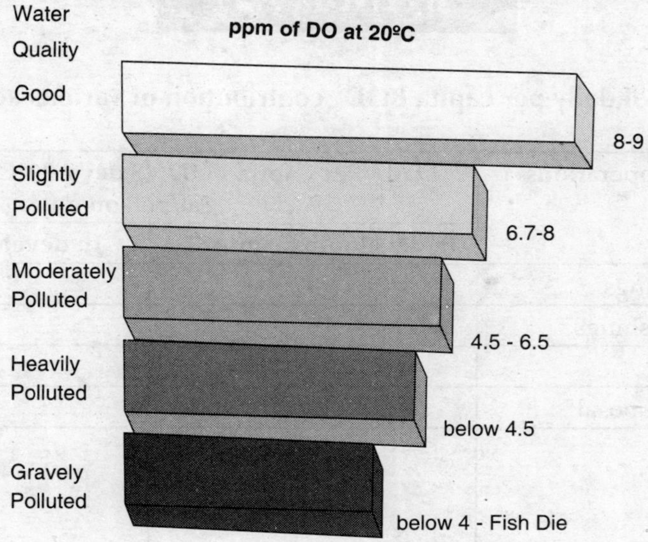

Fig.7.28F$_2$: Water Quality Vs. Dissolved Oxygen in water at 20 °C

APPENDIX 7.29

Average breakdown of daily per capita BOD_5 contribution of various activities/operations

S.N.	Activities/operations	Daily Per Capita BOD_5 (5 days, 20°C) Contribution, gm/person/day	
		In developing countries	In developed countries
1.	Utensils washings	8	6
2.	Individual washings	5	9
3.	Cloth washing	5	9
4.	Solid waste disposal	0	31
5.	Toilets:		
	(i) Faeces	11	11
	(ii) Urine	10	10
	(iii) Others	1	2
	Total average/day/person	40	78

APPENDIX 7.30

Characteristics of well water of coal field area in Eastern India

S.N.	Characteristics	Unit	Average value
1.	pH value	-	7.0
2.	Total dissolved solids (TDS)	mg/litre	672
3.	Total hardness(as $CaCO_3$)	mg/litre	617
4.	BOD, 5 days, 20°C	mg/litre	0.82
5.	COD	mg/litre	4.5
6.	Dissolved solids	mg/litre	6.0
7.	Nitrate	mg/litre	78
8.	Sulfate	mg/litre	47
9.	Chloride	mg/litre	62
10.	Oil & grease	mg/litre	0.038
11.	Iron	mg/litre	0.10
12.	Manganese	mg/litre	0.003
13.	Fluoride	mg/litre	0.23

APPENDIX 7.31

Features of water quality monitoring / measurement instruments

S. N.	Instruments for water quality measurements	Parameters/quality measured	Measurement principles	Remarks/measurement range
1.	Digital pH meter	pH value	Nearnst equation	0-14 pH value
2.	Thermistor Thermometer	Temperature	-	0-50°C
3.	Electrical conductivity meter	Conductivity	Conductivity bridge	(a) 1 mho to 1 micromho (b) 1 micromho to 1 M mho
4.	Dissolved oxygen meter	Dissolved oxygen	-	0-20 ppm
5.	Chloride meter	Chloride	Colourometric technique, Faraday law	-
6.	Chloroscope	Residual chlorine gas	Visual comparison with standard coloured disc (of OTA method)	-
7.	Polarograph	Metals and pesticides	Ilkoive equation, half wave potential	-
8.	Spectrophotometer	Organic and metal pollutants	Beer's Lambert law/Molecular absorption	(a) 350-700 nm (b) 200-2000 nm
9.	Turbidity meter/ comparision tubes	Turbidity	Scattering of radiation by suspended particles	(a) 1-100 JU (b) 0-1000 JU
10.	Nephelometer	Turbidance/ turbidity	Adsorbance of radiation by suspended particles	400-700 nm
11.	Hydrometer	Humidity	-	0-100% humidity
12.	Microscope	Microscopic examination/ biological investigation	-	(a) 1×100 magnification (b) 1×500 magnifiaction
13.	Ion selective electrode	Ionic concentration	-	mg/ng

S. N.	Instruments for water quality measurements	Parameters/quality measured	Measurement principles	Remarks/measurement range
14.	Ion analyser	Ionic concentration	-	10^{-1} mg-ng level
15.	BOD analyser incubator	BOD for 3 days		0.1 mg/litre
16.	BOD and COD monitor	BOD and COD	-	0.1-5 mg/litre
17.	Flame photometer	S-block metals	-	Na, K, Ca fillers
18.	Atomic absorption	Metal pollutants	-	190-800 nm
19.	Gas chromatograph	Volatile organic compounds	FID, ECD, TCD detector	-
20.	IR spectrophotometer	Organic compounds	With KB pellet or Najol technique	-

APPENDIX 7.32

Preservation method and maximum holding period of waste water samples for determination of various parameters

S. N.	Characteristics/parameters of waste water to be analysed/ determined	Water sample preservation method	Maximum holding period of the waste water sample, day
1.	Biological oxygen demand (BOD)	Refrigeration at 4°C	0.25
2.	Color	Refrigeration at 4°C	1
3.	Cyanide	pH of the sample to be raised to 10 or higher with the addition of NaOH	1
4.	Phenol	1 gm $CuSO_4$/litre + H_3PO_4 to lower PH to < 4; refrigeration at 4°C	1
5.	Coliform Bacteria	Sterilised bottle , no specific preservative; refrigeration at 4°C	1.5
6.	Chemical oxygen demand (COD)	20 cc H_2SO_4/litre	7
7.	Fluoride	None required	7
8.	Nitrogen (Kjeldahl,ammonia, nitrite)	Add 40 mg $HgCl_2$/litre; refrigeration at 4°C	7 (except for kjeldahl nitrogen which is unstable)
9.	Phosphorous	40 mg Hg_2Cl_2/litre , refrigeration at 4°C	7
10.	Sulphide	2c.c Zinc acetate/litre	7
11.	Odour	Refrigeration	7
12.	Turbidity	None available	7
13.	Metals	No specific preservation method for sample, but the sample should be acidified.	180

Note: Time allowed between collection & analysis of waste water sample is important to ensure that characteristics to be analysed are not changed between collection & analysis of sample. Preservation method to be followed for samples which are not be analysed immediately after collection of the waste water sample.

APPENDIX 7.33

National (USA) primary drinking water standards as per USEPA with possible adverse health effects of various water contaminants

S.N.	Water Contaminants	Milligram/litre		Potential health effects from ingestion of contaminated water	Sources of contaminant in drinking water
		MCLG	MCL		
1.	Fluoride	4.0	4.0	Skeletal and dental fluorosis	Natural deposits; fertilizer, aluminum industries, water additive
	Volatile Organics:-				
2.	Benzene	Zero	0.005	Cancer	Some foods; gas, drugs, pesticides, paint, plastic industries
3.	Carbon Tetrachloride	Zero	0.005	Cancer	Solvents and their degradation products
4.	P-Dichlorobenzene	0.075	0.075	Cancer	Room and water deodorants and "mothballs"
5.	1,2- Dichloroethane	Zero	0.005	Cancer	Leaded gasoline, fumigants, paints
6.	1,1-Dichloroethylene	0.007	0.007	Cancer	Plastics, dyes, perfumes, paints
7.	Trichloroethylene	Zero	0.005	Cancer	Textiles, adhesives, and metal degreasers
8.	1,1,1-Trichloroethane	0.2	0.2	Liver, Nervous system effects	Adhesives, aerosols, textiles, paints inks, metal degreasers
9.	Vinyl Chloride	Zero	0.002	Cancer	May leach from PVC pipe; formed by solvent break down
	Coliform and Surface Water Treatment:-				

S.N.	Water Contaminants	Milligram/litre		Potential health effects from ingestion of contaminated water	Sources of contaminant in drinking water
		MCLG	MCL		
10.	Giardia lamblia	Zero	TT	Gastroenteric disease	Human and animal fecal waste
11.	Legionella	Zero	TT	Legionnaire's	Indigenous to natural waters; can grow in water heating systems
12.	Standard Plate Count	N/A	TT	Indicates water quality	-
13.	Total Coliform*	Zero	<5%+	Indicates gastroenteric pathogens	Human and animal fecal waste
14.	Turbidity*	N/A	TT	Interferes with disinfection filtration	Soil runoff
15.	Viruses	Zero	TT	Gastroenteric disease	Human and animal fecal waste
	Inorganics:-				
16.	Antimony	0.006	0.006	Cancer	Fire retardants, ceramics, electronics, fireworks, solder
17.	Asbestos(>10m)	7MFL	7MFL	Cancer	Natural deposits; asbestos cement in water systems
18.	Barium*	2	2	Circulatory system effects	Natural deposits, pigments, epoxy sealants, spent coal
19.	Beryllium	0.004	0.004	Bone, lung damage ,Kidney effects	Electrical, aerospace, defense industries
20.	Cadmium*	0.005	0.005	Kidney effects	Galvanized pine corrosion; natural deposits; batteries, paints
21.	Chromium*(total)	0.1	0.1	Liver, kidney, circulatory disorders	Natural deposits; mining, electroplating, pigments
22.	Cyanide	0.2	0.2	Thyroid, nervous	Electroplating, steel, plastics, mining, fertilizer

S.N.	Water Contaminants	Milligram/litre		Potential health effects from ingestion of contaminated water	Sources of contaminant in drinking water
		MCLG	MCL		
23.	Murcury*(inorganic)	0.002	0.002	Kidney, nervous system	Crop runoff; natural deposits; batteries,
24.	Nitrate	10	10	System disorders Methemoglobinemia	Animal waste, fertilizer, natural deposits, septic tanks, sewage
25.	Nitrite	10	10	Methemoglobinemia	Same as nitrate; rapidly converted to nitrate
26.	Selenium	0.05	0.05	Liver damage	Natural deposits; mining, smelting, coal/oil combustion
27.	Thallium	0.0005	0.002	Kidney, liver, brain, intestinal	Electronics, drugs, alloys, glass
	Organics:-				
28.	Acrylamide	Zero	TT	Cancer, nervous system effects	Polymers used in sewage/ wastewater treatment
29.	Adipate, [di(2-ethylhexyl)]	0.4	0.4	Decreased body weight	Synthetic rubber, food packaging, cosmetics
30.	Alachlor	Zero	0.002	Cancer	Runoff from herbicide on corn, soybeans, other crops
31.	Atrazine	0.003	0.003	Mammlary gland tumors	Runoff from use as herbicide on corn and non-cropland
32.	Carbofuran	0.04	0.04	Nervous, reproductive system effects	Soil fumigant on corn and cotton; restricted in some areas
33.	Chlordan*	Zero	0.002	Cancer	Leaching from soil treatment for termites

S.N.	Water Contaminants	Milligram/litre		Potential health effects from ingestion of contaminated water	Sources of contaminant in drinking water
		MCLG	MCL		
34.	Chlorobenzene	0.1	0.1	Nervous system and liver effects	Waste solvent from metal degreasing processes
35.	Dalapon	0.2	0.2	Liver and kidney effects	Herbicide on orchards, beans, coffee, lawns, road/railways
36.	Dibromochl-oropropane	Zero	0.0002	Cancer	Soil fumigant on soybeans, cotton, pineapple, orchards
37.	0-Dichlorobenzene	0.6	0.6	Liver, kidney, blood cell damage	Paints, engine cleaning compounds, dyes, chemical wastes
38.	Cis-1,2- Dichloroeth-ylene	0.07	0.07	Liver, kidney, nervous circulatory	Waste industrial extraction solvents
39.	Trans-1,2-Dichloroethylene	0.1	0.1	Liver, kidney, nervous circulatory	Waste industrial extraction solvents
40.	Dichloromethane	Zero	0.005	Cancer	Paint stripper, metal degreaser, propellant, extraction
41.	1,2-Dichloropropane	Zero	0.005	Liver, kidney effects; cancer	Soil fumigant; waste industrial solvents
42.	Dinoseb	0.007	0.007	Thyroid, reproductive organ damage	Runoff of herbicide from crop and non-crop applications
43.	Dioxin	Zero	0.00000003	Cancer	Chemical production by-product; impurity in herbicides
44.	Diquat	0.02	0.02	Liver, kidney, eye effects	Runoff of herbicide on land and aquatic weeds

S.N.	Water Contaminants	Milligram/litre		Potential health effects from ingestion of contaminated water	Sources of contaminant in drinking water
		MCLG	MCL		
45.	2,4-D*	0.07	0.07	Liver and kidney damage	Runoff from herbicide on wheat, corn, range lands, lawns
46	Endothal	0.1	0.1	Liver, kidney, gastrointestinal	Herbicide on crops, land/ aquatic weeds; rapidly degrade
47	Endrin	0.002	0.002	Liver, kidney, heart damage	Pesticide on insects, rodents, birds; restricted since 1980
48	Epichlorohydrin	0	TT	Cancer	Water treatment chemicals; wastes epoxy resins, coatings
49	Ethylbenzene	0.7	0.7	Liver, kidney, nervous system	Gasoline; insecticides; chemical manufacturing wastes
50	Ethylene dibromide	0	0.00005	Cancer	Leaded gasoline additives; leaching of soil fumigant
51	Glyphosate	0.7	0.7	Liver, kidney damage	Herbicide on grasses, weeds, brush
52	Heptachlor	0	0.0004	Cancer	Leaching of insecticide for termites, very few crops
53	Heptachlor epoxide	0	0.0002	Cancer	Biodegradation of heptachlor
54	Hexchlorobenzene	0	0.001	Cancer	Pesticide production waste by- product
55	Hexachloroyclo-pentadiene	0.05	0.05	Kidney, stomach damage	Pesticide production intermediate

S.N.	Water Contaminants	Milligram/litre		Potential health effects from ingestion of contaminated water	Sources of contaminant in drinking water
		MCLG	MCL		
56	Lindane	0.0002	0.0002	Liver, Kidney, nervous immune, circulatory	Insecticide on cattle, Lumber, gardens; restricted in 1983
57	Methoxychlor	0.04	0.04	Growth, liver, kidney nerve effects	Insecticide for fruits, vegetables, alfalfa livestock, pets
58	Oxamly (Vydate)	0.2	0.2	Kidney damage	Insecticide on apples, potatoes, tomatoes
59	PAHs (benzo(a) pyrene)	0	0.0002	Cancer	Coal tar coatings, burning organic matter volcanoes, fossil fuels
60	PCBs	0	0.0002	Cancer	Coolant oils from electrical transformers; plasticizers
61	Pentachlorophenol	0	0.001	Liver and kidney effects, and cancer	Wood preservatives, herbicide, cooling tower wastes
62	Phthalate [di(2-ethylhexyl)]	0	0.006	Cancer	PVC and other plastics
63	Picloram	0.5	0.5	Kidney, liver damage	Herbicide on broadleaf and woody plants
64	Simazine	0.004	0.004	Cancer	Herbicide on grass sod, some crops, aquatic algae
65	Styrene	0.1	0.1	Liver, nervous system damage	Plastics, rubber, resin, drug industries; leachate from city landfills
66	Trichloroethylene	0	0.005	Cancer	Improper disposal of dry cleaning and other solvents

S.N.	Water Contaminants	Milligram/litre		Potential health effects from ingestion of contaminated water	Sources of contaminant in drinking water
		MCLG	MCL		
67	Toluene	1	1	Liver, Kidney, nervous circulatory	Gasoline additive; manufacturing and solvent operations
68	Toxaphene	0	0.003	Cancer	Insecticide on cattle, cotton, soybeans; canceled in 1982
69	2,4,5-TP	0.05	0.05	Liver and kidney damage	Herbicide on crops, golf courses; canceled in 1982
70	1,2,4-Trichloro-benzene	0.07	0.07	Liver, kidney damage	Herbicide production, dye carrier
71	1,1,2 Trichloro ethane	0.003	0.005	Kidney, liver, nervous system	Solvent in rubber, other organic products; chemical production wastes
72	Xylenes (total)	10	10	Liver, kidney, nervous system	By-product of gasoline refining; paints, links, detergents
73	**Lead and Copper:** Lead*	0	TT++	Kidneys, Nervous system damage	Natural/industrial deposits; plumbing, solder brass alloy faucets
74	Copper	1.3	TT#	Gastrointestinal irritation	Natural/industrial deposits; wood preservatives, plumbing
75	**Other Interim Standards:** Beta/photon emitters	0	4mrem/yr	Cancer	Decay of radionuclides in natural and man-made deposits
76	Alpha emitters	0	15pCi/L	Cancer	Decay of radionuclides in natural deposits
77	Combined radium-226/228	0	5pCi/L	Bone cancer	Natural deposits

S.N.	Water Contaminants	Milligram/litre		Potential health effects from ingestion of contaminated water	Sources of contaminant in drinking water
		MCLG	MCL		
78	Arsenic*	0.05	0.05	Skin, nervous system toxicity	Natural deposits; smeller, glass electronic wastes; orchards
79	Total Trihalo methanes	0	0.10	Cancer	Drinking water chlorination by-products

Notes: TT = special treatment techniques required.

*Contaminants with interim standards which have been revised.

+= Less than 5% positive samples.

MFL= million fibers per liter.

++= Action Level 0.015mg/L.

#= Action Level 1.3 mg/L.

pCi= picocuries

Secondary drinking water standards are established by the USEPA for control of aesthetic qualities relating to public acceptance and include contaminants that affect taste, color, odor & appearance. These standards provide recommended secondary maximum contaminant levels (SMCLs) that are not enforceable maximum values.

Drinking water contamination generally result from:

- Contaminants that find their way into drinking water sources from industrial waste releases, agricultural runoff, atmospheric deposition, and other pollution sources.
- Contaminants formed during the treatment of water supplies including disinfection by-products.
- Materials used for treatment, storage, and distribution of water.

USEPA- United states environmental protection agency

Note:

(a) TT- Special treatment technique required

(b) MLF – Million fibres per litre

MCLG – Maixmum contaminant level goals – They are the maximum level of contaminants in drinking water at which no known or anticipated adverse effects on human health would occur and allow for a margin of safety. MCLG's are not enforceable.

MCL- Maximum contaminant level-They are the maximum allowable concentration of contaminants, which are enforceable and mandatory for all public water supplies.

Contaminants with interim standards which have been revised.

Primary drinking water standards have been established by USEPA (United states environmental protection agency for microbiological and chemical contaminants (MCLG & MCL are regulated contaminants)

APPENDIX 7.34

Glossary of Terms Related to water quality

A

Acid: A compound resulting in a pH less than 7 when in aqueous solution, a molecule that can give up a proton to a base, accept an unshared pair of electrons from a base or react with a base to form a salt, a substance that has more free hydrogen ions, H+, than hydroxyl ions, OH-.

Acute toxicity: As commonly used it is a pronounced effect, severe biological harm or death , produced in an organism by a toxicant, a substance or a mixture of substances within a short period of time, usually 96 hours or less, after exposure. The strict implication is simply one of rapid onset of effects, in comparison with the life span of the organism; severity is not necessarily implied. For algae and bacteria 96 hours would be a chronic exposure since it would last for several generations.

Aeration: The addition, by bubbling, mixing or turbulent exposure, of air or oxygen into water, or by spraying the water into the air, increasing the dissolved oxygen concentration in the water and dissipating or stripping volatile contaminants and other pollutants from the water into the air.

Aerobic: The presence of gaseous or dissolved oxygen, the presence of or utilizing oxygen,

Aesthetic: Pleasing or acceptable to the senses, primarily taste, odor and vision.

Algae: A group of chlorophyll-containing, non-flowering plants, mostly aquatic, although many are planktonic ; some species are extremely large.

Algal bloom: A bloom of algae occurs when their growth is so rapid that they become numerous enough to colour a body of water, a population explosion of phytoplankton in response to changing environmental conditions, including eutrophication from wastewater and non-point sources, blooms can result in oxygen depletion and biological impacts such as fish kills, blooms are often the result of urban runoff of lawn fertilizers.

Alkalinity: The measurement of chemicals in a water supply which determine alkaline conditions, the alkalinity of water is a measure of its capacity to neutralize acids.

Alluvial: Refers to particles carried by running water which are deposited when the flow rate decreases.

Alum: An aluminum sulphate salt, generally potassium or ammonium is the cation, used to coagulate particles in water treatment.

Anaerobe: An organism that can only exist in the absence or near-absence of gaseous or dissolved oxygen.

Anaerobic: Denotes absence of gaseous or dissolved oxygen, submerged sediments below a narrow oxygenated layer may be anaerobic, also refers to metabolic activities, glycolysis, in the absence of oxygen which occurs in some microorganisms.

Annelids: A phylum of segmented marine and freshwater worms, distinguished from non-segmented roundworms and flatworms,

Anoxia: Absence of oxygen,

AOX, Adsorbable organic halides: The total of all halogenated organic compounds particularly those that are fluoridated, chlorinated or brominated,

Aquifer: Water within the soil or rocks beneath the surface of the earth that supplies wells and springs, water in the zone of saturation where all openings in rocks and soil are filled with water, the upper surface of which forms the water table, the streams or pools of water that flow or collect under the surface of the land and not on the surface, these may be confined if there are layers of impermeable material both above & below and it is under pressure so that when the aquifer is penetrated by a well, the water will rise above the top of the aquifer, or unconfined when the upper water surface, the water table, is at atmospheric pressure, and is able to rise and fall.

Any geological formation containing or transmitting water, especially one that supplies the water for wells and springs, use of the term may be restricted to those water-bearing formations capable of yielding water in sufficient quantity to constitute a usable supply,

Artificial recharge: The process where water is put into ground water, or aquifer, storage from surface water supplies such as irrigation water, reclaimed wastewater or induced infiltration from streams or wells

Assimilative capacity: The amount of pollution a water body can receive without noticeable degradation, as a result of the natural ability of the water and its associated chemical and biological systems to dilute or transform contaminants.

Atmospheric deposition: The contribution of atmospheric pollutants or chemical constituents to land or water ecosystems, deposition results from materials in rain or snowfall, combined with dry dust fallout. Atmospheric sources are a significant source of nutrients and contaminant to aquatic systems.

B

Backwashing: Reversing the flow of water through a water treatment filter or membrane to clean and remove deposits.

Bacteria: Small unicellular organisms lacking a nucleus and some other eukaryotic organelles. They may have photosynthetic pigments but lack chloroplasts, the specialized photosynthetic organelles in higher plants, and mitochondria.

Base flow: The volume of flow in a stream or river during dry conditions, as opposed to conditions influenced by storm runoff.

Bay: A relatively small body of water partially isolated from the main portion of the sea, ocean or lake by a relatively narrow opening or channel.

Benthic: Associated with the sediments on the bottom of a water body.

Benthic organism: Any organism that lives in or near the bottom of a water body or in the sediment.

Benthos: The organisms that live in or near the bottom of a water body or in the **sediment.**

Biochemical oxygen demand, BOD: A measure of the amount of **dissolved** oxygen required to completely **oxidize** the available **organic wastes,** a **quantitative** measure of the degree to which **organic** compounds consume oxygen in water, based on a five-day test in which loss of oxygen in a **sample** results from **bacterial** respiration and chemical processes, a traditional water **quality** measurement applied to **wastewater** such as treated **sewage.**

Biosolids: A nutrient-rich organic material resulting from the treatment of wastewater which contains nitrogen and phosphorus along with smaller amounts of other nutrients, such as potassium, sulfur, magnesium, calcium, copper and zinc, soil that is lacking in these substances can be fertilized with biosolids which also improve soil properties and plant productivity reducing dependence on inorganic fertilizers.

Biota: All living organisms including bacteria, plants and animals.

Bioturbation: The disturbance of sediments due to displacement by organisms, bioturbation resulting from burrowing of organisms in the benthic habitat increases sediment aeration and influences contaminant equilibria with the overlying water.

Blackwater: Wastewater from toilets, latrines, privies, water containing feces or body fluids and water from sinks used for food preparation or disposal of chemical or biological ingredients,

Blue-green algae: Prokaryotic organisms with a bacteria-like cell structure, lacking a nucleus and other organelles. These species manufacture photosynthetic pigments but lack chloroplasts, the specialized photosynthetic organelles in higher plants, in some situations an increase in blue-green algae can indicate an environmental stress such as pollution.

Bog: A wetland that is perched above the water table and has no direct hydraulic connection to it, bogs accumulate peat and the vegetation is dominated by sphagnum moss.

Brackish: Water that is neither fresh water nor marine but a mixture of the two or intermediate in salinity, usually found in estuaries where the amount of salinity is constantly fluctuating.

Brine: Highly salty and heavily mineralized water containing heavy metal and organic contaminants

Buffer: A compound or solution capable of resisting a change in **pH.**

C

Carcinogen: A substance capable of causing cancer.

CFU, Colony forming units: A quantitative measure of the concentration of bacteria in a water sample. Bacterial colonies on laboratory media resulting from filtering and culturing bacteria from a water sample, each colony in the laboratory culture is presumed to have arisen from the multiplication of a single bacterium in the original sample.

Chlorination: The addition of chlorine to water primarily for the purpose of **disinfection** but also for other biological or chemical purposes.

Chlorophenols: Broad spectrum pesticides produced when **phenol** rings have a number of chlorine atoms attached, formerly used in cut lumber treatment to prevent discolouration by **fungus,** often associated with pulp mill effluent and wood preservatives.

Chlorophyll: The coloured pigments, often green, red or brown, found in plants & algae which trap and convert light energy to chemically stored energy which is then used to create **organic** molecules from **inorganic** raw materials.

Chloroplasts: The **organelles,** in **eukaryotic** cells that carry out **photosynthesis,** where the **chlorophyll** pigments and related enzymes are located, specialized structures that carry out **photosynthesis** in plants & **algae.**

Chromosomes: The structures in the nucleus of a **eukaryotic** cell which carries the DNA or genetic material in genes.

Chronic toxicity: A long-term **toxic** effect produced in an organism by a **toxicant,** a substance or a mixture of substances.

Coagulation: The use of chemicals to make suspended solids clump together into larger aggregates, flocs, for easier filtration or sedimentation, coagulation in water treatment uses alum to congregate solids in the water into a mass that can be readily trapped by a filter.

Coliform: Non-pathogenic natural gut bacteria monitored when testing water to indicate the possible presence of pathogenic bacteria.

Colloids: Finely divided solids which will not settle out by gravity alone but which may be removed by coagulation or biochemical action.

Combined sewer overflow: A point in **a sewer** collection system where **domestic sewage** mixed in varying proportions with **storm water** overflows to a receiving water body.

Consumptive use: The quantity of water not available for reuse since it is incorporated into a product or in some way at least temporarily removed from the water cycle. Evapotranspiration, evaporation, incorporation into plant tissue, infiltration into ground water and consumption by humans, wildlife or livestock, are some of the reasons water may not be immediately available for reuse.

D

Deionized water: Water free of inorganic chemicals.

Delta: **An alluvial** deposit consisting of rock particles, **sediment** and debris, dropped by a **stream** or **river** as it enters another body of water, the fan-shaped deposit where a **river discharges** to a larger, slower moving water body, important for **wetland habitat** values.

Denitrification: The natural chemical conversion of **dissolved** nitrite nitrogen to nitrate and finally to gaseous nitrogen, removing it from the **aquatic** system.

Desalination: The process of **salt** removal from **sea** or **brackish** water, the removal of **salts** from **saline** water to provide **fresh water,** an increasingly popular way of providing **fresh water** to coastal populations.

Designated water use: **A water use** that is to be protected at a specific location for such purposes as use by **aquatic** life or wildlife, for **irrigation** or stock watering, in **industrial** activities, for recreation or as drinking water.

Detection limit: The lowest **concentration** of a substance in water that can be reproducibly determined by a specific analytical procedure or test method.

Dilution: The process of mixing a liquid, usually water, that has a lower **concentration** of a substance or **pollutant** with **effluent** containing the substance or **pollutant,** or the **pollutant** or substance itself, such that the final **concentration** after mixing is lower than that in the **effluent** or of the pure substance.

Dioxins: Toxic **organic compounds** containing a specific complex **aromatic** ring structure and containing at least one chlorine atom in the molecule.

Discharge: The release of water which may or may not contain **waste** into the environment, often via a pipe or ditch into a **stream,** the volume of water that passes a given point within a given period of time, an all-inclusive outflow term, describing a variety of **flows** such as from a pipe to a **stream** or from a **stream** to a **lake** or **ocean,** usually expressed in **cubic metre per second**

Disinfectant: A chemical usually an **oxidant,** such as chlorine, chloramine, ozone, hydrogen peroxide or potassium permanganate, or radiation, such as ultraviolet light or ionizing radiation which destroys **pathogens** in water, a substance or action used to purify a medium and kill or inactivate infectious organisms, chlorine is currently the most common **disinfectant** used with water.

Disinfection byproducts: **Chlorinated organic** chemicals, including **trihalo methanes,** are formed when water containing **organic** materials is **disinfected** with chlorine, these compounds are **toxic,** the formation of these compounds can be minimized by **filtering** or otherwise removing the **organic** compounds before **chlorination.**

Dispersion: The movement and spreading of **contaminants** from the point of introduction in an **aquifer** or **surface water** body.

Dissolved oxygen deficit, DO deficit: The difference between the oxygen **saturation** value in water as calculated for the measured conditions at the point and time of sampling, and the actual oxygen concentration, the measure is useful because it corrects for temperature, **salinity,** and atmospheric pressure which influence the **saturation** level, a high deficit can be an indicator of a water **quality** problem.

Dissolved solids: **Inorganic** material **dissolved** in water or liquid **wastes,** excessive **dissolved** solids make water unsuitable for drinking or **industrial** uses **,D.**

Dissolved oxygen: The amount of oxygen gas **dissolved** in a given **quantity** of water at a given temperature and atmospheric pressure, usually expressed as a **concentration** in parts per million, **ppm,** or as a percentage of **saturation.**

Domestic water use: Water used for household purposes, such as drinking, food preparation, bathing, washing clothes, dishes & dogs, **flushing** toilets and watering lawns & gardens, most **domestic** water is delivered to homes by a public water supply facility.

Drainage area: **The drainage area** of a **stream** or **river** at a specified location is that area, measured in a horizontal plane, enclosed by a topographic divide from which direct surface **runoff** normally drains by gravity into the **stream** above the specified location.

Drainage basin: The land area where precipitation runs off into **streams, rivers, lakes** and **reservoirs,** a land feature that can be identified by tracing a line along the highest elevations between two areas on a map, often a ridge, large **drainage basins** contain many smaller drainage sub-basins.

Drawdown A lowering of the **ground water** surface caused by pumping from an **aquifer** or lowering the water surface in a **reservoir** by releasing water either through the turbine or outlet pipes or over the spillway.

Drinking water: A water supply, treated or untreated which is intended for human consumption and uses and which is considered to be free of **toxins** & **pathogenic bacteria,** cysts or **viruses, potable** water, fit to drink, **potable** water that has or is to be treated additionally, to enhance **aesthetic quality** and/or reduce mineral content plus other known or unknown, undesirable substances: by one or more point-of-use water processing devices or systems or purified bottled water.

E

Effluent: The out-flow water or waste water from any water processing system or device, softeners, filters or reverse osmosis units, the product water of a given water treatment system, alternatively a general term in waste water treatment for the final water which is discharged from a treatment plant, usually into a natural flowing river or stream, after complete treatment steps which meet current effluent water quality standards, effluent water is often recycled as irrigation water for golf courses, parklands and some agricultural applications, particularly a liquid, that enters the environment from a point source, generally refers to wastewater from a sewage treatment or industrial plant.

Electro dialysis: The salts are extracted from the water by using a membrane with an electrical current separate the ions, positive ions go through one membrane, while the negative ions through a different membrane, leaving the fresh water behind.

Emergent wetlands: **Marshes** in which vegetation is rooted underwater with only the tops exposed, in fully submerged **aquatic** vegetation or upland **habitats** where the roots are also above water surface.

Enteric: Associated with the gut or intestines of animals or gut products such as faeces.

Enteric viruses: A group of **viruses** associated with human feces found in water.

Enterocci: **Bacteria** species which live in the gut of man or other **homeothermic** animals, somi **pathogens,** often used instead of **faecal coliform bacteria** as an **indicator** of water **contamination.**

Epibenthic: Located at the surface of the **sediments,** generally referring to **algae.**

Epilimnion: The upper, warmer, less dense zone of water in a lake, separated by the fairly thin thermocline zone, from the lower, colder, denser zone of water, the hypolimnion.

Erosion: The wearing away of the land surface by wind, water, ice or other geologic agents, occurs naturally from weather or runoff but is often intensified by human land use practices, erosion is a source of sediments, suspended sediments, TDS, total dissolved solids, particulate matter turbidity and solutes in natural waters.

Estuary, Estuaries, Estuarine zone: The lower portion of a river where the ocean and the river mix, the semi-enclosed zone along a coastline where fresh water meets and mixes with the ocean, such as a bay, mouth of a river, salt marsh or lagoon, deep water tidal habitat & tidal wetland, they are usually partially enclosed by land but have free access to the ocean and are occasionally diluted by fresh water runoff from the land.

Euphotic: The surface layer of an ocean, lake, or other body of water into which light can penetrate, also known as the zone of photosynthesis.

Eutrophic: Having a large or excessive supply of plant nutrients, nitrates & phosphates, usually resulting in an increase in biomass and productivity.

Eutrophication: The process of increasing the nutrients, primarily nitrate & phosphate, content of natural waters, usually resulting in an increase in biomass and productivity of algae which may result in the depletion of the oxygen concentration in the water leading to a fish kill, from natural erosion and runoff from the land or from anthropogenic sources

F

Faecal: Refers to **waste** matter, faeces, from the gut or gastrointestinal tract of animals.

Faecal coliform bacteria: Enteric bacteria which ferment lactose with gas and acid formation at a temperature typical of warm-blooded animals, in water, faecal coliforms are commonly used as an indicator of contamination and are measured using filtration and culture on disk media, the portion of the coliform bacteria group which is present in the intestinal tracts and feces of warm-blooded animals, a common pollutant in water,

Fen: A wetland that is at the water table and has a direct hydraulic connection to it, fens accumulate peat and the vegetation is dominated by sphagnum moss & small herbs.

Filtering, Filtration: Passing a solvent with particulate material suspended in it through a material which allows the solvent to pass but retains the particulates, the mechanical process which removes particulate matter by separating water from solid material, by passing it through a filter such as sand in many water filtration plants.

Flocculation: A large scale treatment process involving gentle stirring whereby small particles in flocs are collected into larger particles so that their weight causes them to **settle** to the bottom of the treatment tank,

Flood: An overflow or inundation that comes from a river or other body of water and causes or threatens damage, it can be any relatively high streamflow overtopping the natural or artificial banks in any reach of a stream, also a relatively high flow as measured by either gauge height or discharge quantity.

Flood plain: A strip of relatively flat and normally dry land alongside a **stream, river** or **lake** that is covered by water during a **flood,** land next to a **river** that becomes covered by water when the **river** overflows its banks.

Fluoridated: A compound that has been reacted with the **halide** fluorine and now contains at least one fluorine atom in the molecule.

Flushing: A measure of how often, usually measured in years, water is replaced in a reservoir, bay or other system, based upon **flow** rates into and out of the system,

Fresh water: **Lakes** and **rivers** running off the land to the **sea** and having much lower **solute concentrations** than the **ocean** into which most eventually drain, water containing less than 1,000 parts per million, **ppm** of **dissolved** solids of any type, water that contains less than 1,000 milligrams per liter, mg/l , of **dissolved** solids, generally, more than 500 mg/l of **dissolved** solids is undesirable for drinking and many **industrial** uses.

Fungi, fungus: A major group of multicellular organisms that are non-photosynthetic and often saprophytic, pathogenic or parasitic.

G

Grab sample: A single **sample** taken at a given place and time.

Green Algae: A common algae group, often green in colour, with nucleated cells and photosynthetic pigments contained in organelles called chloroplasts.

Ground water: Water within the earth that supplies wells and springs, water in the zone of saturation where all openings in rocks and soil are filled, the upper surface of which forms the water table, water that flows in aquifers under the surface of the land and not on the surface, water that flows or seeps downward and saturates soil or rock, the upper surface of the saturated zone is called the **water table**, water beneath the surface of the ground, consisting largely of **surface** water that has seeped down, water beneath the earth's surface, occurring in **aquifers** at one or more depth levels.

Ground water hydrology: The branch of **hydrology** that deals with **ground water**, its occurrence & movements, its replenishment & depletion, the properties of rocks that control **ground water** movement & storage and the methods of investigation & utilization of ground water.

Ground water recharge: The inflow of water to a ground water reservoir, primarily from the surface, infiltration of rain & snowfall and its movement to the water table is one form of natural recharge, the volume of water added by this process.

Ground water reservoir: An aquifer or aquifer system in which ground water is stored, water may be placed in the aquifer by artificial or natural means.

H

Hard water: Water containing a high level of calcium, magnesium, and other minerals, **hard water** reduces the cleansing power of soap and produces scale in hot water lines, boilers & appliances.

Hardness (water): A condition caused predominantly by dissolved salts of calcium, magnesium and iron, such as bicarbonates, carbonates, sulfates, chlorides and nitrates, a water-quality indicator of the concentration of alkaline salts in water, hard water requires more soap, detergent or shampoo to raise a lather.

Headwaters: The source and upper reaches of a stream, also the upper reaches of a reservoir, the water upstream from a structure or point on a stream, the small streams that come together to form a river, any and all parts of a river basin except the mainstream river & main tributaries.

Heavy metals: Metallic elements of high molecular weight, typically with **specific gravities** greater than 5, few examples include copper, lead and zinc.

Hydraulic: Related to water and the **flows** and pressures within a connected water-containing system.

Hydrograph: A chart that measures the amount of water **flowing** past a specified point as a function of time.

Hydrologic cycle: The natural pathway water follows as it changes between liquid, solid, & gaseous states; biogeochemical cycle that moves & recycles water in various forms through the environment, evaporation from oceans to the atmosphere, rain and snowfall to the earth's surface, replenishment of ground water, runoff, uptake by plants, and storage in oceans & ice caps, the movement of water from the atmosphere to the earth and its return to the atmosphere through condensation, precipitation, evaporation & transpiration, the cyclic transfer of water from the Earth's surface via evapotranspiration into the atmosphere, from the atmosphere via precipitation back to earth, and through runoff into streams, rivers, & lakes and ultimately into the oceans.

Hydrology: The science that deals with the **hydrologic cycle** or **water cycle** in the environment-land, soil and atmosphere; properties, distribution & circulation of water.

Hypoxia: Depletion of dissolved oxygen in water to low levels, for example less than 2 mg/l, which can result from natural or human introduction of materials with a high BOD or from eutrophication resulting from high nutrient concentrations.

I

Impervious: A material through which water, or other liquids, will not penetrate & thus must run off over the surface or accumulate on the top, surfaces with a low capacity for soil infiltration, paving, roofs, roadways or other human structures, impervious cover increases runoff and affects the quantity and composition of non-point source pollution, the quality or state of being impermeable, resisting penetration by water or plant roots.

Indicator organism: Microorganisms, such as coliform bacteria, that are not in themselves harmful but whose presence is indicative of possible pollution or the presence of other more harmful microorganisms which, through its population size or condition, mirrors environmental condition within an ecosystem.

Indicator tests: Tests for a specific contaminant, organism, group of contaminants or constituent which signals the presence of something else, coliforms indicate the possible presence of other pathogenic bacteria, tests for a specific contaminant, or constituent which signals the possible presence of something else.

Industrial water use: Water used for industrial purposes in such industries as steel, chemical, paper and petroleum refining, primarily from private sources, such as local wells or withdrawal points in a river, but some comes from public sources.

Infiltration: **Flow** of water from the land surface into the subsurface.

Initial dilution zone: Areas immediately adjacent to a **wastewater discharge** in which **chronic water quality objectives** for water or sediment, but not those for fish, may be exceeded; however, they may not exceed the **acute** objectives; they are defined on a **site-specific** basis and may not encroach on water intakes, bathing beaches, shellfish beds, fish spawning and rearing areas, areas of sensitive aquatic vegetation or other specified sensitive areas.

Injection well: A well constructed for the purpose of injecting treated wastewater directly into the ground, wastewater is generally pumped into the well for dispersal or storage into a designated aquifer, one that does not deliver drinking water, an unused aquifer or below the levels of fresh water.

Instream use: Use of water that does not require withdrawal or diversion from its natural watercourse, the use of water for navigation, recreation and habitat for fish and wildlife.

Intake: The place at which a fluid is taken into a channel or pipe, the location where water is withdrawn from a stream.

Invertebrate: An organism without a backbone.

J

Jet: A concentrated, high velocity flow of water capable of causing erosion, used in mining some placer deposits to wash the unconsolidated deposits into sluice boxes.

K

Killifish: Small oviparous Cyprinodontidae or ovoviviparous Poeciliidae, fish used in bioassays and for mosquito control or as bait.

L

Lacustrine: Relating to a **lake** environment.

Lagoon: A shallow **pond** where sunlight, **bacterial** action and oxygen work to purify wastewater, typically used for the storage of **wastewaters, sludges,** liquid **wastes** or spent nuclear fuel, a shallow sound, channel or **pond,** near and generally connected to a larger body of water.

Leaching: Extraction or flushing out of dissolved or suspended materials from the soil, solid waste or another medium by water or other liquids as they percolate down through the medium to ground water or flow laterally through the waste material, the process by which soluble materials in the soil, such as salts, nutrients, pesticide chemicals or contaminants, are washed into a lower layer of soil or are dissolved and carried away by water.

Lentic: Static or standing, non-flowing waters such as **lakes, ponds** and **reservoirs.**

Limnology: The scientific study of physical, chemical & biological conditions and interactions in lentic systems, lakes, ponds and reservoirs.

Littoral zone: Area on or near the shore of a body of water in relatively shallow water.

Livestock water use: **Water used** for livestock watering, feed lots, dairy operations, fish farming, and other on farm needs.

Loading: The rate of introduction of a constituent or contaminant to a receiving water from the environment, significant in relation to the volume and circulation of the receiving water, problems occur when high loadings occur into receiving waters with limited assimilative capacity.

Lotic: A flowing body of fresh water, such as a river or stream.

M

Mainstream: The main course of a **river** or **stream** where most of the water flows most of the time.

Marine: Refers to the **ocean** or to **a sea, saltwater.**

Marsh: A wetland that is usually submerged in shallow water and whose vegetation is dominated by herbs.

Metalloid: Resembling a metal, or having chemical properties similar to metals.

Mg/l , Milligrams per litre: A concentration unit of chemical constituents in solution, the weight of solute per unit volume of solvent, usually water, this measure is equivalent to parts per million (ppm).

Micrograms per litre, µg/l: A concentration unit of chemical constituents in solution; the weight of solute per unit volume of solvent, usually water, one thousand micrograms per liter is equivalent to 1 milligram per litre, this measure is equivalent to parts per billion or ppb.

Micron, µ: A quantitative measure of thickness equal to one millionth of a meter, one thousandth of a millimeter.

Monitoring: To check, measure or examine water quality over a period of time to note any changes which may occur.

Most Probable Number, MPN: The statistically determined number that represents the number of individual bacteria most likely to have been present in a given sample, measurement of fecal coliform indicator bacteria based on gas production in tubes, alternative to the standard fecal coliform test involving filtration and culture on disk media.

Municipal sewage: Sewage from a community which may be composed of domestic sewage, industrial wastes or both.

Muskeg: A wetland or peatland such as a fen or bog that accumulates peat and whose vegetation is dominated by sphagnum moss and small shrubs or herbs.

N

Nekton: Aquatic animals in the water column with sufficient powers of locomotion to overcome currents and go where they want under their own power as opposed to the plankton which, although some are motile, are at the mercy of water currents, fish are examples of the nekton.

Nephelometric turbidity unit, NTU: The unit of measure for the turbidity of water, a measure of the cloudiness of water as measured by a nephelometer, based on the amount of light that is reflected off particles in the water.

Non-consumptive use: Using water in a way that does not reduce the immediate supply such as hunting, fishing, boating, water-skiing, swimming and some power production.

Non-filterable residue: Solids that are not in true solution & that can be removed by filtration they usually contribute directly to turbidity, small particles of solid pollutants that resist separation by conventional methods; operationally greater than 0.45 microns in size; also known as suspended solids, suspended matter or suspended sediment.

Non-point Source Pollution, NPS: Constituents in water, including pollutants, originating from diffuse, land-based sources and generally transported in runoff from precipitation, pollution discharged over a wide land area, not from one specific location, diffuse pollution caused by sediment, nutrients, organic and toxic substances originating from land-use activities, which are carried to lakes and streams by surface runoff, contamination that occurs when rainwater, snowmelt or irrigation washes off ploughed fields, city streets or suburban backyards. As this runoff moves across the land surface, it picks up soil particles and pollutants, such as nutrients & and pesticides, source of pollution in which wastes are not released at one specific, identifiable point but from a number of points that are spread out and difficult to identify & control.

Non-porous: A material which does not allow water to pass through it.

Non-potable: Not suitable for drinking due to **toxins, pathogens** or **aesthetics.**

Nutrient: A substance, element or compound, necessary for the growth, development and reproduction of plants & animals, as a **pollutant** any element or compound, such as phosphorous or nitrogen that encourages abnormally high **organic** growth in ecosystems.

Nutrient cycle: Chemical transformations of nitrogen, phosphorus, silica & other essential elements in continuous cycles between **organic** & living systems and **inorganic** & non-living phases in an ecosystem as organisms grow and die.

O

Ocean:The large, permanent body of **saline** water surrounding the continents and covering most of the surface of the earth.

Oligotrophic:Having **a** low supply of **nutrients** and thus a low **productivity** of **biomass.**

Organelles:The structures in **eukaryotic** cells, notably **chloroplasts** and **mitochondria,** specialized structures that carry out respiration, **photosynthesis** & other functions in the cell.

Organic chemicals or compounds: Usually chemicals or compounds which contain carbon atoms; usually chains of carbon atoms connected directly to each other, chemicals containing carbon, **organic** matter, plant and animal residues, or substances made by living organisms.

Oxygen demand: The need for molecular oxygen to meet the needs of biological and chemical processes in water, even though very little oxygen will dissolve in water, it is extremely important in biological and chemical processes.

P

PAHs, Polycyclic aromatic hydrocarbons: A family of organic compounds with several linked aromatic rings in their structure which are derived from the combustion of fossil fuels, the higher molecular weight PAHs are an environmental concern due to their bioaccumulation in organisms and their toxic & carcinogenic activity.

Palustrine: Relating to a freshwater environment, such as a marsh, fen, lake, pond, river, bog or swamp.

Parasite, parasitic: Organisms that are pathogens and are obliged to live on or in other organisms, often causing disease or death.

Pathogenic: Causing, or capable of causing, disease or death, generally applied to microorganisms.

PCBs, Polychlorinated biphenyls: Man-made liquid chemicals that are stable, non-corroding, fire resistant, toxic and relatively non-biodegradable, once used in electrical transformers because of these properties and in paint, composed of two joined phenol molecules that have chlorine atoms replacing many of the hydrogen atoms, frequently found in industrial wastes, and subsequently in surface water and ground waters, accumulate in the environment, particularly in the sediment where they can remain indefinitely, virtually banned in 1979 but continue to appear in the flesh of fish and other animals.

Peatland: **A wetland** such as a **fen** or **bog** that accumulates peat and whose vegetation is dominated by sphagnum moss & small herbs.

Percolation: The movement of water through the subsurface soil layers, usually continuing downward to the ground water or water table, the oozing or soaking of water through the soil, the movement of water through the openings in rock or soil, the entrance of a portion of the stream flow into the channel materials to contribute to ground water replenishment.

Periphyton: Organisms attached to and growing on structures, sediments or organisms submerged in water.

Photic zone: The upper portion of the water column which admits sufficient light for photosynthesis, the photic zone is reduced with increased turbidity.

Photosynthesis: The process by which the chlorophyll-bearing cells of green plants, in the presence of light, convert carbon dioxide and water into sugar, an organic compound, with the evolution of oxygen, incorporation of solar energy into carbon compounds by green plants, ultimately providing energy and oxygen for the vast majority of life on earth.

Phytoplankton: The photosynthetic portion, primarily algae, of the free-floating community of small, mostl microscopic, organisms in water, collectively called plankton.

Plankton: The total free-floating community of small, mostly microscopic, organisms in water, some are motile but all are at the mercy of water currents.

Plume: A portion of a water body which is distinguishable from the remainder because it is not completely mixed and its characteristics are measurably different, generally downstream from the junction of another stream of water from a tributary or waste discharge, the area taken up by contaminants in an aquifer.

Point source pollution: Source of pollution that involves discharge of wastes from an identifiable point, such as a smokestack or sewage treatment plant, water pollution coming from a single point, such as a sewage outflow pipe.

Polychaetes: Primarily marine annelid worms with paired segmental appendages, separate sexes and free-swimming trochophore larvae, worms of the Class Polychaeta of the invertebrate worm order Annelida, dominant in marine benthos, highly diversified, ranging from detritivores to predators, some species serving as good indicators of environmental stress.

Pond: A relatively small, sometimes ephemeral or seasonal, inland body of **fresh water** occupying a basin or hollow in the earth's surface.

Pore water: The water found in the interstices of submerged **sediments,** the basis of some types of **toxicity** testing, since it is pore water to which benthic organisms are exposed.

Porosity: A measure of the water-bearing capacity of subsurface rock, with respect to water movement, it is not just the total magnitude of porosity that is important, but the size of the voids and the extent to which they are interconnected, as the **pores** in a formation may be open & interconnected, or closed & isolated, clay may have a very high porosity with respect to potential water content, but it constitutes a poor **flow** medium as an **aquifer** because the **pores** are usually so small.

ppt: A **concentration** unit of chemical constituents in **solution;** the weight of **solute** per unit volume of **solvent,** usually applied to **marine, brackish** or **saline water,** this measure is equivalent to parts per thousand.

Primary wastewater treatment: The first stage of the wastewater treatment process consisting of mechanical removal of large settleable solids through filtering, screening and/or settling, primary wastewater treatment is a mechanical treatment in which relatively large solids are removed from the sewage by settling out as sludge, mechanical methods, such as filters and scrapers, are used to remove pollutants, solid material in sewage also settles out in this process.

Pristine: Describes a natural system, water for example, that has not been affected by anthropogenic pollution.

Profoundal zone: The deep-water region of a **lake** that is not penetrated by sunlight.

Prokaryotic: Unicellular organisms with a bacteria-like cell structure, lacking a nucleus and some other eukaryotic organelles, they may have photosynthetic pigments but lack chloroplasts, the specialized photosynthetic organelles in higher plants, and mitochondria.

Protozoan: Single-celled, nucleated, eukaryotic organisms, lacking cell walls, generally microscopic, some are photosynthetic

Q

Qualitative: Refers to what, which chemical or compound or identity regardless of how much.

Quantitative: Refers to a measured value as how much, how fast, how deep, how many or what concentration.

R

Recharge: Water entering an underground **aquifer** through faults, fractures or direct **absorption,** replenishing an **aquifer.**

Reclaimed wastewater: Treated wastewater that can be used for beneficial purposes, such as irrigating certain plants , domestic wastewater which has been treated to a quality suitable for a beneficial use.

Recycled water: Water that is used more than once before it returns to the natural **hydrologic** system.

Red tide: Algal bloom involving dinoflagellate phytoplankton species, such as Gonyaulax monilata and Ptychodiscus brevis, which naturally manufacture biotoxins, can cause fish kills and several types of shellfish poisoning in people.

Reservoir: A natural or artificial basin for collecting and holding a supply of water, tanks, dammed areas, lakes or underground aquifers, where water is collected and used for water storage, regulation and control, large bodies of ground water are called ground water reservoirs, water behind a dam is called a surface reservoir.

Residual chlorine: The unreacted chlorine which remains in **solution** after the reactions with all the **organic** compounds present have occurred.

Return flow: That part of a diverted flow that is not consumptively used and is returned to its original source or to another body of water, drainage water from irrigated farm lands that re-enters the water system to be used further downstream, irrigation water that is applied to an area and which is not consumed in evaporation or transpiration and returns to a surface stream or aquifer.

Reverse osmosis, RO: A water treatment method whereby water is forced through a semi-permeable membrane which filters out impurities, similar in function to a kidney dialysis machine and used in most space programs & navy vessels to turn waste water into potable water, removing salts from water using a membrane, the product water passes through a fine membrane that the salts are unable to pass through, while the salt waste, brine is removed, method of water or wastewater treatment that relies on a semi-permeable membrane to separate waters from pollutants, an external force is used to reverse the normal osmotic process resulting in the solvent moving from a solution of higher concentration to one of lower concentration.

Riparian zone: A stream and all the vegetation on its banks out to the high water mark, associated with the bank of a watercourse, the woodlands bordering a river.

Runoff: Surface flows of water entering rivers, lakes, the ocean or reservoirs, surface water entering rivers, fresh water lakes, or reservoirs, the portion of precipitation that is not absorbed into the soil, but flows into surface streams, that part of the precipitation or irrigation water that appears in uncontrolled surface streams, rivers, drains or sewers, direct runoff or base runoff, storm interflow or ground water runoff, total discharge during a specified period of time, the depth to which a drainage area would be covered if all of the runoff for a given period of time were uniformly distributed over it.

S

Saline: Waters having solute concentrations approaching or exceeding that of seawater, slightly saline water — from 1,000 ppm to 3,000 ppm, moderately saline water — from 3,000 ppm to 10,000 ppm, highly saline water — from 10,000 ppm to 35,000 ppm, water containing more than 1,000 parts per million of dissolved solids of any type, water that contains significant amounts of dissolved solids, water containing dissolved salts, such as the ocean.

Sample: A small portion of water or other substance taken at a given place and time for analysis; it is assumed to be representative of the whole body of water or the rest of the substance within specified statistical limits.

Sanitary sewer: A system of pipes, conduits and pumps used to convey sewage from its source generally to a treatment plant but also to a ground disposal site, the ocean, irrigation site, holding lagoon or a stream, conveys domestic & institutional sewage and blackwater & industrial wastewater, as opposed to storm water.

Saprophyte, saprophytic: Organisms that do not manufacture their own food and use the organic matter of dead, decaying or decayed other organisms as a source of nutrients for growth.

Saturation: A measure of the proportion of the maximum amount of a gas, or solid, that can be **dissolved** in a liquid, usually water, at the specified temperature and pressure, alternatively a measure of the maximum amount of water, or other liquid, that can be **absorbed** or held in a **porous** medium, the condition of a liquid when it has taken into **solution** the maximum possible **quantity** of a given substance at a given temperature and pressure.

Screening: Passing sewage or wastewater through a coarse mesh to remove the larger particles.

Secchi disc: An opaque, black and white disk lowered into water until the demarcation between the black and white portions is no longer visible, this **secchi** depth is a practical, traditional measurement of water clarity, and is correlated with **turbidity** and the depth of the biological **photic** zone.

Secondary sewage treatment: Includes **primary sewage treatment** and provides in addition conditions conducive to the biological **oxidation** of the remaining **organic wastes,** the second step in most **waste** treatment systems, in which **bacteria** break down the **organic** parts of **sewage wastes,** usually accomplished by bringing the **sewage** and **bacteria** together in trickling **filters** or in the activated **sludge** process, involving the biological process of reducing **suspended, colloidal & dissolved organic** matter in **effluent** from **primary sewage treatment** systems and generally removing 80 to 95 % of the **Biochemical oxygen demand, BOD & suspended matter,** accomplished by biological or chemical-physical methods, removes floating and **settleable** solids and about 90 percent of the **suspended solids, disinfection** is the final stage of **secondary treatment.**

Sedimentation: A large scale **water treatment** process where heavy solids **settle** to the bottom of the treatment tank by gravity after **flocculation,** the process by which solids **suspended** in water are allowed to **settle** to the bottom of a basin or container.

Sedimentation tanks: Wastewater tanks in which floating **wastes** are skimmed off and **settled** solids are removed for disposal.

Seepage: The slow movement of water through small cracks, **pores** and interstices of a material, into or out of a body of surface or subsurface water, the loss of water by **infiltration** into the soil from a canal, ditch, lateral, watercourse, **reservoir,** storage facility or other body of water, or from a field.

Separate sewer: A sewer system that carries only **sanitary sewage** or **storm water runoff,** when **sewers** are constructed this way, **wastewater sewage** treatment plants do not have to be sized to handle **storm water flows** avoiding overflows of untreated **sewage.**

Settleable solids: Suspended solids that will **settle** when the **sewage, wastewater** or any other water containing **suspended** material, is held static for a period of time, usually about two hours for **sewage** treatment.

Sewage: Primarily feces, kitchen **waste** and wash water but usually with other **organic & inorganic** materials present.

Sewage treatment: The complete **sewage** treatment process typically involves a three-phase process, in the **primary sewage treatment** process, which incorporates physical aspects, untreated water is passed through a series of **screens** to remove solid **wastes,** in the **secondary sewage treatment** process, typically involving biological & chemical processes, **screened sewage** is then passed through a series of holding & **aeration** tanks and **ponds,** the **tertiary sewage treatment** process consists of **flocculation** basins, clarifiers, **filters** and chlorine basins or **ozone** or **ultraviolet radiation** processes.

Sewage treatment plant: A facility designed to receive the **wastewater** from **domestic** sources and to remove materials that damage water **quality** and threaten public health & safety when **discharged** into receiving **streams** or bodies of water, removes greases & fats, solids from human **waste** & other sources, **dissolved pollutants** from human **waste** & decomposition products, and dangerous microorganisms, facilities employ a combination of mechanical removal steps and **bacterial** decomposition to achieve these results, chlorine is often added to discharge from the plants to reduce the danger of spreading disease by the release of pathogenic bacteria.

Sewer: A system of pipes, conduits and pumps used to convey **sewage** from its source generally to a treatment plant but also to a ground disposal site, the **ocean, irrigation** site, holding **lagoon** or a **stream.**

Side channel: A slow moving body of water usually with two ends open to the main water body.

Siltation: The accumulation of **sediments** transported by water, the deposition of finely divided soil and rock particles on the bottom of **stream** & **river** beds and **lakes & reservoirs.**

Slough: A slow moving body of water usually with one end open to the main water body.

Sludge: Solid matter that **settles** to the bottom of **sedimentation** tanks in a **sewage** treatment plant and must be disposed of by digestion or other methods or **recycled** to the land.

Spawning: Term used to denote the process where fish, and other water dwelling organisms, deposit & **fertilize eggs** during reproduction.

Specific conductance: **A quantitative** measure of the ability of a water to conduct an electrical current, related to the type & **concentration** of **ions** in **solution** and can be used for approximating the total **dissolved** solids **concentration** in water, one can monitor electrical **conductivity** quickly in the field and estimate total **dissolved** solids or **TDS** without doing any lab tests at all using hand-held testers, expressed in units of electrical **conductance,** Siemens per centimeter at 25 ^{0}C, used in **ground water** monitoring as an indicator of the presence of **ions** of chemical substances that may have been released by a leaking landfill or other **waste** storage or disposal facility.

Stomata: The small pores in the epidermis of plants, usually in the leaves, through which atmospheric gas and water exchange takes place and is controlled.

Storm drain: A drainage system designed for **storm water** which is surface **runoff** from streets and other **impervious** surfaces associated with urbanization or other **anthropogenic** activities and distinct from the **sewage** system, a **sewer** that carries only surface **runoff,** street wash and snow melt from the land, **storm drains** are completely separate from those that carry **domestic** and commercial **wastewater.**

Storm sewer: A drainage system designed for **storm water** which is surface **runoff** from streets and other **impervious** surfaces associated with urbanization or other **anthropogenic** activities and distinct from the **sewage** system, a **sewer** that carries only surface **runoff,**

street wash and snow melt from the land, storm **sewers** are completely separate from those that carry **domestic** & commercial **wastewater.**

Storm water: Water which is primarily surface **runoff** from streets and other **impervious** surfaces associated with urbanization or other **anthropogenic** activities.

Storm water discharge: Precipitation that does not **infiltrate** into the ground due to impervious land surfaces, or **evaporate,** but instead **flows** onto adjacent land or water areas and/or is routed into **storm drain** or **sewer** systems.

Stratification: Vertical separation of water masses into layers with different characteristics, dense **salt** water intruding under **fresh water** in a navigation channel can establish **salinity stratification,** temperature differences in **fresh water** can form distinct water layers separated by **density.**

Stream: A relatively small and sometime ephemeral or seasonal **flowing** body of **fresh water,** in a defined channel.

Stream improvement: One or more works, situated in or near a stream, relating to diversion, storing, measuring, conserving, conveying, retarding, confining or using water from the stream.

Supersaturation: **A concentration** of a gas in water above the equilibrium **concentration,** this occurs when the gas enters **solution** more quickly than it can be released from the liquid to gas phase as in extremely high rates of **plankton photosynthesis** or in the tail races of dams and under waterfalls where air is entrained.

Surface water: Water that **flows** in **streams** & **rivers** and in natural **lake** & **ponds,** in **wetlands** and in **reservoirs** constructed by humans but on the surface of the land & not underground, water on the surface on the earth, as distinguished from **ground water.**

Suspended matter: Solids that are not in true **solution** and that can be removed by **filtration** they usually contribute directly to **turbidity,** small particles of solid **pollutants** that resist separation by conventional methods; operationally greater than 0.45 microns in size, also known as **non-filterable residue, suspended solids** or **suspended sediment**

Suspended sediment: Very fine soil particles that remain in **suspension** in water for a considerable period of time due to the upward components of turbulence & currents, can be removed by **filtration** and contribute to **turbidity.**

Suspension: Large particles are kept from **settling** out of a liquid by gravity due to motion or agitation of the liquid.

Swamp: **A wetland** that is permanently or seasonally submerged in shallow water and whose vegetation is dominated by shrubs and trees.

T

TDS, total dissolved solids: A quantitative measure of the total dissolved organic and inorganic solids concentration in water, an indicator test used for water analysis and also a measure of the mineral content of bottled water & ground water, one can monitor electrical conductivity quickly in the field and estimate total dissolved solids or TDS without doing any lab tests at all using hand-held testers since there is a relationship between TDS and conductivity, sum of all dissolved materials such as salts, which are non-filterable and remain following **evaporation** of the water,

Temporal: Of, or relating to, time.

Tertiary sewage treatment: A third step in sewage treatment usually directed towards greatly increasing the removal efficiency of nutrients, removal from wastewater of traces of organic chemicals and dissolved solids that remain after primary sewage treatment and secondary sewage treatment, selected biological, physical, and chemical separation processes to remove organic and inorganic substances that resist conventional treatment practices, consists of flocculation basins, clarifiers, **filters** and chlorine basins or **ozone** and **ultraviolet radiation** processes, the additional treatment of **effluent** beyond that of **primary and secondary sewage treatment** methods to obtain a very high **quality** of **effluent.**

Thermal pollution: An increase in air or water temperature that harms the climate or ecology of an area, a reduction in water **quality** caused by increasing its temperature, often due to disposal of **waste** heat from **industrial** or power generation processes.

Thermocline: The fairly thin zone in a **lake** that separates the upper warmer zone, **epilimnion,** from the lower colder zone, **hypolimnion.**

TOC, Total organic carbon: Sum of all **organic** carbon compounds in water.

Total metals or substances: **A quantitative** measure of the metals or substances both in the **dissolved** state and those ab**sorbed** to **particulate** matter in **suspension.**

Toxicant: An element or compound with a harmful or lethal effect on the physiology, behaviour, reproduction or survival of an organism.

Toxicity: A measure of how poisonous a **toxin** is to an organism.

Toxicity test: A bioassay to determine the toxicity of a chemical or an effluent using living organisms, a toxicity test measures the degree of response of an exposed test organism to a specified concentration of chemical or effluent sample, living organisms are subjected to varying dilutions of polluted water or water containing known amounts of presumed or known toxins or contaminated sediment, mortality, declines in reproductive rates or behavioral changes indicate a toxic response.

Toxin: A compound or element which is **toxic** or poisonous in common usage. More strictly a **toxin** is a natural **toxicant** made by an organism as opposed to poisons manufactured by man.

Transpiration: The loss of water into the atmosphere from living plants, direct transfer of water from the leaves of living plants to the atmosphere, the passage of water vapour from a living body through membranes or pores, process by which water that is absorbed by plants, usually through the roots, is evaporated into the atmosphere from the plant surface, such as leaf pores or stomata,

Tributary: A smaller stream which joins a larger stream, usually, a number of smaller tributaries merge to form a river.

Trophic: Related to nutrition, referring to one of the hierarchial levels in the food web or **food chain** between the many producers at the bottom and the few predators at the top.

TSS, Total suspended solids: The total suspended solids in water removable with a 0.45 micrometer mesh filter.

Turbid: Coloured or opaque due to matter in suspension, rivers and lakes may become turbid after a rainfall due to erosion & surface runoff containing particulate matter, there is sufficient material in suspension that visibility is decreased.The relative lack of clarity or cloudiness, of water, caused by suspended material, sediments, colored materials in solution and plankton, correlates, inversely, with available light for photosynthesis, the quantity of solid particles that are suspended in water and that cause light rays passing through the water to scatter, turbidity makes the water opaque in extreme cases, measured in nephelometric turbidity units, NTU.

U

μg/l , Micrograms per litre: A concentration unit of chemical constituents in solution; the weight of solute per unit volume of solvent, usually water, one thousand micrograms per litre is equivalent to 1 milligram per litre, this measure is equivalent to parts per billion or ppb.

Unsaturated zone: The zone immediately below the land surface where the soil **pores** contain both water and air, the soil is not totally **saturated** with water.

V

Vapourization: The change of state from a liquid to a gas, the change by which any substance is converted from a liquid state and carried off as a vapor, the process of liquid water becoming water vapor from water surfaces, land surfaces and snow fields.

Virus: Small non-cellular **parasitic** organisms consisting of little more than DNA, or RNA in a protein coat, often crystalline, dependent completely on the cellular machinery of their hosts to complete their life cycle and reproduce.

Volatilization: The change of state from a liquid to a gas, the change by which any substance is converted from a liquid state and carried off as a vapor, the process of liquid water becoming water vapor from water surfaces, land surfaces and snow fields.

W

Waste discharge: A pipe or ditch containing **wastewater** which empties into a water course.

Wastewater: Water with waste materials or pollutants dissolved in it, containing waste including greywater, blackwater or water contaminated by contact with waste, including process-generated and contaminated rainfall runoff, water that has been used in homes, industries and businesses that is not suitable for reuse unless it is treated.

Wastewater treatment: The complete wastewater treatment process typically involves a three-phase process, in the primary wastewater treatment process, which incorporates physical aspects, untreated water is passed through a series of screens to remove solid wastes, in the secondary wastewater treatment process, typically involving biological and chemical processes, screened wastewater is then passed a series of holding and aeration tanks and ponds, the tertiary wastewater treatment process consists of flocculation basins, clarifiers, filters and chlorine basins or ozone and ultraviolet radiation processes.

Water column: The portion of an **aquatic** or **marine** environment extending from the water surface to the bottom or the surface of the **sediment.**

Water cycle: The natural pathway water follows as it changes between liquid, solid and gaseous states, biogeochemical cycle that moves and recycles water in various forms through the ecosphere, the circuit of water movement from the **oceans** to the atmosphere and to the Earth and back to the atmosphere through various stages or processes such as precipitation, interception, **runoff, infiltration, percolation,** storage, **evaporation** & transportation.

Water quality: A term used to describe the chemical, physical and biological characteristics of water, usually in respect to its suitability for a particular purpose.

Water quality criteria: Scientifically derived **ambient** numerical values for physical, chemical or biological characteristics of water, **biota** or **sediment** which must not be exceeded to prevent specified detrimental effects from occurring to **water uses,** recommended **concentrations,** levels or narrative statements that should not be exceeded in order to protect the life or health of organisms.

Water quality guideline: Numerical concentration or narrative statement recommended to support and maintain a designated water use.

Water table: The level below the surface of the earth at which the ground becomes saturated with water, the surface of an unconfined aquifer which fluctuates due to seasonal precipitation, the top of the water surface in the saturated part of an aquifer.

Water treatment: A method of cleaning water for a specific purpose, such as drinking water, irrigation water or discharge to a stream

Water use: Water that is used for a specific purpose, such as for domestic use, irrigation, or industrial processing, associated with human influence on the hydrologic cycle, includes water withdrawal from surface water and ground water sources, water delivery to homes and businesses, consumptive use of water, water released from wastewater treatment plants, water returned to the environment and in-stream uses, such as using water to produce hydroelectric power or for navigation.

Watershed: The land area where precipitation runs off into streams, rivers, lakes and reservoirs, a land feature that can be identified by tracing a line along the highest elevations between two areas on a map, often a ridge, large drainage basins contain many smaller drainage sub-basins, land area drained by a river or stream, the natural hydrologic unit associated with numerous ecological and physical processes involving water, the most appropriate geographic unit for management of water quality.

Wetland: Area that is regularly wet or flooded and has a water table that stands at or above the land surface for at least part of the year, an area where saturation with water is the dominant influence on soil parameters and on composition of the plant community, a bog, pond, fen, estuary, swamp, peatland or marsh.

Whole-effluent toxicity: The aggregate toxic effect of an effluent measured directly by a toxicity test.

Withdrawal: Water removed from a ground water or surface water source for anthropogenic use.

Worm: Multicellular invertebrate organisms in several different phyla which are soft-bodied, elongate and often parasitic or pathogenic.

X

Xerophytic: Refers to plants, generally, that are able to grow under very arid conditions where water is scarce, physiological and morphological adaptations to allow growth under arid conditions.

Xeriscaping: Planting vegetation that requires very little water.

Y

Yield: A quantitative measure of how much water can be pumped from a well, for example, either in absolute units or as a percentage of what is actually present in the aquifer, expressed as an amount per unit of time or as an instantaneous, continuous withdrawal rate, the various values will differ depending upon the recharge rate of the aquifer, the increase in biomass or productivity over a season which is not necessarily reflected in the instantaneous standing crop.

Z

Zooplankton: Primarily microscopic animals which swim freely in the water column or are carried about by water currents, many feed on phytoplankton and are in turn a staple diet of small fish.

Zygote: The diploid union of haploid sperm and haploid egg, the start of the next generation.

APPENDIX 7.35

Latest Indian drinking water standards Vs. World Health organization (WHO) guideline values

Characteristic/ parameters	Drinking Water/Potable water Specifications (As per IS 10500:2012)		Standards of Packaged Drinking water		CPHEEO recommended standards for water		WHO, Guidelines Values
	Requirement (Acceptable) Limit)	Permissible Limit in absence of alternate source,	Under FPA Rules, 1955	As per IS 14543:2004	Acceptable mg/l	Cause for rejection	
Organoleptic and Physical Parameters							
Colour (Hz units), max.	5	15	Not more than 2 Hz Units/True Colour Units	2	5	25	Not mentioned
Odour	Agreeable	Agreeable	Agreeable	Agreeable	Unobject-ionable	Unobject-ionable	No guldeline
pH	6.5 to 8.5	No relaxation	6.5-8.5	6.5 to 8.5	7.0 to 8.5	< 6.5 or > 9.2	No guldeline
Taste	Agreeable	Agreeable	Agreeable	do	Unobject-ionable	Unobject-ionable	
Turbidity, NTU, max,	1	5	Not more than 2 nepheloetric turbidity unit (NTU)	2	2.5	10	Not Mentioned
TDS, max.	500	2000	Not more than 500 mg/lite	500	500	12500	Not guideline
General Parameters Concerning Substances undesirable in Excessive Amounts							
Aluminlum (as Al) mg/Max	0.3	0.2	Not more than 0.03 mg/litre	0.03	0.03	0.2	0.2 mg/l
Ammonia (as total ammonia-N), mg/l Max	0.5	No relaxation	-	-	-	-	No guideline
Anionic detergents (as MBAS) mg/l, max	0.2	1.0	Not more than 0.2 mg/liter	0.2	0.2	1.0	
Barium (as Ba), mg/max	0.7	No relaxation	Not more than 1.0 mg/liter	1.0	-	-	0.7
Boron (as B), mg/ max	0.5	1.0	Not more than 5 mg/liter	5	-	-	2.4
Calcium (as Ca), mg/1, max	75	200	Not more than 75 mg/litre	75	75	200	
Chloramines (as CI), mg/litre	4.0	No relaxation					

Chloride (as CI), mg/l, max	250	1000	Not more than 200 mg/litre	200	200	1000	
Copper (as Cu), mg/l, max	0.05	1.5	Not more than 0.05 mg/l, max	0.05	0.05	0.5	2
Fluoride (as F), mg/l, max	1.0	1.5	Not more than 1.0 mg/litre	1.0	1.0	1.5	1.5
Free residual chlorine, mg/l, max	0.2	1	Not more than 0.1 mg/litre	0.2	0.2	< 1.0	5 (c)
Iron (as, Fe), mg/1, max	0.3	No relaxation, Not more than 0.1 mg/ litre	`Not more than 0.1 mg/litre	0.1	0.1	1.0	No guideline
Magnesium (as Mg), mg/1, Max	30	100	Not more than 30 mg/litre	30	30#	150	-
Mangenese (as Mn). mg/1, max	0.1	0.3	Not more than 0.01 mg/litre	0.1	0.05	0.5	0.5 mg/1
Mineral oil, mg/1, max	0.5	No relaxation	Not more than 45 mg/litre	Absent	0.01	0.3	0.5 mg/1
Nitrate (as NO_3) mg/1, Max	45	No ralaxation	Not more than 45 mg/litre	45	45	45	50
Phenolic compounds (as C_6H_5OH) mg/1, Max	0.001	0.002	Not more than 0.001 mg/litre	Absent	0.001	0.002	-
Selenium (as Se), mg/1, Max	0.01	No relaxation	Not more than 0.01 mg/litre	0.01	0.01	0.01	0.04 (P)
Silver , mg/1, Max	0.1	No relaxation	Not more than 0.01 mg/litre	0.01	-	-	-
Sulphate (as SO), mg/1, Max	200	400	Not more than 200 mg/litre	200	200	400	-
Sulphate (H_2S), mg/1, Max	0.05	No realisation	Not more than 0.05	0.05	-	-	-
Total alkalinity as calcium carbonate, mg/1 Max	200	600	Not more than 200 mg/litre	200	200	600	-
Total hardness (as $CaCO_3$) mg/1 Max	200	600	-	-	200	600	guideline
Zinc (as Zn), mg/1, Max	5	15	Not more than 5 mg/litre	5	5.0	15.0	3 mg/1
Nitrite (as NO_2) mg/1, Max	-	-	Not more than 0.02 mg/litre	0.02	-	-	3
Sodium (as Na) mg/1, Max	-	-	Not more than 0.02 mg/litre	200	-	-	40
Antimony (as Sb) mg/1, Max	-	-	Not more than 200 mg/litre	0.005	-	-	0.005 mg/1
Parameters Concering Toxic Substances							

Cadmium (as Cd) mg/1, Max	0.003	No relaxation	Not more than 0.01 mg/litre	0.01	0.01	0.01	0.003
Cyanide (as CN) mg/1, Max	0.05	No relaxation	Not more than 0.05 mg/litre	Absent	0.05	0.05	-
Lead (as Pb) mg/1, Max	0.01	No relaxation	Not more than 0.1 mg/litre	0.01	0.10	0.10	0.01 (A,T)
Mercury (as Hg) mg/1, Max	0.001	No relaxation	Not more than 0.001 mg/litre	0.001	0.001	0.001	0.006
Molybdenum (as Mo) mg/1, Max	0.07	No relaxation	-	-	-	-	
Nickel (as Ni) mg/1, Max	0.2	No relaxation	Not more than 0.02 mg/litre	0.02	-	-	0.07
Pesticides, μg/1. Max	-	No relaxation	Below detectable	-	-	-	-
Polychlorinated biphenyls, mg/1, Max	0.0005	No relaxation	-	-	-	-	-
Poly aromatic hydrocarbons (as PAH), mg/1, Max	0.0001	No relaxation	-	-	-	-	-
Total arsenic (as As), mg /1, Max	0.05	No relaxation	-	-	-	-	0.01 (A, T)
Trhalone-thanes:			-	-	-	-	0.05 (P)
(a) Bromoform, mg/1,Max	0.01	No relaxiation	-	-	-	-	
(b) Dibromocjh-loromethane. mg/1, Max	0.01	No relaxiation	-	-	-	-	100
(c) Bromodich-loromethane, mg/1. Max	0.06	No relaxiation	-	-	-	-	100
(d) Chloroform, mg/1. Max	0.02	No relaxiation	-	-	-	-	60
Parameters Concerning Radioactive Substances							
Radioactive Materials:							
(a) Alpha emitters, Bq/1, Max	0.1	No relaxation	Not more than 0.1 Becquerel/ litre (Bq/l)	0.1	-	-	-
(b) Beta emitters, Bq/1, max	1.0	No relaxation	Not more than 0.1 Becquerel/ litre (Bq/l)	1	-		-
Pesticide Residues Limits							
Pesticides:		Limit in μg/1	-			-	
Alachlor		20	-			-	
Atrazine		0.03	-			-	
Aldrin/Dieldrin		0.1	-		-	-	

Alpha HCH	0.04		-	-	-	-
Beta HCH	125		-	-	-	-
Butachlor	30		-	-	-	-
Chlorpyriphos	1		-	-	-	-
Delta HCH	0,4		-	-	-	-
2,4- Dichloropempxu acetic acid	3		-	-	-	-
DDT (o, p and m, p- isomers of DDT. DDE and DDD)	2		-	-	-	-
Endosulfan (alpha, beta, and sulphate)	9		-	-	-	-
Ethion	190		-	-	-	-
Gamma + HCH (Lindane)	0.3		-	-	-	-
Isoproturan	1		-	-	-	-
Malathion	2		-	-	-	-
Methyl parathion	-		-	Not more than 0.1 mg/l	-	-
Monocrotophos	-		-	-	-	-
Phorate	-		-	-	-	-
(1) Pesticide residues considered individually	-		-	-	-	-
(2) Total pesticide residue	-		-	-	-	-
	-		-	-	-	-
Bacteriological Quality of Drinking Water Organisms	Requirements		-	-	-	-
All water Intended for drinking: (a) E, coli or thermo tolerate coli form bacterias	Shall not be detectable in any 100 ml sample		-	-	-	-
Treated water entering the distribution system: (a) E, coli or themo tolerant coli form bacterias (b) Total coli form bacteria	Shall not be detectable in any 100 ml sample. Shall not be detectable in any 100 ml sample.		-	-	-	-

P- Provisional guideline value, the available on health effects is limited.

T-Provisional guideline value as calculated guideline value is below the level that can the achieved through partial treatment

Note- Fresh water present on the earth is used for drinking, domestic municipal uses, agriculture, irrigation, industries, navigation, recreation...etc. The term 'drinking water' is synonymously used with 'potable water', 'wholesome water', 'safe drinking water', or 'clean drinking water'. Fresh, pure and clean water is tasteless, odorless clear liquid. Pure and healthy drinking water reduces a wide range of contaminants of health concern.

In our country, drinking water norms still remain voluntary in nature. Potable water/ drinking water is still not included under the PFA Act, 1954 as "Food". Groundwater also still remains unlegislated in the country except some states like Goa and other. The Centre, therefore, only 'recommend'. It is the state, and its (state-level) agencies, that have to ensure quality drinking water.

Surface and groundwater both are sources of drinking water after required treatment. Most of water pollution is man-made. It may also occur naturally by addition of soil particles through erosion, animal wastes and leaching of minerals from rocks.

Underground water is considered fairly safe source of drinking water but it is threatened with contamination due to seepage from industrial waste & effluent, municipal waste & sewage and agricultural runoff. The ground water also gets polluted by leaching of salts and minerals due to overuse of ground water sources.

Saltwater encroachment associated with over drafting of aquifers or natural leaching from naturally occurring deposits are natural sources of ground water pollution.

Water is used by every living organism on the earth. In some areas the demand for water already exceeds nature's supplies. Due to rapid urbanization more people move toward urban areas which lead to overuse of water in urban areas. Over exploitation in term is often used when the rate of extraction exceeds the safe yield of any water resources. The most evident problem is a lowering of the water table beyond the reach of existing wells. Well must consequently be deepened to reach the ground water, in some places the water table has dropped hundreds of feet because of extensive well pumping. A lower water table may, in turn, cause other problems such as groundwater-related subsidence and salt water intrusion.

As per United Nations

- In developing countries, 70 percent of industrial wastes are dumped untreated into waters where they pollute the usable water supply;
- Globally, diarrhoea is the leading cause of illness and death, and 88 per cent of diarrhoeal deaths are due to a lack of access to sanitation facilities and unsafe drinking water;
- The provision of improved sanitation and safe drinking water could reduce diarrhoeal diseases by nearly 90 percent.
- Water quality depends on the intended use of water which may be either human consumption or it may be for industries, irrigation, recreation etc. Depending upon the proposed use of water, certain water quality criteria are established and based on these criteria water quality standards are specified by health department and other regulation agencies. Different types of uses require different level of water purity. Drinking water requires highest standards of purity.

In INDIA there are following specifications of water based on physical, chemical and microbiological analysis:

- Potable Water/Drinking Water;
- Packaged Drinking Water
- Packaged Natural Mineral Water;
- The Water for Food Industry;
- Reagent Grade Water;
- Water for Swimming Pools;
- Water for different designated best use

- Water for Pharmaceutical industry;
- Water for Irrigation; and
- Water for Construction....

Drinking water standards

Drinking water/Potable water in urban and rural areas is generally supplied only by the State Govt, undertakings or the local bodies and water is treated & purified by them before it is supplied to the public. Water treated and supplied by local authorities is not within the definition of "food" under PFA Act, 1954. Drinking water remains unlegislated in India. PFA Act prescribes the standards for bottle water only which is implemented by the prescribed authority.

Following standards for drinking water quality are in practice:

- **Standards on Packed Drinking Water notified under The PFA Act, 1954:**
  - For Packaged drinking water; and
  - For Natural Mineral Water.
- **Bureau of Indian Standards (BIS) has formulated following Indian Standards:**
  - Drinking Water Specification: IS 10500: 2012
  - Packaged drinking water (other than natural mineral water) specification IS: 14543: Indian Standard
- Natural Mineral Water, specification IS: 13428:
- Indian Standard For Packing Materials:
  - Bottles, cups, glasses and Jars as per IS:15410; and
  - Pouches and Films as per 1S:15609.
- Drinking water quality standards recommended by CPHEEO.
- Drinking water quality standards recommended by World Health Organization (WHO).

APPENDIX 7.36

General Standards for Discharge of effluents in inland surface water, public sewers, land for irrigation & marine coastal area

S.N.	Parameters	Standards			
		Inland surface water	Public Sewers	Land for Irrigation	Marine Coastal Areas
		(a)	(b)	(c)	(d)
1	Color and odor	All efforts should be made to remove color and unpleasant odor as far as practicable.			
2	Suspended Solids, mg/l (max)	100	600	200	For process Water:100
3	Particle size	Shall pass 850 µ, IS Sieve	-	-	Floatable Solids max. 850 µ
4	pH value	5.5-9.0	5.5-0.9	5.5-9.0	5.5-9.0
5	Temperature	Shall not exceed 5°C above the receiving water	-	-	Shall not exceed 5°C above the receiving water
6	Oil and grease, mg/l max.	10	20	10	20
7	Total Residual Chlorine max. mg/l	1.0	-	-	1.0
8	Ammoniacal Nitrogen, mg/l, max.	50	50	-	50
9	Total kjeldahl N as NH_3 mg/l.max.	5.0	-	-	5.0
10	Free Ammonia (NH_3), mg/l max	5.0	-	-	5.0
11	B.O.D. (3 days , 27°C) mg/ lmax.	0.2	350	100	150
12	C.O.D. mg/l max	250	-	-	250
13	Arsenic (as As), mg/l, max.	0.2	0.2	0.2	0.2
14	Mercury (as Hg), mg/ l,max	0.01	0.01	-	0.01
15	Lead (as Pb), mg/l,max	0.1	1.0	-	2.0

S.N.	Parameters	Standards			
		Inland surface water	Public Sewers	Land for Irrigation	Marine Coastal Areas
16	Cadmium (as Cd), mg/l, max	2.0	2.0	-	2.0
17	Hexavalent Chromium (as Cr^{6+}), mg/l, max	0.1	2.0	-	1.0
18	Total chromium (as Cr), mg/l, max	2.0	2.0	-	3.0
19	Copper (as Cu), mg/l, max	3.0	3.0	-	2.0
20	Zinc (as Zn), mg/l, max	5.0	15.0	-	15.0
21	Selenium (as Se), mg/l max	0.05	0.05	-	0.05
22	Nickel (as Ni) mg/l, max	3.0	3.0	0.2	0.2
23	Cyanide as (CN), mg/l,max	0.2	0.2	-	15.0
24	Fluoride (as F), mg/l, max	2.0	15.0	-	15.0
25	Dissolved Phosphate (as P), mg/l, max	5.0	-	-	-
26	Sulphides (as S), mg/l , Max	2.0	-	-	5.0
27	Phenolic Compounds, (as C_6H_5OH), mg/l, max	1,0	5.0	-	5.0
28	Radio active material prescribed (a) α emitter (b) β emitter	10^{-7} 10^{-6}	10^{-7} 10^{-6}	10^{-8} 10^{-7}	10^{-7} 10^{-6}
29	Bioassay Test	90% Survival of fish after 96 hrs. in100% effluent	90% Survival of fish after 96 hrs. in 100% effluent	90% Survival of fish after 96 hrs, in 100% effluent	90% Survival of fish after 96 hrs, in 100% effluent
30	Manganese (as Mn), mg/l, max	2.0	2.0	-	2.0
31	Iron (as Fe), mg/l, max	3.0	3.0	-	3.0
32	Vanadium (as V), mg/l, max	0.2	0.2	-	0.2
33	Nitrate (as N), mg/l, max	10.0	-	-	20.0

8
Chapter

Water Related Disease

In human body, water acts as a solvent for the secretory & excretory products , is a carrier of nutrient elements to the tissues and removes waste materials from them. It also acts as a regulator of body temperature. Water is more important than food. Deprivation of water brings about death much more quickly than that of food. The total body water constitution is about 70% of adult body weight. Water is an essential constituent of all animals and vegetative matter. Over 80% of the earth's surface is covered by water in the form of relatively pure liquid in lakes, ponds & rivers, as a dilute salt solution in ocean or as solid in snow fields, glaciers & other polar ice caps. Next to air, water is the most important substance for the existence of life on the earth.

About 1.5 billion people in the world still lack access to safe water for drinking and 2.5 billion people have no basic sanitation. The large majority of people are seriously affected by or die from preventable water & sanitation related diseases are rural dwelling and the urban poor in the developing countries. Current global deaths due to water related diseases range from 3 to 6 million annually.

HEALTH HAZARDS & DISEASES ASSOCIATED WITH POLLUTED/CONTAMINATED WATER

There are many diseases which are associated with the contamination of water supplies by animal or human wastes. They include cholera, typhoid fever, paratyphoid fever, dysentery and infectious hepatitis.

The basic sanitation techniques which are essential to the control of water related diseases are the purification of water, and the treatment of sewage. The emphasis in water quality control has shifted from the area of water borne diseases, to control of oxygen consuming wastes, and now to nutrient removal & water reclamation.

The communicable water associated diseases, are caused by the ingestion of biological agents. However, other modes of transmission including simple water contact, as in the case of schistosomiasis, or insect vectors, as in the case of malaria, are also important. The contamination of water by sewage is the principal cause of waterborne diseases. Intestinal discharges of patients or carriers of such diseases contain the biological agents which are responsible for the disease's spread. The disease causing organisms can be transferred from the excrement of an infected person into the mouths of healthy people by means of contaminated food or water. The spread of such infectious diseases can be controlled both by proper treatment & disposal of human wastes, and by purification of drinking water supplies.

The principal biological agents of concern are pathogenic (disease-causing) bacteria, viruses, and parasites. Pathogenic bacteria are the causative agents of the great epidemic diseases cholera & typhoid as well as bacillary dysentery, paratyphoid fever, and tularemia. Fortunately, intestinal bacteria that are discharged into natural waterways usually survive for only a few days.

These bacteria exhibit an initial rapid decrease that results in about a 90% reduction in 2 days. However, thereafter the rate of decrease is considerably slower. This reduction is faster in warm, polluted water than in cold, clean water.

The only viral disease that has been proven to be transmitted through drinking water is infectious hepatitis, although there are others for which transmission through water is theoretically possible. One of the particular difficulties with viruses is that some can survive the normal dosages of chlorine which are applied at water and sewage treatment plants to kill pathogens. Among the parasites that may be ingested is the protozoa *Entamoeba hystolytica,* which causes amebic dysentery. Another parasite is the guinea-worm which causes drantiasis, a common disease among the rural population.

DEFINITION OF WATER RELATED DISEASE

Water related disease encompasses illness resulting from both direct & indirect exposure to water, whether by consumption or by skin exposure during bathing or recreational water use. It includes disease due to waterborne or water-associated pathogens and toxic substances. A broader definition includes illness related to water shortage or water contamination during adverse climate events, such as floods & droughts, diseases related to vectors with part of their life cycle in water habitats; and disease related to inhalation of contaminated water aerosols.

Water-related disease is defined as any significant or widespread adverse effects on human health, such as death, disability, illness or disorders, caused directly or indirectly by the condition, or changes in the quantity or quality of any waters.

The causes of water related disease include micro-organisms, parasites, toxins and chemical contamination of water. Other terms include 'waterborne disease', which implies direct spread and is used mainly to refer to disease caused by microbiological pathogens or chemical contaminants in water.

'Water associated disease' covers the wide range of diseases in which water plays a part, such as legionnaires' disease, as well as diseases related to lack of water for washing and hygiene.

An example of an indirectly related disease is trachoma: the predominant mode of spread is via poor hygiene and flies thriving in conditions of poor sanitation: clean water for hygiene is an important element in prevention, but the disease is not otherwise water-related.

CLASSIFICATION OF WATER RELATED DISEASES

Water related diseases can be grouped into four general classes: water borne diseases , water-washed diseases , water-based diseases and water related insect vectors diseases. The first three classes are closely linked to people's lack of access to safe water supply.

Disease Classification	Causes
Water Borne Diseases	Caused by the consumption of water contaminated by human or animal excreta (feces, urine) containing disease causing organisms such as bacteria, viruses, worms and amoebas
Water Washed Diseases (Water Scared Diseases)	Caused by poor personal hygiene and skin or eye contacts with contaminated water and / or insufficient quantities of water for personal hygiene and washing Ex: Scabies, trachoma (eye infections), flea, lice, typhus
Water based Diseases	Caused by parasite found in intermediate organisms living in contaminated water. These diseases are passed on to humans when they drink / wash with it. Ex: Dracunculiasis, Schistosomiasis, other helminthes
Water Related Insect Vector Diseases	Caused by insects, especially flies and mosquitoes that breed in or feed near contaminated water sources. Ex: Malaria, dengue, blindness, sleeping sickness, yellow fever.

The commonest distinction regarding water exposure is between drinking water and recreational water. For some surveillance systems, disease is also classified according to whether the water supply involved is a public water system, usually regulated by national legislature and local by laws, or other types of supply, such as private wells or untreated water sources. A more detailed classification requires specification of the type of water source, such as groundwater or surface water. For water contact diseases, the classification sub categories include fresh water and marine waters.

Water related disease may also be classified in public health terms, the classification system should also include factors related to intervention, such as the type of water treatment or sanitation measures required to remove the disease-causing factors. Host factors are relevant to water related disease transmission, since many of the agents have more impact on individuals who are malnourished or already suffering from other disease, including diseases associated with immune deficiency.

The classification of water related disease poses particular problems. Agents, diseases and types of exposures are well documented, but the relationship between them is subject to degrees of certainty, timing of exposure and the nature of the exposure. Water in one form or another is a universal exposure for human beings: water is a fundamental requirement for nutrition and overall health, with a corresponding wide range of diseases that can be classified as having a water component. All the agents involved have other means of transmission and the evidence that confirms a water association can be elusive, often depending on whether a large outbreak or documented incident of contamination has been identified. While it has been estimated that the worldwide burden of water-related disease includes millions of cases of illness and deaths, surveillance systems for detecting and confirming water related disease vary considerably.

Water related disease has been a major health problem for both developed and developing countries. Its priority as a surveillance issue has fallen partly as a result of improved water and sanitation in industrialized and northern countries of the world, associated with lower morbidity and mortality from diseases such as cholera and typhoid fever. Food related disease surveillance has overtaken it in importance and in some countries water related disease is classified under food poisoning or gastrointestinal infection, further diminishing its importance in surveillance statistics. The emphasis on the infections affecting the gastrointestinal system is appropriate in terms of numbers of cases, since gastrointestinal infection is the commonest water related illness, worldwide: but a gastrointestinal based classification ignores respiratory infections linked to contaminated water, such as legionellosis, and also skin or systemic disorders. Non infectious water borne disease is increasingly recognized as a public health concern, requiring different types and complexity of surveillance from the traditional systems based on microbiological agents.

While surveillance of illness attributable to treated water is a particular concern in developed or industrialized countries, only a minority of the world's population receives continuous piped supplies. For the 'water poor' populations of the world, the risks associated with untreated water or poor sanitation include both individual risk of illness and public health issues of monitoring, surveillance and prevention. Increased population mobility and travel to different countries has to some extent globalized the risk of water related disease, and the corresponding requirement for national surveillance systems to include the wide range of potential illnesses. Water related disease surveillance is likely to rise in priority with the impact of climate change, including the effects of global warming on water habitats and adverse weather events threatening the security of water supplies. Other threats to safe water include the effect of the world economy on the funds available for public health measures and the possibility of deliberate release of pathogens or toxic agents into water intended for human consumption.

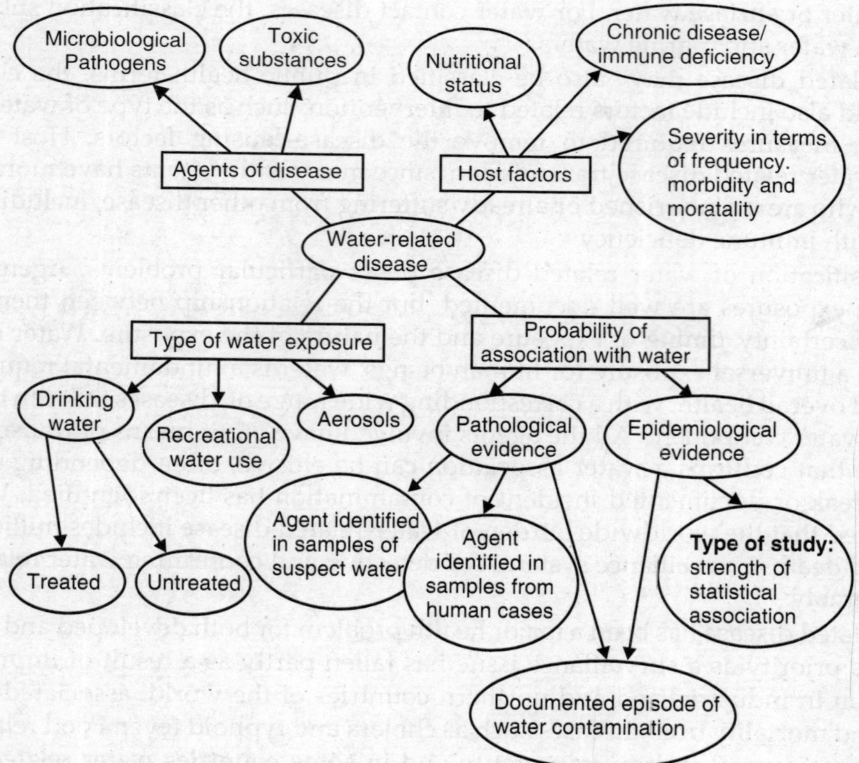

Fig.8.1: Relationships involved in the classification of water-related disease

There are three essential components for the classification: the pathogens & other agents involved in water-related disease; the type of water exposure; and the level of probability of a water cause. Host factors, such as nutritional status, are important in terms of the priority and detail required for surveillance systems in countries with high levels of malnutrition, immune deficiency, or significant mortality from water borne pathogens. Relationships involved in the classification of water-related disease is shown above in figure.

Categories of Water-related Disease

Seven categories of water-related disease are identified: waterborne microbiological disease; waterborne chemical disease; water hygiene disease; water contact disease; water vector habitat disease; excretal disposal disease and water aerosol disease. In addition to the subcategories in table below other subcategories could be included such as those related to duration of exposure (acute/ prolonged).

Table 8.1: Categories , causative agents and examples of of water related disease

Category	Description of disease category	Type of water exposure	Subcategories	Example(s)
Waterborne microbiological disease	Diseases related to consumption of pathogens consumed in water; most due to human or animal faecal contamination of water	Drinking water	Treated or untreated (raw) water Public (municipal) supplies or private supplies	Cholera, Typhoid fever, viral gastroenteritis e.g. due to Norovirus
Waterborne chemical disease	Disease related to ingestion of toxic substances in water	Drinking water	Treated or untreated (raw) water Public (municipal) supplies or private supplies	Arsenicosis
Water hygiene diseases	Diseases whose incidence, prevalence or severity can be reduced by using safe (clean) water to improve personal and domestic hygiene	Any water used for washing/ personal hygiene	Disease related to variations in water quality Disease related to water shortage	Scabies, shigellosis; trachoma
Water contact diseases	Caused by skin contact with pathogen-infested water or with chemical-contaminated water	Recreational water	fresh water sources marine waters	Schistosomiasis (bilharzia); cyanobacteria

Water vector habitat diseases	Diseases where vector lives all or part of its life in or adjacent to a water habitat	Untreated freshwater sources	rivers, streams small collections of stagnant water e.g. water butts	Malaria (mosquitoes); filiariasis (mosquitoes); onchocerciasis (aquatic flies); schistosomiasis (snails); trypanosomiasis (tsetse flies)
Excreta disposal diseases	Diseases related to unsanitary disposal of human waste (faeces and urine)	Drinking water and untreated water sources	diseases related to human/ animal waste in drinking water diseases related to direct/ indirect contact with faeces/ urine	Ascariasis; faecal-oral infections e.g. shigellosis; schistosomiais; trachoma
Water aerosol diseases	Diseases related to respiratory transmission, where a water aerosol containing suspended pathogens enters airway	Drinking or raw water sources	water used in industrial/ residential buildings raw water sources	Legionellosis (legionnaires' disease; humidifier fever); Norwalk-like viral gastroenteritis

Features of Selected Water Borne Diseases are tabulated below.

Table 8.3: Features of Some Water Borne Diseases

Disease	Features
Diarrhea	It is the most common type of water related illness and is caused by drinking water contaminated with disease causing bacteria, viruses and / or tiny parasites like worms / amoebas from human excreta. People who are sick with this have to defecate more often than usual which results in problem of dehydrations and malnutrition.
Dysentery	It is a more serious form of diarrhea and occurs when contaminated water is used for eating / drinking. The persons' feces will frequently contain blood or mucus. It spreads from person to person.

Cholera	It is a highly contagious diarrhea caused by drinking / eating food of water contaminated with the feces or vomit of an infected person. It can also be spread by dirty hands / flies. Cholera outbreaks commonly occur in crowded slums and in the aftermaths of major diseases where water and sanitation facilities are non-existent / damaged / destroyed. In severe cases rapid loss of body fluids leads to dehydration and shock.
Typhoid Fever	Typhoid is a gut infective caused by food / water contaminated with bacteria found in human excreta, and often occurs in epidemics. This disease results in high fever accompanied with diarrhea or vomiting.
Trachoma (Eye Infection)	It is a chronic form of conjunctivitis (pink eye) that get progressively worse and may last for months or many years. The disease is spread by touch of flies. Trachoma is a major cause of blindness in developing countries.
Malaria	Malaria is a disease caused by the micro organisms that are passed onto people who are bitten by malaria infected mosquitoes. People who suffer from malaria suffer from recurring attacks that cause shivers, fevers and aches.
Schistosomiasis	This disease is caused by blood flukes-tiny worms that begin their lives inside fresh water snails. After being released with water as free swimming worms they penetrate the skin who are swimming, bathing or washing in contaminated water. Once in the blood stream, the worm cause victims to suffer from fever, pain in the lower abdomen over time this results in liver damage.
Trypanosomiasis (Sleeping Sickness)	This disease is a dangerous infection spread by infected flies commonly found much in woodland especially around water holes where flies breed. People suffer from fever, ache, fatigue and progressive confusion and difficulty in walking and talking.
Dracunchliasis (Guinea Worm)	This disease is resulted by small worms that enter people in bodies when they drink contaminated water. These worms in body can grow upto 50 cm in length or more just under the skin. The adult worm will form a blister on the skin, normally in the lower parts of the body or legs. When the blister pops, the worm will start to come out of the victim's body.

Features of major water borne diseases are described below:

Cholera. Cholera is an acute, diarrhoeal illness caused by infection of the intestine with the bacterium Vibrio cholerae. A person may get cholera by drinking water or eating food contaminated with the cholera bacterium. This disease spreads very rapidly in areas with inadequate treatment of sewage & drinking water. The cholera bacterium may also live in the environment in brackish rivers and coastal waters.

Approximately 1 in 20 infected people has severe disease characterized by profuse watery diarrhoea, vomiting and leg cramps. In these people rapid loss of body fluids leads to dehydration. Without treatment death can occur within hours.

Dengue. Dengue is an acute infectious disease caused by a virus and transmitted by the bite of the Aedes mosquito, also known as break bone fever and bone-crusher disease. Dengue fever is a severe, flu-like illness that affects infants, young children and adults but rarely causes death.

The symptoms, following an incubation period of five to eight days, are fever, chills, & severe headache, pain in the joints, sweating, and prostration. The clinical features vary according to the age of the patient. Infants and young children may have a feverish illness with rash. Older children and adults may have either a mild feverish illness, or the classical incapacitating disease with abrupt onset and high fever, severe headache, pain behind the eyes, muscle & joint pains and rash.

Dengue hemorrhagic is a potentially lethal complication and is a leading cause of childhood death in several Asian countries. It is characterized by high fever, haemorrhage, enlargement of the liver, and in the most severe cases circulatory failure. Today, dengue is found in tropical and subtropical regions around the world, predominately in urban and semi urban areas, where Aedes mosquitoes are prevalent.

Diarrhoea. Diarrhoea is frequent discharge of watery faeces from the intestines, sometimes containing blood and mucus. Diarrhoea due to infection may last a few days, or several weeks. Persistent diarrhoea results in severe dehydration. It is therefore necessary to replace the fluid lost by the body. Severe diarrhoea may be life threatening due to fluid loss in watery diarrhoea, particularly in infants and young children, the malnourished and people with impaired immunity.

Diarrhoea is a concomitant of many infectious diseases, especially typhoid fever, bacillary or amoebic dysentery, and cholera. Diarrhoea is a symptom of infection by a host of bacterial, viral and parasitic organisms most of which can be spread by contaminated water. Diarrhoea due to infection is widespread throughout the developing world.

Fluorosis. Fluorosis is an abnormal condition caused by excessive intake of fluorine from fluoridated drinking water, characterized by mottling of the teeth. Moderate level chronic exposure (above 1.5 mg/l of water) is quite common. Long-term ingestion of large amounts can lead to potentially severe skeletal problems (skeletal fluorosis). The early symptoms of skeletal fluorosis include stiffness and pain in the joints. In severe case, the bone structure may change and ligaments may calcify, with resulting impairment of muscles and pain. Acute high-level exposure to fluoride causes immediate effects of abdominal pain, excessive saliva, nausea and vomiting. Seizures and muscle spasm may also occur.

Fluoride in water is mostly of geological origin. Waters with high levels of fluoride content are mostly found at the foot of high mountains and in areas where seas has made geological deposits.

Ingestion of excess fluoride, most commonly in drinking water can cause fluorosis which affects the teeth (dental) and bones (skeletal). Moderate amounts lead to dental effects, but long term ingestion of large amounts can lead to potentially severe skeletal problems.

Fluorosis is caused by excessive intake of fluoride. The dental effects of fluorosis develop much earlier than the skeletal effects in people exposed to large amounts of fluoride. Clinical dental fluorosis is characterized by staining and pitting of teeth. In more severe cases all the enamel may be damaged.

High level exposure to fluoride can lead to skeletal fluorosis. Here, fluoride accumulates in the bone progressively over many years. The early symptoms of skeletal fluorosis include stiffness and pain in the joints. In severe cases the bone structure may change and ligaments may calcify resulting in impairment of muscles and pain. Acute high level results in abdominal pain excessive saliva, nausea and vomiting.

Acute high level is very rare and is usually due to accidental contamination of drinking water. Moderate level chronic exposure (>1.5 mg/l) is more common. People affected by fluorosis are often exposed to multiple sources of fluorosis, such as in food, water, air and excessive toothpaste. However, drinking water is typically the most significant source.

Removal of excess fluoride in drinking water is difficult and expensive. The preferred option is to find a supply of safe drinking water with safe fluoride level where access to safe water is already limited, defluoridation may be the only option by use of bone charcoal, contact precipitation, use of Nalgonda activated alumina etc. Since all method produces sludge with very high concentration of fluoride that has to be disposed off ; only water for drinking and cooking purposes should be treated.

Hepatitis. Hepatitis is any disease featuring inflammation of the liver. Two of the viruses that cause hepatitis (hepatitis A and E) can be transmitted through water, food and from person to person. Hygiene is therefore important in their control. The illness starts with an abrupt onset of fever, body weakness, loss of appetite, nausea and abdominal discomfort, followed by jaundice within a few days. The disease may range from mild (lasting 1-2 weeks) to severe disabling disease (lasting several months).

Both hepatitis A and E are found worldwide. Hepatitis A is particularly frequent in countries with poor sanitary and hygienic conditions.

Leptospirosis. Leptospirosis is an infectious disease of domestic animals, especially cattle, swine, and dogs, caused by spirochetes of the genus Leptospira and characterized by jaundice and fever. It is transmissible to humans through direct contact with the urine of infected animals or by contact with a urine contaminated environment, such as surface water, soil and plants.

In humans it causes a range of symptoms, while some infected people may have no symptoms at all. Symptoms of leptospirosis include high fever, severe headache, chills, muscle aches, & vomiting, and may include jaundice, red eyes, abdominal pain, diarrhoea, or a rash. Complications include renal failure, meningitis, liver failure, and respiratory distress. In rare cases death occurs. Leptospirosis occurs worldwide, in both rural & urban areas and in temperate & tropical climates.

Malaria. Malaria is the world's most important parasitic disease transmitted from one person to another through the bite of female Anopheles mosquitoes, which breed in fresh or occasionally brackish water.

Its symptoms are at the onset of malaria, bouts of chills & fever lasting several hours and occurring every three or four days. If the disease is not treated, the spleen & the liver become enlarged, anaemia develops, and jaundice appears. General debility, anaemia, or clogging of the vessels of cerebral tissues by affected red blood cells can be followed by death. Malaria occurs mostly in tropical and subtropical countries.

Scabies. Scabies is a highly contagious parasitic skin disease caused by the itch mite (Sarcoptes scabiei). It is acquired through close contact with an infested individual or contaminated clothing and is most prevalent among those living in crowded and unhygienic conditions. Personal hygiene is an important preventive measure and access to adequate water supply is important in control. The principal sign of the disease is a pimple-like rash that is most commonly found on hands, especially the webbing between the fingers, the skin folds of the wrist, elbow or knee, the penis, the breast or the shoulder. Infestation causes intense itching all over the body, especially at night.

Scabies mites are found worldwide, affecting all socioeconomic classes and in all climates. Epidemics have been linked to poverty, poor water supply, sanitation, and overcrowding.

Typhoid. Typhoid and paratyphoid enteric fever are acute, generalized infections caused by Salmonella typhi and Salmonella paratyphoid respectively. The main sources of infection are contaminated water or milk and, food handlers who are carriers. Their germs are passed in the faeces and urine of infected people. People become infected after eating food or drinking beverages that have been handled by a person who is infected or by drinking water that has been contaminated by sewage containing the bacteria.

Once the bacteria enter the person's body they multiply and spread from the intestines, into the bloodstream. The symptoms of typhoid appear 10 to 14 days after infection; and include high fever, rose colored spots on the abdomen & chest, diarrhoea or constipation, and enlargement of the spleen & liver. In untreated patients , complications may be numerous, affecting practically every body system, and can even include perforation of the intestine with haemorrhage. Typhoid & paratyphoid fevers are common in less industrialized countries, mainly due to the problem of unsafe drinking water, inadequate sewage disposal and flooding.

Other Effects of Water-Bone Diseases

In addition to the immediate and often devastating health effects of water related diseases, affected individuals cannot work. Meager savings are exhausted, people become poor, cannot be productive in turn results in poverty. Water borne infections hamper absorption of food even when intake is sufficient causing malnutrition.

Prevention of water related diseases

Providing wholesome water & good sanitation, constructing sanitary latrines and treating wastewater to allow for biodegradation of human waste will help to curb diseases caused by water pollution.

Water Washed Diseases

They can be controlled effectively with better hygiene for which adequate freshwater is necessary.

Water Based Diseases

Individuals can prevent infection from water-based diseases by washing vegetables in clean water and thoroughly cooking the food. Practicing filtration with nylon gauge clothes to remove guinea worms. Good hygiene, suitable disposal of human waste.

Water Related Insect Vector Diseases

The solution to water related vector diseases would appear to be clear to eliminate the insects that transmit diseases. Putting pesticides, there also have some negative effects. Alternate techniques include using bed nets / introducing predators and sterile insects. Another way is using biological methods and habitat management to reduce / eliminate the natural breeding grounds of the disease vectors. What is important is to have wholesome drinking water to reduce the incidence of diseases and also to reduce malnutrition. Sustainability needs to be addressed by moving away wherever possible from groundwater to surface water resources or ground water recharge.

APPENDIX 8.1

Diseases caused by various microorganisms present in water bodies resulting from sewage disposal

S.N.	Microorganisms	Diseases caused
1	Bacteria	Cholera, typhoid, paratyphoid, dysentery, diarrhea
2	Viruses	Viral hepatitis, poliomyelitis
3	Protozoa	Amoebiasis, giardiasis
4	Helminths	Roundworm, giardiasis

APPENDIX 8.2

Sources and effects of trace metals, present in natural water & waste water

S.N.	Metallic water pollutants	Sources	Effects on man/animal/plants
1	Arsenic	Chemical waste, mining, pesticides, by product	Disturbed peripheral circulation, mental disturbance, liver cirrhosis, lung cancer, hyperkeratosis, ulcer in gastrointestinal tract, kidney damage.
2	Barium	-	Excessive salivation, vomiting diarrhoea, paralysis, colic pain
3	Cadmium	Industrial discharge, mining wastes, metal plating, water pipes	Replaces Zinc biologically, causes destruction of testicular tissue and red blood cells, toxicity to aquatic biota, Growth retardation, diarrhea, bone deformation, kidney damage, testicular atrophy, anemia, injury of central nervous system & liver, hypertension
4	Copper	Metal plating, industrial & domestic waste, mining mineral leaching	Sporadic fever, hypertension, uremia, coma; not very toxic to animals, toxic to plants and algae at moderate levels
5	Cobalt	-	Low blood pressure, lung irritation, diarrhoea, bone deformities, paralysis
6	Hexavalent chromium	-	Diseases in central nervous system, nephritis, gastrointestinal ulceration, cancer
7	Lead	Industry, mining, plumbing, gasoline, coal battery manufacture	Anemia, vomiting, loss of appetite; convulsion; damage of brain, lever & kidney, nervous disorder, wild life destroyed
8	Mercury	Industrial waste, mining, pesticides, coal	Headache, abdominal pain, chest pain, diarrhoea, hemolysis, highly toxic
9	Selenium	Natural geological sources, sulfur, coal	Fever, nervousness; vomiting; low blood pressure, damage of kidney & spleen; blindness and even death; essential at low level but toxic at higher level

S.N.	Metallic water pollutants	Sources	Effects on man/animal/plants
10	Zinc	Metal plating, industrial waste plumbing	Renal damage, vomiting, cramps, essential in many metalloenzymes, toxic to plants at higher level.
12	Beryllium	Coal, nuclear power and space industry	Acute and chronic toxicity, possibly carcinogenic
13	Chromium	Metal plating, cooling tower water	Possibly carcinogenic as Cr (VI). Hexavalent chromium additive (chromate) normally found as Cr (VI) in polluted water
14	Fluorine (fluoride ion)	Natural geological sources, water additive, industrial waste	Possibly tooth decay at about 1 mg/lite, causes mottled teeth and bone damage at about 5 mg/lite.
15	Manganese	Mining, industrial waste, acid mine drainage, microbial action on manganese mineral at low PE.	Relatively non-toxic to animals, toxic to plants at higher levels, stains materials (both room fixtures and clothing)
16	Molybdenum	Natural sources, industrial waste	Possibly toxic to animals, essential for plants

APPENDIX 8.3

Effects of high fluoride ion content in drinking water

Amount of fluoride in drinking water, ppm	Effects
< 0.8	Harmless
1	Mild dental fluorosis to children
1.7-1.8	Mottled teeth to children
2	Molted teeth and tooth disorder
2-3	Tooth disorder
4.5-12	Bone disorder
115	Dangerous to life
2000	Very poisonous, Fatal to life

Note: Fluoride is essential for human being as it helps in normal mineralization of bones and formation of dental enamel. It adversely affects the health of human being when their concentration exceeds the limit of 1.5 mg/1. About 96% of the fluoride in the body is found in bone and teeth. Fluoride is a double edged sword. Ingestion of large amount of fluoride is as harmful as ingestion of its inadequate amount.

Inadequate quantities fluoride causes health problems especially in children. In cold countries like USA, UK etc. problems are related to inadequate consumption of fluoride. In these countries, fluoride is added to water to prevent health hazards. There are areas where dental problems have reduced progressively by adding fluoride in water. Due to inadequacy of fluoride, children suffer from:

- Dental caries
- Lack of formation of dental enamel
- Lack of normal mineralization of bones
- All or a combination of the above
- Fluoride poisoning and the biological response leading to ill effects depend on the following factors:
- Excess concentration of fluoride in drinking water.
- Low Calcium and high alkalinity in drinking water.
- Total daily intake of fluoride
- Duration of exposure to fluoride
- Age of the individual
- Expectant mothers and lactating mothers are most vulnerable groups as, fluoride crosses the placenta hormones for healthy bone formation and bone function are clacitonin, parathormone, vitamin –D and cortisone.

Fluorosis, a disease caused by excess intake of fluoride, is a slow progressive, crippling malady. The tissues affected by fluoride are:

- Dental
- Skeletal
- Non Skeletal

Different fluoride doses (long term ingestion through water) and their effects on human body are given below:

Fluoride (mg/1)	**Effects on human body**
Below 0.5	Dental caries
0.5 to 1.0	Protection against dental caries, Takes care of bone and teeth
1.5 to 3.0	Dental fluorosis
3 to 10	Skeletal fluorosis (adverse changes in bone structure)
10 or more	Crippling skeletal fluorosis and severe ostosclerosis

De-fluorination

Several methods have been suggested for removing excessive fluorides in drinking water. These may be broadly divided into two types.

- Those based upon exchange process or adsorption
- Those based upon addition of chemicals during treatment.

The material used in contact beds includes processed bone. Natural or synthetic tri calcium phosphate, hydroxyl again either alone or in combination with coagulant aid. Other methods include addition to water of material like Magnesia, calcium phosphate, bentonite and fuller's earth, mixing and their separation from water be settling & filtration.

The Nalgonda Technique involves the addition of two simple readily available chemicals Lime and Alum, followed by flocculation, sedimentation & filtration in sequence.

APPENDIX 8.4

Common Diseases Transmitted to Human through Contaminated Drinking Water

S.N.	Type of Organism	Disease	Effects
1	Bacteria	Typhoid fever	Diarrhoea, severe vomiting, enlarged spleen, Inflamed intestine; often fatal if untreated
		Cholera	Diarrhoea, severe vomiting, dehydration; often fatal if untreated
		Bacterial dysentery	rarely fatal except in infants without proper treatment
		Enteritis	Severe stomach pain, nausea, vomiting, rarely fatal
2	Viruses	Infectious hepatitis (Type B)	Fever, severe headache, loss of appetite, abdominal pain, jaundice, enlarged liver; rarely fatal but may cause permanent liver damage
3	Parasitic protozoa	Amoebic dysentery	Severe diarrhoea, headache, abdominal pain, chills, fever; if not treated can cause liver abscess, bowel perforation, and death
		Giardiasis	Diarrhea, abdominal cramps, flatulence, belching, fatigue
		Cryptosporidium	Severe diarrhoea, cramps for up to 3 weeks, and possible death for people with weakened immune systems
4	Parasitic worms	Schistosomiasis	Abdominal pain, skin rash, anemia, chronic fatigue, and chronic general ill health
		Ancylkostomiasis	Severe anemia and possible symptoms of bronchial infection

APPENDIX 8.5

Diseases caused due to Bacterial Pollution of Water

S.N.	Diseases	Bacteria/Virus/Protozoa/Worm
1	**Water borne diseases:**	
	Bacterial:	
	• Typhoid • Cholera • Paratyphoid • Gastroenteritis • Bacterial dysentery	Salmonells typhi Vibrio cholera Salmonella paraypi Enterotoxigenic Escherichia coli Variety of Escherichia coli
	Viral: • Infectious hepatitis • Poliomycetis • Diarrhoea **Other enteric diseases:** • Amoebic dysentery • Other* intestinal illness	 Hepatitis-A virus Polio virus Rota-virus, Norwolk agent, other virus Echono-virus, Coxsachie-virus Ent-amoeba hystolictica Protozoa Glardia sp. And Cryptosporidium sp.
2	**Water-washed diseases:** • Scabies • Trachoma • Bacillary dysentery	 Various skin fungus species Trachoma infecting eyes E.coli*
3	**Water based diseases:** • Schistosomiasis • Guinea worm	 Schistosoma sp. Guinea worm
4	**Infection through water related insect vectors:** • Sleeping sickness • Malaria	 Trypanosoma through tsetse fly Plasmodium through Anaphelis
5	**Infections due to defective sanitation/polluted water:** • Hookworms	 Hookworms Assaris

APPENDIX 8.6

Common disease associated with contaminated water and poor sanitation

S.N.	Disease	Common Vehicle/carrier
1	**Water-borne diseases:** Amoebic dysentery Bacillary dysentery Cholera Cryptosporidiosis Gastroenteritis Gastroenteritis Giardiasis Hepatitis Leptospirosis Paratyphoid fever Salmonellosis Typhoid	Diseases transmitted by ingesting contaminated water and food
2	**Water-washed diseases:** Conjunctivitis Hookworm (Ankylostoma) Leprosy Scabies Skin sepsis and ulcers Truthcuriasis Whipworm (Enterobius) Yews	Lack of adequate quantity of uncontaminated water, and poor personal hygiene create conditions favourable for their spread
3	Water-based diseases: Bitharziosis Dracunculosis Oncholersosis Philariosis Schistosomiasis Threadworm	Diseases causes by infecting agents, by contact with or without ingestion of water. An essential part of the life cycle of the infection agent takes place in an aquatic animal

4	**Fecal-disposal diseases:** Clonorchiasis Diphyttobotheriasis Fasciolopsiasis Paragonimiasis	Diseases caused by infecting agent mostly contracted by eating uncooked fish and other food
5	**Water-related vectors:** Arbpovirus Bancroftian Dengue fever Encephalitis Filariasis Hemorrhagic fever Malaria	Diseases transmitted by insects which live and breed close to water. Infections are spread by mosquitoes, flies, and insect bite

APPENDIX 8.7

Glossary of terms related to effect of water quality on human health

A

Aerobic: A technical word which means in the presence of air, and often refers to metabolic processes that function in the presence of oxygen.

Algae (Singular-Algae)

- Simple single-celled, colonial, or multicelled, mostly aquatic plants, containing chlorophyll and lacking roots, stems & leaves. Aquatic algae are microscopic plants that grow in sunlit water that contains phosphates, nitrates, and other nutrients. Algae, like all aquatic plants, add oxygen to the water and are important in the fish food chain.
- Photosynthetic, almost exclusively aquatic plants of a large and diverse division of the thallophytes, including sea weeds & their fresh water allies. They range in size from simple unicellular forms to giant kelps several metres long, and display varied life cycles and physiological processes with for example different complexes of photosynthetic pigments.
- Nonvascular organisms capable of oxygenic photosynthesis, without sterile cells covering gametangia.

Alkalinity

- Refers to the extent to which water or soils contain soluble mineral salts. Waters with a pH greater than 7.4 are considered alkaline.
- The capacity of water for neutralizing an acid solution. Alkalinity of natural waters is due primarily to the presence of hydroxides, bicarbonates, carbonates and occasionally borates, silicates & phosphates. It is expressed in units of milligrams per litre (mg/1) of $CaCO_3$ (calcium carbonate). A solution having a pH below 6.5 contains no alkalinity.
- Refers to quantity and kinds of compounds present in a lake that collectively shift in pH to the alkaline side of neutrality thereby providing an index to the nature of the rocks within the drainage basin and the degree to which they are weathered.
- Alkalinity is frequently expressed as the total quantity of base, usually in equilibrium with carbonate or bicarbonate that can be determined by titration with a strong acid.

Anaerobic: A technical word which means without air, and often refers to metabolic processes that function in the absence of oxygen,

Anoxia: A state of being in complete absence of oxygen.

Aquifer: An underground layer of water bearing permeable rock or unconsolidated materials from which groundwater can be extracted.

Atrazine: An herbicide that is used globally to stop pre and post emergence broadleaf and grassy weeds in major crops.

Autotrophic: Pertaining to an organism that produces organic compounds from carbon dioxide as a carbon source, using light or the reaction of inorganic chemicals as a source of energy.

B

Bacteria (Singular: Bacterium)

- Microscopic one-celled organisms which live everywhere and perform a variety of functions. While decomposing organic matter in water, bacteria can greatly reduce the amount of oxygen in the water. They also can make unsafe to drink.
- Microscopic unicellular organisms, typically spherical, often clumped into colonies. Some bacteria cause disease, while others perform an essential role in nature in the recycling of materials, for example, decomposing organic matter into a form of bacteria that are used to stabilize organic wastes in wastewater treatment plants, oil spills, or other pollutants. Disease causing form of bacteria are termed 'pathogenic'.
- Some forms of bacteria harmful to man include:
  a. **Total coliform Bacteria -** A particular group of bacteria that are used as indicators of possible sewage pollution. They are characterized as aerobic or facultative anaerobic, gram-negative, non-spore-forming, rod-shaped bacteria which ferment lactose with gas formation within 48 hours. In the laboratory these bacteria are defined as all the organisms that produce colonies with a golden-green metallic sheen within plus or minus 1.0 EC on M-Endo medium (nutrient medium for bacterial growth). Their concentrations are expressed as numbers of colonies per 100 mililitre
  b. **Fecal coliform Bacteria -** Bacteria that are present in the intestine or feces of warm-blooded animals. They are often used as indicators of the sanitary quality of the water. In the laboratory they are defined as all the organisms that produce blue colonisms within 24 hours when incubated at 44.5EC plus or minus 0.2 EC on M-FC medium (nutrient medium for bacterial growth). Their concentrations are expressed as numbers of colonies per 100ml of sample.
  c. **Fecal Straptococal Bacteria –** Bacteria found also in the intestine of warm blooded animals. Their presence in water is considered to verify fecal pollution. They are characterized as gram-positive, cocci bacteria which are capable of growth in brain-heart infusion broth. In the laboratory they are defined as all the organisms which produce red or pink colonies within 24 hours at 35 EC plus or minus 1.0EC on KF-streptococcus medium (nutrient medium for bacterial growth). Their concentrations are expressed as numbers of colonies per 100 ml of sample.

Baseline: The background or starting point from which comparisons can be made.

Bedrock: The native, consolidated rock under the Earth's surface. The chemical composition of bedrock will in part determine the chemical composition of the overlying soil and water.

Benthic: Pertaining to the lowest level of a body of water such as a river, stream or lake. Benthic organis ms inhabit the benthic zone.

- The increase in concentration of a chemical in organisms that reside in environments contaminated with low concentrations of various organic compounds. Also used to describe the progressive increase in the amount of a chemical in an organism resulting from rates of absorption of a substance in excess of its metabolism and excretion.
- Bio concentration plus the accumulation of a compound from food.

Bioavailability

- The fraction of chemical present that is available for uptake by aquatic organisms.
- The Degree to Which toxic substances or other pollutants present in the environment are available to potentially bio degradative microorganisms. Some pollutants might be "bound up" or unavailable because they are attached to clay particles or are buried by sediment. The amount of oxygen, pH, temperature, and other conditions in the water can affect availability.

Biochemical Oxygen Demand (BOD)

- A measure of the quantity of dissolved oxygen, in milligrams per litre, necessary for the decomposition of organic matter by microorganisms, such as bacteria.
- A measure of the amount of oxygen removed from aquatic environments by aerobic micro-organisms for their metabolic requirements. Measurement of BOD is used to determine the level of organic pollution of a stream or lake. The greater the BOD, the greater the degree of water pollution.

Bio magnification: The Process by which a substance is passed up the food chain. A biological process wherein a contaminant's concentration increases at each level up the food chain, including humans, resulting in an especially high level of the substance at upper levels of the food chain. Thus, the availability of such contaminants, even in the seemingly insignificant parts per trillion range, often are ecologically important.

Biomonitoring: A method of inferring ecological condition of an area by examining the organisms that live inhabiting the area. Although biomontoring can occur in any ecosystem, it is most often used to assess water quality of rivers, lakes, streams, and wetlands.

Biota: All living organisms of an ecosystem or region.

C

Carcinogenic: Pertaining to any substance that promotes canner.

Chemical Oxygen Demand (COD): A water quality measure used to indirectly measure the amount of organic compounds in water. This process converts all organic matter into carbon dioxide. It is limited in that it cannot differentiate between levels of biologically active organic substances and those that are biologically inactive.

Chlorophyll-a

- The green pigments of plants. There are seven known types of chlorophyll, Chlorophyll-a and Chlorophyll-b are the two most common forms.
- A green photosynthetic coloring matter of plants found in chloroplasts and made up chiefly of a blue-black ester.
- The primary pigment of photosynthesis in cyanobacteria and eukaryotic autotrophs; often used to indicate biomass; absorbs red and blue light.

Coliform

- A group of bacteria predominantly inhabiting the intestines of man or animals but also found in soil. While typically harmless themselves, Coliform bacteria are commonly used as indicators of the possible presence of pathogenic organisms.
- A group of organisms (Colon bacilli) usually found in the colons of animals and humans; non-pathogenic micro organisms used in testing water to indicate the presence

of pathogenic bacteria. The presence of coliform bacteria in water is an indicator of possible pollution by fecal material. Generally reported as colonies per 100 milliliters (ml) of sample.

Coliform bacteria (total and feacal): Coliform bacteria are a collection of relatively harmless micro-organisms that live in large numbers in the intestines of man and warm & cold blooded animals. Both groups have been widely used as indicators of enteric (intestinal) bacterial pathogens. The total coliform group in not as specific an indicator of faecal contamination as faecal coliforms.

Colour

- One of the most immediately apparent attributes of many natural waters and one that, together with visual clarity, strongly influences human aesthetic perception and recreational use. Colour of waters is a guide to their composition, and remote sensing or water colour is increasingly being used to infer water quality, particularly suspended solids and phytoplankton concentrations.
- The colour of water, with water considered a translucent (i.e. not transparent) material, is commonly associated with transmitted light. However, the colour of natural waters as observed from above is that associated with the upwelling light field that results from back scattering of sunlight illuminating the water volume. In this manner, the colour of natural waters can be objectively specified using their spectral Reflectance, where the reflectance is defined as the ratio of the upwelling light to incident (down welling) light.

Cyanobacteria

- Prokaryotic organisms without organized chloroplasts but having chlorophyll a and oxygen evolving photosynthesis; capable of fixing nitrogen in heterocysts; occurring in lichens both as primary photobionts and as internal or external cephalodia.
- A large and varied group of bacteria which possess chlorophyll a and which carry out photosynthesis in the presence of light and air, producing oxygen. They were formerly regarded as algae and were called ʻblue-green" algae. The group is very old and is believed to have been the first oxygen producing organisms on Earth.

D

Detrital matter: Organic waste material from decomposing plants and animals.

Dissolved Oxygen (DO): The amount of oxygen dissolved in the water. The availability of oxygen is one of the most important indicators of the condition of a water body as DO in necessary for the life of fish and most other aquatic organisms.

E

E coli (Escherichia Coli)

- A bacterial species which inhabits the intestinal tract of man and other warm-blooded animals. Although it poses no threat to human health, its presence in drinking water does indicate the presence of other,, more dangerous bacteria.
- A genus of bacteria that live in the lower intestines of warm-blooded animals, including birds and mammals. They are necessary for the proper digestion of food and are part of the intestinal flora. Its presence in ground and surface waters is a common indicator of fecal contamination.

Estuaries: A semi enclosed coastal body of water that has a free connection with the open sea and within which seawater mixes with freshwater.

Effluent

- Flowing forth or out, especially a stream flowing out of a body of water.
- A surface stream that flows out of a lake or a stream, a branch that flows out of a larger stream (distributary).
- Water Quality - Discharged wastewater such as the treated wastes from municipal sewage plants, brine wastewater from desalting operations, and coolant waters from a nuclear power plant.

Endocrine disrupter: Substances that cause adverse biological effects by interfering with the endocrine system and disrupting the physiological functioning of hormones.

Epilimnion: *The* warm upper layer of a body of water with thermal stratification, which extends down from the surface to the thermocline, which forms the boundary between the warmer upper layers of the epilimnion and the colder waters of the lower depths, or hypolimnion. The epilimnion is less dense than the lower waters, wind circulated, and essentially homothermous. It typically has a higher pH and dissolved oxygen concentration than the hypolimnion.

Eutrophic: Pertaining to water bodies that contain high nutrient concentrations and high levels of primary productivity.

Eutrophication

- *The* degradation of water quality due to enrichment by nutrients, primarily nitrogen (N) and phosphorus (P), which results in excessive plant (principally algae) growth and decay. When levels of N:P are about 7:1, algae will thrive. Low dissolved oxygen (DO) in the water is a common consequence.
- Degrees of Eutrophication typically range from Oligotrophic water (maximum transparency, minimum chlorophyll-o, minimum phosphorus) through Mesotrophic, Eutrophic, to Hypereutrophic water (minimum transparency, maximum chlorophyll-o, maximum phosphorus). Eutrophication of a lake normally contributes to its slow evolution into a bog or marsh and ultimately to dry land. Eutrophication may be accelerated by human activities and thereby speeding up the aging process.

Evapotranspiration (ET)

- Evapotranspiration occurs through evaporation of water from the surface, evaporation from the capillary fringe of the groundwater table, and the transpiration of groundwater by plants (Phreatophytes) whose roots tap the capillary fringe of the groundwater table.
- The process by which plants take in water through their roots and then give it off through the leaves as a by-product of respiration; the loss of water to the atmosphere from the Earth's surface by evaporation and by transpiration through plants.
- The combined processes by which water is transferred from the Earth's surface to the atmosphere; evaporation of liquid or solid water plus transpiration from plants.

F

Facultative bacteria: Bacteria which are capable of switching between aerobic and anaerobic respiration, depending on the oxygen availability of the surrounding environment.

Faecal coliforms: A group of bacteria normally present in large numbers in the intestinal tracts of humans and other warm blooded animals. Specifically, the group includes all of the rod shaped bacteria that are non spore forming, Gram-Negative, lactose-fermenting in 24 hours at 44.5 EC, and which can grow with or without oxygen. In the laboratory, they are defined as all organisms that produce blue colonies with specified time frames. The presence of this type of bacteria in water, beverages, or food is usually taken to indicate that the material is contaminated with solid human waste. Bacteria included in this classification represent a sub group of the larger group termed Coliform. Their concentrations are expressed as number of colonies per 100 ml of sample.

Fungi (Singular - Fungus): Eukaryotic organisms that digest food externally and absorb nutrient molecules into cells. Fungi are the primary decomposers of dead plant and animal matter in many ecosystems.

G

Groundwater

- Water that flows or seeps through water saturated soil or rock, supplying springs and wells. The upper surface of the saturate zone is called the water table.
- Water stored underground in rock crevices and in the pores of geologic materials that make up the Earth's crust.

H

Hardness

- A water-quality indication of the concentration of alkaline salts in water, mainly calcium and magnesium. If the water used is "hard" then more soap, detergent or shampoo is necessary to raise a lather. It also produces scale in hot water pipes, heaters, boilers and other units in which the temperature of water is increased materially.
- A characteristic of water which describes the presence of dissolved minerals. Carbonate hardness is caused by calcium & magnesium bicarbonate; noncarbonate hardness is caused by calcium sulfate, calcium chloride, magnesium sulfate, and magnesium chloride.
- The following scale may assist in appraising water hardness, measured by weight of dissolved salts (in milligrams) per unit (in litres) of water:

  Soft water: 0-60 milligrams/litre (mg/l);

  Moderately Hard water: 61-120 mg/l;

  Hard water: 121 180 mg/l; and

  Very Hard water: over 180 ml

Hydrology: The study of the movement, distribution, and properties of water within the atmosphere and at the Earth's surface.

Hypereutrophic: Pertaining to water bodies that contain extremely high nutrient concentration and extremely high levels of primary productivity.

Hypolimnion

- The lowermost, non-circulating layer of cold water in a thermally stratified lake or reservoir that lies below the Thermocline, remains perpetually cold and is uaually deficient of oxygen.

- The bottom and most dense layer of water in a thermally-stratified lake. Typically, it is non-circulatory and remains cold throughout the year.

I

Impoundment: Body of water created behind a dam or other barrier that obstructs, directs, or retards flow. Also called a reservoir.

L

Lake: A considerable body of inland water or an expanded part of a river. A very slow flowing body of water in a depression in the ground not in contact with the sea.

Leachate: The liquid produced when water percolates through any permeable material, usually containing either or both dissolved and suspended material.

Littoral zone: The area between the high and low water marks in a river, lake, or estuary. However, these areas are difficult to distinguish in non-tidal systems such as rivers & lakes, and so the littoral zone is more commonly referred to as the shallow portion of a body of water, where rooted & floating plants are able to grow and primary production is high. The littoral zone often acts as a nursery for small fish and provides refuge for fish and invertebrates from predators.

M

Macrophyte: A term usually used to refer to rooted or floating aquatic vascular plants.

Macro invertebrate: A term usually used to refer to aquatic invertebrates including insects, crustaceans, molluscs, and worms which inhabit a river channel, pond, lake, wetland, or ocean.

Mesotrophic: Pertaining to water bodies with intermediate nutrient levels and intermediate levels of primary production.

Metabolism: The biochemical modification of chemical compounds in living organisms and cells, include the biosynthesis of complex organic molecules and their breakdown.

Metals

- Metallic materials are normally combinations of metallic elements. They have large numbers of nonlocalized electrons; that is, these electrons are not bound to particular atoms. Many properties of metals are directly attributable to these electrons. Metals are extremely good conductors of electricity & heat and are not transparent to visible light; a polished metal surface has a lustrous appearance. Metals are quite strong, yet deformable, which accounts for their extensive use in structural applications.
- Tend to damage living things at low concentrations and tend to accumulate in the food chain.
- Some of the metals commonly found in water or sediment as pollutants include lead, copper, cadmium, arsenic, silver, zinc, iron, mercury, and nickel.
- Heavy metals: Metallic elements with high atomic weights, e.g., mercury, chromium, cadmium, arsenic, and lead.

Methaemoglobinaemia: Commonly referred to as 'blue baby syndrome', methaemoglobinaemia occurs when methaemoglobin levels in the blood are elevated, limiting the blood's ability to carry sufficient oxygen to cells. High nitrite levels in the blood are usually identified as the root cause, and limiting nitrate or nitrite intake through food, drinking water, and/or drugs can treat the condition. Most commonly detected in infants and in adults that lack the enzyme to convert methaemoglobin back to haemoglobin. Signs and symptoms include a bluish tinge to skin, especially around the mouth, hands & feet.

Methyl mercury: Methyl mercury is shorthand for monomethylmercury, and is more correctly "monomethylmercuric cation". It is composed of a methyl group (CH_3) bonded to a mercury atom; its chemical formula is CH^3Hg^+ (sometimes written as $MeHg^*$). As a positively charged ion it readily combines with such negative ions (anions) as chloride (Cl), hydroxide (OH) and nitrate (NO_3). It also has very high affinity for sulfur, particularly the sulfhydryl (-SH) groups on the amino acid cysteine and hence in proteins containing cysteine, forming a covalent bond. Methyl mercury is a widespread bio accumulative environmental toxin.

Microorganism (microbe): A microorganism or microbe is an organism that is microscopic (too small to be visible to the naked eye). Microorganisms are often described as single-celled, or unicellular organisms; however, some unicellular protists are visible to the naked eye, and some multicellular species are microscopic. The study of microorganisms is called microbiology.

Multicellular: Refers to organisms that contain more than one cell, and having differentiated cells that perform specialized functions.

Mutagenic: Pertaining to an agent that changes the genetic information of an organism and increases the number of mutations above the background level.

Nematode: A phylum of animal composed of the roundworms, most of which are parasitic, that are ubiquitous in freshwater, marine, and terrestrial environments.

Nitrification: Biological oxidation of ammonia with oxygen into nitrite, followed by the oxidation of these nitrites into nitrates.

Nitrogen

- (General) Chemical symbol N, the gaseous, essential element for plant growth, comprising 78 percent of the atmosphere, which is quite inert and unavailable to most plants in its natural form.
- One of the three primary nutrients in a complete fertilizer and the first one listed in the formulation on a fertilizer label: 10-8-6 (nitrogen, phosphorus, potassium).
- (Water Quality) A nutrient present in ammonia, nitrate or nitrite or elemental form in water due possibly to Non-point Source (NPS) pollution or improperly operating wastewater treatment plants.

Non Point Source: Pollution sources which are diffuse and do not have a single point of origin or are not introduced into a receiving stream from a specific outlet. The commonly used categories for non-point sources are; agriculture, forestry, urban, mining, construction, dams & channels, land disposal, and saltwater intrusion.

Nutrients

- An element or compound essential to life, including carbon, oxygen, nitrogen, phosphorus, and many others.
- Animal, vegetable, or mineral substance which sustains individual organisms and ecosystems.
- That portion of any element or compound in the soil that can be readily absorbed and assimilated to nourish growing plants, e.g., nitrogen, phosphorus, potassium, iron

Nutrient Pollution: Contamination of water resources by excessive inputs of nutrients. In surface waters, excess algal production is a major concern. Although natural sources of nutrients exist, major sources are typically Anthropogenic (caused by human activities) and include point sources such as municipal sewage treatment plants and industrial outflows, and non-point sources such as commercial fertilizers, animal waste, and combustion emissions.

O

Oligotrophic: Pertaining to water bodies with very low nutrient levels and low levels of primary production.

Organic matter: Refers to any material that is capable of decay or of being decomposed or is the product of decomposition, and is usually the remains of a recently living organism, and may also include still living organisms.

Organochloride: An organic compound that contains at least one covalently bonded chlorine atom. Most organochlorines have significant biological activity and have been used as powerful pesticides. Many organochlorides are persistent organic pollutants.

Overturn

- The sinking of surface water and rise of bottom water in a lake or sea that results changes in density differences caused by changes in temperature that commonly occur in spring and fall.
- One complete cycle of top to bottom mixing of previously stratified water masses. This phenomenon may occur in the spring or fall, or after storms, and results in uniformity of chemical & physical properties of water at all depths. Also referred to as Turnover.

Oxidation-Reduction State: Also known as the redox potential. Refers to the tendency of a chemical species to acquire electrons and, therefore, be reduced. Many biochemical reactions are oxidation-reduction reactions in which one compound is oxidized and another compound is reduced and the ability of an organism to carry out these reactions is dependent on the oxidation-reduction state of the environment. Redox potential affects the solubility of nutrients and, especially, metals. Oxygen strongly affects redox potential.

P

Parasite: An organism that lives on or in another organism.

Pathogen: A disease-producing agent; usually applied to a living organism. Generally, any viruses, bacteria, or fungi that cause disease.

Pelagic: Pertaining to the portion of a body of water that comprises the water column but does not include the benthic (bottom) or littoral (edge) zones.

Percolation

- The movement, under hydrostatic pressure, of water through the interstices of a rock or soil.
- The movement of water within a porous medium such as soil toward the water table without a definite channel.
- The entrance of a portion of the stream flow into the channel materials to contribute to ground water replenishment.
- Slow seepage of water through a filter.

Persistent Organic Pollutants (POPs)

- Organic-compounds that are resistant to environmental degradation: thus, have been observed to persist in the environment. They are capable of long range transport, bio accumulate in human & animal tissue, and are thought to have serious consequences for the health and well being of humans & wildlife.
- Persistent Organic Pollutants (POPs) are chemicals that remain intact in the environment for long periods, become widely distributed geographically, accumulate in the fatty tissue of living organisms and are toxic to humans & wildlife. POPs circulate globally and can cause damage wherever they travel.
- There is an international convention to combat POPs - Stockholm Convention of Persistent Organic Pollutants.

Pesticide: Any substance or mixture of substances intended for preventing, destroying, repelling, or mitigating any pest. A pesticide may be a chemical substance or biological agent (such as a virus or bacteria) used against pests including insects, plant pathogens, weeds, molluscs, birds, mammals, fish, nematodes (roundworms) and microbes that compete with humans for food, destroy property, spread disease or are a nuisance.

Phosphorus

- An element that is essential to plant life but contributes to an increased trophic level (Eutrophicatuion) of water bodies.
- One of the three primary nutrients in a complete fertilizer and the second one listed in the formulation on a fertilizer label: 10-8-6 (nitrogen, phosphorus, postassium).

Photosynthesis: Biochemical process in which plants, algae, some bacteria, and some protists convert the energy of sunlight to chemical energy. The chemical energy is used to drive synthetic reactions such as the formation of sugars or the fixation of nitrogen into amino acids, the building blocks for protein synthesis. Ultimately, nearly all living things depend on energy produced from photosynthesis for their nourishment, making it vital to life on Earth.

Phytoplankton: The autotrophic component of the plankton that drifts in the water column. The name comes from the Greek terms, phyton or "plant" and planktons , meaning "wanderer" or "drifter". Most phytoplankton are too small to be individually seen with the unaided eye. However, when present in high numbers, their presence may appear as discoloration of the water (the color of which may vary with the phytoplankton present).

Point Source: A stationary location or fixed facility from which pollutants are discharged or emitted. Any single identifiable source of pollution, e.g., a pipe, ditch, ship, ore pit, factory smokestack etc.

Phosphorus

- An element that is essential to plant life but contributes to an increased trophic level (eutrophication) of water bodies.
- One of the three primary nutrients in a complete fertilizer and the second one listed in the formulation on a fertilizer label: 10-8-6 (nitrogen, phosphorus, potassium).

Polychlorinated biphenyls (PCBs): Compounds that are produced by replacing hydrogen atoms in biphenyl with chlorine. PCBs are soluble in most organic solvents, oils, & fats, are very stable and do not degrade easily, thus, considered persistent organic pollutants. They are toxic to and can damage the health and well being of wildlife and humans.

Polycyclic Aromatic Hydrocarbons (PAHs): Chemical compounds that consist of fused aromatic rings and do not contain hetero atoms or carry substituents. Many of them are known or suspected carcinogens. They are formed by incomplete combustion of carbon containing fuels such as wood, coal, diesel, fat, or tobacco.

Precipitation: Any form of water that falls from the sky as part of the weather to the ground. This includes snow, sleet, hail, freezing rain, and rain.

Protozoa: Single-celled eukaryotic (i.e.; containing a nucleus) organisms that have some mobility and are heterotrophic. Grouped in the Kingdom Protista.

R

Respiration: Respiration by individual organisms is the use of O_2 or the production of CO_2 or heat by the organism. Cellular respiration is the process in which the chemical bonds of energy-rich molecules such as glucose are converted into energy usable for life processes.

River

- Surface water finding its way over land from a higher altitude to a lower altitude, all due to gravity.
- The water in a river doesn't all come from surface runoff. Rain falling on the land also seeps into the earth to form ground water. At a certain depth below the land surface, called the water table, the ground becomes saturated with water. If a river bank happens to cut into this saturated layer, as most rivers do, then water will seep out of the ground into the river.
- A natural stream of water of considerable volume, larger than a brook or creek. A river has its stages of development, youth, maturity, and old age. In its earliest stages a river system drains its basin imperfectly; as valleys are deepened, the drainage becomes more perfect, so that in maturity the total drainage area is large and the rate of erosion high. The final stage is reached when wide flats have developed and the bordering lands have been brought low.

Runoff

- A portion of precipitation seeps into the ground to replenish Earth's ground water. Most of it flows downhill as runoff. Runoff keeps rivers and lakes full of water, but it also changes the landscape by the action of erosion.
- That part of the precipitation, snow melt; or irrigation water that appears in uncontrolled surface streams, rivers, drains or sewers. Runoff may be classified according to speed of appearance after rainfall or melting snow as direct runoff or base runoff, and according to source as surface runoff, storm interflow, or groundwater runoff.
- Runoff from agricultural land can carry excess nutrients, such as nitrogen & phosphorus into streams, lakes, and ground water supplies. These excess nutrients have the potential to degrade water quality.

Salinity: Refers to the dissolved salt content of a body of water. The salt content of most natural lakes, rivers, and streams is so small that these waters are termed fresh or even sweet water. The actual amount of salt in fresh water is, by definition, less than 0.05 percent. Otherwise, the water is regarded as brackish, or defined as saline if it contains 3 to 5 percent salt by volume. At well over 5 percent it is considered brine. The ocean is naturally saline at approximately 3.5 percent salt. Some inland salt lakes or seas are even saltier. The Dead Sea, for example, has a surface water salt content of around 15 percent.

Sediments

- Soil, sand, and minerals washed from the land into water, usually after rain. They pile up in reservoirs, rivers, 7 harbours, destroying, fish & wildlife habitat, and clouding the water so that sunlight cannot reach aquatic plants. Careless farming, mining, and building activities will expose sediment materials, allowing them to wash off the land after rainfall.
- Usually applied to material in suspension in water or recently deposited from suspension. In the plural the word is applied to all kinds of deposits from the waters of streams, lakes, or seas.
- The particulate matter settles to the bottom of a liquid.

Sewage: Sewage is the liquid water produced by human society which typically contains washing water, laundry waste, faeces, urine and other liquid or semi liquid wastes. It is one form of wastewater.

Silt: Silt particles fall between 3.9 and 62.5 um in diameter and are larger than clay but smaller than sand. In actuality, silt is chemically distinct from clay, and unlike clay, grains of silt are roughly the same size in all dimensions, and their size ranges overlap.

Spawning ground: Area in lake or river bottom where female fish deposit eggs for to be fertilized.

Stream

- A general term for a body of flowing water; natural water course containing water at least part of the year. In hydrology, it is generally applied to the water flowing in a natural channel as distinct from a canal.
- Some classifications of streams include, in relation to time:
  a. **Ephemeral Streams**–Streams which flow only in direct response to precipitation and whose channel is at all times above the water table.
  b. **Intermittent or Seasonal Streams**–Streams which flow only at certain times of the year when it receives water from springs, rainfall, *or* from surface sources such as melting snow.
  c. **Perennial Streams**–Streams which flow continuously.

And, in relation to ground water:

a. Gaining Streams - Streams or a reach of a stream that receive water from the zone of saturation. Also referred to as an Effluent Stream.

b. Insulated Streams - Streams or a reach of a stream that neither contribute water to the zone of saturation nor receive water from it.

Substrate: The natural environment in which an organism lives, or the surface or medium on which an organism grows or is attached.

Surface water

- An open body of water such as a stream, lake, or reservoir.
- Water that remains on the Earth's surface; all waters whose surface is naturally exposed to the atmosphere, for example, rivers, lakes, reservoirs, ponds, streams, impoundments, seas, estuaries, etc., and all springs, wells, or other collectors directly influenced by surface water. A source of drinking water that originates in rivers, lakes

and run-off from melting snow. It is either drawn directly from a river or captured behind dams and stored in reservoirs.

Suspended Solids: Total suspended solids (TSS) refer to the dry weight of particles trapped by a filter of a specified pore size. The TSS of a water sample is determined by pouring a carefully measured volume of water through a pre weighed filter of a specified pore size, then weighing the filter again after drying to remove all water.

T

Thermal stratification: Separation of water into layers due to temperature gradients set up between surface and deep waters. Stratification usually leads to the formation of the epilimnion, the metalimnion, and the hypolimnion.

Thermocline: The region in a thermally stratified body of water which separates warmer oxygen rich surface water from cold oxygen-poor deep water and in which temperature decreases rapidly with depth. Typically, the temperature decrease reaches 1°C or more for each metre of descent.

Total Dissolved Solids (TDS): Total-dissolved solids is that portion of solids in water that can pass through a filter of a specific pore size. This material can include carbonate, bicarbonate, chloride, sulphate, phosphate, nitrate, calcium, magnesium, sodium, organic ions, and other ions. TDS is measured by filtering a water sample and then collecting the water that passes through the filter. This water is evaporated in a pre weighed dish and then weighed. The increase in weight of the dish represents the total dissolved solids.

Toxic

- Describing a material that can cause acute or chronic damage to biological tissue following physical contact or absorption.
- Substances that even in small quantities may poison, cause injury, or cause death when eaten or ingested through the mouth, absorbed through the skin or inhaled into the lungs.

Toxicity: A measure of the degree to which something is toxic, or poisonous.

Trace metals: Metals in extremely small quantities in animal or plant cells and tissues. The trace metals include iron, magnesium, zinc, copper, chromium, molybdenum, and selenium.

Trophic levels

- Successive stages of nourishment as represented by the links of the food chain. According to a grossly simplified scheme the primary producers (i.e., phytoplankton) constitute the first trophic level, herbivorous zooplankton the second trophic level, and carnivorous organisms the third trophic level.
- Referring to the hierarchy within the food chain. Because energy is lost in the form of heat at each level within the food chain and the quantity of life that can be supported becomes smaller at each level, trophic levels are usually visualized as a pyramid.

  Trophic state (eutrophic, mesotrophic, oligotrophic, hypereutrophic, hyper oligotrophic
- State Index (TSI) - A measure of Eutrophication of a body of water using a combination of measures of water transparency or turbidity (using Secchi Disk depth recordings), Chlorophyll-a concentrations, and total phosphorus levels. TSI measures range from a scale 20-80.

- (referred to as Carlson's Trophic State Index).
- Degrees of eutrophication typically range from Oligotrophic water (maximum transparency, minimum chlorophyll-a, minimum phosphorus) through Mesotrophic, Eutrophic, to hypereutrophic water (minimum transparency, maximum chlorophyll-a, maximum phosphorus).
  a. Eutrophic Reservoirs and lakes which are rich in nutrients and very productive in terms of aquatic animal & plant life.
  b. Mesotrophic Reservoirs and lakes which contain moderate quantities of nutrients and are moderately productive in terms of aquatic animal & plant life.
  c. Oligotrophic Reservoirs and lakes which are nutrient poor and contain little aquatic plant or animal life.

Turbidity

- The amount of solid particles that are suspended in water and that cause light rays shining through the water to scatter. Thus, turbidity makes the water cloudy or even opaque in extreme cases. Turbidity is measured in nephelometric turbidity units (NTU) or Formazin turbidity units (FTU) depending on the method and equipment used. The turbidity may be caused by a wide variety of suspended materials, such as clay, silt, finely divided organic & inorganic matter, soluble coloured organic compounds, plankton and other microscopic organisms & similar substances.
- Turbidity in water has public health implications due to the possibilities of pathogenic bacteria encased in the particles and thus escaping disinfection processes. Turbidity interferes with water treatment (filtration), and affects aquatic life. Excessive amounts of turbidity also make water aesthetically objectionable. The degree of the turbidity of water is measured by a Turbidimeter.

U

Ultra oligotrophic: Pertaining to water bodies that contain extremely low nutrient concentrations and extremely low levels of primary productivity.

Ultraviolet radiation: Electromagnetic radiation with a wavelength shorter than visible light (1 - 380 nm).

V

Vascular plants: Refers to plants in the Kingdom Plantae (also called Viridiplantae) that have specialized tissues for conducting water. Vascular plants include the ferns, clubmosses, horsetails, flowering plants, conifers and other gymnosperms.

Vertical gradient: Commonly found in lakes and deep water systems, vertical gradients are set up as a result of density differences in the water where lighter, less dense water tends to float to the surface and heavier, denser water settles to the bottom. Gradients in salinity, pH, dissolved oxygen, and temperature can often be detected during relatively stable periods.

Virus: Refers to a submicroscopic particle that can infect the cells of a biological organism. At the most basic level viruses consist of genetic material contained within a protective protein shell called a capsid, which distinguishes them from other virus like particles such as prions and viroids. The study of viruses is known as virology.

W

Wastewater treatment: Any of the mechanical or chemical processes used to modify the quality of waste water in order to make it more compatible or acceptable to man and his environment.

a. **Primary wastewater treatment-** The first stage of the wastewater treatment process where mechanical methods, such as filters and scrapers, are used to remove pollutants. Solid material in sewage also settles out in this process.

b. **Secondary wastewater treatment-** The treatment (following primary wastewater treatment) involving the biological process of reducing suspended, colloidal, & dissolved organic matter in effluent from primary treatment systems and which generally removes 80 to 95 percent of the Biochemical Oxygen Demand (BOD) & suspended matter. Secondary wastewater treatment may be accomplished by biological or chemical-physical methods. Activated sludge and trickling filters are two of the most common means of secondary treatment. It is accomplished by bringing together waste, bacteria, and oxygen in trickling filters or in the activated sludge process. This treatment removes floating & settleable solids and about 90 percent of the oxygen demanding substances & suspended solids. Disinfection is the final stage of secondary treatment.

c. **Tertiary wastewater treatment-** Selected biological, physical, and chemical separation processes to remove organic & inorganic substances that resist conventional treatment practices; the additional treatment of effluent beyond that of primary and secondary treatment methods to obtain a very high quality of effluent.

Water quality: A term used to describe the chemical, physical, and biological characteristics of water, usually in respect to its suitability for a particular purpose. A neutral term that relates to the composition of water as affected by natural processes and human activities. It depends not only on water's chemical condition, but also its biological, physical and radiological condition. The quality of water is also related to specific use, and is usually measured in terms of constituent concentrations. The level of water quality is based upon the evaluation of measured quantities and parameters, which then are compared to water quality standards, objectives or criteria.

Water treatment

- Processes undertaken to purify water acceptable to some specific use, e.g., drinking. Most water treatment processes include some form, or combination of forms, of sedimentation, filtration, and chlorination.

Water Treatment, Combined Technique: A relatively new water disinfection technique greatly reducing the need for chlorination while effectively destroying up to 99.9 percent of coli phage (intestinal bacteria) in raw water. The method combines two purification techniques that have been previously used separately for water purification - potassium permanganate and copper/silver ions - but in combination the processes kill bacteria up to 10 times faster than metal ions alone and up to five times faster than potassium permanganate alone.

Watershed

- The land area that drains water to a particular stream, river, or lake. It is a land feature that can be identified by tracing a line along the highest elevations between two areas on a map, often a ridge, by topographic divides that drains a given stream or river.

- The natural or disturbed unit of land on which all of the water that falls (or emanates from springs or melts from snow packs), collects by gravity, and fails to evaporate, runs off via a common outlet. All lands enclosed by a continuous hydrologic drainage divide and lying upslope from a specified point on a stream; a region or area bounded peripherally by a water parting and draining ultimately to a particular water course or body of water. Also referred to as Water Basin or Drainage Basin.

Z

Zooplankton: Small protozoans or metazoans (e.g, invertebrates and other animals) that feed on other planktonic organisms that inhabit the water column of bodies of water.

APPENDIX 8.8

Some Water Related diseases and their causative organisms

S.N.	Disease	Causative organisms	Mode of Spread	Symptoms
1	Typhoid	Salmonella typhi	Ingestion of contaminated food, water, milk, unwashed raw vegetables and flies	Continuous fever which progressively increases day by day, the temperature being higher in the evening than in the morning accompanied by body aches, headache and constipation, Hemorrhage from an ulceration in the small intestine
2	Cholera	Vibrio Cholerae	Ingestion of water or food contaminated by the bacteria from the stool of a cholera patient	Painless diarrhea followed by vomiting; patient may pass 30 to 40 stools per day which soon becomes typically watery and colourless with flakes of mucous floating in them
3	Bacterial dysentery	Shigella spp	Through contaminated food, water and by direct personal contact	Diarrhoea, with presence of blood and mucous in the stools accompanied by severe griping pain in the abdomen, Stools are not too frequent (4-10 per day) and the faecal matter is scanty, Patient looks ill.
4	Leptospirosis	Laptospira	Primary hosts are rodents, which carry the organisms in their kidenys and the patient may become infected by bathing or swimming in water contaminated with the rodent™S urine	Fever, pain in legs,nausea, vomiting are common, congestion of the conjunctival blood vessels around the corneas of the eyes
VIRUSES				
5	Infective hepatitis	Hepatitis virus	Stools that contain virus contaminating the water and food	Loss of appetite, nausea, vomiting and diarrhea accompanied by fever. The urine is dark coloured. Eye and skin have yellow colouration

S.N.	Disease	Causative organisms	Mode of Spread	Symptoms
			PROTOZOA	
6	Amoebic dysentery	Entamoeba histolytic a	Ingestion of cysts in food and water	Abdominal discomfort to diarrhoea, with or without the presence of blood or mucous in the stools, accompanied by fever, chills and griping pain in the abdomen
7	Diarrhoea	Gistfis (= Lamblia) intestinalis	Cysts which are voided with the faeces and enter the new hosts in food or water	Intestinal disorders leading to epigastric pain , abdominal discomfort, loss of appetite, headache and loose bowels
			HELMINTHS	
8	Bilharzia	Schistosoma spp.	Eggs of the flukes pass out with human faeces or urine and if they reach fresh water, develop into miracidia larvae which infect snails. The cercaria stage develops in the snails and on leaving the host, cercaria penetrate the skin of humans wading in the water.	Allergy like itch, rash, aches, fever, eosinophilia, etc. When infection is heavy, the eggs may be deposited in the arterioles of the lunds causing cardio pulmonary schistosomiasis or corpulmonale or ayerza disease, which may lead to congestive heart failure
9	Guinea worm	Dracunculus medinensis	Unfiltered water containing the infected copepods	Blister near the ankle, burns around the blister, allergy and aches

Note: In the developed world, water associated disease are rare, due essentially to the presence of efficient water supply and waste water disposal systems. However, in the developing world, the majority of people without a safe water supply and adequate sanitation are prone to water associated diseases.

A WHO survey has highlighted the following facts:

- Each day, 30,000 people die from water associated diseases.
- In developing countries, 80% of all illnesses are water associated.

Safe adequate and accessible supplies of water, combined with proper sanitation, are basic need and can help to reduce many of the disease affecting under privileged populations.

A wide range of natural and human influences affects water quality. The most important of the natural influence are geological hydrological and climatic conditions, since these affect the quantity & the quality of water available.

There are two main types of water pollution:

Chemical pollution of water and diseases

Almost every known element existing normally in the environment can become poisonous when introduced into the human system in larger than normal quantities. One major way of introducing these elements into the environment and later into the human system is through the discharge of industrial effluents into water sources such as rivers.

These pollutants include detergents, solvents, nitrogenous substances, dyes, ammonia etc. This can affect human health directly or indirectly by accumulating in aquatic life.

Some elements or chemicals may be found in water in excessive or inadequate amounts. For instance, excess fluorine in water causes dental fluorosis or mottled enamels. While lack of fluorine in water causes dental cavities (decay), therefore, maintaining the optimal level (1 mg/1) is essential.

It has been observed that surface water is often low in fluoride content but the concentration is high in underground water sources.

A concentration of 10 ppm of nitrogen is thought to be harmful. It causes infant methaemoglobinemia (blue baby syndrome).

Biological pollution of water and diseases

Water may contain numerous pathogenic organisms and thereby become a means of transmission for many diseases.

All water associated disease require an infectious agent, a transmission route and the exposure of a susceptible living organisms for their spread.

Water associated disease can be defined as a disease in relation to water supply and sanitation. There are four categories:

- Water borne diseases – Several infections enteric or intestinal diseases of man are transmitted through water contamination by fecal matter. Pathogens excreted in water by an infected person include all major categories such as bacteria, viruses, protozoa and parasitic worms. Water acts only as a passive vehicle for the infectious agent. Table below shows examples of waterborne diseases.

Table 8.8T_1: Waterborne diseases with their etiologies

S.N.	Types of organism	Disease Types
1	Bacteria	• Typhoid and paratyphoid fever • Cholera • Diarrhoeas (caused by salmonella, Yersinia) • Compylobacter dysentery • Bacillary dysentery (caused by shigella)
2	Virus	• Hepatitis-A • Poliomyelitis • Viral gastroenteritis
3	Protozoa	• Amoebic dysentery • Giardia (Imbliasis) • Balantidiasis
4	Helminthes	Helminthiasis caused by Ascaris and Trichinas

The cycle of infection due to waterborne diseases is explained in figure below.

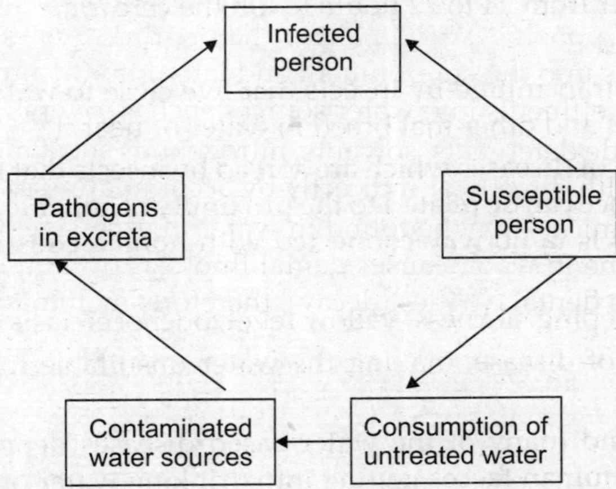

Fig.8.8F$_1$: Cycle of infection due to waterborne diseases

It should be noted that waterborne diseases form part of the group of water washed diseases as well. They may also be transmitted by any of the faeco-oral routes: dirty hands, dirty food, dirty water. etc.

- **Water washed diseases**

  These comprise diseases linked to a back of water for personal hygiene.

  Examples of water washed diseases are:

- Dermatological disease such as scabies.
- Ophthalmic disease such as trachoma and conjunctivitis.
- House borne diseases such as house borne typhus and relapsing fever. Lack of good personal hygiene and inability to wash clothes encourages the proliferation of lice and the problems associated of lice and the problems associated with their presence (itching, scratching, skin sores).

To prevent this type of disease, provision of an ample amount of water and personal hygiene are very essential.

- **Water based diseases**

  These are diseases caused by infectious agents that are spread by contact with water. The essential part of the life cycle of the infecting agent takes place from an aquatic animal.

  A number of diseases depend upon the pathogenic organisms spending part of their life cycle in water or in an intermediate host which lives in water. Thus, infection of humans cannot occur by immediate ingestion of, or contact with, the organism excreted by sufferers.

  Many of the diseases in this class are caused by worms, which infest the sufferer and produce eggs. Which are then discharged in feces or urine. Typical examples are schistosomisasis and dracunculiasis (guinea worm).

  To prevent this group of diseases, the following methods may be implemented:

  – Avoidance of contact with and ingestion of contaminated water.

- Reduction of intermediate hosts (snail) by using ' endod' or Lemma toxin.
- Storage of water from 24 to 72 hours to kill the cercaria.

- **Water-related diseases**

  These are diseases transmitted by insects that live close to water. Infections are spread by mosquitoes, flies and other that breed in water or near it.

  There are a number of diseases which are spread by insects that breed of feed near water so that their incidence can be related to the proximity of suitable water sources. Infection with these diseases is in no way connected with human consumption or contact with the water.

Example: Malaria, sleeping sickness, yellow fever, onchocerciasis etc.

To prevent this type of disease, making the water unsuitable for breeding of insects is essential.

All the waterborne and many of the water-based diseases depend for their dispersion on infecting agents from human faeces getting into drinking water or into food. The chain of disease transmission may be broken effectively by sanitary disposal of faeces and the provision of safe & adequate water supplies.

Improvement in the reality of community water supplies will basically only affect the waterborne such as bacillary dysentery, cholera and typhoid. Many of the diarrhoeal diseases are due more to a lack of safe and adequate quantities of water. Skin and eye infections are in this group of water associated diseases.

When water supplies are developed without complementary improvements in personal hygiene, food handing and preparation, and in general health care, they are unlikely to produce the expected health benefits. Table below gives in comparative form, the four categories of water-associated diseases, and gives examples of typical diseases in each category, the causative agents and the preventive strategies.

Table 8.8T$_2$: The four mechanisms of water associated diseases and preventive strategies

Water-associated diseases	Example	Agent	Preventive strategies
1. Waterborne (feacal-oral) (a) Low infectious dose (b) High infectious dose	Typhoid cholera Bacillary dysentery	A A	• Improve water quality • Prevent usual use of other contaminated source • Health education
2. Water-washed (a) Skin and eye (b) Other	Scabies, trachoma Louse-borne fever	F E	• Improve water quantity • Improve water access. • Health education
3. Water-based (a) Penetrating skin (b) Ingested	Schistosomiasis Guineas worms	D D	• Decrease need for untreated water contact • Control swail population • Improve quality of water • Filter out cyclops • Health education

Water-associated diseases	Example	Agent	Preventive strategies
4. Water-related (a) Biting near water (b) Breeding in water	Sleeping sickness, Malaria	C C	• Proper site selection of house • Using personal protection materials • Destroy breeding sites of insects • Decrease need to visit breeding site • Health education

Key: A- Bacteria B- Virus C- Protozoa
D- Helminths E- Spirochetes F- Other agents

APPENDIX 8.9

Major water pollutants and their effect on human health

S.N.	Pollutant	Origin	Human Health Effects
1	Arsenic	Pesticide industries and agriculture	Toxic, eruption on skin, carcinogenic
2	Cadmium	Industrial waste discharge	May substitute calcium in bone structure, causes a disease-Itai-Itai in Japan
3	Chlorine	Industries using chlorine bleach	Physiological
4	Chromium	Industries using chlorine bleach	Toxic and carcinogenic
5	Cyanide	Pesticides	Toxic and fatal
6	Lead	Industrial, mine and smelter discharges, water pipes	Causes a disease called plumbism
7	Nitrates and Nitrites	Fertilizer industries	Cause methemoglobinaemia in infants
8	Sulphates	Leaching of gypsum or industrial waste discharges	Have laxative, effect in combination with magnesium or sodium in drinking water, may form hydrogen sulphide gas which is fatal for workers in sewers
9	Selenium	Pesticides, industrial waste	Carcinogenic, causes caries
10	Mercury	Pesticides brain damage	Toxic, causes fatal
11	Bacteria	Sewage	Cholera,typhoid, hepatitis, dysentery

Note: Water-Borne Diseases are listed below:

1. **Caused by the presence of an infective agent:**
   - a. Viral – Infective hepatitis, poliomyelitis, rotavirus diarrhea in infants
   - b. Bacterial – Typhoid & paratyphoid fever, bacillary dysentery, cholera
   - c. Protozoal – Amoebiasis, giardiasis
   - d. Helminthic – Round worm, thread worm
   - e. Leptosiral – weil's disease
2. **Due to the presence of an aquatic host:**
   - a. Cyclops – Guinea worm
3. **Water washed Diseases:**
   - a. skin, conjunctivitis
4. **Water related Diseases (vector-borne):**
   - a. Malaria, Filaria, Dengue
5. **Water related diseases-** Fluorosis

APPENDIX 8.10

Chemical contaminants in drinking water and their adverse effects on human body

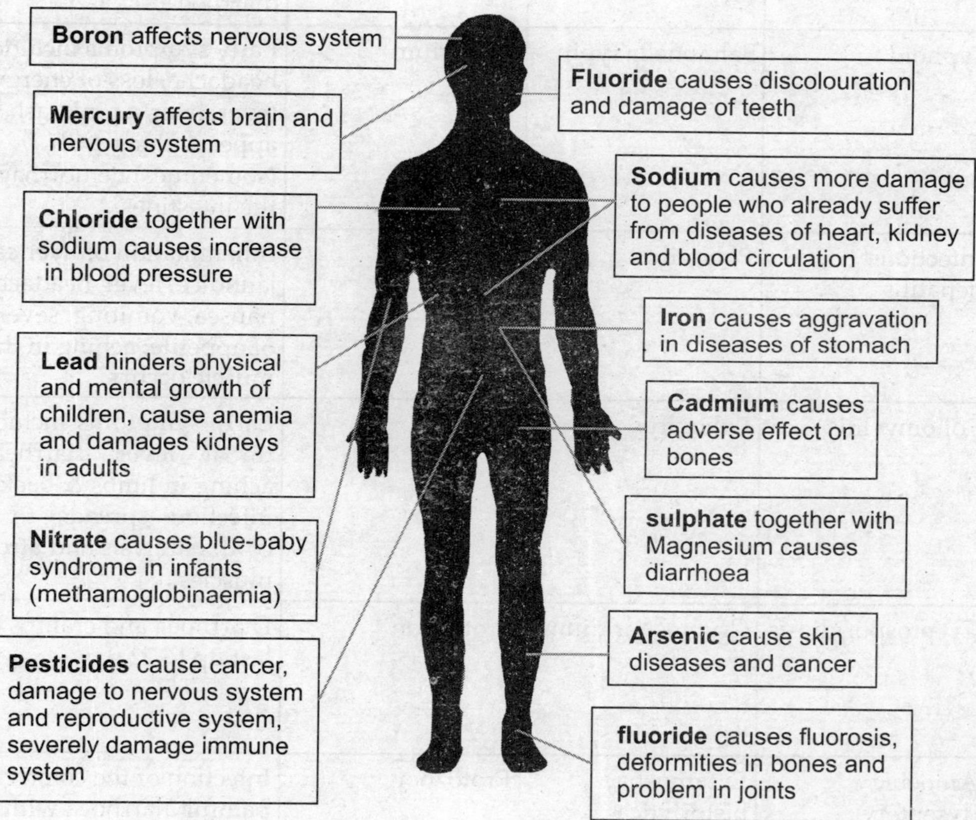

Fig.8.10F$_1$: Adverse effects of Chemical contaminants in drinking water on human body

Note: *Causative Agents And Symptoms Of Some Human Diseases Transmitted By Polluted Water Is Tabluated Below.*

S.N.	Disease	Infectious Agent	Type of Organism	Symptoms
1	Cholera	Vibrio cholerae	Bacterium	Severe diarrhoea, vomiting; fluid loss of as much as 20 quarts per day causes cramps and collapse.
2	Dysentery	Shiugella dysenteries	Bacterium	Infection of the colon causes painful diarrhoea with mucus and blood in the stools; abdominal pain

S.N.	Disease	Infectious Agent	Type of Organism	Symptoms
3	Enteritis	Clostridium perfringens, other bacteria	Bacterium	Inflammation of the small intestine causes general discomfort, loss of appetite, abdominal cramps, and diarrhoea
4	Typhoid	Salmonella typhi	Bacterium	Early symptoms include headache, loss of energy, fever; later, a pink rash appears along with (sometimes) hemorrhaging in the intestines.
5	Infectious hepatitis	Poliovirus	Virus	Inflammation of liver causes jaundice, fever, headache, nausea, vomiting, severe loss of appetite; aching in the muscle occurs.
6	Poliomyelitis	Poliovirus	Virus	Early symptoms include sore throat, fever, diarrhea, and aching in limbs & back: when infection spreads to spinal cord, paralysis and atrophy of muscles.
7	Cryptosporidiosis	Cryptosporidium sp.	Protozoon	Diarrhoea and cramps that last up to 22 days
8	Amoebic dysentery	Entamoeba bistorlytica	Protozoon	Infection of the colon causes painful diarrhoea with mucous and blood in the stools; abdominal pain.
9	Schistosomiasis	Schistosoma sp.	Fluke	Tropical disorder of the liver and bladder causes blood in urine, diarrhea, weakness, lack of energy, repeated attacks of abdominal pain.

9

Chapter

Thermal Pollution of Water

Thermal pollution of water means the addition of excess of undesirable heat to water thereby making it harmful to aquatic life in river bodies, lakes, ponds, oceans etc.

Waste water/ cooling water generated by the thermal power plants, nuclear plants etc. has higher temperature than the normal temperature of the surface water (More than 8-10°C). This causes thermal pollution of water.

Thermal pollution is defined as sudden increase or decrease in temperature of natural body of water (e.g., ocean, lake, river, pond) by human influence. This normally occurs when a plant or facility takes in water from a natural resource and puts back with an altered (high or low) temperature.

Thermal Pollution occurs when an industry draws in water from a natural source and either cools it down or heat up. Then it ejects this hot/cold water back into the natural source (river/ ocean/lake/pond etc.) which changes the dissolved oxygen levels and have disastrous effects on local ecosystem and community.

- The basic definition of thermal water pollution is the deterioration of water quality as a result of any process that changes ambient water temperature
- It is usually associated with the increases of water temperatures in a stream, lake, or ocean due to the discharge of heated water from industrial processes, such as the generation of electricity from the thermal power plants
- However this rise in the ambient water temperature can also occur in streams where shading vegetation along the banks is removed or where sediments have made the water more turbid. Both of these effects allow more energy from the sun to be absorbed by the water and thereby increase its temperature
- Instances in which the effects of colder-than-normal water temperatures can be also observed. For example, the discharge of cold bottom water from deep water reservoirs behind large dams changed the downstream biological communities in systems such as the River.

CAUSES OF THERMAL POLLUTION

- **Water as Cooling Agent in Power, Manufacturing and Industrial plants:** Production and Manufacturing plants are biggest source of thermal pollution. These plants draw water from nearby source to keep machines cool and then release back to the source with higher temperature. When heated water returns to the river or ocean, the water temperature rises sharply. When oxygen levels are altered in the water, this can also degrade the quality and longevity of life in wildlife that lives underwater. This process can also wipe away stream side vegetation, which constantly depends on constant levels of oxygen and temperature. By altering these natural environments, industries are essentially helping decrease the quality of life for these marines based life forms and can ultimately destroy habitats if they are not controlled and careful about their practices.

- **Soil Erosion:** Soil erosion is another major factor that causes thermal pollution. Consistent soil erosion causes water bodies to rise, making them more exposed to sunlight. The high temperature could prove fatal for aquatic biomes as it may give rise to anaerobic conditions.
- **Deforestation:** Trees and plants prevent sunlight from falling directly on lakes, ponds or rivers. When deforestation takes place, these water bodies are directly exposed to sunlight, thus absorbing more heat and raising its temperature. Deforestation is also a main cause of the higher concentrations of greenhouse gases i.e. global warming in the atmosphere.
- **Runoff from Paved Surfaces:** Urban runoff discharged to surface waters from paved surfaces like roads and parking lots can make water warmer. During summer seasons, the pavement gets quite hot, which creates warm runoff that gets into the sewer systems and water bodies.
- **Natural Causes:** Natural causes like volcanoes and geothermal activity under the oceans and seas can trigger warm lava to raise the temperature of water bodies. Lightening can also introduce massive amount of heat into the oceans. This means that the overall temperature of the water source will rise, having significant impacts on the environment.

SOURCES OF THERMAL POLLUTION

Heat producing industries i.e.,

- Thermal power plants, nuclear power plants, refineries, steel mills etc. are the major sources of thermal pollution. Power plants utilize only 1/3 of the energy provided by fossil fuels for their operations. Remaining 2/3 is lost in the form of heat generally to cooling water. Cold water, generally, is drawn from some nearby water-body. passed through the plant and returned to the same water body with temperature about 10°C higher than the initial temperature. Excess of heat reaching such water bodies causes thermal pollution of water.

The following sources contribute to thermal pollution:

- Nuclear power plants
- Coal fired plants
- Industrial effluents
- Domestic sewage
- Hydro-electric power

Nuclear power plants: Nuclear power plants including drainage from hospitals, research institutions, nuclear experiments and explosions, discharge a lot of heat that is not utilized along with traces of toxic radio nuclides into nearby water streams. Emissions from nuclear reactors and processing installations are also responsible for increasing the temperatures of water bodies. The operations of power reactors ant) nuclear fuel processing units constitutes the major contributor of heat in the aquatic environment. Heated effluents from power plants are discharged at 10 °C higher than the receiving waters that, affects the aquatic flora and fauna.

Coal fired power plants: Coal fired power plants constitute a major source of thermal pollution. The condenser coils in such plants are cooled with water from nearby lakes or rivers. The resulting heated water is discharged into streams thereby raising the water temperature by 15 °C. Heated effluent decreases the dissolved content of water resulting in death of fish and other aquatic organisms. The sudden fluctuation of temperature also leads to *'thermal shock'* killing aquatic life that have become acclimatized to living in a steady temperature.

Industrial effluents: Industries like textile, paper, pulp and sugar manufacturing release huge amounts of cooling water along with effluents into nearby natural water bodies. The waters polluted by sudden and heavy organic loads result in severe drop in levels of dissolved oxygen leading to death of several aquatic organisms.

Domestic Sewage: Domestic sewage is discharged into rivers, lakes, canals or streams with minimal treatment or without any treatment. These wastes have a higher organic temperature and organic load. This leads to decrease in dissolved oxygen content in the receiving waters resulting in the set-up of anaerobic conditions causing release of foul and offensive gases in water. Eventually, this leads to development of anoxic conditions resulting in rapid death of aquatic organisms.

Hydro electric power: Generation of hydroelectric power sometimes leads to negative thermal loading in water systems. Apart from electric power industries, various factories with coding requirementcontribute to thermal loading.

THERMAL POLLUTION IN STREAMS BY HUMAN ACTIVITIES

- Industries and power plants use water to cool machinery and discharge the warm water into a stream
- Stream temperature rises when trees and tall vegetation providing shade are cut.
- Soil erosion caused due to construction also leads to thermal pollution
- Removal of stream side vegetation
- Poor farming Practices also lead to thermal pollution

Effects of Thermal Pollution

- With rise in water temperature in river bodies, density & viscosity of water reduces and its vapor pressure in creases increases the evaporation loss.
- Warmer temperatures of the discharged effluents from the thermal power plants lower the dissolved oxygen content of the lake or stream by decreasing the solubility of oxygen in water. High temperature becomes barrier for oxygen penetration in deep cold water. The toxicity of pesticides, detergents and chemical in the effluent increases with increases in temperature. Warning of air reduces its air saturation value and its capacity to dissolve oxygen.
- These warmer water also enhance the respiration rates to the aquatic organisms thus forcing them to consume oxygen faster. Warning of water accelerates the decomposition

process taking palace in water. If water is chemically polluted, then its warning increases the toxicity to fish warmer water allows algae to flourish on surface of water and over the long term, growing algae can decrease oxygen level in the water.

- With the decrease in the dissolved oxygen level and increase in the respiration rate of the aquatic species, berating breathing problems arise in the aquatic organisms which may prove to be final. Temperature rise of water affects physiology controlling reproductive cycle, digestion rates, respiration rate and many chemical activities. Dissolved oxygen in water for fish to survive is 5-7 mg/litre (ppm). In fishes, several activities like nest building, spawning, hatching and reproduction depend on optimum temperature.
- It also increase their susceptibility to disease, parasites, and toxic chemicals.
- Temperature rise of water causes rise in metabolic rates which implies that an aquatic animal will require more food to maintain body weight. With reduction in temperature to 5°C fishes require less food compared to 15°C (natural food). Higher temperature implies faster growth rate and shorter life span temperature is of vital signification to physical metabolic and biochemical processes that control respiratory rates, digestion, excretion and overall development of aquatic organisms. Temperature changes cause total disruption to ecosystem.
- Water temperature of 35°C is dangerous for fishes survival at higher temperature (a) bacterial growth rate and rate of physiological process increases, (b) decomposition of organic and other oxygen demanding waste is speeded up and (c) rate of oxygen depletion increases and dissolved oxygen content decrease.
- Thermal discharge is favorable to bacteria and pathogens.
- Discharge of heated water into shallow water near the shore of a lake may also damage spawning and kill young fish. Fishes do not survive above 30°C. Salmon fry sockeye fish die at 22.5°C while large mouth bass fish die at 26°C., Thermal death of fish is due to thermal shock of water.
- The adaptability of the fishes and other aquatic organisms do not safeguard them form sudden and sharp changes in temperature of the effluent water from the thermal power plants.
- These effect of sharp changes in water temperature when new power plants open up or when plants shut down for repair is known as thermal shock. Abrupt changes in temperature alters the seasonal variation in the type and abundance of lower organisms leading to shortage of right food for fish at the right time.
- Fish and other organisms adapted to a particular temperature range can also be killed from thermal shock. Thermal pollution is directly responsible for mortality of aquatic organisms. Increase in water temperature leads to exhaustion of micro-organisms thereby shortening the life span of fishes. Above a certain temperature, fishes die due to failure of respiratory system and nervous system failure.
- Many fish do die on the intake screens which are employed to prevent the fishes from clogging the heat exchanger pipes.
- Growth of green algae is good at 30 – 35°C and that of blue green algae at 35 – 40°C. Thermal discharges in rivers/ lakes/ ocean favors their resulting in damage to eco-system as blue green algae is poorer food source and toxic to fish.

- The composition of flora and fauna changes because the species sensitive to increased temperature due to thermal shock will be replaced by temperature tolerant species.
- Metabolic activities of aquatic organisms increase requiting more oxygen and food.
- Fish migration is affected due to formation of various thermal zones.
- Warning of earth due to nuclear plants/fossil fuel fired power plants and geothermal energy increases heat load on the earth added by heat radiation from earth due to green house effect, the average earth temperature increase causing rise in water body temperature. Increase in per capita energy consumption increases the thermal load on earth.
- **Increase in Toxins:** With the constant flow of high temperature discharge from industries, there is a huge increase in toxins that are being regurgitated into the natural body of water. These toxins may contain chemicals or radiation that may have harsh impact on the local ecology and make them susceptible to various diseases.
- **Loss of Biodiversity:** A dent in the biological activity in the water may cause significant loss of biodiversity. Changes in the environment may cause certain species of organisms to shift their base to some other place while their could be significant number of species that may shift in because of warmer waters. Organisms that can adapt easily may have an advantage over organisms that are not used to the warmer temperatures.
- **Ecological Impact:** A sudden thermal shock can result in mass killings of fish, insects, plants or amphibians. Hotter water may prove favorable for some species while it could be lethal for other species. Small water temperature increases the level of activity while higher temperature decreases the level of activity. Many aquatic species are sensitive to small temperature changes such as one degree Celsius that can cause significant changes in organism metabolism and other adverse cellular biology effects.
- **Affects Reproductive Systems:** A significant halt in the reproduction of marine wildlife (although this may be true, reproduction can still occur between fish but the likelihood of defects in newborns is significantly higher) can happen due to increasing temperatures as reproduction can happen with in certain range of temperature. Excessive temperature can cause the release of immature eggs or can prevent normal development of certain eggs.
- **Increases Metabolic Rate:** Thermal pollution increases the metabolic rate of organisms as increasing enzyme activity occurs that causes organisms to consume more food than what is normally required, if their environment were not changed. It disrupts the stability of food chain and alter the balance of species composition.
- **Migration of organisms:** Warm water may also cause particular species of organisms to migrate to suitable environment that would cater to its requirements for survival resulting is loss for those species that depend on them for their daily food as the food chain is interrupted.
- A one megawatt nuclear power plant, may require five million m^3 cooling water each day. The water used in such a plant has its temperature increased by about 17°C during the cooling process. For this reason, such plants are usually built very close to an abundant water supply such as a lake, a large river, or the ocean.

When thermal pollution drives water temperatures up, most aquatic and marine wildlife cannot survive. Immobile organisms, such as plants and shellfish, simply die. The amount of any gas that can be dissolved in water varies inversely with the temperature. As water is warmed, therefore, it is capable of dissolving less and less oxygen. Organisms that need oxygen to survive will, in such cases, cant be able tosurvive.

When heated water is released from a plant or factory, it does not readily mix with the cooler water around it. Instead, it forms a stream like mass known as a **thermal plume** that spreads out from the outflow pipes. It is in this thermal plume that the most severe effects of thermal pollution are likely to occur. Only over an extended period of time does the plume gradually mix with surrounding water, producing a mass of homogenous temperature.

Invasion of Destructive Organism

Water temperatures can have other, less expected effects also. As an example, trout can swim less rapidly in water above 19°C making them less efficient predators. Organisms may become more subject to disease in warmer water too. The bacterium *Chondrococcus columnaris* is harmless to fish at temperatures of less than 10°C. Between temperatures of 10 – 21°C, however, it is able to invade through wounds in a fish's body and at temperatures above 21°C it can even attack healthy tissue.

Urban Heat dome

Another example of thermal pollution is the development of urban heat islands. An urban heat island consists of a dome of warm air over an urban area caused by the release of heat in the region. Since more human activity occurs in an urban area than in the surrounding rural areas, the atmosphere over the urban area becomes warmer than it is over the rural areas.

It is not uncommon for urban heat islands to produce measurable climate changes. For example, the levels of pollutants trapped in an urban heat island can reach 5 to 25 percent greater than the levels over rural areas. Fog and clouds may reach twice the level of comparable rural areas, wind speeds may be produced by up to 30 per cent, and temperatures may be 0.5 - 2 °C higher than in surrounding rural areas. Such differences may cause both personal discomfort and, in some cases, actual health problems for those living within an urban heat island.

Undesirable changes in Algal population: Excess Nutrients from the washout water from farm lands combined with thermal pollution cause an excessive algal growth with consequent changes. High Temperature promotes blue green algal blooms which disrupts the aquatic food chain

Thermal Enrichment of Water

- It is the concept of using heated water released from the thermal power plants for beneficial purposes
- Warm water released from the power plants can be used for irrigation purpose to prolong the growing season of some species in frost-prone areas.
- Cycled through aquaculture pens it can be also used to enhance the growth of commercially valuable fish and shellfish. For example, waste hot water is used to cultivate oysters in aquaculture lagoons in Japan and in New York's Long Island Sound and to cultivate catfish & redfish in Texas.
- Heated water can be also used to heat buildings and greenhouses.
- Can be run under sidewalks to melt snow in the cold countries.
- To desalinate ocean water.

Control Measures for thermal pollution

The following methods can be adapted to control high temperature of water caused by thermal discharges:

- **Cooling towers**: Use of water from water systems for cooling systems for cooling purposes, with subsequent return to the water way after passage through a condenser, is called cooling process. Cooling towers transfer heat from hot water to the atmosphere by evaporation. Cooling towers are of two types:
  i. **Wetcooling tower**: Hot water coming out from the condenser (reactor) is allowed to spray over baffles. Cool air. with high velocity, is passed from sides, which takes away the heat and cools the water.
  ii. **Dry cooling tower:** Here, hot water is allowed to flow In long spiral pipes. Cool air with the help of a fan is passed over' these hot pipes, which cools down hot water. This cool water can be recycled.
- **Cooling ponds:** Cooling ponds are the best way to coot thermal discharges. Heated effluents on the surface of the water in cooling panels maximise dissipation of heat to the atmosphere and minimise the water area and volume. The warm water wedge acts like a cooling pond.
- **Spray ponds:** The water coming out from condensers is allowed to pass into the ponds through sprayers. Here water is sprayed through nozzles as fine droplets. Heat from the fine droplets gets dissipated to the atmosphere.
- **Artificial lakes:** Artificial takes are man made water bodies that offer once-through cooling. The heated effluents can be discharged into the lake at one end and water for cooling purposes may be withdrawn from the other end. The heat is eventually dissipated through evaporation. Working principles and the diagram of the above methods of thermal pollution control is given below.

Cooling Tower

a. **Wet cooling tower:** Hot water is sprayed over baffles. Cool air entering from sides takes away the heat and cools water.

This cool water can be recycled or discharged. Large amount (2%)of water is lost through evaporation and in the vicinity of wet cooling tower extensive fog is formed which is not good for environment and causes damage to vegetation.

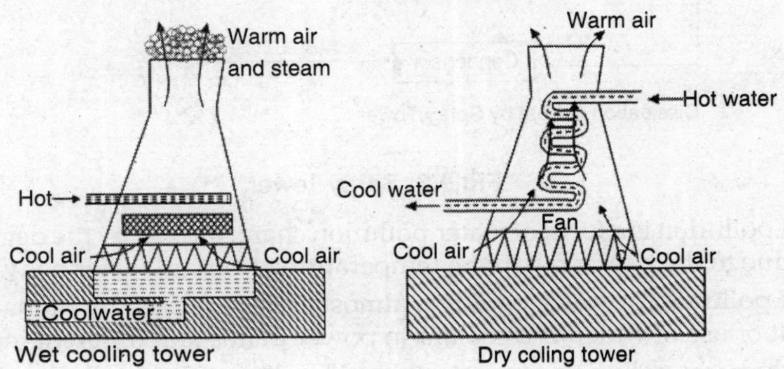

Fig.9.1: Wet and Dry Cooling Tower

b. **Dry cooling tower:** The heated water flows in a system of spiral pipes. Air is passed over these hot pipes with fans. There is no water loss in this method and installation and operatizn of dry cooling tower is many times higher than wet cooling tower.

Cooling Ponds/ Artificial lakes

Outgoing hot water from power plant condensers is stored in cooling ponds or artificial lakes natural evaporation cools water which can be recirculated or discharged in nearby water body.

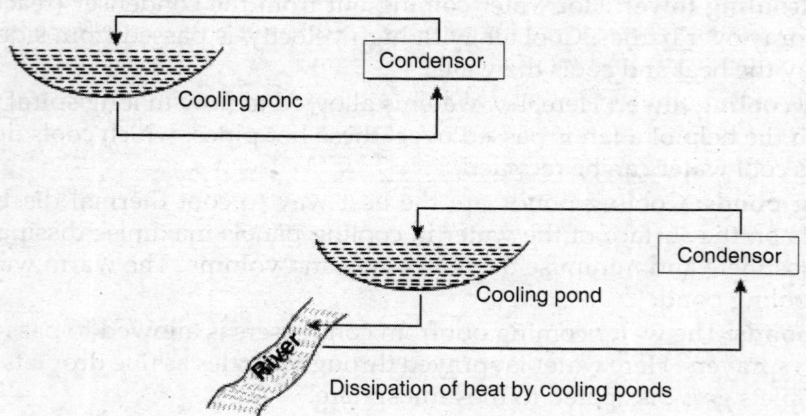

Fig.9.2: Cooling Pond

Spray Towers

The water form power plant condensers is received in spray ponds. Here the water is sprayed through nozzles where fine droplets are formed. Heat from these fine droplets is dissipated to the atmosphere.

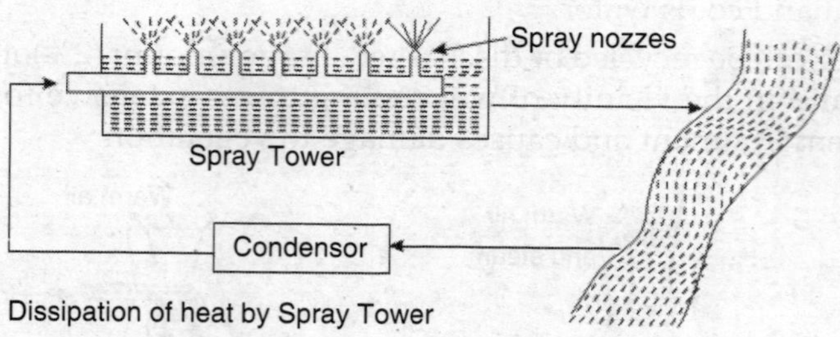

Fig.9.3: Spray Tower

- Thermal pollution is a form of water pollution characterized by the degradation of water quality due to the rise or fall in the temperature of certain water body.
- Thermal pollution of water bodies is almost always caused by humans, and is mostly the result of use of water as a coolant in power plants and different industrial facilities.
- Unlike chemical pollution of water, thermal pollution primarily results in a change in the physical properties of water.

- Thermal water pollution can cause huge damage to many aquatic ecosystems because there are many fish, and other aquatic organisms that have adapted to particular temperature of the water and once this temperature changes they often fail to adapt and die in the process as a result of thermal shock.
- Thermal water pollution is mostly associated with elevated water temperatures. The higher temperatures of water cause oxygen level to decrease and this can lead to major decline of fish species. The higher water temperatures can also lead to increased bacteria levels.
- The less common is the case of thermal pollution by the release of very cold water from the base of reservoirs and into warmer water bodies such as rivers.
- Thermal water pollution can lead to reduced biodiversity of the affected area and cause migration of many fish species.
- Even small temperature changes of one to two degrees Celsius can cause significant changes to many aquatic species and lead to loss in biodiversity.
- Thermal pollution can in some very rare cases have beneficial effect and may even, lead to positive impact on certain aquatic species such as manatee.
- More than 75% of thermal pollution in United States is attributed to power plants.
- In order to control the levels of thermal pollution many industrial facilities in developed world' are equipped with cooling ponds and cooling towers'.
- Urban runoff from roads and parking lots can also become major source of thermal pollution, particularly during warm weather.
- Thermal pollution of water can also lead to increased levels of chemical pollution because toxicity of many harmful chemicals often increases with the increase in temperature.
- One of the best ways to reduce thermal pollution Is to primarily focus on energy facilities that do hot have large water requirements for cooling purposes.

APPENDIX 9.1

Saturation value of dissolved oxygen (in mg/litre) for different sources/types of waters at different temperature and atmosphere pressure

S.No.	Sources/types of water	Temperature in °C						
		0	5	10	15	20	25	30
1	Fresh water/river water (Chlorides=0ppm	14.6	12.8	11.3	10.2	9.2	8.4	7.6
2	Brackish water/lake water (Chlorides = 10000 ppm	12	11.4	10.1	9.1	8.3	7.6	6.9
3	Sea water	11.3	10.1	9.0	8.1	7.4	6.7	6.1

APPENDIX 9.2

Temperature Limit For Discharge Of Condenser Cooling Water From Thermal Power Plant(TPP)

A. New thermal power plants commissioned after June 1, 1999.

New thermal power plants, which will be using water from rivers/lakes/reservoirs, shall install cooling towers irrespective of location and capacity. Thermal power plants which will use sea water for cooling purposes, the condition below will apply.

B. New projects in coastal areas using sea water.

The thermal power plants using sea water should adopt suitable system to reduce water temperature at the final discharge points so that the resultant rise in the temperature of receiving water does not exceed 7°C over and above the ambient temperature of the receiving water bodies.

C. Existing thermal power plants.

Rise in temperature of condenser cooling water from inlet to the outlet of condenser shall not be more than 10°C.

D. Guidelines for discharge points:

1. The discharge point shall preferably be located at the bottom of the water body at mid stream for proper dispersion of thermal discharge.
2. In case of discharge of cooling water into sea, proper marine outfall shall be designed to achieve the prescribed standards. The point of discharge may be selected in consultation with concerned State Authorities/NIO.
3. No cooling water discharge shall be permitted in estuaries or near ecologically sensitive areas such as mangroves, coral reefs/Spanish and breeding grounds of aquatic flora & fauna.

Chapter 10 Marine Pollution

The Ocean covers approximately 70% of the earth's total surface area. In the total water content of earth, 97% is present in the Oceans.

Oceans are the main regulatory agent of earth's climate. About 60% of the world's population live within 60 kms of coastline and use the coastline for their livelihood. Environmental pollution of the coast, inshore water and deep ocean is one of the important topical issue in the context of human health and global warming.

The major marine pollutants are oil, sewage, garbage, toxic chemicals, pesticides, heavy metal, radioactive waste, thermal pollution, and eutrophication.

Marine pollution means introduction by man, directly/ indirectly of substances / energy into the marine environment resulting in deleterious effects such as harm to living resources, hazards to quality for use of seawater and reduction of amenities.

The continuing environmental degradation of our oceans and coastal areas is a matter of concern because it is detrimental to human health, economic development, climate and our planet's store of biodiversity. It is interfering with the sustainability of environment and its resources.

"Biodiversity" means the variability among living organisms from all sources including terrestrial, marine and other aquatic ecosystems and the ecological complexes of which they are part; this includes diversity within species, between species and of ecosystems".

Sustainability is defined as the development that meet the need of the present, without compromising the ability of future generation to meet their own needs.

SOURCES OF MARINE POLLUTION

The main sources are atmosphere, river runoff, agriculture, livestocks, urban runoff, automobiles, land clearing, sewage outfall, industrial waste etc. In general the coastal zone receives pollutants from major ways viz., atmosphere, reverine and glacier. Also anthropogenic activities like discharging the effluents through piped outfalls, direct dumping, operation of ships etc also cause marine pollution.

In the earth's surface, 18% is represented by coastal zone, which provide space for 60% of the world's human population, because, 70% of the world cities with population exceeding 1.6 million are located in this zone. 90% of the world fish catch is obtained from coastal zone. The coastal zone is only 8% in the hydrosphere, but responsible for 18-33% of the total primary production. The biological wealth of this zone is very high as it served as feeding, nursery and spawning grounds with rich biodiversity.

Pollution may arise from point source like a single sewer pipe or factory wastewater outfall, or it can arise from a variety of geographic points called as non point source.

In the point source pollution the concentration of the substance or the intensity of the effect (e.g. temperature near cooling water outfalls of power plants) declines with increasing distance from the point source.

The non-point source effects cannot be attributed to any single. Runoff arises due to rain is an example of a non point source. The toxic substance and other fertilizer based nutrients may spread over the coast.

INDICATORS OF MARINE POLLUTION

Many organisms serve as indicators of marine pollution. For example, the ecological shift in the succession of phytoplankton communities from diatoms to dinoflagellates due to metal pollution and thermal pollution. The poor hatching of eggs and soft calcium shell of birds and mass mortalities of birds due to failure in the earlier metabolites by the interference of pesticides,

Large scale mass killing of birds due to casting, asphyxiation and smothering by oil pollution, reproductive failure in mammal are also some of the indicators of severe pollution.

Usually the pollutants exceeding threshold limits combined with environmental variables such as temperature, salinity, pH, dissolved oxygen, hydrogen sulphide etc., give stress to the organisms.

In most cases the migratory organisms such as fishes could avoid the polluted area. Whereas the residential or sedentary organisms are exposed to the pollutants experience the stress conditions.

EFFECT OF POLLUTANTS ON MARINE ORGANISMS

Pollutants enters into to the oceans are mostly diluted, however the organisms living in the oceans tend to concentrate the pollutants into their body by various mechanisms, like adsorption, absorption, ingestion etc. The concentration of pollutant Increases with higher trophic levels, i.e. from primary producers to the tertiary consumer. The threshold limit of organisms to the pollutant concentration increase with increasing trophic levels. Various mode of pollution transport in organisms includes bioconcentration, bioaccumulation and biomagnifications.

Bioconcentration

The bioconcentration is the process by which a contaminant such as oil is directly taken up (i.e., by absorption only) from water and is accumulated to levels greater than those found in the surrounding water.

Bioaccumulation

Marine organisms can accumulate chemical species in amounts far exceeding their surrounding (sea water or sediment) concentrations. The bioaccumulation is the process by which a contaminant is taken up by organisms directly through the physical exposure pathway or through consumption fo food containing the contaminated substance,

Bioaccumulation incorporates the concepts of bioconcentration and biomagnifications. The term '**concentrator factor**' has been used to quantify the bioaccumulation and is defined as the concentration of the chemical species in the Organisms or one of its components divided by the seawater or surrounding concentration the environment from which the organism was living. Organisms may take up the food, concentration factors can vary from one environment to another for a specific organisms with a given chemical.

Biomagnification

Biomagnification is the increase in tissue concentrations of a bioaccumulated chemical substance such as oil as it passes up though the trophic levels.

Biodegradation

Biodegradation is the process of degradation of a contaminant such as oil because of its use as a food a source for certain microorganisms. This process is limited to a great extent by temperature, nutrients and oxygen availability.

TYPES AND EFFECTS OF VARIOUS MARINE POLLUTANTS

There are two types of marine pollutants–biodegradable and non-biodegradable pollutants. Most of the materials, which reach the sea, disintegrate either through simple chemical reactions or because of the activities of bacteria and some larger organisms. There are some substances which are either extremely stable or else they have a very slow rate of degradation. The biodegradable pollutants are not persistent in the environment (e.g. oil, organic compound in sewage). But the non-biodegradable pollutants are persisting in environment for longer time (e.g. heavy metals, plastics, nuclear wastes etc.).

Sewage

Sewage is discharged into the oceans all over the world mostly from urban settlement. Sewage adds to the amount of small particles suspended in the water and contributes large amounts of nutrients.

The raw sewage contains higher concentrations of organic particulate matter, home washings, detergents , small amount of oil, higher concentration of bacteria and viruses, which includes harmful pathogenic forms. Sewage has bad odour, irritant and harmful to human & other organisms.

The main sources of sewage are the coastal outfall located near the cities. Also the increasing shipping activities also add higher concentration of sewage to the harbor and shipping routes. Moreover many rivers transport sewage from the inner regions of the land.

In open coast, the effect of sewage is difficult to detect due to higher mixing up of tidal, wave and current actions. However in semi-enclosed areas like small bays, harbours etc., the effects are devastating such as higher Biochemical Oxygen Demand(BOD) which expresses the organic load concentration. BOD is the amount of oxygen required by the microorganisms for the decomposition of the organic compounds dissolved in a given volume of sewage.

Consequently high BOD values indicate high concentrations of organic material and therefore low quantities of dissolved oxygen (DO). Near sewage outflow areas in temperate waters, the benthic invertebrate communities have degraded, kelp debs have disappeared and diseased fish have become more prevalent. Sewage input in the coastal environment normally increases the nutrient concentration of the outfall area. In tropical waters, outflows near coral reefs have caused a bloom of algal species that grows over the coral, and eventually it leads to the death of coral and its associated forms. This some times leads to the change in the ecological succession, that is the coral reef may be replaced by an algal reef.

Oil

Oil is discharged in to the sea in various forms as crude oil and as separate fractions. Most of the oil fractions are biodegradable. Oil and its fractions are used in various ways from household needs to automobiles and industries. The spilled oil cause more devastation in the marine environment. Petroleum oils are very complex mixtures of large numbers of hydrocarbons, containing also small amounts of sulphur, nitrogen, oxygen and different metals combined with the hydrocarbons.

The main ways in which oil enters in to the sea are:

- Natural release
- Tanker operations
- Oil tanker and other ship accidents
- Operation of ships other then tankers
- Offshore oil drilling and production platforms
- Ship-shore oil terminal operation
- Refinery operations
- Discharges of oil products on land and subsequent discharge by runoff

As the oil enters into sea it is exposed to processes, which modify it both physically and chemically. This process is called weathering. As the oil tends to float on the surface of sea, the prevailing wind, wave, tide and current conditions make the oil to spread to a wider area from the point of release. The majority of crude oil forms sticky layers on the surface, which prevents free diffusion of gasses, clogs adult organisms feeding structures and decreases the sunlight available for photosynthesis. Volatile components of an oil spill eventually evaporate into the air, leaving heavier tars behind which forms into tar balls, which fall to the bottom and may be assimilated by bottom organisms or incorporated into sediments.

The alteration of physical and chemical properties of discharged oil through a series of natural processes, which begin when the discharge occurs and continue until the oil is removed. Major processes, which contribute to weathering, include evaporation, dissolution, photo oxidation and emulsification.

Very little fraction of miscible fraction of oil is dissolved in the seawater and other fractions are immiscible. Evaporation is the single most important weathering process for the first several days of an oils discharge. Evaporation results in the loss of lighter fractions through volatilization from petroleum products (e.g. benzene, naphthalene).

The process whereby oil is incorporated into water in the form of small droplets is called emulsification. Emulsions in water can either be oil-in-water or water-in-oil, formed as a result of wave actions. There is some solution of hydrocarbons in the water and this also is more marked in the more volatile constituents. The oil-in-water leading to the wide dispersion of fine oil droplets in the sea. The water-in-oil composition leading to the formation of semi solid masses known as '**chocolate mousse**'.

If silt or sediment is present, oil may be absorbed on to it and may fall with it to the seabed. This effect can be seen as tar ball formation, which is mostly observed over the seabed of oil spoiled area.

Some of the oil constituents are naturally attached by bacteria present in the sea. The simpler compounds, the normal paraffin, are most easily broken down and consumed. The other higher and complex molecular weight fractions are broken down very slowly.

When there is a major oil spill the degree of effect depends on many factors such as:

- Type of oil
- Quantity of oil
- Physiography of the area
- Weather conditions at the time of spill
- Biodiversity of the area
- Season of the spill
- Previous exposure of the area to oil
- Exposure to other pollutants and
- Treatment of oil spill.

As the oil is of floating type, it tends to spread and eventually reach the coastline. In many coastal areas, animals and plants situated between high & low water marks are heavily coated with oil and smothered. The exposure of organism to oil adds foreign smell in its body and tissues and the process is called tainting. The shellfish can be tainted either by deposition of oil on their shell or by ingestion of naturally emulsified oil during normal feeding. The local fish population in the immediate vicinity of oil discharge can acquire a marked taint from oil components dissolved in the water.

Almost all groups of organisms like invertebrates, vertebrates like fishes, reptiles, mammals etc., are affected by oil. When the plumage of a diving birds get oiled, the water repellent properties of capacity. This leads to a condition called hypothermia, subsequently the bird may die.

Birds may also ingest the oil. All these cause large scale death of marine birds.

The tainting can be cleared by rearing the affected organisms in clean water for a considerable period. In shellfishes, during the process of moulting, the tainted shells are removed. Also, the birds's feathers and some animal's body can be cleaned with detergent. In this type of clean up, the within about five years, as crude oil is not as highly toxic as it is biodegradable.

The polynuclear aromatic hydrocarbons (PAH) in particular 3, 4 benzopyrene are well known carcinogens (cancer producing). These compounds are present in crude oil in small amounts. The PAH assimilated by marine organisms can be passed upwards through food chain towards man.This will lead to serious concern of human health.

The environmental degradation of oil includes the destruction of algal beds, spreading of oil over rocky shores, incorporation in beach sediment and subsequent loss on scenic beauty.

Heavy Metals

Normally, the metals having the atomic weight more than twenty is considered as heavy metals. The term 'heavy metals' is used to denote elements with specific gravity higher than those of Iron (Fe) and mainly include Lead (Pb), Mercury(Hg), Copper(Cu), Cadmium(Cd) and Chromium (Cr).

Coal combustion, electric utilities, steel and iron manufacturing, fuel oils, fuel additives, and incineration of urban refuse are the more sources of oceanic and atmospheric contamination by heavy metals. Heavy metal contaminated runoff from the land, rain of pollutants from the air, and fallout from ship wrecks pollute the ocean with dangerous metals.

In general, the effects of heavy metals toxicity can be classified into three main categories:

- The blocking of essential functional groups of the biomolecule (e.g. proteins, enzymes).
- The displacement of a metal ion from a biomolecule.
- The modification of the structure of biomolecules in space which is very important for their function.

SOME METAL POLLUTANTS AND THEIR EFFECT ON HUMAN BEINGS & MARINE ORGANISMS

Lead

Affects human central nervous system. A poison by ingestion and moderately irritating. A common air contaminant due to use of bad fuel in the auto industry, now being phased out through the introduction of lead-free petrol; also in air in the vicinity of industrial plants using lead. It is a carcinogen of the lungs and kidneys. Lead is flammable in the form of dust when exposed to heat or flame. From the point of view of industrial poisoning, inhalation of lead is much more important than is ingestion. Lead is a cumulative poison. Increasing amounts build up in the body and eventually a point is reached where symptoms and disability occur. Lead can cause irreversible behavioral disturbances, neurological damages and other developmental problems in young children & babies.

Cadmium

This metal is toxic to humans by inhalation and other routes. It can enter through ingestion, intraperitoneal, subcutaneous, intramuscular and intravenous routes. Has no known biochemical or nutritional function. Also cadmium is highly toxic to freshwater and marine organism. It is bioaccumulative through the food chain. Increased exposure can increase risk of lung cancer, cadmium was responsible for the 'Itai-itai' disease in japan.

Chromium

The metal exists in two forms, i.e. trivalent and hexavalent. Hexavalent chromium in high dosages has been implicated as the cause of digestive tract cancers, cutaneous & nasal mucous membrane ulcers and dermatitis. Certain chromate salts, i.e. calcium chromate are carcinogenic, at least when inhaled.

Zinc

Zinc is human skin irritant and produces pulmonary system effects. The difficulty arises from oxidation of zinc fumes prior to inhalation or presence of impurities such as cadmium, antimony, arsenic and lead. Zinc however is a needed micronutrient for humans. It is needed in agriculture. It is used in galvanizing industry. The real problem is of its association with heavy metals like lead.

Arsenic

Toxic by subcutaneous, intramuscular & intraperitoneal routes and is reported to produce systemic, skin & gastrointestinal effects. It is also a human carcinogen and experimental teratogen.

Mercury

The silvery white liquid metal is a potent neurotoxin, capable of causing severe brain damage in developing fetuses and mild tremors & emotional disturbances in exposed adults.

Implicated in the occurrence of Minamata disease in Japan, largest source of mercury contamination is through aquatic animals in whom it accumulates and gets bio transferred to organocompounds, e.g. methyl mercury. Mercury and lead poisoning cause brain damage & behavioral disturbances in children.

Copper

It is also dangerous to marine organisms and it has been used in marine anti-corrosive paints.

Pesticides

Pesticides are toxins (chemical compounds) used in agricultural farm to kill the unwanted pest organisms attaching the cultivates plants. A pesticide is any substance or mixture of substances intended for preventing, destroying, repelling, or mitigating any pest. Pests can be insects, mice & other animals, unwanted plants (weeds), fungi, or microorganisms like bacteria & viruses.

There are natural and artificial pesticides. The natural pesticides are biodegradable and are mostly derived from the extracts of plants & animals. However, artificial pesticides are man made and produced extensively only after the industrial revolution. These pesticides belong to four main categories.

1. Organochlorine compounds (e g, DDT- Dichloro diphenyl trichloroethane)
    - Persistent in the environment or decompose very slowly.
    - Half life is 9-116 years.
    - Accumulate in many organisms
    - Higher concentrations on higher organisms-mammals like dolphins, whales, tortoise etc.
    - Detected in Antarctic penguins.
2. Organophosphate esters (e.g., parathion)
    - Very toxic
    - Much toxic than previous one.

3. Carbamates (eg.)
   - Decompose rapidly through hydrolysis.
   - Half-life period approximately one week.
4. Chlorophenyl-acils 2, 4, 5-T
   - Hazardous compounds.
   - Used in Vietnam War for defoliation.

Pesticides are the examples of non-point source pollutants. These toxic materials used in the highland for agricultural and other purposes end up into the sea either through runoff into streams & rivers or having been transported by the wind(dust and aerial spray).

Depending on the type of the chemical, its partial and vapour pressure, evaporation, condensation and deposition occur at various places of the globe leading to their global partitioning. Because of such conditions these chemicals, which originate from lower latitudes, move to higher latitudes.

Depending on various conditions, these chemicals migrate in a series of relatively short jumps termed as "grasshoffer effect". Because of this grasshopper effect, concentration of such chemicals even in the remote parts of the world (Arctic and Antarctic) may rise even if the use of these chemicals are curtailed in source countries.

Existence of contamination by these chemicals in the environmental media leads to potential impact on almost all species including human. All the organisms, obtain their first dosage of these chemical from their food and water, wile the mammalian species derive these toxins from theirs mothers through milk. While the plants and lower trophic level organisms in not openly seen. Organo chlorines have been found to cause microsomal enzyme induction, accelerating body steroids such as estrogens and also to have an effect on the normal reproductive activities of many experimental animals. It is generally known that 5% eggshell thinning occurs at a concentration 5 mg/gm of DDT in the bird egg. It is also found that concentrations of 20-1000 mg/gm of DDT in the liver of birds are considered to pose a threat to individual bird reproduction and therefore on the population as a whole.

POLYCHLORINATED BIPHENYLS (PCBs)

PCBs are a series of technical mixtures consisting of many isomers and compounds that vary from mobile oily liquid to white crystalline solids and hard noncrystalline resins.

PCBs usually come from older electrical equipment like transformers, insulations etc.

Toxic by ingestion, inhalation and skin contact. A suspected human carcinogen, affects skin and liver. Usual signs of systemic intoxication include nausea, vomiting, weight loss, jaundice, edema and abdominal pain. Severe liver damage may cause coma and death. PCBs typically cause reproduction problems in most marine organisms. The reduction in sperm motility in human with increasing PCB concentration was observed indicating the possible effect of these chemicals on the reproductive capability of mammalian species.

RADIOACTIVE MATERIALS

The world's oceans have been a sink for radioactive waste from the production waste that enters the ocean from nuclear weapon testing (this is banned by international conventions), the releasing or dumping of wastes from nuclear fuel cycle systems, nuclear accidents (Chernobyl Accident in USSR in May 1985), ongoing pollution from smaller but more or less continuous leaks from nuclear power stations (Sellafield) or nuclear submarines and carries etc.

Dumping of high-level radioactive waste is no longer permitted in the ocean, but dumping of low-level wastes is still permitted. Low-level waste contains less radioactivity per gram than high-level waste. High- level wastes usually have longer half-lives. For example, one common high-level waste that is produced by spent nuclear fuel has a half-life of 24, 100 years.

It has been suggested that contained nuclear waste should be disposed in the deep sea.

The time required for half the atoms in a given quantity of a radioactive element to disintegrate is called its half life.

So little is known about the deep sea environment or the consequences of containment leakage and failure, that the effects could be devastating. The bioaccumulation process is not associated with the radioactivity properties of a molecule but with its chemical properties. If, however, a bioacumulated molecule happens to be radioactive, its effects on the organism are much more severe.

Radioactive materials bring about two principal sorts of consequences at the organism level:

- Direct toxicity caused by the ionization atoms and molecules of living materials (and especially water) resulting in the production of strong oxidizing agents, which are harmful to living tissues.
- The mutagen activity: here radiation is a factor, which causes errors during the DNA transcription leading to genome alteration. These errors, called mutations, are in most cases lethal or else lead to teratogenesis when they occur during reproduction. They can also result in cancer when they take place during the division of somatic cells.

NUTRIENTS

The nutrients are mainly nitrogen, phosphorus and silicates. Nutrients are used by plants & animals in a complicated cycle, returned to the environment and then reused. They are essential for the growth of marine primary producers like phytoplankton, seaweed and seagrass. Studies have shown that too much nutrient in the sea and abnormal algal blooms can upset the balance of plants & animals living near the seabed.

Agriculture in the land, aquaculture practices in the coastal areas and release of sewage as outfall are some the major sources. Fertilizers used on land are washed into the ocean via rivers and streams. Also the aquaculture farms locates near the coastal zone also release the wastewater in to the streams and estuaries or directly to the marine environment. Apart from this the sewage released into the coastal areas also contains higher concentration of nutrients.

The higher nutrient concentrations are always utilized by phytoplankton. The release of excess nutrients into coastal waters leads to the condition called Eutrophication. Eutrophication leads to higher production of phytoplankton called bloom. The death and decay of phytoplankton lead to higher oxygen consumption, which is detrimental to the organisms.

High nutrient concentration cause phytoplankton blooms such as, red tides, various yellow & green foams, slimes, and slicks. Although algal blooms are natural, a higher frequency of their occurrence in the past twenty years indicates an unhealthy ecosystem. The toxicity of recent blooms is increasing, which can have a direct effect on the organisms that feeds on them. Also, phytoplankton naturally contains DMS (Dimethyl Sulfoxide), which is released from dead phytoplankton into the atmosphere and can be changed to sulfuric acid, which eventually contribute to acid rain.

THERMAL POLLUTION OF SEA

In this type, the normal temperature of marine source water is increased to 7-8 ^{0}C from the ambient temperature. Mostly the power generating plants along the ocean coastlines use the marine waters for cooling purpose, which leads to heated water expelled into the marine environment.

The excess temperature usually gets reduced by various ways. As the water has higher specific heat, it takes prolonged time for cooling unless there is an external mechanism to accelerate this, The dilution by surrounding water, evaporation and precipitation may accelerate cooling within a reasonable time.

Thermal pollution seems to only affects the communities immediately adjacent to the discharge. In continuous discharge the exposure time is also continuous, which generally result in community shift that is more heat tolerant species may replace the normal biota. The longtime exposure to the heat conditions may be detrimental to the normal biota. Many studies have been done on the effects of thermal pollution on the marine environment. Thermal discharge is most noted in the tropical areas, where organisms are near their thermal maximum. For example, mangrove trees in a thermal heated bay no longer reproduce and no new seedlings can be found in the lagoon.

Solid Wastes

Plastic is produced in large amount and has danger of solid waste into the marine environment.

Plastic is not biodegradable and therefore effecting the oceans for long periods of time. When compared to other types of pollutants the biological consequences are low.

Solid wastes are dumped into the sea mainly by the anthropogenic activities.

Mostly of no biodegradable plastics made materials such as used bottles, cups, plastic bags, plastic fishing nets, metal pieces, abandoned wreckage ships, etc. are dumped into the sea. This is also called Ocean dumping.

- Six-pack holders will not decompose for 400 years.
- 100, 000 marine mammals and 2 million sea birds die every year after ingesting or being caught in plastic debris.
- Adding ingredients to plastics that would hasten decomposition would add 5-7% to costs, but this price increase in unacceptable to industries.

The solid waste on the other hand acts as a substratum for many fouling marine organisms. The initial biofilm formation on a new substratum and subsequent macro fouling is useful in some ways.

If the solid waste is not releasing any toxic substances then it can be used to build artificial reefs and has been considered as a very good way to increase the environmental heterogeneity (biodiversity) in areas which function as nursery or feeding grounds for important fishing stocks.

There have been cases that fish having been trapped inside plastic bags and other plastic objects are dying of suffocation. Also some cases have been reported of fish getting entangled in plastic fishing nets which have been torn off from fishing boats and drift in the water column, causing a phenomenon known as "ghost fishing". Sea turtles mistake plastic bags for jellyfish, eat and die from internal blockages. Seals and sea lions starve after being muzzled by six-pack rings or entangled by nets.

Birds get entangled in discarded fishing line and drowned, while lost drift nets are particularly dangerous to marine forms.

Marine oil pollution sources

Land based sources-15 %

Natural sources-8%

Oil Industry-32 %

Marine transport-45%

Control Measures for Oil Pollution

- Natural process of emulsification of oil by use of chemical dispersants: can be sprayed on the oil.
- Slick-lickers: continuous belt of absorbent material dips through the oil slick and is passed through rollers to extract oil.
- Rocks can be cleaned with high pressure steam.

Share of Various Sources of Marine Pollution

S.No.	Sources	Share of total(percent)
1.	Runoff and discharges from land	44
2.	Airborne emissions from land	33
3.	Shipping and accidental spills	12
4.	Ocean dumping	10
5.	Offshore mining, oil and gas drilling	1
	All sources	100

Control Measures for Marine Pollution

- Local communities near sea must protect sea.
- Management by government by enforcing the laws.
- Incentives must be offered for conservation.
- Industrial units should be equipped with pollution control equipments.
- Oil floating on the sea can be removed by using:-
  a. Skimmers
  b. Absorbents
  c. Coagulants

Protective Measures

- Municipal & industrial wastes should be treated well and then disposed to sea.
- Pollution levels should be detected/ checked periodically.
- Soil erosion in the coast has to be arrested.
- Recreational beaches should be maintained to meet hygienic standards.

Sources of hydrocarbons inputs into the seas by various sources is tabulated below.

Table 10.1: Sources of hydrocarbons inputs into the seas

S.No.	Sources of hydrocarbon inputs	Amount introduced (metric tons/year)
1.	**Natural Sources:-** • Maine seeps • Sediment erosion	**0.25** (0.2) (0.05)
2.	**Offshore Production**	**0.05**
3.	**Transportation:-** • Tanker operations • Dry- docking • Marine terminals • Bilge and fuel oils • Tanker accidents • Non tanker accidents	**1.47** (0.7) (0.03) (0.02) (0.3) (0.4) (0.02)
4.	**Atmospheric Transport**	**0.3**
5.	**Municipal and Industrial:-** • Municipal wastes • Refineries • Non refining industrial wastes • Urban runoff • River runoff • Ocean dumping	**1.18** (0.7) (0.1) (0.2) (0.12) (0.04) (0.02)
	Total	**3.2**

Pollutants and Their Effect on the Marine Environment is Tabulated below.

Table 10.2: Impact of various pollutant sources on marine environment

S.No.	Pollutants Sources	Impact on marine environment
1.	Municipal and domestic waste	Reduced dissolved Oxygen(DO); increased hydrogen sulphide levels; incidence of faecal coliform and faecal streptococci; high biological oxygen demand (BOD)
2.	Industrial waste	Affect DO, temperature, turbidity, PH, ammonia values; increases BOD, suspended solids
3.	Toxic metals	Causes change in chemical and biochemical processes, increase in turbidity, lethal and sub lethal effects on marine life
4.	Oil pollution	Cause smothering, clogging and toxicity
5.	Fertilisers	Affect nutrient levels and may cause eutrophication
6.	Dredging and reclamation siltation	Affects habitats of marine organisms; lethal and sublethal effects; affects flushing capacity of the water body increases in nutrient levels and can cause excessive algal bloom; may also cause damage to coral reefs and coastal nurseries
7.	Discharge of coolant waters	Raises the temperature of the water; can cause the growth of the blue-green algae
8.	Toxic chemicals	Causes lethal and sub lethal effects on marine organisms
9.	Offshore mining	Increases particulate loading which can lead to loss of light and reduced primary productivity; smothering and clogging of benthic communities
10.	Radionuclides	Bioaccumulation in fish and other benthic community

APPENDIX 10.1

Effects of Oil pollution on marine Organisms

S.N.	Marine Organisms affected	Immediate effect of oil pollution	Effect on the periphery of the spillage area	Possible long range damage to the marine organisms
1.	Fish	Premature hatch if in larval stage, may also be killed	Nutrition is upset	Unknown, possibility of stored hydrocarbons building up in the food chain.
2.	Birds	Blindness, birds may also freeze	Resistance is lowered	Do
3.	Mammals	May freeze	Resistance is lowered	Do
4.	Algae	Killed	Replacement of native species with resistant foreign species	Do
5.	Phytoplankton	Killed	Retardation of cell division, damage to cell membranes	Do
6.	Molluscs	Smoothered	Shell fish are sterilized hydrocarbons are stored in the muscle.	Do
7.	Benthic organisms	Smoothered	Disorder in marine communities	Do
8.	zooplankton	killed	Increased death rate	Unknown
9.	Higher plants	Killed	Gradual deterioration of the environment due to storage of oil in the sediment	Unknown

Note: The dangers of oil pollution for marine animals
One metric ton of oil pollutes an area of 12 square kilometers

Effects of Oil Spills

- **Volatile Organics** Compounds immediately kill many of the aquatic organisms (especially plankton & larvae) and contaminate fish
- **Floating oil** coats birds and ocean mammal; destroys natural insulation & buoyancy and causes deaths.
- **Heavy oil** sinks to ocean bottom and washes into estuaries where it contaminates crabs, oysters, mussels, clams, etc.
- **Oil slicks** on beaches harm intertidal life and cause economic losses to tourism and fishing industries.

- **Oil spills** can be caused by:
  - Tanker accidents
  - Intentional dumping
  - Drilling/pumping operations
- Petroleum is biodegradable.
- Many pollution experts consider oil to be among the least damaging ocean pollutants.

Various processes act to break up and degrade oil in the ocean environment as shown in figure.

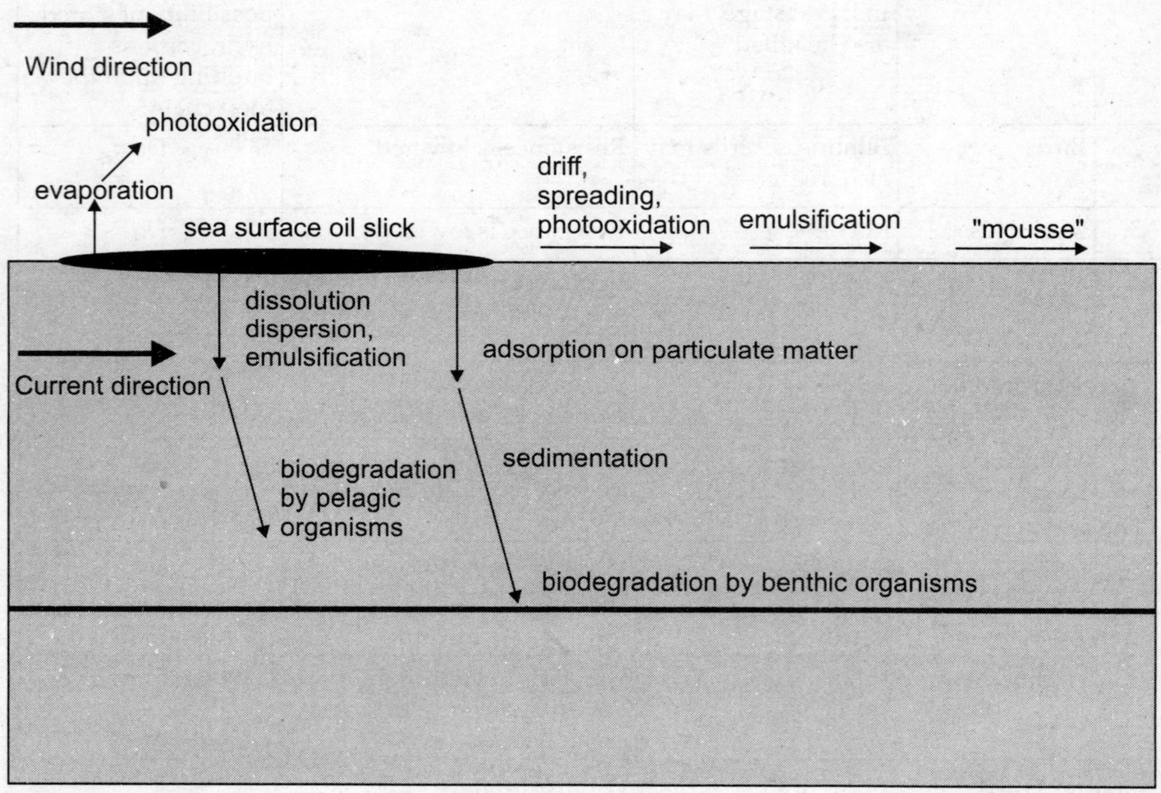

Fig.10.1F$_1$: Break up & degradation of oil in the ocean environment by various processes

Industries, residential areas, construction sites, farms etc. contribute to pollution of coastal water as shown below in figure.

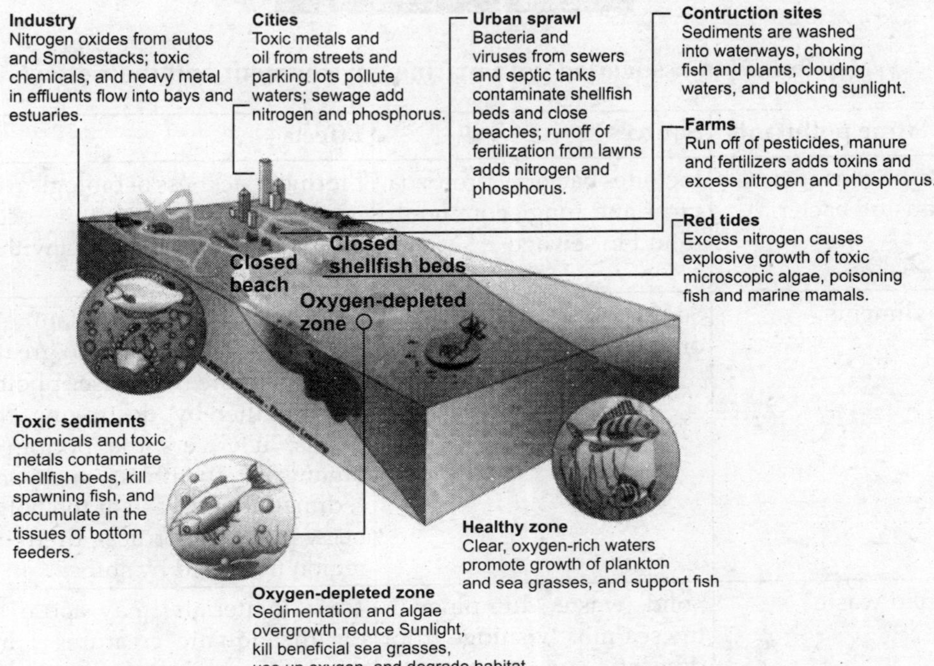

Fig.10.1F$_2$: Sources of pollution of coastal water

The world's largest oil spills is shown below in table.

Table 10.1T$_1$: Details of world's largest oil spills

Rank	Date on which oil spill occurred	Location of oil spill	Sources of oil spill	Size of oil spill	
				Million litres	Million gallons
1.	1/1991	Kuwait, Saudi Arabia	Oil terminals, tankers	908	240
2.	6/1979	Gulf of Mexico	Ixtoc oil well	530	140
3.	3/1992	Uzbekistan	Oil well	333	88
4.	2/1983	Iran	Oil well	303	80
5.	8/1983	Near coast of South Africa	Castillo de Bellver tanker	299	79
6.	3/1978	Near coast of France	Amoco Cadiz, tanker	261	69
53	3/1989	PrinceWilliam Sound, Alaska	Exxon Valdez, tanker	44.0	11.6

APPENDIX 10.2

Water Pollution associated with marine environment and their effects

S. N.	Marine pollutants	Sources	Effects
1.	Pathogens (disease causing bacteria)	Includes bacteria, protozoa, virus and fungi; commonly found in sewage	Produces sickness or biological unbalance in plants/animals within ocean or in humans coming in contact with sea water or eat sea fish/crab etc.
2.	Sediments	Sediments get deposited on the bottom of the sea.	Sediments affect the plant growth by blocking light reaching to greater depths and therefore decrease photosynthetic activity thereby destroying the rooted plants; it covers the bottom dwelling organisms (benthos) such as oysters sediment deposition at the bottom of sea necessitates periodical dredging of the region traversed by ships.
3.	Solid waste	Solid waste dumping in the sea mostly sludge from domestic sewage treatment plant	Toxic materials may pose problems for the aquatic creatures ; inert solid materials are beneficial to the marine environment; artificial fishing reefs enhances the habitat area for small fishes.
4.	Excess heat/ Thermal pollution	Fossil Fuel fire or nuclear power plants discharges	Alters ambient conditions and these changes may be detrimental to the organisms present
5.	Fresh water and brine	Fresh water sources for discharge into sea include storm drainage system, water diversion network associated with dams and effluent of some industrial processes. Brine is a by-product of desalination plant	Discharge of too much of fresh water in the oceans produces environmental change within a small area, though excessive fresh water discharge is not a critical problem; Discharge of brine into fresh water sources may cause permanent damage to organisms which are acclimated to a particular salinity
6.	Toxic substances (inorganic & organic)	Toxic inorganics emitted from industries, Toxic organics includes pesticides, insecticides halogenated hydrocarbons, petroleum and industrial chemicals. It does deteriorate very easily over a long period of time.	Inorganics in small quantity are relatively harmless; Pesticides like DDT is more soluble in oil than in water, so it collects within the fatty tissues of marine organisms. Toxic organics are discharged into ocean as manufacturing effluent or as run off after utilization.

7.	Petroleum oil/ products	Enters into marine environment due to accidents such as ship tanker damage, transfer loss, natural seepage, off shore production losses associated with oil refineries, run off originating as drippings or disposal of used automobile lubricants and unburned hydrocarbons from automobiles.	It has the effect of toxic organics on marine organisms, though it is biodegradable given enough time.
8.	Nutrients (fertilisers)	Nutrients(fertilisers) are present in domestic sewage effluents, agricultural run off & non-point source run off from urban area.	These chemicals/nutrient adds to the total load of nutrient in coastal area and does not allow it to decay as a result dissolved oxygen content of water is rapidly used and depleted causing danger to aquatic creatures
9.	Radioactive discharges-	Discharged by nuclear power plant, nuclear fuel production & reprocessing plant, uranium ore mining & concentration plants, coal burning, nuclear weapon testing, mine drainage, accidental spillage, long term storage & disposal of nuclear wastes	Hazardous for human beings depending on the concentration.
10.	Oxygen demanding materials	Sewage sludge, organic waste materials, excessive plant growth due to over supply of nutrients	Need oxygen for degeneration thereby depriving the marine animals of oxygen, decreases marine animals population
11.	Acids and bases	Industrial discharges, accidental discharges, rupturing of tankers	Disturbs the natural ecological balance of the system, toxicity of toxic materials increases
12.	Aesthetically displeasing materials	All materials having foul smell or which is unpleasant to look at. Tar balls, floatables, gas (often H_2S) producing materials, coloring agents	They offend the senses, poses threat to the ecology of area.

Solutions to Coastal Water Pollution

Prevention	Cleanup
Reduce input of toxic pollutants. Separate sewage and storm water lines. Ban dumping of wastes and sewage by ships in coastal waters. Ban dumping of hazardous material. Strictly regulate coastal development, oil drilling, and oil shipping. Require double hulls for oil tankers.	Improve oil-spill cleanup capabilities. Use nanoparticles on sewage and oil spills to dissolve the oil or sewage (still under development). Require secondary treatment of coastal sewage. Use wetlands, Solar-aquatic, or other methods to treat sewage.

APPENDIX 10.3

Elemental Analysis of sea water

S.N.	Elemental Constituents of sea water	Concentration in sea water, mg/kg sea water
1.	Chlorine	18980
2.	Sodium	10561
3.	Magnesium	1277
4.	Sulphur	884
5.	Calcium	400
6.	Potassium	380
7.	Bromine	65
8.	Carbon	28
9.	Strontium	13
10.	Boron	4.6
11.	Fluorine	4.6
12.	Silicium	4.0
13.	Other 28 elements	<1

Chapter

Water Management

Watershed is defined as land area bounded by divide line from which water drains under influence of gravity in to stream, lakes, reservoir. eg. pits, dams, farm, ponds etc.Watershed Management is the management of rainfall and resultant run off.

The watershed is the land area from which water drains under gravity to a common drainage channel. Thus watershed is a delineated area with a well defined topographic boundary and one water outlet. The watershed can range from a few square kilometers to few thousand square kilometers in size.

Necessity of water shed management

Sustainable management of water resources requires systemic, integrated decision- making that recognises the interdependence in three areas. First, decisions on land use also affect water, and decisions on water also affect the environment and land use. Second, decisions on our economic and social future currently organised by socioeconomic sectors and fragmented, affect the hydrology and ecosystems in which humans live. Third, decisions at the international, national, and local levels are interrelated.

OBJECTIVES AND APPROACHES TO WATERSHED MANAGEMENT

Watershed management focuses on how to manage the river basin for optimum utilization of water resource and thereby sustainable development of the society. As the development goals have widened from increasing food production to include poverty reduction & environmental sustainability, protecting the environment is a big challenge and greater emphasis should be given to check the exploitation of the natural resources base. Soil erosion & land degradation coupled with declining per capita availability of land & freshwater are posing serious threat to environment. This is becoming more intense with the increasing population causing food security problems. Hence efforts are needed for management of natural resources for increased productivity of the soils. Several government and nongovernment agencies have launched watershed development and management projects to tackle some of these problems with the objectives of soil conservation, improving the land productivity and promoting appropriate technologies for efficient & sustainable use of natural resources.However, many watershed projects around the world have not performed well because of the poor community participation.

To ensure their sustainability, water resources must be viewed holistically, both in their natural state and in balancing competing demands on them—domestic, agricultural, industrial (including energy), & environmental. Sustainable management of water resources requires systemic, integrated decision-making that recognises the interdependence in three areas. First, decisions on land use also affect water, and decisions on water also affect the environment and land use. Second, decisions on our economic and social future currently organised by socioeconomic sectors and fragmented, affect the hydrology and ecosystems in which humans live. Third, decisions at the international, national, and local levels are interrelated.

Watershed management practices

In watershed management the aspects of development are considered with regard to availability of the resources. The practices of conservation and development of land & water are taken up with respect to their suitability for people's benefit as well as sustainability. Various measures taken up for watershed management include the following:

Water harvesting: Proper storage of water is done with provision for use in dry seasons in low rainfall areas. It also helps in moderation of flood.

Afforestation and agro-forestry: In watershed development, afforestation & crop plantation play a very important role. They help to prevent soil erosion and retention of moisture. In high rainfall areas, woody trees are grown in between crops to substantially reduce the runoff and loss of fertile soil. In Dehradun tree like Eucalyptus, Leucaena and grasses like chrysopogon are grown along with maize or wheat to achieve the objectives. Woody trees grown successfully in such agro-forestry programs include Sheesham, Teak and Keekar which have been used in watershed areas of river Yamuna.

Mechanical measures for reducing soil erosion and runoff losses: Several mechanical measures like terracing, bunding, bench terracing, no till farming, contour cropping, strip cropping etc. are used to minimize runoff and soil erosion particularly on the slopes of watersheds. Bundling has proved to be a very useful method in reducing runoff, peak discharge and soil loss.

Scientific mining and quarrying: Due to improper mining, the hills lose stability and get disturbed resulting in landslides, rapid erosion etc. Contour trenching at an interval of one meter on overburdened dump, planting some soil binding plants land draining of water courses in the mined area are recommended for minimizing the destructive effects of mining in watershed areas.

Public participation: People's involvement including the farmers & tribal's is the key to the success of any watershed management program, particularly the soil & water conservation. The communities are to be motivated for protecting a freshly planted areas and maintaining a water harvesting structure implemented by the government or some external agency (NGO) independently or by involving the locale people. Properly educating the people about the campaign and its benefits or sometimes paying certain incentives to them can help effective people's participation. Watershed management in Himalayan region is of vital importance since most of the watersheds of our country lie there. Several anthropogenic activities accelerate its slope instability which should be prevented and efforts should be made to project the watershed by preventing overgrazing, terracing & contour farming to check runoff and erosion etc. On steeper slopes with sliding faces, straw mulching tied with thin wires and ropes helps in establishing the vegetation and stabilizing the slopes.

Enrichment of nutrients (Eutrophication) from surrounding watershed affects the penetration of light through the water, causing damage to the characteristic of water and aquatic life.

WATERSHED DEVELOPMENT IN INDIA

India began to look at the watershed development programs in the 1970s for increasing land, controlling land degradation and increasing the productivity of soils. In the 1970s, watershed development held no special significance for the development community in India, however by the end of the 1980s the situation changed radically. Initially watershed projects concentrated on soil and water conservation issues. A decade later, it was felt that technical & physical works alone would not lead to the desired objectives of watershed development and it must also take into account the social, financial & institutional aspects of rural development. Watershed programs have been established over a diverse range of rain fed agro eco regions in India.

Watersheds in India are broadly grouped into five agro climatic zones:

(i) Trans-Gangetic Plain zone,
(ii) Western Himalayan zone,
(iii) Western Plateau and Hill zone,
(iv) Gujarat Plains and Hill zone, and
(v) Southern zone.

Due to inherent heterogeneity of agro-climatic characteristics over different regions, they have divergent potentials and opportunities. Depending upon the size of the watersheds, these are broadly divided into micro and macro watersheds. Watersheds with areas up to 1250 hectares were classified as micro-watersheds, whereas the macro-watersheds were those, of greater area.

In 1994, the Ministry of Rural Development (MORD) produced a set of guidelines for implementing its watershed programmes, which aimed to tackle the concerns related to the realisation of the full benefits of watershed work. This progressive policy was essentially people centered and it incorporated good practice from NGO and government policy, such as awareness raising, bottom-up planning, partnerships with NGOs, and community participation. Since 1994-95 Ministry of rural areas and employment, government of India has implemented nearly 15000 watersheds.

ADVANTAGES OF WATERSHED PROJECTS

Advantages include:

- Improved access to drinking water in project areas during drought.
- Increase in cultivation area leading to increase in employment.
- Increase in crop yield, resulting better income to rural population.
- Improved availability of fodder for animals and increase in milk yield.
- Increase in employment & involvement of women.
- Increase in net returns from all crops.
- Decrease in soil erosion.
- Restoration of ecological balance.

CURRENT STATUS OF WATERSHED MANAGEMENT IN INDIA

Food production increased markedly in many parts of the world during the twentieth century. In India, the rate of growth of food production surpassed the rate of population growth, resulting in increased food per capita. Increased food production has required rapid expansion of irrigated area and water diversion to agriculture, which has resulted in groundwater overdraft and stream flow depletion in some basins. Water scarcity resulting from basin closure has numerous implications for water management, including increased competition and an imperative to improve water productivity in order to maintain growth in the agricultural economy. Basin closure also occurs in the context of global climate change, which may impact water availability and introduce additional uncertainties. Water resources analysis in regions experiencing closure should be carried out in a basin context, because uses in one area affect water availability in downstream areas.

In India, water management has been a government priority for irrigation and drinking water supplies. Today the scope for expanding surface & ground water sources is limited. Thus better management of existing water resources, through a judicious mix of delegation of responsibility & authority to local institutions and large-scale investment in re-directing surplus water to deficit areas is needed.

The new paradigm shift in watershed development focused on achieving the overall goal of enhancing sustainable rural livelihoods for reducing the incidence of rural poverty. Most of the watershed projects were launched for the following purposes:

- Raising farm income
- Enhancing agricultural productivity
- Soil and water conservation
- Generating rural employment
- Reducing risk by diversifying crops in rain fed areas.

  Several useful studies have been conducted to assess the contribution of participatory watershed programs and the results from these studies are discussed here. Many studies revealed that participatory watershed projects had a positive impact on crop productivity.
- The participatory watershed management is a critical area of rural development that could support rural people in many ways. India's watershed development project is seen as flagship project of Ministry of Rural development, Government of India.
- Watershed management in India has undergone dramatic change to include greater stakeholder's participation for management of natural resources in a sustainable way. It is increasingly recognized that community participation was central to watershed development. More participatory approaches have achieved greater success in enhancing livelihoods in an equitable fashion.
- It has been noted that participatory watershed management projects have been raising income, agricultural productivity, generating employment and conserving soil and water resources.
- Evidence from the three case studies and other general impact studies suggests that watershed development brought several positive trends including diversification of the rural economy, development of new institutions, increasing cropping intensity

, improved fodder production, increased availability of drinking water with rising ground water table, capacity development of the community etc.

- Based on the evidence found, it has been suggested that participatory watershed management could be a viable strategy of rural development for achieving sustainable rural livelihoods in India.

A FRAMEWORK FOR IMPROVING WATER RESOURCE MANAGEMENT

The adoption of a comprehensive framework for analyzing policies and options would help guide decisions about managing water resources in countries where significant problems exist, or are emerging, concerning the scarcity of water, the efficiency of service, the allocation of water, or environmental damage. The complexity of the analysis would vary according to the country's capacity and circumstances, but relatively simple frameworks can often clarify priority issues. The framework would facilitate the consideration of relationships between the ecosystem and socioeconomic activities in river basins. The analysis should take account of social, environmental, and economic objectives; evaluate the status of water resources within each basin; and assess the level and composition of projected demand. Special attention should be given to the views of all stakeholders. The results of the analyses at a river basin level would become part of the national strategy for water resource management. The analytical framework would provide the supporting structure for formulating public policies on regulations, incentives, public investment plans, and environmental protection and on the inter- linkages among them. It would establish the parameters, ground rules, and price signals for decentralized implementation by government agencies and the private sector. Decentralizing the delivery of water services and adopting pricing that induces efficient use of water are key elements of sound water resource management. But, for decentralized management to be effective, a supportive legal framework and adequate regulatory capacity are required, as well as a system of water charges to endow water entities with operational and financial autonomy for efficient and sustainable delivery of services.

WATER POLICY OBJECTIVES

Differences among countries notwithstanding, water resource management that follows the principles of comprehensive analysis, opportunity cost pricing, decentralization, stakeholder participation, and environmental protection will yield more coherent policies and investments across sectors, promote conservation, and improve the efficiency of water allocation. The objective is to achieve, over time, the following improvements:

- ***For industry,*** extensive water conservation and protection of groundwater sources. Experience in industrial countries suggests that controlling pollution will also substantially reduce the quantity of water used per unit of industrial output.
- ***For water supply and sanitation,*** more efficient and accessible delivery of water services & sewage collection, treatment, and disposal, with the ultimate goal of providing universal coverage. This will be achieved by extending existing supplies through water conservation & reuse and by using other sustainable methods. Greater involvement of the private sector, non governmental organizations, and user groups will be required, as will cost recovery to ensure financial viability while applying graduated fees to assist the poor.

- ***For irrigation and hydropower,*** modernized irrigation practices, greater attention to cost recovery, drainage & salinity control, measures to reduce pollution from agricultural activities, improvements in operation & maintenance of existing systems, and investments in small scale irrigation & various water harvesting methods. This calls for the development of institutions & technologies that respond to the needs of farmers for higher quality services, including greater participation of community groups and user associations, while reinforcing the efficient management of demand. Particular attention will be given to the needs of small scale farmers, who comprise most of the agricultural community. Greater priority should be given to managing the demand for energy, identifying small scale and renewable energy alternatives, promoting watershed conservation practices, and retrofitting and enhancing dam facilities.
- ***For the environment and poverty alleviation,*** more rigorous attention to minimizing resettlement, maintaining biodiversity, and protecting ecosystems in the design & implementation of water projects. Water & energy supplies gained through conservation and improved efficiency can be used instead of developing new supplies to extend service to the poor and maintain water dependent ecosystems. Low cost and environmentally benign methods of developing new water supplies for agriculture, rural drinking water, and industry will be pursued. The water supply needs of rivers, wetlands, & fisheries will be considered in decisions concerning the operation of reservoirs and the allocation of water.

POLICY IMPLICATIONS

For the interventions to have a positive impact on watershed conditions, the general principles are:

- Concentrate on contiguous sites defined by the threats to the landscape, chances of success and cost-effectiveness of the investment, where landscape and economic improvement will be self-evident.
- Include all stakeholders in watershed management rather than only the poor farmers in the target areas, as is the current practice among most development organisations.
- Select preventive rather than curative activities, and base them on land use capacity and income generating potential for maximum cost-effectiveness.
- Treat farmers, large and small, as informed clients to whom development organisations are accountable and who are capable of deciding what is good for them in the light of their resources, priorities and values.

Water-conserving technology

An important element in any strategy to conserve water will be incentives for adopting technologies and management approaches that increase the efficient use, allocation, & distribution of water. Such technologies and management approaches will make it easier to conserve water, to increase the efficiency of water use & conveyance, and to reuse wastewater. As water scarcity and waste disposal problems become more acute, adopting & improving water conservation practices, wastewater reuse systems, and overall approaches to reduce pollution will become increasingly important.

Environmental protection

Preservation of the environment and the resource base are essential for sustainable development. The protection, enhancement, and restoration of water quality and the abatement of water pollution will therefore be a focus of World Bank supported operations, particularly since providing safe drinking water is so critical to maintaining & improving human health. Accordingly, the Bank will increase its support of government efforts to improve and expand sanitation and the collection and treatment of wastewater. Similarly, the Bank will promote the use of efficiency pricing and "**the polluter-pays**" principle through the imposition of pollution charges to encourage water conservation and reduce pollution. For industrial waste, mining runoff, and wastewater discharges, a balanced strategy involving economic incentives, effective legislation and regulatory systems, and guidelines for levels of pollution control will be used to reduce effluents at the source especially toxic substances- and to stimulate reuse. For pollution originating from agricultural activities, the Bank will support initiatives that restore and protect surface and subsurface waters degraded by agricultural pollutants and that minimize soil erosion. The Bank will assist governments in developing strategies and cost-effective mechanisms for the ecologically sustainable management, protection, & restoration of recharge areas and water-dependent ecosystems, such as wetlands, riverine floodplain areas, estuaries, & coastal zones. Investments that involve resettlement should be avoided or minimized, and, where resettlement is necessary, former incomes and living standards should be restored or improved. Given the increasing importance of groundwater, especially in arid and semiarid areas, the Bank will pay attention to the linkages between ground and surface water in managing river basins and will support the establishment of government programs and policies, including land use policies, which restore and protect the quality of groundwater and preserve groundwater recharge areas.

WATER MANAGEMENT AND WATER SERVICES IN INDIA

The Indian Constitution also enshrines the right to adequate potable water, although it does not specify quantities.

Central Government

The Ministry of Water Resources is incharge of overall planning, coordination and guidance in the sector of water resources. Department of Drinking Water Supply is in the Ministry of Rural Development (MORD). In addition, the Department of Land Resources (DOLR) in the MORD is in charge of watershed-based rural development programmes such as the Desert Development Programme (DDP), the Drought-Prone Areas Programme (DPAP) and the Integrated Wasteland Development Programme (IWDP), which carries out water resource development activities, including building check dams and water harvesting structures.

In addition, the Ministry of Agriculture and Cooperation (MOAC) also funds and implements watershed based development programmes such as the National Watershed Development Project for Rainfed Areas (NWDPRA) and the Watershed Development Project in Shifting Cultivation Areas (WDPSCA), besides externally aided projects like the Integrated Watershed Development Project (IWDP-Hills-Phase II), the Karnataka Watershed Development Project, and Comprehensive Watershed Development Projects in Tirunelveli (Tamil Nadu), Ramanathapuram (Tamil Nadu), Koraput (Orissa), Madhya Pradesh and Karnataka.

State Government

In addition to centrally sponsored schemes, which are implemented by state governments with a 100% grant from the central government, the State government also implements some schemes where they share the costs with the Central Government. Within States, the Department of Irrigation is in charge of developing and maintaining major, medium and minor irrigation projects as well as groundwater development, while the Department of Panchayati Raj and Rural Development, the Department of Environment, Forests, Science and Technology and the Department of Agriculture implement watershed- based development programmes. In addition, the Department of Finance and Planning overseas the work of the state remote sensing agency, which is in charge of investigating and proposing areas in the state for water management, afforestation, etc.

WATER HARVESTING

Water harvesting is the capture, diversion, & storage of rainwater for plant irrigation and other uses. It is appropriate for large scale landscapes such as parks, schools, commercial sites, parking lots, and apartment complexes, as well as small scale residential landscapes. The problem of water shortage in arid and semi-arid regions is one of low rainfall & uneven distribution through out the season, which makes rainfed agriculture a risky enterprise. The irrigation water are often at the break of overuse and therefore untapped sources of (irrigation) water have to be sought for increasing agricultural productivity. Water harvesting for dry-land agriculture is a traditional water management technology to ease future water scarcity in many arid and semi-arid regions of world.

Historically, harvested rain water provided water for drinking, landscape watering, and for agricultural uses. Once urban areas started to develop, centralized water supply systems replaced the need to harvest water. Harvesting rainwater can reduce the use of drinking water for land scape irrigation. It is also an effective water conservation tool and proves more beneficial when coupled with the use of native, low-water-use and desert-adapted plants. Additionally, rainwater is available free of charge and puts no added strain on the municipal supply or private wells.

As the appropriate choice of technique depends on the amount of rainfall and its distribution, land topography, soil type & soil depth and local socio-economic factors, these systems are very site specific. The water harvesting methods applied strongly depend on local conditions and include such widely differing practices as bunding, pitting, micro-catchments water harvesting, flood water & ground water harvesting.

THE OBJECTIVES AND METHODS OF RAINWATER HARVESTING

Rainwater harvesting is a technique of increasing the recharge of groundwater by capturing and storing rainwater. This is done by constructing special water-harvesting structures like dug wells, percolation pits, lagoons, check dams etc. Rainwater, wherever it falls, is captured and pollution of this water is prevented.

The annual average rainfall in India is 1200 mm; however, in most places it is concentrated over the rainy season, from June to September. In fact Cherapunji in Assam (the place receiving the second highest annual rainfall as11000 mm) still suffers from water scarcity. The water flows with runoff and there is little vegetation to check the runoff & allow infiltration. Till now there is hardly any rain water harvesting being done in this region, thereby losing all the water that comes through rainfall.

Rainwater harvesting has the following objectives:

- To reduce run off loss.
- To avoid flooding of roads.
- To meet the increasing demands of water.
- To raise the water table by recharging ground water.
- To reduce ground water contamination.
- To supplement ground water supplies during lean seasons.

Rainwater can be mainly harvested by anyone of the following methods:

- By storing in tanks or reservoirs above or below ground.
- By constructing pits, dug wells, lagoons, trench or check dams on small rivulets.
- By recharging the ground water.

VARIOUS TYPES OF WATER HARVESTING

Three groups of water harvesting techniques are described below:

Rainwater harvesting: Rainwater harvesting is defined as a method for inducing, collecting, storing and conserving local surface runoff for agriculture in arid & semi-arid regions. Three types of water harvesting are covered by rainwater harvesting.

(a) Water collected from roof tops, courtyards and similar compacted or treated surfaces is used for domestic purpose or garden crops.

(b) Micro catchment water harvesting is a method of collecting surface runoff from a small catchment area and storing it in the root zone of an adjacent infiltration basin. The basin is planted with a tree, a bush or with annual crops.

(c) Macro-catchment water harvesting, also called harvesting from external catchments, is the case where runoff from hill-slope catchments is conveyed to the cropping area located at hill foot on flat terrain.

Flood water harvesting can be defined as the collection and storage of creek flow for irrigation use. Flood water harvesting, also known as 'large catchment water harvesting' or 'Spate Irrigation', may be classified into following two forms:

(a) In case of '**floodwater harvesting within stream bed**', the water flow is dammed and as a result, inundates the valley bottom of the flood plain. The water is forced to infiltrate and the wetted area can be used for agriculture or pasture improvement.

(b) In case of '**floodwater diversion**', the flood water is forced to leave its natural course and conveyed to nearby cropping fields. It is difficult to give exact figure on the present area under various forms of floodwater harvesting systems.

Groundwater harvesting is employed to cover traditional as well as unconventional ways of ground water extraction. Qanat systems, underground dams and special types of wells are few examples of the groundwater harvesting techniques. For example,

Qanats, widely used in Iran, Pakistan, North Africa and in Spain, consists of a horizontal tunnel that taps underground water in an alluvial fan, brings it to the surface due to gravitational effect. Qanat tunnels have an inclination of 1-2% and a length of up to 30 km. Many are still maintained and deliver steadily water to fields for agriculture production and villages for drinking water supply.

Groundwater dams like 'Subsurface Dams' and **'Sand Storage Dams'** are other examples of groundwater harvesting. They obstruct the flow of ephemeral streams in a river bed; the water is stored in the sediment below ground surface and can be used for aquifer recharge. Sand filled reservoirs have the following advantages:

- Evaporation losses are reduced,
- No reduction in storage volume due to siltation,
- Stored water is less susceptible to pollution, and
- Health hazards due to mosquito breeding are avoided.

Groundwater harvesting does not play the same role globally as rain and floodwater harvesting.

Traditional Method of Rainwater Harvesting:

In India, it is an old practice in high rainfall areas to collect rainwater from roof tops into storage tanks. In foot-hills water flowing from springs are collected by embankment type water storage. In Himalayan foot-hills people use the hollow bamboos as pipelines to transport the water of natural springs. Rajasthan is known for its "Tankas" (underground tanks) and "Khadins" (Embankments) for harvesting rainwater. In our ancient times, we had adequate Tellabs, Baawaris,Johars, Hauz etc. in every cities, village and capital cities of our Kings which were used to collect rainwater and ensure adequate water supply in dry periods.

Modern Techniques of Rainwater Harvesting:

In arid and semi-arid regions artificial ground water recharging is done by constructing shallow percolation tanks. Check-dams made of any suitable native material (brush, polls, rocks, plants, loose rocks, wire nets, stones, slabs, sacks etc.) are constructed for harvesting runoff from large catchment areas. Rajendra singh of Rajasthan popularly known as "Waterman" has been doing a commendable job for harvesting rainwater by building check-dams in Rajasthan and he has been honored with the prestigious Magsaysay Award for his work.

Ground water flow can be intercepted by building ground water dams or storing water underground. As compared to surface dams, ground water dams have several advantages like minimum evaporation loss, reduced chances of contamination etc.

In roof top rainwater harvesting which is a low cost and effective technique for urban houses & buildings, the rainwater from the top of the roofs is diverted to some surface tank or pit through a delivery system which can be later used for several purposes. Also it can be used to recharge underground aquifers by diverting the stored water to some abandoned dug well or by using a hand pump.

All the above techniques of rainwater harvesting are low cost methods with little maintenance expenses. Rainwater harvesting helps in recharging the aquifers, improves ground water quality by dilution, improves soil moisture and reduces soil erosion by minimizing run off water.

BENEFITS OF HARVESTING RAINWATER

- Water harvesting not only reduces potable water use and related costs, but also reduces off-site flooding and erosion by holding rainwater on the site.
- If large amounts of water are held in pervious areas where water penetrates easily,some of the water may percolate to the water table.
- Rainwater is a clean, salt-free source of water for plants.
- Rainwater harvesting can reduce salt accumulation in the soil which can be harmful to root growth. When collected, rainwater percolates into the soil, forcing salts down and away from the root zone area. This allows for greater root growth and water uptake, which increases the drought tolerance of plants.
- Limitations of water harvesting are few and are easily met by good planning and design.

KIND OF STORAGE

Above-ground water storage: In most rain and floodwater harvesting schemes the water delivered by surface runoff & overland flow is stored (only) in the soil matrix. This means, that its application is limited to the rainy season. To allow cropping outside the rainy season, a number of storage media are employed, ranging from Ferro-cement tanks of little metre square content to large reservoirs, storing millions of cubic metre.

Underground Storage: As several disadvantages are connected with surface storage of water eg. large evaporation losses, loss of storage caused by siltation, pollution problems & loss of agricultural land, underground storage of water may be an interesting alternative. This storage can be done in near surface **aquifers** (e.g. in wadi beds), calling for a conjunctive management of water resources, or in cisterns. **Cisterns** are man made caves or underground constructions to store water. Often the walls of these cisterns are plastered; their water losses by deep percolation or by evaporation can be minimal. The construction of cisterns was already practiced several thousand years ago; chalky rocks were preferred. Traditionally, in Mediterranean houses, one cellar room was specifically designed to store rainwater. Similar in-house cisterns are known from Rajasthan, NW India. In the same region, 'Kunds', covered underground tanks with plastered catchments, are found. Nowadays cisterns are often constructed using concrete.

Parameters for Identification of Suitable Rain and Floodwater Harvesting Areas

The most important parameters to be considered in identifying areas suitable for rain and floodwater harvesting are as follows:

(a) Rainfall: The knowledge of rainfall characteristics (*intensity and distribution*) for a given area is one of the pre-requisites for designing a water harvesting system. The availability of rainfall data series in space and time and rainfall distribution is important for rainfall-runoff process and also for determination of available soil moisture. A threshold rainfall events (*e.g. of 5 mm/event*) is used in many rainfall runoff models as a start value for runoff to occur. The intensity of rainfall is a good indicator of which rainfall is likely to produce runoff. Useful rainfall factors for the design of a rain or floodwater harvesting system include:

- Number of days in which the rain exceeds the threshold rainfall of the catchment, on a weekly or monthly basis.

- Probability and occurrence (in years) for the mean monthly rainfall.
- Probability and reoccurrence for the minimum and maximum monthly rainfall and
- Frequency distribution of storms of different specific intensities.

(b) **Land use or vegetation cover:** Vegetation is another important parameter that affects the surface runoff. An increase in the vegetation density results in a corresponding increase in interception losses, retention and infiltration rates which consequently decrease the volume of runoff. Vegetation density can be characterized by the size of the area covered under vegetation. There is a high degree of congruence between density of vegetation and suitability of the soil to be used for cropping.

(c) **Topography and terrain profile:** The land form along with slope gradient and relief intensity are other parameters to determine the type of water harvesting. The terrain analysis can be used for determination of the length of slope, a parameter regarded of very high importance for the suitability of an area for macro catchment water harvesting. With a given inclination, the runoff volume increases with the length of slope. The slope length can be used to determine the suitability for macro or micro or mixed water harvesting systems decision making.

(d) **Soil type & soil depth:** The suitability of a certain area either as catchment or as cropping area in water harvesting depend strongly on its soils characteristics viz.

- Surface structure; which influence the rainfall-runoff process.
- The infiltration & percolation rate; which determine water movement into the soil and within the soil matrix.
- The soil depth including soil texture; which determines the quantity of water which can be stored in the soil.

(e) **Hydrology and water resources:** The hydrological processes relevant to water harvesting practices are those involved in the production, flow and storage of runoff from rainfall within a particular project area. The rain falling on a particular catchment area can be effective (as direct runoff) or ineffective (as evaporation, deep percolation). The quantity of rainfall which produces runoff is a good indicator of the suitability of the area for water harvesting.

(f) **Socio-economic and infrastructure conditions:** The socio-economic conditions of a region being considered for any water harvesting scheme are very important for planning, designing and implementation. The chances for success are much greater if resource users and community groups are involved from early planning stage onwards. The farming systems of the community, the financial capabilities of the average farmer, the cultural behaviour together with religious belief of the people, attitude of farmers towards the introduction of new farming methods, the farmers knowledge about irrigated agriculture, land tenure & property rights and the role of women and minorities in the communities are crucial issues. The existing or planned infrastructure as well as regional development plans have to be duly taken into account when planning a water harvesting scheme.

(g) **Environmental and ecological impacts:** Dry area ecosystems are generally fragile and have a limited capacity to adjust to change. If the use of natural resources (land and water), is suddenly changed by water harvesting, the environmental consequences are often far greater than foreseen. Consideration should be given to the possible effect on natural wetlands as on other water users, both in terms of water quality and quantity. New water harvesting systems may intercept runoff at the upstream part of the catchment, thus depriving potential down stream users of their share of the resources. Water harvesting technology should be seen as one component of a regional water management improvement project. Components of such integrated plans should be the improvement of agronomic practices, including the use of good plant material, plant protection measures and soil fertility management.

Rainwater & Flood water Harvesting have the potential to increase the productivity of arable & grazing land by increasing the yields and by reducing the risk of crop failure. They also facilitate afforestation, fruit tree planting or agro forestry. With regard to tree establishment, rainwater and floodwater harvesting can contribute to the fight against desertification. Most of these techniques are relatively cheap and can therefore be a viable alternative where irrigation water from other sources is not readily available or too costly. Unlike pumping water, water harvesting saves energy and maintenance costs. Using harvested rainwater helps in decreasing the use of other valuable water sources like groundwater. Remote sensing and Geographical Information Systems can help in the determination of areas suitable for water harvesting.

Rainwater harvesting should suit its purpose, be accepted by local population, and be sustainable in local environment. In dry areas (and without storage facilities), field crops with deep rooting and drought resistant trees constitute the most promising application.

The decision making process concerning the best method applicable in particular environmental and geophysical conditions depends on kind of crop to be grown and prevalent socio-economic and cultural factors. Local availability of labour and material are the most important factors.

The accessibility of the site has also to be considered for construction of water harvesting structures and distance from village.

One of the crucial social aspects for the success is the involvement/ participation of the stakeholders or beneficiaries. All stakeholders have to get involved in planning, designing and implementation of water harvesting structure. A consensus is necessary for operation and maintenance of water harvesting structures. Involvement of local NGOs may also benefit the community for collective action.

Water management problems can only be tackled in a holistic way, integrating land, water and labour management. Finally a comparison between water harvesting techniques and the construction of large or medium dams shows that:

- Through the introduction of water harvesting, water resources in upstream watershed can be managed more efficiently.
- Water harvesting can supplement irrigation water supply during water scarcity or low water availability periods. Its proximity to cropping area can be an important point in improving water use efficiency and avoiding field losses.
- Water harvesting may be of small scale but certainly have edge over dams due to its suitability for immediate local environment, they are labour intensive (*local employment generating*), democratic and participatory in nature.
- With the small scale of water harvesting technology, no foreign investment is needed (but banking facilities are sometimes needed).

- Some of the benefits of large dams like generating hydropower energy, supplying drinking water for big cities etc, can not be offered by water harvesting and
- Water harvesting to be successful requires local capacity building and agriculture extension services, training & credit facilities for resources users, co-operation and extensive participation.

WATER RECYCLING

Though recycling is a term generally applied to aluminum cans, glass bottles, and newspapers, water can be recycled as well. *Water recycling is reusing treated wastewater for purposes such as agricultural & landscape irrigation, industrial processes, toilet flushing, and replenishing a ground water basin (referred to as ground water recharge).*Water is sometimes recycled & reused onsite; for example, when an industrial facility recycles water used for cooling processes. A common type of recycled water is water that has been reclaimed from municipal wastewater, or sewage. The term water recycling is generally used synonymously with water reclamation and water reuse.

A common example of unplanned water recycling occurs when cities draw their water supplies from rivers, which receive wastewater discharges upstream from those cities. Water from these rivers has been reused, treated, and piped into the water supply a number of times before the last downstream user withdraws the water. Planned projects are those that are developed with the goal of beneficially reusing a recycled water supply.

The need for alternative water resources, coupled with increasingly stringent water quality discharge requirements, are the driving forces for developing water reuse strategies in the world today. Water reuse enables practitioners to manipulate the water cycle, thereby creating needed alternative water resources and reducing effluent discharge to the environment. The growing trend is to consider water reuse as an essential component of integrated water resources management and sustainable development, not only in dry & water deficient areas, but in water abundant regions as well. In areas with high precipitation where water supply may be costly due to extensive transportation and/or pumping, water reuse has become an important economic alternative to developing new sources of water.

The health of the river systems and aquifers is also forcing its way to the forefront of public consciousness as whole landscapes lose their ability to absorb, provide and purify water. This not only threatens water supplies but also increases risks and impacts associated with pest species, disease vectors and catastrophic weather events. The environment, we now know to our cost, must also have its share of available water. Water reclamation and reuse have almost become necessary for conserving and extending available water supplies.

Reuse of wastewater for agricultural irrigation is practiced today in almost all arid areas of the world. Numerous countries have established water resources planning policies based on maximum reuse of urban wastewater. In many dry regions, particularly in developing countries in Asia, Africa, and Latin America, unplanned use of inadequately treated wastewater for irrigation of crops continues today and is often confused with planned & regulated reuse. This major health concern makes it imperative to governments and the global community to implement proper reuse planning and practices, emphasizing public health and environmental protection, during this era of rapid development of wastewater collection and treatment. Within the next 2 decades, 60 percent of the world's population will live in cities. As increasingly ambitious targets for sewage collection are pursued, massive and growing volumes of wastewater will be disposed of without treatment to rivers and natural water bodies. The challenges will be particularly acute in mega-cities (cities with a population of 10 million or more), over 80 percent of which will be located in developing countries.

Water reuse may also present communities with an alternate wastewater disposal method as well as provide pollution abatement by diverting effluent discharge away from sensitive surface waters. Already accepted and endorsed by the public in many urban & agricultural areas, properly implemented non-potable reuse projects can help communities meet water demand & supply challenges without any known significant health risks.

BENEFITS OF RECYCLED WATER

Recycled water can satisfy most water demands, as long as it is adequately treated to ensure water quality appropriate for the use. As for any water source that is not properly treated, health problems arises from drinking or being exposed to recycled water, if it contains disease causing organisms or other contaminants.

Recycled water is most commonly used for non-potable (not for drinking) purposes, such as agriculture, landscape, public parks, and golf course irrigation. Other non-potable applications include cooling water for power plants & oil refineries, industrial process water for such facilities as paper mills & carpet dyers, toilet flushing, dust control, construction activities, concrete mixing, and artificial lakes.

Although most water recycling projects have been developed to meet non-potable water demands, a number of projects use recycled water indirectly for potable purposes. These projects include recharging ground water aquifers and augmenting surface water reservoirs with recycled water. In ground water recharge projects, recycled water can be spread or injected into ground water aquifers to augment ground water supplies, and to prevent salt water intrusion in coastal areas.

Indirect potable reuse refers to projects that discharge recycled water to a water body before reuse. Direct potable reuse is the use of recycled water for drinking purposes directly after treatment. While direct potable reuse has been safely used in Namibia (Africa), it is not a generally accepted practice elsewhere.

ENVIRONMENTAL BENEFITS OF WATER RECYCLING

In addition to providing a dependable, locally-controlled water supply, water recycling provides tremendous environmental benefits. By providing an additional source of water, water recycling can help find ways to decrease the diversion of water from sensitive ecosystems.

Other benefits include decreasing wastewater discharges and reducing & preventing pollution. Recycled water can also be used to create or enhance wetlands and riparian habitats. Environmental benefits of water recycling are listed below:

- ***Water recycling can decrease diversion of freshwater from sensitive ecosystems:***

  Plants, wildlife, and fish depend on sufficient water flows to their habitats to live and reproduce. The lack of adequate flow, as a result of diversion for agricultural, urban, & industrial purposes, can cause deterioration of water quality & ecosystem health. Water users can supplement their demands by using recycled water, which can free considerable amounts of water for the environment and increase flows to vital ecosystems.

- ***Water recycling decreases discharge to sensitive water bodies:*** In some cases, the impetus for water recycling comes not from a water supply need, but from a need to eliminate or decrease wastewater discharge to the ocean, an estuary, or a stream..By avoiding the conversion of salt water marsh to brackish marsh, the habitat for endangered species can be protected.
- ***Recycled water may be used to create or enhance wetlands and riparian (stream) habitats:*** Wetlands provide many benefits, which include wildlife & wildfowl habitat, water quality improvement, flood diminishment, and fisheries breeding grounds. For streams that have been impaired or dried from water diversion, water flow can be augmented with recycled water to sustain and improve the aquatic & wildlife habitat.
- Water recycling can reduce and prevent pollution: When pollutant discharges to oceans, rivers, and other water bodies are curtailed, the pollutant loadings to these bodies are decreased. Moreover, in some cases, substances that can be pollutants when discharged to a body of water can be beneficially reused for irrigation. For example,recycled water may contain higher levels of nutrients, such as nitrogen, than potable water. Application of recycled water for agricultural and landscape irrigation can provide an additional source of nutrients and lessen the need to apply synthetic fertilizers.

WATER REUSE DRIVERS

The main drivers for water reuse development worldwide are:

- **Increasing water demands** to sustain industrial and population growth. This is the most common and important driver for dry and water-abundant regions in developed, developing, and transitional countries.
- **Water scarcity and droughts**, particularly in arid and semi-arid regions. In this case, reclaimed water is a vital and drought-proof water source to ensure economic & agricultural activities.
- **Environmental protection and enhancement** in combination with wastewater management needs represent an emerging driver, in a number of industrialized countries, coastal areas, and tourist regions. In areas with more stringent wastewater discharge standards, such as in Europe, Australia, & South Africa, wastewater reuse becomes a competitive alternative to advanced water treatment from both economic and environmental points of view.
- **Socio-economic factors** such as new regulations, health concerns, public policies, and economic incentives are becoming increasingly important to the implementation of water reuse projects. For example, the increase in the cost of potable water will help promote the implementation of wastewater reuse.
- **Public health protection** is the major driver in developing countries where lack of access to fresh water supplies coupled with high market access in urban and peri- urban areas, drives untreated reuse in agriculture. Public health protection and environmental risk mitigation are key components of any reuse program under these conditions.

In spite of the economic and ecological advantages associated with wastewater reuse, the key issue remains public health safety. The reuse of raw wastewater, still widely practiced in several regions in China, India, Morocco, Egypt, Pakistan, Nepal, Vietnam and most of South America, leads to enteric diseases, helminthic infections, & dangerous epidemics. In addition to public health risks, insufficiently treated effluent may also have detrimental effects on the environment. For example, high salinity levels in effluent can lead to a decrease in productivity for certain crops and destabilization of the soil structure. Another possible adverse effect is groundwater pollution.

Water Reuse Applications – Urban and Agriculture

Agriculture is the largest user of water, accounting for approximately 80 percent of the global demand. Consequently, agricultural irrigation is the major water reuse application worldwide. In a number of arid and semiarid countries - Israel, Jordan, and Tunisia – water reuse provides the greatest share of irrigation water. Israel is the world's leader in this area, with over 70 percent of collected and treated wastewater reused for agricultural purposes.

Urban water reuse is developing rapidly, particularly in large cities, coastal, and tourist areas. Japan is the leader in urban water reuse, with 8 percent of the total reclaimed water used for urban purposes. The most common urban uses are for the irrigation of green areas (parks, golf courses, & sports fields), urban development (waterfalls, fountains, & lakes), road cleaning, car washing, and firefighting. Another major type of reuse is on-site water reuse within commercial and residential buildings. For example, Australia, Canada, Japan, and the United Kingdom use treated domestic wastewater for toilet flushing. Golf course irrigation is reported as the most rapidly growing application of urban water reuse in Europe, while replenishment of river flows for recreational uses is becoming increasingly popular in Spain and Japan. There are several advantages to implementing urban reuse versus agricultural reuse:

- Most urban reuse, such as toilet flushing, vehicle washing, stack gas cleaning, and industrial processing is non-consumptive; therefore, the water can be reused again for subsequent consumptive uses in agriculture or industry.
- The urban markets for water reuse are generally closer to the points of origin of the reclaimed water than are the agricultural markets.
- Urban reuse water generally holds a higher value than agricultural reuse because it can be metered and appropriate charges levied.

Wastewater treatment for reuse may have a lower cost than developing new water supply sources, particularly for low-quality reuse in toilet flushing and similar non-potable urban uses. Agricultural irrigation will probably continue to dominate water reuse practices for many years into the future, especially in developing countries. However, reclamation projects are not likely to be built to serve agriculture. Over recent years, there has been increasing interest in indirect potable reuse in a number of industrialized countries (Australia, Belgium, France, Spain, South Africa, Singapore, and the U.S.) for water supply augmentation through the replenishment of surface reservoirs, aquifers, & salt intrusion barriers in coastal areas.

Untreated reuse water is a large and rapidly growing problem practiced in both low and middle income countries around the world.

BENEFIT OF RECYCLED WATER

Recycled water can satisfy most water demands, as long as it is adequately treated to ensure water quality appropriate for the use. Figure below shows types of treatment processes and suggested uses at each level of treatment. In uses where there is a greater chance of human exposure to the water, more treatment is required. As for any water source that is not properly treated, health problems could arise from drinking or being exposed to recycled water if it contains disease-causing organisms or other contaminants.

Recycled water is most commonly used for non-potable (not for drinking) purposes, such as agriculture, landscape, public parks, and golf course irrigation. Other non-potable applications include cooling water for power plants and oil refineries, industrial process water for such facilities as paper mills and carpet dyers, toilet flushing, dust control, construction activities, concrete mixing, and artificial lakes.

Although most water recycling projects have been developed to meet non-potable water demands, a number of projects use recycled water indirectly1 for potable purposes. These projects include recharging ground water aquifers and augmenting surface water reservoirs with recycled water. In ground water recharge projects, recycled water can be spread or injected into ground water aquifers to augment ground water supplies, and to prevent salt water intrusion in coastal areas. For example, since 1976, the Water Factory Direct Injection Project, located in Orange County, California, has been injecting highly treated recycled water into the aquifer to prevent salt water intrusion, while augmenting the potable ground water supply.

Indirect potable reuse refers to projects that discharge recycled water to a water body before reuse. Direct potable reuse is the use of recycled water for drinking purposes directly after treatment.

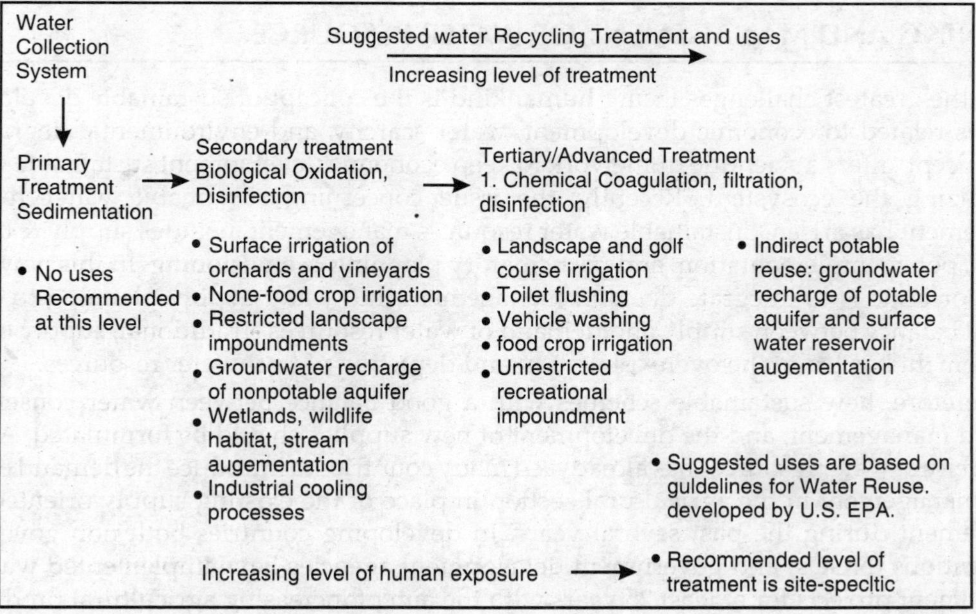

Fig.11.1: Suggested water Recycling Treatment and uses*

WATER REUSE PROGRAMS IN INDIA

The principal reuse application remains agricultural irrigation, especially in developing countries. Urban, non-potable reuse, such as reuse for, landscape irrigation, road cleaning, car washing, toilet flushing, and river flow augmentation, is developing rapidly in high density urban and tourist areas. Indirect potable reuse and the use of reclaimed water for industrial purposes have also been receiving increased attention in several industrialized countries.

Water recycling has proven to be effective and successful in creating a new and reliable water supply, while not compromising public health. Non-potable reuse is a widely accepted practice that will continue to grow. However, in many parts of the United States, the uses of recycled water are expanding in order to accommodate the needs of the environment and growing water supply demands. Advances in wastewater treatment technology and health studies of indirect potable reuse have led many to predict that planned indirect potable reuse will soon become more common.

While water recycling is a sustainable approach and can be cost-effective in the long term, the treatment of wastewater for reuse and the installation of distribution systems can be initially expensive compared to such water supply alternatives as imported water or ground water. Institutional barriers, as well as varying agency priorities, can make it difficult to implement water recycling projects. Finally, early in the planning process, agencies must implement public outreach to address any concerns and to keep the public involved in the planning process.

As water demands and environmental needs grow, water recycling will play a greater role in our overall water supply. By working together to overcome obstacles, water recycling, along with water conservation, can help us to conserve and sustainable manage our vital water resources.

PLANNING AND MANAGEMENT OF WATER RESOURCE

One of the greatest challenges facing humankind is the concept of sustainable development, which is related to economic development, water scarcity, and environmental degradation. This concept offers a scientific framework to keep economic development sustainable without jeopardizing the ecosystem. Recently, the issue concerning sustainable water resources management has arisen. Sustainable water resources management includes supply & demand aspects, policy implementation, and participatory planning & envisioning. In this new vision, the major issue is to integrate demand management into water supply planning to achieve rational balance between supply and demand of water resources. Traditional, supply oriented management has led to the overexploitation and depletion of freshwater resources.

Therefore, new sustainable schemes with a good balance between water conservation, demand management, and the development of new supplies should be formulated. Actually, the scarcity of water resources has already led many countries to introduce the demand oriented water management in the agricultural section in place of the existing supply oriented water management during the past several years. In developing countries both non government organizations (NGOs) and government development agencies have implemented watershed management projects for at least 25 years with the aim of increasing agricultural productivity and reducing poverty on hillsides in rural areas. In the last seven years or so, biodiversity conservation organizations have also set up a few watershed management projects as a way to intensify production and reduce farmer encroachment on neighboring forested areas with high biological diversity.

Many of the watershed management projects throughout the world have not taken into account land use capacity and its restoration & prevention potential. They have centered on activities that although important at the plot level do not add up to transformations at the landscape level. Additionally, they have been top down, have not accommodated the interests of resource users nor motivated their interests, and have not incorporated all stakeholders nor learned from their feedback. They have been fixated on rigid technology solutions geared to replace instead of complement local conservation practices. With these approaches, projects have not been able to foster activities that strongly reinforce both economic development and long term management of natural resources.

INTEGRATED WATER RESOURCES MANAGEMENT(IWRM)

At its simplest, integrated water resources management is a logical and appealing concept. Its basis is that the many different uses of water resources are interdependent. High irrigation demands and polluted drainage flows from agriculture mean less freshwater for drinking or industrial use; contaminated municipal & industrial wastewater pollutes rivers and threatens ecosystems; if water has to be left in a river to protect fisheries & ecosystems, less can be diverted to grow crops. There are plenty more examples of the basic theme that unregulated use of scarce water resources are wasteful and inherently unsustainable.

Integrated management means that all the different uses of water resources are considered together. Water allocations and management decisions consider the effects of each use on the others. They are able to take account of overall social and economic goals, including the achievement of sustainable development. This also means ensuring coherent policy making related to all sectors. As we shall see, the basic IWRM concept has been extended to incorporate participatory decision- making. Different user groups (farmers, communities, environmentalists…) can influence strategies for water resource development and management. That brings additional benefits, as informed users apply local self regulation in relation to issues such as water conservation and catchment protection far more effectively than central regulation & surveillance can achieve.

Management emphasises that we must not only focus on development of water resources but that we must consciously manage water development in a way that ensures long term sustainable use for future generations.

"Integrated water resources management is therefore a systematic process for the sustainable development, allocation and monitoring of water resource use in the context of social, economic and environmental objectives."

The primary objectives of integrated water resources management are:

- Empower women, men, & communities to decide on their level of access to safe water and hygienic living conditions and on the types of water-using economic activities they desire and to organise to achieve them.
- Produce more food and create more sustainable livelihoods per unit of water applied (more crops & jobs per drop) and ensure access for all to the food required to sustain healthy & productive lives.

- Manage human water use so as to conserve quantity and quality of freshwater and terrestrial ecosystems that provide services to humans and living things.

  Five primary actions are required to achieve these objectives:

- Involve all stakeholders in integrated management.
- Move to full-cost pricing of water services for all human uses.
- Increase public funding for research and innovation in the public interest.
- Recognise the need for cooperation on integrated water resource management in international river basins.
- Massively increase investments in water.

Water is vital for human survival, health and dignity and a fundamental resource for human development. The world's freshwater resources are under increasing pressure yet many still lack access to adequate water supply for basic needs. Growth in population, increased economic activity and improved standards of living lead to increased competition for, and conflicts over, the limited freshwater resource. Here are a few reasons why people argue that the world faces impending water crisis:

- Water resources are increasingly under pressure from population growth, economic activity and intensifying competition for the water among users;
- Water withdrawals have increased more than twice as fast as population growth and currently one third of the world's population live in countries that experience medium to high water stress;
- Pollution is further enhancing water scarcity by reducing water usability downstream;
- Shortcomings in the management of water, a focus on developing new sources rather than managing existing ones better and top down sector approaches to water management result in uncoordinated development and management of the resource.
- More and more development means greater impacts on the environment.
- Current concerns about climate variability and climate change demand improved management of water resources to cope with more intense floods and droughts.

Water Crisis Facts are given below.

- Only 0.4% of total of global water in the world is available for humans.
- Today more than 2 billion people are affected by water shortages in over 40 countries.
- 263 river basins are shared by two or more nations.
- 2 million tonnes per day of human waste are deposited in water courses.
- Half the populations of the developing world are exposed to polluted sources of water that increase disease incidence.
- 90% of natural disasters in the 1990s were water related.
- The increase in numbers of people from 6 billion to 9 billion will be the main driver of water resources management for the next 50 years.

KEY ISSUES IN WATER MANAGEMENT

Water Governance Crisis

Sectoral approaches to water resources management have dominated in the past and are still prevailing. This leads to fragmented and uncoordinated development & management of the resource. Moreover, water management is usually in the hands of top down institutions, the legitimacy and effectiveness of which have increasingly been questioned. Thus, weak governance aggravates increased competition for the finite resource. IWRM brings coordination and collaboration among the individual sectors, plus a fostering of stakeholder participation, transparency and cost-effective local management.

Securing Water for People

Although most countries give first priority to satisfying basic human needs for water, one fifth of the world's population is without access to safe drinking water and half of the population is without access to adequate sanitation. These service deficiencies primarily affect the poorest segments of the population in developing countries like India, China, etc. In these countries, meeting water supply and sanitation needs for urban and rural areas represents one of the most serious challenges in the years ahead. Halving the proportion of the population lacking water and sanitation services by 2020 is one of the Millennium Development Goals. Doing so will require a substantial reorientation of investment priorities, which will be very much more readily achieved in those countries that are also implementing IWRM.

Securing water for food production

Population projections indicate that over the next 25 years food will be required for another 3 billion people. Water is increasingly seen as a key constraint on food production, equivalent to if not more crucial than land scarcity. Irrigated agriculture is already responsible for more than 70% of all water withdrawals (more than 90% of all consumptive use of water). Even with an estimated need for an additional 20% of irrigation water over the next 25 years, serious conflicts are likely to arise between water for irrigated agriculture and water for other human & ecosystem uses. IWRM offers the prospect of greater efficiencies, water conservation and demand management equitably shared among water users, and of increased recycling & reuse of wastewater to supplement new resource development.

Integrated Water Resource Management(IWRM) Cycle is shown below in figure.

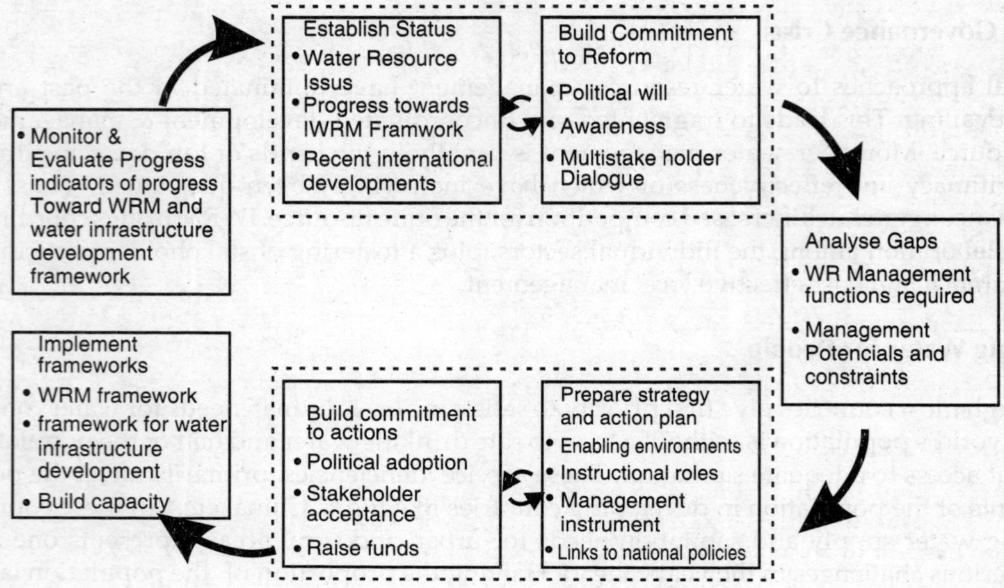

Fig.11.2: IWRM cycle

Protecting vital ecosystems

Terrestrial ecosystems in the upstream areas of a basin are important for rainwater infiltration, groundwater recharge and river flow regimes. Aquatic ecosystems produce a range of economic benefits, including such products as timber, fuel wood & medicinal plants, and they also provide wildlife habitats & spawning grounds. The ecosystems depend on water flows, seasonality and water table fluctuations and are threatened by poor water quality. Land & water resources management must ensure that vital ecosystems are maintained and that adverse effects on other natural resources are considered and where possible reduced when development & management decisions are made. IWRM can help to safeguard an "environmental reserve" of water commensurate with the value of ecosystems to human development.

Gender disparities

Formal water management is male dominated. Though their numbers are starting to grow, the representation of women in water sector institutions is still very low. That is important because the way that water resources are managed affects women and men differently. As custodians of family health and hygiene and providers of domestic water and food, women are the primary stakeholders in household water and sanitation. Yet, decisions on water supply and sanitation technologies, locations of water points and operation and maintenance systems are mostly made by men. The Gender and Water Alliance cites the example of a well meaning NGO that helped villagers to install pour flush latrines to improve their sanitation and hygiene, without first asking the women about the extra two liters of water they would have to carry from distant sources for every flush. A crucial element of the IWRM philosophy is that water users, rich and poor, male and female, are able to influence decisions that affect their daily lives.

WATER MANAGEMENT PRINCIPLES

A meeting in Dublin in 1992 gave rise to four principles that have been the basis for much of the subsequent water sector reform.

Principle 1. Fresh water is a finite and vulnerable resource, essential to sustain life, development and the environment.

The notion that freshwater is a finite resource arises as the hydrological cycle on average yields a fixed quantity of water per time period. This overall quantity cannot yet be altered significantly by human actions, though it can be, and frequently is, depleted by man made pollution. The freshwater resource is a natural asset that needs to be maintained to ensure that the desired services it provides are sustained. This principle recognises that water is required for many different purposes, functions and services; management therefore, has to be holistic (integrated) and involve consideration of the demands placed on the resource and the threats to it. The integrated approach to management of water resources necessitates co-ordination of the range of human activities which create the demands for water, determine land uses and generate waterborne waste products. The principle also recognises the catchment area or river basin as the logical unit for water resources management.

Principle 2. Water development and management should be based on a participatory approach, involving users, planners and policymakers at all levels.

Water is a subject in which everyone is a stakeholder. Real participation only takes place when stakeholders are part of the decision making process. The type of participation will depend upon the spatial scale relevant to particular water management and investment decisions. It will be affected too by the nature of the political environment in which such decisions take place. A participatory approach is the best means for achieving long lasting consensus and common agreement.

Participation is about taking responsibility, recognizing the effect of sectoral actions on other water users and aquatic ecosystems and accepting the need for change to improve the efficiency of water use and allow the sustainable development of the resource. Participation does not always achieve consensus, arbitration processes or other conflict resolution mechanisms also need to be put in place. Governments have to help create the opportunity and capacity to participate, particularly among women and other marginalised social groups. It has to be recognised that simply creating participatory opportunities will do nothing for currently disadvantaged groups unless their capacity to participate is enhanced. Decentralising decision making to the lowest appropriate level is one strategy for increasing participation.

Principle 3. Women play a central part in the provision, management and safeguarding of water.

The pivotal role of women as providers and users of water and guardians of the living environment has seldom been reflected in institutional arrangements for the development and management of water resources. It is widely acknowledged that women play a key role in the collection & safeguarding of water for domestic and in many cases agricultural use, but that they have a much less influential role than men in management, problem analysis and the decision making processes related to water resources. IWRM requires gender awareness. In developing the full and effective participation of women at all levels of decision making, consideration has to be given to the way different societies assign particular social, economic and cultural roles to men and women. There is an important synergy between gender equity and sustainable water management. Involving men and women in influential roles at all levels of water management can speed up the achievement of sustainability; and managing water in an integrated and sustainable way contributes significantly to gender equity by improving the access of women & men to water and water related services to meet their essential needs.

Principle 4. Water has an economic value in all its competing uses and should be recognised as an economic good as well as a social good.

Within this principle, it is vital to recognise first the basic right of all human beings to have access to clean water and sanitation at an affordable price. Managing water as an economic good is an important way of achieving social objectives such as efficient and equitable use, and of encouraging conservation and protection of water resources. Water has a value as an economic good as well as a social good. Many past failures in water resources management are attributable to the fact that the full value of water has not been recognised.

Value and charges are two different things and we have to distinguish clearly between them. The *value* of water in alternative uses is important for the rational allocation of water as a scarce resource, whether by regulatory or economic means. *Charging (or not charging)* for water is applying an economic instrument to support disadvantaged groups, affect behavior towards conservation and efficient water usage, provide incentives for demand management, ensure cost recovery and signal consumers' willingness to pay for additional investments in water services.

Treating water as an economic good is an important means for decision making on the allocation of water between different water use sectors and between different uses within a sector. This is particularly important when extending supply is no longer a feasible option.

WATER USE, IMPACTS AND BENEFITS

Impacts

Most uses of water bring benefits to society but most also have negative impacts which may be made worse by poor management practices, lack of regulation or lack of motivation due to the water governance regimes in place.

Each country has its priority developmental and economic goals set according to environmental, social and political realities. Problems and constraints arise in each water use area, but the willingness and ability to address these issues in a coordinated way is affected by the governance structure of water. Recognising he inter-related nature of different sources of water and thus also the inter-related nature and impacts of the differing water uses is a major step to the introduction of IWRM.

Impacts of the water use sectors on water resources are tabulated below.

Table 11.1: Impacts of the water use sectors on water resources

	Positive impacts	Negative impacts
Environment	# Purification # Storage # Hydrological cycle	-
Agriculture	# Return flows # Increased infiltration # Decreased erosion # Groundwater recharge # Nutrient recycling	# Depletion # Pollution # Salinisation # Water logging # Erosion
Water supply & sanitation	# Nutrient recycling	# High level of water security required # Surface and groundwater pollution

Benefits from IWRM

Environment benefits

(a) Ecosystems can benefit from applying an integrated approach to water management by giving environmental needs a voice in the water allocation debate. At present these needs are often not represented at the negotiating table.

(b) IWRM can assist the sector by raising awareness among other users of the needs of ecosystems and the benefits these generate for them. Often these are undervalued and not incorporated into planning and decision making.

(c) The ecosystem approach provides a new framework for IWRM that focuses more attention on a system approach to water management: protecting upper catchments (e.g. reforestation, good land husbandry, soil erosion control), pollution control (e.g. point source reduction, non point source incentives, groundwater protection) and environmental flows. It provides an alternative to a sub-sector competition perspective that can join stakeholders in developing a shared view and joint action.

Agriculture benefits

(a) As the single largest user of water and the major non-point source polluter of surface and groundwater resources, agriculture has a poor image. Taken alongside the low value added in agricultural production, this frequently means that, especially under conditions of water scarcity, water is diverted from agriculture to other water uses. However, indiscriminate reduction in water allocation for agriculture may have far reaching economic and social consequences. With IWRM, planners are encouraged to look beyond the sector economics and take account of the implications of water management decisions on employment, the environment and social equity.

(b) By bringing all sectors and all stakeholders into the decision- making process, IWRM is able to reflect the combined "value" of water to society as a whole in difficult decisions on water allocations. This may mean that the contribution of food production to health, poverty reduction and gender equity, for example, could over-ride strict economic comparisons of rates of return on each cubic metre of water. Equally, IWRM can bring into the equation the reuse potential of agricultural return flows for other sectors and the scope for agricultural reuse of municipal and industrial wastewaters.

(c) IWRM calls for integrated planning so that water, land and other resources are utilised in a sustainable manner. For the agricultural sector IWRM seeks to increase water productivity (i.e. more crops per drop) within the constraints imposed by the economic, social and ecological context of a particular region or country.

Water supply and sanitation benefits

(a) Above all, properly applied IWRM would lead to the water security of the world's poor and unserved being assured. The implementation of IWRM based policies should mean increased security of domestic water supplies, as well as reduced costs of treatment as pollution is tackled more effectively.

(b) Recognizing the rights of people, and particularly women and the poor, to fair share of water resources for both domestic and household-based productive uses, leads inevitably to the need to ensure proper representation of these groups on the bodies that make water resource allocation decisions.

(c) The focus on integrated management and efficient use should be a stimulus to the sector to push for recycling, reuse and waste reduction. High pollution charges backed by rigid enforcement have led to impressive improvements in industrial water-use efficiencies in the industrialised countries, with benefits for domestic water supplies and the environment.

(d) Past sanitation systems often focused on removing the waste problem from the areas of human occupation, thus keeping the human territories clean and healthy, but merely replacing the waste problem, with often detrimental environmental effects elsewhere. Introduction of IWRM will improve the opportunity for introduction of sustainable sanitation solutions that aim to minimise waste-generating inputs, and reduction of waste outputs, and to solve sanitation problems as close as possible to where they occur.

(e) At a practical local level, improved integration of water resource management could lead to greatly reduced costs of providing domestic water services, if for instance more irrigation schemes were designed with a domestic water component explicitly involved from the start.

WATER CONSERVATION

Water sources include:

- Fresh water
- River
- Stream
- Pond
- Ocean

Need for water conservation:

- Population increases water requirement also increases.
- Due to deforestation annual rainfall decreases.
- Over exploitation of ground water.

Ways of water conservation:

- Reducing evaporation loss.
- Reducing irrigation loss.
- Reuse water.
- Avoid sewage discharge.

Water conservation method:

- Rain water harvesting
- Watershed management

Water being one of the most precious and indispensable resources needs to be conserved. The following strategies can be adopted for conservation of water.

- **Decreasing run-off losses:** Huge water-loss occurs due to run-off on most of the soils, which can be reduced by allowing most of the water to infiltrate into the soil. This can be achieved by using contour cultivation, terrace farming, water spreading, chemical treatment or improved water-storage system.

  (a) **Contour cultivation**: on small furrows & ridges across the slopes trap rainwater and allow more time for infiltration. Terracing constructed on deep soils have large water-storage capacity. On gentle slopes trapped run off is spread over a large area for better infiltration.

  (b) **Conservation bench terracing**: It involves construction of a series of benches for catching the runoff water.

  (c) **Water spreading is done by channeling or lagoon leveling**, In channeling, the water flow is controlled by a series of diversions with vertical intervals. In lagoon leveling, small depressions are dug in the area so that there is temporary storage of water.

  (d) **Chemical wetting agents (Surfactants)**: These seem to increase the water intake rates when added to normal irrigated soil.

  (e) **Surface crop residues, tillage, mulch, animal residues etc**. help in reducing run-off by allowing more time for water to penetrate into the land.

  (f) **Chemical conditioners like gypsum ($CaSO_4.2H_2O$)** when applied to sodic soils improve soil permeability and reduce run off. Another useful conditioner is HPAN (hydrolyzed polyacrylonitrile)

  (g) **Water-storage structures like farm ponds**, dug-outs etc. build by individual farmers can be useful measures for conserving water through reduction of runoff.

- **Reducing evaporation losses:** This is more relevant in humid regions. Horizontal barriers of asphalt placed below the soil surface increase water availability and increase crop yield This is more effective on sandy soil but less effective on loamy sand soils. A co-polymer of starch and acrylonitrile called 'super slumper' has been reported to absorb water up to 1400 times its weight. This chemical has been found to be useful for sandy soils.

- **Storing water in soil:** Storage of water takes place in the soil root zone in humid regions when the soil is wetted to field capacity. By leaving the soil fallow for one season water can be made available for the crop grown in next season.
- **Reducing irrigation losses**:
  - (a) Use of lined or covered canals to reduce seepage.
  - (b) Irrigation in early morning or late evening to reduce evaporation losses.
  - (c) Sprinkling irrigation and drip irrigation to conserve water.
  - (d) Growing hybrid crop varieties with less water requirements and tolerance to saline water help conserve water.
- **Reuse of water:**
  - (a) Treated wastewater can be used for ferti-irrigation.
  - (b) Using grey water from washings, bath-tubs etc. for watering gardens, washing cars or paths help in saving fresh water.
- **Preventing wastage of water:** This can be done in households, commercial buildings and public places.
  - (a) Closing taps when not in use.
  - (b) Repairing any leakage from pipes.
  - (c) Uszing small capacity flush in toilets.
- **Increasing block pricing:** The consumer has to pay a proportionately higher bill with higher use of water. This helps in economic use of water by the consumers.

The adoption of a watershed wide approach will necessarily require some institutional adaptations.

Our framework is based on the assumption that watershed management will work provided that the right biophysical context & potential are matched with the proper socioeconomic incentive and the most supportive policy context.

The framework also assumes the need to adopt a flexible and learning approach to watershed management. This requires active learning from feedback, keeping track of what really is happening, and particularly what the stakeholders are doing & thinking.

The most important change needed from both implementing organizations and watershed stakeholders is the adoption of a holistic and 'systems' approach to watershed management.

A systems view will require engaging all stakeholders in a watershed. Part of the goal of watershed management will be to resolve conflicts of land use, which requires that organizations facilitate a dialogue between residents of the watershed and those downstream as well as the active involvement of the relevant local governments & institutions.

APPENDIX 11.1

Role of various Government Organizations Involved in Water Management in India

The drinking water supply is the responsibility of the state Governments to provide safe drinking water to the people. In our country , the following agencies are connected with regulating, monitoring and laying down standards of water:

- Ministry of Health & Family Welfare;
- Bureau of Indian Standards (BIS) under the Ministry of Consumer Affairs, Food and Public Distribution;
- Ministry of Rural Development;
- Ministry of Urban Development;
- Ministry of Environment and Forests;
- Ministry of Water Resources; and
- Local Bodies

Central Government

- **The ministry of Health and Family Welfare**: The ministry administers the Prevention of Food Adulteration Act, 1954 along with the prevention of Food Adulteration Rules, 1955, drinking water was not included in the preview of this act. Only packaged water("bottled drinking water" or "packed drinking water") has been included under the definition of food vide notification GSR 202(E), dated 21st March, 2001. Thereafter, it had been brought under the compulsory certification scheme of the BIS. Potable water/ drinking water is still not included under the PFA as "Food".

  The PFA Act covers food standards, general procedures for sampling, analysis of food, power of authorized officers, nature of penalties and other parameters related to food. Then this act was repealed by the Food Safety and Standards Act, 2006 which is related to:-

  (i) Procedure of Licensing and Registration of Food Business Operators;
  (ii) Operationalization of Food safety and Standards Act by the State Government; and
  (iii) To deal with other business of food safety etc.

Food Safety and Standards Authority of India (FSSAI)

It has been established under Food Safety and Standards Act, 2006 (FSS Act) which consolidates central acts & order [like Prevention of food Adulteration Act, 1954, Fruit Products Order, 1955, Meat Food Product Order, 1973, Vegetable Oil Products (Control) Order, 1947, Edible Oils Packing (Regulation) Order, 1988, Solvent Extracted Oil, De-Oiled Meal and Edible Flour (Control) Order, 1967, Milk and Milk Product Order, 1992 etc.] that have hitherto handle food related issues in various Ministries and Department. FSSAI has been created for laying down science based standards for articles of food and to regulate their manufacture, storage, distribution, sale and import to ensure availability for safe and wholesome food for human consumption.

The FSS Act also aims to establish a single reference point for all matters relating to food safety and standards, by moving from multi-level, multi departmental control to a single line to command. To this effect, the Act established an independent statutory Authority – The Food Safety and Standards Authority of India with head office at Delhi. Food Safety and standards Authority of India (FSSAI) and the State Food Safety Authorities enforce various provisions of the Act. Ministry of Health & Family Welfare, Government of India is the Administrative Ministry for the implementation of FSSAI.

- **The Bureau of Indian Standards (BIS):** The Bureau of Indian Standards (BIS) an autonomous national standards body of the country under the administrative control of the Department of Consumer Affairs is mandated to prepare and implement standards. BIS bottle water norms have been notified under the Prevention of Food Adulteration Act, 1954, whereas the drinking water/potable water norms still remain voluntary in nature. BIS ensures the quality of Packaged Drinking Water manufactured by its licensees through a well-defined certification scheme wherein regular surveillance is done of the licensees by factory inspections and drawing of samples from factory & market and their independent testing to check the conformity of the product to relevant Indian Standards. If such licensees are found not meeting the norms prescribed in the relevant Indian Standards, actions like warning, stop marking, deferment of renewal of licence and cancellation of licence are taken.
- **The Ministry of Rural Development and The Ministry of Urban Development:** They are concerned with the supply of drinking water in the country. It is the Department of Drinking Water Supply under the Ministry of Rural Development in regard to rural areas and Central Public Health and Environmental Engineering Organization (CPHEEO) under the Ministry of Urban Development for urban areas. Though the norms for quality of drinking water have been laid down by both these agencies, these are only recommendatory in nature. The implementation part vests with the State Governments.
- **The Ministry of Environment and Forests (MOEF):** Ministry is concerned with implementation of policies and programmes related to conservation of the country's natural resources including lakes & rivers and the prevention & abatement of pollution. There are many Acts/ Legislation available in the country for control/abatement of water pollution like: The Water (Prevention and Control of Pollution) Act,1974, The Water Cess (Prevention and Control of Pollution) Act, 1977 and The Environment (Protection) Act,1986, which monitors surface water quality, chiefly rivers. The Central Pollution Control Board and State Pollution Control Board under the Ministry of Environment and Forests implement these legislations.
- **The Ministry of Water Resources:** The Ministry is responsible for laying down policy guidelines and programmes for the development and regulation of country's water resources.
  - **Central Ground Water Authority:** It has been constituted under Section 3 (3) of the Environment (Protection) Act, 1986 to regulate and control the development and management of ground water resources in the country and function under the administrative control of the Ministry of Water Resources.
  - **Central Ground Water Board:** Central Ground Water Board (CGWB), a subordinate office of the Ministry of Water Resources, Government of India, is the National Apex Agency entrusted with the responsibilities of providing scientific inputs for

management, exploration, monitoring, assessment, augmentation and regulation of ground water resources of the country. Central Ground Water Board was established in 1970 by renaming the Exploratory Tube wells Organization under the Ministry of Agriculture, Government of India. It was merged with the Ground Water Wing of the Geological Survey of India during 1972.

– **Central Water Commission:** Central Water Commission is a premier Technical Organization of India in the field of Water Resources and is presently functioning as an attached office of the Ministry of Water Resources, Government of India. The Commission is entrusted with the general responsibilities of initiating, coordinating and furthering in consultation of the State Governments concerned, schemes for control, conservation and utilization of water resources throughout the country, for purpose of Flood Control, Irrigation, Navigation, Drinking Water Supply and Water Power Development. It also undertakes the investigations, construction and execution of any such schemes as required.
– **The Water Quality Assessment Authority:** There are several agencies, which are operating water quality network in the country. These include the Central Ground Water Board/Authority, Central Water Commission, Water Supply Authorities, Industries and Educational Research Institutes. All these agencies are working more or less independent of each other and there is hardly any co-ordination among these. The result was that, total confusion as one agency did not know what the other was doing and very often there was a great deal of overlapping. In order to address this problem of multiplicity and with a view to bringing the various agencies on a single interactive platform, the Central Government, Ministry of Environment and Forests constituted **The Water Quality Assessment Authority** in May, 2001 which functions under the administrative control of the Ministry of water Resources, Govt, of India.
– **Functions of Water Quality Assessment Authority at Central level**:To exercise powers under Section 5 of Environment (Protection) Act 1986 for issuing direction and for taking measures for investigations and research; establishment or recognition of environmental laboratories and institutes; collection & dissemination of information and preparation of manuals, codes or guides relating to the prevention, control and abatement of water pollution.

- **National Water Development Agency:** National Water Development Agency(NWDA) was set up in July, 1982 as Autonomous Society under the Societies Registration Act, 1860, to carry out the water balance and other studies on a scientific & realistic basis for optimum utilisation of Water Resources of the Peninsular rivers system for preparation of feasibility reports and thus to give concrete shape to Peninsular Rivers Development Component of National Perspective. In 1990, NWDA was also entrusted with the task of Himalayan Rivers Development Component of National Perspectives. Recently, the functions of NWDA have been further modified and the work of preparation of detailed Project Reports (DPR) of various link proposals and Pre-feasibility Reports and feasibility reports of intra-State links as proposed by the States have been included in the functions of NWDA.
- **National Ganga River Basin Authority:** To take measures for abatement of pollution and conservation of the river Ganga to maintain its ecology, water quality with other measures.

State Government

Drinking water is a state subject and role of the Union Ministry is merely recommendatory in nature. It is the state government that must adopt standards and enforce them. CPHEEO has recommended drinking water quality standards in its Manual on Water Supply and Treatment, but the standards are not legally binding on local governments. BIS has also developed drinking water quality standards but they are not mandatory. Local bodies such as Municipal Corporation/Nagar Parishad/Nagar Panchayat and Public Health Engineering Department in urban areas are expected to follow these guidelines.

Functions of various state level agencies

- Line Department of State Government: Implementation of provisions of PFA Rules and water quality management programmes and schemes;
- Water Quality Review Committee at state under the Water Quality Assessment Authority at Central level: Generation of reliable water quality data; facilitation of activities for prevention and control of pollution of water bodies and improve co-ordination between central & state agencies;
- State Ganga River Conservation Authority: To take measures for abatement of pollution and conservation of the river Ganga in the State to maintain its ecology, water quality with other measures;
- State Pollution Control Board: Enforcement of provisions of various acts & rules for prevention and control of water pollution; and Rural & Urban Local Bodies: Implementation of water supply & quality management programmes formulated by the Govt. in the field of Drinking Water.

APPENDIX 11.2

Water harvesting- A tool for water conservation

Water scarcity is a serious problem throughout the world for both urban and rural communities. More and more water is required for domestic, construction and industrial use. The rate of withdrawal is far in excess of the rate of recharging the water table. Pollution and overexploitation have shrunk the availability of clean and potable water to a trickle.

Urbanization has resulted in over exploitation of ground water, reduction in open soil surface & water infiltration rate and a resultant deterioration in water quality.

Water covers about 75 % of the earth's surface, but only about 2% is fresh water, and a larger portion of it is polar ice. 86% of Asia's fresh water is used for agriculture, 8%for industry and 6%for domestic purposes. Our country uses 83%of fresh water for agriculture.

Fresh water, once considered an inexhaustible resource, is now fast becoming a rare and scarce commodity. The water policy of the Government of India puts a norm of 180 litres per capita for our domestic needs. But it is not possible to supply even 100 litres per day even in urban areas.

At present one third of the global population is facing water stress.

How are we going to increase the ground water level? The main source is precipitation, in the form of rain or snow. This rain water can be used to recharge the ground water, by adopting a simple technique called rainwater harvesting. Rainwater harvesting means making use of each and every drop of rainwater to recharge the groundwater, by directing the rainfall to go into the well or under the ground, without wasting the rainwater.

India uses only 10-20 percent of its annual rainfall when it percolates and reaches the ground water aquifers, while the major portion goes into the ocean.

Artificial Recharging

Artificial recharging to augment ground water resources has become a necessity and we should therefore develop and popularize some of the cost effective rainwater harvesting methods in urban & rural areas. The main objectives of rainwater harvesting are:

- To conserve the surface run-off during monsoons.
- To recharge the aquifers and increase availability of ground water.
- To improve the quality of ground water where required.
- To overcome the problem of flooding and stagnation of water during the monsoon season.
- To arrest salt-water intrusion.

Various methods are available to recharge the water table by increasing the rate of infiltration. The artificial recharge can be achieved by obstructing the flow of water, by storing the water, by spreading the water or by injection through wells and bore wells.

The following are some methods used to recharge the ground water:

- Pebble bed method
- Percolation pits
- Recharge wells
- Ridges and furrows
- Check dams
- Gully control/stone wall structures
- Contour bunding, trenching
- Land flooding

The technology of water harvesting

Rainwater, runoff and the floodwaters from rivers are all harvested. Water harvesting systems are located:

- in the open to capture rainwater where it fell
- in the path of a stream or its runoff
- beside rivers to catch the flood waters

The design and structure of each system is decided by the terrain and rainfall pattern of the region. Hence each eco-zone of India has unique techniques for harvesting water.

a. **In the hills and mountainous regions** where there are plenty of streams, simple engineering structures are used to divert the water into channels that fed the fields. The structures became more sophisticated and much bigger when the streams turned to rivers.

b. **In the arid and semi-arid regions,** where the streams are more seasonal, the diversion channels first led the water to a storage structure like a tank for later use. Storage systems to collect just runoff from the watershed are also built.

c. **In the flood plains,** several unique systems to control and harness the floodwaters are devised.

d. **In the coastal areas** where there is danger of river water turning saline, several ingenious ways came up to regulate the flow of saline water.

e. **In regions with good groundwater aquifers,** dugwells with innovative methods to lift the water are in use. Deep wells were dug in the beds of tanks & rivers, both to serve as a source of good water when the water recedes and also to recharge the groundwater when they are fully submerged.

f. **In areas where rainfall is the only option,** people devise methods to literally "catch rainwater where it fell".

Advantages of harvesting water

When a small water body or source is created, there are several benefits:

- The standing water percolates into the ground and recharges the water table
- Wells in the surrounding areas have plenty of good water
- Green cover increases in the surrounding areas
- Soil erosion is reduced'

- Silting of rivers is reduced
- Floods and runoff get controlled

But today, even with the large water projects, the problem of water famine continues. For several decades different parts of India have been in the grip of drought. Even as the serious problems accompanying changes in the natural course of a river are being realized, the big dams continue to be built.

Slowly, there is a growing awareness that the traditional water harvesting systems will be a better and cheaper alternative to big or small dams. Groups of villages have revived small water harvesting systems and are enjoying the benefits.

Some of the contemporary practices being adopted to harvest rain are...

Check dams are constructed across small streams with high banks.

- The velocity of the runoff is slowed down and reduces soil erosion
- Soil moisture is improved by standing water and percolation recharges the aquifers

Contour trenches are dug on hill slopes and on barren wastelands for soil conservation through moisture retention. The trenches break the slope and reduce the velocity of surface runoff. The slopes can then be used for afforestation.

Bunds or small earthen barriers are built on sloping agricultural land. A long slope is converted into several short ones in order to:

- increase the standing time of rainwater where it falls, thereby allowing it to percolate into the soil
- minimise velocity, thereby reducing the erosion caused by runoff
- divert runoff for water harvesting purposes

Contour stonewalls made of stones are built across a hill slope to intercept the runoff. Soil moisture is thus conserved and erosion is reduced.

Sub-surface dams or groundwater dams are constructed to obstruct the natural flow of groundwater and store it. They are used where groundwater flows vary from very high flows after the rain to very low flows in the dry season. When water is stored underground , it reduces loss due to evaporation and is protected from contamination & parasites.

Percolation ponds like irrigation tanks have a bund to restrain the water flowing through a watershed, and a waste weir for the surplus to flow through. Water percolating into the bund is lead downstream by drains under the bund.

Artificial Glaciers

In Ladakh the water from melting glaciers reaches the fields only in summer. The people have hence channelised water through metal pipes to the shadow area of a mountain. This water freezes and being at a lower altitude & closer to the villages, it melts and reaches them in spring.

Nadis

In Rajasmand district of western Rajasthan, nadis (ponds) once served as the principal drinking water source. They received water from erratic, torrential rainfall. As large quantities of sediments were regularly deposited in them, they were destroyed by quick siltation. 20-30 nadis with a command area in excess of 500 hectares have now been constructed. Spillways have been added to prevent their damage by siltation. Afforestation of the drainage basin and construction of silt traps have helped further.

Polymer kunds

To deal with acute water scarcity in Churu District in Rajasthan, villagers are encouraged to build and renovate kunds (well-shaped tanks). The traditional materials like clay, silt, lime, ash and gravel used to make the catchment area of a kund are not completely impermeable, and hence some water is lost due to seepage. A water-based non-toxic polymer solution that permeates the highly porous sandy soil is now being used to increase runoff into the kunds. These materials reduce loss due to seepage of water into the soil.

Chauka System

In Jaipur, Rajasthan, degraded pastures have been dyked to form chaukas to harvest rain. Chaukas are rectangular plots arranged in a zig zag pattern and lie along a small gradient. 1.5 m high dykes are built along the three sides that lie towards the lower part of the gradient. Trees are also planted on the dykes to withstand rain. When it rains, water collects in the dyked lower half of the chauka. As the water level rises in one chauka, it spills into the next and so on. The entire pasture hence receives water. The chaukas do not get flooded. The excess water from the last chauka flows into a drain. The chauka system also promotes recharge of groundwater.

Jaldhar Model

In the Eastern Highlands, rainwater is harvested in a portion of the farmland. Pits are dug in each plot (which is bunded) to accumulate water. This pit also helps the subsurface flow of water to lower plots and improves the soil moisture of the area as a whole.

Networking of farm ponds

In Adihalli watershed, in Arasikere taluka of Hassan district in Karnataka, 330 farm ponds have been excavated. A series of ponds, constructed along contour lines and connected to one another, allow easy access to water & better soil retention.

Traditional water harvesting in India

Throughout India, several ingenious ways have been devised to catch and store rainwater for future use. They are known as traditional water harvesting systems. It is traditional water harvesting systems that have made life possible even in the Thar Desert.

The technology and engineering of the traditional water harvesting systems differed, depending on whether they were to provide drinking water or to be used for irrigation. Those meant for drinking water were generally smaller, sometimes covered and with steps leading down to the water. This ensured that people could only collect water manually in small quantities to meet their individual or family's needs. Irrigation systems on the other hand spread over large open and had a complicated network of pipes & channels for collecting & distributing.

Traditional water harvesting systems used in various eco-Zones of India is tabulated below

Table 11.2T_1: Traditional water harvesting systems used in various eco-Zones of India

Eco-zone	Traditional harvesting systems	Description	Found in
1. Trans-Himalayan Region	Zing	Tanks for collecting water from melted ice	Ladakh
2. Western Himalayas	Kul	Water channels in mountain areas	Jammu, Himachal Pradesh
	Naula	Small ponds	Uttaranchal
	Kuhl	Headwall across a ravine to divert water from a natural stream for irrigation	Himachal Pradesh
	Khatri	Chambers carved in hard rock for storing water	Himachal Pradesh
3. Eastern Himalayas	Apatani	Terraced plots connected by inlet and outlet channels	Arunachal Pradesh
4. Northeastern Hill Ranges	Zabo	Impounding runoff	Nagaland
	Cheo-oziihi	Channels from rivers	Nagaland
	Bamboo drip irrigation	Water from streams in the hills is brought to the plains via bamboo pipes for drip irrigation	Meghalaya
5. Brahmaputra Valley	Dongs	Ponds	Assam
	Dungs/jampois	Small irrigation canals kinking rice fields and a stream	W.Bengal
6. Ingo-Gangetic Plain	Ahar-pynes	Embanked catchment basin and channels	S.Bihar
	Bengal's inundation Channels	Inundation canals	W.Bengal
	Dighis	Small square of circular reservoir fed by canals from rivers	Delhi
	Baolis	Stepwells	Delhi

Eco-zone	Traditional harvesting systems	Description	Found in
7. Thar Desert	Kunds/kundis	Underground storage	W. Rajasthan
	Kuis/beris	Deep pits near tanks	W. Rajasthan
	Jhalaras	Tank	Rajasthan, Gujarat
	Nadi	Village ponds	Jodhpur, Rajasthan
	Tankas	Underground tank	Bikaner, Rajasthan
	Khadins	Embankment across lower hill slopes	Jaisalmer, W. Rajasthan
	Vav/Vavdi/Baoli/Bavadi	Stepwells	Gujarat, Rajasthan
	Virdas	Shallow wells	Rann of Kutch, Gujarat
	Paar	Area where water has percolated, accessed by kuis	-
8. Central Highlands	Talab/Bandhis	Reservoirs	Bundelkhand, Madhya Pradesh
	Saza Kuva	Open well	Mewar, E. Rajasthan
	Johads	Earthen check dams	Alwar district, Rajasthan
	Naada/bandh	Stone check dam	Mewar, Thar desert
	Pat	Diversion bund across stream	Jhabua district, Madhya Pradesh
	Rapat	Percolation tank	Rajasthan
	Chandela tank	Tank	Rajasthan
9. Eastern Highlands	Katas/Mundas/Bandhas	Earthen embankments across drainage lines	Orissa & Madhya Pradesh
10. Deccan Plateau	Cheruvu	Resrvoirs to store runoff	Chitoor, Cuddapah districts of Andhra Prades
	Kohli tanks	Tanks	Maharashtra
	Bhandaras	Check dams	Maharashtra
	Phad	Check dams and canals	North western
			Maharastra

Eco-zone	Traditional harvesting systems	Description	Found in
	Kere	Series of tanks	Central Karnataka
	Ramtek Model	Intricate network of groundwater and surface water bodies, connected through surface and underground canals	Ramtek, Maharashtra
11. Western Ghats	Surangam	Horizontal well	Kasasrgode, Kerala
12. Western Coastal Plains	Virdas	Shallow wells	Rann of Kutch, Gujarat
13. Eastern Ghats	Korambu	Temporary wall of brushwood, grass and mud laid across channels to raise the level of water	Kerala
14. Eastern Coastal Plains	Yeri	Tank	Tamilnadu
15. The Islands	Jackwells	Bamboo pipes are used to lead water into shallow pits	Great Nicobar Island

Quantity of water that can be harvested

The following is an illustrative theoretical calculation that highlights the enormous potential for rainwater harvesting. Consider a building with a flat terrace area of 100 m^2. The average annual rainfall in Chennai is approximately 1100 mm. In simple terms, this means that if the terrace floor is assumed to be impermeable, and all the rain that falls on it is retained without evaporation, then, in one year, there will be rainwater on the terrace floor to a height of 1100 mm.

Area of plot = 100 m^2

Height of rainfall = 1.1 m (1100 mm)

Volume of rainfall over the plot = Area of plot x Height of rainfall

100 m^2 x 1.1 m = 110 m^3(1,10,000 litres).

Assuming that only 60% of the total rainfall is effectively harvested, volume of water harvested = 66,000 litres

This volume is about four times the annual drinking water requirement of a five member family. The average daily water requirement per person is 10 litres according to IS 1172:

Rainwater harvesting Methods

The main aim of water harvesting is to conserve rainwater in the areas where it falls so that it may satisfy local needs. Harvesting rainwater and recharging the water table can be achieved through the following methods.

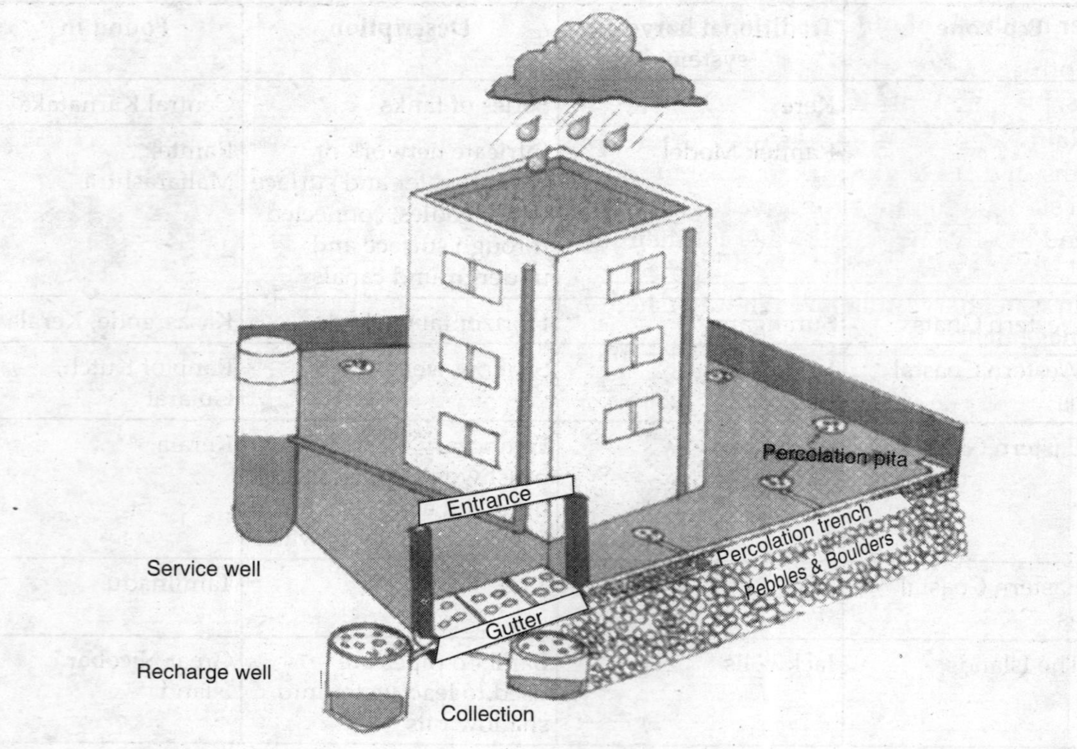

Fig.11.2F$_1$: Scheme of rain water harvesting

By using the existing well

The run-off water from roof tops can be led into the existing well through pipes and a small settling pit to filter the turbidity and other pollutants. In this cost effective process we not only conserve the precious rainwater but also help to increase the local ground water table. Even an abandoned well can be used for this purpose.

Through percolation pits

Percolation pits (1m x 1m x 3m) may be dug a little away from the building. The pit is then filled with brick jelly/pebbles followed by river sand for the purpose of better percolation. The top layer of sand may be cleaned and replaced once in two years to remove the settled silt for improving the percolation.

Decentralized percolation trenches

This method can be adopted by those who reside in houses with large open areas. Run-off water from the roof top can be diverted into bare soil/ garden in the premises. Apart from this a longitudinal trench of 1.0 m depth and 0.5 m width may be dug at the periphery of the plot and filled with pebble stones & sand in order to store the excess run-off during rainy season that will eventually percolate into the ground.

Other methods of utilizing rain water

Run-off should be diverted into suitably designed structures near pavements, parking lots, parks, play grounds, etc.

Rainwater can also be stored in underground storage tanks (water sumps) for a few months and used directly for washing, flushing and other domestic purposes. Industries, multi storied buildings, various offices, etc., can implement this cost effective method, as they depend mostly on ground water for their needs.

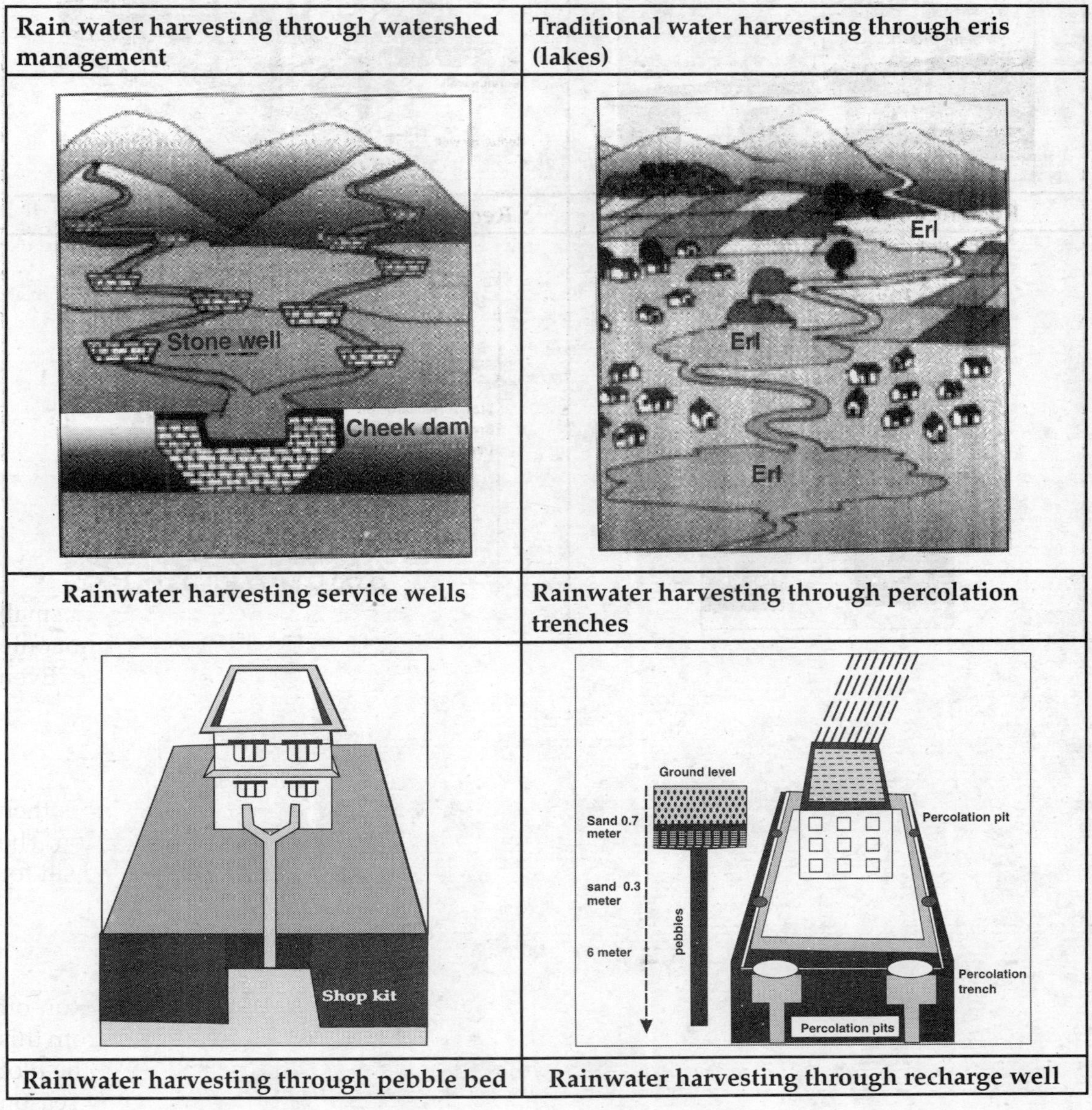

Rain water harvesting through watershed management | **Traditional water harvesting through eris (lakes)**

Rainwater harvesting service wells | **Rainwater harvesting through percolation trenches**

Rainwater harvesting through pebble bed | **Rainwater harvesting through recharge well**

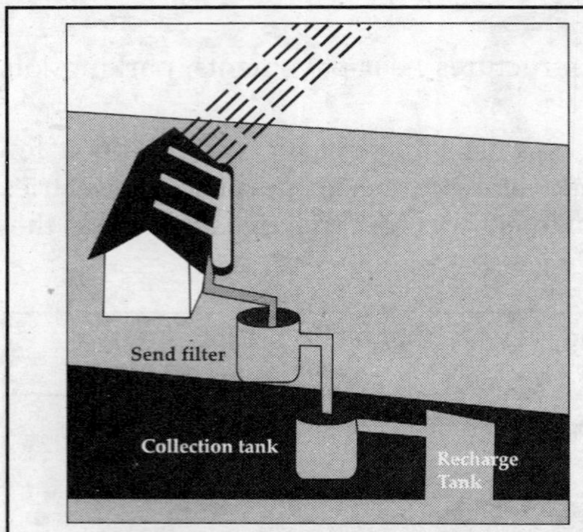

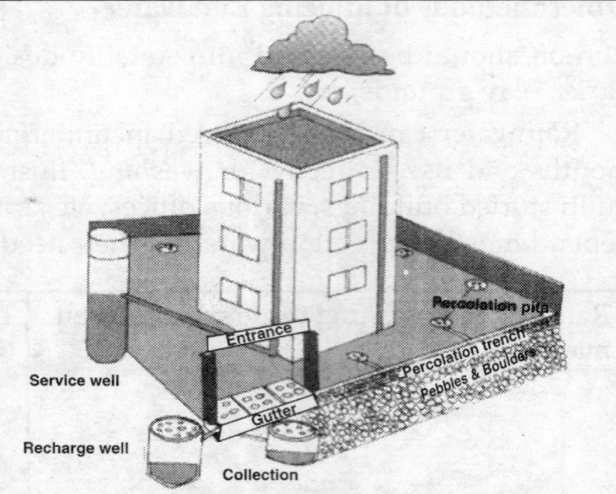

Recharge water through soak pits

Recharge ground water by pebble beds

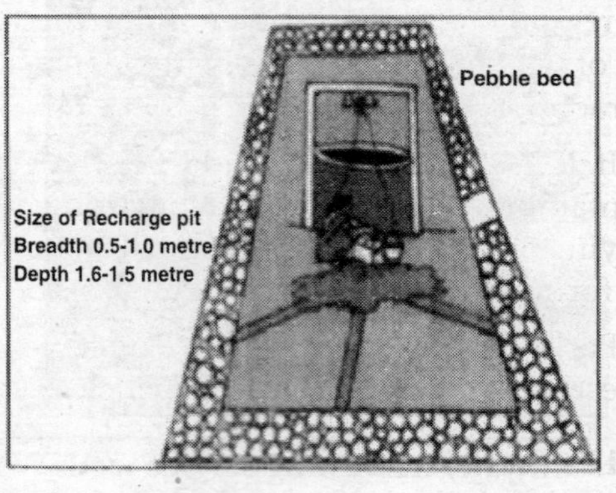

12

Chapter

Water Pollution Control Acts & Rules

WATER (PREVENTION AND CONTROL OF POLLUTION) ACT, 1974

The regulatory regimes for environmental conservation comprise a legislative framework, and a set of regulatory institutions. Inadequacies in each have resulted in accelerated environmental degradation.

It is expedient to provide for the prevention and control of water pollution and the maintaining or restoring of wholesomeness of water, for the establishment, with a view to carrying out the purposes aforesaid, of Boards for the prevention and control of water pollution and for conferring on and assigning to such Boards powers and functions relating thereto.

Parliament has no power to make laws for the States with respect to any of the matters aforesaid except as provided in articles 249 and 250 of the Constitution;

PRELIMINARY

Short title, application and commencement.

(1) Short title, application and commencement. This Act may be called the Water (Prevention and Control of s Pollution) Act, 1974.

(2) It applies in the first instance to the whole of the States of Assam, Bihar, Gujarat, Haryana, Himachal Pradesh, Jammu and Kashmir, Karnataka, Kerala, Madhya Pradesh, Rajasthan, Tripura and West Bengal and the Union territories; and it shall apply to such other State which adopts this Act by resolution passed in that behalf under clause (1) of article 252 of the Constitution.

(3) It shall come into force, at once in the States of Bihar, Gujarat, Haryana, Himachal Pradesh, Jammu and Kashmir, Karnataka, Kerala, Madhya Pradesh, Rajasthan, Tripura and West Bengal and in the Union territories, and in any other State which adopts this Act under clause (1) of article 252 of the Constitution on the date of such adoption and any reference in this Act to the commencement of this Act shall, in relation to any State or Union territory, mean the date on which this Act comes into force in such State or Union territory.

Definitions

In this Act, unless the context otherwise requires,-

(a) "Board" means the Central Board or State Board; " Central Board" means the Central Pollution Control Board Constituted under section 3;

(b) "members" means a member of a Board and includes the chairman thereof;

(c) "occupier", in relation to any factory or premises, means the person who has control over the affairs of the factory or the premises, and includes, in relation to any substance, the person in possession of the substance. "outlet" includes any conduit pipe or channel, open or closed carrying sewage or trade effluent or any other holding arrangement which causes, or is likely to cause, pollution;

(d) "pollution" means such contamination of water or such alteration of the physical, chemical or biological properties of water or such discharge of any sewage or trade effluent or of any other liquid, gaseous or solid substance into water (whether directly or indirectly) as may, or is likely to, create a nuisance or render such water harmful or injurious to public health or safety, or to domestic, commercial, industrial, agricultural or other legitimate uses, or to the life and health of animals or plants or of aquatic organisers;

(e) "prescribed" means prescribed by rules made under this Act by the Central Government or, as the case may be, the State Government,

(f) "sewage effluent" means effluent from any sewerage system or sewage disposal works and includes sullage from open drains; "sewer" means any conduit pipe or channel, open or closed, carrying sewage or trade effluent;

(g) "State Board" means a State Pollution Control Board constituted under section 4";

(h) "State Government" in relation to a Union territory means the Administrator thereof appointed under article 239 of the Constitution;

(i) "stream" includes-

(i) river;

(ii) Water course (whether flowing or for the time being dry);

(iii) inland water (whether natural or artificial);

(iv) subterranean waters;

(v) sea or tidal waters to such extent or, as the case may be, to such point as the State Government may, by notification in the Official Gazette, specify in this behalf ;

(j) "trade effluent" includes any liquid, gaseous or solid substance which is discharged from any premises used for carrying on any "Industry, operation or process, or treatment and disposal system" other than domestic sewage.

THE CENTRAL AND STATE BOARDS FOR PREVENTION AND CONTROL OF WATER POLLUTION

Constitutions of Central Board.

(1) The Central Government shall, with effect from such date (being a date not later than six months of the commencement of this Act in the States of Assam, Bihar, Gujarat, Haryana, Himachal Pradesh, Jammu and Kashmir, Karnataka, Kerala, Madhya Pradesh, Rajasthan, Tripura and West Bengal and in the Union territories) as it may,

by notification in the Official Gazette, appoint, constitute a Central Board to be called the Central Pollution Control Board powers conferred on and perform the functions assigned to that Board under this Act.

(2) The Central Board shall consist of the following members, namely

(a) a full- time chairman, being a person having special knowledge or practical experience in respect of matters relating to environmental protections or a person having knowledge and experience in administering institutions dealing with the matters aforesaid, to be nominated by the Central Government;

(b) such number of officials, not exceeding five, to be nominated by the Central Government to represent that Government;

(c) such number of persons, not exceeding five, to be nominated by the Central Government, from amongst the members of the State Boards, of whom not exceeding two shall be from those referred to in clause (c) of sub-section (2) of section 4.

(d) such number of non-officials, not exceeding three, to be nominated by the Central Government, to represent the interests of agriculture, fishery or industry or trade or any other interest which, in the opinion of the Central Government, ought to be represented.

(e) two persons to represent the companies or corporations owned, controlled or managed by the Central Government, to be nominated by that Government.

(f) a full-time member-secretary, possessing qualifications, knowledge and experience of scientific, engineering or management aspects of pollution control to be appointed by the Central Government.

(3) The Central Board shall be a body corporate with the name aforesaid having perpetual succession and a common seal with power, subject to the provisions of this Act, to acquire, hold and dispose of property and to contact and may by the aforesaid name, sue or be used.

Constitutions of State Board.

(4) The State Government shall, with effect from such date as it may by notification in the Official Gazette, appoint, constitute a State Pollution Control Board under such name as may be specified in the notification, to exercise the powers conferred on and perform the functions assigned to that Board under this Act.

(5) A State Board shall consist of the following members, namely:-

(a) A chairman, being a person having special knowledge or practical experience in respect of matters relating to environmental protection or a person having knowledge and experience in administering institutions dealing with the matters aforesaid, to be nominated by the State Government Provided that the chairman may be either whole-time or part time as the State Government may think fit;

(b) such number of officials, not exceeding five, to be nominated by the State Government to represent that Government;

(c) such number of persons, not exceeding five, to be nominated by the State Government from amongst the members of the local authorities functioning within the State;

(d) such number of non-officials, not exceeding three, to be nominated by the State Government to represent the interests of agriculture, fishery or industry or trade or any other interest which, in the opinion of the State Government, ought to be represented.

(e) two persons to represent the companies or corporations owned, controlled or managed by the State Government, to be nominated by that Government.

(f) "a full-time member-secretary, possessing qualifications, knowledge and experience of scientific, engineering or management aspects of pollution control, to be appointed by the State Government."

(6) Every State Board shall be a body corporate with the name specified by the State Government in the notification under sub-section (1), having perpetual succession and a common seal with power, subject to the provisions of this Act, to acquire, hold and dispose of property and to contract, and may, by the said name, use or be used.

(7) Notwithstanding anything contained in this section, no State Board shall be constituted for a Union territory and in relation to a Union territory, the Central Board shall exercise the powers and perform the functions of a State Board for that Union territory.

Provided that in relation to any Union territory the Central Board may delegate all or any of its powers and functions under this sub-section to such person or body of persons as the Central Government may specify.

Terms and conditions of service of members

(1) Save as otherwise provided by or under this Act, a member of a Board, other than a member-secretary, shall hold office for a term of three years from the date of is nomination:

Provided that a member shall, notwithstanding the expiration of his term, continue to hold office until his successor enters upon his office.

(2) The term of office of a member of a Board nominated under clause (b) or clause (e) of sub-section (2) of section 3 or clause or. 176. cause (e) of sub-section (2) of section 4 shall come to an end as soon as he ceases to hold the office under the Central Government or the State Government or, as the case may be, the company or corporation owned, controlled or managed by the Central Government or the State Government, by virtue of which he was nominated.

(3) The Central Government or, as the case may be, the State Government may, if it thinks fit, remove any member of a Board before the expiry of his term of office, after giving him a reasonable opportunity of showing cause against the same.

(4) A member of a Board, other than the member-secretary, may at any time resign his office by writing under his hand addressed

(a) in the case of the chairman, to the Central Government or, as the case may be, the State Government

(b) in any other case, to the chairman of the Board; and the seat of the chairman or such other member shall there upon become vacant.

(5) A member of a Board, other than the member-secretary, shall be deemed to have vacated his seat if he is absent without reason, sufficient in the opinion of the Board, from three consecutive meetings of the Board, or where he is nominated under clause (c) or clause (e) of sub-section (2) of section 3 or under clause (c) or clause (e) of sub-section (2) of section 4, if he ceases to be a member of the State Board or of the local authority or, as the case may be, of the company or corporation owned, controlled or managed

by the Central Government or the State Government and such vacation of seat shall, in either case, take effect from such date as the Central Government or, as the case may be, the State Government may, by notification in the Official Gazette,specify.

(6) A casual vacancy in a Board shall be filled by a fresh nomination and the person nominated to fill the vacancy shall hold office only for the remainder of the term for which the member in whose place he was nominated.

(7) A member of a Board shall be eligible for renomination.

(8) The other terms and conditions of service of a member of a Board, other than the chairman and member-secretary, shall be such as may be prescribed.

(9) The other terms and conditions of service of the chairman shall be, such as may be prescribed.

Disqualifications

(1) No person shall be a member of a Board, who;

(a) is, or at any time has been adjudged insolvent or has suspended payment of his debts or has compounded with his creditors, or

(b) is of unsound mind and stands so declared by a competent court, or

(c) is, or has been, convicted of an offence which, in the opinion of the Central Government or, as the case may be, of the State Government. Involves moral turpitude, or

(d) is, or at any time has been, convicted of an offence under this Act,

(e) is, or has directly or indirectly by himself or by any partner, any share or interest in any firm or company carrying on the business of manufacture, sale or hire of machinery, plant, equipment, apparatus or fittings for the treatment of a sewage or trade effluents, or

(f) is a director or a secretary, manager or other salaried officer or employee of any company or firm having any contract with the Board, or with the Government constituting the Board, or with a local authority in the State, or with a company or corporation owned, controlled or managed by the Government, for the carrying out of sewerage schemes or for the installation of plants for the treatment of sewage or trade effluents,

(g) has so abused, in the opinion of the Central Government or as the case may be, of the State Government, his position as a member, as to render his continuance on the Board detrimental to the interest of the general public.

(2) No order of removal shall be made by the Central Government or the State Government, as the case may be, under this section unless the member concerned has been given a reasonable opportunity of showing cause against the same.

(3) Notwithstanding anything contained in sub-sections (1) and (7) of section 5, a member who has been removed under this section shall not be eligible for renomination as a member.

Vacation of seats by members

If a member of a Board becomes subject to any of the disqualifications specified in section 6, his seat shall become vacant.

Meetings of Board

A Board shall meet at least once in every three months and shall observe such rules of procedure in regard to the transaction of business at its meetings as may be prescribed:

Provided that if, in the opinion of the chairman, any business of an urgent nature is to be transacted, he may convene a meeting of the Board at such time as he thinks fit for the aforesaid purpose.

Constitutions of committees

(1) A Board may constitute as many committees consisting wholly of members or wholly of other persons or partly of members and partly of other persons, and for such purpose or purposes as it may think fit.

(2) A committee constituted under this section shall meet at such time and at such place, and shall observe such rules of procedure in regard to the transaction of business at its meetings, as may be prescribed.

(3) The members of a committee (other than the members of the Board) shall be paid such fees and allowances, for attending it, meetings and for attending to any other work of the Board as may be prescribed.

Temporary association of persons with Board for particular purpose

(1) A Board may associate with itself in such manner, and for such purposes, as may be prescribed any person whose assistance or advice it may desire to obtain in performing any of its functions under this Act.

(2) A person associated with the Board under sub-section (1) for any purpose shall have a right to take part in the discussions of the Board relevant to that purpose, but shall not have a right to vote at a meeting of the Board, and shall not be a member for any other purpose.

(3) A person associated with the Board under sub-section (1) for any purpose shall be paid such fees and allowances for attending its meetings and for attending to any other work of the Board, as may be prescribed.

Vacancy in Board not invalidate acts or proceedings

No act or proceeding of a Board or any committee thereof shall be called in question on the ground merely of the existence of any vacancy in, or any defect in the constitution of, the Board or such committee, as the case may be.

A Delegation of powers to chairman

The chairman of a Board shall exercise such powers and perform such duties as may be prescribed or as may, from time to time, be delegated to him by the Board.

Member secretary and officers and other employees of Board

(1) The terms and conditions of service of the member secretary shall be such as may be prescribed.

(2) The member-secretary shall exercise such powers and perform such duties as may be prescribed or as may, from time to time, be delegated to him by the Board or its chairman.

(3) Subject to such rules as may be made by the Central Government or, as the case may be, the State Government in this behalf, a Board may appoint such officers and employees as it considers necessary for the efficient performance of its functions.

(3A) The method of recruitment and the terms and conditions of service (including the scales of pay) of the officers (other than the member secretary) and other employees of the Central Board or a State Board shall be such as may be determined by regulations made by the Central Board or, as the case may be, by-the State Board:

Provided that no regulation made under this sub-section shall take effect unless,-

(a) in the case of a regulation made by the Central Board, it is approved by the Central Government; and

(b) in the case of a regulation made by a State Board, it is approved by the State Government. (3B) The Board may, by general or special order, and subject to such conditions and limitations, if any, as may be specified in the order, delegate to any officer of the Board such of its powers and functions under this Act as it may deem necessary.

(4) Subject to such conditions as may be prescribed, a Board may from time to time appoint any qualified person to be a consulting engineer to the Board and pay him such salaries and allowances and subject him to such other terms and conditions of service as it thinks fit.

JOINT BOARDS

Constitutions of Joint Boards

(1) Notwithstanding anything contained in this Act, an agreement may be entered into-

(a) by two or more Governments of contiguous States, or

(b) by the Central Government (in respect of one or more Union territories) and one or more Governments of States contiguous to such Union territory or Union territories, to be in force for such period and to be subject to renewal for such further period, if any, as may be specified in the agreement to provide for the constitution of a Joint Board,-

(i) in a case referred to in clause (a), for all the participating States, and

(ii) in a case referred to in clause (b), for the participating Union territory or Union territories and the State or States.

(2) An agreement under this section may

(a) provide, in a case referred to in clause (a) of sub- section (1), for the apportionment between the participating States and in a case referred to in clause (b) of that sub- section, for the apportionments between the Central Government and the participating State Government or State Governments, of the expenditure in connection with the Joint Board; (b) determine, in a case referred to in clause (a) of sub- section (1), which of the participating State Governments and in a case referred to in clause (b) of that sub-section, whether the Central Government or the participating State Government (if there are more than one participating State, also which of the participating State Governments) shall exercise and perform the several powers and functions of the State Government under this Act and the references in this Act to the State Government shall be construed accordingly;

(c) provide for consultation, in a case referred to in clause (a) of sub-section (1), between the participating State Governments and in a case referred to in clause (b) of that sub-section, between the Central Government and the participating State Government or State Governments either generally or with reference to particular matters arising under this Act ;

(d) make such incidental and ancillary provisions, not inconsistent with this Act, as may be deemed necessary or expedient for giving effect to the agreement.

(3) An agreement under this section shall be published, in a case referred to in clause (a) of sub-section (1), in the Official Gazette of the participating States and in a case referred to in clause (b) of that sub-section, in the Official Gazette of the participating Union territory or Union territories and the participating State or States.

Composition of Joint Boards

(1) A Joint Board constituted in pursuance of an agreement entered into under clause (a) of sub-section (1) of section 13 shall consist of the following members, namely:-

(a) a full time chairman, being a person having special knowledge or practical experience in respect of matters relating to environmental protection] or a person having knowledge and experience in administering institutions dealing with the matters aforesaid, to be nominated by the Central Government;

(b) two officials from each of the participating States to be nominated by the concerned participating State Government to represent that Government;

(c) one person to be nominated by each of the participating State Governments from amongst the members of the local authorities functioning within the State concerned;

(d) one non-official to be nominated by each of the participating State Governments to represent the interests of agriculture, fishery or industry or trade in the State concerned or any other interest which, in the opinion of the participating State Government, is to be represented ;

(e) two persons to be nominated by the Central Government to represent the companies or corporations owned, controlled or managed by the participating State Governments;

(f) a full time member secretary, possessing qualifications, knowledge and experience of scientific, engineering or management aspects of pollution control, to be appointed by the Central Government.

(2) A Joint Board constituted in pursuance of an agreement entered into under clause (b) of sub-section (1) of section 13 shall consist of the following members, namely:-

(a) a full- time chairman, being a person having special knowledge or practical experience in respect of matters relating to environmental protection] or a person having knowledge and experience in administering institutions dealing with the matters aforesaid, to be nominated by the Central Government;

(b) two officials to be nominated by the Central Government from the participating Union territory or each of the participating Union territories as the case may be, and two officials to be nominated, from the participating State or each of the participating States, as the case may be, by the concerned participating State Government;

(c) one person to be nominated by the Central Government from amongst the members of the local authorities functioning within the participating Union territory or each of the participating Union territories, as the case may be, and one person to be nominated, from amongst the members of the local authorities functioning within the participating State or each of the participating States, as the case may be, by the concerned participating State Government;

(d) one non-official to be nominated by the Central Government and one person to be nominated by the participating State Government or State Governments to represent the interests of agriculture, fishery or industry or trade in the Union territory or in each of the Union territories or the State or in each of the States, as the case may be, or any other interest which in the opinion of the Central Government or, as the case may be, of the State Government is to be represented;

(e) two persons to be nominated by the Central Government to represent the companies or corporations owned, controlled or managed by the Central Government and situate in the participating Union territory or territories and two persons to be nominated by the Central Government to represent the companies or corporations owned, controlled or managed by the participating State Governments;

(f) a full-time member-secretary, possessing qualifications knowledge and experience of scientific, engineering or management aspects of pollution control, to be appointed by the Central Government.

(3) When a Joint Board is constituted in pursuance of an agreement under clause (b) of sub-section (1) of section 13, the provisions of sub-section (4) of section 4 shall cease to apply in relation to the Union territory for which the Joint Board is tituted.

(4) Subject to the provisions of sub-section (3), the provisions of subsection (3) of section 4 and sections 5 to 12 (inclusive) shall apply in relation to the Joint Board and its member secretary as they apply In relation to a State Board and its member-secretary.

(5) Any reference in this Act to the State Board shall, unless the context otherwise requires, be construed as including a Joint Board.

Special provision relating to giving of directions

Notwithstanding anything contained in this Act where any Joint Board is constituted under Section – 13.

(a) The Government of the State for which the Joint Board is constituted shall be competent to give any direction under this Act only in cases where such direction relates to matter within the exclusive territorial jurisdiction of the State;

(b) The Central Government alone shall be competent to give any direction under this Act where such direction relates to a matter within the territorial jurisdiction of two or more States or pertaining to a Union territory. A Board shall meet at least once in every three months and shall observe such rules of procedure in regard to the transaction of business at its meetings as may be prescribed:

Provided that if, in the opinion of the chairman, any business of an urgent nature is to be transacted, he may convene a meeting of the Board at such time as he thinks fit for the aforesaid purpose.

POWERS AND FUNCTIONS OF BOARDS

Functions of central board

(1) Subject to the provisions of this Act, the main function of the Central Board shall be to promote cleanliness of streams and wells in different areas of the States.

(2) In particular and without prejudice to the generality of the foregoing function, the Central Board may perform all or any of the following functions, namely:-

(a) advise the Central Government on any matter concerning the prevention and control of water pollution.

(b) co-ordinate the activities of the State Boards and resolve disputes among them;

(c) provide technical assistance and guidance to the State Boards, carry out and sponsor investigations and research relating to problems of water pollution and prevention, control or abatement of water pollution

(d) plan and organise the training of persons engaged or to be engaged in programs for the prevention, control or abatement of water pollution on such terms and conditions as the Central Board may specify.

(e) organise through mass media a comprehensive programme regarding the prevention and control of water pollution.

(f) collect, compile and publish technical and statistical data relating to water pollution and the measures devised for its effective prevention and control and prepare manuals, codes or guides relating to treatment and disposal of sewage and trade effluents and disseminate information connected therewith.

(g) lay down, modify or annul, in consultation with the State Government concerned, the standards for a stream or well: Provided that different standards may be laid down for the same stream or well or for different streams or wells, having regard to the quality of water, flow characteristics of the stream or well and the nature of the use of the water in such stream or well or streams or wells.

(h) plan and cause to be executed a nation-wide programme for the prevention, control or abatement of water pollution.

(i) perform such other functions as may be prescribed.

(3) The Board may establish or recognise a laboratory or laboratories to enable the Board to perform its functions under this section efficiently, including the analysis of samples of water from any stream or well or of samples of any sewage or trade effluents.

Functions of state board

(1) Subject to the provisions of this Act, the functions of a State Board shall be –

(a) to plan a comprehensive programme for the prevention, control or abatement of pollution of streams and wells in the State and to secure the execution thereof;

(b) to advise the State Government on any matter concerning the prevention, control or abatement of water pollution

(c) to collect and disseminate information relating to water pollution and the prevention, control or abatement thereof

(d) to encourage, conduct and participate in investigations and research relating to problems of water pollution and prevention, control or abatement of water pollution;

(e) to collaborate with the Central Board in organising the training of persons engaged or to be engaged in programmes relating to prevention, control or abatement of water pollution and to organise mass education programmes relating thereto

(f) to inspect sewage or trade effluents, works and plants for the treatment or sewage and trade effluents and to review plans, specifications or other data relating to plants set up for the treatment of water, works for the purification thereof and the system for the disposal of sewage or trade effluents or in connection with the grant of any consent as required by this Act

(g) lay down, modify or annul effluent standards for the sewage and trade effluents and for the quality of receiving waters (not being water in an inter-State stream) resulting from the discharge of effluents and to classify waters of the State

(h) to evolve economical and reliable methods of treatment of sewage and trade effluents, having regard to the peculiar conditions of soils, climate and water resources of different regions and more specially the prevailing flow characteristics of water in streams and wells which render it impossible to attain even the minimum degree of dilution

(i) to evolve methods of utilisation of sewage and suitable trade effluents in agriculture

(j) to evolve efficient methods of disposal of sewage and trade effluents on land, as are necessary on account of the predominant conditions of scant stream flows that do not provide for major part of the year the minimum degree of dilution

(k) to lay down standards of treatment of sewage and trade effluents to be discharged into any particular stream taking into account the minimum fair weather dilution available in that stream and the tolerance limits of pollution permissible in the water of the stream, after the discharge of such effluents

(l) to make, vary or revoke any order—

(i) for the prevention, control or abatement of discharge of waste into streams or wells

(ii) requiring any person concerned to construct new systems for the disposal of sewage and trade effluents or to modify, alter or extend any such existing system or to adopt such remedial measures as are necessary to prevent control or abate water pollution

(m) to lay down effluent standards to be complied with by persons while causing discharge of sewage or sullage or both and to lay down, modify or annul effluent standards for the sewage and trade effluents

(n) to advice the State Government with respect to the location of any industry the carrying on of which is likely to pollute a stream or well

(o) to perform such other functions as may be prescribed or as may, from time to time be entrusted to it by the Central Board or the State Government.

(2) The Board may establish or recognise a laboratory or laboratories to enable the Board to perform its functions under this section efficiently, including the analysis of samples of water from any stream or well or of samples of any sewage or trade effluents.

Powers to give directions

(1) In the performance of its functions under this Act --

(a) the Central Board shall be bound by such directions in writing the Central Government may give to it; and

(b) every State Board shall be bound by such directions in writing as the Central Government or the State Government may give to it:

Provided that where a direction given by the State Government is inconsistent with the direction given by the Central Board, the matter shall be referred to the Central Government for its decision.

(2) Where the Central Government is of the opinion that and State Board has defaulted in complying with any directions given by the Central Government under sub-section (1) and as a result of such default a grave emergency has arisen and it is necessary or expedient so to do in the public interest, it may, by order, direct the Central Board to perform any of the functions of the State Board in relation to such area for such period and for such purposes, as may be specified in the order.

(3) Where the Central Board performs any of the functions of the State Board in pursuance of a direction under sub-section (2), the expenses, if any, incurred by the Central Board with respect to performance of such functions may, if the State Board is empowered to recover such expenses, be recovered by the Central Board with interest (at such reasonable rate as the Central Government may, by order, fix) from the date when a demand for such expenses is made until it is paid from the person or persons concerned as arrears of land revenue or of public demand.

(4) For the removal of doubts, it is hereby declared that any directions to perform the functions of any State Board given under sub-section (2) in respect of any area would not preclude the State Board from performing such functions in any other area in the State or any of its other functions in that area.

PREVENTION AND CONTROL OF WATER POLLUTION

Power of State Government to Restrict the Application of the Act to Certain Areas

(1) Notwithstanding contained in this Act, if the State Government, after consultation with, or on the recommendation of, the State Board, is of opinion that the provisions of this Act need not apply to the entire State, it may, by notification in the Official Gazette, restrict the application of this Act to such area or areas as may be declared therein as water pollution, prevention and control area or areas and thereupon the provisions of this Act shall apply only to such area or areas.

(2) Each water pollution, prevention and control area may be declared either by reference to a map or by reference to the line of any watershed or the boundary of any district or partly by one method and partly by another.

(3) The State Government may, by notification in the Official Gazette-

(a) alter any water pollution prevention and control area whether by way of extension or reduction; or

(b) define a new water pollution, prevention and control area in which may be merged one or more water pollution, prevention and control areas, or any part or parts thereof.

Power to Obtain Information

(1) For the purpose of enabling a State Board to perform the function conferred on it by or under this Act, the State Board or any officer empowered by it in that behalf, may make surveys of any area and gauge and keep records of the flow or volume and other characteristics of an stream or well in such area, and may take steps for the measurement and recording of the rainfall in such area or any part thereof and for the installation and maintenance for those purposes of gauges or other apparatus and works connected therewith, and carry out stream surveys and may take such other steps as may be necessary in order to obtain any information required for the purposes aforesaid.

(2) A State Board may give directions requiring any person who in its opinion is abstracting water from any such stream or well in the area in quantities which are substantial in relation to the flow or volume of that stream or well or is discharging sewage or trade effluent into any such stream or well, to give such information as to the abstraction or the discharge at such times and in such form as may be specified in the directions.

(3) Without prejudice to the provisions of sub-section (2), a State Board may, with a view to preventing or controlling pollution of water, give directions requiring any person in charge of any establishment where any industry, operation or process, or treatment and disposal system is carried on, to furnish to it information regarding the construction, installation or operation of such establishment or of any disposal system or of any extension or addition thereto in such establishment and such other particulars as may be prescribed.

Power to Take Samples of Effluents and Procedure to be Followed In Connection Therewith

(1) A State Board or any officer empowered by it in this behalf shall have power to take for the purpose of analysis samples of water from any stream or well or samples of any sewage of trade effluent which is passing from any plant or vessel or from or over any place into any such stream or well.

(2) The result of any analysis of a sample of any sewage or trade effluent taken under sub- section (1) shall not be admissible in evidence in an legal proceeding unless the provisions of sub-sections (3), (4) and (5) are complied with.

(3) Subject to the provisions of sub-sections (4) and (5), when a sample (composite or otherwise as may be warranted by the process used) of any sewage or trade effluent is taken for analysis under sub-section (1), the person taking the sample shall --

- (a) serve on the person in charge of, or having control over, the plant or vessel or in occupation of the place (which person is hereinafter referred to as the occupier) or any agent of such occupier, a notice, then and there in such form as may be prescribed of his intention to have it so analysed;
- (b) in the presence of the occupier or his agent, divided the sample into two parts;
- (c) cause each part to be placed in a container which shall be marked and sealed and shall also be signed both by the person taking the sample and the occupier or his agent;
- (d) send one container forthwith,--
  - (i) in a case where such sample is taken from any area situated in a Union territory, to the laboratory established or recognised by the Central Board under section; and

(ii) in any other case, to the laboratory established or recognised by the State Board under section 17;

(e) on the request of the occupier or his agent, send the second container --

(i) in a case where such sample is taken from any area situated in a Union territory, to the laboratory established or specified under sub-section (1) of section 51; and

(ii) in an other case, to the laboratory established or specified under sub- section (1) of section 52.

(4) When a sample of any sewage of trade effluent is taken for analysis under sub-section (1) and the person taking the sample serves on the occupier or his agent, a notice under clause (a) of sub-section (3) and the occupier or his agent willfully absents himself, then-

(a) the sample so taken shall be placed in a container which shall be marked and sealed and shall also be signed by the person taking the sample and the same shall be sent forthwith by such person for analysis to the laboratory referred to in sub-clause (i) or sub-clause (ii), as the case may be, of clause (e) of sub-section (3) and such person shall inform the Government analyst appointed under sub-section (1) or sub- section (2), as the case may be, of section 53, in writing about the wilful absence of the occupier or his agent; and

(b) the cost incurred in getting such sample analysed shall be payable by the occupier or his agent and in case of default of such payment, the same shall be recoverable from the occupier or his agent, as the case may be, as an arrear of land revenue or of public demand:

Provided that no such recovery shall be made unless the occupier or, as the case may be, his agent has been given a reasonable opportunity of being heard in the matter.

(5) When a sample of any sewage or trade effluent is taken for analysis under sub-section (1) and the person taking the sample serves on the occupier or his agent a notice under clause (a) of sub-section (3) and the occupier or his agent who is present at the time of taking the sample does not make a request for dividing the sample into two parts as provided in clause

(b) of sub-section (3), then, the sample so taken shall be placed in a container which shall be marked and sealed and shall also be signed by the person taking the sample and the same shall be sent forthwith by such person for analysis to the laboratory referred to in sub-clause

(i) or sub-clause (ii), as the case may be, of clause (d) of sub-section (3).

Reports of Results of Analysis on Samples Taken Under Section 21

(1) Where a sample of any sewage or trade effluent has been sent for analysis to the laboratory established or recognised by the Central Board or, as the case may be, the State Board, the concerned Board analyst appointed under sub-section (3) of section 53 shall analyse the sample and submit a report in the prescribed form of the result of such analysis in triplicate to the Central Board or the State Board, as the case may be.

(2) On receipt of the report under sub-section (1), one copy of the report shall be sent by the Central Board or the State Board, as the case may be, to the occupier or his agent referred to in section 21, another copy shall be preserved for production before the

court in case any legal proceedings are taken against him and the other copy shall be kept by the concerned Board.

(3) Where a sample has been sent for analysis under clause (e) of sub-section (3) or sub-section (4) of section 21 to any laboratory mentioned therein, the Government analyst referred to in that sub-section shall analyse the sample and submit a report in the prescribed form of the result of the analysis in triplicate to the Central Board or, as the case may be, the State Board which shall comply with the provisions of sub-section (2).

(4) If there is any inconsistency or discrepancy between, or variation in the results of, the analysis carried out by the laboratory established or recognised by the Central Board or the State Board, as the case may be, and that of the laboratory established or specified under section 51 or section 52, as the case may be, the report of the latter shall prevail.

(5) Any cost incurred in getting any sample analysed at the request of the occupier or his agent shall be payable by such occupier or his agent and in case of default the same shall be recoverable from him as arrears of land revenue or of public demand.

Power of Entry and Inspection

(1) Subject to the provisions of this section, any person empowered by a State Board in this behalf shall have a right at any time to enter, with such assistance as he considers necessary, any place—

(a) for the purpose of performing any of the functions of the Board entrusted to him;

(b) for the purpose of determining whether and if so in what manner, any such functions are to be performed or whether any provisions of this Act or the rules made thereunder of an notice, order, direction or authorisation served, made, given, or granted under this Act is being or has been complied with;

(c) for the purpose of examining any plant, record, register, document or any other material object or for conducting a search of any place in which he has reason to believe that an offence under this Act or the rules made thereunder has been or is being or is about to be committed and for seizing any such plant, record, register, document or other material object, if he has reason to believe that it may furnish evidence of the commission of an offence punishable under this Act or the rules made thereunder:

Provided that the right to enter under this sub-section for the inspection of a well shall be exercised only at reasonable hours in a case where such well is situated in any premises used for residential purposes and the water thereof is used exclusively for domestic purposes.

(2) The provisions of the Code of Criminal Procedure, 1973, or, in relation to the State of Jammu and Kashmir, the provisions of any corresponding law in force in that State, shall, so far as may be, apply to an search or seizure under this section as they apply to any search or seizure made under the authority of a warrant issued under section 94 of the said Code, or, as the case may be, under the corresponding provisions of the said law.

Explanation – For the purposes of this section, "place" includes vessel.

Prohibition on use of stream or well for disposal of polluting matter, etc

(1) Subject to the provisions of this section—

(a) no person shall knowingly cause or permit any poisonous, noxious or polluting matter determined in accordance with such standards as may be laid down by the State Board to enter (whether directly or indirectly) into any stream or well or sewer or on land; or

(b) no person shall knowingly cause or permit to enter into any stream any other matter which may tend, either directly or in combination with similar matters, to impede the proper flow of the water of the stream in a manner leading or likely to lead to a substantial aggravation of pollution due to other causes or of its consequences.

(2) A person shall not be guilty of an offence under sub-section (1), by reason only of having done any of the following acts, namely;--

(a) constructing, improving a maintaining in or across or on the bank or bed of any stream any building, bridge, weir, dam, sluice, dock, pier, drain or sewer or other permanent works which he has a right to construct, improve or maintain;

(b) depositing any materials on the bank or in the bed of any stream for the purpose of reclaiming land or for supporting, repairing or protecting the bank or bed of such stream provided such materials are not capable of polluting such stream;

(c) putting into an stream any sand or gravel or other natural deposit which has flowed from or been deposited by the current of such stream;

(d) causing or permitting, with the consent of the State Board, the deposit accumulated in a well, pond or reservoir to enter into any stream.

(3) The State Government may, after consultation with, or on the recommendation of, the State Board, exempt, by notification in the Official Gazette, any person from the operation of sub-section (1) subject to such conditions, if any, as may be specified in the notification and any conditions so specified may by a like notification and be altered, varied or amended.

Restrictions on New Outlets and New Discharges

(1) Subject to the provisions of this section, no person shall, without the previous consent of the State Board,--

(a) establish or take any steps to establish any industry, operation or process, or any treatment and disposal system or an extension or addition thereto, which is likely to discharge sewage or trade effluent into a stream or well or sewer or on land (such discharge being hereafter in this section referred to as discharge of sewage); or

(b) bring into use any new or altered outlets for the discharge of sewage; or

(c) begin to make any new discharge of sewage;

Provided that a person in the process of taking any steps to establish any industry, operation or process immediately before the commencement of the Water (Prevention and Control of Pollution) Amendment Act, 1988, for which no consent was necessary prior to such commencement or, if he has made an application for such consent, within the said period of three months, till the disposal of such application.

(2) An application for consent of the State Board under sub-section (1) shall be made in such form, contain such particulars and shall be accompanied by such fees as may be prescribed.

(3) The State Board may make such inquiry as it may deem fit in respect of the application for consent referred to in sub-section (1) and in making any such inquiry shall follow such procedure as may be prescribed.

(4) The State Board may —

(a) grant its consent referred to in sub-section (1), subject to such conditions as it may impose, being—

(i) in cases referred to in clauses (a) and (b) of sub-section (1) of section 25, conditions as to the point of discharge of sewage or as to the use of that outlet or any other outlet for discharge of sewage;

(ii) in the case of a new discharge, conditions as to the nature and composition, temperature, volume or rate of discharge of the effluent from the land or premises from which the discharge or new discharge is to be made; and

(iii) that the consent will be valid only for such period as may be specified in the order, and any such conditions imposed shall be binding on any person establishing or taking any steps to establish any industry, operation or process, or treatment and disposal system or extension or addition thereto, or using the new or altered outlet, or discharging the effluent from the land or premises aforesaid; or

(b) refuse such consent for reasons to be recorded in writing.

(5) Where, without the consent of the State Board, any industry operation or process, or any treatment and disposal system or any extension or addition thereto, is established, or any steps for such establishment have been taken or a new or altered outlet is brought into use for the discharge of sewage or a new discharge of sewage is made, the State Board may serve on the person who has established or taken steps to establish any industry, operation or process, or any treatment and disposal system or any extension or addition thereto, or using the outlet, or making the discharge, as the case may be, a notice imposing any such conditions as it might have imposed on an application for its consent in respect of such establishment, such outlet or discharge.

(6) Every State Board shall maintain a register containing particulars or conditions imposed under this section and so much of the register as relates to any outlet, or to any effluent, from any land or premises shall be open to inspection at all reasonable hours by any person interested in, or affected by such outlet, land or premises, as the case may be, or by any person authorised by him in this behalf and the conditions so contained in such register shall be conclusive proof that the consent was granted subject such conditions]

(7) The consent referred to in sub-section (1) shall, unless given or refused earlier, be deemed to have been given unconditionally on the expiry of a period of four months of the making of an application in this behalf complete in all respects to the State Board.

(8) For the purposes of this section and sections 27 and 30 –

(a) the expression "new or altered outlet" means any outlet which is wholly or partly constructed on or after the commencement of this Act or which (whether so constructed or not) is substantially altered after such commencement;

(b) the expression "new discharge" means a discharge which is not, as respects the nature and composition, temperature, volume, and rate of discharge of the effluent substantially a continuation of a discharge made within the preceding twelve months (whether by the same or different outlet), so however that a discharge which is in other respects a continuation of previous discharge made as aforesaid shall not be deemed to be a new discharge by reason of any reduction of the temperature or volume or rate of discharge of the effluent as compared with the previous discharge.

Provision regarding existing discharge of sewage or trade effluent

Where immediately before the commencement of this Act any person was discharging any sewage or trade effluent into a stream or well or sewer or on land, the provisions of section 25 shall, so far as may be, apply in relation to such person as they apply in relation to the person referred to in that section subject to the modification that the application for consent to be made under sub-section (2) of that section shall be made on or before such date as may be specified by the State Government by notification in this behalf in the Official Gazette.

Refusal or Withdrawal of Consent by State Board

(1) A State Board shall not grant its consent under sub-section (4) of section 25 for the establishment of any industry, operation or process, or treatment and disposal system or extension or addition thereto, or to the bringing into use of a new or altered outlet unless the industry, operation or process, or treatment and disposal system or extension or addition thereto, or the outlet is so established as to comply with an conditions imposed by the Board to enable it to exercise its right to take samples of the effluent.

(2) A State Board may from time to time review --

(a) any condition imposed under section 25 or section 26 and may serve on the person to whom a consent under section 25 or section 26 is granted a notice making any reasonable variation of or revoking any such condition.

(b) the refusal of any consent referred to in sub-section (1) of section 25 or section 26 or the grant of such consent without any condition, and may make such orders as it deemed fit.

(3) Any conditions imposed under section 25 or section 26 shall be subject to any variation made under sub-section (2) and shall continue in force until revoked under that sub section.

Appeals

(1) Any person aggrieved by an order made by the State Board under Section 25, section 26 or section 27 may within thirty days from the date on which the order is communicated to him, prefer an appeal to such authority (hereinafter referred to as the appellate authority) as the State Government may think fit to constitute:

Provided that the appellate authority may entertain the appeal after the expiry of the said period of thirty days if such authority is satisfied that the appellant was prevented by sufficient cause from filing the appeal in time.

(2) An appellate authority shall consist of a single person or three persons as the State Government may think fit, to be appointed by that Government.

(3) The form and manner in which an appeal may be preferred under sub-section (1), the fees payable for such appeal and the procedure to be followed by the appellate authority shall be such as may be prescribed.

(4) On receipt of an appeal preferred under sub-section (1), the appellate authority shall, after giving the appellant and the State Board an opportunity of being heard, dispose of the appeal as expeditiously as possible.

(5) If the appellate authority determines that any condition imposed, or the variation of any condition, as the case may be, was unreasonable, then,--

(a) where the appeal is in respect of the unreasonableness of any condition imposed, such authority may direct either that the condition shall be treated as annulled or that there shall be substituted for it such condition as appears to it to be reasonable;

(b) where the appeal is in respect of the unreasonableness of any variation of a condition, such authority may direct either that the condition shall be treated as continuing in force unvaried or that it shall be varied in such manner as appears to it to be reasonable.

Revision

(1) The State Government may at any time either of its own motion or on an application made to it in this behalf, call for the records of any case where an order has been made by the State Board under section 25, section 26 or section 27 for the purpose of satisfying itself as to the legality or propriety of any such order and may pass such order in relation thereto as it may think fit:

Provided that the State Government shall not pass any order under this sub-section without affording the State Board and the person who may be affected by such order a reasonable opportunity of being heard in the matter.

(2) The State Government shall not revise any order made under section 25, section 26 or section 27 where an appeal against that order lies to the appellate authority, but has not been preferred or where an appeal has been preferred such appeal is pending before the appellate authority.

Power of State Board to Carry Out Certain Works

(1) Where under this Act, any conditions have been imposed on any person while granting consent under section 25 or section 26 and such conditions require such person to execute any work in connection therewith and such work has not been executed within such time as may be specified in this behalf, the State Board may serve on the person concerned a notice requiring him within such time (not being less than thirty days) as may be specified in the notice to execute the work specified therein.

(2) If the person concerned fails to execute the work as required in the notice referred to in sub section (1), then, after the expiration of the time specified in the said notice, the State Board may itself execute or cause to be executed such work.

(3) All expenses incurred by the State Board for the execution of the aforesaid work, together with interest, at such rate as the State Government may, by order, fix, from the date when a demand for the expenses is made until it is paid, may be recovered by that Board from the person concerned, as arrears of land revenue, or of public demand.

Furnishing of Information to State Board and Other Agencies in Certain Cases

(1) If at any place where any industry, operation or process, or any treatment and disposal

system or any extension or addition thereto is being carried on, due to accident or other unforeseen act or event, any poisonous, noxious or polluting matter is being discharged, or is likely to be discharged into a stream or well or sewer or on land and, as a result of such discharge, the water in any stream or well is being polluted, or is likely to be polluted, then the person incharge of such place shall forthwith intimate the occurrence of such accident, act or event to the State Board and such other authorities or agencies as may be prescribed.

(2) Where any local authority operates any sewerage system or sewage works the provisions of sub-section (1) shall apply to such local authority as they apply in relation to the person in charge of the place where an industry or trade is being carried on.

Emergency Measures In Case of Pollution of Stream or Well

(1) Where it appears to the State Board that any poisonous, noxious or polluting matter is present in any stream or well or on land by reason of the discharge of such matter in such stream or well or on such land or has entered into that stream or well due to any accident or other unforeseen act or event, and if the Board is of opinion that it is necessary or expedient to take immediate action, it may for reasons to be recorded in writing, carry out such operations as it may consider necessary for all or any of the following purposes, that is to say-

(a) removing that matter from the stream or well or on land and disposing it of in such manner as the Board considers appropriate;

(b) remedying or mitigating any pollution caused by its presence in the stream or well;

(c) issuing orders immediately restraining or prohibiting the persons concerned from discharging any poisonous, noxious or polluting matter into the steam or well or on land] or from making insanitary use of the stream or well.

(2) The power conferred by sub-section (I) does not include the power to construct any works other than works of a temporary character which are removed on or before the completion of the operations.

Power of board to make application to courts for restraining apprehended pollution of water in streams of wells.

(1) Where it is apprehended by a Board that the water in any stream or well is likely to be polluted by reason of the disposal or likely disposal of any matter in such stream or well or in any sewer, or on any land, or otherwise, the Board may make an application to a court, not inferior to that of a Metropolitan Magistrate or a Judicial Magistrate of the first class, for restraining the person who is likely to cause such pollution from so causing.

(2) On receipt of an application under sub-section (I) the court make such order as it deems fit.

(3) Where under sub-section (2) the court makes an order restraining any person from polluting the water in any stream or well, it may in that order-

(i) direct the person who is likely to cause or has caused the pollution of the water in the stream or well, to desist from taking such action as is likely to cause pollution or, as the case may be, to remove such stream or well, such matter, and

(ii) authorise the Board, if the direction under clause (i) (being a direction for the removal of any matter from such stream or well) is not complied with by the person to whom such direction is issued, to undertake the removal and disposal of the matter in such manner as may be specified by the court.

(4) All expenses incurred by the Board in removing any matter in pursuance of the authorisation under clause (ii) of sub-section (3) or in the disposal of any such matter may be defrayed out of any money obtained by the Board from such disposal and any balance outstanding shall be recoverable from the person concerned as arrears of land revenue or of public demand.

A Power to Give Directions

Notwithstanding anything contained in any other law, but subject to the provisions of this Act, and to any directions that the Central Government may give in this behalf, a Board may, in the exercise of its powers and performance of its functions under this Act, issue any directions in writing to any person, officer or authority, and such person, officer or authority shall be bound to comply with such directions.

FUNDS, ACCOUNTS AND AUDIT & MISCELLANEOUS

FUNDS ACCOUNTS AND AUDIT

Contributions by Central Government

The Central Government may, after due appropriation made by Parliament by law in this behalf, make in each financial year such contribution to the Central Board as it may think necessary to enable the Board to perform its functions under this Act.

Contributions by State Government

The State Government may, after due appropriation made by the Legislature of the State by law in this behalf, make in each financial year such contributions to the State Board as it may think necessary to enable that Board to perform its functions under this Act.

Fund of Central Board

(1) The Central Board shall have its own fund, and all sums which may, from time to time, be paid to it by the Central Government and all other receipts (by way of gifts, grants, donations, benefactions fees or (otherwise) of that Board shall be carried to the fund of the Board and all payments by the Board shall be made therefrom.

(2) The Central Board may expend such sums as it thinks fit for performing its functions under this Act, and, where any law for the time being in force relating to the prevention, control of abatement or air pollution provides for the performance of any function under such law by the Central Board, also for performing its functions under such law and such sums shall be treated as expenditure payable out of the funds of that Board.

Fund of State Board

(3) The State Board shall have its own fund, and the sums which may, from time to time, be paid to it by the State Government and all other receipts (by way of gifts, grants, donations, benefactions fees or (otherwise) of that Board shall be carried to the fund of the Board and all payments by the Board shall be made therefrom.

(4) The State Board may expend such sums as it thinks fit for performing its functions under this Act, and, where any law for the time being in force relating to the prevention, control or abatement of air pollution provides for the performance of any function under such law by the State Board, also for performing its functions under such law] and such sums shall be treated as expenditure payable out of the fund of that Board.

A Borrowing Powers of Board

A Board may, with the consent of, or in accordance with, the terms of any general or special authority given to it by the Central Government or, as the case may be, the State Government, borrow money from any source by way of loans or issue of bonds, debentures or such other instruments, as it may deem fit, for the performance of all or any of its functions under this Act.

Budget

The Central Board or, as the case may be, the State Board shall, during each financial year, prepare, in such form and at such time as may be prescribed, a budget in respect of the financial year next ensuing showing the estimated receipt and expenditure, and copies thereof shall be forwarded to the Central Government, or as the case may be, the State Government.

Annual Report

(1) The Central Board shall, during each financial year, prepare, in such form as may be prescribed, an annual report giving full account of its activities under this Act during the previous financial year and copies thereof shall be forwarded to the Central Government within four months from the last date of the previous financial year and that Government shall cause every such report to be laid before both Houses of Parliament within nine months from the last date of the previous financial year.

(2) Every State Board shall, during each financial year, prepare, in such form as may be prescribed, an annual report giving full account of its activities under this Act during the previous financial ear and copies thereof shall be forwarded to the State Government within four months from the last date of the previous financial year and that Government shall cause every such report to be laid before the State legislature within a period of nine months from the last date of the previous financial year.

Account and Audit

(1) Every Board shall maintain proper accounts and other relevant records and prepare an annual statement of accounts in such form as may be prescribed by the Central Government or, as the case may be, the State Government.

(2) The accounts of the Board shall be audited by an auditor duly qualified to act as an auditor of companies under section 226 of the Companies Act, 1956. (1 of 1956).

(3) The said auditor shall be appointed by the Central Government or, as the case may be, the State Government on the advice of the Comptroller and Auditor-General of India.

(4) Every auditor appointed to audit the accounts of the Board under this Act shall have the right to demand the production of books, accounts, connected vouchers and other documents and papers and to inspect any of the offices of the Board.

(5) Every such auditor shall send a copy of his report together with an audited copy of the accounts to the Central Government or, as the case may be, the State Government.

(6) The Central Government shall, as soon as may be after the receipt of the audit report under sub-section (5), cause the same to be laid before both Houses of Parliament.

(7) The State Government shall, as soon as may be after the receipt of the audit report under sub-section (5), cause the same to be laid before the State Legislature.

PENALTIES AND PROCEDURE

Failure to comply with directions under sub-section (2) or sub-section (3) of section of 20, or orders issued under clause (c) of sub-section (1) of 32 or directions issued under sub- section (2) of section 33 or section 33A.

(1) Whoever fails to comply with any direction given under sub-section (2) or sub- section (3) of section 20 within such time as may be specified in the direction shall, or conviction, be punishable with imprisonment for a term which may extend to three months or with fine which may extend to ten thousand rupees or with both and in case the failure continues, with an additional fine which may extend to five thousand rupees for every day during which such failure continues after the conviction for the first such failure.

(2) Whoever fails to comply with any order issued under clause (c) of sub-section (1) of section 32 or any direction issued by a court under sub-section (2) of section 33 or any direction issued under section 33A shall, in respect of each such failure and on conviction, be punishable with imprisonment for a term which shall not be less than one year and six months but which may extend to six years and with fine, and case the failure continues, with an additional fine which may extend to five thousand rupees for every day during which such failure continues after the conviction for the first such failure.

(3) If the failure referred to in sub-section (2) continues beyond a period of one year after the date of conviction, the offender shall, on conviction, be punishable with imprisonment for a term which shall not be less than two years but which may extend to seven years and with fine.

Penalty for Certain Acts

(1) Whoever

(a) destroys, pulls down, removes, injures or defaces any pillar, post or stake fixed in the ground or any notice or other matter put up, inscribed or placed, by or under the authority of the Board, or

(b) obstructs any person acting under the orders or directions of the Board from exercising his powers and performing his functions under this Act, or

(c) damages any works or property belonging to the Board, or

(d) fails to furnish to any officer or other employees of the Board any information required by him for the purpose of this Act, or

(e) fails to intimate the occurrence of an accident or other unforeseen act or even under section 31 to the Board and other authorities or agencies as required by that section, or

(f) in giving any information which he is required to give under this Act, knowingly or willfully makes a statement which is false in any material particular, or

(g) for the purpose of obtaining any consent under section 25 or section 26, knowingly or willfully makes a statement which is false in any material particular shall be punishable with imprisonment for a term which may extend to three months or with fine which may extend to ten thousand rupees or with both.

(2) Where for the grant of a consent in pursuance of the provisions of section 25 or section 26 the use of a meter or gauge or other measure or monitoring device is required and such device is used for the purposes of those provision, any person who knowingly or willfully alters or interferes with that device so as to prevent it from monitoring

or measuring correctly shall be punishable with imprisonment for a term which may extend to three months or with fine which may extend to ten thousand rupees or with both.

Penalty for Contravention of Provisions of Section 24

Whoever contravenes the provisions of section 24 shall be punishable with imprisonment for a term which shall not be less than one year and six months but which may extend to six years and with fine.

Penalty for Contravention of Section 25 or Section 26

Whoever contravenes the provision of section 25 or section 26 shall be punishable with imprisonment for a term which shall not be less than two years but which may extend to six years and with fine.

Enhanced penalty after previous conviction

If any person who has been convicted of any offence under section 24 or 25 or section 26 is again found guilty of an offence involving a contravention of the same provision, he shall, on the second and on every subsequent conviction, **be punishable with imprisonment for a term which shall not be less than one and half years but which may extend to seven years and with fine:**

Provided that for the purpose of this section no cognizance shall be taken of any conviction made more than two years before the commission of the offence which is being punished.

A Penalty for contravention of certain provisions of the act

Whoever contravenes any of the provisions of this Act or fails to comply with any order or direction given under this Act, for which no penalty has been elsewhere provided in this Act, shall be punishable with imprisonment which may extend to three months or with fine which may extend to ten thousand rupees or with both and in the case of a continuing contravention or failure, with an additional fine which may extend to five thousand rupees for every day during which such contravention or failure continues after conviction for the first such contravention or failure.

Publication of names of offenders

If any person convicted of an offence under this Act commits a like offence afterwards it shall be lawful for the court before which the second or subsequent conviction takes place to cause the offender's name and place of residence, the offence and the penalty imposed to be published at the offender's expense in such newspapers or in such other manner as the court may direct and the expenses of such publication shall be deemed to be part of the cost attending the conviction and shall be recoverable in the same manner as a fine.

Offences by companies

(1) Where an offence under this Act has been committed by a company, every person who at the time the offence was committed was in charge of, and was responsible to the company for the conduct of, the business of the company, as well as the company, shall

be deemed to the guilty of the offence and shall be liable to be proceeded against and punished accordingly:

Provided that nothing contained in this sub-section shall render any such person liable to any punishment provided in this Act if he proves that the offence was committed without his knowledge for that he exercised all due diligence to prevent the commission of such offence.

(2) Notwithstanding anything contained in sub-section (1), where an offence under this Act has been committed by a company and it is proved that the offence has been committed with the consent or connivance of, or is attributable to any neglect on the part of, any director, manager, secretary or other officer of the company, such director, manager, secretary or other officer shall also be deemed to be guilty of that offence and shall be liable to be proceeded against and punished accordingly.

Explanation For the purpose of this section,--

(a) "company" means any body corporate, and includes a firm or other association of individuals; and

(b) "director" in relation to a firm means a partner in the firm.

Offences by government departments

Where an offence under this Act has been committed by any Department of Government, the Head of the Department shall be deemed to be guilty of the offence and shall be liable to be proceeded against and punished accordingly

Provided that nothing contained in this section shall render such Head of the Department liable to any punishment if he proves that the offence was committed without his knowledge or that he exercised all due diligence to prevent the commission of such offence.

Cognizance of offences

(1) No court shall take cognizance of any offence under this Act except on a complaint made by--

(a) a Board or any officer authorised in this behalf by it; or

(b) any person who has given notice of not less than sixty days, in the manner prescribed, of the alleged offence and of his intention to make a complaint, to the Board or officer authorised as aforesaid, and no court inferior to that of a Metropolitan Magistrate or a Judicial Magistrate of the first class shall try any offence punishable under this Act.

(4) Where a complaint has been made under clause (b) of sub section (1), the Board shall, on demand by such person, make available the relevant reports in its possession to that person:

Provided that the Board may refuse to make any such report available to such person if the same is, in its opinion, against the public interest

(5) Notwithstanding anything contained in section 29 of the Code of Criminal Procedure, 1973", it shall be lawful for any Judicial Magistrate of the first class or for any Metropolitan Magistrate" to pass a sentence of imprisonment for a term exceeding two years or of fine exceeding two thousand rupees on any person convicted of an offence punishable under this Act.

Members, officers and servants of board to be public servants

All members, officers and servants of a Board when acting or purporting to act in pursuance of any of the provisions of this Act (45 of 1860) and the rules made thereunder shall be deemed to be public servants within the meaning of section 21 of the Indian Penal Code.

MISCELLANEOUS

Central water laboratory

(1) The Central Government may, by notification in the Official Gazette --
   (a) establish a Central Water Laboratory; or
   (b) specify any laboratory or institute as a Central Water Laboratory, to carry out the functions entrusted to the Central Water Laboratory under this Act.

(2) The Central Government may, after consultation with the Central Board, make rules prescribing--
   (a) the functions of the Central Water Laboratory;
   (b) the procedure for the submission to the said laboratory of samples of water or of sewage or trade effluent for analysis or tests, the form of the laboratory's report thereunder and the fees payable in respect of such report;
   (c) such other matters as may be necessary or expedient to enable that laboratory to carry out its functions.

State water laboratory

(1) The State Government may, by notification in the Official Gazette --
   (a) establish a State Water Laboratory; or
   (b) specify any State laboratory or institute as a State Water Laboratory, to carry out the functions entrusted to the State Water Laboratory under this Act.

(2) The State Government may, after consultation with the State Board, make rules prescribing --
   (a) the functions of the State Water Laboratory;
   (b) the procedure for the submission to the said laboratory of samples of water or of sewage or trade effluent for analysis or rests, the form of the laboratory's report thereon and the fees payable in respect of such report;
   (c) such other matters as may be necessary or expedient to enable that laboratory to carry out its functions.

Analysts

(1) The Central Government may, by notification in the Official Gazette appoint such persons as it thinks fit and having the prescribed qualifications to be Government analysts for the purpose of analysis of samples of water or of sewage or trade effluent sent for analysis to any laboratory established or specified under sub-section (1) of section 51.

(2) The State Government may, by notification in the Official Gazette, appoint such persons as it thinks fit and having the prescribed qualifications to be Government analysts for

the purpose of analysis of water or of sewage or trade effluent sent for analysis to any laboratory established or specified under sub-section (1) of section 52.

(3) Without prejudice to the provisions of sub-section (3) of section 12, the Central Board or, as the case may be, the State Board may, by notification in the Official Gazette, and with the approval of the Central Government or the State Government, as the case may be, appoint such persons as it thinks fit and having the prescribed qualifications to be Board analysts for the purpose of analysis of samples of water or of sewage or trade effluent sent for analysis to any laboratory established or recognised under section 16, as the case may be, under section 17.

Reports of analysts

Any document purporting to be report signed by a Government analyst or, as the case may be, a Board analyst may be used as evidence of the facts stated therein in any proceeding under this Act.

Local authorities to assist

All local authorities shall render such help and assistance and furnish such information to the Board as it may require for the discharge of its functions, and shall make available to the Board for inspection and examination such records, maps, plans and other documents as may be necessary for the discharge of its functions.

Compulsory acquisition of land for the state board

Any land required by a State Board for the efficient performance of its function under this Act shall be deemed to be needed for a public purpose and such land shall be acquired for the State Board under the provisions of the Land Acquisition Act, 1894 (1 of 1984), or under any other corresponding law for the time being in force.

Returns and reports

The Central Board shall furnish to the Central Government, and a State Board shall furnish to the State Government and to the Central Board such reports, returns, statistics, accounts and other information with respect to its fund or activities as that government, or, as the case may be, the Central Board may, from time to time, require.

Bar of jurisdiction

No civil court shall have jurisdiction to entertain any suit or proceeding in respect of any matter which an appellate authority constituted under this Act is empowered by or under this Act to determine, and no injunction shall be granted by any court or other authority in respect of any action taken or to be taken in pursuance of any power conferred by or under this Act.

Protection of action taken in good faith

No suit or other legal proceedings shall lie against the Government or any officer of Government or any member or officer of a Board in respect of anything which is in good faith done or intended to be done in pursuance of this Act or the rules made thereunder.

Overriding effect

The provisions of this Act shall have effect notwithstanding anything inconsistent therewith contained in any enactment other than this Act.

Power of central government to supersede the central board and joint boards

(1) If at any time the Central Government is of opinion --

(a) that the Central Board or any Joint Board has persistently made default in the performance of the functions imposed on it by or under this Act; or

(b) that circumstances exist which render it necessary in the public interest so to do, the Central Government may, by notification in the Official Gazette, supersede the Central Board or such Joint Board, as the case may be, for such period not exceeding one year, as may be specified in the notification:

Provided that before issuing a notification under this sub-section for the reasons mentioned in clause (a), the Central Government shall give a reasonable opportunity to the Central Board or such Joint Board, as the case may be, to show cause why it should not be superseded and shall consider the explanations and objections if any, of the Central Board or such Joint Board, as the case may be.

(2) Upon the publication of notification under sub-section (1) superseding the Central Board or any Joint Board,

(a) all the members shall, as from the date of supersession vacate their offices as such;

(b) all the powers, functions and duties which may, by or under this Act, be exercised, performed or discharged by the Central Board or such Joint Board, shall, until the Central Board or the Joint Board, as the case may be, is reconstituted under sub- section (3) be exercised, performed or discharged by such person or persons as the Central Government may direct;

(c) all property owned or controlled by the Central Board or such Joint Board shall, until the Central Board or the Joint Board, as the case may be, is reconstituted under sub-section (3) vest in the Central Government.

(3) On the expiration of the period of supersession specified in the notification issued under sub-section (1), the Central Government may -

(a) extend the period of supersession for such further term, not exceeding six months, as it may consider necessary; or

(b) reconstitute the Central Board or the Joint Board, as the case may be, by fresh nomination or appointment, as the case may be, and in such case any person who vacated his office under clause (a) of sub-section (2) shall not be deemed disqualified for nomination or appointment:

Provided that the Central Government may at any time before the expiration of the period of supersession, whether originally specified under sub-section (1) or as extended under this sub-section, take action under clause (b) of this sub-section.

Power of state government to supersede state board

(1) If at any time the State Government is of opinion --

(a) that the State Board has persistently made default in the performance of the functions imposed on it by or under this Act; or

(b) that circumstances exist which render it necessary in the public interest so to do, the State Government may, by notification in the Official Gazette, supersede the State Board for such period, not exceeding one year, as may be specified in the notification:

Provided that before issuing a notification under this sub-section for the reasons mentioned in clause (a), the State Government shall give a reasonable opportunity to the State Board to show cause why it should not be superseded and shall consider the explanations and objections, if any, of the State Board.

(2) Upon the publication of a notification under sub-section (1) superseding the State Board, the provisions of sub sections (2) and (3) of section 61 shall apply in relation to the supersession of the State Board as they apply in relation to the supersession of the Central Board or a Joint Board by the Central Government.

Power of central government to make rules

(1) The Central Government may, simultaneously with the constitution of the Central Board, make rules in respect of the matters specified in sub-section (2):

Provided that when the Central Board has been constituted, no such rule shall be made, varied, amended or repealed without consulting the Board.

(2) In particular, and without prejudice to the generality of the foregoing power, such rules may provide for all or any of the following matters namely:-

(a) the terms and conditions of service of the members (other than the chairman and member secretary) of the Central Board under sub-section (8) of section 5;

(b) the intervals and the time and place at which meetings of the Central Board or of any committee thereof constituted under this Act, shall be held and the procedure to be followed at such meetings, including the quorum necessary for the transaction of business under section 8, and under sub section (2) of section 9;

(c) the fees and allowances to be paid to such members of a committee of the Central Board as are not members of the Board under sub-section (3) of section 9;

(d) the manner in which and the purposes for which persons may be associated with the Central Board under sub section (1) of section 10 and the fees and allowances payable to such persons;

(e) the terms and conditions of service of the chairman and the member-secretary of the Central Board under sub section (9) of section 5 and under sub-section (1) of section 12;

(f) conditions subject to which a person may be appointed as a consulting engineer to the Central Board under sub-section (4) of section 12;

(g) the powers and duties to be exercised and performed by the chairman and member secretary of the Central Board;

(h) the form of the report of the Central Board analyst under sub-section (1) of section 22;

(i) the form of the report of the Government analyst under sub-section (1) of section 22;

(j) the form in which the time within which the budget of the Central Board may be prepared and forwarded to the Central Government under section 38;

(k) the form in which the annual report of the Central Board may be prepared under section 39;

(l) the form in which the accounts of the Central Board may be maintained under section 40;

(m) the manner relating to the Central Board, including the powers and functions of that Board in relation to Union territories;

(n) any other matter which has to be, or may be, prescribed.

(3) Every rule made by the Central Government under this Act shall be laid, as soon as may be after it is made, before each House of Parliament while it is in session for a total period of thirty days which may be comprised in one session or in two or more successive sessions, and if, before the expiry of the session immediately following the session or the successive sessions aforesaid, both Houses agree in making any modification in the rule of both Houses agree that the rule should not be made, the rule shall thereafter have effect only in such modified form or be of no effect, as the case may be; so, however, that any such modification or annulment shall be without prejudice to the validity of anything previously done under that rule.

Power of state government to make rules

(1) The State Government may, simultaneously with the constitution of the State Board, make rules to carry out the purposes of this Act in respect of matters not falling within the purview of section 63;

Provided that when the State Board has been constituted, no such rule shall be made, varied, amended or repealed without consulting that Board.

(2) In particular, and without prejudice to the generality of the foregoing power, such rules may provide for all or any of the following matters, namely:--

(a) the terms and conditions of service of the members (other than the chairman and the member-secretary) of the State Board under sub-section (8) of section 5;

(b) the time and place of meetings of the State Board or of any committee of that Board constituted under this Act and the procedure to be followed at such meeting, including in quorum necessary for the transaction of business under section 8 and under sub-section (2) of section 9;

(c) the fees and allowances to be paid to such members of a committee of the State Board as are not members of the Board under sub-section (3) of section 9;

(d) the manner in which and the purpose for which persons may be associated with the State Board under sub-section (1) of section 10 6[and the fees and allowances payable to such persons;

(e) the terms and conditions of service of the chairman and the member secretary of the State Board under sub-section (9) of section 5 and under sub-section (1) of section 12;

(f) the conditions subject to which a person may be appointed as a consulting engineer to the State Board under sub-section (4) of section 12;

(g) the powers and duties to be exercised and discharged by the chairman and the member-secretary of the State Board;

(h) the form of the notice referred to in section 21;

(i) the form of the report of the State Board analyst under sub section (3) of section 22.

(j) the form of the report of the Government analyst under sub- section (3) of section 22;

(k) the form of application for the consent of the State Board under sub-section (2) of section 25 and the particular it may contain;

(l) the manner in which inquiry under sub section (3) of section 25 may be made in respect of an application for obtaining consent of the State Board and the matters to be taken in to account in granting or refusing such consent;

(m) the form and manner in which appeals may be filed, the fees payable in respect of such appeals and the procedure to be followed by the appellate authority in disposing of the appeals under sub-section (3) of section 28;

(n) the form in which and the time within which the budget of the State Board may be prepared and forwarded to the State Government under section 38;

(nn) the form in which the annual report of the State Board may be prepared under section 39;

(o) the form in which the accounts of the State Board may be maintained under sub- section (l) of section 40;

(p) the manner in which notice of intention to make a complaint shall be given to the State Board or officer authorised by it under section 49; any other matter which has to be, or may be, prescribed.

THE WATER (PREVENTION AND CONTROL OF POLLUTION) ACT, 1974 (AS AMENDED IN 1978 AND 1988).

The Water Act is a comprehensive legislation providing for the Prevention and Control of Water Pollution and for maintaining or restoring the wholesomeness of water in streams or wells. The Act provides for the establishment of the Central Pollution Control Board at the Centre and State Pollution Control Boards in the respective States.

(A) The functions of the Central Board at the national level are to:

(i) Advise the Central Govt. on matters relating to prevention and control of water pollution.

(ii) Coordinate the activities of the State Board and resolve disputes among them.

(iii) Provide technical assistance and guidance to the State Boards.

(iv) Carry out and sponsor research and investigation in the problems of water pollution.

(v) Set the standards for streams and wells.

(vi) Create environmental awareness and

(vii) To act as State Board for the Union Territories.

(B) State Board has executive and territorial functions which include:

(i) Planning for prevention, control or abatement of pollution of streams and wells.

(ii) Advise the State Govt. on matters relating to water pollution-

(iii) Inspection of sewage or industrial effluent, including municipal wastewater treatment plants for the treatment of sewage or trade effluents.

(iv) Setting standards for the sewage and industrial effluents discharge.

There is a provision of joint boards for two or more contiguous States. In case of dispute between two State Boards, the Central Board has authority to arbitrate.

Important provisions in the Water (Prevention and Control of Pollution) Act, 1974 (As amended In 1978 And 1988) are:

(i) Pollution Control Board (PCB) has the right

- To obtain any information regarding the construction, installation or operation of an industrial establishment or treatment and disposal system
- To take samples of trade effluent for the purpose of analysis in the prescribed manner
- To enter and inspect any industrial establishment, record, register, document or any other material object.
- To prohibit use of stream or sewer or land for disposal system without prior consent of the PCB.

(ii) Restriction on establishment and the operation of any industry process or any treatment and disposal system without prior consent of the PCB.

(iii) PCB's right to refuse or withdraw consent, for discharge of effluents.

(iv) Industry to comply with the conditions stipulated in the consent.

(v) PCB's to grant consent within four months after the date of receipt of the application complete in all respects.

(vi) Industry to appeal to the Appellate Authority, in case of grievances against the order passed by the PCB regarding grant, refusal or withdrawal of the consent within the specified time in the prescribed manner.

(vii) Industry to furnish information to the PCB and other specified agency (ies) in case of discharge of poisonous, noxious or polluting matter into a stream, sewer or land, occurred or likely to occur resulting in pollution due to an accident or any other unforeseen event.

(viii) PCB's right to issue orders restraining or prohibiting an industry from discharging any poisonous, noxious or polluting matter in case of emergencies, warranting immediate action.

(ix) PCB's have power to make an application to the court for restraining likely disposal of polluting matter in a stream or on land.

(x) Bar of jurisdiction in civil court in respect of any matter under purview of the Appellate Authority constituted under the Act and no grant of injunction in respect of any action taken or proposed in pursuance of the Act.

(xi) Bar on filing of any suit or legal proceedings against the Government or Board officials, for action taken in good faith in pursuance of the Act.

(xii) PCB's to make inquiries, in the prescribed manner, for grant of consent for discharge of effluents.

(xiii) PCB's power to issue directions for:
- The closure, prohibition or regulation of any industry, operation or process or,
- The stoppage or regulation of supply electricity, water or any other service to industry in the prescribed manner

(xiv) Industry to comply with the directions of the PCB within the specified time.

(xv) PCB's to maintain a consent register containing particulars of the consent issued and to provide access to industry at all reasonable hours.

THE WATER PREVENTION AND CONTROL OF POLLUTION CESS ACT, 1977

Water (prevention and control of pollution) Cess Act of 1977 provides funds for the Central and state pollution control boards. It also talks about this Act which empowers the Central Government to impose a cess on water consumed by industries listed in Schedule I of the act.

This Act of Parliament received the assent of the President on the 7th December, 1977, and is hereby published for general information:–

An Act to provide for the levy and collection of a cess on water consumed by persons carrying on certain industries and by local authorities, with a view to augment the resources of the Central Board and the State Boards for the prevention and control of water pollution constituted under the Water (Prevention and Control of Pollution) Act, 1974.

Be it enacted by Parliament in the Twenty eighth Year of the Republic of India as follows:–

SHORT TITLE, EXTENT, APPLICATION AND COMMENCEMENT

(1) This Act may be called the Water (Prevention and Control of Pollution) Cess Act, 1977.

(2) It extends to the whole of India except the State of Jammu and Kashmir.

(3) Subject to the provisions of sub-section (2), it applies to all the States to which the Water (Prevention and Control of Pollution) Act, 1974 applies and the Union territories.

(4) It shall come into force on such date as the Central Government may, by notification in the Official Gazette, appoint.

DEFINITIONS

In this Act, unless the context otherwise requires:

(a) "local authority" means a municipal corporation or a municipal council (by whatever name called) or a cantonment board or any other body, entrusted with the duty of supplying the water under the law by or under which it is constituted;

(b) "prescribed" means prescribed by rules made under the Act;

(c) "specified industry" means any industry specified in Schedule I;

(d) words and expressions used but not defined in this Act and defined in the Water (Prevention and Control of Pollution) Act, 1974 (6 of 1974) shall have the meanings respectively assigned to them in that Act.

LEVY AND COLLECTION OF CESS

(1) There shall be levied and collected a cess for the purpose of the Water (Prevention and Control of Pollution) Act, 1974 (6 of 1974) and utilisation there under.

(2) The cess under sub-section (1) shall be payable by--

(a) every person carrying on any specified industry; and

(b) every local authority,

And shall be calculated on the basis of water consumed by such person or local authority, as the case may be, for any of the purposes specified in column (1) of Schedule II, at such rate, not exceeding the rate specified in the corresponding entry in column (2) thereof, as the Central Government may, by notification in the Official Gazette, from time to time, specify.

(2A) Where any person carrying on any specified industry or any local authority consuming water for domestic purpose liable to pay cess fails to comply with any of the provisions of section 25 of the Water (Prevention and Control of Pollution) Act, 1974 (6 of 1974) or an of the standards laid so down by the Central Government under the Environment (Protection) Act, 1986, cess shall be and payable at such rate, not exceeding the rate specified in column (3) of Schedule II, as the Central Government may, by notification in the Official Gazette, from time to time specify.

(3) Where any local authority supplies water to any person carrying on any specified industry or to any other local authority and such person or other local authority is liable to pay cess under sub-section (2) or sub-section (2A) in respect of the water so supplied, then, notwithstanding anything contained in that sub-section, the local authority first mentioned shall not be liable to pay such cess in respect of such water.

Explanation--For the purpose of this section and section 4, "consumption of water" includes supply of water.

AFFIXING OF METERS

(1) For the purpose of measuring and recording the quantity of water consumed, every person carrying on any specified industry and every local authority shall affix meters of such standards and at such places as may be prescribed and it shall be presumed that the quantity indicated by the meter has been consumed by such person or local authority, as the case may be, until the contrary is proved.

(2) Where any person or local authority fails to affix any meter as required by sub-section (1), the Central Government shall after notice to such person or local authority, as the case may be, cause such meter to be affixed and the cost of such meter together with the cost for affixing the meter may be recovered from such person or local authority by the Central Government in the same manner as an arrear of land revenue.

FURNISHING OF RETURNS

(1) Every person carrying on any specified industry and every local authority, liable to pay the cess under section 3, shall furnish such returns, in such form at such intervals and containing such particulars to such officer or authority, as may be prescribed.

(2) If a person carrying on any specified industry or a local authority, liable to pay the cess under section 3, fails to furnish any return under sub-section (1), the officer or the authority shall give a notice requiring such person or local authority to furnish such return before such date as may be specified in the notice.

ASSESSMENT OF CESS

(1) The officer of authority to whom or which the return has been furnished under section 5 shall, after making or causing to be made such inquiry as he or it thinks fit and after satisfying himself or itself that the particulars stated in the return are correct, by order, assess the amount of cess payable by the concerned person carrying or any specified industry or local authority, as the case may be.

(1A) If the return has not been furnished to the officer or authority under sub-section (2) of section 5, he or it shall, after making or causing to be made such inquiry as he or it thinks fit, by order, assess the amount of cess payable by the concerned person carrying on any specified industry or local authority, as the case may be.

(2) An order of assessment made under sub-section (1) or sub-section (1A) shall specify the date within which the cess shall be paid to the State Government.

(3) A copy each of the order of assessment made under sub-section (1) or sub-section (1A) shall be sent to the person or, as the case may be, to the local authority concerned and to the State Government.

(4) The State government shall, through such of its officers or authorities as may be specified by it in this behalf by notification in the Official Gazette, collect the cess from the person or local authority liable to pay the same and pay the amount so collected to the Central Government in such manner and within such time as may be prescribed.

REBATE

Where any person or local authority, liable to pay the cess under this Act, instals any plant for the treatment of sewage or trade effluent, such person or local authority shall from such date as may be prescribed, be entitled to a rebate of twenty five per cent of the cess payable by such person or, as the case may be, local authority.

Provided that a person or local authority shall not be entitled to a rebate, if he or it--

(a) consumes water in excess of the maximum quantity as may be prescribed in this behalf for any specified industry or local authority; or

(b) fails to comply with any of the provisions of section 25 of the Water (Prevention and Control of Pollution) Act, 1974 (6 of 1974) or any of the standards laid down by the Central Government under the Environment (Protection) Act, 1986 (29 of 1986).

CREDITING PROCEEDS OF CESS TO CONSOLIDATED FUNDS OF INDIA AND APPLICATION THEREOF

The proceeds of the cess levied under section 3 shall first be credited to the Consolidated Fund of India and the Central Government may, if Parliament by appropriation made by law in this behalf, so provides, pay to the Central Board and every State Board, from time to time, from out of such proceeds, after deducting the expenses on collection, such sums of money as it may think fit for being utilised under the Water (Prevention and Control of Pollution) Act, 1974 (6 of 1974)

Provided that while determining the sum of money to be paid to any State Board under this section, the Central Government shall have regard to the amount of cess collected by the State Government concerned under sub-section (4) of section 6.

Explanation-For the purpose of this section, "State Board" includes a Joint Board, if any, constituted under section 13 of the Water (Prevention and Control of Pollution) Act, 1974 (6 of 19743).

POWER OF ENTRY

Any officer or authority of the State Government specially empowered in this behalf by that Government may,-

(a) with such assistance, if any, as he or it may think fit, enter at an) reasonable time any place which he or it considers it necessary to enter for carrying out the purposes of this Act including the testing of the correctness of the meters affixed under section 4

(b) do within such place anything necessary for the proper discharge of his or its duties under this Act; and

(c) exercise such other powers as may be prescribed.

INTEREST PAYABLE FOR DELAY IN PAYMENT OF CESS

If any person carrying on any specified industry or any local authority fails to pay any amount of cess payable under section 3 to the State government within the date specified in the order of assessment made under section 6, such person or local authority, as the case may be, shall be liable to pay interest on the amount to be paid at the rate of two per cent for every month or part of a month comprised in the period from the date on which such payment is due till such amount is actually paid.

PENALTY OF AMOUNT DUE UNDER THE ACT

If any amount of cess payable by any person carrying on any specified industry or any Local authority under section 3 is not paid to the State (government within the date specified in the order of assessment made under section 6, it shall be deemed to be in arrears and the authority prescribed in this behalf may, after such inquiry as it deems fit, impose on such person Of, as the case may be, Local authority, a penalty not exceeding the amount of cess in arrears:

Provided that before imposing any such penalty, such person or, as the case may be, the local authority shall be given a reasonable opportunity of being heard and if after such hearing the said authority is satisfied that the default was for any good and sufficient reason, no penalty shall be imposed under this section.

RECOVERY OF AMOUNT DUE UNDER THE ACT

Any amount due under this Act (including any interest or penalty payable under section 10 or section 11, as the case may be) from any person carrying on any specified industry or from any local authority may be recovered by the Central Government in the same manner as an arrear of land revenue.

APPEALS

(1) Any person or local authority aggrieved by an order of assessment made under section 6 or by an order imposing penalty made under section 11 may, within such time as may be prescribed, appeal to such authority in such form and in such manner as may be prescribed.

(2) Every appeal preferred under sub-section (I) shall be accompanied by such fees as may be prescribed.

(3) After the receipt of an appeal under sub-section (1), the appellate authority shall, after giving the appellant an opportunity of being heard in the matter, dispose of the appeal as expeditiously as possible.

(4) Every order passed in appeal under this section shall be final and shall not be called in question in any court of law.

PENALTY

(1) Whoever, being under an obligation to furnish a return under this Act, furnishes any return knowing, or having reason to believe, the same to be false shall be punishable with imprisonment which may extend to six months or with fine which may extend to one thousand rupees or with both.

(2) Whoever, being liable to pay cess under this Act willfully or intentionally evades or attempts to evade the payment of such cess shall be punishable with imprisonment which may extend to six months or with fine which may extend to one thousand rupees or with both.

(3) No court shall take cognizance of an offence punishable under this section save on a complaint made by or under the authority of the Central Government.

OFFENCES BY COMPANIES

(1) Where an offence under this Act has been committed by a company, every person who, at the time the offence was committed, was in charge of and was responsible to, the company for the conduct of the business of the company as well as the company, shall be deemed to be guilty of the offence and shall be liable to be proceeded against and punished accordingly: Provided that nothing contained in this sub-section shall render any such person liable to any punishment, if he proves that the offence was committed without his knowledge or that he exercised all due diligence to prevent the commission of such offence.

(2) Notwithstanding anything contained in sub-section (1), where an offence under this Act has been committed by a company and it is proved that the offence has been committed with the consent or connivance of, or is attributable to any neglect on the part of, any director, manager, secretary or other officer of the company, such director, manager, secretary or other officer shall also be deemed to be guilty of that offence and shall be liable to be proceeded against and punished accordingly.

Explanation-For the purpose of this section,-

(a) "company" means any body corporate and includes a firm or other association of individuals; and

(b) "director", in relation to firm, means a partner in the firm.

POWER TO AMEND SCHEDULE I

(1) The Central Government may, by notification in the Official Gazette, add to Schedule I any industry having regard to the consumption of water in the carrying on of such industry and the consequent discharge thereof resulting in pollution of any stream and

thereupon Schedule I shall, subject to the provisions of sub-section (2), be deemed to be amended accordingly.

(2) Every such notification shall be laid before each House of Parliament, if it is sitting, as soon as may be after the issue of the notification and is it is not sitting, within seven days of its re-assembly and the Central Government shall seek the approval of Parliament to notification by a resolution moved within a period of fifteen days beginning with the day on which the notification is so laid before the House of the People, and if Parliament makes any modification in the notification or directs that the notification should cease to have effect, the notification shall thereafter have effect only in such modified form pr be of no effect, as the case may be, but without prejudice to the validity of anything previously done thereunder.

POWER TO MAKE RULES

(1) The Central Government-may make rules for carrying out the purposes of this Act.

(2) Without prejudice to the generality of the foregoing power, such rules may provide for all or any of the following matters, namely:-

(a) the standards of the meters to be affixed and the places at which such meters are to be affixed under sub-section (I) of section 4;

(b) the returns to be furnished under section 5, the form in which and the intervals at which such returns are to be furnished, the particulars which such returns contain and the officer or authority to who or which such returns shall be furnished;

(c) the manner in which and the time within which the cess collected shall be paid to the Central Government under sub-section (4) of section 6;

(d) the date from which any person or local authority liable to pay cess shall be entitled to the rebate and the maximum quantity of water in excess ofconsumption whereof any person or local authority shall not be entitled to the rebate under section 7.

(e) the powers which may be exercised by the officer or authority under section 9;

(f) the authority which may impose penalty under section 11;

(g) the authority to which an appeal may be filed under sub-section (I) of section 13 and the time within which and the form and manner in which such appeal may be filed;

(h) the fees which shall accompany an appeal under sub-section (2) of section 13; and

(i) any other matter which has to be or may be prescribed.

(3) Every rule made under this Act shall be laid, as soon as may be after it is made, before each House of Parliament while it is in session for a total period of thirty days which may be comprised in one session or in two successive sessions and if, before the expiry of the session immediately following the session or the successive sessions aforesaid, both Houses agree in making any modification in the rule or both Houses agree that the rule should not be made, the rule shall thereafter have effect only in such modified form or be of no effect, as the case may be; so, however, that any such modification or annulment shall be without prejudice to the validity of anything previously done under that rule.

SCHEDULE I

1. Ferrous metallurgical industry.
2. Non-ferrous metallurgical industry.
3. Mining industry.
4. Ore processing industry.
5. Petroleum industry.
6. Petro-chemical industry
7. Chemical industry.
8. Ceramic industry.
9. Cement industry.
10. Textile industry. (including cotton synthetic and semi-synthetic fibres manufactured from these fibres);
11. Paper industry.
12. Fertilizer industry.
13. Coal (including coke) industry.
14. Power (thermal, diesel) and Hydel generating industry
15. Processing of animal or vegetable products industry including processing of milk, meat, hides and skins, all agricultural products and their wastes.
16. Engineering industry

SCHEDULE II

Table 12.1: Water Cess

Purpose for which water is consumed	Maximum rate under sub-section (2) of section 3	Maximum rate under sub-section (2A) of section 3
1. Industrial cooling, spraying in mine pits or boiler feeds	One and a half paise per kilo litre	Two and one-fourth paise per kilo litre.
2. Domestic purpose	Two paise per kilo litre	Three paise per kilo litre
3. Processing whereby water gets polluted and the pollutants are easily biodegradable and are toxic.	Four paise per kilo litre	Seven and a half paise per kilo litre.
4. Processing whereby water gets polluted and the pollutants are not easily biodegradable and are toxic.	Five paise per kilo litre	Nine and a half paise per kilo litre

NOTIFICATION

New Delhi, the 16th January, 1980

G.S.R. 190.-In exercise of the powers conferred by clause (1) of article 258 of the Constitution, the President, with the consent of the State Governments concerned hereby entrusts to the Governments of each of the States of Andhra Pradesh, Bihar, Gujarat, Haryana, Himachal Pradesh, Kerala, Madhya Pradesh, Punjab, Rajasthan, Uttar Pradesh and West Bengal, the functions of the Central Government under sub-section (2) of section 4, section 12 and sub-section (3) of section 14 of the Water (Prevention and Control of Pollution) Cess Act, 1977 (36 of 1977) subject to the conditions that not withstanding this entrustment the Central Government may itself exercise any of the said functions should deem fit to do so in any case.

MINISTRY OF ENVIRONMENT AND FORESTS NOTIFICATION

New Delhi, the 26th January, 1992

S.0.78(E).-In exercise of the powers conferred by sub- section (2) of section 1 of the Water (Prevention and Control of Pollution) Cess (Amendment) Act, 1991 (53 of 1991), the Central Government hereby appoints the 26th day of January,1992 as the date on which the said Act shall come into force.

THE WATER PREVENTION AND CONTROL OF POLLUTION CESS ACT 1977 (AMENDED IN 1991)

The Water Cess Act provides for the levy of a cess on water consumed by persons carrying on specified industries given in Schedule-I of the Act and also local authorities entrusted with the duty of supplying water under the laws by or under which they are constituted at the rates specified in Schedule-II of the Act. The Cess is levied and collected by the State Government concerned and credited to the consolidated Fund of India. An industry which installs and operates its effluent treatment plant is entitled to a rebate of 25% on the cess payable. The cess has been introduced mainly to augment the resources of the Central and the State Pollution ControlBoards.

APPENDIX 12.1

Schedule of specified industries for water cess

S.N.	Industry
1	Ferrous metallurgical industries
2	Non-ferrous metallurgical industries
3	Mining industry
4	Ore processing industry
5	Petroleum industry
6	Petro chemical industry
7	Chemical industry
8	Ceramic industry
9	Cement industry
10	Textile industry (including cotton synthetic and semi-synthetic fibers manufactured from these fibers)
11	Paper industry
12	Fertilizer industry
13	Coal industry
14	Power (thermal, diesel and hydel) generating industry
15	Processing of animal or vegetable products (including processing of milk, meat, hides and skins, all agricultural products their wastes)
16	Engineering industry

APPENDIX 12.2

Prescribed rate of water cess in India

S.N.	Processes for which water is consumed	Rate of cess under sub-section (2) of section, 3, Paise/Kilolitre	Rate of cess under sub-section (2A) of Section 3, paise/Kilolitre
1	Industrial cooling, spraying in mine pits or boiler feeds	1.5	2.25
2	Domestic purpose	2	3
3	Processing whereby water gets polluted and the pollutants are easily biodegradable	4	7.5
4	Processing whereby water gets polluted and the pollutants are easily biodegradable and are toxic.	5	9.5

*The rates shall be applicable only in case of non-compliance with any one of the prescribed consent conditions or any one of the standards for discharge of effluent prescribed by the central government under environment protection act, 1986, otherwise as specified under column 2.

APPENDIX 12.3

List of projects requiring environmental clearance from the Central government of India

S.N.	Projects requiring environmental clearance from central government
1	Nuclear power and related projects such as heavy water plants, nuclear fuel complex, rare earths
2	River valley projects including hydel power, major Irrigation and their combination including flood control
3	Ports, harbors, airports (except minor ports and harbours)
4	Petroleum refineries including crude & product pipelines
5	Chemical fertilizers (nitrogenous and phosphatic) other than single superphosphate (SSP)
6	Pesticides (technical)
7	Petrochemical complexes (both olefinic and chromatic) and petrochemical intermediates such as DMT, caprolactun, LAB etc. and product of basic such as LLDPE, HDPE, PP, PVC etc..
8	Bulk drugs and pharmaceuticals
9	Exploration for oil & gas and their production, transportation & storage.
10	Synthetic rubber
11	Asbestos and asbestos products
12	Hydrocyanic acids and its derivatives
13	Primary metallurgical industries (such as production of iron & steel, aluminum, Copper zinc, lead and ferro alloys)
14	Chlor-alkali industries
15	Integrated paint complex including manufacture of resins and basic raw materials required in the manufacture of paints.
16	Viscose staple fibre and filament yarn
17	Storage batteries integrated with manufacture of oxides and lead antimony alloys
18	All tourism projects between 200-500 metres of high water line and at locations within elevation of more than 1000 metres with investments of > 5 crores
19	Thermal Power plants
20	Mining projects (with leases more than 5 hectares)
21	Highway Projects

S.N.	Projects requiring environmental clearance from central government
22	Tarred roads in the Himalayas and /or forest areas.
23	Distilleries
24	Raw skins and hides
25	Pulp, paper and Newsprint
26	Dyes
27	Cement
28	Foundries (individual)
29	Electroplating

APPENDIX 12.4

Depriciation table under the income tax act 1961 for pollution control equipments/systems

S.N.	Pollution control equipments	Depreciation
1	**Air pollution control equipments:** (i) Electrostatic Precipitation (ii) Felt filter systems (iii) Dust collector systems (iv) Scrubber-counter current/venture/packed bed/cyclonic	100%
2	**Water pollution control equipment:** (i) Mechanical screen systems (ii) Aerated detritus chambers (including air compressor) (iii) Mechanically skimmed oil & grease removal systems (iv) Chemical feed systems and flash arising equipments (v) Mechanical flocculators and mechanical reactors (vi) Diffused air / mechanically aerated activated sludge systems (vii) Aerated lagoon systems (viii) Bio-filters (ix) Methane recovery anaerobic digester system (x) Air floatation systems (xi) Air/steam stripping systems (xii) Urea hydrolysis systems (xiii) Marine outfall systems (xiv) Centrifuge for dewatering sludge (xv) Rotating biological contactor or bio-disc (xvi) Ion-exchange resin columns (xvii) Activated carbon column	100%
3	(i) Solid waste control equipment -caustic/lime/chrome/crème/ cryolite recovery system (ii) Solid waste recycling and resource recovery system	100%
	Income tax notification no.9374 dated 16.9.1993 (Income tax seventeenth amendment rules, 1993, They will come into force w.e.f. 1.4.1994	

APPENDIX 12.5

Parameters under National Water Quality Monitoring Programme(INDIA)

Field Observations	Core Parameters	General Parameters	Bio-Monitoring	Trace Metals	Pesticides
• Depth of main stream/ depth of water table • Colour and intensity • Odour • Visible effluent discharge • Human activities around station • Station detail	• pH • Temperature • Conductivity, μmhos/cm • Dissolved Oxygen, mg/l • BOD, mg/l • Nitrate –N, mg/l • Nitrite –N, mg/l • Faecal Coliform, MPN/100 ml • Total Coliform, MPN/100 ml	• Turbidity, NTU • Phenolphthalein Alkalinity, as $CaCO_3$ • Total Alkalinity , as $CaCO_3$ • Chlorides, mg/l • COD, mg/l • Total Kjeldahl-N, as N mg/l • Ammonia-N, as N mg/l • Hardness, as $CaCO_3$ • Calcium, as $CaCO_3$ • Suphate, mg/l • Sodium, mg/l • Total Dissolved Solids, mg/l • Total fixed Dissolved Solids, mg/l • Total Suspended Solid, mg/l • Phosphate, mg/l • Boron, mg/l • Magnesium, as $CaCO_3$ • Potassium, mg/l • Fluoride, mg/l	• Saprobity index • Diversity index • P/R Ratio	• Arsenic, μg/l • Cadmium,mg/l • Copper, μg/l • Lead, μg/l • Chromium (Total), μg/l • Nickel, μg/l • Zinc, μg/l • Mercury, μg/l • Iron (Total), μg/l	• Alpha BHC, μg/l. • Beta BHC, μg/l • Gamma BHC(Lindane), μg/l • O P DDT, μg/l • P P DDT, μg/l • Alpha Endosulphan, μg/l • Beta Endosulphan, μg/l • Aldrin, μg/l • Dieldrin, μg/l • Carbory(Carbam), μg/l • 2-4, μg/l • Malathian, μg/l • Methyl Parathian, μg/l • Anilophos, μg/l • Chloropyriphos, μg/l

Note—Significance of various parameters involved in monitoring of water quality are described below.

pH Value

The pH of a solution is measured as negative logarithm of hydrogen ion concentration. At a given temperature, the intensity of the acidic or basic character of a solution is indicated by pH or hydrogen ion concentration. pH values from 0 to 7 are diminishing acidic, 7 to 14 increasingly alkaline and 7 is neutral.Measurement of pH is one of the most important and frequently used tests, as every phase of water and waste water treatment & waste quality management is pH dependent.

The pH of natural water usually lies in the range of 4 to 9 and mostly it is slightly basic because of the presence of bicarbonates & carbonates of alkali and alkaline earth metals. pH value is governed largely by the carbon dioxide/ bicarbonate/ carbonate equilibrium. It may be affected by humic substances, by changes in the carbonate equilibriums due to the bioactivity of plants and in some cases by hydrolysable salts. The effect of pH on the chemical and biological properties of liquid makes its determination very important. It is used in several calculations in analytical work and its adjustment to an appropriate value is absolutely necessary in many of analytical procedures.

In dilute solution, the hydrogen ion activity is approximately equal to the concentration of hydrogen ion.

Colour

The method is useful in the field by comparing the colour of sample with a comparator. When viewed by transmitted light through a depth of several feet, pure water exhibits a light blue colour which may be altered by the presence of organic matter to greenish blue, green, greenish yellow, yellow or brown. Colour is removed to make water suitable for general and industrial applications. Coloured industrial wastewater may require colour removal before discharge into water sources. The visual comparison method is applicable to nearly all samples of potable water. Pollution by certain industrial wastes may produce unusual colours that cannot be matched. In such a case, an instrumental method is useful. Colour is expressed in terms of Pt/Co standard unit.

Hardness

Water hardness is a traditional measure of the capacity of water to precipitate soap. Hardness of water is not a specific constituent but is a variable and complex mixture of cations and anions. It is caused by dissolved polyvalent metallic ions. In fresh water, the principal hardness causing ions are calcium and magnesium which precipitate soap. Other polyvalent cations also may precipitate soap, but often are in complex form, frequently with organic constituents, and their role in water hardness may be minimal and difficult to define. Total hardness is defined as the sum of the calcium and magnesium concentration, both expressed as $CaCO_3$, in mg/1. The degree of hardness of drinking water has been classified in terms of the equivalent $CaCO_3$ concentration as follows:

Soft 0-60 mg/l , Medium 60-120 mg/l , Hard 120-180 mg/l , Very hard >180 mg/l

Although hardness is caused by cation, it may also be described in terms of carbonate (temporary) and non-carbonate (permanent) hardness. Carbonate hardness refers to the amount of carbonates and bicarbonates in solution that can be removed or precipitated by boiling. This type of hardness is responsible for the deposition of scale in hot water pipes and kettles. When total hardness is numerically greater than that of total alkalinity expressed as $CaCO_3$ the amount of hardness equivalent to alkalinity is called carbonate hardness. When the hardness is numerically equal to less than total alkalinity, all hardness is carbonate hardness. The amount of hardness in excess of total alkalinity expressed as $CaCO_3$ is non-carbonate hardness. Non-carbonate hardness is caused by the association of the hardness-causing cation with sulphate, chloride or nitrate and is referred to as "permanent hardness". This type of hardness cannot be removed by boiling. Public acceptability of the degree may vary considerably from community depending on local conditions, and the association.

Total Solids

The term 'solid' refers to the matter either filterable or non-filterable that remains as residue upon evaporation and subsequent drying at a defined temperature. Further categorization depends upon the temperature employed for drying and ignition. Different forms of solids are defined on the basis of method applied for their determination. Solids may affect water or effluent quality adversely in number of ways. Water with high dissolved solids may include an unfavorable physiological reaction in the transient consumer and generally are of inferior palatability. Highly mineralized waters are unsuitable for many industrial applications. High suspended solids in waters may be aesthetically unsatisfactory for such purposes as bathing. Analysis of total solids are important to decide upon the various unit operations & processes in physical & biological wastewater treatment and to asses its performance evaluation. For assessing compliance with regulatory agency, wastewater effluent limitations for various forms of solids act as indicating parameters.

Turbidity

Turbidity can be measured by its effect on the scattering light, which is termed as Nephelometry. Turbidimeter can be used for sample with moderate turbidity and nephelometer for sample with low turbidity. Higher the intensity of scattered lights higher the turbidity.

Turbidity is an expression of the optical property that causes light to be scattered and absorbed rather than transmitted in straight lines through the sample. The standard method for the determination of turbidity has been based on the Jackson candle turbidity meter. However, the lowest turbidity value that can be measured directly on this instrument is 25 units. An indirect method is necessary to estimate the turbidity in the range of 0-5 units; the turbidities of treated water generally fall in this range. Most commercial turbidimeters available for measuring low turbidities give comparatively good indicators of the intensity of light scattered in one particular direction, predominantly at right angle to the incident light. These nephelometers are relatively unaffected by small changes in design parameters and are therefore specified as the standard instrument for measurement of low turbidities. Results from nephelometric measurements are expressed as nephelometric turbidity units (NTU).

Alkalinity

Alkalinity of sample can be estimated by titrating with standard sulpuric acid (0.02N) at room temperature using phenolphthalein and methyl orange indicator. Titration to discoloration of phenolphthalein indicator will indicate complete neutralization of OH and ½ of CO_3, while sharp change from yellow to orange of methyl orange indicator will indicate total alkalinity (complete neutralization of OH, CO_3, HO_3).

Conductivity

Conductivity is the capacity of water to carry an electrical current and varies both with number and types of ions in the solutions, which in turn is related to the concentration of ionized substances in the water. Most dissolved inorganic substances in water are in the ionized form and hence contribute to conductance.

Oil and Grease

Oil and grease is any material recovered as a substance soluble in petroleum ether, hexane or n-hexane. It includes other materials extracted by the solvent from an acidified sample such as sulphur compounds, certain organic dyes and chlorophyll. Oil and grease are defined by the method used for their determination. The oil and grease content of domestic industrial wastes and of sludges, is an important consideration in the handling and treatment of these materials for ultimate disposal. When treated effluents are discharged in water body, it leads to environmental degradation. Hydrocarbons, esters, oils, fats, waxes and high molecular weight fatty acids are the major materials dissolved by hexane. All these materials have a 'greasy feel'. Three methods for oil and grease estimations are (i) the partition-gravimetric method, (ii) the partition infrared method and (iii) the Soxhlet extraction method. Though methods-(i) does not provide the needed precision, it is widely used for routine analysis of samples because of its simplicity and it needs no special instrumentation and (ii) is identical to hydrocarbons. In method (iii) adequate instrumentation allows for the measurement of as little as 0.2 mg oil and grease.

Chlorine (Residual)

The chlorination of water supplies and polluted waters serves primarily to destroy or deactivate disease-producing micro-organisms. A secondary benefit, particularly in treating drinking water, is the overall improvement in water quality resulting from the reaction of chlorine with ammonia, iron, manganese, sulphide and some organic substances.

Chlorination may produce adverse effects. Taste & odour characteristics of phenols and other organic compounds present in a water supply may be intensified. Potentially carcinogenic chloro-organic compounds such as chloroform may be formed. Combined chlorine formed on chlorination of ammonia or amine bearing waters adversely affects some aquatic life. To fulfil the primary purpose of chlorination and to minimise any adverse effects, it is essential that proper testing procedures be used with a before knowledge of the limitations of the analytical determination.

Chlorine applied to water in its molecular or hypochlorite form initially undergoes hydrolysis to form free chlorine consisting of aqueous molecular chlorine, hypochlorous acid and hypochlorite ion. The relative proportion of these free chlorine forms is pH and temperature dependent. At the pH of most waters, hypochlorous acid and hypochlorite ion will predominate.

Free chlorine reacts readily with ammonia and certain nitrogenous compounds to form combined chlorine. With ammonia, chlorine reacts to form the chloramines: monochloramine, dichloramine and nitrogen trichloride. The presence and concentrations of these combined forms depend chiefly on pH, temperature, initial chlorine-to-nitrogen ratio, absolute chlorine demand and reaction time. Both free and combined chlorine may be present simultaneously. Combined chlorine in water supplies may be formed in the treatment of raw waters containing ammonia or by the addition of ammonia or ammonium salts. Chlorinated wastewater effluents, as well as certain chlorinated industrial effluents, normally contain only combined chlorine. Historically the principal analytical problem has been to distinguish between free and combined forms of chlorine.

Phenols

The presence of organic compounds in water and wastewater is a matter of increasing concern to the water industry, environmentalists and general public. The impact of such compounds may differ. For example, biologically oxidisable compounds will deplete oxygen in the water body and petroleum products. Oil and grease will adversely affect biological activity. Phenols, detergents and other organic materials may be toxic to phyto and zoo-plankton beyond certain levels. Organic compounds may enter water environment through human waste disposal and industrial discharges. A variety of sensitive instruments have been developed to quantify the organic pollutants. BOD, COD, TOC, oil & grease, phenols, detergents-surfactants, and TFPs, are certain demand parameters considered to evaluate levels of contamination.

Phenols are defined as hydroxyl derivatives of benzene, and its condensed nuclei occur in domestic & industrial wastewaters, natural wastes and potable water supplies. Odoriferous and objectionable tasting chlorophenols are formed as a result of chlorination of water containing phenol. Phenols may be present in raw water owing to the discharge of wastewaters from coke distillation plants, the petrochemical industry and numerous other industries where phenols serve as intermediates. They are also present in municipal wastewaters.

Presence of phenols may lead to objectionable taste in chlorinated drinking water and hence its monitoring is essential. Phenols can be removed from drinking water by super-chlorination (chlorine-dioxide or chloramines treatment) ozonation and activated carbon adsorption.

Dissolved Oxygen (DO)

All living organisms are dependent upon oxygen in one form or the other to maintain the metabolic processes that produce energy for growth and reproduction. Aerobic processes are of great interest, which need free oxygen for wastewater treatment. Dissolved Oxygen (DO) is also important in precipitation and dissolution of inorganic substances in water.

DO levels in natural waters and wastewaters depend on physical, chemical and biological activities in water body. The solubility of atmospheric oxygen in fresh water ranges from 14.6 mg/l at 0°C to about 7.0 mg/l at 35°C under normal atmospheric pressure. Since it is poorly soluble gas, its solubility directly varies with the atmospheric pressure at any given temperature.

Analysis of DO is a key test in water pollution control and wastewater treatment processes. The following illustrations reveal importance of DO as a parameter:

- It is necessary to know DO levels to assess quality of raw water and to keep a check on stream pollution.
- In wastewaters, dissolved oxygen is the factor that determines whether the biological changes are brought out by aerobic or anaerobic organisms.
- DO test is the basis of BOD test which is an important parameter to evaluate pollution potential of Wastes.
- DO is necessary for all aerobic biological wastewater treatment processes.
- Oxygen is an important factor in corrosion. DO test is used to control the amount of oxygen in boiler feed waters either by chemical or physical methods.

Biochemical Oxygen Demand (BOD)

The Biochemical Oxygen Demand (BOD) is an empirical standardized laboratory test which measures oxygen requirement for aerobic oxidation of decomposable organic matter & certain inorganic materials in water, polluted waters and wastewater under controlled conditions of temperature & incubation period. The quantity of oxygen required for above oxidation processes is a measure of the test. The test is applied for fresh water sources (rivers, lakes), wastewater (domestic, industrial), polluted receiving water bodies, marine water (estuaries, coastal water) and also for finding out the level of pollution, assimilative capacity of water body & also performance of waste treatment plants.

Guideline BOD values for classification of raw untreated water is given below.

Table 12.5T$_1$: Guideline BOD values for classification of raw untreated water

Raw Water Quality Class	Designated best use	BOD value	Remarks
A	Drinking water source without conventional treatment but with chlorination.	2 or less	Could cause problems in treatment, larger Cl_2 demand and residual taste/odour problem.
B	Drinking water source with conventional treatment.	3 or less	

Chemical Oxygen Demand (COD)

Chemical Oxygen Demand (COD) test determines the oxygen requirement equivalent of organic matter that is susceptible to oxidation with the help of a strong chemical oxidant. It is an important, rapidly measured parameters as a means of measuring organic strength for streams and polluted water bodies. The test can be related empirically to BOD, organic carbon or organic matter in samples from a specific source taking into account its limitations. The test is useful in studying performance evaluation of wastewater treatment plants and monitoring relatively polluted water bodies. COD determination has advantage over BOD determination. COD results can be obtained in 3-4 hrs as compared to 3-5days required for BOD test. Further, the test is relatively easy, precise, and is unaffected by interferences as in the BOD test. The intrinsic limitation of the test lies in its inability to differentiate between the biologically oxidisable & biologically inert material and to find out the system rate constant of aerobic biological stabilization.

Total Organic Carbon (TOC)

The Total Organic Carbon (TOC) estimate provides speedy way of estimating organic carbon but does not permit to differentiate between biologically and chemically oxidisable portion of organic compounds.

The organic carbon in water and wastewater is composed of a variety of organic compounds in various oxidation states. Some of these carbon compounds can be oxidized by biological or chemical processes. Therefore, biochemical oxygen demand (BOD) and chemical oxygen demand (COD) are used conventionally. TOC is a more convenient and direct expression of total organic content. If a repeatable empirical relationship is established between TOC and BOD or COD, then TOC can be used to estimate the equivalent BOD or COD. The relationship must be established independently, for each set of matrix conditions, such as various points in a treatment process. Unlike BOD or COD, TOC is independent of the oxidation state of the organic matter and does not measure other organically bound elements, such as nitrogen & hydrogen, and inorganics that can contribute to the oxygen demand measured by BOD & COD. Hence, TOC measurement does not replace BOD and COD testing.

Phosphate (PO_4^{-3})

Phosphorous occurs in natural waters and in wastewater almost solely in the form of various types of phosphates. These forms are commonly classified into orthophosphates and total phosphates. These may occur in the soluble form, in particles of detritus or in the bodies of aquatic organisms. The various forms of phosphates find their way into wastewater, effluents and polluted water from a variety of sources. Larger quantities of the same compounds may be added when the water is used for laundering or other cleaning, since these materials are major constituents of many commercial cleaning preparations. Orthophosphates applied to agricultural or residential cultivated land as fertilizers are carried into surface waters with storm runoff and to a lesser extent with melting snow. Organic phosphates are formed primarily by biological processes. They are contributed to sewage by body wastes and food residues.

Presence of phosphates in water and wastewater analysis has a great significance. Phosphate in small concentration are used in water supplies to reduce scale formation, to increase carrying capacity of mains, to avoid corrosion in water mains, to remove iron & manganese in micro quantities and in coagulation especially in acid conditions. The presence of phosphate in large quantities in fresh waters indicates pollution through sewage and industrial wastes. It promotes growth of nuisance causing micro-organisms. Though phosphate posseses problems in surface waters, its presence is necessary for biological degradation of wastewaters. Phosphorus is an essential nutrient for the growth of organisms and helps for the primary productivity of a body of water.

Nitrogen (Ammonia)

Ammonia is produced by the microbiological degradation of organic nitrogenous matter. It appears, therefore, in many ground waters as well as surface waters. Concentrations of ammonia above a certain level in water polluted either due to sewage or industrial waste is toxic to fish. The proportions of the two forms of ammonia nitrogen in surface water depend on pH.

pH	**6**	**7**	**8**	**9**	**10**	**11**
%NH_3	0	1	4	25	78	96
%NH_4	100	99	96	75	22	4

For accurate results, it is generally preferable to distill off ammonia from the sample, and absorb in boric acid. It is then determined either by titration or colorimetrically using Nessler reagent. Direct nesslerization of the sample is quicker depending upon interference.

Nitrogen (Nitrate) (NO_3)

Determination of nitrate (NO_3) is difficult because of the relatively complex procedures required, the high probability that interfering constituents will be present and the limited concentration ranges of the various techniques. Nitrate is the most highly oxidised form of nitrogen compounds commonly present in natural waters. Significant sources of nitrate are chemical fertilizers, decayed vegetable & animal matter, domestic effluents, sewage sludge disposal to land, industrial discharge, leachates from refuse dumps and atmospheric washout. Depending on the situation, these sources can contaminate streams, rivers, lakes and ground water. Unpolluted natural water contains minute amounts of nitrate. Excessive concentration in drinking water is considered hazardous for infants because of its reduction to nitrite in intestinal track causing methemoglobinaemia. In surface water, nitrate is a nutrient taken up by plants and converted into cell protein. The growth stimulation of plants, especially of algae may cause objectionable eutrophication.

Nitrogen (Nitrite) (NO_2)

Nitrite in water is either due to oxidation of ammonium compounds or due to reduction of nitrate. As an intermediate stage in the nitrogen cycle, it is unstable. A usual concentration in natural water is in the range of some tenths of mg/1. Higher concentrations are present in industrial wastes, sewage and in biologically purified effluents and in polluted streams. In chlorinated supplies, levels of nitrite are often less than the limit of detection, i.e. 0.005 mg/l , but high levels may occur in unchlorinated water. Very high nitrite levels are usually associated with water of unsatisfactory microbiological activity.

Sulphate (SO_4^-)

Sulphate ions usually occur in natural waters. Many sulphate compounds are readily soluble in water. Most of them originate from the oxidation of sulphate ores, the solution of gypsum & anhydrite, the presence of shales, particularly those rich in organic compounds, and the existence of industrial wastes. Atmospheric sulphur dioxide formed by the combustion of fossil fuels and emitted by the metallurgical roasting processes may also contribute to the sulphate compounds of water. Sulphur trioxide (SO_3) produced by the photolytic oxidation of sulphur dioxide comes with water vapours to form sulphuric acid which is precipitated as acid rain or snow. Sulphur bearing mineral are common in most sedimentary rocks. In the weathering process , gypsum (calcium sulphate) is dissolved and sulphide minerals are partly oxidised, giving rise to a soluble form of sulphate that is carried away by water. In humid region, sulphate is readily leached from the zone of weathering by infiltrating waters and surface run off but in semiarid & arid regions the soluble salts may accumulate within a few tens of feet of land surface. Where this occurs, sulphate concentration in shallow ground water may exceed 5000 mg/l and gradually decrease with depth.

Ingestion of water containing high concentration of sulphate can have a laxative effect, which is enhanced when sulphate is consumed in combination with magnesium. Water containing magnesium sulphate at levels about 1000 mg/l acts as a purgative in human adults. Taste threshold concentrations for the most prevalent sulphate salts are 200-500 mg/l for sodium sulphate , 250-900 mg/l for calcium sulphate, and 400-600mg/l for magnesium sulphate. Essentially on the basis of above values, which are also allied to the cathartic effect of sulphate, a guidelines value of 400 mg/l is proposed. Sulphates cause scaling in water supplies, and problem of odour & corrosion in wastewater treatment due to its reduction to H_2S.

Boron (B)

Boron is an essential element for plant growth. However, in excess of 2 mg/l in irrigation water, it is deleterious to certain plants and some plants may be affected adversely by concentrations as low as 1mg/l (or even less in commercial green-houses). Drinking water rarely contain more than 1 mg/l and generally less than 0.1 mg/l , concentrations considered innocuous for human consumption. Boron may occur naturally in some water or may find its way into a water course through cleaning compounds and industrial effluents. Seawater contains approximately 5 mg/l B. The ingestion of large amounts of boron can affect the central nervous system. Protracted ingestion may result in a clinical syndrome known as **"Borism"**.

Sodium (Na)

Sodium ranks sixth among the elements in order of abundance and is present in most natural water. The levels may vary from less than 1 mg Na/1 to more than 500 mg Na/1. Relatively high concentrations may be found in brines and hard water softened by the sodium exchange process. The ratio of sodium to total cations is important in agriculture and human pathology. Soil permeability can be harmed by a high sodium ratio. Persons afflicted with certain diseases require water with low sodium concentration. A limiting concentration of 2 to 3 mg/1 is recommended in feed water destined for high-pressure boilers. When necessary, sodium can be removed by the hydrogen-exchange process or by distillation. Sodium compounds are used in many applications, including caustic soda , fertilizers and water treatment chemicals.

Potassium (K)

Potassium ranks seventh among the elements in order of abundance, yet its concentration in most drinking water seldom reaches 100 mg/l. Potassium is an essential element in both plant & human nutrition and occurs in ground water as a result of mineral dissolution.

Do not store water samples in soft-glass bottles because of the possibility of contamination from leaching of the glass. Use acid washed polyethylene or borosilicate glass bottles. Adjust sample to pH < 2 with nitric acid. This will dissolve potassium salts and reduce adsorption on vessel walls.

Fluoride (F^-)

Fluoride ions have dual significant in water supplies. High concentration of F causes dental fluorosis (disfigurement of the teeth). At the same time, a concentration less than 0.8 mg/l results in "dental caries". Hence, it is essential to maintain the F concentration between 0.8 to 1.0 mg/l in drinking water. Among the many methods suggested for the determination of fluoride ion in water, the colorimetric method and the ion selective electrode method are the most satisfactory and applicable to variety of samples. Because all of the colorimetric methods are subject to errors due to presence of interfering ions, it may be necessary to distill the sample before making the fluoride estimation.

Chloride (Cl^-)

The presence of chloride in natural waters can be attributed to dissolution of salt deposits, discharges of effluents from chemical industries, oil well operations and seawater intrusion in coastal areas. Each of these sources may result in local contamination of both surface water and groundwater. The salty taste produced by chloride depends on the chemical composition of the water. A concentration of 250 mg/l may be detectable in some waters containing sodium ions. On the other hand, the typical salty taste may be absent in water containing 1000 mg/l chloride when calcium and magnesium ions are predominant. High chloride content may harm pipes and structures as well as agricultural plants.

Aluminium (Al)

Aluminium is the third most abundant element on the earth's crust. The presence of aluminium in all natural water is in the form of soluble salt, a colloid, or an insoluble compound. Soluble, colloidal, and insoluble aluminium may be present in treated water in residual form of coagulation with aluminium containing material. The USEPA water standard for Al is 0.05 mg/l max. BIS desirable limit is 0.03 mg Al/1.

Arsenic (As)

Severe poisoning can arise from the ingestion of as little as 100 mg of Arsenic trioxide. Chronic effects may result from the accumulation of arsenic compounds in the body at low intake levels. Carcinogenic properties also have been observed in arsenic compounds. For the protection of aquatic life, the average concentration of As in water should not exceed 72 ug/l and maximum should not exceed 140 ug/1. FAO-recommended max level for irrigation water is 100 µg/l. USEPA water standard is 0.05 mg/l. BIS desirable limit is 0.05 mg As/1

Cadmium (Cd)

Cadmium occurs in sulphide minerals that also contain zinc, lead or copper. The metal is used in electroplating, batteries, paint pigments and in alloys with various other metals. Cadmium is usually associated with zinc. Cadmium is highly toxic and has been implicated in some cases of poisoning through food. Minute quantities of cadmium are suspected of being responsible for adverse changes in arteries of human kidneys. Cadmium also causes generally cancers in laboratory animals and has been linked epidemiologically with certain human cancers. A cadmium concentration of 200 ug/l is toxic to certain fishes. Cadmium may enter water as a result of industrial discharges or the deterioration of galvanized pipe. The FAO recommended maximum level for cadmium for irrigation water is 10 ug/l. USEPA drinking water standard for cadmium is 0.005 mg/l. BIS desirable limit is l mg/1.

Chromium (Cr)

Chromium is found chiefly in chrome-iron ore. Chromium is used in alloys, in electroplating and in pigment. Chromium salts are used extensively in industrial processes and may enter a water supply through the discharge of wastes. Chromate compounds frequently are added to cooling water for corrosion control. Chromium may exist in water supplies in both the hexavalent and the trivalent state although the trivalent form rarely occurs in potable water. FAO recommended the maximum limit for irrigation water is 100 u.g/l. USEPA drinking water standard for chromium is 0.1 mg/1.

Copper (Cu)

Copper occurs in nature in its native state, but also found in many minerals, the most important of which are those containing sulphide compounds, but also those with oxides or carbonates. Copper salts are used in water supply systems to control biological growth of reservoirs & distribution pipes and to catalyse in oxidation of manganese. Corrosion of copper-containing alloys in pipe fitting may introduce measurable amounts of copper into the water in a pipe system. FAO recommended maximum level for irrigation water is 200 ug/l. USEPA drinking water standard for chromium and copper is 1 mg/1. BIS limit for copper is 0.05 mg/l.

Iron (Fe)

Iron occurs in the minerals as hematite, taconite and pyrite. It is widely used in steel and other alloys. Elevated iron levels in water can cause stains in plumbing, laundry & cooking utensils and can impart objectionable taste & colour to foods. The United Nations FAO recommended level for irrigation water is 5mg/l. The USEPA secondary drinking water standard MCL is 0.3 mg/l. The BIS standard desirable limit is 0.3 mg/l.

Lead (Pb)

Lead in a water supply may come from industrial, smelter discharges and mine or from the dissolution of plumbing & plumbing fixture. Tap water that are inherently not corrosive or not suitably treated may contain lead resulting from an attack on lead service pipes, lead interior plumbing, brass fixtures and fittings on solder pipe joints chiefly from galena (PbS). It is used in batteries, ammunition, solder, piping, pigments, insecticides and alloys. Lead also was used in gasoline for many years as an anti-knock agent in the form of tetraethyl lead. The Food and drug administration regulates lead content in food and in house paints. USEPA drinking water standard for lead is 15 ug/l. The BIS desirable limit is 0.05 mg/l.

Manganese (Mn)

Manganese is associated with iron minerals and occurs in nodules in ocean, fresh water and soils. The common ores are pyrolusite and psilomelane. Manganese is used in steel alloys, batteries and food additives. The FAO recommended maximum level for manganese is 50 ug/l. BIS desirable limit is 0.1 mg/1.

Nickel (Ni)

The average abundance of Ni in the earth's crust is 1.2 ppm; in soils it is 2.5 ppm; in streams it is l ug/1, and in ground water it is <0.1 mg/1. Nickel is obtained chiefly from pyrrhotite and garnierite. Nickel is used in alloys, magnets, protective coating, catalysts, and batteries. The common aqueous species is Ni^{2+}. In reducing conditions insoluble sulphide can form, while in aerobic conditions insoluble sulphide can form, while in anerobic conditions ligands can from. It is suspected to be an essential trace element for some plants and animals. The standards of United Nations -FAO for irrigation water is 200 ug/l. The USEPA primary drinking water standard MCL is 0.1 mg/1.

Zinc (Zn)

Zinc is an essential and beneficial element for human growth. Concentration above 5 mg/l can cause a bitter and stringent taste in water. The zinc concentration in water varies from

0.06 to 7.0 mg/l with a mean concentration of 1.33 mg/l. Zinc most commonly enters the domestic water supply from deterioration of galvanized iron and dezincification of brass. In such cases lead and cadmium also may be present because they are impurities of the zinc used in galvanising. Zinc in water also may result from industrial waste pollution. The FAO recommended maximum level for zinc in irrigation water is 2 mg/l. The USEPA primary drinking water standard MCL is 5 mg/l.

Cyanide (CN)

Cyanide refers to all of the CN groups in cyanide compounds that can be determined as the cyanide ion, CN" by the methods used. The cyanide compounds in which cyanide can be obtained as CN are classed as simple and complex cyanide.

MICROBIOLOGICAL CONTAMINANTS OF WATER

Water borne pathogens can enter the human body through skin, inhalation, ingestion, aspiration, or by direct contact with mucosa of the eye, ear, nose, mouth & genitals, and cause disease. Besides failure of potable water systems, which usually contribute to the greatest number of outbreaks of waterborne diseases, common outdoor recreational activities such as swimming, boating, bathing, camping and hiking, all place humans at risk of waterborne diseases from ingestion or direct contact with contaminated water. A risk analysis approach is required for overcoming the problems with waterborne diseases and water quality monitoring. While this could be achieved relatively easily with respect to chemical contaminants, and standards could be set up for chemical contaminants, setting standards for microbial contaminants has proved more difficult, because, (a) methods for detection of many pathogens are not yet available; (b) days to weeks are sometimes required to obtain results; and (c) expensive and time consuming methods are often required to detect & quantify the presence of specific pathogens.

To overcome these difficulties, the concept of testing water samples for indicator organisms has been developed for assessing faecal contamination of water and the concept relies on the fact that certain non-pathogenic bacteria present in water can be easily isolated & quantified by simple bacteriological methods; detection of these bacteria would essentially mean that faecal contamination has occurred and also suggest that enteric pathogens also may be present.

No single indicator provides assurance that water is pathogen-free. Pathogens such as *Psedomonas, Aeromonas, Plesidomnas, Yersinia, Vibrio, Legionella and Mycobacterium* may not correlate with the most commonly used coliform indicators, and also, traditional bacteria indicators may not also correlate with viruses & parasites in pristine waters or ground waters.

Coliform bacteria

Coliform bacteria include all aerobic and facultative anaerobic germ negative, non sore forming, rod-shaped bacteria that produce gas upon fermentation in prescribed culture media within 48 hr at 35°C. In recent years, new criteria have been added to traditional definition of coliform bacteria and E. *coli.* using the presence of characteristic enzymatic activities. Definition of coliform or faecal coliform basically relies on the activity of a single enzyme B-*galactosidase.* The new enzymatic definition of total coliform bacteria is based on the presence of B- *galactosidase;* and that of E. *coli* is based on the enzymatic action of B- *glucuronidase.*

Deficiencies with the use of coliform bacteria as indicator of water quality include:

- Regrowth in aquatic environments.
- Regrowth in distribution systems, including biofilm colonization.
- Suppression by high background bacterial growth.
- Not indicative of health threat.
- No relationship with enteric protozoan and viral concentration.

Coliform bacteria had been used historically to assess the microbial quality of drinking water. However, coliform bacteria does not indicate the presence of pathogenic waterborne *Giardia* or enteric viruses. Numerous outbreaks have occurred in which coliform standards were met, because of greater resistance of viruses and *Giardia* to chlorination. A new approach is needed to ensure the microbial safety of drinking water.

Most Probable Number (MPN), the Membrane filter (MF) and presence-absence (P/A) tests are used to detect the total coliforms.

Faecal coliform bacteria

Faecal coliform bacteria are differentiated in the laboratory by their ability to ferment lactose, with production of acid and gas at 44.5°C within 24 hours.

Faecal coliforms pose some of the same limitations as those posed by coliforms (Regrowth in distribution system, less resistance to water treatment than viruses and protozoa, etc.) Faecal coliforms are also detected by similar methods (MPN, MF and P/A) used for total coliforms.

Faecal streptococci

Faecal streptococci include the gram-positive, Lancefield Group-D streptococci, belonging to the genera *Enterococcus* and *Streptococcus*.

The genus *Enterococcus* is differentiated from other streptococci by ability to grow in 6.5% NaCl, pH 9.6 and 45°C (includes spp. E. *avivum,* E. *faecium,* E. *faecalis* and E. *gallinarium).* E. faecium and E. faecalis are more specific to human gut. Both MPN and MF methods may be used for isolation and enumeration of faecal streptococci.

Clostridium perfringens

They are sulphite-reducing anaerobic spore-formers, gram-positive, rod-shaped and exclusively of faecal origin. The hardly spores (resistant to 75°C for 15 min.) limit its usefulness as an indicator; it could be used as an indicator of past pollution and an indicator of removal of protozoan parasites and viruses during treatment of drinking water & wastewater.

Pseudomonas aeruginosa

They are gram-negative, non-sporulating, rod-shaped bacteria, producing characteristics pyocyanin (blue-green) and the green fluorescent pigment fluorescein. One of the most common opportunistic pathogens. Numerous cases of folliculitis, dermatitis, ear and urinary tract infections are due to P. *aeruginosa,* associated with swimming in contaminated waters. Because of this association and also its consistent presence in high numbers in sewage, this organism has been suggested as an indicator for water quality in swimming pools, hot tubs and other recreational waters. But its value as an indicator is very little, because it is ubiquitous in nature and can multiply under natural conditions (can grow even in distilled water).

Hydrogen sulphide (H_2S) producing bacteria

H_2S producing bacteria are invariably present in faeces. There is a very strong correlation between presence of H_2S producing bacteria and faecal pollution of water. By detecting the presence of H_2S producing bacteria in water samples, faecal pollution of such waters can be deduced.

Coli phages and other bacteriophages

Their use as indicators of faecal pollution is based on the assumption that their presence in water samples denotes the presence of bacteria capable of supporting replication of the phages. Advantages of using Coli phages is that they can be detected by simple and inexpensive techniques, in 8-18 hrs. Both plating and MPN method can be used to detect coliphages, in volumes ranging from 1-100 ml of water.

Another bacteriophages of interest is the bacteriophages that infect *Bacleriodes fragilis.* These bacteria (Bacteriodes spp.) are strict anaerobes and are a major component of the human faecal microflora. The bacteriophages that infect *Bacteriodes fragilis* appear to be exclusively of human origin and can serve as an ideal indicator organism. The main disadvantage is that host organism (B. *fragilis)* is an anaerobic organism, so the detection of the indicator organism would involve tedious and complicated methodologies.

The Coli phages and other bacteriophages have also been suggested as indicators of viral pollution. This is because the structure, morphology and size, as well as their behaviour in the aquatic environment closely resemble those of enteric viruses. These bacteriophages are quite useful in evaluation of virus resistance to disinfectants.

An essential goal for the provision of safe drinking water is that it be essentially free of (at low risk of containing) disease-causing micro-organisms. Since the beginning of the 20th century, the detection of faecal indicator bacteria in drinking water has been used as the bases of criteria, guidelines & standards for acceptable limits of faecal contamination and as the basis for judging or predicting the possible presence or absence or pathogenic (disease-causing) micro-organisms. The WHO Guidelines for Drinking Water Quality & many other authorities continue to support the use of bacterial indicator levels and their measurement as a basis for judging & verifying drinking water quality. However, such faecal indicator analysis of drinking water as a measure of end-product quality and determinant of microbial disease risk is only one of many measures and activities in an overall system for providing safe drinking water.

Water borne pathogens enter the drinking water systems through faecal contamination, and the list of such organisms (which include bacteria, viruses, protozoans and parasites) is a very long one. It would neither be practicable, nor is it necessary, to test a sample of drinking water for all the pathogens that might possible be present in it, as that would involve many different procedures, a multitude of media and other materials, incubation conditions, etc. It would, therefore, be sufficient if we can determine whether the water sample was contaminated with faces, such water is not fit for human consumption, as there would be a possibility that such water might harbour pathogenic organisms. Therefore, by establishing faecal contamination of water, we will be able to infer upon its potabilility.

In the laboratory there are three simple procedures of testing water samples for faecal contamination:

H_2S Strip method: This is a simple, reliable and easy to perform. 'Presence / Absence" test for bacteriological quality, which works on the principle that there is a close correlation between faecal contamination and the presence of hydrogen sulphide (H_2S) producing bacteria and, the faecal pollution of water can be deduced by demonstration of H_2S production. H_2S strip method shows >90% agreement with the conventional MPN test.

Most Probable Number (MPN) method for coliform bacteria-using the multiple tube fermentation technique: In this method the MPN of total coliform bacteria, faecal coliform bacteria (or the thermos tolerant coliforms) present in the water sample is determined, along with the presence / absence *of Escherichia coli.*

APPENDIX – A

Short Q & A On Water Pollution Control Engineering

Q. What are the main sources of water ?

Ans. The main sources of water include rainwater, surface water and groundwater. The distribution of water varies as per its locations. Water also exists on earth as a solid, liquid or gas, oceans, rivers, clouds, and rain, all of which contain water, are in a frequent state of change.

Water Cycle recycles the earth's water supply. The sun is the energy that powers this process. Its energy in the form of light, & heat causes water to evaporate from oceans, rivers, lakes etc. Warm air currents rising from the earth's surface lift this water vapour up into the atmosphere.

When the air currents reach the cooler layers of the atmosphere, the water vapour condenses around and clings on to fine particles in the air. This step is called condensation. When enough vapours attach itself to tiny pieces of dust, pollen or pollutants, it forms a cloud. As the air gets more and more moist, the droplets that form the clouds grow larger and larger. Eventually they will get so big that the swirling atmospheric winds can no longer hold them up. The droplets then fall from the sky as precipitation.

Precipitation can be in the form of rain, snow, sleet or hail depending on other atmospheric conditions such as temperature. Some of the precipitation will be absorbed into the ground. This is called infiltration. Once in the ground, the water can join the earth's groundwater supply. This is one of the world's largest storehouses of water. The water could also be absorbed from the ground by the roots of plants.

Q. Describe the water quality classification system followed in India.

Ans. In India, five water quality classes have been designated (A-E) on the basis of the water quality requirements for a particular use:

***Class A* waters**: for use as drinking water source without conventional treatment but after disinfection.

***Class B* waters**: for use for organised outdoor bathing.

***Class C* waters:** for use as drinking water source with conventional treatment followed by disinfection.

***Class D* waters:** to maintain aquatic life (i.e. propagation of wildlife and fisheries).

Class E **waters:** for use for irrigation, industrial cooling and controlled waste disposal.

The five classes have been used to set quality objectives for stretches of the Yamuna and Ganga rivers, and surveys have been carried out to compare the actual river-quality classification with that required to sustain the designated best use. Where a river has multiple uses, the quality objectives are set for the most stringent (best) use requirements. After comparing ambient water quality with the designated water quality objective, any deficiencies will require appropriate pollution control measures on the discharges, including discharges in upstream stretches. This system is also helpful for the planning and siting of industry.

No industries are permitted to discharge any effluent in stretches of rivers classified in Class A.

Q. Describe the features & properties of water.

Ans. Features & properties of water are described below:

- **Water has a high specific heat-**Since water has a high specific heat, it can absorb large amounts of heat energy before it begins to get hot. It also means that water releases heat energy slowly when situations cause it to cool. Water's high specific heat allows for the moderation of the Earth's climate and helps organisms regulate their body temperature more effectively.
- **Water in a pure state has a neutral pH.-**Pure water is neither acidic nor basic. Water changes its pH when substances are dissolved in it. Rain has a naturally acidic pH of about 5.6 because it contains natural derived carbon dioxide ,nitrogen oxides and sulfur dioxide.
- **Water conducts heat more easily than any other liquid except mercury. -**This feature causes large bodies of liquid water like lakes and oceans to have a uniform vertical temperature profile.
- **Water molecules exist in liquid form over a wide range of temperature from 0 to 100 ^{0}C -** This range allows water molecules to exist as a liquid in most places on our planet.
- **Water is a universal solvent.-** It is able to dissolve a large number of different chemical compounds. This feature also enables water to carry solvent nutrients in runoff, infiltration, groundwater flow, and also in living organisms.
- **Water has a high surface tension-** Water is adhesive & elastic, and tends to aggregate in drops rather than spread out over a surface as a thin film. This phenomenon also causes water to stick to the sides of vertical structures despite gravity's downward pull. Water's high surface tension allows for the formation of water droplets & waves, allows plants to move water (and dissolved nutrients) from their roots to their leaves, and the movement of blood through tiny vessels in the bodies of some animals.
- Water molecules are the only substance on Earth that exists in all three physical **states of matter: solid, liquid, and gas-**This feature plays an important role in the redistribution of heat energy in the Earth's atmosphere. In terms of heat being transferred into the atmosphere, approximately 75 % of this process is accomplished by the evaporation and condensation of water.

- **The freezing of water molecules causes their mass to occupy a larger volume-** When water freezes it expands rapidly adding about 9% by volume. Fresh water has a maximum density at around 4 ^{0}C. Water is the only substance on this planet where the maximum density of its mass

Q. What are five main physical actions involved in hydrological/water cycle ?

Ans. The water cycle - technically known as the hydrologic cycle - is the circulation of water within the earth's hydrosphere, involving changes in the physical state of water between liquid, solid, and gas phases. The hydrologic cycle refers to the continuous exchange of water between atmosphere, land, surface and subsurface waters, & organisms. In addition to storage in various compartments (the ocean is one such compartment); the multiple cycles that make up the earth's water cycle involve five main physical actions: ***evaporation, precipitation, infiltration, runoff, and subsurface flow.***

Evaporation – It occurs when radiant energy from the sun heats water, causing the water molecules to become so active that some of them rise into the atmosphere as vapour. It is the transfer of water from bodies of surface water into the atmosphere. This transfer entails a change in the physical nature of water from liquid to gaseous phases. Along with evaporation can be counted transpiration from plants. Thus, this transfer is referred to as evapotranspiration. About 90% of atmospheric water comes from evaporation, while remaining 10% is from transpiration. Transpiration occurs when plants take in water through roots and release it through the leaves, a process that can clean water by removing contaminants and pollution. Evapotranspiration is water *evaporating* from the ground and *transpiration* by plants. Evapotranspiration is also the way water vapour re-enters the atmosphere.

Precipitation. In cold air way up in the sky, rain clouds form. Rising warm air carries water vapor high into the sky where it cools, forming water droplets around tiny bits of dust in the air. Some vapor freezes into tiny ice crystals which attract cooled water drops. The drops freeze to the ice crystals, forming larger crystals called snowflakes. When the snowflakes become heavy, they fall. When the snowflakes meet warmer air on the way down, they melt into rain drops. In tropical climates, cloud droplets combine together around dust or sea salt particles. They bang together and grow in size until they're heavy enough to fall. Sometimes there is a layer of air in the clouds that is above freezing, or 0^{0}C. Then closer to the ground the air temperature is once again below freezing. Snowflakes partially melt in the layer of warmer air, but then freeze again in the cold air near the ground. This kind of precipitation is called sleet. It bounces when it hits the ground. If snowflakes completely melt in warmer air, but temperatures are below freezing near the ground, rain may freeze on contact with the ground or the streets. This is called freezing rain, and a significant freezing rain is called an ice storm which are extremely dangerous because the layer of ice on the streets can cause traffic accidents. Ice can also build up on tree branches & power lines, causing them to break and our lights to go out.

There is another kind of precipitation that comes from thunder storms called hail.

Infiltration. Under some circumstances precipitation actually evaporates before it reaches the surface. More often, though, precipitation reaches the Earth's surface, adding to the surface water in streams and lakes, or infiltrating a portion of the precipitation that reaches the earth's surface seeps into the ground through the process called infiltration. Infiltration into the ground is the transition from surface water to groundwater. The infiltration rate will depend upon soil or rock permeability as well as other factors. Infiltrated water may reach another compartment known as groundwater (i.e., an aquifer). Groundwater tend to move slowly, so the water may return as surface water after storage within an aquifer for a period of time that can amount to thousands of years in some cases. Water returns to the land surface at lower elevation than where it infiltrated, under the force of gravity or gravity induced pressures.

Runoff, The amount of water that infiltrates the soil varies with the degree of land slope, the amount & type of vegetation, soil & rock type, and whether the soil is already saturated by water. The more is openings in the surface (cracks, pores, joints), the more infiltration occurs. Water that doesn't infiltrate the soil flows on the surface as runoff. Precipitation that reaches the surface of the earth but does not infiltrate the soil is called **runoff.** Runoff can also come from melted snow and ice. It also includes the variety of ways by which land surface water moves down slope to the oceans. Water flowing in streams and rivers may be delayed for a time in lakes. Not all precipitated water returns to the sea as runoff; much of it evaporates before reaching the ocean or reaching an aquifer.

Subsurface Flow. Surface flow incorporates movement of water within the earth, either within the recharge zone or aquifers. After infiltrating, subsurface water may return to the surface or eventually seep into the ocean.

Q. What are ground water resources ?

Ans. About 70% of the rainfall is lost to evaporation. Most of the rest runs off the surface into streams & lakes, and eventually returns to the sea. Some, however, slowly percolates down through the soil until it reaches a layer of impermeable rock. The water which builds up on the top of the impermeable rock fills all the openings, crevices, & pores in the overlying sand and rock, to a level called the *water table.* It then very slowly flows to the oceans. When the ground level drops below the water table, the ground water helps supply streams and lakes giving a continuity of discharge they would not otherwise have. Groundwater resources are extremely important because they supply to water withdrawal. Difficulties exist with this in that the rate of replenishment of these supplies is so slow that they must be considered to be essentially nonrenewable resources.

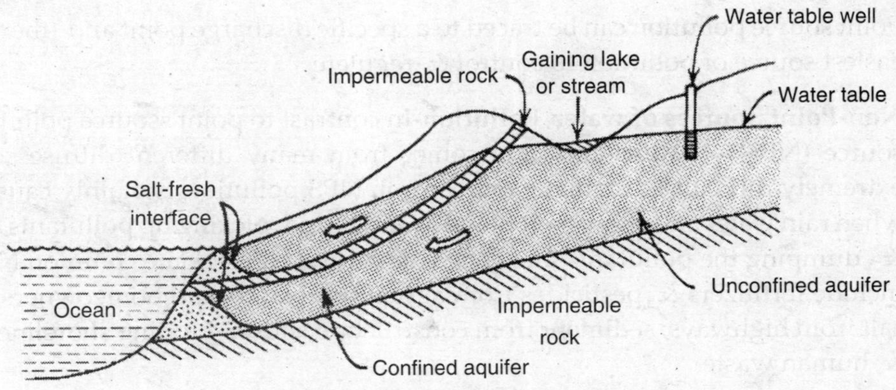

Fig.1: Groundwater resources

One problem that can result from a drop in the water table is a corresponding sinking of the ground level in a process called subsidence. Another problem that can result from overexploitation of groundwater is that the water table can drop so low that it becomes uneconomical to pump it out. When this happens, the area involved may lose its principal water supply. Coastal areas face another sort of groundwater problem because of the close proximity of saline and fresh water. Under natural conditions there is a balance between the inflow of salt water and the outflow of fresh water, which establishes a boundary between the two regimes. However, as freshwater is pumped out of the area, the water table drops and the boundary moves inland. After a period of time, a coastal freshwater well may need to be abandoned because of this intrusion.

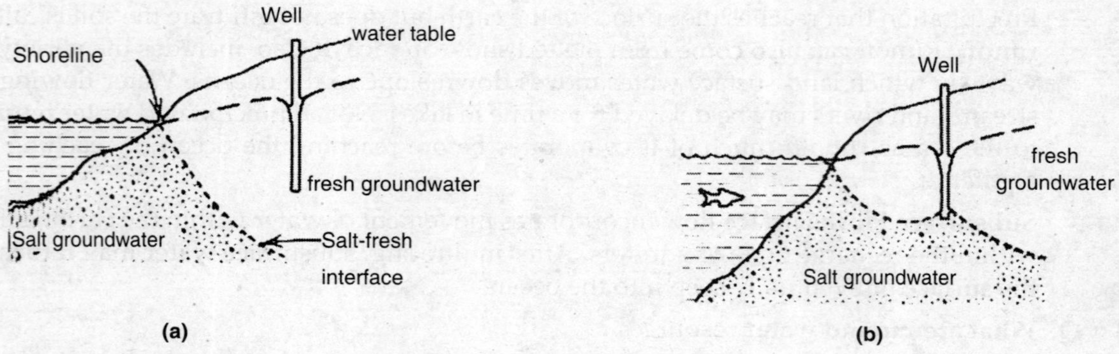

Fig.2: Saltwater intrusion into a freshwater well due to over-pumping. (a) before, (b) after.

Q. What do you mean by point sources and non-point sources of water pollution ?

Ans. Point Sources of water Pollution-When pollutants enter the river through a specific entry point, such as a drain pipe draining directly into a river, it's called point source pollution. Industrial water discharges and sewage treatment plants are the main sources of this type of pollution. Point source pollutants can include many different organic & inorganic substances, including human waste and toxic metals.

Point source pollution can be traced to a specific discharge point and therefore, it is the easiest source of pollution to control & regulate.

Non-Point Sources of water Pollution-In contrast to point ssource pollution, nonpoint source (NPS) of water pollution comes from many different diffuse sources and is extremely difficult to regulate and control. NPS pollution is mainly caused by runoff, when rain water and snow melt move over the land, picking up pollutants along the way & dumping the pollutants into rivers, tanks and lakes. Some common NPS pollutants include fertilizers & pesticides from agricultural lands and homeowners; oil, grease & salt from highways; sediment from construction sites & eroding shorelines; and animal & human waste.

Q. What are various categories of water pollutants ?

Ans. Various Types of water Pollutants are as follows:

- ***Oxygen-demanding wastes*** are biodegradable organic compounds contained in domestic sewage or certain industrial effluents. When these compounds are decomposed by bacteria, oxygen is removed from the water. If the oxygen level drops low enough, the fish will die.
- ***Disease-causing agents*** are various pathogenic microorganisms which usually enter the water with human sewage. Contact with these microbes can be made by drinking the water or through various water-contact activities.
- ***Synthetic organic compounds*** include detergents & other household aids, pesticides, and various synthetic industrial chemicals. Many of these compounds are toxic to aquatic life and may be harmful to humans.
- ***Plant nutrients*** such as the nitrogen & phosphorus which drains from fertilized lands, as well as the effluent from most sewage treatment plants, stimulate the growth of algae and water weeds.
- ***Inorganic chemicals and mineral substances*** includes the acids which form when water drains from abandoned mines, as well as the heavy metals such as mercury & cadmium.
- ***Sediments*** are particles of soils, sands, and minerals washed from the land. They can smother bottom life such as shellfish & coral, as well as fill in reservoirs and harbors. Improper soil management leading to erosion is a major contributor.
- ***Radioactive substances*** can enter the water from the mining & processing of radioactive ores, from various nuclear power operations, from medical facilities, and from nuclear weapons testing.
- ***Thermal discharges of hot water*** from steam-electric power plants raise the temperature of the receiving water by 7-10 ^{0}C resulting in various changes in local eco systems.

Q. What is the importance & implication of Dissolved Oxygen in water ?

Ans. The amount of dissolved oxygen (DO) in water is an important parameter of water quality. Fish, for example, require certain minimum amounts of DO for their survival depending upon their species, stage of development, level of activity, and the water temperature. For example, for a well-rounded, warm water fish population, it is recommended that the DO remains above 5 ppm for at least 16 hours of the day and during the other 8 hours it should not drop below 3 ppm.

There are four processes which affect the amount of oxygen in the water: **Reaeration, photosynthesis, respiration, and the oxidation of wastes**.

Reaeration is the process wherein oxygen enters the water through the contact the water surface makes with the atmosphere. The solubility of oxygen in fresh water, at a pressure of 1 atmosphere, decreases with temperature as shown in Figure below.

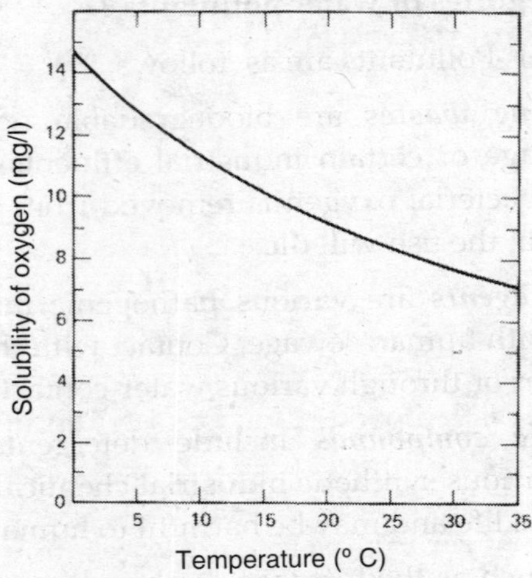

Fig.3: Solubility of oxygen in fresh water exposed to water saturated air at 1 atmosphere pressure.

When the actual amount of oxygen in the water is less than the saturation value given in the figure, atmospheric oxygen passes into the water at a rate which is proportional to the deficit. By increasing the surface area in contact with the atmosphere the transfer of oxygen in contact with the atmosphere is increased, so a bubbling stream will pick up oxygen easier than a stagnant pond.

During photosynthesis, which of course occurs only during the daylight hours, oxygen is liberated, thereby increasing the DO level in the water. Respiration, however, is a process which continuously removes oxygen from the water. Combining the three effects of photosynthesis, respiration, and reaeration produces a diurnal variation of DO as shown in Figure below. In this figure, photosynthesis is assumed to occur from 6 AM to 6 PM, bringing the DO level to slightly above saturation in the afternoon. While the water is supersaturated, oxygen diffuses out of the water instead of into it.

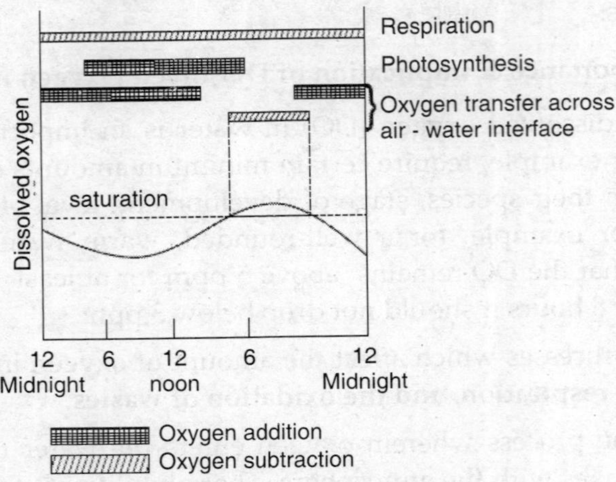

Fig.4: Diurnal variation of DO assuming photosynthesis from 6 AM to 6 PM.

The fourth process which affects the amount of oxygen in the water is the oxidation of wastes. Microorganisms, especially bacteria, use organic wastes as food and in the process break down the complex organics into simple organic & inorganic materials. This decomposition may occur in the presence of oxygen, in which case it is called *aerobic* decomposition, or it may occur in the absence of oxygen in which case it is called *anaerobic* decomposition.

The general form of the equation for aerobic decomposition of organic material is

Organic matter + bacteria + $O_2 \rightarrow CO_2 + H_2O$ + new bacterial cells

The reaction products from aerobic decomposition are not obnoxious, being simply carbon dioxide & water along with some sulfates and nitrates. During aerobic decomposition oxygen is removed from the water, resulting in a drop in DO.

If too much organic matter must be decomposed, the DO supply in the water may fall to zero. If this happens, not only will the oxygen-dependent aquatic life be killed, but the end products from the resulting anaerobic decomposition will be toxic and malodorous. Typical anaerobic decomposition reaction products include ammonia, methane, hydrogen sulfide, carbon dioxide, and water.

Q. What is Biochemical Oxygen Demand (BOD) and its importance in water pollution ?

Ans. One of the most widely used measurements for estimating the water pollution potential of a given amount of organic waste is called the *biochemical oxygen demand*, or BOD. The BOD is a measure of the amount of oxygen required by bacteria to oxidize waste aerobically to carbon dioxide and water. If we were to introduce an amount of organic matter into some water and monitor its decomposition, we would observe that the amount of organic material remaining would decrease exponentially with time. The amount of BOD which remains after some period of time is proportional to the remaining organic material, which results in a curve as shown in Figure below.

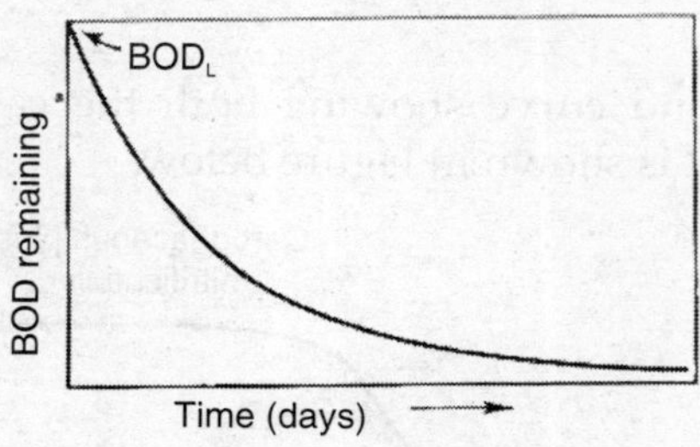

Fig.5: BOD remaining as a function of time with initial value BOD_L

The initial value of BOD will of course be the total oxygen requirement to oxidize the organic material. This quantity is called the *ultimate BOD*, or BOD_L.

Usually when curves of BOD versus time are drawn, they are actually made for BOD utilized and not BOD remaining. BOD utilized will be proportional to the amount of organic matter which has been oxidized and hence will have the shape as shown in Figure below.

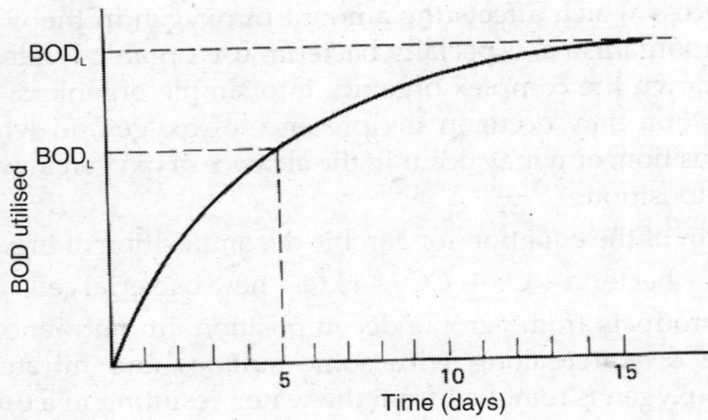

Fig.6: This is the usual curve for BOD, and is actually BOD utilized. Also shown is the 5 day BOD.

A standard way to measure BOD is to determine the amount of oxygen required by thebacteria during the first 5 days of decomposition (at 20°C). This is the 5-day BOD called BOD_5, which is shown in Figure above.

The importance of the concept of BOD in water quality management must be emphasized. Consider that normal domestic sewage may have a 5-day BOD of around 200 mg/1 and that industrial sewage may have a BOD of several thousands of milligrams per liter. Compare these requirements for oxygen to the amounts actually dissolved in water.

It is thus easy to anticipate the quick depletion of oxygen in any receiving water unless the dilution factor is quite high. Depleting the water of oxygen, will kill the fish and the anaerobic decay products will be extremely objectionable. There is a secondary effect which causes the demand for oxygen to suddenly increase after about 8-10 days. The organic nitrogen in the wastes is converted to ammonia during decomposition and the subsequent oxidation of the ammonia to nitrite (NO_2) and then nitrate (NO_3) requires oxygen.

The total oxygen demand curve showing both the carbonaceous demand and the nitrification requirement is shown in Figure below.

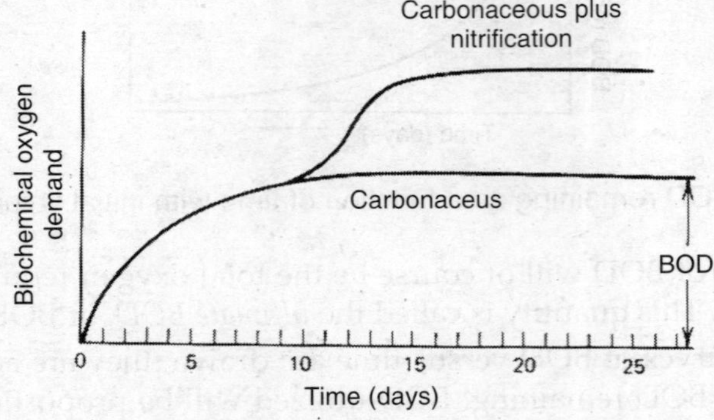

Fig.7: The complete BOD curve includes the carbonaceous and nitrification demands

Q. What is oxygen sag curve and its importance in water pollution ?

Ans. When wastes are discharged into a body of water, the amount of dissolved oxygen will decrease due to oxidation by bacteria. Opposing this drop in dissolved oxygen is reaeration which replaces oxygen through the surface, at a rate which is proportional to the depletion of oxygen below the saturation value. The simultaneous action of deoxygenation and reaeration produces what is called the oxygen sag curve, as shown in Figure below.

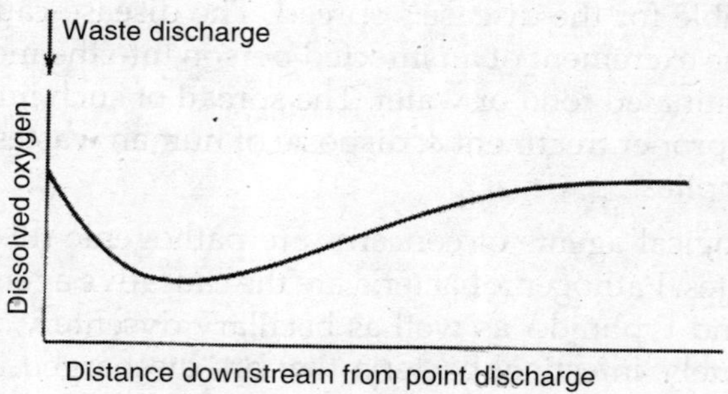

Fig.8: The oxygen sag curve

The DO curve initially drops as the wastes deplete the oxygen faster that it can be replaced. At the point where the DO is a minimum, the rate of reaeration becomes equal to the rate of oxygen utilization. Beyond that point the rate of reaeration exceeds the rate of utilization and the DO level eventually returns to normal. This sequence is referred to as the natural self-purification ability of water.

The horizontal axis of the oxygen sag curve may be either time or distance. If for example, a certain amount of waste is released all at one time into some impounded water, the DO level will be a function of time. If, however, there is a continuous discharge of wastes into a stream, then the oxygen sag curve will be a function of the distance downstream from the point of discharge. As we move downstream from the point of discharge, and the dissolved oxygen begins to drop, there will be a corresponding change in the biota. Normal game fish will be replaced by fishes tolerant of the turbid, low DO waters. At the point downstream where the DO reaches its lowest value, conditions will be at their worst. If the stream goes anaerobic, there will be no fish and the only organisms present will be those able to obtain their oxygen from the surface, or those which are tolerant of low oxygen conditions.

Q. what are the health hazards &diseases associated with polluted/contaminated water ?

Ans. There are many diseases which are associated with the contamination of water supplies by animal or human wastes. They include cholera, typhoid fever, paratyphoid fever, dysentery and infectious hepatitis.

The basic sanitation techniques which are essential to the control of water related diseases are the purification of water, and the treatment of sewage. The emphasis in water-quality control has shifted from the area of water-borne diseases, to control of oxygen consuming wastes, and now to nutrient removal & water reclamation.

The communicable water-associated diseases, are caused by the ingestion of biological agents.

However, other modes of transmission including simple water contact, as in the case of schistosomiasis, or insect vectors, as in the case of malaria, are also important. The contamination of water by sewage is the principal cause of waterborne diseases. Intestinal discharges of patients or carriers of such diseases contain the biological agents which are responsible for the disease's spread. The disease-causing organisms can be transferred from the excrement of an infected person into the mouths of healthy people by means of contaminated food or water. The spread of such infectious diseases can be controlled both by proper treatment & disposal of human wastes, and by purification of drinking water supplies.

The principal biological agents of concern are pathogenic (disease-causing) bacteria, viruses, and parasites. Pathogenic bacteria are the causative agents of the great epidemic diseases cholera and typhoid - as well as bacillary dysentery, paratyphoid fever, and tularemia. Fortunately, intestinal bacteria that are discharged into natural waterways usually survive for only a few days.

These bacteria exhibit an initial rapid decrease that results in about a 90% reduction in 2 days. However, thereafter the rate of decrease is considerably slower. This reduction is faster in warm, polluted water than in cold, clean water.

The only viral disease that has been proven to be transmitted through drinking water is infectious hepatitis, although there are others for which transmission through water is theoretically possible. One of the particular difficulties with viruses is that some can survive the normal dosages of chlorine which are applied at water and sewage treatment plants to kill pathogens. Among the parasites that may be ingested is the protozoa *Entamoeba hystolytica*, which causes amebic dysentery. Another parasite is the guinea-worm which causes drantiasis, a common disease among the rural population.

Q. What is Water Quality? Discuss important water quality parameters and their regulatory limits.

Ans. "The water quality is that which a body of water should be able to support its designated uses, which include drinking, industrial uses, shell fishing, swimming, other body contact, recreational boating, or fishing".

Any physical, chemical, or biological condition that prevents the designated use(s) of a body of water represents poor water quality.

Some of the important water quality parameters and regulatory limits for them are as follows:

Dissolved oxygen

Dissolved oxygen (DO) is critical for aquatic life and also controls chemical processes. DO is important to aquatic organisms to perform their decomposition function ("aerobic" decomposers). DO enter the water from the air and from aquatic plant photosynthesis. Oxygen is consumed by respiration (Biological Oxygen Demand) and some chemical reactions. Values of DO less than a daily mean of 5.0 mg/liter (ppm) or instantaneous value of 4.0 mg/l is considered to be low and less than 2.0 mg/l to be dangerous to aquatic life.

Turbidity

Turbidity is caused by suspended material that blocks light. High levels of turbidity can slow the growth of aquatic plants and reduce visibility for aquatic animals that need to see their prey. Turbidity is a measure of how much light is scattered by materials suspended in the water. Scattered light is much less available for aquatic plants, so high levels of turbidity can limit the growth of aquatic plants as well as restrict the visibility of prey and enemies to aquatic animals. Consequently, there are standards for turbidity in fresh (50 NTU) & salt (25 NTU) waters. An NTU is a Nephelometric Turbidity Unit.

Fecal coliform bacteria

Fecal Coliform Bacteria are indicators of recent contamination by fecal matter from warm blooded creatures, including humans, other mammals and birds. The MFC membrane filtration technique/ method, enumerate fecal coliforms in water samples. Most of the bacteria detected by this method are E. coli, the most abundant bacterium in the guts of humans. While E. coli itself is usually not a pathogen, its presence is an indicator of the possible presence of real pathogens that can cause disease.

Coliform bacteria themselves are not necessarily pathogenic, but they indicate the likelihood that more dangerous organisms, such as the organisms that cause salmonellosis, cholera or hepatitis, are present. Although contamination of waters by human wastes is especially serious, animal wastes can also transmit disease. Consequently, high levels of fecal coliform bacteria, regardless of the source, imply a high risk of disease transmission. Two additional warnings apply: a) some pathogens live longer in surface waters than fecal coliform bacteria, and b) many bacteria are alive in the water but cannot be cultured and detected by the standard techniques. Owing to the risk of disease transmission by contact with waters contaminated by fecal wastes, standards for this water quality parameter are quite stringent.

As guidelines, waters designated for shell fishing should not have average fecal coliform levels above 14 colony forming units (CFU) per 100 ml. Waters designated for most other uses, e.g., swimming, should not average above 200 CFU/100 ml.

Total Nitrogen and Phosphorus

Total Nitrogen is a measure of the quantity of the important plant nutrient, nitrogen, present in the water in dissolved and particulate forms. It is one measure of the degree of nitrogen loading to surface waters.

Total Phosphorus is a measure of the important plant nutrient, phosphorus, which not only stimulates aquatic weed growth, but also blooms of nitrogen-fixing blue-green algae (cyanobacteria) and bacteria.

These plant nutrients occur at variable but typically low levels in natural waters. Elevated levels of nitrogen and/or phosphorus can stimulate excessive plant growth ("eutrophication"), which can be very harmful to aquatic ecosystems. There are no general standards for nitrogen or phosphorus. Non-point sources of nitrogen and phosphorus are not currently regulated. Some countries have banned high-phosphate detergents, in order to control algae blooms in rivers receiving sewage effluents.

Total suspended solids

Sediment loading to surface waters raises turbidity & suspended solids levels and causes siltation, other pollutant loadings, and problems for aquatic life. Sedimentation is a serious problem in areas downstream of improperly managed land disturbing activity. Suspended solids cause siltation, clog the gills of aquatic animals, and carry toxic chemicals and nutrients with them when they enter the water. High levels of suspended solids are considered to be one of the most widespread water quality problems, and are frequently caused by human activities.

Chlorophyll

Chlorophyll a is a measure of phytoplankton biomass. All plants have the photosynthetic pigment chlorophyll, so its concentration in the water is a measure of plant abundance. It is a good measure for phytoplankton, which is otherwise hard to quantify. Standard for chlorophyll is 40 micrograms per liter (μg/l), above which water quality is considered to be impaired by excess phytoplankton. However, chlorophyll measures only the biomass of small phytoplankton. Larger aquatic plants that grow on surfaces or the big aquatic weeds that fill many ponds are not measured by this technique. There is no accepted standard method for the measurement of these larger plants, so measures of chlorophyll in a body of water may grossly underestimate the actual amount of plant growth there. High levels of plant growth indicate excessive nutrient loading and can cause problems with low dissolved oxygen at night.

Because shallow waters usually have enough light for plants to grow, unusually high levels of chlorophyll a usually indicate excessive nutrient loading.

Q. What are primary water quality standards in India for various water uses ?

Ans. Indian water quality standards are tabulated below.

A. Drinking water source without conventional treatment but after disinfections:

- Total coliform organism MPN/100ml shall be 50 or less
- pH between 6.5 and 8.5
- Dissolved Oxygen 6 mg/l or more
- Biochemical Oxygen Demand 5 days 20°C 2 mg/l or less

B. Outdoor bathing (Organised)

- Total coliforms organism MPN/100ml shall be 500 or less
- pH between 6.5 and 8.5
- Dissolved Oxygen 5 mg/l or more

C. Drinking water source after conventional treatment and disinfection:

- Total coliforms organism MPN/100ml shall be 5000 or less
- pH between 6 and 9
- Dissolved Oxygen 4 mg/l or more
- Biochemical Oxygen Demand 5 days 20°C 3 mg/l or less

D. Propagation of wild life and fisheries:

- pH between 6.5 to 8.5
- Dissolved Oxygen 4mg/l or more
- Free Ammonia (as N) 1.2 mg/l or less

E. Irrigation, industrial cooling, controlled waste disposal:

- Electrical Conductivity at 25°C micro mhos/cm Max. 2250
- Sodium absorption Ratio Max. 26
- Boron Max. 2mg/l

Q. What are deleterious effects of industrial effluents ?

Ans. Industries need a wide variety of raw materials and chemicals which are later discharged as effluents. Acids, alkalis, toxic metals, pesticides and other poisonous substances such as cyanide, dyes, oils, detergents, resins, rubbers are a few to mention. Heated effluents that impart thermal loading on receiving waters and effluents containing radioactive materials are also of prime concern. Some of the effluents such as from tanning and meat packing may also contain pathogenic bacteria. The nature and extent of pollution depends on the materials present in the effluent and on the quantity discharged.

Effects on water courses

Color: The effluents contain dyes in higher concentrations which impart color to the receiving streams and they persist for longer distances. Photosynthesis of phytoplankton is affected by these colors.

pH value: The extreme alkalinity makes the receiving water unfit for any purpose. Further, it is deleterious to most of the aquatic life.

Suspended impurities: The colloidal and suspended impurities produce turbidity in the receiving waters. The turbidity and color along with the oil & scum create an unsightly appearance.

Depletion of oxygen: Natural substances such as starch and dextrin and inorganic substances such as sulfide and nitrite present in the effluent exert an immediate oxygen demand. The stream will then be devoid of oxygen and the aquatic life are affected adversely.

Toxic substances: Chromium, sulfide, chlorine and aniline dyes present in these wastes are directly toxic to fish and microbial organisms which carry out purification. Thus the self purification of the water body is affected.

Oils: Various oils (mineral) in the effluent interfere with the oxygenation of stream as they form a blanket on the surface and prevents the entry of oxygen at air/water interface.

Dissolved minerals: The mineral materials, mostly sodium salts increase the salinity of the water and consequently it becomes unfit for irrigation.

Effects on land

- The excess content of sodium (60%) and boron (2 mg/l) are deleterious to crops.
- The high sodium alkalinity combined with salinity impairs the growth of plants.
- Texture of the soil is affected by sodium and penetration of roots is prevented.

- Soil permeability is also affected by sodium and ultimately the soil will lose its productivity.
- Suspended and colloidal impurities clog the pores and form a mat on the surface of soil preventing the passage of air, water etc.

Q. What Are The Effects Of Various Water Pollutants ?

Ans. Effects of various water pollutants are tabulated below.

Water Pollutants	Effects
Detergent	Complex formation between DNA and mercury or cadmium results in birth defects. Although ,detergents are not highly toxic to fishes ,they do cause damage to gills and remove the protective mucous from the skin and the intestine.
Domestic waste	Water borne diseases like cholera (Vibrio cholera) typhoid, dysentery and various health problems , depletion of dissolved oxygen, objectionable odour.
Industrial effluents	It causes deleterious effects on living things and may bring death or sub lethal pathology of kidneys, liver, lungs, brain and reproductive system.
Agricultural waste	Excessive fertilizer leads to accumulation of nitrates in children called methemoglobinaemia. Richness of nutrients results in eutrophication.
Eutrophication	During Eutrophication, algal bloom release toxic chemicals into the aquatic system. Algal bloom leads to oxygen depletion and an increase in CO_2 level. Thus aquatic organisms begin to die which leads to succession.
Bioaccumulation (or) Biological magnification	Aquatic plants and animals can accumulate certain pesticides in their body tissues in greater concentration than in water. This phenomenon is commonly referred to as biological magnification or biological amplification eg.,DDT. It is more threatening as its concentration continuously increases in successive trophic levels in a food chain which results in many health hazards. DDT absorbed by fish eating birds 25 ppm ↑ DDT in large fish 2 ppm ↑
	DDT in small fish 0.5 ppm ↑ DDT in zooplankton 0.003 ppm ↑ DDT in water 0.000003 ppm or 0.003 ppb
Lead (pb)	Anaemia, vomiting, damage of liver, brain and kidney

Arsenic(A)	Mental disturbance, lung cancer, ulcer, kidney damage
Mercury(Hg)	Abdominal pain, headache, diarrhea, chest pain
Cadmium(Cd)	Growth retardation, diarrhea, bone deformation, kidney damage, anemia, damage to lever
Barium(Ba)	Excessive salvation, diarrhea, paralysis
Chromium(Cr)	Gastro intestinal ulceration, diseases of central nervous system, cancer, nephritis
Zinc(Zn)	Vomiting, renal damage
Copper(Cu)	Hypertension, uremia, coma
Temperature	Reduction of dissolved oxygen , Increase in Biological Oxygen Demand Early hatching of fish eggs and fish mortality, Mitigation of aquatic biota
Radioactivity	Serious skin cancer, carcinoma, melanoma, breast cancer, leukemia, DNA breakage and cataract
Siltation	Reduced visibility, Reduction in direct light penetration, Decrease in photosynthetic rate, Chances of anaerobic digestion in the benthos zone.
Oil	

Q. Describe In Detail Major Types Of Water Pollutants And Their Effects.

Ans. Human activities give rise to water pollution by introducing various categories of substances or waste heat into a water body. The more common types of polluting substances include pathogenic organisms, oxygen demanding organic substances, plant nutrients which stimulate algal blooms, inorganic & organic toxic substances and oil & grease.

Pathogenic Organisms. Many serious human diseases such as cholera, typhoid, bacterial & amoebic dysentery, enteritis, polio and infectious hepatitis are caused by water-borne pathogens. In addition, malaria, yellow fever and filariasis are transmitted by insects that have aquatic larvae. Faecal pollution of water resources by untreated or improperly treated sewage is a major cause for the spread of water-borne diseases. To a lesser extent, disease causing organisms may also be derived from animal rearing operations & food processing factories with inadequate waste water treatment facilities. The spread of water-borne infectious diseases can be controlled through the introduction of water & sewage treatment facilities and through improved hygiene.

Biodegradable Organic Substances. Human & animal wastes as well as effluents from industries processing plant or animal products contain a mixture of complex organic substances such as carbohydrates, proteins & fats as their major pollution load. These substances are readily biodegradable and when introduced into the environment are quickly decomposed through the action of natural microbial populations. Some of the organic matter is oxidised to carbon dioxide & water while the rest is assimilated and used for the synthesis of new microbial cells. In due course, these organisms will also die and become food for other decomposers. Eventually virtually all of the organic carbon will be oxidised.

When a biodegradable organic waste is discharged into an aquatic ecosystem such as a stream, estuary or lake, oxygen dissolved in the water is consumed due to respiration of micro-organisms that oxidise the organic matter. The more biodegradable a waste, the more rapid is the rate of its oxidation &the corresponding consumption of oxygen. In an aquatic ecosystem, a greater number of species of organisms are supported when the dissolved oxygen concentration is high. Oxygen depletion due to waste discharge has the effect of increasing the numbers of decomposer organisms at the expense of others.

When oxygen demand of a waste is so high as to eliminate all or most of the dissolved oxygen from a stretch of a water body, organic matter degradation occurs through the activities of anaerobic organisms which do not require oxygen. Not only does the water then become devoid of aerobic organisms, but anaerobic decomposition also results in the formation of a variety of foul smelling volatile organic acids & gases such as hydrogen sulfide & mercaptans. The stench from these is quite unpleasant.

Plant Nutrients. The availability of plant nutrients, eg, nitrogen & phosphorus are important determinants of the biological productivity of aquatic ecosystems. Nutrient deficient aquatic environments are called "oligotrophic" and those rich in nutrients, "eutrophic". Young lakes are generally oligotrophic, but they naturally accumulate nutrients over time, derived from drainage and sediment run off from its catchment. When human activities greatly accelerate nutrient enrichment of water bodies, the process is called "cultural eutrophication". Sewage, animal wastes and many industrial effluents contain high levels of nitrogen & phosphorus. Another major source is fertiliser run off from urban and agricultural catchments.

While in the long term, cultural eutrophication accelerates the natural successional progress of aquatic ecosystems towards a terrestrial system,in the short term problems arise due to cyclic occurrences of algal blooms & decay. In warm weather, nutrients stimulate rapid growth of algae and floating aquatic weeds. The water often becomes opaque and has unpleasant tastes & odours. When these organisms die they become food for decomposer bacteria. Depletion of dissolved oxygen leads to anaerobic conditions and a general decline in the ecological & aesthetic qualities of the water body. Algal blooms also reduce light penetration into the water making it impossible for sea grasses and other bottom anchored plants to survive.

Toxic Inorganic Pollutants. Many inorganic substances are released by natural weathering of rocks and are washed into water courses. However, human activities such as mining and mineral processing as well as wastage are responsible for far greater quantities of toxic inorganic pollutants entering water supplies & aquatic ecosystems.Of particular concern among these are arsenic, an ingredient of some pesticides, and heavy metals such as mercury, lead, tin & cadmium as they tend to accumulate in tissues. Mine drainage and leaching of mine tailings as well as metal finishing & inorganic chemical industries are major sources of metal pollution in the water environment.

MERCURY poisoning causes birth defects and permanent brain damage. The worst case of mercury poisoning of a community to date occurred in Minamata in Japan, due to the consumption of contaminated fish & shell fish from the bay which received discharges of chemical industry effluents over a long period of time. Insoluble metallic mercury is converted by bacteria in the marine sediments into water soluble methyl mercury. It is then concentrated through the trophic levels of the food chain due to selective retention in tissues.

LEAD causes damage to the nerve system, and experts recommend a tolerance limit of less than 10 parts per billion in drinking water. Prior to the introduction of PVC water pipes, lead piping and cast iron pipes using lead solder were widely used for public water supplies. The main source of environmental lead is automobile exhausts. Lead has long been an additive to petrol to boost its octane rating. Exhaust fumes from cars & other vehicles are eventually washed down by rain and enter watercourses. Some inorganic industrial effluents also contain lead.

Other heavy metals such as **CHROMIUM** and **ZINC** are present in effluents from metal finishing industries. Discharge of waste waters from many small metal finishing industries causes ground water pollution. Chromium salts are added to cooling waters of air conditioning systems for corrosion prevention. Significant contribution of this metal therefore comes from this source as well.

Production of alumina from bauxite gives rise to highly caustic effluents which also contain dissolved aluminium and other metals. **ALUMINIUM** has recently been implicated in the Alzheimer's disease, although there is still uncertainty about the link. **CAUSTIC EFFLUENTS** can raise the pH of receiving waters to levels unsuitable for many organisms. **ACIDIC EFFLUENTS** are produced from mine drainage and by many industries such as ore smelting, metal finishing, leather tanning and petroleum processing. Many lakes are acidified due to acid precipitation. Increased levels of sulfur & nitrogen oxides in the atmosphere from the combustion of fossil fuels such as coal and petroleum are the principal causes of acid rain.

Acidification lowers the pH of the water, especially when there is little buffering capacity in the form of alkalinity to neutralise it. Fish, amphibians and many insects will be killed by increased acid levels and in severe cases only a few resistant species such as fungi may survive.

Toxic Organic Chemicals. Many thousands of natural and synthetic organic chemicals are in use today for the manufacture of a variety of products ranging pesticides, pigments, pharmaceuticals & plastics. Several of these are known to cause birth abnormalities, genetic defects and cancer. Some chemicals like DDT and PCB's (Poly Chloro Biphenyls) are concentrated in tissues to dangerous levels. Many are only very slowly biodegradable and persist in the environment for long periods of time.

The major causes of organic chemical pollution are improper disposal of domestic & industrial wastes and herbicide & pesticide run off from farming areas, where they are used in substantial quantities. In addition, large quantities of hazardous wastes have, in the past, been disposed of in landfills with inadequate containment from where they are slowly leached into surface & ground water supplies, eventually finding their way into the food chain.

Oil Pollution. Petroleum is one of the major energy sources today and huge volumes of oil are transported between points of production & consumption around the globe.

All along these major transportation routes oil spills happen regularly and oil slicks are ever present. With serious spills, many marine birds and other animals are choked to death by the oil slick. Even when dispersed, many hydrocarbons in the oil are toxic to aquatic organisms. Some are thought to be carcinogenic. Oil being lighter than water floats on the surface as a thin film which can interfere with the transfer of gases such as oxygen & carbon dioxide, as well as heat, between the water and the atmosphere.

Routine petroleum refining, storage & use also results in pollution by leaking oil, oily waste water and sludge.Uunderground fuel storage tanks in service also leak oil into the ground water.

Thermal Pollution. Tremendous quantities of waste heat is produced by power plants and to a lesser extent by a broad spectrum of other industries. Cooling water drawn from the ocean, river, lake or aquifer is passed through heat exchangers where it absorbs the waste heat and is subsequently discharged back into the water environment.

Significant rises in water temperature can be caused in the receiving water in the vicinity of cooling water disposal sites. Such increases in temperature alter the species composition in ecosystems as organisms normally tolerate temperature variations only over a very small range. At higher temperatures, oxygen solubility is reduced, but bacterial respiration rate will increase, making the water more prone to deoxygenation. Temperature variations will also cause alterations in pH due to changes in the degree of ionisation and increased solubility or precipitation of bottom deposits.

Q. Describe Various Methods Of Control Of Water Pollution.

Ans. With increasing urbanisation and expanding agricultural & industrial production, water pollution problems have become serious necessitating the adoption of suitable control measures for reducing water pollution.

Approaches to controlling sources of water pollution may be grouped into three broad categories: (1) minimisation of waste or pollutant generation, (2) Treatment prior to disposal of waste water streams at source, and (3) "in-situ" reduction or elimination of pollution.

Minimisation of Pollutant Generation. Reduction of quantity of waste /pollutants generated is the most desirable approach to pollution control. Since it conserves resources that would otherwise be wasted, and at the same time eliminates the cost of removing pollutant after they are produced, it is the cheapest & the most effective alternative. For non-point pollution sources, this is the only practicable method of pollution control. Yet, this approach has not been exploited to its fullest extent.

As a general rule, a resource becomes a waste when it can no longer be economically utilised or recovered. It is then disposed of in the environment in the cheapest manner possible. Availability of economical technology for resource processing & usage has been a main determinant of when the resource is discarded as waste.

Minimising soil erosion by improved agricultural practices (e.g. by minimising surface runoff & leaving crop residues in the ground), more efficient use of nutrients (e.g., though the use of slow release fertilisers) and the development & use of biological pest control techniques in preference to the use of non-biodegradable toxic chemicals are some of the measures for minimising water pollution from agriculture.

Considerable potential also exists in many industries to reduce waste generation. Development and use of non-polluting technology to modify or replace existing manufacturing processes, and recycling or recovering materials that would otherwise be wasted are two approaches which not only reduce pollutant generation, but can even result in a saving for the industry by minimising or eliminating the need for waste treatment for pollutant removal.

In some cases, it may be more practical to segregate strong & weak waste streams to facilitate materials or energy recovery. Good housekeeping practices, like minimising spillage & materials wastage, also leads to waste reduction and savings in production cost.

Wastewater Treatment at Source. In nature, a variety of different mechanisms operate to degrade & transform waste materials into stable, harmless end products such as carbon dioxide. This cleansing ability is referred to as the "self-purification" or "assimilative" capacity. When the quantities of wastes to be disposed of are large, however, the natural purification processes become overloaded and can no longer assimilate the wastes without adversely affecting environmental quality. Man-made treatment systems are then needed to reduce pollutant loads to acceptable levels for discharge. For the most part, these purification systems make use of the same mechanisms as in the natural environment to bringing about waste stabilisation.

The wastewater treatment technologies are classified as physical, chemical & biological processes, depending on the nature of the purification mechanism employed. The character of the pollutants and the form (suspended or dissolved) in which they are present usually determine the most suitable process for their removal. For example, total suspended solids and floatable materials such as oil & fat are readily removed by physical processes such as sedimentation or flotation respectively.

BIOLOGICAL METHODS are effective & economical when the waste water contains mostly biodegradable pollutants such as organic matter. An advantage of biological processes is that the microorganisms involved in waste stabilisation are themselves produced in the process. For dilute wastes - including general domestic wastewaters, "aerobic" biological processes (activated sludge, oxidation ponds and aerobic biofilter) are usually favoured since they are capable of producing an effluent with very low residual pollutant concentrations. These processes, require oxygen, in proportion to the pollutant load present. Oxygen is supplied through aeration, which is a significant cost component. For strong wastes, "anaerobic" biological treatment in enclosed vessels is generally preferred as they proceed in the absence of oxygen, and in addition produce a useful, energy-rich by-product in the form of methane. The effluent from anaerobic processes, however, contain higher levels of residual organic materials and may require further polishing treatment (often in aerobic processes).

CHEMICAL TREATMENT is used when the pollutant is non- biodegradable and is not amenable to removal by simple physical means (e.g. when it occurs in dissolved form). Heavy metals are typically removed by chemical precipitation, while toxic substances such as cyanide may be chemically oxidised. An important disadvantage of chemical treatment methods is that they generally require dosing with a chemical which can be quite expensive. In addition, disposal of the chemical sludge produced in these processes may also pose some problems.

When a community based treatment system is impractical, it is possible to provide a degree of treatment prior to discharging sewage into the environment. A popular method used for individual homes & small groups of residences is the SEPTIC TANK. It consists of a simple baffled tank which traps most of the solids in the waste water and also affords some decomposition of soluble organic matter. The effluent is disposed of into the ground through a system of leach drains. As solids progressively accumulate in the tank, it is necessary to periodically desludge the system, typically every 3 to 7 years. As deep sewering in built-up areas is very expensive, other more efficient alternatives to the septic tank are also desirable for on- site use.

In-situ Pollution Control. Waste minimisation & treatment help prevent pollution from occurring and should be the principal approaches to water quality maintenance. Occasionally, however, when a water body is already adversely affected, it will be necessary to consider action aimed at helping the ecosystem recover from the impact of pollution. Methods to facilitate this are collectively grouped under in-situ control techniques.

Aeration of lakes & reservoirs, especially when they are thermally stratified (in summer), has been used to prevent anaerobic conditions from occurring. Forced circulation of water in stratified lakes is an alternative method. Dredging nutrient rich superficial sediments from highly eutrophic lakes, while very expensive, helps reduce occurrence of severe algal blooms. Addition of aluminium or iron salts to assist the precipitation of phosphorus has also been practiced in some lakes to control dissolved phosphorus levels in the water.

Q. Discuss the Composition of water in Rivers and Lakes.

Ans. The composition of a river, and a lake, is a reflection of four inputs: atmospheric input of gases & solutes, biologic processes, discharge of groundwater, and local interactions with mineral components in the soil or stream bed.

At base flow, the water in a gaining river is derived from groundwater, so the composition of the river is a reflection of the aquifer, while for a losing stream the reverse holds good - the composition of the shallow groundwater reflects the river, for example an alluvial aquifer.

The composition of a river is a dynamic thing, with concentrations generally (but not always) increasing downstream, but pH generally increasing.

Q. What is hard water ?

Ans. Hard water occurs when excess minerals in the water create certain problems but water hardness is not a safety issue. Hard water is safe for drinking, cooking, & other household uses.

Hard water can cause problems for like decreased life of household plumbing & water-using appliances, increased difficulty in cleaning & laundering , decreased efficiency of water heaters, and white/chalky deposits on items such as plumbing, tubs, sinks, pots & pans.

Q. Describe the Causes of Hard Water.

Ans. Approximately 22 % of the earth's fresh water is groundwater, and naturally, as it flows through soil & rock, it picks up minerals. Hard water results when an excessive amount of calcium & magnesium are present. Total hardness is measured in milligrams per litre (mg/l) or in parts per million (ppm). Parts per million (ppm) measures the unit(s) of a substance for every one million units of water. Milligrams per litre (mg/l) and parts per million (ppm) are roughly equal in water analysis. When conducting chemical analysis, laboratories usually measure hardness minerals in milligrams per liter of water (mg/l).

Identifying Hard Water

- Decreased cleaning capabilities of soaps & detergents, resulting in dingy laundry and reduced life of fabrics.
- Increased buildup of scale on plumbing fixtures and cooking utensils.
- Film left on the body resulting in dry skin and dull, limp hair.
- Soap scum on bathtubs, shower tiles, and basins.
- Increased water heating costs due to scale buildup and mineral deposits, and more frequent replacement of hot water heating elements.
- Clogged pipes or appliances resulting in reduced water flow and increased repairs.

Q. What are different Types of Water Hardness ?

Ans. There are two types of water hardness-temporary & permanent hardness.

Temporary Hardness-This refers to hardness whose effects can be removed by boiling the water. Such waters have usually percolated though limestone formations and contain bicarbonate HCO_3 along with small amounts of carbonate CO_3 as the principal negative ions. Boiling the water promotes the reaction: $2\,HCO_3 \rightarrow CO_3 + CO_2$ by driving off the carbon dioxide gas. The CO_3 reacts with Ca_2+ or Mg_2+ ions, to form insoluble calcium & magnesium carbonates which precipitate out. By tying up the metal ions in this way, the amounts available to form soap scum are greatly reduced.

Permanent Hardness-Waters that contain other anions such as chloride or sulphate cannot be remediated by boiling, and are said to be 'permanently' hard. The only practical treatment is to remove all the ions by various methods.

Q. What do you mean by water quality ? What is its significance ?

Ans. Water quality is a term used to express the suitability of water for various uses or processes. Any particular use of water will have certain requirements for the physical,chemical or biological characteristics of water; for example limits on the concentrations of toxic substances for drinking water use, or restrictions on temperature & pH ranges for water supporting animals. Consequently, water quality can be defined by a range of variables which limit the water. Although many uses have some common requirements for certain variables, each use will have its own demands and influences on water quality. It should be free from colour, turbidity, taste and odour, of moderate temperature and aerated.

The quality of water may be described in terms of the concentration and state (dissolved or particulate) of some or all of the organic & inorganic material present in the water, together with certain physical characteristics of the water.

Water temperature, pH, specific conductance, turbidity, dissolved oxygen and suspended sediments are important parameters if we discuss about water quality. Potable water is an important parameter for human beings, because it is useful for maintaining human health.

Q. What are the main water quality parameters ?

Ans. Water quality is determined by various water parameters like temperature,pH,specific conductance,turbidity,dissolved oxygen,hardness ,suspended sediments etc.

Water temperature- It is important to fisherman , industries, and growth of fish & algae. A lot of water is used for cooling purposes in power plants that generate electricity. They need cool water to start with, and they generally release warmer water back to the environment. The temperature of the released water can affect downstream habitats. Temperature also can affect the ability of water to hold oxygen as well as the ability of organisms to resist certain pollutants.

pH of water-pH is a measure of how acidic/basic a water sample is. The range goes from 0 - 14, with 7 being neutral. The pH of less than 7 indicates acidity, whereas a pH of greater than 7 indicates a base. Since pH can be affected by chemicals in the water, pH is an important indicator of water that is changing chemically. The pH is reported in logarithmic units. Each number represents a 10-fold change in the acidity/basicity of the water.Water with a pH of 5 is ten times more acidic than water having a pH of 6. Pollution can change a water's pH, which in turn can harm animals and plants living in the water. For instance, water coming out of an abandoned coal mine can have a pH of 2, which is very acidic and would definitely affect any fish and other microorganisms.

SPECIFIC CONDUCTANCE-Specific conductance is a measure of the ability of water to conduct an electrical current. It is highly dependent on the amount of dissolved solids (such as salt) in the water. Pure water, such as distilled water, will have a very low specific conductance, and sea water will have a high specific conductance. Rainwater often dissolves airborne gases & dust while it is in the air, and thus often has a higher specific conductance than distilled water. Specific conductance is an important water-quality measurement because it gives a good idea of the amount of dissolved material in the water.

TURBIDITY-Turbidity is the amount of particulate matter that is suspended in water. Turbidity measures the scattering effect that suspended solids have on light: the higher the intensity of scattered light, the higher the turbidity. Material that causes water to be turbid include:

- clay
- silt
- finely divided organic & inorganic matter
- soluble coloured organic compounds
- plankton
- microscopic organisms

Turbidity makes the water cloudy or opaque. Turbidity is measured by shining a light through the water and is reported in Nephelometric Turbidity Units (NTU). During periods of low flow (base flow), many rivers are a clear green colour, and turbidities are low, usually less than 10 NTU. During a rainstorm, particles from the surrounding land are washed into the river making the water a muddy brown colour, indicating water that has higher turbidity values. Also, during high flows, water velocities are faster and water volumes are higher, which can more easily stir up and suspend material from the stream bed, causing higher turbidities.

DISSOLVED OXYGEN-Although water molecules contain oxygen atom, this oxygen is not what is needed by aquatic organisms living in natural waters. A small amount of oxygen, upto about ten molecules of oxygen per million of water, is actually dissolved in water. This dissolved oxygen is breathed by fish & zooplankton and is needed by them to survive. Rapidly moving water, such as in a mountain stream or large river, tends to contain a lot of dissolved oxygen, while stagnant water contains little. Bacteria in water can consume oxygen as organic matter decays. Thus, excess organic material in the tanks, lakes and rivers can cause an oxygen-deficient situation to occur. Aquatic life can have a hard time in stagnant water that has a lot of rotting, organic material in it, especially in summer, when dissolved oxygen levels are at a seasonal low.

HARDNESS-The amount of dissolved calcium & magnesium in water determines its 'hardness.Where the water is relatively hard, it is difficult to get lather up when washing hands or clothes. And, industries in these area have to spend money to soften their water, as hard water can damage equipment. Hard water can even shorten the life of fabrics and clothes.

SUSPENDED SEDIMENT- It is the amount of soil moving along in a stream. It is highly dependent on the speed of the water flow, as fast-flowing water can pick up and suspend more soil than calm water. During storms, soil is washed from the stream banks into the stream. The amount that washes into a stream depends on the type of land in the river's watershed and the vegetation surrounding the river. If land is disturbed along a stream and protection measures are not taken, then excess sediment can harm the water quality of a stream. These bunds and fences are supposed to trap sediment during a rainstorm and keep it from washing into a stream, as excess sediment can harm the creeks, rivers, lakes, and reservoirs. Sediment coming into a reservoir is always a concern; once it enters it cannot get out - most of it will settle to the bottom. Reservoirs can 'silt in' if too much sediment enters them. The volume of the reservoir is reduced, resulting in less area for water storage for agriculture as well as reducing the power-generation capability of the hydel power plant in the dam.

Q. What are various Impurities found in Water ?

Ans. As water goes through the hydrologic cycle, it gathers many impurities. Dust, smoke, & gases in the air contaminate rain, snow, hail, and sleet. Rain water picks up silt, chemicals, and disease organisms. As it enters the earth through seepage and infiltration, some of the suspended impurities may be filtered out. However, other minerals & chemicals are dissolved and carried along. As groundwater it may contain disease organisms and harmful chemicals. In addition to the impurities in water resulting from infiltration, many are contributed by industries. The garbage, sewage, industrial waste, pesticides etc., are all possible contaminants of raw water. Impurities in raw water are either suspended or dissolved. Suspended impurities include diseases organisms, silt, bacteria,& algae.They must be removed/destroyed before the water is consumed. Dissolved impurities include salts, (calcium, magnesium, & sodium), iron, manganese, and gases (oxygen, carbon dioxide, hydrogen sulphide, pH & nitrogen). These impurities must be reduced to levels acceptable for human consumption.

Q. How are Raw Water Classified ?

Ans. Raw water is classified as fresh, brackish, or salt water (seawater) based on the concentration of TDS (Total Dissolved Solids). Fresh water has a TDS concentration of less that 1,500 ppm. Brackish water is high in minerals and has a TDS concentration between 1,500 ppm and 15,000 ppm. Salt water has a TDS concentration greater than 15,000 ppm. Generally, groundwater (subsurface) has less chemical or biological contaminants than surface water, provided reasonable care is exercised in the selection of the well site. Harmful microorganisms are usually reduced to tolerable levels by passage through the soil.

Q. How are treated water classified ?

Ans. Treated water is classified as palatable water and potable water.

Palatable Water. Palatable water is water that is pleasing in appearance and taste. It is significantly free from colour, turbidity, taste, and odour. It should also be cool and aerated.

Potable Water. Potable water means water that is fit for drinking. Quality standards for treated water reflect the values of substances allowed in potable water. Standards exist to measure the physical, chemical, microbiological , & radiological quality of water and to test for the presence of chemical agents.

Q. What are Physical Quality characteristics of water ?

Ans. The main physical quality characteristics of water are colour, odour , taste, turbidity, and temperature. These physical quality characteristics and their related quality standards are described below.

Colour - Colour in water is derived from coloured substances, such as vegetable matter, dissolved from roots & leaves, from humus, or from inorganic compounds such as iron & manganese salts. The colour standard is designed to make drinking water more palatable.

Odour and Taste-There is no set standards for odour & taste as there are no specific tests for these. Odour and taste found in water are most commonly caused by algae, decomposed organic matter, dissolved gases, or industrial waste.

Turbidity-Turbidity refers to a muddy or unclear condition of water caused by suspended clay, silt, organic & inorganic matter, plankton and other microorganisms. The turbidity standard was established to improve the efficiency of disinfection by reducing particles to which microorganisms could attach.

Temperature - Warm water tastes flat. Cooling water suppresses odours & tastes and makes it more palatable. Temperature also effects the chlorination & purification of water. Disinfection takes longer when water is colder, and purification capacity is reduced with reverse osmosis treatment equipment. Water having physical characteristics exceeding the limits or making it less palatable should not be used for drinking. Otherwise, reduced consumption and increased risk of dehydration may result. High water temperature in rivers and stream reduces dissolved oxygen content posing threat to survival of aquatic creatures,fishes etc.

Q. What are Chemical Quality characteristics of water ?

Ans. The chemical quality of water depends on the chemical substances it contains. These substances include TDS, chlorides, sulphates and other ions. The chemical quality of water involves its hardness, alkalinity, acidity, and corrosiveness. Chemical substances having an adverse health effect have established standards that should not be exceeded.

Potential Hydrogen (pH) - The pH is a measure of the acidic or alkaline nature of water. It is defined as the negative logarithm of the hydrogen ion concentration. It ranges from 0 to 14 and pH of 7 is neutral. The pH influences the corrosiveness of the water, the amount of chemicals needed for proper disinfection, and the ability of an analyst to detect contaminants. Water with a pH below 7 is regarded as acidic while that with a pH above 7 is regarded as alkaline. The pH standard was established to ensure effective purification and disinfection.

Arsenic: Arsenic can be present in natural water sources in a wide range of concentrations. It can come from either natural or industrial sources. Ingestion of low concentrations of arsenic can cause nausea, vomiting, abdominal pain, or nerve damage.

Chloride: Chloride exists in most natural waters. It is the main anion found in seawater. Chloride comes from natural salt deposits, domestic & industrial waste, and agricultural runoff. Even in low concentrations, chloride can produce an objectionable taste in water.

Calcium, Magnesium & Hardness: Calcium & magnesium in drinking water, is an important parameter is to be taken into account Water hardness is a measure of the capacity of water to precipitate soap, which is in practice the sum of concentrations of all polyvalent cations present in water (Ca, Mg, Sr, Ba, Fe, Al, Mn, etc.); nevertheless, since the other ions (apart from Ca & Mg) play a minor role , generally hardness is defined as the sum of the Ca & Mg concentrations.

Water calcium & magnesium result from decomposition of calcium & magnesium aluminosilicates and, at higher concentrations, from dissolution of limestone, magnesium limestone, magnesite, gypsum & other minerals. Drinking water may be intentionally supplemented with these elements while treated, as happens with deacidification of underground waters by means of calcium hydroxide or filtration through different compounds counteracting acidity such as $CaCO_3$, $MgCO_3$ and MgO, and also with stabilization of low-mineralized waters by addition of CaO and CO_2. In low & medium mineralized underground & surface waters (as drinking waters are), calcium & magnesium are mainly present as simple ions Ca^{2+} and Mg^{2+} , the Ca levels varying from tens to hundreds of mg/l and the Mg concentrations varying from units to tens of mg/l.

Magnesium is usually less abundant in waters than calcium, which is easy to understand since magnesium is found in the Earth's crust in much lower amounts as compared with calcium. In common underground and surface waters the weight concentration of Ca is usually several times higher compared to that of Mg, the Ca to Mg ratio reaching up to 10. However , a common Ca to Mg ratio is about 4.

Sulphates: Sulphates occur naturally in water as the result of dissolution of sulfur bearing minerals. High concentrations also result from industry sources, such as coal mine drainage, pulp paper mills, tanneries, textile mills, and domestic waste water.

When ingested, sulphates have a laxative effect. They also can produce a bad taste in water. The sulphate standard was established to prevent chemically induced diarrhea.

Total Dissolved Solids(TDS): The TDS of water is composed of mineral salts and small amounts of other inorganic & organic substances. The proportion of each constituent is the result of weathering of rocks found in the drainage basin and of any industrial contributions. Since TDS is composed of chloride, magnesium, sulfate, and other ions, its ingestion in water has the same effects. Therefore, the TDS standard was established to prevent chemically induced diarrhea.Chemical water quality standards are based on the effect the water will have on the health of living organisms. The effect of a particular chemical substance determines if a limit is established for that substance. Chemical substances having a negative physical effect will have a mandatory limit that should not be exceeded. Some substances, such as iron and manganese, have no significant negative physical effect, but may restrict the use of the water, such as for the laundering of clothes.

Q. What are Microbiological Quality characteristics of water ?

Ans. The microbiological quality of potable water shows its potential for transmitting waterborne diseases which are caused by viruses, bacteria, protozoa, or higher organisms. The testing for microorganisms in water is extremely difficult. The number of these organisms is usually very low, even in a badly polluted water supply, and the test used to find them is difficult. For these reasons, indicator organisms are used to detect the presence of contamination. The bacterial organisms used as an indicator of possible contamination are total coliform. These organisms occur in large quantities in the intestines of warm-blooded animals. The presence of coliform organism in potable water is an indication of either inadequate treatment or the introduction of undesirable materials to the water after treatment. While the detection of many disease-causing microbes is difficult, the test to detect a surrogate organism, E. coli, is simple and effective for field use. Because of its relative simplicity and field adaptability, the membrane filter technique has gained wide acceptance as the preferred technique for the presumptive determination of the presence of coliform organisms in potable water. The microbiological standard was established to ensure infectious microorganisms would not cause diseases.

Q. Describe the acceptable water quality standards.

Ans. Water quality management should ensure that water supplied is free from pathogenic organisms, clear, potable, free from undesirable taste & odour, of reasonable temperature, neither corrosive nor scale forming and free from minerals which could produce undesirable physiological effect.

Physical & chemical standards for Potable and Safe water is tabulated below.

S.N.	Characteristics	* Acceptable	** Cause for Rejection
1.	Turbidity (Units on JTU scale)	2.5	10
2.	Colour (units on platinum-cobalt scale	5.0	25
3.	Taste and odour	Unobjectionable	Unobjectionable
4.	pH	7.0 to 8.5	6.5 to 9.2
5	Total dissolved solids(mg/l)	500	1500
6	Total hardness (mg/l as $CaCO_3$)	200	600
7	Chlorides (mg/l as Cl)	200	1000
8	Sulphates (as SO_4, mg/l)	200	1000
9	Fluorides (as F, mg/l)	1.0	1.5
10	Nitrates (as NO_3, mg/l)	45	100
11	Calcium (as Ca, mg/l)	75	200
12	Magnesium (as Mg, mg/l)	30	150
13	Iron (as Fe, mg/l)	0.1	1.0
14	Manganese (as Mn, mg/l)	0.05	0.5

Bacteriological Standards

(a) Coliform count in any water sample of 100 ml should be zero. A sample of water entering the distribution system that does not conform to this standard calls for an immediate investigation into both efficacy of the purification and the method of sampling.

(b) Water in the distribution system should satisfy all the three criteria indicated below:

- *E,Coli* count in 100 ml of any sample should be zero.
- Coliform organisms not more than 10 per 100 ml shall be present in any sample.
- Coliform organisms should not be detectable in 100 ml of any two consecutive samples or more than 50% of the samples collected for the year.

World Health Organisation (WHO) Guideline Values for chemicals (in **mg/l**) that are of health significance in drinking water is as follows ;

Arsenic - 0.01

Fluoride - 1.5

Manganese- 0.4

Nitrate- 50

Nitrite- 3

Chlorine - 5

Copper - 2000

Lead - 10

Nickel - 20

Q. What is water pollution ? What causes it ?

Ans. Water pollution occurs when a body of water is adversely affected due to the addition of large amounts of foreign materials to the water. When it is unfit for its intended use, water is considered polluted. Two types of water pollutants exist; point source and non-point source. Point sources of pollution occur when harmful substances are emitted directly into a body of water.

Many causes of pollution including sewage and fertilizers contain nutrients such as nitrates & phosphates. In excess levels, nutrients over stimulate the growth of aquatic plants & algae. Excessive growth of these types of organisms consequently clogs waterways, use up dissolved oxygen as they decompose, and block light to deeper waters.

Q. How are different types of water pollution sources classified ?

Ans. Water pollution can be classified into three types as domestic pollution, industrial pollution and agricultural pollution.

Domestic pollution. Important source of water pollution is domestic sewage system, which pollutes well & rivers, which are important source of drinking water. The solution to excreta disposal & its proper management is required to prevent the spread of pathogens in the environment which directly influences disease transmission through person to person contact, water and the food chain. Inadequate system for disposing of faeces, faeces deposited near the homes threaten household hygiene, by direct contact with people or by being carried into homes & kitchens by children, domestic animals or insects.

Domestic neighbourhood or district water supplies such as wells, tanks & reservoirs may be contaminated by poorly designed or maintained sewage disposal systems. The potable water contaminated with faeces is the chief cause of some important diseases of human beings. Use of open field for defecation results in faeces causing pollution of potable water as they consist primarily of intestinal bacteria.

Since a primary cause of contamination of water is inadequate or improper disposal of human & animal excreta, better water quality only improves health when sanitation is improved as well and when the quantity of water is sufficient. Deteriorating water quality is a particular threat in developing countries, where people lack access to clean drinking water and the majority of sewage are discharged into surface waters without waste water treatment. When untreated or inadequately treated wastewater or excreta (faecal sludge) is applied to soil & crops, disease transmission can occur. The persons at risk are the farmers, farm workers and their families as well as consumers of crops produced in such a way.

Industrial Pollution. Sewage is not the only the cause of water pollution, industrial waste is also a significant polluter – giving rise to contamination with heavy metals. The chemicals discharged by the factories are more harmful than the sewerage that flows into the river. The effluent discharged by the factories contains detergents, non-biodegradable materials and toxic chemicals hazardous to health & hygiene. Other sources of water pollution are distillery ,potassic fertilizers, electroplating plant, which contains harmful heavy metals, and cyanides which causes excessive acidity in water, which is harmful for aquatic life. Water pollution ranged from moderate to very high in case of textiles, jute, leather, vegetables fats and brewery. Moderate pollution results from dairy and canning establishments.

Agricultural Pollution. In addition to problems of waterlogging, desertification, salinization, erosion, etc., that affect irrigated areas; the problem of downstream degradation of water quality by salts, agrochemicals and toxic leachates is a serious environmental problem. 'Intensive use of pesticides and fertilisers has increased the level of nitrates in water. This can lead to brain damage and even death among children. Salinisation of water resources is a major concern to the sustainability of irrigation than is that of the salinisation of soils. Trace toxic constituents, such as Se, Mo and As in agricultural drainage waters may cause pollution problems that threaten the continuation of irrigation in some projects.

Q. Describe various physical parameters of water pollution.

Ans. Various water pollution physical parameters are described below. **pH.** pH is a measurement of how **acidic** or how **basic (alkaline)** a solution is. When substances dissolve in water they produce charged molecules called ions. Acidic water contains extra hydrogen ions (H+) and basic water contains extra hydroxyl (OH-) ions.

Pollution can change pH in water, which in turn can harm animals and plants living in the water. For instance, water coming out of an abandoned coal mine can have a pH of 2, which is very acidic and would definitely affect any fish that live in it. The mine-drainage water would be 100,000 times more acidic than neutral water.

TOTAL SUSPENDED SOLIDS(TSS). The TSS is solid materials, including organic & inorganic, that are suspended in the water. These would include silt, plankton and industrial wastes. High concentrations of suspended solids can lower water quality by absorbing light.

Waters then become warmer which lessen the ability of the water to hold oxygen necessary for aquatic life. Because aquatic plants also receive less light, photosynthesis decreases and less oxygen is produced. The combination of warmer water, less light and less oxygen makes it impossible for some forms of aquatic life to exist.

Suspended solids affect aquatic life in other ways. They can clog fish gills, reduce growth rates, decrease resistance to disease, and prevent egg & larval development. Particles that settle out can smother fish eggs and those of aquatic insects, as well as suffocate newly-hatched larvae. The material that settles also fills the spaces between rocks and makes these microhabitats unsuitable for various aquatic insects. Suspended solids can result from erosion from urban runoff and agricultural land, industrial wastes, river bank erosion, bottom feeders (such as carp), algae growth or wastewater discharges.

CONDUCTIVITY. Conductivity is a measurement of the ability of an aqueous solution to carry an electrical current. An ion is an atom of an element that has gained or lost an electron which will create a negative or positive state. For example, sodium chloride consists of sodium ions

(Na+) and chloride ions (Cl-) held together in a crystal. In water it breaks apart into an aqueous solution of sodium & chloride ions. This solution will conduct an electrical current. Degree to which water will carry an electrical current depends upon:

- the concentration or number of ions;
- mobility of the ion;
- oxidation state (valency) and;
- temperature of the water.

Resistance, expressed in ohms, is the opposite of conductivity. Conductivity is then expressed in reciprocal ohms. A more convenient unit of measurement in the chemical analysis of water is micromhos. The specific conductance or conductivity measurement is related to ionic strength and does not indicate what specific ions are present. Conductivity measurement is used to determine a number of applications related to water quality. These are as follows:

- Determ**ining mineralization:** This is commonly called total dissolved solids(TDS) Total dissolved solids information is used to determine the overall ionic effect in a water source. Certain physiological effects on plants & animals are affected by the number of available ions in the water.
- Noting variation or changes in natural water and waste waters quickly.
- Estimating the sample size necessary for other chemical analyses.
- Determining amounts of chemical reagents or treatment chemicals to be added to water sample.

Elevated dissolved solids can cause 'mineral tastes' in drinking water. Corrosion or encrustation of metallic surfaces by waters high in dissolved solids causes problems with industrial equipment and boilers as well as domestic plumbing, hot water heaters, toilet flushing mechanisms & washing machines. Indirect effects of excess dissolved solids are primarily the elimination of desirable food plants and habitat forming plant species. Agricultural uses of water for livestock watering are limited by excessive dissolved solids and high dissolved solids can be a problem in water used for irrigation.

HARDNESS. Hardness is due to the presence of multivalent metal ions which come from minerals dissolved in the water. Hardness is based on the ability of these ions to react with soap to form a precipitate or soap scum. In fresh water the primary ions are calcium & magnesium; however iron & manganese may also contribute. Carbonate hardness is equal to alkalinity but a non-carbonate fraction may include nitrates & chlorides.. Hardness is expressed in mg/l of $CaCO_3$ (even though all the hardness may not be due to $CaCO_3$).

Classification of Water by Hardness Content (Hardness Unit: mg/l) is as follows:

Concentration, mg/l	Description
0 - 75	Soft water
75 - 150	Moderately hard water
150 – 300	Hard water
300 and more	Very hard water

The most important impact of hardness on fish & other aquatic life appears to be the affect the presence of these ions has on the other more toxic metals such as lead, cadmium, chromium & zinc. Generally, the harder the water, the lower the toxicity of other metals to aquatic life. In hard water some of the metal ions form insoluble precipitates and drop out of solution and are not available to be taken in by the organism. Large amounts of hardness are undesirable mostly for economic or aesthetic reasons. If a stream or river is a drinking water source, hardness can present problems in the water treatment process. Hardness must also be removed before certain industries can use the water.

TOTAL DISSOLVED SOLIDS(TDS). The TDS comprise inorganic salts (principally calcium, magnesium, potassium, sodium, bicarbonates, chlorides and sulphates) and small amounts of organic matter that are dissolved in water. The TDS in drinking-water originate from natural sources, sewage, urban runoff and industrial wastewater. Concentrations of TDS in water vary considerably in different geological regions owing to differences in the solubilities of minerals. The presence of high levels of TDS in drinking-water may be objectionable to consumers.

Q. Describe various chemical parameters of water pollution.

Ans. Various chemical parameters of water pollution is described below.

BIOLOGICAL OXYGEN DEMAND (BOD). Biochemical Oxygen Demand (BOD) refers to the amount of oxygen that would be consumed if all the organics in one litre of water were oxidised by bacteria and protozoa. The first step in measuring BOD is to obtain equal volumes of water from the area to be tested and dilute each specimen with a known volume of distilled water which has been thoroughly shaken to insure oxygen saturation.

After this, an oxygen meter is used to determine the concentration of oxygen within one of the vials. The remaining vial is than sealed and placed in darkness and tested five days later. BOD is then determined by subtracting the second meter reading from the first. The range of possible readings can vary considerably: water from an exceptionally clear lake might show a BOD of less than 2 ml/l of water. Raw sewage may give readings in the hundreds and food processing wastes may be in the thousands.

Generally, when BOD levels are high, there is a decline in DO (Dissolved Oxygen) levels. This is because the demand for oxygen by the bacteria is high and they are taking that oxygen from the oxygen dissolved in the water. If there is no organic waste present in the water, there would not be as many bacteria present to decompose it and thus the BOD will tend to be lower and the dissolve oxygen level will tend to be higher..BOD Vs. water quality is shown below.

BOD(ppm)	Water Quality
1 - 2	**Very Good-**There will not be much organic waste present in the water supply.
3 - 5	**Fair:** Moderately Clean water
6 - 9	**Poor: Somewhat Polluted water.** Usually indicates organic matter is present and bacteria are decomposing this waste.
100 or greater	**Very Poor: Highly Polluted water.** Contains organic waste.

CHEMICAL OXYGEN DEMAND (COD). The COD test will give a good estimate of the first stage oxygen demand for most wastewaters. An advantage of the COD test over the biochemical oxygen demand (BOD) test is 2 to 3 hours versus 5 days. The COD test also is used to measure the strength of wastes that are too toxic for the BOD test. The COD test should be considered an independent measurement and not a quick substitute for the BOD test. The COD is usually higher than the BOD, but the amount will vary from waste to waste. Chemical Oxygen Demand measures the ability of hot chromic acid solution to oxidize organic matter. This analyzes both biodegradable & non-biodegradable organic matter. expressed as O_2. The results of the COD tests are usually higher than the corresponding BOD test for several reasons. Many organic compounds which are dichromate oxidizable are not biochemically oxidizable; Certain inorganic substances, such as sulphides, sulphates, thiosulphates, nitrites and ferrous iron are oxidized by dichromate, creating an inorganic COD, which is misleading when estimating the organic content of the wastewater.

The BOD to COD ratio is nothing more than the BOD concentration divided by the COD concentration for the same sample (e.g., if BOD is 60 mg/L, and COD is 100 mg/L for a sample, the ratio is 60/100, or 0.60). There are three fairly reliable 'rules of thumb' correlations between COD/BOD. Ratios for COD to BOD of 1 to 2 are found in potable water or clean surface or groundwater. Ratios of COD to BOD of 2 to 4 are usually seen in routine domestic/municipal sewage wastes. Ratios of COD to BOD of 4 to 6 are usually indicative of industrial type wastes.

In some industrial effluents (pretreatment program), BODs *can* be higher than CODs (for example, some effluents which are high in sugars, as can be found in the bakery industry, or soda bottling. Some industrial effluents will have higher demand because of the higher quantities of chemicals that demand oxygen).

Most applications of COD determine the amount of organic pollutants found in surface water (e.g. lakes and rivers), making COD a useful measure of water quality. It is expressed in milligrams per liter (mg/ l) or ppm , which indicates the mass of oxygen consumed per liter of solution.

CHLORIDE. Chloride in drinking water originates from natural sources, sewage & industrial effluents, urban runoff containing deicing salt and saline intrusion. The main source of human exposure to chloride is the addition of salt to food, and the intake from this source is usually greatly in excess of that from drinking water. Excessive chloride concentrations increase rates of corrosion of metals in the distribution system, depending on the alkalinity of the water. This can lead to increased concentrations of metals in the water supply. Chloride concentrations in excess of about 250 mg/litre can give rise to detectable taste in water.

SULPHATE. Sulphates occur naturally in numerous minerals and are used commercially, principally in the chemical industry. They are discharged into water in industrial wastes and through atmospheric deposition; however, the highest levels usually occur in groundwater and are from natural sources. In general, the average daily intake of sulphate from drinking water, air and food is approximately 500 mg, food being the major source. However, in areas with drinking water supplies containing high levels of sulphate, drinking-water may constitute the principal source of intake.

Q. What are the biological parameters of water pollution.

Ans. Various biological parameters are discussed below.

COLIFORMS: Coliforms are bacteria that are always present in the digestive tracts of animals, including humans, and are found in their wastes. They are also found in plant and soil material.

Total coliforms, Faecal coliforms, and E. coli: The most basic test for bacterial contamination of a water supply is the test for total coliform bacteria. Total coliform counts give a general indication of the sanitary condition of a water supply. Total coliforms include bacteria that are found in the soil, in water that has been influenced by surface water, and in human or animal waste.

Fecal coliforms are the group of the total coliforms that are considered to be present specifically in the gut and feces of warm blooded animals. Because the origins of fecal coliforms are more specific than the origins of the more general total coliform group of bacteria, fecal coliforms are considered a better indication of animal or human waste than the total coliforms.

The major species in the fecal coliform group is *Escherichia coli (E. coli)*. Of the five general groups of bacteria that comprise the total coliforms, only *E. coli* cannot be found growing and reproducing in the environment. Consequently, *E. coli* is considered to be the species of coliform bacteria that is the best indicator of fecal pollution and the possible presence of pathogens.

Coliform bacteria may harm us by causing diseases and food spoilage. Of environmental concern are the many types of coliform bacteria. Fecal coliform bacteria are a group of bacteria that are passed through the fecal excrement of humans, livestock and wildlife. They aid in the digestion of food. A specific subgroup of this collection is the fecal coliform bacteria, the most common member being Escherichia coli. These organisms may be separated from the total coliform group by their ability to grow at elevated temperatures and are associated only with the fecal material of warm-blooded animals. Bacteria reproduce rapidly if conditions are right for growth. Most bacteria grow best in dark, warm, moist environments with food. Some bacteria form colonies as they multiply which may grow large enough to be seen. By growing and counting colonies of fecal coliform bacteria from a sample of stream water, we can determine approximately how many bacteria were originally present.Most coliform bacteria do not cause disease. However, some rare strains of *E. coli*, are pathogenic and can cause serious illness. Recent outbreaks of disease caused by E. coli 0157:H7 have generated much public concern about this organism. *E. coli* 0157:H7 has been isolated from cattle, chickens, swine, and sheep. Most of the reported cases have been attributed to the consumption of partially cooked hamburger. Cases of E. coli 0157:H7 caused by contaminated drinking water supplies are extremely rare.

Standards

- The first standard is a mandatory one that must not be exceeded; there should be no more than 10,000 total coliforms per 100ml of water or 2,000 faecal coliforms per 100ml of water.
- The second standard is a guideline that should be achieved where possible; within 100ml of water there should be no more than 500 total coliforms, 100 faecal coliforms and 100 faecal streptococci.

Q. What are the effects of water pollution ?

Ans. Water pollution affects the health of the waterway, the health of the organisms living in & around the water bodies, and, eventually, the health of humans. The effects of water pollution can range from aquatic deformities to contaminated fish to 'dead' lakes. Toxic pollutants can also alter the genetic makeup of an organism, resulting in either death or extreme deformities.

Persistent Organic Pollutants (POPs), such as dioxin, PCBs and DDT, are chemical substances that persist in the environment & bio accumulate through the food web. Therefore, POPs can also cause sickness and disease in humans, who are at the end of the food chain. People who regularly consume a lot of fish will have larger levels of toxic chemicals in their bodies than those who only eat fish occasionally. Toxic chemicals can lead to reproductive problems, cancer and neurological disorders. People who are most at risk of health problems due to contaminated fish consumption are those with weakened immune systems, including children, pregnant women & the elderly.

Water-borne diseases are caused by drinking water contaminated by human or animal faeces, which contain pathogenic microorganisms.There are many types of water-borne diseases, like Cholera, Dengue, Diarrhoea, Leptospirosis, Malaria, Hepatitis, Scabies, Typoid and Fluorosis. **Fluorosis** is a dental problem which attacks the children by making the brownish tinch on their teeth.The amount of nutrients entering the water bodies has intensified due to increased urbanization & agriculture, leading to increased biological growth, or eutrophication. Under eutrophic conditions, nutrient loading (more nutrients than the water body can handle) stimulates excessive plant growth, which in turn decreases the amount of oxygen in the water and kills off certain species of animal life.

Freshwater resources all over the world are threatened not only by over exploitation and poor management but also by ecological degradation. The main source of freshwater pollution can be attributed to discharge of untreated waste, dumping of industrial effluent, and run-off from agricultural fields. Industrial growth, urbanization and the increasing use of synthetic organic substances have serious and adverse impacts on freshwater bodiesThe developed countries suffer from problems of chemical discharge into water sources mainly groundwater,while developing countries face problems of agricultural run-off in water sources.

Q. Describe the Causes, sources , effects and control measures of water pollution.

Ans. Water pollution is defined as alteration in physical, chemical or biological characteristics of water making it unsuitable for designated use in its natural state.

Causes of water pollution: Water is an essential commodity needed for drinking, cooking, bathing, washing, irrigation, and for industrial operations. Most water for such uses comes from rivers, lakes or ground water sources. Water dissolves many substances in it , therefore, it can easily get polluted. Pollution of water can be caused by point sources or non-point sources. Point sources are specific sites near water bodies which directly discharge effluents into them. Major point sources of water pollution are industries, power plants, underground coal mines, offshore oil wells etc. The discharge from non-point sources is not at any particular site, rather, these sources are scattered, which individually or collectively pollute water. Surface run-off from agriculture fields, overflowing small drains, rain water sweeping roads & fields, atmospheric deposition etc. are the non-point sources of water pollution.

Sources of water pollution

Ground water pollution: Ground water forms about 6 % of the total water available on earth and is about 30 times more than surface water. Ground water seems to be less prone to pollution as the soil mantle through which water passes helps to retain various contaminants due to its caption exchange capacity. However, there are a number of potential sources of ground water pollution. Septic tanks, industry, deep well injection, mining etc. are mainly responsible for ground water pollution, which is irreversible. Ground water pollution with arsenic, fluoride & nitrate are posing serious health hazard.

Surface water pollution: The major sources of surface water pollution are:

Sewage: Pouring the drains and sewers in fresh water bodies causes water pollution. The problem is severe in cities.

Industrial effluents: Industrial wastes containing toxic chemicals, acids, alkalis, metallic salts, phenols, cyanides, ammonia, radioactive substances, etc. are sources of water pollution. They also cause thermal pollution of water.

Synthetic detergents: Synthetic detergents used in washing and cleaning products foam and pollute water.

Agrochemicals: Agrochemicals like fertilizers and pesticides washed by rain water and surface run off pollute water.

Oil: Oil spillage into sea-water during drilling and shipment pollute it.

Waste heat: Waste heat from industrial discharges increases the temperature of water bodies and affects distribution and survival of sensitive species.

Effects of Water Pollution. Organic matter which reaches water bodies is decomposed by micro-organisms present in water. For this degradation oxygen dissolved in water is consumed. Dissolved oxygen (DO) is the amount of oxygen dissolved in a given quantity of water at a particular temperature and atmospheric pressure. Amount of dissolved oxygen depends on aeration, photosynthetic activity in water, respiration of animals and plants and ambient temperature.

The saturation value of DO varies from 8-15 mg/l. For active fish species 5-8 mg/l of DO is required whereas less desirable species like carp can survive at 3.0 mg/l of DO.

Lower DO may be harmful to animals especially fish population. Oxygen depletion helps in release of phosphates from bottom sediments and causes eutrophication.

Control of water pollution. Following steps will help reduce water pollution:

(i) Prevent run-off of manure. Divert such run off to basin for settlement. The nutrient rich water can be used as fertilizer in the fields.

(ii) Separate drainage of sewage and rain water should be provided to prevent overflow of sewage with rain water.

(iii) Planting trees would reduce water pollution by sediments and will also prevent soil erosion.

For controlling water pollution from point sources, treatment of waste waters is essential before being discharged. Parameters which are considered for reduction in such water are-Total solids, biological oxygen demand (BOD), chemical oxygen demand (COD), nitrates & phosphates, oil & grease, toxic metals etc.

Wastewaters should be properly treated by primary and secondary treatments to reduce the BOD, COD levels up to the permissible levels for discharge.

Advanced treatment for removal of nitrates and phosphates will prevent eutrophication. Before the discharge of waste water, it should be disinfected to kill the disease-causing organisms like bacteria.

Proper chlorination should be done to prevent the formation of chlorinated hydrocarbons or disinfection should be done by ozone or ultra violet radiations.

Q. Describe the causes, effects and control of thermal pollution of water.

Ans. Thermal pollution is defined as presence of waste heat in the water which increases its temperature can cause undesirable changes in the natural water environment.

Causes of thermal pollution: Heat producing industries i.e., thermal power plants, nuclear power plants, refineries, steel plants etc. are the major sources of thermal pollution. Power plants utilize only 1/3 of the energy provided by fossil fuels for their operations. Remaining 2/3 is generally lost in the form of heat to the water used for cooling. Cold water, generally, is drawn from some nearby water-body, passed through the plant and returned to the same water-body, with temperature 10-12^{0}C higher than the initial temperature. Excess of heat reaching such water bodies causes thermal pollution of water.

Effects of Thermal Pollution:

(i) The dissolved oxygen content of water is decreased as the solubility of oxygen in water is decreased at high temperature. High temperature becomes a barrier for oxygen penetration into deep cold waters.

(ii) Toxicity of pesticides, detergents and chemicals in the effluents increases with increase in temperature.

(iii) The composition of flora & fauna changes because the species sensitive to increased temperature due to thermal shock will be replaced by temperature tolerant species.

(iv) Metabolic activities of aquatic organisms increase at high temperature and require more oxygen, whereas oxygen level falls under thermal pollution.

(v) Discharge of heated water near the shores can disturb spawning and can even kill young fishes.

(vi) Fish migration is affected due to formation of various thermal zones.

Control of thermal Pollution: Following methods can be employed for control of thermal pollution:

(i) Cooling ponds

(ii) Spray Ponds

(iii) Cooling towers

Water from condensers is stored in ponds where natural evaporation cools the water which can then be re circulated or discharged in nearby water body. Spray ponds and cooling towers can be used to cool heated water before being discharged into water bodies eg, rivers,sea etc.

Q. Describe the sources , effects and control of marine pollution.

Ans. Sources: The main sources of marine pollution are (i)rivers, which bring pollutants from their drainage basins, (ii) Catchment area i.e. coastline where human settlements in the form of hotels, industry, agricultural practices have been established, and (iii) oil drilling and shipment.

Most of the rivers ultimately join the ocean. The pollutants which these rivers carry from their drainage basins are finally poured into the sea. These include sewage sludge, industrial effluents, synthetic detergents, agrochemicals, solid wastes, plastics, metals and waste heat released by industries. In the sea the pollutants get diluted and the organic matter is further broken down as in river water. Still many pollutants specially the recalcitrant once remain unchanged or are partially degraded causing marine pollution. These pollutants get biomagnified and affect fisheries and other marine life. Another important source of marine pollution is the leaking toxic substances, radioactive wastes etc. which are stored in large containers and dumped in deep sea considering sea to be a better disposal site than land.

Tankers and other shipping means, industries, lubricating oil using industry, metal industry, paint industry, automotive wastes, refineries, ship-accidents and off shore production add to marine pollution. Tankers transporting oil contribute to oil pollution significantly. After delivering the oil through sea-route, earlier empty tankers used to be filled with water called ballast water to maintain balance. The ballast-water containing residual oil from tankers are released into the sea on completion of return journey. Now-a-days the oil floating on the ballast water is removed in the newly designated 'load on top tankers' before ballast water is let off.

Effects: Oil in sea water can spread over a large area of the sea, remain dispersed or get adsorbed on sediments. It can cause adverse effects on marine life.

Oil in sea water affects sensitive flora and fauna. Phytoplankton, zooplankton, algal species, and various species of invertebrates, coral reefs, fish, birds & mammals are after the slimy mucus and gills is affected. Oil disrupts the installing capacity of feathers. Death occurs due to loss of buoyancy and subsequent drowning of birds. Leakage from oil tanker near Alaska in 1989 caused damage to coral reefs and resulted in death of lakhs of birds. Some important incidents of bird mortality due to oil have been reported at Brittany, France where 20 thousand birds died due to more than 220 tons of oil spillage in 1978. At Elbe, Germany 500 thousand birds in 1955. During the 1991 Gulf War 200 million gallons of oil spread in the Persian Gulf badly affected the marine ecosystem adversely.

Control:

(i) Toxic pollutants from industries and sewage treatment plants should not be discharged in coastal waters.

(ii) Run off from non-point sources should be prevented to reach coastal areas.

(iii) Sewer overflows should be prevented by having separate sewer and rain water pipes.

(iv) Dumping of toxic, hazardous wastes and sewage sludge should be banned.

(v) Development activities on coastal areas should be minimized.

(vi) Oil and grease from service stations should be processed for reuse.

(vii) Oil ballast should not be dumped into sea.

(viii) Ecologically sensitive coastal areas should be protected by not allowing drilling.

Q. What Are Major Water Borne Diseases ?

Ans. Major water borne diseases are described below:

Cholera. Cholera is an acute, diarrhoeal illness caused by infection of the intestine with the bacterium Vibrio cholerae. A person may get cholera by drinking water or eating food contaminated with the cholera bacterium. This disease spreads very rapidly in areas with inadequate treatment of sewage & drinking water. The cholera bacterium may also live in the environment in brackish rivers and coastal waters.

Approximately 1 in 20 infected people has severe disease characterized by profuse watery diarrhoea, vomiting and leg cramps. In these people rapid loss of body fluids leads to dehydration. Without treatment death can occur within hours.

Dengue. Dengue is an acute infectious disease caused by a virus and transmitted by the bite of the *Aedes* mosquito, also known as *break bone fever* and *bone-crusher disease*. Dengue fever is a severe, flu-like illness that affects infants, young children and adults but rarely causes death.

The symptoms, following an incubation period of five to eight days, are fever, chills, & severe headache, pain in the joints, sweating, and prostration. The clinical features vary according to the age of the patient. Infants and young children may have a feverish illness with rash. Older children and adults may have either a mild feverish illness, or the classical incapacitating disease with abrupt onset and high fever, severe headache, pain behind the eyes, muscle & joint pains and rash.

Dengue hemorrhagic is a potentially lethal complication and is a leading cause of childhood death in several Asian countries. It is characterized by high fever, hemorrhage, enlargement of the liver, and in the most severe cases circulatory failure. Today, dengue is found in tropical and sub tropical regions around the world, predominately in urban and per urban areas, where Aedes mosquitoes are prevalent.

Diarrhoea. Diarrhoea is frequent discharge of watery faeces from the intestines, sometimes containing blood and mucus. Diarrhoea due to infection may last a few days, or several weeks. Persistent diarrhoea results in severe dehydration. It is therefore necessary to replace the fluid lost by the body. Severe diarrhoea may be life threatening due to fluid loss in watery diarrhoea, particularly in infants and young children, the malnourished and people with impaired immunity.

Diarrhoea is a concomitant of many infectious diseases, especially typhoid fever, bacillary or amoebic dysentery, and cholera. Diarrhoea is a symptom of infection by a host of bacterial, viral and parasitic organisms most of which can be spread by contaminated water. Diarrhoea due to infection is widespread throughout the developing world.

Fluorosis. Fluorosis is an abnormal condition caused by excessive intake of fluorine from fluoridated drinking water, characterized by mottling of the teeth. Moderate level chronic exposure (above 1.5 mg/l of water) is quite common. Long-term ingestion of large amounts can lead to potentially severe skeletal problems (skeletal fluorosis). The early symptoms of skeletal fluorosis include stiffness and pain in the joints. In severe case, the bone structure may change and ligaments may calcify, with resulting impairment of muscles and pain. Acute high-level exposure to fluoride causes immediate effects of abdominal pain, excessive saliva, nausea and vomiting. Seizures and muscle spasm may also occur.

Fluorosis affects millions of people around the world, but as regards dental fluorosis the very mild forms are the most frequent. Waters with high levels of fluoride content are mostly found at the foot of high mountains and in areas where the sea has made geological deposits.

Hepatitis. Hepatitis is any disease featuring inflammation of the liver. Two of the viruses that cause hepatitis (hepatitis A and E) can be transmitted through water, food and from person to person. Hygiene is therefore important in their control. The illness starts with an abrupt onset of fever, body weakness, loss of appetite, nausea and abdominal discomfort, followed by jaundice within a few days. The disease may range from mild (lasting 1-2 weeks) to severe disabling disease (lasting several months).

Both hepatitis A and E are found worldwide. Hepatitis A is particularly frequent in countries with poor sanitary and hygienic conditions.

Leptospirosis. Leptospirosis is an infectious disease of domestic animals, especially cattle, swine, and dogs, caused by spirochetes of the genus Leptospira and characterized by jaundice and fever. It is transmissible to humans through direct contact with the urine of infected animals or by contact with a urine-contaminated environment, such as surface water, soil and plants.

In humans it causes a range of symptoms, while some infected people may have no symptoms at all. Symptoms of leptospirosis include high fever, severe headache, chills, muscle aches, & vomiting, and may include jaundice, red eyes, abdominal pain, diarrhoea, or a rash. Complications include renal failure, meningitis, liver failure, and respiratory distress. In rare cases death occurs. Leptospirosis occurs worldwide, in both rural and urban areas and in temperate and tropical climates.

Malaria. Malaria is the world's most important parasitic disease transmitted from one person to another through the bite of female Anopheles mosquitoes, which breed in fresh or occasionally brackish water.

Its symptoms are at the onset of malaria, bouts of chills & fever lasting several hours and occurring every three or four days. If the disease is not treated, the spleen and the liver become enlarged, anemia develops, and jaundice appears. General debility, anaemia, or clogging of the vessels of cerebral tissues by affected red blood cells can be followed by death. Malaria occurs mostly in tropical and subtropical countries.

Scabies. Scabies is a highly contagious parasitic skin disease caused by the itch mite (*Sarcoptes*

scabiei). It is acquired through close contact with an infested individual or contaminated clothing and is most prevalent among those living in crowded and unhygienic conditions. Personal hygiene is an important preventive measure and access to adequate water supply is important in control. The principal sign of the disease is a pimple-like rash that is most commonly found on hands, especially the webbing between the fingers, the skin folds of the wrist, elbow or knee, the penis, the breast or the shoulder. Infestation causes intense itching all over the body, especially at night.

Scabies mites are found worldwide, affecting all socioeconomic classes and in all climates. Epidemics have been linked to poverty, poor water supply, sanitation, and overcrowding.

Typhoid. Typhoid and paratyphoid enteric fever are acute, generalized infections caused by

Salmonella typhi and *Salmonella paratyphoid* respectively. The main sources of infection are contaminated water or milk and, food handlers who are carriers. Their germs are passed in the faeces and urine of infected people. People become infected after eating food or drinking beverages that have been handled by a person who is infected or by drinking water that has been contaminated by sewage containing the bacteria.

Once the bacteria enter the person's body they multiply and spread from the intestines, into the bloodstream. The symptoms of typhoid appear 10 to 14 days after infection; and include high fever, rose-colored spots on the abdomen & chest, diarrhoea or constipation, and enlargement of the spleen & liver. In untreated patients , complications may be numerous, affecting practically every body system, and can even include perforation of the intestine with hemorrhage. Typhoid & paratyphoid fevers are common in less-industrialized countries, mainly due to the problem of unsafe drinking water, inadequate sewage disposal and flooding.

Q. What are the Major Sources of Water Pollution ?

Ans. Principal sources of water pollution are:

- industrial discharge of chemical wastes and byproducts.
- discharge of poorly-treated or untreated sewage.
- surface runoff containing pesticides.
- slash and burn farming practice, which is an element within shifting cultivation
- agricultural systems.
- surface runoff containing spilled petroleum products.
- surface runoff from construction sites, farms, or paved and other impervious surfaces e.g. silt.
- discharge of contaminated and/or heated water used for industrial processes.
- acid rain caused by industrial discharge of sulfur dioxide (by burning high-sulfur fossil fuels).
- excess nutrients added by runoff containing detergents or fertilizers.
- underground storage tank leakage, leading to soil contamination, thence aquifer Contamination.

Q. What Are The Organic & Inorganic Pollutants Present In Polluted Water ?

Ans. Some organic water pollutants are:

- insecticides & herbicides, a range of organohalide and other chemicals.
- bacteria, often is from sewage or livestock operations;
- food processing waste, including pathogens.
- tree and brush debris from logging operations.

- VOCs (Volatile Organic Compounds, industrial solvents) from improper storage.

Some inorganic water pollutants are:

- heavy metals including acid mine drainage.
- acidity caused by industrial discharges (especially sulfur dioxide from power plants).
- chemical waste as industrial by products.
- fertilizers, in runoff from agriculture including nitrates & phosphates.
- Silt in surface runoff from construction sites, logging, slash and burn practices or land clearing sites.

Q. What Are Main Point Sources And Non - Point Sources Of Water Pollution ?

Ans. Point Sources. They include transfer of pollutants from municipal & industrial liquid waste disposal sites and from municipal & household hazardous waste and refuse disposal sites.

Non Point Sources. They include:

Diffuse agricultural sources:

Wash off and soil erosion from agricultural lands carrying materials applied during agricultural use, mainly fertilizers, herbicides & pesticides.

Diffuse urban sources:

Run off from city streets, from horticultural, gardening & commercial activities in the urban environment and from industrial sites & storage areas.

Q. Describe The Transport And Chemical Reactions Of Water Pollutants ?

Ans. Most water pollutants are eventually carried by the rivers into the oceans. There are areas of oxygen depletion, caused by chemicals using up oxygen and by algae blooms, caused by excess nutrients from algal cell death & decomposition. Fish and shellfish kills have been reported, because toxins climb the food chain after small fish consume copepods, then large fish eat smaller fish, etc. Each step up the food chain concentrates certain toxins like heavy metals and DDT by approximately a factor of ten.

Many chemicals undergo reactive decay / change over long periods of time in groundwater reservoirs. Such chemicals include chlorinated hydrocarbons such as trichloroethylene (used in industrial metal degreasing) and tetrachloroethylene used in the dry cleaning industry. Both of these chemicals, which are carcinogens , undergo partial decomposition reactions leading to new hazardous chemicals.

Groundwater pollution is much more difficult to abate than surface pollution because ground water can move great distances through unseen aquifers. Non-porous aquifers such as clays partially purify water of bacteria by simple filtration (adsorption & absorption), dilution, and, in some cases, chemical reactions & biological activity. However, in some cases, the pollutants merely transform to soil contaminants. Ground water that moves through cracks and caverns is not filtered and can be transported as easily as surface water.

There are a variety of secondary effects stemming not from the original pollutant, but a derivative condition. Some of these secondary impacts are:

- Silt bearing surface runoff can inhibit the penetration of sunlight through the water column, hampering photosynthesis in aquatic plants.
- Thermal pollution of water can induce fish kills and invasion by new thermophyllic species.

Q. What Are The Effects Of Water Pollution On Aquatic System ?

Ans. Water pollution affects the health of the waterway, the health of the organisms living in & around the waterway, and, eventually, the health of humans. The effects of water pollution on aquatic systems can range from aquatic deformities to contaminated fish to 'dead' lakes. Heavy metals like mercury & lead, and human-made organic chemicals such as pesticides, biomagnify as they move up the food chain, resulting in tumors and death for predatory animals, such as lake trout, herring gulls, & even humans.

Toxic pollutants can also alter the genetic makeup of an organism, resulting in either death or extreme deformities. Cormorants suffer from cross-billed syndrome , while terns exhibit birth defects from dioxin, PCBs (Poly Chlorinated Biphenyls) and furan exposure Other examples of deformities include large fish tumors and three-legged frogs.

Q. What Is Eutrophication And Its Effects ?

Ans. Before the introduction of chemical fertilizers the water contained little plant nutrients and were clear due to their size & depth. These lakes support high levels of animal life and receive proper amounts of nutrients, mainly phosphorous & nitrogen, from natural sources, such as decomposing plant matter. Urbanisation , agricultural revolution and industrialization has changed the situation. The amount of nutrients entering the water bodies has increased greatly, mainly due to increased urbanization and agriculture, leading to increased **biological growth, or eutrophication**.

Under eutrophic conditions, nutrient loading (more nutrients than the waterbody can handle) stimulates excessive plant growth, which in turn decreases the amount of dissolved oxygen in the water and eventually kills off certain species of animal life. Other pollution-tolerant species, such as worms and carp, grow more rapidly; thus, the ecological balance of the lake is altered.

Q. What are various water treatment techniques employed for potable (drinking) water?

Ans. Any physical, biological, or chemical change in water quality that adversely affects living organisms or makes water unsuitable for desired uses is considered pollution. The most serious water pollutants for human health are pathogenic organisms from human and animal wastes. We have traditionally taken advantage of the capacity of ecosystems to destroy these organisms, but as the population density has increased , these systems have become overloaded & ineffective necessitating water treatment to make it usable.

In industrialized nations, toxic chemical wastes have become an increasing problem. Agricultural and industrial chemicals have been released or spilled into surface waters and are seeping into groundwater supplies.

The water treatment processes are classified as:

- **eration;**
- **sedimentation;**
- **coagulation and flocculation;**
- **filtration;**
- **disinfection.**

Potential industrial and agricultural contaminants (chemicals such as oil, phosphates, sulphates, heavy metals, etc.) end up in water resources in increasing amounts causing water pollution. These contaminants must be removed if the water is to be made potable. Since water treatment presents the most demanding component of a water supply system, it must be examined whether alternative methods exist that yield a measure of water quality improvement: protection of the water source from contamination, as well as rehabilitation, upgrading & systematic monitoring of already existing works, and construction of efficient sanitation facilities.

Q. What is the basic purpose and range of applications of aeration in water treatment ?

Ans. The basic purpose of aeration is the reduction of substances which cause unpleasant tastes & odors as well as discoloration in water. Aeration is used for treatment of groundwater where it also has additional positive side effects (precipitation of iron and manganese). When treating surface water aeration is useful in adding oxygen to the raw water. Aeration always precedes some other treatment process.

Aeration equipment is used to intensively mix air and water to facilitate the transfer of gases into or out of the water. The following effects can be achieved by aeration of water:

- Addition of oxygen; this may be necessary for surface water where the natural oxygen content was depleted due to the presence of large amounts of organic substances. Aeration contributes positively to subsequent biological treatment (e.g. slow sand filtration).
- **Removal of dissolved iron & manganese.** -These are oxidized and form nearly insoluble hydroxide sludges. They can be removed in a settling tank or by means of a coarse filter.
- **Removal of excess carbon dioxide** (CO_2) to prevent corrosion of metal and concrete surfaces.
- **Reduction of H_2S, CH_4 and other volatile compounds** which produce objectionable taste & odor.
- **Temperature reduction.**

Q. Describe the role and area of applications of SEDIMENTATION in water treatment.

Ans. Sedimentation occurs in nature perpetually and aids the natural purification of lakes & rivers. Use is made of this physical process in the treatment of water by passing it through settling basins or storage tanks at low and uniform velocities. This is a simple means of reducing the contents of suspended matter and partially of bacteria. It can be combined by preceeding it with coagulation & flocculation, and succeeding it with slow sand filtration. Following these procedures, disinfection is required for high bacteria contents.

Areas of Application of sedimentation

Turbidity control. Under the influence of gravity, suspended matter in rain water settles out if it as it has a density greater than that of the water itself. The efficiency of a settling basin depends on the nature (shape, size, density) of the particles that are accountable for the turbidity; gravity, sand & silt, which pollute surface waters heavily and settle easily, especially during the rainy season. Colloidal matter which contributes much to turbidity is held in suspension mainly by electrostatic forces & because of its low density. Colloidal particles,when brought in contact with coagulants, form flocculent material that can be settled or filtered out.

Pathogenic Organisms removal. Simple sedimentation by passing water through a settling tank does not achieve a significant removal of pathogens. Two to four weeks storage, though, can reduce bacteria populations considerably (50- 90%) by means of biological processes. Storage in excess of one month can reduce the viral count. The degree of purification depends on the severity of pollution and on the presence of other pollutants. Storage induced contamination (mosquito breeding due to algal growth) must be avoided by covering tanks. Schistosoma larvae, infectious agents of Bilharzia, usually cannot survive more than two days of protected storage, provided suitable hosts (snails) are not present.

Color removal. Removal of color without the use of chemical procedures can only be achieved by very long storage times.

Q. What Is The Purpose Of Coagulation And Flocculation In Water Treatment ?

Ans. Finely dispersed , suspended and colloidal particles causing turbidity & color of the water cannot be removed sufficiently by the ordinary sedimentation process. Adding a coagulant and mixing & stirring the water causes the formation of settleable particles. These flocs are large enough to settle rapidly under the influence of gravity, and may be removed from suspension by filtration. The choice and dose rates of coagulants depend on the characteristics of the water to be treated. The coagulation chemicals like aluminium sulfate should be readily available and their application must be closely monitored.On the household level, coagulation by means of natural coagulants of plant & soil origin and simple devices has been practiced traditionally in developing countries.

Q. Describe the Mechanisms of Coagulation.

Ans. Colloidal particles (dia:10^{-4} to 10^{-6} mm) generally carry a negative electrical charge and are surrounded by an electrical double layer (due to attachment of positively charged ions from the ambient solution) and thus inhibit the close approach of each other. They remain finely divided & don't agglomerate. Due to their low specific gravity, they don't settle out. A coagulant (generally positively charged) causes compression of the double layer and thus the neutralization of the electrostatic surface potential of the particles. The resulting destabilized particles stick sufficiently together when contact is made. Rapid mixing (a few seconds) is important at this stage to obtain uniform dispersion of the chemical and to increase the opportunity for particle-to-particle contact. Subsequent gentle and prolonged (several minutes) mixing cements the still microscopic coagulated particles into larger flocs. These flocs then are able to aggregate with suspended polluting matter. When increased sufficiently in size and weight, the particles settle to the bottom.

Q. What are various coagulants commonly used in water treatment ?

Ans. Coagulants used include chemicals , materials of soil origin, materials of plant origins & natural coagulants.

Chemicals. It is most common practice to use aluminium and iron salts as coagulant in water treatment. Both salts hydrolyze when added to water and form insoluble materials -aluminium & ferric hydroxides -when reacting with calcium and magnesium carbonates, which are almost always present in water. If these carbonates are not present in sufficient **quantity** (as in soft water), hydrated lime $Ca(OH)_2$ or sodium carbonate Na_2CO_3 may be added also. The formation of the insoluble hydroxides depends on the pH. Aluminium sulphate coagulates best in a pH range between 4.4 and 6. At higher pH values, higher rates of soluble aluminate ions form. Sodium aluminate is generally used at medium pH values (6.5 to 8). Irons salts are effective over a wide range of pH values (except for values between 7 and 8.5). Whereas turbidity is best removed within a pH range of 5.7 to 8.0, color removal is generally obtained at acid pH of about 4.4 to 6.0. To improve the coagulation and flocculation process and to reduce the dose of coagulants, **flocculation aids** may be used. The most commonly used material is activated silica. Yet diatoms (kieselgur), activated carbon in powder form, bentonite & certain other types of adsorptive clays, organic substances and derived materials are also used.

Materials of Soil Origin. Mineral substances are used as flocculation aids in modern water treatment. A dose of 10 mg/l of bentonite, for instance, together with 10 mg/l of aluminium sulphate , yield significantly better results than a higher dose of aluminum sulphate alone. In rural households in developing countries, however, various naturally occurring materials are traditionally used as coagulants: e.g., fluvial clays from rivers , clarifying rock material from desert regions, and earth from termite hills. Their main constituents are quartz, montmorillonite, kaolinite, calcite and feldspar. Their coagulating mechanisms differ greatly from those of metal salts.

Application of-clay as a coagulant yields the following results:

- reduction of turbidity;
- no effect on pH value;
- an initial mineral taste, later on normal;
- no effect on bacteria count.
- Clays contain traces of heavy metals (mostly chromium and manganese). High intakes of these metals may have toxic effects;
- viruses survive in the settled sludge.

Coagulants of Plant Origin. Such substances are widely used in developing countries to purify water. Usually the plants are not cultivated. Rather, according to passed on experience, certain substances are gathered, prepared and added to the water that is to be purified. Some examples of traditionally used coagulants and coagulant aids are:

- seeds from the Indian Nirmali tree (strychnos potatorum);
- seeds of the trees of the family of the Moringaceae: Moringa Olifeira, occurring in India, Senegal, Sudan (Behenus tree) and Moringa Stenopetala, Kenya;
- sap from the stem of the tuna cactus (opuntia ficus indica) occurring in Peru and Chile.

- the bark of the south American tree Schinopsis Quebracho-Colorado which contains tannin: it is known commercially as "Floccatan ;"
- potato starch.

For most of these plant materials, it is not known which particular substance actually triggers the coagulation. Neither is it known whether there are toxic side effects from frequent use. To obtain the optimal dose for various substances and raw water qualities, coagulation experiments must be carried out; generally this dose is smaller than that of aluminum sulphate.

For coagulation with Moringa Olifeira seeds, the following effects have been achieved.

- significant reduction of turbidity;
- pleasant taste;
- unchanged pH value;
- initial reduction of the bacteria count, followed by a secondary rise after only 24 hours, reaching or even surpassing the initial concentration;
- antibiotic effect on various bacteria and fungi.

Nirmali seeds and Tunaflex as natural coagulants and aid substances combined with alum salts have been successfully used in municipal water treatment. Substantial savings in primary coagulants can be achieved which, reduces the overall cost considerably.

Other Natural Coagulants include:

- Algae-derived substances;
- Chitosan, acting faster than any known coagulant from plant materials (produced from the shells of shrimp and lobster);
- Dough from millet bread (Sudan) or curds (thin layers).

Q. What is filtration and types of filters used in water treatment ?

Ans. Filtration is the passage of polluted water through a porous medium, thus utilizing the principle of natural cleansing of the soil. This technique used in water treatment is based on several simultaneously occurring phenomena:

- mechanical straining of undissolved suspended particles (screening effect);
- charge exchange, flocculation , adsorption of colloidal matter ;
- bacteriological-biological processes within the filter.

Filters may be divided into two principally different types:

- slow sand (or biological) filtration (v = 0.1 to 0.3 m/h),
- rapid filtration (v = 4 to 15 m/h).

In-between types also exist. Depending on the filtration rate, different mechanisms are operative within the filter. Generally, a filter consists of the following components:

- filter. medium (inert medium: quartz sand; or chemically activated medium: burnt material),
- support bed (gravel) and under-drain system,
- influent and effluent pipes, -wash and drain lines, -control and monitoring appurtenances.

Q. Describe the principle, process and mechanism of Rapid Filtration ?

Ans. Rapid filtration is based on the principle of mechanical straining of suspended matter due to the screening effect of the filter bed (sand, gravel, etc.). The particles in the water pass into the filter bed and lodge in the voids between grains of the filter medium. The cleaning of the rapid filter is facilitated by backwashing i.e., by reversing the flow direction; a backwash may be done simply with water or by use of a water-air mix (upward air scour). The impurities are thus dislodged and removed from the filter bed.

The performance of a rapid filter regarding the removal of suspended matter is determined by the following filtration process variables and parameters:

- filtration rate (v),
- influent characteristics, i.e., particle size, distribution, etc.,
- filter medium characteristics which control the removal of the particles and their release upon backwashing, respectively.

Generally, the treatment effect can be improved by:

- reduced filtration rates,
- smaller granulation size of the filter medium,
- increasing depth of the filter bed,
- decreasing concentration of particles to be retained.

Q. What are various types of Rapid Filters used in water treatment ?

Ans. Various types of rapid filters used in water treatment are described below:

a. **Conventional (Down flow) Filters.** Rapid filtration is complex, demanding & expensive process. This is due to the need for frequent filter washing which requires elaborate back washing systems. Additional complexities associated with the generation of pressure arise for pressure filters. Combined with coagulation, flocculation and sedimentation, rapid filtration is a very efficient treatment process for the removal of impurities. However, it should only be used in larger plants. For smaller plants in rural areas, simple rapid filters -without backwashing facilities are recommended. A number of filter types operating at filtration rates lower than those for conventional filers are described below. Generally, they serve as pretreatment units to reduce the turbidity of water. The removal of pathogens requires, in addition, either slow sand filtration and/or disinfection.

b. **Up flow Filter.** In up flow filters, the direction of flow of the raw water is upwards through the filter bed. Backwashing is done by reversal of the flow direction. The effect of the filter depends on the type of the filter medium, the filtration rate, and preceding aeration or addition of a coagulant. For coarse organic & inorganic substances, the filter may act as a simple screen. Or else it may retain precipitated iron compounds. At low filtration rates and sufficient oxygen content of the raw water, biological activity can be observed.

   The advantages of up flow filters as compared with gravity down flow rapid filters are:

- can be constructed from locally available materials,
- quality requirements (uniformity & gradation) and volume of the filter medium are lower. Instead of sand, gravel, crushed bricks, coconut and other type fibers can be used,
- longer filter runs,
- better turbidity removal.

Up flow filters can be constructed at a variety of degrees of complexity. A rather simple type can be built from a drum equipped with a raw water inlet pipe, a somewhat larger size drain at the bottom, and an outlet pipe for the clarified water near the top of the drum

Filtration effect: Reduction of between 50 and 70% of organic and inorganic coarse and fine particles, slight reduction of bacteria.

Filtration rate: 0.5 to 1.5 m/in.

Filter medium: Coarse sand, grain size between 3 and 4 mm diameter.

Filter bed depth: 0.3 m.

Support layer and under drain: gravel covered by perforated metal tray.

Cleaning: Shut off of the inlet. Quick removal of drain stopper so that supernatant as well as water in the filter bed drain out together with retained particles.

Cleaning of the filter which takes ten minutes should be done every day for preventing the filter bed from clogging.. As bacteria cannot be sufficiently removed, subsequent disinfection is indispensible in case of bacterial water contamination.Higher filtration rates result in higher buoyancy forces on the filter medium. The top layer of the sand may be spewn up. This can be avoided by covering the filter bed with a metal grate or by raising the depth of the filter bed. In the latter case though, backwashing by means of simply draining the water in a reversed direction may become increasingly impossible. Conventional backwashing facility have to be added. Better results may be obtained by using smaller grains and stratified filter beds with decreasing grain size from bottom to top (e.g., 0.7 to 2 mm over a depth of 1 to 1.5 m).

c. **Coarse Filters:** Rapid filters preceding slow sand filters are frequently used to retain coarse particles and to sufficiently reduce turbidity. Coarse sand, gravel or plant fibres are used as a filter medium. It can be replaced upon cleaning. Such pre filtration can be done either horizontally or vertically. The filtration rates for a coarse filter are lower than those for a conventional rapid filter.

Filtration effect: Reduction of turbidity by between 50 and 80% (max. load 250 NTU)

Filtration rate: 0.5 to 1.0 m/h

Filterbox: same as slow sand filter

Operating head: 1.0 to 1.5 m.

Filter medium: coarse sand, gravel, shredded coconut fibres. Two or more layers of different material possible (coarser material up top and finer material below).

Filter bed depth: 0.8 to 1.4 m.

Drainage system: same as slow sand filter.

Cleaning: replace medium completely when head loss exceeds certain value, i.e., when too big (approximately once every 3-4 months).

d. **Household Size Rapid Filter.** Household filters can be made from sand or gravel of different grain sizes, from ceramics, porcelain or other fine porosity materials. They operate on the principle of mechanical straining of the particles contained in the water. The filter performance depends on the porosity of the filter medium. Through additives in the filter material, additional effects can be obtained (eg, adsorption, disinfection).

e. **Multiple layer filter.** Using metal drums, plastic containers or clay vessels and filling them with several layers of sand, gravel or charcoal, simple household filters can be put together. They do not perform well at removing pathogens, though. After filtration, the water therefore needs to be disinfected. Charcoal adsorbs organic substances which cause disagreeable color & taste. This effect can only be sustained, however, if the charcoal is frequently renewed. If this is not possible, or if the filter (empty or filled with water) is left unused for some time, the charcoal can become a breeding ground for bacteria. The result is that the filtered water exhibits a higher bacteria count than the raw water. Monitoring of the filter condition is rendered more difficult by the fact that there is no visual indication given for the point when charcoal should be replaced. Charcoal cannot be regenerated. It is for these reasons that the use of filters with charcoal media is not recommended.

f. **Ceramics filter.** For the household purpose ceramics filters may be used for the purification of drinking water The purifying agent is a filter element, also called candle, through which the water is passed. Suspended particles are thus mechanically retained, and, depending on the size of the pores, also pathogens. Ceramics filters should only be used if the water is not too turbid, as the pores clog quickly. Ceramics filter elements can be made from various different material compositions (e g., diatomaceous earth, porcelain); they have pore sizes of between 0.3 and 50 μ. If the pore size is smaller than or equal to 1.5 μ all pathogens get removed with certainty. Post treatment of the water prior to consumption is rendered unnecessary. Filters with larger pores only retain macro organisms such as cysts and worm eggs. The filtered water must be boiled subsequently or otherwise disinfected.

   The impurities held back by the candle deposit on the candle's surface. At regular intervals, this coating can be brushed off under running water. After the cleaning, candle should be boiled. Candles made from diatomaceous earth which contain silver, avoid recontamination of purified water due to infestation of the filter material with bacteria laden washing water can be avoided.

   Filters operating at atmospheric pressure exhibit a very slow rate of percolation. This can be increased considerably by forcing the water through the medium. Ceramics filters must be checked from time to time for fissures so as to prevent the water from passing through the medium without being filtered.

g. **Cartridge microfilter.** Besides ceramics filters, other microfilters made from fine porosity materials are also available:eg,synthetics, paper, felt like material (pore size between 25 and 50 μ). They are inserted into a bell-like filter device, which is mounted on the top of a water pipe. When the filter material becomes clogged, i.e., used up, it must be discarded and replaced new. Even though these filters are cheaper than ceramics filters, their use is more expensive, since the filter material cannot be regenerated.

Q. What is Slow Sand Filtration & its mechanism ?

Ans. Slow sand filtration is accomplished by passing raw water slowly - driven by gravity through a medium of fine sand. On the surface of the sand bed, a thin biological film develops after some time of ripening (different from the rapid filter). This film consists of active microorganisms and is called filter skin. It is responsible for the bacteriological purification effect. The slow sand filter is therefore also called "surface filter" or biological filter.The principle purification processes taking place during slow sand filtration are:

Sedimentation: The water body sitting on top of the filter bed acts as a settling reservoir. Settleable particles sink to the sand surface.

Mechanical straining: The sand acts as a strainer. Particles too big to pass through the interstices between the sand grains are retained.

Adsorption: The suspended particles and colloids that come in contact with the surface of the sand grains by following the passage of the water are retained by:

- adhesion to the biological layer ,
- physical mass attraction (Van der Waals force), and
- electrostatic & electrokinetic attractive forces (Coulomb forces).

On account of these forces, an agglomerate of opposite charged particles forms within the top layer of sand. This process needs some time of ripening to fully develop.

Biochemical processes in the biological layer are:

- partial oxidation and breakdown of organic substances forming water, CO_2 & inorganic salts,
- conversion of soluble iron & manganese compounds into insoluble hydroxides which attach themselves to the grain surfaces,
- killing of E. Coli and of pathogens.

Organic substances are deposited on the upper layer of sand, where they serve as a breeding ground and food for bacteria & other types of microorganisms. These produce a slimy, sticky, gelatinous film which consists of active bacteria, their wastes & dead cells and partly assimilated organic materials. The dissimilation products are carried away by the water to greater depth. Similar processes occur there. The bacterial activity gradually decreases with depth. Different types of bacteria are normally found at various depths.

Algae can contribute to the breakdown of organic material and bacteria. They can improve the formation of the biological layer (filter skin). In uncovered filters, growth of algae is driven by photosynthesis. The presence of large amounts of algae in the supernatant reservoir of a filter generally impedes the functioning of the filter. Dead cell material may clog the filter. Increased consumption of oxygen due to the presence of dead cell material increases the possibility that anaerobic conditions will occur. There is always a diurnal variation in the oxygen content due to growth and decay of the algae mass. When algae growth is strong, the algae must be either removed regularly or the filter must be covered.

The conditions necessary for those biochemical processes are:

- sufficient ripening of the biological layers,
- uniform and slow flow of water through the filter ,

- a depth of the filter bed of 1 m (0.5 m is needed solely for the biochemical process) of specific grain sizes,
- sufficient oxygen in the raw water (at least 3 mg/l) to induce biological activity.

At MPN (Most Probable Number) Contents Greater than 1000 E. Coli/100 ml, Raw Water Should Subsequently Be Disinfected. Microbiological processes and chemical activity are very sensitive to changes in temperature. Both slow down under conditions of low temperature. A reduction in filtration rate can compensate for this effect. Under prolonged cold conditions, the filter should be covered to prevent heat loss, and subsequent disinfection should be provided.

Q. What is the purpose of disinfection of water in water treatment ?

Ans. Water disinfection processes are designed to destroy disease producing organisms by means of disinfectants. Drinking water should be free of pathogenic organisms. Storage, sedimentation, coagulation, flocculation and filtration of water both individually & jointly reduce the contents of bacteria in water to a certain extent. None of these methods can guarantee the complete removal of germs. Disinfection is needed at the end. Water with low turbidity may even be disinfected without any additional treatment for bacteria removal. Groundwater abstracted from deep wells is usually free of bacteria. Surface water and water obtained from shallow wells and open dug wells generally need to be disinfected.

The degree or efficiency of disinfection depends on the method employed and on the following factors influencing the process:

- kind and concentration of microorganisms in the water,
- other constituents of the water which may impede disinfection or render it impossible,
- contact time provided (important for chemical disinfectants, since their effect is not instantaneous, a minimum time of contact is necessary),
- temperature of the water (higher temperatures speed up chemical reactions).

Q. What are various means/agents of water disinfection ?

Ans. Water disinfection can be accomplished by several means:

- physical treatment: removal of bacteria through slow sand filtration, straining of macro organisms by means of micro screening , application of heat (boiling), storage, etc.
- irradiation, such as UV (Ultra Violet) light,
- metal ions, such as silver (and copper),
- chemical treatment, use of oxidants (halogens & halogen compounds -chlorine, iodine, bromine , ozone, potassium permanganate, hydrogen peroxide, etc.).

Q. What are the requisites of a good chemical disinfectant ?

Ans. A good chemical disinfectant should have the following capabilities:

- destroy all organisms present in the water within reasonable contact time, the range of water temperature encountered, and the fluctuation in composition, concentration & condition of the water to be treated;

- accomplish disinfection without rendering the water toxic or carcinogenic;
- permit simple and quick measurement of strength & concentration in the water,
- persist in residual concentration as a safeguard against recontamination;
- allow safe & simple handling, application and monitoring;
- ready and dependable availability at reasonable cost.

Though there are potential hazards for the human organism associated with prolonged ingestion of chemicals , the application of chlorine & its compounds for the purpose of water disinfection is the best and most tested compromise when evaluated according to the aforementioned criteria. The other methods differ significantly from each other in terms of their effect, the technological level and particularly in their applicability. They are introduced only briefly.

Q. Describe pros and cons of chlorination as a means of water disinfection.

Ans. Chlorination is the addition of chlorine gas or some other oxidizing chlorine compound (sodium or calcium hypochlorite, chlorinated lime, chlorine dioxide etc.) to the water to be treated. The actual disinfection agent is hypochlorous acid (HOCl) which forms when chlorine is added to water: Hypochlorous acid also forms subsequent to dissociation, when chlorinated lime or

hypochlorites are added. At pH levels between 3 and 6, hypochlorous acid dissociates poorly. Chlorination is most effective in that range of pH. At pH levels greater than 8, hypochlorite ions predominate or exist almost exclusively. Hence the disinfecting effect drops off rapidly as the pH level increases. Simultaneously with the dissociation, hypochlorous acid partly breaks up, forming monatomic oxygen, which contributes to the oxidizing effect:

Chlorination is the most widely used method for drinking water disinfection. It is effective and economical. The fraction that becomes effective as an oxidizing agent when chlorine or some of its compounds is added to raw water is called "free available" or "active" chlorine. Small amounts of chlorine, due to its ability to penetrate cells of microorganisms, are sufficient to destroy many different strains of bacteria. Many types of viruses & macro-organisms like schistosoma larvae can be killed. A contact time of at least 30 minutes is required, at the end of which the residual chlorine concentration in the water must still be between 0.1 and 0.5 mg/l (= ppm). Amoebic cysts and spores with resistant cell membranes require higher doses and longer contact times.

Chlorine also reacts with many other oxidizable water constituents such as iron & manganese compounds, ammonia, and compounds thereof (forming chloramines), as well as numerous types of organic particles. The presence of these substances reduces the germicidal effect considerably. Sufficient chlorine must be added to the water to make sure that there is a residual concentration to prevent recontamination.

It is advisable to remove or reduce prior to chlorination, those substances by sedimentation and/or filtration which would impede disinfection. Through such pretreatment,helminth eggs (parasitic worms) can be removed which are insensitive to chlorination. Through chlorination, certain undesirable side effects may occur.

Particularly in industrialized areas, synthetic organic compounds may enter the hydrologic cycle in high concentrations. The presence of chlorine enhances the danger of the formation of carcinogenic compounds (e.g., chloroform and other trihalomethanes).

Q. Describe the merits and demerits of various chemicals used for chlorination of water for its disinfection.

Ans. **Chlorine gas &chlorine dioxide** are the most widely used in water treatment on account of their high efficiency and ease of application. Handling & transport, however, are considered too demanding and hazardous (explosive, toxic). Several other chlorine compounds which have various active chlorine contents are more easily applicable. In some form or another they are available virtually anywhere.

Sodium hypochlorite (NaOCl), commonly known as bleach is generally available in dissolved form. Its commercial strength in terms of active chlorine is between 1 and 15%. It is stored in dark glass or plastic bottles. The solution loses some of its strength during storage. Prior to use, active chlorine content should be tested. Sunlight and high temperatures accelerate deterioration of the solution. The containers therefore should be stored in cool darkened areas. The stability of the solution decreases with increasing contents of available chlorine. A 1% solution is relatively stable. But it is not economical to store. Even though hypochlorite solutions are less hazardous than chlorine gas, precaution should be taken to avoid skin contact and to protect containers against physical damage.

Chlorinated Lime or Bleaching Powder ($CaO \cdot 3\ CaOCl_2 \cdot 3\ H_2O$) in the powder form is readily available and inexpensive. It is stored in corrosion resistant cans. When fresh, it contains 35% active chlorine. When exposed to air, it quickly loses its effectiveness. It is usually applied in solution form which is prepared by adding the powder to a small amount of water to form a soft cream. Stirring prevents lumping when more water is added. When the desired volume of the solution has been prepared, it is allowed to settle before decanting. Solutions should have concentrations between 5 and 1% of free chlorine, the latter being the most stable solution. Some 10% of the chlorine remains in the settled sludge.

High Test Hypochlorite (HTH) is a stabilized version of calcium hypochlorite ($Ca(OCl)_2$) containing between 60 and 70% available chlorine. Under normal storage conditions, commercial preparations will maintain their initial strength with little loss. Even though HTH is expensive, it may be economical, due to its properties. It is available in tablet or granular form.

These chemicals must be handled with great caution. They are caustic, corrosive and sensitive to light. They should be stored in tightly closed containers and in darkened spaces. When handling the material, contacts with skin, eyes and other body tissues must be avoided. Chlorine corrodes metal and to a less extent, wood and some synthetic materials. Metal parts which come in contact with the chemicals should be resistant.

Q. What are the factors for Determination of Chlorine Dose during disinfection by chlorination ?

Ans. Chlorine of any type must be added to water in closely controlled concentrations which depend on the characteristics of the water. As the use of dry chemicals doesn't always permit sufficient accuracy of dosing, solutions are preferred. Chlorine is usually added to the water for disinfection at the end of the treatment process. This allows the most effective treatment at the lowest level of chlorine application. Measurements of the chlorine demand & residual chlorine must be taken to assure that sufficient free chlorine is available to accomplish disinfection. Water characteristics and, hence, the chlorine demand may vary due to external influences (e.g., rainy season, etc.). It is therefore necessary to monitor the water quality from time to time, at the points of consumption in cases where the chlorine dosage is fixed. The objective of disinfection via chlorination can only be obtained if the chlorine dosage is adjusted to the changed water characteristics.

The chlorine demand of water of a given quality can be determined as follows: One litre samples of the water are taken. Chlorine solution of a known concentration is added and mixed with the water. After 30 minutes of contact time, the residual chlorine content is measured. The difference to the amount added then yields the chlorine consumption.

Chlorine demand = chlorine consumption + desired residual chlorine

Usually 1% chlorine solutions are applied. The chlorine flow is set such that a chlorine residual level of between 0.1 and 0.3 mg/l is obtained. Higher levels are used if rapid recontamination is likely.

Colorimetric tests are employed to determine total chlorine residuals. Chemical agents are used which are oxidized by chlorine to produce a colored complex, intensity of which is proportional to the amount of chlorine present. Reading the colors and matching color standards by means of a comparator and disks, gives the amount of free, available & residual chlorine. Various simple test kits are available, using permanent glass and containing DPD reagents in liquid /compressed tablet form.

Q. Discuss various factors to be considered for Practical Application of chlorination for disinfection of water.

Ans. Chlorination should never be performed prior to slow sand filtration (residual chlorine destroys biological agents). Sedimentation and filtration preceding chlorination enhance the disinfection effect. The lower the turbidity, the smaller the amount of chlorine necessary for effective disinfection. The chlorine solution can either be added to a batch of water (non-continuous or diffusion chlorination) or alternatively, it can be fed continuously to a constant flow of water.

Batch Chlorination. Where tanks are used for storage of drinking water, the required amount of chlorine can be added to the tank periodically. It is advantageous to alternate between two tanks

While one tank is in use, the other one is refilled and treated with chlorine. The water can be used after a minimum of 30 minutes contact time. This procedure allows uninterrupted supply.

If the water quality of a given source varies, the chlorine demand must be reevaluated from time to time. Before a tank is used the first time for storing water, it must be cleaned carefully and disinfected (application of between 50 and 100 ppm active chlorine). Once the water has been disinfected, recontamination must be carefully prevented. Tanks should be covered. A tap should be used to release the water so as to avoid scooping out the water with unclean jars and the like. If the water is not used immediately but left in the house for a while, only well cleaned and covered jars should be used.

Chlorine Tablets: In certain situations, e.g., while traveling, chlorine tablets can be used. They are used for periodic chlorination of small batches of water.

Diffusion Chlorination. Open wells are often bacteriologically contaminated because of non hygienic methods for lifting the water, or due to careless use of the surroundings of the well.

Various devices for diffusion chlorination include Type I which is a clay jar (12 to 15 l volume), filled nearly half-way with a mix of 1.5 kg bleach powder and 3 kg coarse sand (grain size 1.4 to 1.6 mm). It has two holes above the sand surface. The jar is covered with a plastic foil. The jar is suspended approximately 1 m below the water surface in the well. The chlorine can thus diffuse through the two holes into the well water.

Type II also consists of a clay jar (volume 7 to 10 l). It has 6 to 8 holes in the bottom. These are covered with stones on top of which a layer of gravel is placed. On top of that is put a mix of 1.5 kg bleaching powder and 3 kg of coarse sand. Stones are filled to the rim of the jar, which is then lowered into the water.For larger wells & higher rates of water use, two jars should be used which are interchangeably refilled.

In **Type III** used for small household wells, a double jar is recommended which releases less chlorine per time unit. The inner jar contains a mix of 1 kg bleaching powder and 2 kg coarse sand. The diffusion openings are provided. As these devices are not fit for large variations in water use, insufficient chlorination may occur at higher rates of water use.

Continuous Chlorination. Simple chlorine dosing instruments can be installed in piped water supply systems. Chlorine is fed to the water in proportion to the flow rate. A pipeline transmits the water from the source to the reservoir, passing through some sort of pretreatment (e.g., coagulation/flocculation and settling). Before entering the reservoir, the water is passed through a mixing chamber where a dosing apparatus introduces droplets of chlorine into the water.

A variety of types of chlorine dosing instruments are commercially available. They range from manually controlled types to fully automated ones. Usually a unit consists of a storage tank and a diaphragm pump for feeding the hypochlorite solution. The feed rate is proportional to the water flow rate and, thus depends on the consumption. The use of these devices is limited to piped water supply systems.

Q. Describe the use of Iodine as a water disinfectant.

Ans. Iodine is an excellent water disinfectant, effective against bacteria, amoeba cysts, cercerea and some viruses. It is added to the water mostly in the form of an aqueous solution @ 2 droplets per liter of water of a 2% iodine tincture. Iodine preparations are also available in tablet form.

In comparison to chlorination, the use of iodine has the following advantages:

- effectiveness over a wider range of pH values (up to pH 10), except at very low temperatures;
- ammonia and organic nitrogenous compounds have little effect on germicidal efficiency because they do not form substitution compounds with iodine;

- action depends less on contact time and temperature;
- effectiveness against more pathogenic organisms within short times;
- use and handling is simpler.

Since operating costs are too high, the use of iodine is not expected to ever become an important widely applied disinfectant.

The applicability is limited due to the following disadvantages:

- higher concentrations than chlorine (on a ppm basis) are necessary for effective action; -muddy or turbid water substantially affect germicidal action;
- iodine is about 20 times as expensive as chlorine per unit of germicidal effectiveness;
- taste & slight color produced by the iodine affect palatability and aesthetic quality;
- physiological effect of prolonged use of iodine (especially in children) is suspected. Allergies were ascertained.

In view of these economic and health implications the use of iodine for disinfection is recommended only for occasional application (e.g., in case of catastrophe or while traveling). Aside from that, iodine is a highly effective and technically widely applicable disinfectant.

Q. Describe Ozonation (application of ozone) as a disinfectant in water treatment.

Ans. Ozone (O_3) is one of the most effective water disinfectants. As a powerful oxidant, it reduces the contents of iron, manganese, & lead, and eliminates most of the objectionable taste & odor present in water. Its effectiveness does not depend on pH value, temperature or ammonia content of the water. Since ozone is relatively unstable, it is generated almost invariably at the point of use. Ozone is obtained by passing a current of dried and filtered air (or oxygen) through between two electrodes (plates or tubes) subjected to an alternating current potential difference. A portion of the oxygen is then converted into ozone.

This principle of ozone production has been used in Europe for a long time, since it has the advantage of being applicable under a wide range of conditions. It leaves no chemical residuals behind in the treated water. On the other hand, no lasting protection against recontamination is provided either. Capital costs for the instrumentation of ozone production & feeding, as well as operating costs due to the electrical energy requirements, are very high. Moreover, operation of ozonizers requires continuous and skilled monitoring.

Q. Describe the use of Potassium Permanganate as a disinfectant for water.

Ans. Potassium permanganate ($KMnO_4$), a powerful oxidant, is rarely applied in water treatment for disinfection. It is effective against cholera bacteria, but not against other pathogenic germs. A dose of 1 to 5 ppm $KMnO_4$ is recommended for application. It creates a purple brown precipitate which coats the walls of the storage tank. It cannot be removed easily. In recent years, potassium permanganate has gained steadily in the application in pretreatment since it has proved effective at:

- removing objectionable odor & taste by means of oxidation of organic material,& hydrogen sulfide;

- preventing algal growth;
- removing iron & manganese compounds by means of oxidation and subsequent separation by filtration.

Q. Discuss the role of silver in disinfection of water ?

Ans. Preparations containing silver (available as liquid/powder) may be used to reduce the germ count of water. They are readily soluble in water and can be dosed easily. Effectiveness of silver can be explained by the oligo-dynamic properties of silver ions (silver nitrate or salt compounds). Even minute concentrations (0.03 to 0.04 ppm) are notably effective. The silver ions curb the growth of germs. After contact of between 30 minutes and 6 hours, depending on the level of bacteriological contamination, water of a very low germ content may be obtained. Odor and taste of the water are not affected by the application of silver. Disinfection by silver is a simple and very effective method. Its major advantage is that it provides already treated water with long lasting protection against recontamination by germs.

The effect of silver and other metals has been known to many peoples for a long time. The tradition of storing drinking water in silver vessels is still maintained in wealthier families. Although there is a tendency at present to exchange the metallic containers for plastic ones, even Indian villagers can still be seen fetching water from the well in brass or copper vessels. The metallic vessels are believed to have antiseptic qualities. Silver preparations are also used in ceramic filters. Major disadvantages of silver for the purpose of disinfection are the costs of treatment (about 200 times higher than gaseous chlorine), relatively long contact periods are required, organic substances and iron, sulphur, etc., inhibit action, thus limiting the applicability of the technique. If combined with chlorine,silver preparations are more widely applicable (direct disinfection and protection against recontamination).

Q. Describe the role of water boiling as a means of water disinfection.

Ans. Boiling water is a very effective though energy consuming method to destroy pathogenic germs: bacteria, viruses, spores, cercerea and amoeba cysts, worm eggs, etc. The presence of turbidity or other impurities has little effect on germicidal effectiveness. If boiling is the only type of treatment available, it is recommended to let the water settle before, and decant it or filter it through a fine-meshed cloth so as to remove coarse impurities and suspended particles. The water is then brought to a strong boil which is maintained for at least five, preferably twenty minutes. For storing, it must not be transferred to a different vessel, but left in the former one and covered, so as to protect it from recontamination.

Boiling, together with the associated release of gases, especially CO2, alters the taste of water. But through stirring while boiling and by letting the water sit in the partially filled vessel for a few hours afterward, the water picks up air and loses its bland taste. To improve the taste of the water, flavoring plant materials may be added during boiling. If done properly, boiling is a very effective and simple disinfection method. Since it requires a significant amount of energy, this method is only recommended in exceptional cases

Q. Describe the use of Ultra Violet (UV) Radiation for disinfection of water.

Ans. The germicidal effect of UV rays is well known to harness it for water disinfection. The effect of sunlight on surface water is initiated in a more intense and controllable way. The most commonly used source of UV radiation is a low pressure quartz mercury vapor lamp which emits invisible light at a wavelength in the range of 200 to 300 nm (nanometer). The germicidal effect depends on the electric power of the lamp and on the time of exposure of the water to the UV radiation. It decreases with increasing distance between water and lamp. Also, many substances present even in pretreated water (e.g., small amounts of dissolved iron) absorb UV light. Other constituents (turbidity, suspended matter) inhibit or prevent the transmission of radiation. A disinfection unit is built such that the water is made to flow through a pipe in a thin film around the lamp, which is located at the pipe's center, emitting radiation. The flow rate is adjusted as required. The water must be pre-filtered.Disinfection by UV radiation is a "clean" process, since no chemical additives are used. Residual matter doesn't occur, and tastes & odors are neither produced in the water nor altered. Automatic devices are available which indicate when the lamp's output is not sufficient.

Some of disadvantages as given below, forbid its widespread use for water disinfection:

- commercially available devices are relatively expensive,
- there is a dependence on steady power supply ,
- the lamp's powers of penetration are limited; thin water films are necessary,
- turbidity, and impurities reduce the effectiveness notably,
- the lamps gradually lose their radiation power, accelerated by a coating of dirt. The lamp's average life is 1000 to 5000 hours,
- disinfection occurs rather quickly and effectively (up to 99.9%), though no protection for recontamination is provided.

Q. Why waste water treatment is necessary. Discuss its importance.

Ans. Wastewater treatment aims at removal of unwanted components in waste stream. Its objective is to produce a waste stream (or treated effluent) and a solid waste or sludge also suitable for discharge or reuse back into the environment. This material is contaminated with toxic organic & inorganic compounds.Waste water treatment can be done by physical, chemical and biological methods either alone or in combination. However, most of the treatment processes are not completely effective and also do not offer complete solutions for the removal of water contaminants as it transfers the contaminant to a different component. The wastewater generated by residential, commercial, & industrial sources is collected by sewers and transported to the Wastewater Treatment Plant.. A lot of sewage also includes some surface water from roofs or hard-standing areas. Municipal wastewater includes residential, commercial, & industrial liquid waste discharges, and may include storm water runoff.

The wastewater contains many of the physical, chemical & biological impurities which may impair quality of the receiving water body as well as interfere with the aquatic ecosystem in many ways. In order to render it safe ,it would be appropriate to treat the wastewater prior to disposal by adopting a systematic logical sequence of treatment.

Sewage systems capable of handling storm water are known as combined systems. Such systems are usually avoided since they complicate and thereby reduce the efficiency of sewage treatment plants owing to their seasonality. Besides, heavy storms may overwhelm the sewage treatment system, causing a spill or overflow. It is preferable to have a separate storm drain system for storm water. Sewerage systems that transport liquid waste discharges and storm water together to a common treatment facility are called combined sewer systems.

As rainfall runs over the surface of roofs and the ground, it may pick up various contaminants including soil particles, (sediment), heavy metals, organic compounds, animal waste, and oil & grease. Some jurisdictions require storm water to receive some level of treatment before being discharged directly into waterways.

Q. What are the major constituents of preliminary, primary and secondary treatment of waste water ?

Ans. The wastewater contains many floatable materials that may interfere with the downstream processes or be detrimental to the plant equipments. To increase the waste water treatment efficiency of the secondary treatment units it is essential to have **preliminary and primary treatment** of wastewater.

Materials removed may include rags, plastic, lumber, and grit. This is done by bar screens which remove large objects, rags, and plastic that would cause plugging problems in the downstream lines. The grit basins remove the grit by sedimentation. Grit is abrasive materials such as sand and gravel that must be removed to minimize wear on the downstream equipment. Materials removed by the bar screens and grit removal system are de-watered, discharged to a dump trailer, and transported to the landfill for disposal.

Secondary treatment helps in reducing the organic load present in wastewater. The commonly used units in this treatment are Trickling filter, Activated sludge, Water treatment ponds etc.

The secondary treatment of water is based on the natural self-purification capacity of natural waters.This treatment is designed to substantially degrade the biological content of the sewage such as are derived from human waste, food waste, soaps and detergent. The majority of the municipal & industrial water treatment plants treat the settled sewage liquor using aerobic biological processes. For this to be effective, the biota require nutrients and a substrate on which to live. There are number of ways in which this is done. In all these methods, the bacteria and protozoa consume biodegradable soluble organic contaminants (e.g. sugars, fats, organic short-chain carbon molecules, etc.) and bind much of the less soluble fractions into flocs.

Secondary treatment systems are classified as **fixed film** or suspended growth, such as rock filters where the biomass grows on media and the sewage passes over its surface. In **suspended growth systems** - such as activated sludge - the biomass is well mixed with the sewage and be operated in a smaller space than fixed film systems that treat same amount of water.However, fixed film systems are more able to cope better with drastic changes in the amount of biological material and can provide higher removal rates for organic material & suspended solids than suspended growth systems. This is reflected in two different possibilities for water treatment. The first type is a so-called fixed bed bio-reactor, usually referred to as a trickling filter.

Q. What are different stages in preliminary treatment of waste water ? What are its uses and necessity ?

Ans. It is used to reduce Solids(50-60%), and BOD (30- 50%). The commonly used units in the preliminary treatment of water are bar screens, Comminutor,and Grit chamber (Settling tank).

Preliminary treatment, the first treatment process, consists of the removal of substances that may interfere with the downstream processes or be detrimental to the plant equipment. Primary treatment is to reduce oils, grease, fats, sand, grit, and coarse (settleable) solids. This step is done entirely with machinery, hence the name mechanical treatment.

Influx (influent) and removal of large objects: In the mechanical treatment, the influx (influent) of sewage water is strained to remove all large objects that are deposited in the sewer system, such as rags, sticks, condoms, sanitary towels (sanitary napkins) or tampons, cans, fruit, plastic, lumber, grit., etc. This is most commonly done with a manual or automated mechanically raked screen. This type of waste is removed because it can damage the sensitive equipment in the sewage treatment plant.

Sand and grit removal: This stage includes a sand or grit channel where the velocity of the incoming waste water is carefully controlled to allow sand grit and stones to settle but still maintain the majority of the organic material within the flow. This equipment is called a detritor or sand catcher. Sand grit and stones need to be removed early in the process to avoid damage to pumps and other equipment in the remaining treatment stages. Sometimes there is a sand washer (grit classifier) followed by a conveyor that transports the sand to a container for disposal. The contents from the sand catcher may be fed into the incinerator in a sludge processing plant, but in many cases, the sand and grit is sent to a land-fill. Materials removed may include rags, plastic, lumber, and grit. This is done by bar screens which remove large objects, rags, & plastic that may cause plugging problems in the downstream lines. The grit basins remove the grit by sedimentation. Grit is abrasive materials such as sand and gravel that must be removed to minimize wear on the downstream equipments. Materials removed by the bar screens and grit removal system are de-watered, discharged to a dump trailer, and transported to the landfill for disposal.

Grit and Screen Facility: The Grit and Screen Facility is where sewage first enters the plant. The speed of the wastewater is reduced to 0.3 metre per second and air is pumped in to set up a rolling motion.This movement separates the heavier grit and sand materials from the suspended organic matter. The larger sand particles settle to the bottom of the tanks, thus preventing harm to equipment within the Plant or clogging of pipes during the treatment process. The settled sand is removed once a month and buried.Wastewater continues through the grit tanks and passes through bar screens to remove any material larger than 12.5mm in size. The wastewater then flows by gravity to the Primary Settling Basins.

Q. What Are Different Stages In Primary Treatment Of Waste Water ? What Is Its Purpose?

Ans. The second step in the waste water treatment process is primary treatment. The waste water enters two primary clarifiers (sedimentation basins) which remove suspended and floating materials.The primary clarifiers remove about 60% of the Total Suspended Solids and about 30% of the Biochemical Oxygen Demand (BOD) in the incoming wastewater. BOD is a measure of the amount of oxygen needed to biochemically degrade the organic matter in the wastewater. Primary treatment does not remove colloidal or dissolved solids.

Sedimentation: Many plants have a sedimentation stage where the sewage is allowed to pass slowly through large tanks, commonly called "primary clarifiers" or "primary sedimentation tanks". The tanks are large enough that faecal solids can settle and floating material such as grease & plastics can rise to the surface and be skimmed off. The main purpose of the primary stage is to produce a generally homogeneous liquid capable of being treated biologically and a sludge that can be separately treated or processed. Primary settlement tanks are usually equipped with mechanically driven scrapers that continually drive the collected sludge towards a hopper in the base of the tank from where it can be pumped to further sludge treatment stages.

Primary Settling Basins: This is where suspended solids settle out and floating scum is removed for further treatment. Upon entering the Primary Settling Basins, the velocity of the wastewater is reduced to 3 metres per minute, allowing fine particles to settle to the bottom. Settled sludge on the bottom of the basins is continuously scraped into hoppers at the end of tanks. It takes about 4 hours for waste water to flow through the Primary Settling Basins. Upon completion, the primary effluent is pumped to the Bioreactors and the settled sludge is pumped to the Fermenters. The scum from the top of the basins is collected in hoppers and pumped to the Digesters.

Q. What Are Different Stages And Purpose Of Secondary Treatment Of Waste Water ?

Ans. Secondary treatment usually consists of two steps which remove the dissolved & colloidal organic material not removed by the primary treatment. There are various secondary treatment processes in use. This treatment use microorganisms to convert organic wastes into stabilized compounds - similar to self-purification process in streams.

It is used to reduce Solids (85-95%), BOD (80-95%), and Coliform (90-95%). The commonly used units in this treatment are Trickling filter, Activated sludge,and Water treatment ponds.

The biological treatment of water is based on the natural self-purification capacity of natural waters.This treatment is designed to substantially degrade the biological content of the sewage which are derived from human waste, food waste, soaps and detergent. The majority of the municipal & industrial water treatment plants treat the settled sewage liquor using aerobic biological processes. For this to be effective, the biota require nutrients and a substrate on which to live. There are number of ways in which this is done. In all these methods, the bacteria and protozoa consume biodegradable soluble organic contaminants (e.g. sugars, fats, organic short-chain carbon molecules, etc.) and bind much of the less soluble fractions into floc.

Secondary treatment systems are classified as **fixed film** or **suspended growth**, such as rock filters where the biomass grows on media and the sewage passes over its surface. In **suspended growth systems** - such as activated sludge - the biomass is well mixed with the sewage & can be operated in a smaller space than fixed film systems that treat same amount of water. However, fixed film systems cope better with drastic changes in the amount of biological material and can provide higher removal rates for organic material & suspended solids than suspended growth systems. This is reflected in two different possibilities for water treatment. The first type is a so-called fixed bed bio-reactor, usually referred to as a trickling filter.

Roughing filters are intended to treat particularly strong or variable organic loads, typically industrial, to allow them to then be treated by conventional secondary treatment processes. Characteristics include typically tall, circular filters filled with open synthetic filter media to which sewage is applied at a relatively high rate. Designed to allow high hydraulic loading and a high flow-through of air. On larger installations, air is forced through the media using blowers. The resultant liquor is usually within the normal range for conventional treatment processes.

Q. What do you mean by biological treatment of waste water ?

Ans. Treatment of industrial effluents using microorganisms is is called biological treatment of waste water. Secondary treatment typically utilizes biological treatment processes, in which microorganisms convert non settleable solids to settleable solids. Biological treatment processes followed by settling tanks removes approximately 85% of the BOD & TSS in wastewater. Secondary treatment for municipal wastewater is the minimum level of treatment required. Its objective is to produce a stream (or treated effluent) and a solid waste or sludge also suitable for discharge or reuse back into the environment.

Q. Describe FLOCCULATION technique of waste water treatment.

Ans. Generally aluminium compounds are used as flocculents. The aluminium ion, when released into the water, will form aluminium hydroxide, $Al(OH)_3$, flocs, which will partly block the pores of a sand filter and thus enhance filtering action. A flocculent should in principle also be able to destabilize colloid solutions. A colloid in water containing a low concentration on some ions, will be surrounded by charged particles, that create an electrical double layer around it. The electrical potential of that layer is called the zeta-potential. The higher this zeta potential is, the stronger the repulsion between the particles and thus the smaller the chance that two particles will collide and then coagulate. If to such a solution aluminium ions are added, the zeta-potential of the colloid will be strongly reduced, the colloid solution becomes unstable and flocculation will occur. This process is called salt flocculation.

Q. Describe The Activated Sludge Process Of Waste Water Treatment ?

Ans. The activated sludge process is a biological process in which air or oxygen is forced into sewage liquor to develop a biological floc utilizing an active biomass. The activated sludge water treatment plants use a variety of mechanisms and processes to use dissolved oxygen to promote the growth of biological floc that substantially removes organic material. It also traps particulate material and can, under ideal conditions, convert ammonia to nitrite & nitrate and ultimately to nitrogen gas. The biomass is largely composed off saprophytic bacteria but also has an important protozoan flora mainly composed of amoebae, Spirotrichs, Peritrichs including Vorticellids and a range of other filter feeding species. In poorly managed activated sludge,a range of mucilaginous filamentous bacteria can develop including *Sphaerotilus natans* which produces a sludge that is difficult to settle and can result in the sludge blanket decanting over the weirs in the settlement tank to severely contaminate the final effluent quality. This material is often described as sewage fungus but true fungal communities are relatively uncommon to neutralize the organic matter in the wastewater.

The active biomass (called return activated sludge) is mixed with the wastewater leaving the primary clarifiers and with air in the aeration basins. Part of the settled material, the sludge, is returned to the head of the aeration system to re-seed the new sewage entering the tank. This fraction of the floc is called R.A.S - Return Activated Sludge. The remaining sludge, also called W.A.S - Waste Activated Sludge, is further treated prior to disposal. (W.A.S is also sometimes called S.A.S - Surplus Activated Sludge) The microorganisms, in the presence of oxygen, use organic matter & nutrients in wastewater to sustain their life processes, including reproduction. Effluent from the aeration basins contains large quantities of suspended biological solids (called activated sludge) that must be removed prior to discharge. Therefore, the second step of the secondary treatment process is sedimentation in the secondary clarifiers. Most of the solids removed in the secondary clarifiers are returned to the aeration basins as returned activated sludge to maintain the appropriate population of microorganisms needed to assimilate the organic matter entering the aeration basins.

Q. Explain The Trickling Filter Method Of Waste Water Treatment.

Ans. In the trickling filter method of waste disposal, a bed of crushed rock or synthetic media supports a film of aerobic microorganisms. The liquid waste water is sprayed over the bed and the microorganisms act on the contaminants as they pass through the filter. This is a common method used in sewage treatment plants to treat acetaldehyde, acetic acid, benzene, chlorinated hydrocarbons, cyanides, formaldehyde, ketones, and resins. Trickling filterization is generally effective for aqueous wastes where concentrations are < 1%. Like all biological treatment, it requires stable, consistent operating conditions. It is not suitable for removal of highly chlorinated organics, aliphatics, amines, and aromatic compounds. Heavy metals and organic chemicals may kill the microorganisms. Heavy metals and non-biodegradable organics may also concentrate in the sludge. Hydrogen sulfide gas may also be released.

In a trickling filter, water is divided over the top and is allowed to flow slowly over the filter bed. In the course of that flow organic matter, both dissolved & suspended, is attacked by micro-organisms. In the first part of the filter (the top) mainly carbon decomposition takes place. The depth of that carbon decomposition zone depends on the load: if the water contains much organic carbon the zone will be very deep. In heavily loaded filters carbon decomposition may be the only process taking place. In the lower part of the trickling filter nitrification can occur. It will take rather long before a population of nitrifying bacteria has been established, since they grow slowly. It will take several weeks before a filter is properly conditioned.

If a trickling filter is just put into operation the first organisms that will grow in it will be bacteria and fungi. These will coat the filling material: a bio-film is formed. On this film the secondary colonizers will start growing. To this group belong among others protozoa and nematodes. The growth of organisms in and on the bio-film will increase its thickness. If a film becomes too thick it will start sloughing. This sloughed film will appear in the water as flocs. These flocs settle easily, so if water from the trickling filter is led through a settling area first, nearly all organic matter can be removed.

Processes happening in a trickling filter are the following:

- dissolved organic matter is mineralized.
- particulate organic matter & colloids are either converted into dissolved organic matter and then mineralized, or are adsorbed onto the bio-film and removed from the water as settleable flocs.
- inorganic matter, particulate, colloidal or dissolved can react with or be adsorbed onto the bio-film and can thus be removed with the flocs.

The organisms that can be found in a trickling filter do not differ much from the ones found in activated sludge. No differences have been found between the bacterial flora of activated sludge and that of trickling filters. In activated sludge most are attached to suspended flocs, whereas in a trickling filter they are attached to the filter bed.

Q. Describe The Process Of Anaerobic Digestion Of Sewage Sludge.

Ans. Waste treatment in this fashion uses the same process as that which naturally occurs in landfills. In an oxygen free environment it reduces organic waste to a relatively stable solid residue (digestate) similar to compost. Anaerobic digestion is particularly suited to wet, organic material and as such has been used for the treatment of sewage sludge.

The organic portion is separated from the waste to remove plastic, glass & metals and then placed in a sealed reactor. The conditions necessary for biological degradation are created which produces biogas, a mixture of methane and carbon dioxide. This can then be used as a fuel. The methane is generally burnt on site for heating or to produce electricity on a small scale. The digester will require an energy input to retain the material at elevated temperatures (between 20 and 40°C - the mesophilic range, or up to 60°C - the thermophilic range) in order to speed digestion. Some of the energy available from the methane generated can be recycled for this purpose, typically 20-40% of the total energy produced. Alternatively the process can produce solid fuel, a fibre product for board manufacture, or more commonly the residue is used as a soil conditioner.

The wastewater exiting the anaerobic digestion facility have elevated levels of BOD and COD. Some of this material is termed as 'hard COD' meaning it cannot be accessed by the anaerobic bacteria for conversion into biogas. If this effluent was put directly into water courses it would negatively affect them by causing eutrophication. As such further treatment of the wastewater is often required. This treatment will typically be an oxidation stage where air is passed through the water in a sequencing batch reactors or reverse osmosis unit.

Q. Describe Various Pond Treatment Processes For Waste Water Treatment.

Ans. The treatment of wastewater could also be accomplished by various biological pond treatment processes which are listed below:

Aerobic Stabilization Ponds. Treatment is done with aerobic bacteria; oxygen is supplied by algal photosynthesis and natural surface reaeration; depth of 0.15 to 1.5 m is maintained ,

Carbonaceous BOD removal is achieved.

Maturation (tertiary) Ponds. Use aerobic treatment; applied loadings are low to preserve aerobic conditions , Secondary effluent polishing and seasonal nitrification.

Facultative Ponds. Treatment with aerobic, anaerobic and facultative bacteria; the pond has 3 zones: a surface aerobic zone, a bottom anaerobic zone, and an intermediate zone partly aerobic-anaerobic , Carbonaceous BOD removal is achieved.

Anaerobic Ponds. Treatment done with anaerobic bacteria; depths of up to 9.1 m to conserve anaerobic conditions , Carbonaceous BOD removal (waste stabilization) is achieved.

Q Discuss The Role Of Fermenters , Bioreactors And Secondary Clarifiers In Biological Treatment Of Waste Water.

Ans. Fermenters are where the primary sludge settling out from the Primary Settling Basins is broken down. The fermentation process converts the organic material into volatile acids called Volatile Fatty Acids (VFA's). An example of a VFA is acetic acid or vinegar. The VFA's are sent to the Bioreactors and utilized by the nutrient consuming microorganisms.

Bioreactors are where the Biological Nutrient Removal (BNR) process takes place. This natural process greatly reduces the concentration of dissolved organic compounds in the effluent and removes unwanted carbon, phosphorus, & nitrogen from the wastewater without the addition of chemicals. In the Bioreactor, the effluent from the Primary Settling Basins is mixed with micro-organisms and Volatile Fatty Acids (VFA's). The microorganisms come from the secondary clarifiers , where they have settled out and are then returned to the Bioreactors. The VFA's are created in the Fermenter. The micro-organisms naturally break down excess carbon and nutrients present in the wastewater. The effluent is moved through carefully controlled anaerobic (absence of available oxygen), anoxic (chemically available oxygen only) & aerobic (abundance of free oxygen) zones of the Bioreactors.Each of these areas remove specific organic compounds as the micro-organisms continue to grow and flourish, consuming impurities in the wastewater. After approximately 9 hours in the Bioreactors, the effluent flows to the Secondary Clarifiers.

Secondary Clarifiers are where any remaining solids along with the microorganisms from the Bioreactors settle to the bottom and the clear effluent flows out the top of the basins. Of the settled micro-organisms, about 97% are returned to the Bioreactors to be used again in the BNR process and 3% are sent to the DAF Thickener and then to the digesters for further treatment. The clarified final effluent flows from the Secondary Clarifiers to the Chlorine Contact Chamber before being released to the river.

Q. Describe The Working Of Rotating Biological Contactors (RBCs) Used In Secondary Waste Water Treatment Systems.

Ans. Rotating Biological Contactors (RBC's) , used in waste water treatment plants are mechanical secondary treatment systems, which are robust and capable of withstanding surges in organic load. The rotating disks support the growth of bacteria and microorganisms present in the sewage, which breakdown and stabilise organic pollutants.To be successful, micro-organisms need both oxygen to live and food to grow. Oxygen is obtained from the atmosphere as the disks rotate. As the micro-organisms grow, they build up on the media until they are sloughed off due to shear forces provided by the rotating discs in the sewage.

The primary function of these bio-reactors is the reduction of organic matter. It consists of series of closely spaced discs mounted on a horizontal shaft and rotated while 40% of each disc is submerged in wastewater. The discs are made up of light-weight plastic and are covered with a biofilm. The slime is 1-3 mm in thickness on disc. As the disk rotates the biofilm is exposed to the wastewater only part of the time then with atmosphere for adsorption of oxygen. The rotation in and out of the wastewater serves to vary the feeding cycle of the microorganisms that make up the biofilm. The shaft rotates at 1-2 rpm (slowly). Excess solids are removed by shearing forces created by the rotation mechanism. Effluent from the RBC is then passed through final clarifiers where the micro-organisms in suspension settle as a sludge. The sludge is withdrawn from the clarifier for further treatment.

The primary and secondary treatment processes generally remove at least 85% of the total suspended solids and biochemical oxygen demand. The secondary process is a very sensitive biological process and can be adversely impacted by the discharge of incompatible or toxic wastes into the sewer system. Precise control of this process is necessary to effectively treat the wastewater.

Q. Define advanced waste water treatment. Why is it needed ?

Ans. Advanced Wastewater Treatment is defined as any process designed to produce an effluent of higher quality than normally achieved by secondary treatment processes or containing unit operations not normally found in Secondary Treatment units. Primary and secondary waste water treatment processes removes the majority of BOD & Suspended Solids found in waste waters. However, in many cases this level of treatment is insufficient to protect the receiving waters or to provide reusable water for industrial and/or domestic recycle. Thus, additional treatment steps have been added to wastewater treatment plants to provide for further organic & solids removals or to provide for removal of nutrients and toxic materials.

Once freshwater has been used for an economic / beneficial purpose, it is generally discarded as waste. These wastewaters are discharged, either as untreated waste or as treated effluent, into natural watercourses, from which they are abstracted for further use after undergoing "self purification" within the stream. Through this system of indirect reuse, wastewater may be reused up to a dozen times or more before being discharged to the sea. Such indirect reuse is common in the larger river systems. However, more direct reuse is also possible: the technology to reclaim wastewaters as potable or process waters is a technically feasible option for agricultural & some industrial purposes (such as for cooling water or sanitary flushing), and a largely experimental option for the supply of domestic water. Waste water reuse for drinking raises public health, and religious, concerns among consumers. Due to stringent requirement of the quality of potable and process water, tertiary treatment of water is not a luxury any more rather it has emerged out as the only way to provide adequate safe guard for community health.

Q. What Are Different Types , Purpose And Stages Of Advanced Wastewater Treatment.

Ans. Advanced Wastewater Treatment is categorised by the type of process flow scheme as:

- Tertiary Treatment
- Physical-Chemical Treatment
- Combined Biological-Physical Treatment

Tertiary treatment is defined as any treatment process in which unit operations are added to the flow scheme following conventional secondary treatment. Additions to conventional secondary treatment could be the addition of a filter for suspended solids removal or the addition of many other unit processes for organic, suspended solids, nitrogen and phosphorous removal.

Physical-chemical treatment is defined as a treatment process in which biological and physical-chemical processes are intermixed to achieve the desired effluent characteristics.

Combined biological-physical-chemical treatment is differentiated from tertiary treatment in that in tertiary treatment any unit processes are added after conventional biological secondary treatment, while in combined treatment, biological and physical –chemical treatment are mixed.

Another way to classify advanced wastewater treatment is to differentiate on the basis of desired treatment goals. Advanced wastewater treatment is used for:

- Additional organic and suspended solids removal
- Removal of nitrogenous oxygen demand (NOD)
- Nutrient removal
- Removal of toxic materials

The first three reasons for additional organic removal through advanced wastewater treatment are simple. The fourth requires explanation. The performance of secondary treatment plants is measured in terms of BOD and SS removals. A well designed & operated secondary water treatment plant will remove from 85 to 95% of the influent BOD & SS. However, the BOD test does not measure all of the organic material present in the wastewater. An average secondary effluent may have a BOD of 20 mg/L and a COD of 60 to 100 mg/L. The average secondary plant removes approximately 65% of the influent COD. Thus, when high quality effluents are required, additional organic removals must be accomplished. In addition to the organic materials remaining in most secondary effluents, there is an additional oxygen demand resulting from the nitrogen present in the wastewater.

In wastewaters, much of the nitrogen is found in the form of ammonia. When secondary treatment is used, a great deal of this ammonia is discharged in the effluent. Bacteria can utilize this ammonia as an energy source and convert ammonia to nitrite & nitrate.

Another reason for advanced wastewater treatment may be to remove nutrients contained in discharges from secondary treatment plants. The effluents from secondary treatment plants contain both nitrogen (N) & phosphorous (P). which are ingredients of fertilizers. When excess amounts of N and P are discharged, plant growth in the receiving waters may be accelerated. Algae growth may be stimulated causing blooms which are toxic to fish life as well as aesthetically unpleasing. Fixed plant growth may also be accelerated causing the eventual process of a lake becoming a swamp to be speeded up. Therefore, is necessary to remove nitrogen and phosphorous prior to discharge in some cases.

Toxic materials, both organic & inorganic are discharged into many sewage collections systems. When these materials are present in sufficient quantities to be toxic to bacteria, it is necessary to remove them prior to biological treatment. In other cases, it is necessary to remove even small amounts of these materials prior to discharge to protect receiving waters or drinking water supplies. Thus, advanced wastewater treatment processes have been used in cases where secondary treatment was not possible due to materials toxic to bacteria entering the plant as well as in cases where even trace amounts of toxic materials are unacceptable in plant effluents.

Q. Describe Ion Exchange Method Of Water Purification.

Ans. **Ion exchange** is a process for water purification in which ions are exchanged between a solution and an **ion exchanger**, a non-aqueous solid or gel. Typical ion exchangers are ion exchange resins, zeolite, montmorillonite, clay, and humus. Ion exchangers are either **cation exchangers** for positively charged cations or **anion exchangers** for negatively charged anions. There are also ion exchangers called mixed beds which have both anion & cation exchanging resins within them. Ion exchange is a reversible process and the ion exchanger can be *regenerated* or *loaded* by washing with an excess of the ions to be exchanged.

Ion exchange method is widely used in household and industrial water purifications to produce soft water. This is accomplished by exchanging calcium Ca_2+ and magnesium Mg_2+ cations against sodium Na+ or hydrogen H+ cations. Ion exchange is also applied in water softeners.Ion Exchange can be used in the food & beverage, hydrometallurgical, metals finishing, chemical & petrochemical, pharmaceutical, sugar & sweeteners, ground & potable water, nuclear, softening & industrial water, semiconductor, power, and a host of other industries.

Q. Describe the features of Ion exchange resins used in water purification.

Ans. An **ion exchange resin** is an insoluble matrix (on support structure) normally in the form of small (1-2 mm diameter) beads, usually white or yellowish, fabricated from an organic polymer substrate on the surface of which are sites with easily trapped and released ions in a process called ion exchange. There are different types of ion exchange resin which are fabricated to selectively prefer one or several different types of ions. This resin is used extensively for water softening during water purification. They were developed as a more flexible alternative to the use of natural or artificial zeolites.

Most ion exchange resins are based on crosslinked polystyrene. The required active groups can be introduced after polymerization, or substituted monomers can be used. The crosslinking is usually achieved by adding a small proportion of divinyl benzene to styrene. Non-crosslinked polymers are used only rarely because of their tendency to change dimensions in dependence on the ions bonded. However, crosslinking somewhat decreases the capacity of the resin and prolong the time for reaching the equilibrium for the ions in solution and in the resin. Particle size also influences the resin parameters; smaller particles have larger surface, but cause larger head loss in the column. Membranes made of highly cross-linked ion exchange resins that allow passage of ions but not of water are used for electrodialysis.

There are four main types of ion exchange resins differing in their functional groups:

- strongly acidic (sulfonic acid groups, eg. sodium polystyrene sulfonate or polyAMPS)
- strongly basic, (trimethylammonium groups, eg. polyAPTAC)
- weakly acidic (carboxylic acid groups)
- weakly basic (amino groups, eg. polyethylene amine)

There are also specialised types:

- chelating resin

Q. What is the difference between normal osmosis process and reverse osmosis process?

Ans: In the normal osmosis process, more dilution of concentrated solution occurs by the movement of molecules from the less concentrated side. But in case of reverse osmosis process, due to the application of high external pressure, the molecules of solution move from the more concentrated side to the less concentrated side through the semi-permeable membrane.

Q. Describe The Reverse Osmosis Process Used For Desalination Of Brackish Water & Sea Water.

Ans. To desalinate water, we must create a flow through a membrane, causing the water to leave the salty side of the membrane, flowing into the unsalted side. To achieve this, pressure must be created upon the water column on the salt side of the membrane; firstly, to remove the natural osmotic pressure and secondly, to create extra pressure on the water column, in order to push the water through the membrane. For the desalination of seawater, the pressure must be about 50-60 bars.

There are several different techniques that can be applied for water desalination. Examples are reverse osmosis, electro dialysis, distillation and ion exchange: **Reverse osmosis** is the most economic process for the **desalination** of brackish water and seawater. When we compare this process to the traditional thermic process of distillation, the capital investments and the energy use are much lower. Membranes occupy through a selective separation wall. Certain substances can pass through the membrane, while other substances are caught.

Membrane filtration: This can be used as an alternative for flocculation, sediment purification techniques, adsorption (sand filters & active carbon filters, ion exchangers), extraction and **distillation.** Membrane filtration can be divided up between micro & ultra filtration on the one hand and nano filtration & Reverse Osmosis (RO or hyper filtration) on the other hand. When membrane filtration is used for the removal of larger particles, micro filtration and ultra filtration are applied. Because of the open character of the membranes the productivity is high while the pressure differences are low. When salts need to be removed from water, nano filtration and Reverse Osmosis are applied. Nano filtration and RO membranes do not work according to the principle of pores; separation takes place by diffusion through the membrane. The pressure that is required to perform nano filtration and Reverse Osmosis is much higher than the pressure required for micro & ultra filtration, while productivity is much lower.

Q. Describe the role of Electrodialysis in treating wastewater.

Ans. **Electrodialysis** is used to transport salt ions from one solution through ion exchange membranes to another solution under the influence of an applied electric potential difference. This is done in an electrodialysis cell. The cell consists of a feed (diluate) compartment and a concentrate (brine) compartment formed by an anion exchange membrane and a cation exchange membrane placed between two electrodes. In practical electrodialysis processes, multiple electrodialysis cells are arranged into a configuration called an electrodialysis stack, with alternating anion and cation exchange membranes forming the multiple electrodialysis cells.

Electrodialysis processes are unique compared to distillation techniques and other membrane based processes (such as reverse osmosis) in that dissolved species are moved away from the feed stream rather than the reverse. Because the quantity of dissolved species in the feed stream is far less than that of the fluid, electrodialysis offers the practical advantage of much higher feed recovery in many applications.

Q. Describe the working principle of electrodialysis cell used for water *purification*.

Ans. In an electrodialysis stack, the diluate (D) feed stream, brine or concentrate (C) stream, and electrode (E) stream are allowed to flow through the appropriate cell compartments formed by the ion exchange membranes. Under the influence of an electrical potential difference, the negatively charged ions (e.g., chloride) in the diluate stream migrate toward the positively charged anode. These ions pass through the positively charged anion exchange membrane, but are prevented from further migration toward the anode by the negatively charged cation exchange membrane and therefore stay in the C stream, which becomes concentrated with the anions. The positively charged species (e.g., sodium) in the D stream migrate toward the negatively charged cathode and pass through the negatively charged cation exchange membrane. These cations also stay in the C stream, prevented from further migration toward the cathode by the positively charged anion exchange membrane. As a result of the anion and cation migration, electric current flows between the cathode and anode. Only an equal number of anion and cation charge equivalents are transferred from the D stream into the C stream and so the charge balance is maintained in each stream.

The result of the electrodialysis process is an ion concentration increase in the concentrate stream with a depletion of ions in the diluate solution feed stream.

The E stream is the electrode stream that flows past each electrode in the stack. This stream may consist of the same composition as the feed stream (eg,sodium chloride) or may be a separate solution containing a different species (e.g., sodium sulfate). Depending on stack configuration, anions & cations from the electrode stream may be transported into the C stream, or anions & cations from the D stream may be transported into the E stream. In each case, this transport is necessary to carry current across the stack and maintain electrically neutral stack solutions.

Anode and Cathode Reactions

Reactions take place at each electrode.

At the cathode,

$$2e^- + 2\,H_2O \rightarrow H_2\,(g) + 2\,OH$$

While at the anode

$$H_2O \rightarrow 2\,H+ + ½\,O_2\,(g) + 2e\text{-} \text{ or, } 2\,Cl \rightarrow Cl2\,(g) + 2e\text{-}$$

Small amounts of hydrogen gas are generated at the cathode and small amounts of either oxygen or chlorine gas (depending on composition of the E stream and end ion exchange membrane arrangement) at the anode. These gases subsequently dissipated as the E stream effluent from each electrode compartment is combined to maintain a neutral pH and discharged or re-circulated to a separate E tank.

Q. Discuss various practical applications of electrodialysis.

Ans. The major application of electrodialysis is for desalination of brackish water / seawater as an alternative to RO for potable water production & seawater concentration for salt production. In normal potable water production without the requirement of high recoveries, RO is generally believed to be more cost effective when total dissolved solids (TDS) are 3,000 parts per million (ppm) or greater, while electrodialysis is more cost-effective for TDS feed concentrations less that 3,000 ppm or when high recoveries of the feed are required.

Another important application for electrodialysis is the production of pure water and ultrapure water by electro deionization (EDI). In EDI, the compartments (diluate, concentrate, or both) of the electrodialysis stack are filled with ion exchange resin. When fed with low TDS feed (e.g., feed purified by RO), the diluate can reach very low levels (e.g., 18 Megaohms). The ion exchange resins act to retain the ions, allowing these to be transported across the ion exchange membranes. The main usage of EDI systems are in electronics, pharmaceutical, and cooling tower applications.

Some practical applications of electrodialysis include:

- Large scale brackish and seawater desalination and salt production.
- Small and medium scale drinking water production (e.g., towns & villages, construction & military camps, nitrate reduction, hotels and hospitals).
- Water reuse (e.g., industrial laundry wastewater, produced water from oil/gas production, cooling tower makeup & blowdown, metals industry fluids, wash-rack water).
- Pre-demineralization (e.g ,boiler makeup and pre treatment, ultrapure water pre treatment, process water desalination, power generation, semiconductor, chemical manufacturing, food & beverage).
- Food processing.
- Agricultural water (e.g., water for greenhouses, hydroponics, irrigation, livestock).
- Glycol desalting (e.g., antifreeze or engine-coolants, capacitor electrolyte fluids, oil & gas dehydration, conditioning & processing solutions, industrial heat transfer fluids, secondary coolants from heating, venting,& air conditioning (HVAC).
- Glycerene Purification.

In application, electrodialysis systems can be operated as continuous production or batch production processes. In a continuous process, feed is passed through a sufficient number of stacks placed in series to produce the final desired product quality. In batch processes, the diluate or concentrate streams are re-circulated through the electrodialysis systems until the final product or concentrate quality is achieved. Electrodialysis is usually applied to deionization of aqueous solutions. However, desalting of sparingly conductive aqueous organic and organic solutions is also possible.

Q. What are the limitations of electrodialysis ?

Ans. Electrodialysis has inherent limitations, working best at removing low molecular weight ionic components from a feed stream. Non-charged, higher molecular weight, and less mobile ionic species will not typically be significantly removed. Also, in contrast to RO, electrodialysis becomes less economical when extremely low salt concentrations in the product are required and with sparingly conductive feeds: current density becomes limited & current utilization efficiency typically decreases as the feed salt concentration becomes lower, and with fewer ions in solution to carry current, both ion transport and energy efficiency greatly declines. Hence, comparatively large membrane areas are required to satisfy capacity requirements for low concentration (and sparingly conductive) feed solutions.

As with RO, electrodialysis systems require feed pretreatment to remove species that coat, precipitate onto, or otherwise "foul" the surface of the ion exchange membranes. This fouling decreases the efficiency of the electrodialysis system. Species of concern include calcium & magnesium hardness, suspended solids, silica, and organic compounds. Water softening can be used to remove hardness, and micron or multimedia filtration can be used to remove suspended solids. Hardness in particular is a concern since scaling can build up on the membranes. Various chemicals are also available to help prevent scaling. Also, electrodialysis reversal systems seek to minimize scaling by periodically reversing the flows of diluate & concentrate and polarity of the electrodes.

Q. What is sewage ? Describe various sewage treatment methods.

Ans. Sewage is created by residences, institutions, and commercial & industrial establishments. It can be treated close to where it is created (in septic tanks, bio filters or aerobic treatment systems), or collected and transported via a network of pipes and pump stations to a municipal treatment plant. Sewage collection and treatment is typically subject to local, state and federal regulations and standards.

Industrial sources of wastewater often require specialized treatment processes. Raw influent (sewage) is the liquid waste from toilets, baths, showers, kitchens, sinks etc. Household waste is that which is disposed of via sewers. Municipal wastewater therefore includes residential, commercial, and industrial liquid waste discharges, and may include stormwater runoff. Sewage systems capable of handling storm water are known as combined systems.

Typically , sewage treatment involves three stages, called *primary*, *secondary* and *tertiary treatment*. Sewage Treatment Methods include:

- **Mechanical treatment**;

  Removal of large objects

  Removal of sand and grit

  Pre-precipitation
- **Biological treatment**;

Oxidation bed (oxidizing bed) or aeration system

Post precipitation

Effluent

- **Chemical treatment** (this step is usually combined with settling and other processes to remove solids, such as filtration.)

Primary treatment. It involves removal of large objects, sand and grit. Primary treatment removes the materials that can be easily collected from the raw wastewater and disposed of. The typical materials that are removed during primary treatment include fats, oils, & greases , sand, gravels and rocks (also referred to as grit), larger settleable solids including human waste and floating materials. This step is done entirely with machinery, hence the name mechanical treatment.

Secondary treatment. **It** is designed to substantially degrade the biological content of the sewage such as are derived from human waste, food waste, soaps and detergent. The majority of municipal and industrial plants treat the settled sewage liquor using aerobic biological processes. For this to be effective, the biota require both oxygen and a substrate on which to live. There are number of ways in which this is done. In all these methods, the bacteria and protozoa consume bio degradable soluble organic contaminants (e.g. sugars, fats, organic short chain carbon molecules, etc.) and bind much of the less soluble fractions into floc.

Secondary treatment systems are classified as **fixed film** or suspended growth. Fixed-film treatment process including trickling filter and rotating biological contactors where the biomass grows on media and the sewage passes over its surface. In **suspended growth**

systems—such as activated sludge—the biomass is well mixed with the sewage and can be operated in a smaller space than fixed-film systems that treat the same amount of water.

However, fixed-film systems are more able to cope with drastic changes in the amount of biological material and can provide higher removal rates for organic material and suspended solids than suspended growth systems.

Tertiary treatment. This treatment provides a final stage to raise the effluent quality before it is discharged to the receiving environment (sea, river, lake, ground, etc.). More than one tertiary treatment process may be used at any treatment plant. If disinfection is practiced, it is always the final process. It is also called "effluent polishing".

Q. Describe various primary sewage treatment units.

Ans. Primary treatment removes the materials that can be easily collected from the raw wastewater and disposed off eg; fats, oils, & greases , sand, gravels and rocks , larger settleable solids including human waste & floating materials.

Influx (influent) and removal of large objects: In the mechanical treatment, the influx (influent) of sewage water is strained to remove all large objects that are deposited in the sewer system, such as rags, sticks, condoms, sanitary towels / napkins / tampons, cans, fruit, etc. This is most commonly done with a manual or automated mechanically raked screen. This type of waste is removed because it can damage or clog the equipment in the sewage treatment plant.

Sand & grit removal: Primary treatment typically includes a sand or grit channel or chamber where the velocity of the incoming wastewater is carefully controlled to allow sand , grit & stones to settle, while keeping the majority of the suspended organic material in the water column. This equipment is called a detritor or sand catcher. Sand grit and stones need to be removed early in the process to avoid damage to pumps and other equipment in the remaining sewage treatment stages. Sometimes there is a sand washer (grit classifier) followed by a conveyor that transports the sand to a container for disposal. The contents from the sand catcher may be fed into the incinerator in a sludge processing plant, but in many cases, the sand and grit is sent to a landfill.

Sedimentation: Many plants have a sedimentation stage where the sewage is allowed to pass slowly through large tanks, commonly called "primary clarifiers" or "primary sedimentation tanks". The tanks are large enough that fecal solids can settle and floating material such as grease & oils can rise to the surface and be skimmed off. The main purpose of the primary stage is to produce a generally homogeneous liquid capable of being treated biologically and a sludge that can be separately treated or processed. Primary settlement tanks are usually equipped with mechanically driven scrapers that continually drive the collected sludge towards a hopper in the base of the tank from where it can be pumped to further sludge treatment stages.

Q. How does secondary treatment help in treating sewage water? What are various treatment units involved in secondary treatment of sewage ?

Ans. Secondary treatment degrades the biological content of the sewage such as are derived from human waste, food waste, soaps and detergent. The majority of municipal and industrial plants treat the settled sewage liquor using aerobic biological process. For this to be effective, the biota require both oxygen and a substrate on which to live. There are number of ways in which this is done. In all these methods, the bacteria and protozoa consume biodegradable soluble organic contaminants (e.g. sugars, fats, organic short-chain carbon molecules, etc.) and bind much of the less soluble fractions into floc.

Secondary treatment systems are classified as **fixed film** or **suspended growth**. Fixed-film treatment process including trickling filter and rotating biological contactors where the biomass grows on media and the sewage passes over its surface.

In **suspended growth** such as activated sludge ,the biomass is well mixed with the sewage and can be operated in a smaller space than fixed-film systems that treat the same amount of water.

However, fixed-film systems are more able to cope with drastic changes in the amount of biological material and can provide higher removal rates for organic material and suspended solids than suspended growth systems.

Roughing filters: These are intended to treat particularly strong or variable organic loads, typically industrial, to allow them to then be treated by conventional secondary treatment processes. Characteristics include typically tall, circular filters filled with open synthetic filter media to which wastewater is applied at a relatively high rate. They are designed to allow high hydraulic loading and a high flow-through of air. On larger installations, air is forced through the media using blowers. The resultant wastewater is usually within the normal range for conventional treatment processes.

Activated sludge: Activated sludge plants encompasses a variety of mechanisms and processes that use dissolved oxygen to promote the growth of biological floc that substantially removes organic material. It also traps particulate material and can, under ideal conditions, convert ammonia to nitrite and nitrate ultimately to nitrogen gas.

Fluidized bed reactors: The carbon adsorption following biological treatment are particularly effective in reducing both the BOD and COD to low levels. A fluidized bed reactor is a combination of the most common stirred tank packed bed, continuous flow reactors and has excellent heat and mass transfer characteristics.

In a fluidized bed reactor, the substrate is passed upward through the immobilized enzyme bed at

a high velocity to lift the particles. However the velocity must not be so high that the enzymes are swept away from the reactor entirely. This causes low mixing; these type of reactors are suitable for the exothermic reactions. It is most often applied in immobilized enzyme catalysis.

Filter beds (oxidising beds): In older plants and plants receiving more variable loads, trickling filter beds are used where the settled sewage liquor is spread onto the surface of a deep bed made up of coke (carbonised coal), limestone chips or specially fabricated plastic media. Such media must have high surface areas to support the biofilms to that form.

The liquor is distributed through perforated rotating arms radiating from a central pivot. The distributed liquor trickles through this bed and is collected in drains at the base. These drains also provide a source of air which percolates up through the bed, keeping it aerobic. Biological films of bacteria, protozoa & fungi form on the media's surfaces and eat or reduce the organic content. This biofilm is grazed by insect larvae and worms which help maintain an optimal thickness.

Overloading of beds increases the thickness of the film leading to clogging of the filter media and ponding on the surface.

Biological aerated filters: Biological Aerated (or Anoxic) Filter (BAF) or Bio filters combine filtration with biological carbon reduction, nitrification or denitrification. BAF usually includes a reactor filled with a filter media. The media is either in suspension or supported by a gravel layer at the foot of the filter. The dual purpose of this media is to support highly active biomass that is attached to it and to filter suspended solids. Carbon reduction and ammonia conversion occurs in aerobic mode and sometime achieved in a single reactor while nitrate conversion occurs in anoxic mode. BAF is operated either in upflow or downflow configuration depending on design.

Membrane biological reactors: Membrane biological reactors (MBR) combines activated sludge treatment with a membrane liquid-solid separation process. The membrane component utilizes low pressure microfiltration or ultra filtration membranes and eliminates the need for clarification and tertiary filtration. The membranes are typically immersed in the aeration tank (however, some applications utilize a separate membrane tank). One of the key benefits of a membrane bioreactor system is that it effectively overcomes the limitations associated with poor settling of sludge in conventional activated sludge (CAS) processes.

This technology permits bioreactor operation with considerably higher mixed liquor suspended solids (MLSS) concentration than CAS systems, which are limited by sludge settling. The process is typically operated at MLSS in the range of 8,000–12,000 mg/l, while CAS are operated in the range of 2,000–3,000 mg/l. The elevated biomass concentration in the membrane bioreactor process allows for very effective removal of both soluble and particulate biodegradable materials at higher loading rates. Thus increased Sludge Retention Times, usually exceeding 15 days ensure complete nitrification even under extreme cold weather operating conditions.

The cost of building and operating a MBR is usually higher than conventional wastewater treatment, however, as the technology has become increasingly popular and has gained wider acceptance throughout the industry, the life-cycle costs have been steadily decreasing.

Secondary sedimentation: The final step in the secondary treatment stage is to settle out the biological floc or filter material and produce sewage water containing very low levels of organic material and suspended matter.

Rotating biological contactors: Rotating biological contactors (RBCs) are mechanical secondary treatment systems, which are robust and capable of withstanding surges in organic load. The rotating disks support the growth of bacteria and micro-organisms present in the sewage, which breakdown and stabilise organic pollutants. Microorganisms need both oxygen to live and food to grow. Oxygen is obtained from the atmosphere as the disks rotate. As the micro-organisms grow, they build up on the media until they are sloughed off due to shear forces provided by the rotating discs in the sewage. Effluent from the RBC is then passed through final clarifiers where the micro-organisms in suspension settle as a sludge. The sludge is withdrawn from the clarifier for further treatment.

Q. Why is it essential to go for tertiary treatment of sewage ?

Tertiary treatment provides a final stage to raise the effluent quality before it is discharged to the receiving environment (sea, river, lake, ground, etc.). More than one tertiary treatment process may be used at any treatment plant. If disinfection is practiced, it is always the final process. It is also called "effluent polishing".

Filtration: Sand filtration removes much of the residual suspended matter. Filtration over activated carbon removes residual toxins.

Lagooning: Lagooning provides settlement and further biological improvement through storage in large man made ponds or lagoons. These lagoons are highly aerobic & colonization by native macrophytes, especially reeds, is often encouraged. Small filter feeding invertebrates such as Daphnia and species of Rotifera greatly assist in treatment by removing fine particulates.

Constructed wetlands: Constructed wetlands include engineered reed beds and a range of similar methodologies, all of which provide a high degree of aerobic biological improvement and can often be used instead of secondary treatment for small communities.

Waste removal: Wastewater may contain high levels of the nutrients ; nitrogen and phosphorus whose excessive release to the environment can lead to a build up of nutrients, called eutrophication, which can in turn encourage the overgrowth of weeds, algae, and cyanobacteria (blue-green algae). This may cause an algal bloom, a rapid growth in the population of algae. The algae numbers are unsustainable and eventually most of them die. The decomposition of the algae by bacteria uses up so much of oxygen in the water that most or all of the animals die,which creates more organic matter for the bacteria to decompose. In addition to causing deoxygenation, some algal species produce toxins that contaminate drinking water supplies.

Different treatment processes are required to remove nitrogen and phosphorus.

Nitrogen removal: The removal of nitrogen is effected through the biological oxidation of nitrogen from ammonia (nitrification) to nitrate, followed by denitrification, the reduction of nitrate to nitrogen gas. Nitrogen gas is released to the atmosphere and thus removed from the water.

Phosphorus removal: Phosphorus can be removed biologically in a process called enhanced biological phosphorus removal. In this process, specific bacteria, called polyphosphate accumulating organisms, are selectively enriched and accumulate large quantities of phosphorus within their cells (up to 20% of their mass). When the biomass enriched in these bacteria is separated from the treated water, these bio solids have a high fertilizer value.

Disinfection: The purpose of disinfection in the treatment of wastewater is to substantially reduce the number of microorganisms in the water to be discharged back into the environment. The effectiveness of disinfection depends on the quality of the water being treated (e.g., cloudiness, pH, etc.), type of disinfection being used, the disinfectant dosage (concentration & time), and other environmental variables. Cloudy water will be treated less successfully since solid matter can shield organisms, especially from ultraviolet light or if contact times are low. Generally, short contact times, low doses and high flows all militate against effective disinfection. Common methods of disinfection include ozone, chlorine, or ultraviolet light. Chloramine, which is used for drinking water, is not used in wastewater treatment because of its persistence.

Q. Briefly explain the overall method of sewage water treatment.

Ans. Sewage can be treated close to where the sewage is created, in a decentralised system (in septic tanks, biofilters or aerobic treatment systems), or be collected & transported by a network of pipes & pump stations to a municipal water treatment plant, a centralised system.Sewage treatment generally involves three stages, called primary, secondary and tertiary treatment.

Primary treatment consists of temporarily holding the sewage in a quiescent basin where heavy solids can settle to the bottom while oil, grease and lighter solids float to the surface. The settled and floating materials are removed.

Secondary treatment removes dissolved and suspended biological matter. Secondary treatment is typically performed by indigenous, water-borne micro-organisms in a managed habitat.

Tertiary treatment ,Treated water is sometimes disinfected chemically or physically (for example, by lagoons and microfiltration) prior to discharge into a stream, river, bay, lagoon or wetland, or it can be used for the irrigation of a golf course, green way or park.

Primary treatment includes Screening , Flow Equalization, Neutralization, Coagulation , Primary sedimentation.

Secondary treatment includes Biological treatment like Activated sludge process, Trickling filter, Membrane Bio Reactor,Secondary clarifier

Tertiary treatment includes Heavy metal removal ,Disinfection (UV treatment).

Pretreatment

Pretreatment of sewage water removes materials that can be easily collected from the raw sewage before they damage or clog the pumps and sewage lines of primary treatment clarifiers.

Screening. The influent sewage water passes through a bar screen to remove all large objects like cans, rags, sticks, plastic packets etc. carried in the sewage stream.This is done with an automated mechanically raked bar screen in modern plants serving large populations, whilst in smaller or less modern plants, a manually cleaned screen may be used.

Grit removal. Pretreatment may include a sand or grit channel or chamber, where the velocity of the incoming sewage is adjusted to allow the settlement of sand, grit, stones, and broken glass.

Flow equalization. Clarifiers and mechanized secondary treatment are more efficient under uniform flow conditions. Equalization basins may be used for temporary storage of diurnal or wet weather flow peaks. Basins provide a place to temporarily hold incoming sewage during plant maintenance and a means of diluting.

Fat & grease removal. In some larger plants, fat & grease are removed by passing the sewage through a small tank where skimmers collect the fat floating on the surface. Air blowers in the base of the tank may also be used to help recover the fat as a froth.

Primary clarifier: In the primary sedimentation stage, sewage flows through large tanks, the tanks are used to settle sludge while grease & oils rise to the surface and are skimmed off.

Secondary treatment: Secondary treatment systems are classified as *fixed-film* or *suspended-growth* systems.

Surface-aerated basins (lagoons). Most biological oxidation processes for treating industrial wastewaters have in common the use of oxygen (or air) and microbial action. Surface-aerated basins achieve 80 to 90 percent removal of BOD with retention times of 1 to 10 days.

Filter beds (oxidizing beds). In older plants and those receiving variable sewage loadings, trickling filter beds are used where settled sewage liquor is spread onto the surface of a bed made up of coke limestone chips or specially fabricated plastic media. The liquor is typically distributed through perforated spray arms. The distributed liquor trickles through the bed and is collected in drains at the base.Biological films of bacteria, protozoa and fungi form on the media's surfaces and eat or otherwise reduce the organic content.

Membrane bioreactors combine activated sludge treatment with a membrane liquid-solid separation process. The membrane component uses low pressure microfiltration or ultrafiltration membranes and eliminates the need for clarification and tertiary filtration. The membranes are immersed in the aeration tank.

Secondary sedimentation. The final step in the secondary treatment stage is to settle out the biological floc or filter material through a secondary clarifier and to produce sewage water containing low levels of organic material and suspended matter.

Tertiary treatment: The purpose of tertiary treatment is to provide a final treatment stage to further improve the effluent quality before it is discharged to the receiving environment (sea, river, lake, ground, etc.).

Filtration: Sand filtration removes much of the residual suspended matter. Filtration over activated carbon, also called *carbon adsorption*, removes residual toxins.

Disinfection: The purpose of disinfection in the treatment of waste water is to substantially reduce the number of microorganisms in the water to be discharged back into the environment for the later use of drinking, bathing, irrigation, etc. The type of disinfectant used, the disinfectant dosage (concentration & time), and other environmental variables affect the disinfection. Cloudy water will be treated less successfully, since solid matter can shield organisms, especially from ultraviolet light or if contact times are low. Generally, short contact times, low doses and high flows all act against effective disinfection. Common disinfectant include ozone, chlorine, ultraviolet light, or sodium hypochlorite. Chloramine, which is used for disinfecting drinking water, is not used in the treatment of waste water because of its persistence.

Overall process of sewage water treatment is shown below in the figure.

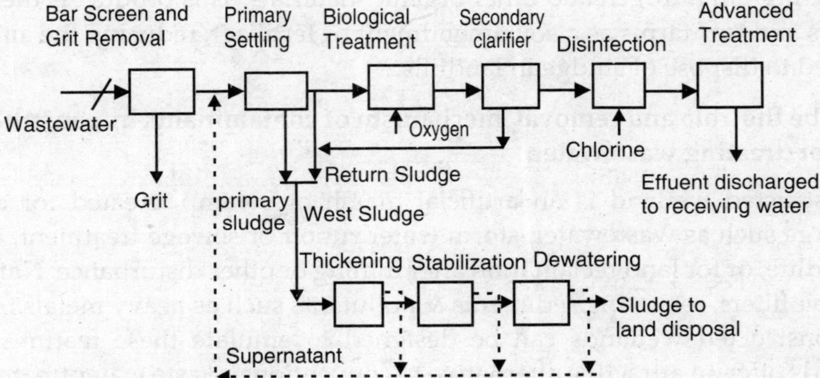

Q. Briefly Describe Sewage Sludge Treatment & Disposal Method.

Ans. The sludge accumulated in a wastewater treatment process must be treated and disposed of in a safe & effective manner. The purpose of digestion is to reduce the amount of organic matter and the number of disease causing microorganisms present in the solids. The most common treatment options include anaerobic digestion, aerobic digestion, and composting. The choice of a wastewater solid treatment method depends on the amount of solids generated and other site-specific conditions. However, in general, composting is most often applied to smaller-scale applications followed by aerobic digestion and then lastly anaerobic digestion for the larger-scale municipal applications.

Anaerobic digestion: Anaerobic digestion is a bacterial process carried out in the absence of oxygen. The process can either be *thermophilic* digestion, in which sludge is fermented in tanks at a temperature of 55°C, or *mesophilic*, at a temperature of around 36°C. Though allowing shorter retention time (and thus smaller tanks), thermophilic digestion is more expensive in terms of energy consumption for heating the sludge.

One major feature of anaerobic digestion is the production of biogas, which can be used in generators for electricity production and/or in boilers for heating purposes.

Aerobic digestion: Aerobic digestion is a bacterial process occurring in the presence of oxygen. Under aerobic conditions, bacteria rapidly consume organic matter and convert it into carbon dioxide. The operating costs are much greater than for anaerobic digestion because of the energy costs needed to add oxygen to the process.

Composting: Composting is also an aerobic process that involves mixing the wastewater solids with sources of carbon such as sawdust, straw or wood chips. In the presence of oxygen, bacteria digest both the wastewater solids and the added carbon source and, in doing so, produce a large amount of heat.

Sludge disposal: When a liquid sludge is produced, further treatment may be required to make it suitable for final disposal. Typically, sludges are thickened (dewatered) to reduce the volumes and transported off site for disposal. No process completely eliminates the need to dispose off bio solids. However, an additional step some cities are taking to superheat the wastewater sludge and convert it into small pelletized granules that are high in nitrogen and other organic materials. This product is then sold to local farmers and turf farms as a soil amendment or fertilizer, reducing the amount of space required to dispose of sludge in landfills.

Q. Describe the role and removal mechanism of contaminants in constructed wetlands used for treating wastewater.

Ans. A constructed wetland is an artificial marsh or swamp, created for anthropogenic discharge such as wastewater, storm water runoff or sewage treatment, and as habitat for wildlife, or for land reclamation after mining or other disturbance. Natural wetlands act as biofilters, removing sediments & pollutants such as heavy metals from the water, and constructed wetlands can be designed to emulate these features. Constructed wetlands offer an attractive alternative to conventional waste water treatment in certain circumstances but removal of phosphorus is strongly dependent on the bed medium. Calcium, iron and aluminium rich solid media are recommended.

Process of removing contaminants. Physical, chemical, and biological processes combine in wetlands to remove contaminants from wastewater. A thin aerobic film around each root hair is aerobic due to the leakage of oxygen from the rhizomes, roots, and rootlets. Decomposition of organic matter is facilitated by aerobic and anaerobic micro-organisms present. Microbial nitrification & subsequent denitrification releases nitrogen as gas to the atmosphere.Phosphorus is coprecipitated with iron, aluminum, and calcium compounds located in the root-bed medium. Suspended solids are filtered out as they settle in the water column in surface flow wetlands or are physically filtered out by the medium within subsurface flow wetland cells. Harmful bacteria & viruses are reduced by filtration and adsorption by biological films on the rock media in subsurface flow and vertical flow systems.

Q Define watershed management.What are its types ?

Ans. Watershed is defined as land area bounded by divide line from which water drains under influence of gravity in to stream, lakes, reservoir. Eg. Pits, dams, Farm, ponds etc. Watershed Management is the management of rainfall and resultant run off.

Its various types are micro,mini and macro.

Q. What is the aim, objectives and approaches to water shed management ?

Ans. Watershed management focuses on how to manage the river basin for optimum utilization of water resource and thereby sustainable development of the society. As the development goals have widened from increasing food production to include poverty reduction & environmental sustainability, protecting the environment is a big challenge and greater emphasis should be given to check the exploitation of the natural resources base. Soil erosion & land degradation coupled with declining per capita availability of land & freshwater are posing serious threat to environment. This is becoming more intense with the increasing population causing food security problems. Hence efforts are needed for management of natural resources for increased productivity of the soils. Several government and nongovernment agencies have launched watershed development and management projects to tackle some of these problems with the objectives of soil conservation, improving the land productivity and promoting appropriate technologies for efficient and sustainable use of natural resources.However, many watershed projects around the world have not performed well because of the poor community participation.

To ensure their sustainability, water resources must be viewed holistically, both in their natural state and in balancing competing demands on them—domestic, agricultural, industrial (including energy), and environmental. Sustainable management of water resources requires systemic, integrated decision-making that recognises the interdependence in three areas. First, decisions on land use also affect water, and decisions on water also affect the environment and land use. Second, decisions on our economic and social future currently organised by socioeconomic sectors and fragmented, affect the hydrology and ecosystems in which humans live. Third, decisions at the international, national, and local levels are interrelated.

Q. Discuss in detail the water shed management practices.

Ans. The watershed is defined as the land area from which water drains under gravity to a common drainage channel. Thus watershed is a delineated area with a well defined topographic boundary and one water outlet. The watershed can range from a few square kilometers to few thousand square kilometers in size.

Watershed management practices: In watershed management the aspects of development are considered with regard to availability of the resources. The practices of conservation and development of land & water are taken up with respect to their suitability for people's benefit as well as sustainability. Various measures taken up for watershed management include the following:

Water harvesting: Proper storage of water is done with provision for use in dry seasons in low rainfall areas. It also helps in moderation of flood.

Afforestation and agro-forestry: In watershed development, afforestation & crop plantation play a very important role. They help to prevent soil erosion and retention of moisture. In high rainfall areas, woody trees are grown in between crops to substantially reduce the runoff and loss of fertile soil. In Dehradun tree like Eucalyptus, Leucaena and grasses like chrysopogon are grown along with maize or wheat to achieve the objectives. Woody trees grown successfully in such agro-forestry programs include Sheesham, Teak and Keekar which have been used in watershed areas of river Yamuna.

Mechanical measures for reducing soil erosion and runoff losses: Several mechanical measures like terracing, bunding, bench terracing, no till farming, contour cropping, strip cropping etc. are used to minimize runoff and soil erosion particularly on the slopes of watersheds. Bundling has proved to be a very useful method in reducing runoff, peak discharge and soil loss in Dehradun.

Scientific mining and quarrying: Due to improper mining, the hills lose stability and get disturbed resulting in landslides, rapid erosion etc. Contour trenching at an interval of one meter on overburdened dump, planting some soil binding plants land draining of water courses in the mined area are recommended for minimizing the destructive effects of mining in watershed areas.

Public participation: People's involvement including the farmers& tribal's is the key to the success of any watershed management program, particularly the soil & water conservation. The communities are to be motivated for protecting a freshly planted areas and maintaining a water harvesting structure implemented by the government or some external agency (NGO) independently or by involving the locale people. Properly educating the people about the campaign and its benefits or sometimes paying certain incentives to them can help effective people's participation. Watershed management in Himalayan region is of vital importance since most of the watersheds of our country lie there. Several anthropogenic activities accelerate its slope instability which should be prevented and efforts should be made to project the watershed by preventing overgrazing, terracing & contour farming to check runoff and erosion etc. On steeper slopes with sliding faces, straw mulching tied with thin wires and ropes helps in establishing the vegetation and stabilizing the slopes.

Q. How do the nutrients from agricultural fields affect the watershed?

Ans. Enrichment of nutrients (Eutrophication) from surrounding watershed affects the penetration of light through the water, causing damage to the characteristic of water and aquatic life.

Q. Describe The History Of Watershed Development In India ?

Ans. A watershed is a logical, natural planning unit for sustainable agricultural research and development particularly when environmental considerations are emphasized.

Hydrologically, watershed could be defined as an area from which the runoff drains through a particular point in the drainage system.

India began to look at the watershed development programs in the 1970s for increasing land controlling land degradation and increasing the productivity of soils. In the 1970s, watershed development held no special significance for the development community in India, however by the end of the 1980s the situation changed radically. Initially watershed projects concentrated on soil and water conservation issues. A decade later, it was felt that technical and physical works alone would not lead to the desired objectives of watershed development and it must also take into account the social, financial and institutional aspects of rural development. Watershed programs have been established over a diverse range of rain-fed agro eco regions in India.

Watersheds in India are broadly grouped into five agro climatic zones:

(i) Trans-Gangetic Plain zone,

(ii) Western Himalayan zone,

(iii) Western Plateau and Hill zone,

(iv) Gujarat Plains and Hill zone, and

(v) Southern zone.

Due to inherent heterogeneity of agro-climatic characteristics over different regions, they have divergent potentials and opportunities. Depending upon the size of the watersheds, these are broadly divided into micro and macro watersheds. Watersheds with areas up to 1250 hectares were classified as micro-watersheds, whereas the macro-watersheds were those, of greater area.

In 1994, the Ministry of Rural Development (MORD) produced a set of guidelines for implementing its watershed programmes, which aimed to tackle the concerns related to the realisation of the full benefits of watershed work. This progressive policy was essentially people-centred and it incorporated good practice from NGO and government policy, such as awareness raising, bottom-up planning, partnerships with NGOs, and community participation. Since 1994-95 Ministry of rural areas and employment, government of India has implemented nearly 10000 watersheds.

Q. What are the Advantages of Watershed projects ?

Ans. Advantages include:

– Improved access to drinking water in project areas during drought.

– Increase in cultivation area leading to increase in employment.

– Increase in crop yield, resulting better income to rural population.

– Improved availability of fodder for animals and increase in milk yield.

– Increase in employment & involvement of women.

– Increase in net returns from all crops.

– Decrease in soil erosion.

– Restoration of ecological balance.

Q. Describe Current Status Of Watershed Management In India.

Ans. Food production increased markedly in many parts of the world during the twentieth century. In India, the rate of growth of food production surpassed the rate of population growth, resulting in increased food per capita. Increased food production has required rapid expansion of irrigated area and water diversion to agriculture, which has resulted in groundwater overdraft and stream flow depletion in some basins. Water scarcity resulting from basin closure has numerous implications for water management, including increased competition and an imperative to improve water productivity in order to maintain growth in the agricultural economy. Basin closure also occurs in the context of global climate change, which may impact water availability and introduce additional uncertainties. Water resources analysis in regions experiencing closure should be carried out in a basin context, because uses in one area affect water availability in downstream areas.

In India, water management has been a government priority for irrigation and drinking water supplies. Today the scope for expanding surface & ground water sources is limited. Thus better management of existing water resources, through a judicious mix of delegation of responsibility & authority to local institutions and large-scale investment in re-directing surplus water to deficit areas is needed.

Q. What is water harvesting and its necessity ?

Ans. Water harvesting is the capture, diversion, and storage of rainwater for plant irrigation and other uses. It is appropriate for large scale landscapes such as parks, schools, commercial sites, parking lots, and apartment complexes, as well as small scale residential landscapes.The problem of water shortage in arid and semi-arid regions is one of low rainfall and uneven distribution through out the season, which makes rainfed agriculture a risky enterprise. The irrigation water are often at the break of overuse and therefore untapped sources of (irrigation) water have to be sought for increasing agricultural productivity. Water harvesting for dry-land agriculture is a traditional water management technology to ease future water scarcity in many arid and semi-arid regions of world.

Historically, harvested rain water provided water for drinking, landscape watering, and for agricultural uses. Once urban areas started to develop, centralized water supply systems replaced the need to harvest water. Harvesting rainwater can reduce the use of drinking water for land scape irrigation. It is also an effective water conservation tool and proves more beneficial when coupled with the use of native, low-water-use and desert-adapted plants. Additionally, rainwater is available free of charge and puts no added strain on the municipal supply or private wells.

As the appropriate choice of technique depends on the amount of rainfall and its distribution, land topography, soil type and soil depth and local socio-economic factors, these systems are very site specific. The water harvesting methods applied strongly depend on local conditions and include such widely differing practices as bunding, pitting, micro-catchments water harvesting, flood water and ground water harvesting.

Q. What Are Various Types Of Water Harvesting ?

Ans. Three groups of water harvesting techniques is described below:

Rainwater harvesting: Rainwater harvesting is defined as a method for inducing, collecting, storing and conserving local surface runoff for agriculture in arid and semi-arid regions. Three types of water harvesting are covered by rainwater harvesting.

(a) Water collected from roof tops, courtyards and similar compacted or treated surfaces is used for domestic purpose or garden crops.

(b) Micro catchment water harvesting is a method of collecting surface runoff from a small catchment area and storing it in the root zone of an adjacent infiltration basin. The basin is planted with a tree, a bush or with annual crops.

(c) Macro-catchment water harvesting, also called harvesting from external catchments, is the case where runoff from hill-slope catchments is conveyed to the cropping area located at hill foot on flat terrain.

Flood water harvesting can be defined as the collection and storage of creek flow for irrigation use. Flood water harvesting, also known as 'large catchment water harvesting' or 'Spate Irrigation', may be classified into following two forms:

(a) In case of '**floodwater harvesting within stream bed**', the water flow is dammed and as a result, inundates the valley bottom of the flood plain. The water is forced to infiltrate and the wetted area can be used for agriculture or pasture improvement.

(b) In case of '**floodwater diversion**', the wadi water is forced to leave its natural course and conveyed to nearby cropping fields. It is difficult to give exact figure on the present area under various forms of floodwater harvesting systems.

Groundwater harvesting is employed to cover traditional as well as unconventional ways of ground water extraction. Qanat systems, underground dams and special types of wells are few examples of the groundwater harvesting techniques. For example, **Qanats**, widely used in Iran, Pakistan, North Africa and in Spain, consists of a horizontal tunnel that taps underground water in an alluvial fan, brings it to the surface due to gravitational effect. Qanat tunnels have an inclination of 1-2% and a length of up to 30 km. Many are still maintained and deliver steadily water to fields for agriculture production and villages for drinking water supply.

Groundwater dams like 'Subsurface Dams' and **'Sand Storage Dams'** are other examples of groundwater harvesting. They obstruct the flow of ephemeral streams in a river bed; the water is stored in the sediment below ground surface and can be used for aquifer recharge. Sand filled reservoirs have the following advantages:

- Evaporation losses are reduced,
- No reduction in storage volume due to siltation,
- Stored water is less susceptible to pollution, and
- Health hazards due to mosquito breeding are avoided.

Groundwater harvesting does not play the same role globally as rain and floodwater harvesting.

Q. Describe The Objectives And Methods Of Rainwater Harvesting.

Ans. Rainwater harvesting is a technique of increasing the recharge of groundwater by capturing and storing rainwater. This is done by constructing special water-harvesting structures like dug wells, percolation pits, lagoons, check dams etc. Rainwater, wherever it falls, is captured and pollution of this water is prevented.

The annual average rainfall in India is 1200 mm; however, in most places it is concentrated over the rainy season, from June to September. In fact Cherapunji in

Assam (the place receiving the second highest annual rainfall as11000 mm) still suffers from water scarcity. The water flows with runoff and there is little vegetation to check the runoff and allow infiltration. Till now there is hardly any rain-water harvesting being done in this region, thereby losing all the water that comes through rainfall.

Rainwater harvesting has the following objectives:

- To reduce run off loss.
- To avoid flooding of roads.
- To meet the increasing demands of water.
- To raise the water table by recharging ground water.
- To reduce ground water contamination.
- To supplement ground water supplies during lean seasons.

Rainwater can be mainly harvested by anyone of the following methods:

- By storing in tanks or reservoirs above or below ground.
- By constructing pits, dug wells, lagoons, trench or check dams on small rivulets.
- By recharging the ground water.

Q. Describe the modern techniques of rain water harvesting.

Ans. Traditional Method of Rainwater Harvesting: In India, it is an old practice in high rainfall areas to collect rainwater from roof tops into storage tanks. In foot-hills water flowing from springs are collected by embankment type water storage. In Himalayan foot-hills people use the hollow bamboos as pipelines to transport the water of natural springs. Rajasthan is known for its "Tankas" (underground tanks) and "Khadins" (Embankments) for harvesting rainwater. In our ancient times, we had adequate Tellabs, Baawaris,Johars, Hauz etc. in every cities, village and capital cities of our Kings which were used to collect rainwater and ensure adequate water supply in dry periods.

Modern Techniques of Rainwater Harvesting: In arid and semi-arid regions artificial ground water recharging is done by constructing shallow percolation tanks. Check-dams made of any suitable native material (brush, polls, rocks, plants, loose rocks, wire nets, stones, slabs, sacks etc.) are constructed for harvesting runoff from large catchment areas. Rajendra singh of Rajasthan popularly known as "Waterman" has been doing a commendable job for harvesting rainwater by building check-dams in Rajasthan and he has been honored with the prestigious Magsaysay Award for his work.

Ground water flow can be intercepted by building ground water dams or storing water underground. As compared to surface dams, ground water dams have several advantages like minimum evaporation loss, reduced chances of contamination etc.

In roof top rainwater harvesting which is a low cost and effective technique for urban houses and buildings, the rainwater from the top of the roofs is diverted to some surface tank or pit through a delivery system which can be later used for several purposes. Also it can be used to recharge underground aquifers by diverting the stored water to some abandoned dug well or by using a hand pump.

All the above techniques of rainwater harvesting are low cost methods with little maintenance expenses. Rainwater harvesting helps in recharging the aquifers, improves ground water quality by dilution, improves soil moisture and reduces soil erosion by minimizing run-off water.

Q. What Are The Benefits Of Harvesting Rainwater ?

Ans.

- Water harvesting not only reduces potable water use and related costs, but also reduces off-site flooding and erosion by holding rainwater on the site.
- If large amounts of water are held in pervious areas where water penetrates easily,some of the water may percolate to the water table.
- Rainwater is a clean, salt-free source of water for plants.
- Rainwater harvesting can reduce salt accumulation in the soil which can be harmful to root growth. When collected, rainwater percolates into the soil, forcing salts down and away from the root zone area. This allows for greater root growth and water uptake, which increases the drought tolerance of plants.
- Limitations of water harvesting are few and are easily met by good planning and design.

Q. What Is Water Recycling ?

Ans. Though recycling is a term generally applied to aluminum cans, glass bottles, and newspapers, water can be recycled as well.*Water recycling is reusing treated wastewater for purposes such as agricultural & landscape irrigation, industrial processes, toilet flushing, and replenishing a ground water basin (referred to as ground water recharge).*Water is sometimes recycled & reused onsite; for example, when an industrial facility recycles water used for cooling processes. A common type of recycled water is water that has been reclaimed from municipal wastewater, or sewage. The term water recycling is generally used synonymously with water reclamation and water reuse.A common example of unplanned water recycling occurs when cities draw their water supplies from rivers, which receive wastewater discharges upstream from those cities. Water from these rivers has been reused, treated, and piped into the water supply a number of times before the last downstream user withdraws the water. Planned projects are those that are developed with the goal of beneficially reusing a recycled water supply.

Q. What Are The Benefits Of Recycled Water ?

Ans. Recycled water can satisfy most water demands, as long as it is adequately treated to ensure water quality appropriate for the use. As for any water source that is not properly treated, health problems arises from drinking or being exposed to recycled water if it contains disease-causing organisms or other contaminants.

Recycled water is most commonly used for non-potable (not for drinking) purposes, such as agriculture, landscape, public parks, and golf course irrigation. Other non-potable applications include cooling water for power plants & oil refineries, industrial process water for such facilities as paper mills & carpet dyers, toilet flushing, dust control, construction activities, concrete mixing, and artificial lakes.

Although most water recycling projects have been developed to meet non-potable water demands, a number of projects use recycled water indirectly1 for potable purposes. These projects include recharging ground water aquifers and augmenting surface water reservoirs with recycled water. In ground water recharge projects, recycled water can be spread or injected into ground water aquifers to augment ground water supplies, and to prevent salt water intrusion in coastal areas.

Indirect potable reuse refers to projects that discharge recycled water to a water body before reuse. Direct potable reuse is the use of recycled water for drinking purposes directly after treatment. While direct potable reuse has been safely used in Namibia (Africa), it is not a generally accepted practice elsewhere.

Q. Describe the Environmental Benefits of Water Recycling.

Ans. In addition to providing a dependable, locally-controlled water supply, water recycling provides tremendous environmental benefits. By providing an additional source of water, water recycling can help find ways to decrease the diversion of water from sensitive ecosystems.

Other benefits include decreasing wastewater discharges and reducing & preventing pollution. Recycled water can also be used to create or enhance wetlands and riparian habitats. Environmental benefits of water recycling is listed below:

- ***Water recycling can decrease diversion of freshwater from sensitive ecosystems:*** Plants, wildlife, and fish depend on sufficient water flows to their habitats to live and reproduce. The lack of adequate flow, as a result of diversion for agricultural, urban, & industrial purposes, can cause deterioration of water quality & ecosystem health.

  Water users can supplement their demands by using recycled water, which can free considerable amounts of water for the environment and increase flows to vital ecosystems.
- ***Water recycling decreases discharge to sensitive water bodies:*** In some cases, the impetus for water recycling comes not from a water supply need, but from a need to eliminate or decrease wastewater discharge to the ocean, an estuary, or a stream.. By avoiding the conversion of salt water marsh to brackish marsh, the habitat for endangered species can be protected.
- ***Recycled water may be used to create or enhance wetlands and riparian (stream) habitats:*** Wetlands provide many benefits, which include wildlife & wildfowl habitat, water quality improvement, flood diminishment, and fisheries breeding grounds. For streams that have been impaired or dried from water diversion, water flow can be augmented with recycled water to sustain and improve the aquatic & wildlife habitat.
- ***Water recycling can reduce and prevent pollution:*** When pollutant discharges to oceans, rivers, and other water bodies are curtailed, the pollutant loadings to these bodies are decreased. Moreover, in some cases, substances that can be pollutants when discharged to a body of water can be beneficially reused for irrigation. For example,recycled water may contain higher levels of nutrients, such as nitrogen, than potable water. Application of recycled water for agricultural and landscape irrigation can provide an additional source of nutrients and lessen the need to apply synthetic fertilizers.

Q. Discuss various water sources and need & ways/methods of water conservation.

Ans. Water sources include:

- Fresh water
- River

- Stream
- Pond
- Ocean

Need for water conservation:

- Population increases water requirement also increases.
- Due to deforestation annual rainfall decreases.
- Over exploitation of ground water.

Ways of water conservation:

- Reducing evaporation loss.
- Reducing irrigation loss.
- Reuse water.
- Avoid sewage discharge.

Water conservation method:

- Rain water harvesting

§ **Watershed management**

Q. Describe in detail about the water conservation methods.

Ans. Water being one of the most precious and indispensable resources needs to be conserved. The following strategies can be adopted for conservation of water.

- **Decreasing run-off losses:** Huge water-loss occurs due to run-off on most of the soils, which can be reduced by allowing most of the water to infiltrate into the soil. This can be achieved by using contour cultivation, terrace farming, water spreading, chemical treatment or improved water-storage system.

(a) **Contour cultivation**: on small furrows & ridges across the slopes trap rainwater and allow more time for infiltration. Terracing constructed on deep soils have large water-storage capacity. On gentle slopes trapped run off is spread over a large area for better infiltration.

(b) **Conservation-bench terracing**: It involves construction of a series of benches for catching the runoff water.

(c) **Water spreading is done by channeling or lagoon-leveling**, In channeling, the water flow is controlled by a series of diversions with vertical intervals. In lagoon leveling, small depressions are dug in the area so that there is temporary storage of water.

(d) **Chemical wetting agents (Surfactants)**: These seem to increase the water intake rates when added to normal irrigated soil.

(e) **Surface crop residues, tillage, mulch, animal residues etc**. help in reducing run-off by allowing more time for water to penetrate into the land.

(f) **Chemical conditioners like gypsum ($CaSO_4.2H_2O$)** when applied to sodic soils improve soil permeability and reduce run off. Another useful conditioner is HPAN (hydrolyzed polyacrylonitrile)

(g) **Water-storage structures like farm ponds**, dug-outs etc. build by individual farmers can be useful measures for conserving water through reduction of runoff.

- **Reducing evaporation losses:** This is more relevant in humid regions. Horizontal barriers of asphalt placed below the soil surface increase water availability and increase crop yield This is more effective on sandy soil but less effective on loamy sand soils. A co-polymer of starch and acrylonitrile called 'super slumper' has been reported to absorb water up to 1400 times its weight. This chemical has been found to be useful for sandy soils.
- **Storing water in soil:** Storage of water takes place in the soil root zone in humid regions when the soil is wetted to field capacity. By leaving the soil fallow for one season water can be made available for the crop grown in next season.
- **Reducing irrigation losses**:
  (a) Use of lined or covered canals to reduce seepage.
  (b) Irrigation in early morning or late evening to reduce evaporation losses.
  (c) Sprinkling irrigation and drip irrigation to conserve water.
  (d) Growing hybrid crop varieties with less water requirements and tolerance to saline water help conserve water.
- **Reuse of water:**
  (a) treated wastewater can be used for ferti-irrigation.
  (b) using grey water from washings, bath-tubs etc. for watering gardens, washing cars or paths help in saving fresh water.
- **Preventing wastage of water:** This can be done in households, commercial buildings and public places.
  (a) Closing taps when not in use.
  (b) Repairing any leakage from pipes.
  (c) Using small capacity flush in toilets.
- **Increasing block pricing:** The consumer has to pay a proportionately higher bill with higher use of water. This helps in economic use of water by the consumers.

Q. What is waste water ?

Ans. Wastewater is "used" water that is deposited into drains by residents and businesses. Flushing the toilet, taking a shower, and running the washing machine are all examples of everyday activities that produce wastewater. This wastewater drains down the sink or toilet and travels through a network of underground pipes known as the sanitary sewer system. It includes substances such as human waste, food scraps, oils, soaps and chemicals. In homes, this includes water from sinks, showers, bathtubs, toilets, washing machines and dishwashers. Businesses and industries also contribute their share of used water that must be cleaned.

Q. What are the effects of wastewater pollutants ?

Ans. If wastewater is not properly treated, then the environment and human health can be negatively impacted. These impacts can include harm to fish & wildlife populations, oxygen depletion, beach closures and other restrictions on recreational water use, restrictions on fish & shellfish harvesting and contamination of drinking water.

Q. Give some examples of waste water pollutants and their harmful effects

Ans. Some examples of pollutants found in wastewater and the potentially harmful effects these substances can have on ecosystems and human health care listed below:

- decaying organic matter and debris can use up the dissolved oxygen in a lake so fish arrange all the star bulleted items as shown in word file and other aquatic biota cannot survive;
- excessive nutrients, such as phosphorus & nitrogen (including ammonia), can cause eutrophication, or over-fertilization of receiving waters, which can be toxic to aquatic organisms, promote excessive plant growth, reduce available oxygen, harm spawning grounds, alter habitat and lead to a decline in certain species;
- chlorine compounds and inorganic chloramines can be toxic to aquatic invertebrates, algae & fish;
- bacteria, viruses & disease causing pathogens can pollute beaches and contaminate shellfish populations, leading to restrictions on human recreation, drinking water consumption and shellfish consumption;
- metals, such as mercury, lead, cadmium, chromium & arsenic can have acute and chronic toxic effects on species.
- other substances such as some pharmaceutical & personal care products, primarily entering the environment in wastewater effluents, may also pose threats to human health, aquatic life and wildlife.

Q. What is The Need for Wastewater Treatment ?

Ans. Wastewater treatment is needed so that we can use our rivers & streams for fishing, swimming and drinking water. Pollution in the urban waterways results in frequent occurrences of low dissolved oxygen, fish kills, algal blooms and bacterial contamination. Early efforts in water pollution control prevented human waste from reaching water supplies or reduced floating debris that obstructed shipping. Pollution problems and their control were primarily local, not national, concerns.

Since then, population and industrial growth have increased demands on our natural resources, altering the situation dramatically. Progress in abating pollution has barely kept ahead of population growth, changes in industrial processes, technological developments, changes in land use, business innovations, and many other factors.

Increases in both the quantity & variety of goods produced can greatly alter the amount & complexity of industrial wastes and challenge traditional wastewater treatment technology. The application of commercial fertilizers & pesticides, combined with sediment from growing development activities, continues to be a source of significant pollution as runoff washes off the land.

The basic function of the wastewater treatment plant is to speed up the natural processes by which water purifies itself. In earlier years, the natural treatment process in streams & lakes was adequate to perform basic wastewater treatment. As our population and industry grew to their present size, increased levels of treatment prior to discharging domestic wastewater became necessary.

Q. How wastewater is collected for treating ?

Ans. The most common form of pollution control consists of a system of sewers and wastewater treatment plants. The sewers collect municipal wastewater from homes, businesses, & industries and deliver it to facilities for treatment before it is discharged to water bodies or land, or reused.

During the early days people living in both the cities & the countryside used cesspools to dispose of domestic wastewater. Cities began to install wastewater collection systems later because of an increasing awareness of water borne disease.

The use of sewage collection systems has brought improvements to public health.

Q. Briefly describe the Conventional Wastewater Treatment Processes.

Ans. Conventional wastewater treatment process is described below.

1. **Preliminary treatment: Screening:** Upon arrival via the sewer system, the wastewater is sent through a bar screen, which removes large solid objects such as sticks and rags. Leaving the bar screen, the wastewater flow is slowed down entering the grit tank. This allows sand, gravel, and other heavy material that was small enough not to be caught by the bar screen to settle to the bottom. All the collected debris from the grit tank and bar screen is disposed of at a sanitary landfill or recycled.

   Wastewater entering the treatment plant includes items like wood, rocks, and even dead animals. Unless they are removed, they could cause problems later in the treatment process. Most of these materials are sent to a landfill.

2. **Pumping:** The wastewater system relies on the force of gravity to move sewage from home to the treatment plant. So wastewater-treatment plants are located on low ground, often near a river into which treated water can be released. If the plant is built above the ground level, the wastewater has to be pumped up to the aeration tanks. From here on, gravity takes over to move the wastewater through the treatment process.

3. **Aerating:** One of the first steps that a water treatment facility can do is to just shake up the sewage and expose it to air. This causes some of the dissolved gases (such as hydrogen sulfide, which smells like rotten eggs) that taste & smell bad to be released from the water. Wastewater enters a series of long, parallel concrete tanks. Each tank is divided into two sections. In the first section, air is pumped through the water.

4. **Primary treatment:** Primary treatment is the second step in wastewater treatment. It allows for the physical separation of solids and greases from the wastewater. The screened wastewater flows into a primary settling tank where it is held for several hours. This allows solid particles to settle to the bottom of the tank and oils & greases to float to the top. As organic matter decays, it uses up oxygen. Aeration replenishes the oxygen. Bubbling air through the water also keeps the organic material suspended while it forces 'grit' (sand and other small, dense particles) to settle out. Grit is pumped out of the tanks and taken to landfills.

   Wastewater then enters the second section or sedimentation tanks. Here, the sludge (the organic portion of the sewage) settles out of the wastewater and is pumped out of the tanks. Some of the water is removed in a step called thickening and then the sludge is processed in large tanks called digesters.

5. **Secondary treatment**: Secondary treatment is a biological treatment process that removes dissolved organic material from wastewater. The partially treated wastewater from the settling tank flows by gravity into an aeration tank. Here it is mixed with solids containing micro-organisms that use oxygen to consume the remaining organic matter in the wastewater as their food supply. The aeration tank uses air bubbles to provide the

mixing and the oxygen, both of which are needed for the micro-organisms to multiply.

From here the liquid mixture, composed of solids with micro-organisms and water, is sent to the final clarifier. Here the solids settle to the bottom where some of the material is sent to the solids handling process, and some is recirculated to replenish the population of micro-organisms in the aeration tank to treat incoming wastewater.

As sludge is settling to the bottom of the sedimentation tanks, lighter materials are floating to the surface. This 'scum' includes grease, oils, plastics, and soap. Slow-moving rakes skim the scum off the surface of the wastewater. Scum is thickened and pumped to the digesters along with the sludge.

Many plants also use filtration in sewage treatment. After the solids are removed, the liquid sewage is filtered through a substance, usually sand, by the action of gravity. This method gets rid of almost all bacteria, reduces turbidity & color, removes odors, reduces the amount of iron, and removes most other solid particles that remained in the water. Water is sometimes filtered through carbon particles, which removes organic particles.

6. **Final treatment: Killing pathogens**: Wastewater that remains is disinfected to kill harmful micro-organisms before being released into receiving waters. Although there are many methods available to kill these micro-organisms, chlorine and ultraviolet disinfection are the most widely used.

   Dechlorination occurs in the final wastewater treatment step. A solution of sodium bisulfite is added to the chlorinated effluent to remove residual chlorine.

   Following disinfection and dechlorination, the treated wastewater (now called final effluent) can be returned to the receiving waters from which it came. The flow is conveyed to an outfall and discharged through a series of diffusers into a surface water body or stream.

   Finally, the wastewater flows into a 'chlorine contact' tank, where the chemical chlorine is added to kill bacteria, which could pose a health risk, just as is done in swimming pools. The chlorine is mostly eliminated as the bacteria are destroyed, but sometimes it must be neutralized by adding other chemicals. This protects fish and other marine organisms, which can be harmed by the smallest amounts of chlorine.

   The production of water of satisfactory quality is dependent upon the elimination or destruction of two groups of living organisms (1) the pathogenic microorganisms, which may infect humans through the use of contaminated water and (2) the algae and other aquatic growths, which may render water aesthetically unfit for human consumption. The treated water (called effluent) would then be discharged to a local river or the ocean.

   Wastewater Residuals: Solids processing: Primary solids from the primary settling tank and secondary solids from the clarifier are sent to the digester. During this process, micro-organisms use the organic material present in the solids as a food source and convert it to by-products such as methane gas and water. Digestion results in a 90% reduction in pathogens and the production of a wet soil-like material called "bio solids" that contain 95-97% water. To remove some of this water and reduce the volume, mechanical equipment such as filter presses or centrifuges are used to squeeze water from the bio solids. The bio solids are then sent to landfills, incinerated, or beneficially used as a fertilizer or for soil amendment.

Another part of treating wastewater is dealing with the solid-waste material. These solids are kept for 20 to 30 days in large, heated and enclosed tanks called 'digesters.' Here, bacteria break down (digest) the material, reducing its volume, odors, and getting rid of organisms that can cause disease. The finished product is mainly sent to landfills, but sometimes can be used as fertilizer.

Our home is serviced by either a sanitary and a storm sewer or a combined sewer, a single pipe which conveys both sanitary sewage and storm water. It is 99.90 percent water, with only 0.01 percent of the wastewater being dissolved & suspended solid material. The cloudiness of sewage is caused by suspended particles which, in untreated sewage, range from 100 to 350 mg/l. A measure of the strength of the wastewater is biochemical oxygen demand, or BOD5. The BOD5 measures the amount of oxygen microorganisms require in five days to break down sewage. Untreated sewage has a BOD5 ranging from 100 mg/l to 300 mg/l. Pathogens or disease causing organisms are present in sewage. Coliform bacteria are used as an indicator of disease causing organisms. Sewage also contains nutrients (such as ammonia & phosphorus), minerals, and metals. Ammonia can range from 12 to 50 mg/l and phosphorus can range from 6 to 20 mg/l in untreated sewage.

The sewer systems are built to follow the natural slope of the land. This design allows gravity to do most of the work transporting the wastewater to treatment facilities. In some situations, pump stations are constructed in low-lying areas to force the wastewater uphill to the treatment facility.

The organic & inorganic material found in this wastewater is too abundant to be handled by the environment. Therefore, treatment of wastewater is an essential process that prevents contamination & destruction of waterways, drinking water sources, and natural water resources.

Sewage treatment is a multi-stage process to renovate wastewater before it reenters a body of water, is applied to the land, or is reused. The goal is to reduce or remove organic matter, solids, nutrients, disease causing organisms, and other pollutants from wastewater. Each receiving body of water has limits to the amount of pollutants it can receive without degradation. Therefore, each sewage treatment plant must treat it to allowable levels of BOD_5, suspended solids, coliform bacteria, and other pollutants.

Q. Differentiate between Domestic sewage and Industrial Wastewater.

Ans. The term wastewater is commonly used to describe liquid wastes that are collected and transported to a treatment facility through a system of sewers. Wastewater is generally divided into two broad classifications: domestic wastewater and industrial wastewater.

DOMESTIC WASTEWATER: Domestic sewage is wastewater discharged from sanitary conveniences in residential, office, commercial, factories and various institutional properties. It is a complex mixture containing primarily water (approximately 99%) together with organic and inorganic constituents. These constituents or contaminants comprise of suspended, colloidal and dissolved materials. Domestic sewage, since it contains human wastes, also contains large numbers of micro-organisms and some of these can be pathogenic(disease causing). Waterborne bacterial diseases that can be present in sewage include cholera, typhoid and tuberculosis. Viral diseases can include infectious hepatitis. Inorganic constituents include chlorides & sulphates, various forms of nitrogen & phosphorous, as well as carbonates &bicarbonates. Proteins and carbohydrates constitute about 90% of the organic matter in domestic sewage. These arise from the excreta, urine, food wastes, and wastewater from bathing, washing, & laundering, and because of the latter, soaps, detergents, & other cleaning products can be found as well.

Domestic sewage has a flow pattern which typically shows two peaks — in the morning before the start of working hours and in the evening after the population has returned from work. Typically these hydraulic peaks would also become more distinct as the sewage flows considered come from smaller populations and consequently smaller sewer networks. Variations in sewage characteristics across a given community tend to be relatively small although variation across communities can be more readily detected. Notwithstanding these variations, the composition of domestic sewage is such that it lends itself well to biological treatment in terms of the availability of and balance between carbonaceous components and nutrients.

The biodegradability of sewage can be estimated by considering its Chemical Oxygen Demand (COD) and the corresponding BOD_5 (5 day BOD), and is indicated by its COD:BOD_5 and BOD_5:N:P ratios. This would typically be about 1.5:1 and 25:4:1 respectively. The nitrogen, N, would typically be in the form of organic nitrogen and ammonia nitrogen (Amm-N). Nitrates (NO_3-N) would not be expected to be present as conditions in the sewers would be such that nitrate formation is unlikely while nitrate degradation because of anoxic reactions is likely. The phosphorous (P) would be a combination of organic and phosphate (PO_4) forms. The pH of sewage would be within the range of 6–9 and this is generally considered suitable for biological processes. Examples of values of BOD_5, TSS (Total Suspended Solids), and TKN (Total Kjeldhal Nitrogen) which have been used for purposes of plant design are 250, 300 and 40 mg/l respectively. As indicated earlier in this paragraph, sewage characteristics can vary across communities and a raw sewage BOD_5 of 500mg/ l has been encountered.

INDUSTRIAL WASTEWATER: Industrial wastewater characteristics vary with each type of discharge. Food processing, dairy operations and meat packing all have wastes that are very high in BOD, sometimes over 1000 mg/l. This BOD is high in dissolved sugars, fats, & proteins and low in suspended material. It will dramatically increase the loading on the secondary treatment processes. Some food process wastes can also have a high pH, like potato processing and some vegetable canning, or a low pH, like green chili processing or fruit canning.

Toxins, like heavy metals coming from metal plating, battery shops,& heavy equipment manufacturing, and solvents from industries like body shops, dry cleaners, & furniture manufacturing can kill the microorganisms that are needed to treat the waste. Copper, chrome, lead, and cyanide are all chemicals that are commonly found in heavy industrial manufacturing discharges.

Hospitals and other medical facilities have the potential to discharge wastes that are radioactive or represent a biohazard.

In some cases, pretreatment of these industrial discharges may be required. Systems that deal with industrial discharges should adopt a pretreatment ordinance to cover conditions that the system may need to impose on the industrial dischargers in order to maintain their NPDES requirements on the treatment plant effluent. If local ordinances are different from specific categorical limits imposed in the Federal Regulations, the most stringent limits of the two must be followed. Pretreatment ordinances can require actual treatment of the industrial waste prior to discharge to the system, or it may levy a monetary surcharge based on the addition loading that results from the discharge.

Any industrial discharger is considered to be a "Significant Industrial User" or SIU if:

(A) The discharge requires pretreatment.

(B) The flow is over 25,000 gpd (gallon per day) , or

(C) The flow is more than 5% of total average plant BOD or hydraulic load

Q. What is the impact of industrial wastewater on water bodies ?

Ans. The impacts of of industrial waste water on water bodies is as follows:

(a) Physical effects — These include impact on clarity of the water and interference to oxygen dissolution in it. Water clarity is affected by turbidity which may be caused by inorganic suspended Solids and/or organic particulates suspended in the water (Volatile Suspended Solids). The latter may undergo biodegradation and thereby also have oxidation effects. Turbidity reduces light penetration and this reduces photosynthesis while the resultant loss in clarity would adversely affect the food gathering capacity of aquatic animals because these may not be able to see their prey. Very fine particulates may also clog the gill surfaces of fishes and thereby affecting respiration & eventually killing them. Settleable particulates may accumulate on plant foliage and bed of the water body forming sludge layers which would eventually smother benthic organisms. As the sludge layers accumulate, they may eventually become sludge banks and if the material in these is organic then its decomposition would give rise to malodours.

In contrast to the settleable material, particulates lighter than water eventually float to the surface and form a scum layer. The latter also interferes with the passage of light and oxygen dissolution. Because of the former, these scum layers affect photosynthesis. Discharge limits on wastewater or treated wastewater discharges typically have a value for TSS such as 30- 50 mg/l. Many industrial wastewaters contain oil and grease (O&G). While some of the latter may be organic in nature, there are many which are mineral oils. Notwithstanding their organic or mineral nature, both types cause interference at the air-water interface and inhibit the transfer of oxygen. Apart from their interference to the transfer of oxygen from atmosphere to water, the O&G (particularly the mineral oils) may also be inhibitory. Unlike domestic sewage, industrial discharges can have temperatures substantially above ambient temperatures. These raise the temperatures of the receiving water and reduce the solubility of oxygen. Apart from this, rapid changes in temperature may result in thermal shock and this may be lethal to the more sensitive species. Heat, however, does not always have a negative impact on organisms as it may positively affect growth rates although there are limits here too , since the condition may favor certain species within the population more than others and over time biodiversity may be negatively affected.

(b) Oxidation and residual dissolved oxygen — Waterbodies have the capacity to oxygenate themselves through dissolution of oxygen from the atmosphere and photosynthetic activity by aquatic plants. Of the latter, algae often plays an important role. However, there is a finite capacity to this re-oxygenation and if oxygen depletion, as a result of biological or chemical processes induced by the presence of organic or inorganic substances which exert an oxygen demand (i.e. as indicated by the BOD or COD), exceeded this capacity then the dissolved oxygen (DO) levels would decline. The latter may eventually decline to such an extent that septic conditions occur. A manifestation of such conditions would be the presence of malodours released by facultative and anaerobic organisms. An example of this is the reduction of substances with combined oxygen such as sulphates by facultative bacteria and resulting in the release of hydrogen sulphide. The depletion of free oxygen would affect the survival of aerobic organisms. DO levels do not, however, need to drop to zero before adverse impacts are felt. A decline to 3–4 mg/l, which still means the water contains substantial quantities of oxygen, may already adversely affect higher organisms like some species of fish. If inhibitory substances are also present, then the DO level at which adverse effects may be felt can be even higher than before. The case of elevated water temperatures due to warm discharges is somewhat different. The elevated temperatures can affect metabolic rates positively (possibly two fold for each 10^0C rise in temperature) but elevated temperatures also reduce the solubility of oxygen in water. This would mean increasing demand for oxygen while its availability declines. Because of the impact of DO levels on aquatic life, much importance has been placed on determining the BOD value of a discharge. Typical BOD5 limits set values are 20 and 50m

(c) Inhibition or toxicity and persistence — These effects may be caused by organic or inorganic substances and can be acute or chronic. Examples of these include the pesticides and heavy metals. Many industrial wastewaters do contain such potentially inhibitory or toxic substances. The presence of such substances in an ecosystem may bias a population towards members of the community which are more tolerant to the substances while eliminating those which are less tolerant and resulting in a loss of biodiversity. For similar reasons, an awareness of the impact such substances have on biological systems is not only relevant in terms of protection of the environment but is of no less importance in terms of their impact on the biological systems used to treat industrial wastewaters. Even successful treatment of such a wastewater may not necessarily mean that the potability of water in a receiving waterbody would not be affected. For example small quantities of residual phenol in water can react with chlorine during the potable water treatment process giving rise to chlorophenols which can cause objectionable tastes & odors in the treated water. Apart from the organic pollutants which are potentially inhibitory or toxic, there are those which are resistant to biological degradation. Such persistent compounds can be bio accumulated in organisms resulting in concentrations in tissues being significantly higher than concentrations in the environment and thereby making these organisms unsuitable as prey/food for organisms (including Man) higher up the food chain. While some organic compounds may be persistent, metals are practically non-degradable in the environment.

(d) Eutrophication—The discharge of nitrogenous and phosphorous compounds into receiving waterbodies may alter their fertility. Enhanced fertility can lead to excessive plant growth. The latter may include algal growth. The subsequent impact of such growth on a waterbody can include increased turbidity, oxygen depletion, and toxicity issues. Algal growth in unpolluted water bodies is usually limited because the water is nutrient limiting. While nutrients would include marco-nutrients like nitrogen, phosphorous, and carbon, and micro-nutrients like cobalt, manganese, calcium, potassium, magnesium, copper, & iron which are required only in very small quantities, the focus in concerns over eutrophication would be on phosphorous and nitrogen as quantities of the other nutrients in the natural environment are often inherently adequate. In freshwaters ,the limiting nutrient is usually phosphorous while in estuarine and marine waters it would be nitrogen. Treatment of industrial wastewater (or domestic sewage for that matter) can then target the removal of either phosphorous or nitrogen, depending on the receiving waterbody, to ensure that the nutrient limiting condition is maintained. When the nutrient limiting condition is no longer present in the water body, and when other conditions such as ambient temperature are appropriate, excessive algal growth or algal blooms (e.g. the red tide) may occur. Apart from aesthetic issues, such algal blooms may affect the productivity of the fisheries in the locale. It should be noted that not all industrial wastewaters contain excessive quantities of nutrients, macro and micro. This deficiency, if there is, results in process instability and/or the proliferation of inappropriate microbial species during biological treatment of the wastewaters.

Bulking sludge is a manifestation of such an occurrence. To address this deficiency, nutrients supplementation is required. The quantities used should be carefully regulated so that an excessive nutrients condition is not inadvertently created and these excess nutrients subsequently discharged with the treated effluent. In terms of BOD:N:P, the optimal ratio for bio treatment is often taken as 100:5:1 while the minimum acceptable condition can be 150:5:1.

(e) Pathogenic effects — Pathogens are disease-causing organisms and an infection occurs when these organisms gain entry into a host (e.g. man or an animal) and multiply therein. These pathogens include bacteria, viruses, protozoa, and helminthes. While domestic and medical related wastewaters may typically be linked to such micro-organisms (and especially the bacteria and viruses), industrial wastewaters are not typically associated with this category of effects. The exception to this is wastewaters associated with the sectors in the agro-industry dealing with animals. The concern here would be the presence of such organisms in the wastewater which is discharged into a receiving water body and diseases, if any, are then transmitted through the water. While many of these organisms can be satisfactorily addressed with adequate disinfection of the treated effluent and raw potable water supplies during the water treatment process, there are those which cannot be dealt with so easily. Two examples of such organisms, Cryptosporidum and Giardia, belong to the protozoa family. The difficulty is that the infected host does not necessarily shed the organism but is likely also to shed its eggs or oocysts. The latter can unfortunately be resistant to the usual disinfection processes.

Q. What is the impact of industrial wastewater on municipal waste water treatment plants ?

Ans. On some occasions industrial wastewaters are discharged into a sewerage system serving commercial and residential premises. Such a combination of wastewater streams is known as municipal wastewater and the quality of such a mixture of waste waters can vary depending on the proportion of industrial wastewaters in it and the type of industries contributing the industrial wastewater streams. Usually the domestic & commercial components in municipal wastewater can be expected to provide some buffering in terms of the characteristics of the combined flow. This is then expected to enable the combined wastewater to be treated easily compared to the treatment of the industrial wastewater on its own. However, even where the option of discharging into a sewerage system is available, some degree of pretreatment is frequently required at the factory before such discharge is permitted. Such pretreatment may include pH adjustment to 6–9 and BOD5 reduction to 400 mg/l. This is to protect the receiving sewers from corrosion and also protect the performance of the receiving treatment plant from an organic substrate overload.

All treatment plants should be planned giving serious consideration for the reduction of waste volume & strength, through process of chemical substitution, chemical recovery & recycling of water. The pollutional load is dealt with operations like segregation, neutralization, equalization, chemical precipitation, chemical oxidation & biological oxidation. Several chemicals are used to reduce the BOD by chemical coagulation such as alum, ferric sulphate, ferrous sulphate & ferric chloride. Lime or H_2SO_4 is used to adjust pH in this process. The dye waste may be economically treated by biological methods prior to equalization, neutralization & chemical oxidation.

The industrial waste when discharged into sewer not only chokes the sewer due to the deposition of solids but also reduces the cross section of the sewer arising out of the lime encrustature.

Chromium compounds in excess of 10-20mg disrupt the operation of the trickling filter. Sulphides are toxic to the micro organisms and are removed along with the sludge. The sludge is dried over sand drying beds and can be used as good manure. Chemical coagulation (Alum, ferric chloride, and ferrous sulphate) with or without prior neutralization followed by biological treatment is necessary for better quality of effluent.

Q. Give a brief account of the following with respect to industrial waste water.

(a) Volume reduction

(b) Strength reduction

(c) Neutralization

(d) Equalization

Ans. a. Volume reduction

In general the first step in minimizing the effects of industrial wastes on relieving streams & treatment plants is to reduce the volume of such wastes. This may be accomplished by:

(i) Classification of wastes
(ii) Conservation of wastewater
(iii) Change in production to decrease wastes.
(iv) Re using both industrial & municipal effluent as raw water supplies.

Classification of wastes

If wastes are classified so that manufacturing process waters are separated from cooling waters, the volume of water requiring intensive treatment may be reduced considerably. Sometimes it is possible to classify & separate the process waters themselves, so that only the most polluted once are treated & the relatively uncontaminated are discharged without treatment.

The 3 main classes of wastes are:

(A) Wastes from manufacturing process --These include waters used in forming paper, discharge from plating solutions in metal fabrication, those discharge from washing of milk cans in dairy plants etc.

(B) Waters used as cooling agents in industrial process - The volume of these wastes varies from one industry to another. Although cooling water can be become contaminated by small leaks, corrosion products or the effect of heat, these wastes contain little , if any organic matter & are classed as non pollution.

(C) Wastes from sanitary uses - The volume depends on many factors including size of the plant, amount of waste producing materials washed from the floors & the degree of cleanliness required & number of workers in the unit.

Unfortunately, in most old plants process, cooling & sanitary waste water are mixed in one pipeline & many industries are paying little attention to segregating wastes to avoid stream pollution.

b. Strength Reduction

Waste Strength reduction is the second major objective for an industrial plant concerned with waste treatment. The strength of wastes may be reduced by:

1. Process Changes
2. Equipment Modifications
3. Segregation of Wastes
4. Equalization of Wastes
5. By-Product Recovery
6. Proportioning of Wastes and
7. Monitoring Waste Streams

Process Changes

In reducing the strength of wastes through process changes, the sanitary engineer is concerned with wastes that are most troublesome from a pollution standpoint.

Equipment Modification

Changes in equipment can effect a reduction in the strength of the waste, usually by reducing the amounts of contaminants entering the waste stream. An outstanding example of waste strength reduction occurred in the dairy industry. The new cans were constructed with smooth necks so that they could be drained faster and more completely. This prevented a large amount of milk waste from entering streams and sewage plants.

Segregation of Wastes

Segregation of Wastes reduces the strength and/or the difficulty of treating the final waste from industrial plant. It usually results in two wastes: one strong and small in volume and the other weaker with almost the same volume as the original unsegregated waste. The small- volume strong waste can then be handled with methods specific to the problem it presents. In terms of volume reduction alone, segregation of cooling waters and storm waters from process waste will mean a saving in the size of the final treatment plant.

Equalization of Wastes

Plants, which have many products, from a diversity of processes, prefer to equalize their wastes. This requires holding wastes for a certain period of time, depending on the time taken for the repetitive process in the plant. For example, if a manufactured item requires a series of operations that take eight hours, the plant needs an equalization basin designed to hold the wastes for that eight hours period. The effluent from an equalization basin is much more consistent in its characteristics than each separate influent to that same basin.

Stabilization of pH & B.O.D and settling of Solids & Heavy Metals are among the objectives of equalization. Stable effluents are treated more easily and efficiently, than unstable ones by industrial& municipal treatment plants.

By-Product Recovery

All wastes contain by products, the exhausted materials used in the process. Since some wastes are very difficult to treat at low cost, it is advisable to consider the possibility of building a recovery plant which will produce a marketable By-Product and at the same time solve a troublesome wastes problem.

Proportioning Wastes

By Proportioning its discharge of concentrated wastes into the main sewer , a plant can often reduce the strength of its total waste to the point where it will need a minimum of final treatment or will cause the least damage to the stream or treatment plant.

It may prove less costly to proportion one small but concentrated waste into the main flow. According to the rate of the main flow, than to equalize the entire waste of the plant in order to reduce the strength.

Monitoring Waste Streams

Accidental spills are often the sole cause of stream pollution or malfunctioning of treatment plants and these can be controlled & often eliminated completely, if all significant sources of wastes are monitored.

c. Neutralization

Excessively acidic or alkaline wastes should not be discharged without treatment into a receiving stream. A stream is adversely affected by low or high pH values. This adverse condition is even more critical when sudden sludge of acids or alkalis are imposed upon the stream.

Acceptable Methods of Neutralization are:

1. Mixing wastes so that the net effect is a neutral pH.
2. Passing acid wastes through beds of limestone.
3. Mixing acid wastes with lime slurries.
4. Adding the proper proportions of concentrated solutions of caustic soda (NaOH) or soda ash (Na_2CO_3) to acid wastes.
5. Adding compressed CO_2 to alkaline wastes.
6. Adding sulfuric acid to alkaline wastes.

The material and method used should be selected on the basis of the overall cost, since material costs vary widely and equipment for utilizing various agents will differ with the method selected. The volume, kind and quality of acid or alkali to be neutralized are also factors in deciding which neutralizing agent to use.

d. Equalization

Equalization is a method of retaining wastes in a basin so that the effluent discharged is fairly uniform in its characteristics (pH, color, turbidity, alkalinity, B.O.D etc). A secondary but significant effect is that of lowering the concentration of effluent contaminants. A retention pond serves to level out the effects of peak loadings on the plant while substantially lowering the B.O.D and suspended solids load to the aeration unit.

Air is sometimes injected into these basins to provide:

1. Better mixing
2. Chemical oxidation of reduced compounds
3. Some degree of biological oxidation
4. Agitation to prevent suspended solids from settling.

The size and shape of the basins vary with the quantity of waste and the pattern of its discharge from the industry. The capacity should be adequate to hold and render homogeneous, all the wastes from the plant. Almost all industrial plants operate on a cycle basis; thus if the cycle of operations is repeated for every two hours, an equalization tank which can hold a two -hour flow will usually be sufficient.

The mere holding of waste, however is not sufficient to equalizing it. Each unit volume of waste discharged must be adequately mixed with other unit volumes of waste discharged many hours previously.

This mixing may be brought about in the following ways:

1. Proper distribution and baffling
2. Mechanical agitation
3. Aeration and
4. Combination of all three.

Q. In what different ways the neutralization of industrial waste water is achieved?

Ans. Excessively acidic or alkaline wastes should not be discharged without treatment into a receiving stream. A stream is adversely affected by low or high pH values. This adverse condition is even more critical when sudden sludge of acids or alkalis are imposed upon the stream.

Acceptable Methods of Neutralization are:

1. Mixing wastes so that the net effect is a neutral pH.
2. Passing acid wastes through beds of limestone.
3. Mixing acid wastes with lime slurries.
4. Adding the proper proportions of concentrated solutions of caustic soda (NaOH) or soda ash (Na_2CO_3) to acid wastes.
5. Adding compressed CO_2 to alkaline wastes.
6. Adding sulfuric acid to alkaline wastes.

The material and method used should be selected on the basis of the overall cost, since material costs vary widely and equipment for utilizing various agents will differ with the method selected. The volume, kind and quality of acid or alkali to be neutralized are also factors in deciding which neutralizing agent to use.

Q. Why equalization and proportioning is of much importance in industrial wastewater treatment? How is it carried out?

Ans.

Equalization: Equalization is a method of retaining wastes in a basin so that the effluent discharged is fairly uniform in its characteristics (pH, color, turbidity, alkalinity, B.O.D etc). A secondary but significant effect is that of lowering the concentration of effluent contaminants. A retention pond serves to level out the effects of peak loadings on the plant while substantially lowering the B.O.D and suspended solids load to the aeration unit.

Air is sometimes injected into these basins to provide:

1. Better mixing
2. Chemical oxidation of reduced compounds
3. Some degree of biological oxidation
4. Agitation to prevent suspended solids from settling.

The size and shape of the basins vary with the quantity of waste and the pattern of its discharge from the industry. The capacity should be adequate to hold and render homogeneous, all the wastes from the plant. Almost all industrial plants operate on a cycle basis; thus if the cycle of operations is repeated for every two hours, an equalization tank which can hold a two -hour flow will usually be sufficient.

The mere holding of waste, however is not sufficient to equalizing it. Each unit volume of waste discharged must be adequately mixed with other unit volumes of waste discharged many hours previously.

This mixing may be brought about in the following ways:

1. Proper distribution and baffling
2. Mechanical agitation
3. Aeration and
4. Combination of all three.

Proportioning: Proportioning means the discharge of industrial wastes in proportion to the flow of municipal sewage in the sewers or to the stream flow in the receiving river. In most cases it is possible to combine equalization and proportioning in the same basin. The effluent from the equalization basin is metered into the sewer or stream according to a predetermined schedule. The objective of proportioning in sewers is to keep constant the percentage of industrial wastes to domestic sewage flow entering the municipal sewage plant. This procedure has several purposes:

1. To protect municipal sewage treatment using chemicals from being impaired by a sudden overdose of chemicals contained in the industrial waste.
2. To protect biological treatment devices from strong loads of industrial wastes which may inactivate the bacteria.
3. To minimize fluctuations of sanitary standards in the treated effluent
4. The rate of flow of industrial waste varies from instant to instant, as does the flow of domestic sewage system. Therefore the industrial waste must be equalized and retained, then proportioned to the sewer or stream according to the volume of domestic sewage or stream.

Air is sometimes injected into these basins to provide:

1. Better mixing
2. Chemical oxidation of reduced compounds
3. Some degree of biological oxidation
4. Agitation to prevent suspended solids from settling.

   The size and shape of the basins vary with the quantity of waste and the pattern of its discharge from the industry. The capacity should be adequate to hold and render homogeneous, all the wastes from the plant. Almost all industrial plants operate on a cycle basis; thus if the cycle of operations is repeated for every two hours, an equalization tank which can hold a two -hour flow will usually be sufficient.

   The mere holding of waste, however is not sufficient to equalizing it. Each unit volume of waste discharged must be adequately mixed with other unit volumes of waste discharged many hours previously.

This mixing may be brought about in the following ways:

1. Proper distribution and baffling
2. Mechanical agitation
3. Aeration and
5. Combination of all three.

Q. Discuss the following with reference to removal of dissolved inorganic solids from industrial wastewater: (a) Dialysis (b) Reverse osmosis (c) Ion exchange

Ans. a. Dialysis-It is the separation of solute by means of their unequal diffusion through membranes. It is most useful in recovering pure solutions for reuse in manufacturing processes for e.g. caustic soda in textile industry. Dialysis work on the simple principle of passing a concentrated, impure caustic solution upward, counter current to a downstream water supply, from which it is separated by a semi-permeable membrane. The caustic soda permeates the membranes and goes into water more rapidly than the other impurities contained in the water The quantity of NaOH diffusing through the membrane diaphragm depends upon the time, the area of the dialyzing surface, the mean concentration difference and the temperature. Dialysis is an operation requiring very little operator attention and although its main role is to conserve raw materials and to reduce plant waste, at the same time it aids in waste treatment. With the introduction of acid resistant membranes, dialysis has been used successfully in the recovery of sulphuric acid in the copper, stainless steel and other industries. Some operation can recover as much as the 70 to 75% of the acid, but recovery as little as 20% may be justifying the process. In dialysis, the driving force of separation is natural diffusion because of concentration gradient. Electro dialysis is another form of dialysis in which the natural driving force is enhanced by the application of electrical energy. Electro-dialysis can achieve as much as 44 percent reduction in concentration of dissolved solids in industrial effluents

b. **Reverse osmosis** -It is process for separating relatively pure water or some other solvent from a less pure solution. The solution is passed over the surface of a specific semi-permeable membrane at a pressure in excess of the effective osmotic pressure of the feed solution. The permeating liquid is collected as the product and the concentrated feed solution is generally discarded. The membrane must be (1) highly permeable to water (2) highly impermeable to solutes (3) capable of withstanding the applied pressure without failure, (4) chemically inert, mechanically strong. Although several types of membranes have been developed, two types of membranes are generally used in commercial equipment.(1). The first is a symmetric of skinned cellulose acetate membranes made in flat or Tubular forms.(2). The second type of membrane is an aromatic polyamide or polyamide hydrazide. Reverse osmosis is an advanced unit operation in water treatment. Reverse osmosis membranes are capable of removing at least 90 percent of the dissolved solids in water as well as organics, bacteria and others impurities. Osmosis can be defined as the spontaneous passage of a liquid from dilute to a more concentrated solution across an ideal semipermeable membranes which allows the passage of the solvent (water) but not the dissolved solids (solutes). The transport of the water from one side of the membrane to the other continues until the pressure (p) is large enough to prevent any net transfer of the solvent (water) to the more concentrated solution. At equilibrium, the quantity of water passing in the either direction is equal, and pressure (p) is then defined as the osmotic pressure of the solution having that particular concentration and make up of dissolved solids. If a piston is placed on the more concentrated solution side of a semipermeable membrane and a pressure (P) is applied to the solution, the following conditions can be realized:

(i) P is less them the osmotic pressure of the solution and solvent flows (in the same direction) spontaneously toward the more concentrated solution. (ii) P equals the osmotic pressure of the solution and solvent flows at the same rate in both directions that is no net change in water levels. (iii) P is greater than the osmotic pressure of the solution and solvent flows from the more concentrated solution to the pure solvent side of the membrane leaving a higher concentration behind. This condition represents the phenomenon of reverse osmosis. Substantial energy is needed to constantly pressurized the incoming salty water is one of the drawback of the reverse osmosis.

c. **Ion exchange**-Ion exchange is basically a process of exchanging certain undesirable cations and anions of the wastewater for sodium, hydrogen or other ions in resinous materials. The resins, both natural and artificial are commonly referred to as Zeolites. When the ability of the zeolite bed to produce soft water is exhausted, the softener is temporarily cut out of service. It is then backwash to clean and hydraulically regrade the bed, regenerated with a solution of common salt; which remove the calcium and magnesium in the form of their soluble chlorides and restores the zeolite to its original conditions, rinsed free of these and the excess salt, and finally returned to service.

Q. List the techniques applied in removal of suspended solids.

Ans. There are three methods for the removal of suspended solids, which are as follows:

1. Screening.
2. Flotation
3. Sedimentation

Screening: Screening of industrial wastes is generally practiced on waste containing larger suspended solids of variable sizes. For example from tanneries, pulp and paper mills, or poultry processing plants. It is an economical and effective means of rapid separation of these larger suspended solids from the remaining waste material. In many cases screening alone will reduce the suspended solids to a low enough concentration to be acceptable for discharge into a municipal sewer or a nearby stream.

Screens are available in sizes ranging from coarse (10 or 20 mesh) to fine (120 to 320mesh). In the industry, generally two types of screens are used:

(i) North Screen
(ii) Sweco Screen

North screen: They are generally rotary, self-cleaning gravity-type, units. It has been used successfully in treating wastes from meat-packing, grain-washing, tanning, making, woolen and sea food plants.

Sweco screen: They are circular, vibratory screens. They have been quite effectively used for screening wastes from food-packing processes such as meat and poultry packing or fruit and vegetable canning.

Floatation: The floatation process consists of creation of air bubbles in the waste body or tank by introducing air to the system. These air bubbles increase the buoyancy of the finer particles which could be easily lifted by skimming (i.e. the process of removing floating grease or scum from the surface of the effluent in a tank).

Advantages of floatation as wastewater process are:

(i) Grease, and light solids rising to the tops and grit and heavy solids settling to the bottom are all removed in one unit.
(ii) Odor nuisance are minimized, because of short detention period.
(iii) High overflow rates and short detention periods means smaller tank sizes, resulting in decreased space requirements and possible saving in construction costs

Disadvantages of floatation are:

(i) The additional equipment required results in higher operating costs.
(ii) Floatation units generally do not give as effective treatment as gravity settling units, although the efficiency varies with the waste.

(iii) More skilled maintenance is required for a flotation unit than for a gravity settling unit.

Sedimentation: Sedimentation processes are very effective in removing suspended solids in industrial wastewater. The efficiency of sedimentation tank depends, in general on the following factors:

* Detention periods.
* Wastewater characteristics
* Tank design.
* Particle size.
* Velocity & density of particles.

The clarifiers/tanks either rectangular or circular in shape are most commonly used in the application of sedimentation in wastewater treatment. The design of the clarifiers are based on several factors, which are as follows:

* Influent TSS concentration.
* Effluent TSS concentration.
* Surface loading
* Detention time.
* Sludge generation

Q. Explain any three methods of sludge disposal.

Ans. Both primary & secondary sewage treatments involve settling of particulate matter, and thus produce sludge. The concentration of solids in the primary sewage sludge is about 5%; the activated sludge contains about 1%; and the sludge from trickling filters has about 2% solids.

Thus the sludge is composed of almost entirely of water and volume reduction is the key to economic disposal. In addition to reducing its high water content, the sludge must be stabilized so that its biological activity and tendency towards putrefaction are reduced drastically.

The sludge is concentrated by gravity settling and floatation. After concentration , the sludge is subjected to anaerobic digestion in a digester in which the organic content of the sludge decomposes to give mainly methane and carbon dioxide and at the same time the bound water is released from the sludge.

The sludge is then conditioned to improve its dewatering characteristics by adding chemicals like iron salts and polyelectrolytes. These chemicals bind the sludge particles together and encourage the release of water. The sludge is then heated under pressure and after a period of time the gel structure of the sludge breaks down so that the water is released.

The thickened sludge is then dewatered for efficient handling and disposal. The dewatering is accomplished by mechanical methods, the most common being centrifugation and filtration.

The dewatered sludge is then subjected to oxidation to reduce the organic content, with the consequent destruction of bacteria and significant reduction in their volume.

Incineration and oxidation are the two common methods employed for oxidation. Several methods are employed for the ultimate disposal of sludge. The wet digested sludge may be sprayed on to a cropland where it functions as fertilized Dried sludge may be used a land fill or soil conditioner. Wet or partially dewatered sludge or ash from incineration may be transported from the shore to dumping grounds at sea. The potential drawback to the use of sewage sludge as fertilizer in agricultural fields is the presence of both organic and inorganic toxic compounds. The former compounds are oxidation-resistant organic substances, such as organochlorine species which become bound in the organic matrix of the sludge. The inorganic toxicants are represented by heavy metals, mainly arsenic, cadmium, lead, mercury and zinc. These metals can be taken up by crops and introduced into the food chains or leached to the ground water.

Drying beds / sand bed drying

For small industrial waste treatment plants, sludge can be dewatered on open or covered sand beds. Drying of the sludge occurs by percolation and evaporation. The proportion of the water removed by percolation may vary from 20- 55%. Depending on the initial solids contents of the sludge & on the characteristics of the solids. The design and use of drying beds are affected by climatic conditions (rainfall & evaporation)

Sludge drying beds usually consists of 10-23cms of sand, 20-46 cms of gravel or stone. The sand has an effective size of 0.3 to 1.2mm & a uniformity coefficient < 5.0. Gravel is graded from 0.32 to 2.54cm. The beds are provided with under drains spaced from 2.7 to 6.1m apart. The under drain piping having a mini dia of 10cm & minimum slope of 1% The filtration is returned to the treatment plants. Wet sludge is usually applied to the drying beds at depths of 20 -30cms. Removal; of the dried sludge varies with both individual judgment and final disposal means many cases, the bed turn over can be substantially increased by the use of chemicals. Slum treatment can be reduce the sludge drying time by 50%.The use of polymers can increase the rate of bed dewatering and also increase the depth of application.

Drying and incineration

A large volume of sludge can be reduced to a small volume of ash, which is free from organic matter & therefore easily disposable, by combination of heat drying & incineration. Incineration involves drying & combustion. Various types of incineration units are available to accomplish these reactions in single or combined units. In the incineration process, a sludge temperature is raised to 100 °C, at which point moisture is evaporated from the sludge. The water vapour & air temperature are increased to ignition point.Some excess air is required for complete combustion of the sludge. Self sustaining combustion is often possible with dewatered waste sludges ones the burning auxiliary fuel raises the incineration temperature to ignition point. The primary end products of combustion are CO_2, SO_2 & ash.

Incineration can be accomplished in multiple hearth furnaces in which the sludge passes vertically through a series of hearths. In fluidized bed sludge particle are fed into a bed of sand fluidized by upward moving air. A temperature of 760-815 °C is maintained in the beds, resulting in rapid drying and burning of sludge. Ash is removed from the bed by the upward flowing combustion gases.

Sanitary landfill

Sanitary landfill is used to burry garbage, refuse & sludge in a planned and methodical manner. It is a simple, effective & inexpensive method for disposing of dry matter such as refuse, but sludge is usually too liquid for this procedure. However, vacuum filter or sand bed dried sludge can be disposed off in this manner. The area proposed for the sanitary land fill should be easily accessible, yet removed from sources of water supply & recreational areas at the same time on the land which is not too costly. The suitability of soil & possible future use of the property are also imp. Consideration sanitary landfills should be located above the ground water level and 152m away from water sources, particularly when the soil is sandy, gravely or of limestone. At the end of each days dumping, the sludge should be covered and compacted by a bulldozer or tractor.

Sludge barging

Sludge barging is a means of final disposal of sludge to a deep body of water usually an ocean. There is little theory involved in this method of treatment. Raw precipitated, digested or filtered sludge solids are pumped into a waiting ship barge. When the barge is fully loaded it transports the sludge to a suitability site far from the shore, where it is discharged usually by being pumped out deep under the surface of the water. The sludge should be somewhat concentrated before being loaded on the barge (to conserve space) but not too much that pumping difficulties arise.

Pumping raw or digested sludge directed into the ocean & disposing it by dilution technique is usually practiced in coastal cities. These sludges however, should not be contaminated by oil & grease, odor high coliform counts, toxic materials discolouration or be likely to create sludge banks or deposits.

Q. What is sludge thickening? Enumerate various methods and describe the gravity sludge thickener.

Ans. Thickening is a procedure used to increase the solids content of sludge by removing a portion of the liquid fraction. To illustrate, if waste activated sludge, which is typically pumped from secondary tanks with a content of 0.8% solids, can be thickened to a content of 4% solids, then a fivefold decrease in sludge volume is achieved. Thickening is generally accomplished by physical means including co-settling, gravity settling, flotation, centrifugation, gravity belt, and rotary drum. The volume reduction obtained by sludge concentration is beneficial to subsequent treatment processes, such as digestion, dewatering, drying and combustion from the following standpoints:

1. Capacity of tanks and equipment required
2. Quantity of chemicals required for sludge conditioning
3. Amount of heat required by digesters and amount of auxiliary fuel required for heat drying or incineration, or both.

Most common techniques of thickening in sludge processing are:

1. Gravitational settling
2. Dissolved Air Flotation(DAF)
3. Basket centrifuge
4. Solid-bowl centrifuge

5. Gravity belt filter
6. Rotary drum

Thickening is the process by which bio solids are condensed to produce a concentrated solids product and a relatively solids-free supernatant. Thickening wastewater solids reduces the volume of residuals, improves operation, and reduces costs for subsequent storage, processing, transfer, end use, or disposal. For example, thickening liquid-solids (slurry) from 3 to 6 percent will reduce the volume by 50 percent.

There are several different methods for thickening bio solids, including dissolved air floatation (DAF), centrifugal thickening, gravity belt thickening, and gravity thickening. Gravity thickening uses the natural tendency of higher-density solids to settle out of liquid to concentrate the solids.

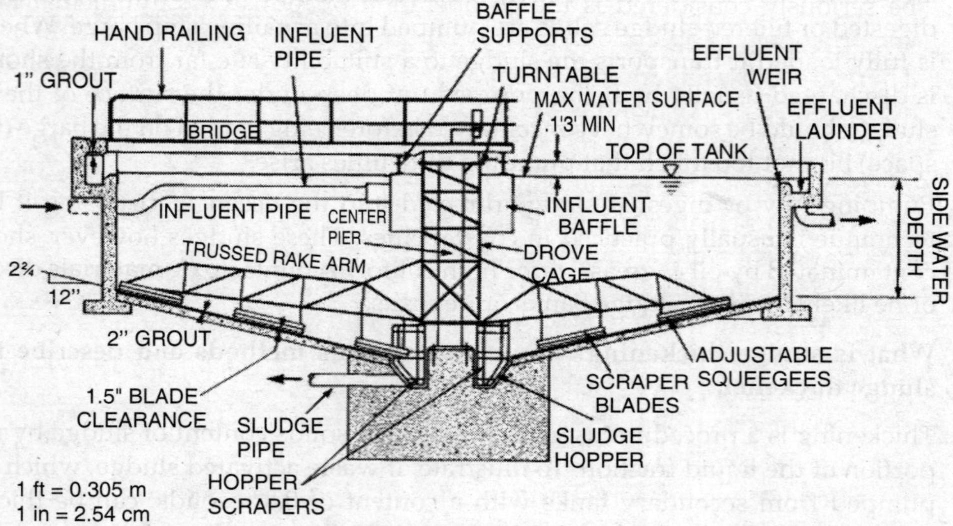

Fig.9: Gravity Thickener

Gravity thickeners consist of a circular tank (usually with a conical bottom) that is fitted with collectors or scrapers at the bottom. Primary and/or secondary solids are fed into the tank through a center well, which releases the solids at a low velocity near the surface of the tank. The solids settle to the bottom of the tank by gravity, and the scrapers slowly move the settled, thickened solids to a discharge pipe at the bottom of the tank. A v notch weir located at the top of the tank allows the supernatant to return to a clarifier. In addition, many systems also use a skimmer to collect and remove any floatables and grease that have accumulated at the top of the tank.

The bio solids concentration and thickening that occurs in the tank is achieved through three different settling processes, which include gravity settling, hindered settling, and compaction settling. Gravity settling occurs when solid particles travel downward due to their weight. Settlement continues as solids begin to concentrate near the bottom of the tank, but the settlement rate decreases as the solids concentrations increase. This is known as hindered settling.

Compaction settling occurs when bottom solids are further concentrated due to the pressure of solids on top of them.

Solids at the bottom of the tank can reach as high as 15 percent total solids (TS). A more typical result is 4 to 6 percent TS. Liquid at the surface of the tank is nearly clear, with suspended solids concentrations as low as 200 mg/l.

Q. What are the advantages and disadvantages of combined treatment of Industrial raw waste with the domestic waste?

Ans. It is often possible & advisable for an industry to discharge its waste water directly into a municipal sewage treatment plant, where a certain portion of the pollution can be removed. A municipal sewage plant, if designed & operated properly can be handle almost any type & quantity of industrial waste. Hence one possibility that should be seriously considered is the co operation of industry & municipality in the joint construction and operation of a municipal waste water treatment plant.

There are many advantages to be gained from such a joint venture:

1. Here the responsibility is placed with one owner, while at the same time, the cooperative spirit between industry & municipality increases, particularly if the division of costs is mutually satisfactory.
2. Only one chief operator is required, whose sole obligation is the management of the treatment plant i.e. he is not burden by the miscellaneous duties often given to the industrial employee in charge of waste disposal & the chances of mismanagement and neglect which may result if industrial production men operate waste treatment plants, are eliminated.
3. Since the operator of such a large treatment plant usually receives higher pay than separate domestic plant operators, better trained people are available.
4. Even if identical equipment is required construction costs are less for a single plant than for two or more. Furthermore, municipalities can apply for state & or federal aid for plant construction, which private industry is not eligible to receive.
5. The land required for plant construction & for disposal of waste products is obtained more easily by the municipality.
6. Operating costs are lower, since more waste is treated at a lower rate per unit of volume.
7. Possible cost advantages resulting from lower municipal financing cost & federal grants.
8. Some wastes may add valuable nutrient for biological activity to counter act other industrial wastes that are nutrient deficient. Thus bacteria in the sewage are added to organic industrial wastes as seeding material. These microorganisms are vital to biological treatment. Also, acids from 1 industry may help to neutralization alkaline wastes from another industry.
9. The treatment of all waste water generated in the community in a municipal plant, enables the municipality to assure a uniform level of treatment to all the users of the river & even to increase the degree of treatment given to all waste water to the maximum level obtainable with technological advance.

Q. What is meant by 'Effluent standards and Stream standards'? What are the salient features of them?

Ans.

EFFLUENT STANDARDS

Effluent standards pertain to the quality of the discharge water itself.

They are based on economics than on absolute protection of the stream.

Easy to control.

Detailed stream analysis are not required.

They do not establish an overall level of pollutant loading for a given water body.

Ratio of wastewater to stream flow are not considered.

Treatment is obligatory irrespective of the size of industry.

For effective protection of an overloaded stream, the effluent standards are required to be upgraded.

Large industries have an edge over small industry.

STREAM STANDARDS

Stream standards refer to the quality of the receiving water downstream from the origin of the wastewater discharge.

- They are based on establishing classification of quality for a stream.
- The quality of the receiving water is regulated to maintain established stream classification.
- Prevention of excessive pollution/ Loading is limited to what the stream can assimilate.
- No consideration of type and location of industry.
- Allows public to establish goals for present and future water quality.
- Confusion of zone of different classification.
- Controversy over proportion of stream to be reserved for future usage (municipal, industrial, agriculture etc.).
- Opposition from industry/ public to change the established classification.
- A detailed stream analysis is required to determine the level of wastewater treatment required to maintain the health of the ecosystem.
- Cost of treatment may affect the survival of industry.

Q. Enumerate and explain in detail the factors to be considered for stream sampling.

Ans. Before embarking on designing of any industrial wastewater treatment facility a comprehensive stream-sampling program should be conducted. This will in turn help considerably for efficient operation of facility.

The factors to be considered before starting a sampling program are:

1. Overall objectives of the program
2. Total number of samples
3. Points of collection
4. Method of collection
5. Data to be obtained

6. Frequency of sample collection
7. Time of the year for sampling
8. Statistical handling of data
9. Care of samples, prior to analysis

Overall objectives of the program: There are various objectives of stream sampling. In one instance the engineer may be concerned with the effect of an upstream industry on the water quality downstream, of special interest in this case might be the color of the water in the receiving stream. In another instance, he may be attempting to ascertain the dissolved oxygen sag characteristics of the stream during the summer season only. In another case, he could concern that the stream characteristics comply with the standards stipulated by the state pollution control board. The importance of the 8 factors will depend to some degree on the overall objectives of the sampling program.

Total number of samples: The no. of samples required depends on the objectives of the programme & the amount of time alloted for completion of the same. The use of the few locations & more samples to define the results is usually much more reliable than using many stations with only a few samples from each. Also, samples are frequently taken over a long time interval, during which the condition of the water course is subject to variation. In many instances, an attempt to test all condition by infrequent, random sampling produces no definite pattern & in fact may be misleading. It may be better to concentrate on well defined frequent & intensive sampling

A well planned survey with a specific objective will require a minimum samples. If for ex, someone wishes to determine the river characteristics, especially the DO profile during low flow periods, 2 or 3 samples collected at the proper time will suffice. These samples must be collected, however during extreme drought periods, if factors influencing the character of the stream to be clearly established

However, since the industrial processing is so varied & usually unpredictable, that many samples of the receiving water under all conditions must be collected in order to evaluate the effect of waste upon the stream.

Points of collection: Sampling points should be selected with great care & due consideration should be given to the sources of pollution, dilution by branch streams, changes in surrounding topography & slope of the river significant river side features also influent the choice of sampling points such as a municipal water intake, a park, an industrial area, a good fishing spot, a hotel or a camping site would each have a definite bearing on the usage of the stream. Since the acceptable pollution limits for water vary according to usage, samples should to collected & the record made of the condition of the stream just above & just below all such points of stream use are changed.

A minimum of 4 stream stations is recommended:

(a) An upstream site where the water is uncontaminated.

(b) Just below the source of pollution or dilution.

(c) Where the stream is in worst condition due to the source of pollution.

(d) A point midway between bottom of oxygen sag & the recovery of oxygen level.

Method of collection: Samples should be taken from 0.6 ft depth in streams less than 2 ft. The volume of samples depends on the number and type of analysis to be carried out. A standard type dissolved oxygen sampler is recommended for collecting most samples. Glass bottles with glass caps or polyethylene containers are most widely used. Any doubt about the cleanliness of the sample bottle can usually be dispelled by rinsing it first with some of the actual stream water, but special bottles are required for bacteriological samples.

Data to be collected: The scientific collection of data for stream analysis may be divided into three major categories i.e., hydrologic factors, sources of pollution and water course sampling. The type of data to be obtained depends on the objectives of sampling programme and the amount of time and money available for the investigation. For example, if the oxygen resources of a section of a stream is required then, DO, water temperature and stream flow should be measured. If the sampling programme is of general nature, the stream analyst should undertake as many chemical, physical and biological tests as possible. For design of wastewater treatment plants, the rate of flow, temperature, BOD and DO can supply sufficient information. In addition, data on pH, color and turbidity may indicate the general physical condition of the stream. Biological analysis is required when the stream water is used for drinking, bathing or flushing. In this case, the coliform count is usually determined.

Frequency of sample collection: Samples should be collected as frequently as possible to provide a representative of total sample. The master sample should contain individual constituents of variation expected. For ex, if the pH is known to vary from 4 to 10, individual samples with pH values of 4,5,6,7,8,9 and 10 should appear in the composite sample at least once during each sampling period. If the situation requires instantaneous analysis, more individual samples with little or no composting are collected.

Time of the year for sampling: Time of the year is of at most importance when there is a deadline for producing results. In stream studies, dealing with industrial waste treatment is concerned primarily with the critical condition of pollution which generally exists when the environment is at its warmest & its stream flow slowest, & the man made pollution greatest. But many studies must be undertaken in the spring & autumn seasons due to stream conditions and man power problems. However, many times due to the urgency of the problem or unusual conditions of pollution, the investigations may have to be carried in any of the seasons. The objective of every stream analysis should however to be collect the data during critical stream conditions of temperature, flow and pollution load. It may sometime be necessary to project stream analysis to conditions which might obtain during future critical points. This is sometimes requiring even when a survey is made during the summer. Because the intensity of the problem varies to some degree from year to year & even from one period of the year to another.

Statistical handling of data: It is a well known fact that data can be manipulated to emphasize that aspect of the survey which the analyst fill the most important. For this the engineer must have a working knowledge of statistic and mathematics in order to convey this information in the best form to the layman.

The rate of flow during critical conditions is of importance to the industrial waste engineering. If he incorporates the minimum flow ever recorded in the stream, he may end up with unrealistic evaluation. On the other hand the use of the mean summer or low flow value can be dangerous, since lower than this often occurs. Therefore, some regulatory agencies make use of minimum of 7 days to review once in 10 years as the criteria for designing waste water treatment facilities.

Q. Explain the different zones of pollution in a stream.

Ans. Zones of Purification: When sewage is discharged into water, a succession of changes in water quality takes place. If the sewage is emptied into a lake in which currents about the outfall are sluggish and shift their direction with the wind, the changes occur in close proximity to each other and, as a result, the pattern of changes is not crisply distinguished. If, on the other hand, the water moves steadily away from the outfall, as in a stream, the successive changes occur in different river reaches and establish a profile of pollution which is well defined. However, in most streams, this pattern is by no means static. It shifts longitudinally along the stream and is modified in intensity with changes in season and hydrography.

When a single large charge of sewage is poured into a clean stream, the water becomes turbid, sunlight is shut out of the depths, and green plants, which by photosynthesis remove carbon dioxide from the water and release oxygen to it, die off. Depending on the stream velocity, the water soon turns nearly black. Odorous sulfur compounds are formed and solids settle to the bottom, forming a sludge. The settled solids soon decompose, forming gases such as ammonia, carbon dioxide, and methane or marsh gas. Scavenging organisms increase in number until they match the food supply.

The oxygen resources are drawn upon heavily and, when overloaded, become exhausted. Life in such waters is confined to anaerobic bacteria (which exist when no oxygen is available), larvae of certain insects such as mosquitoes, and a few worms. There are no fish; turtles are generally the only forms of higher life present. This condition is known as the zone of degradation.

In a second zone, or **zone of decomposition**, more solids settle out, the water becomes somewhat clearer, and sunlight penetrates the surface. Oxygen is absorbed from the atmosphere at the air-water interface permitting the establishment of aerobic (oxygen available) conditions. The aerobic bacteria continue the conversion of organic matter into nitrates, sulfates. and carbonates. These, together with the carbon dioxide produced by decomposition as well as by bacteria and plant life, are food sources. With sunlight now penetrating the water, and with abundant food, algae begin to flourish and form a green scum over the surface.

In the third zone, or **zone of recovery**, algae become more numerous and self-purification proceeds more rapidly. Green plants utilizing carbon dioxide and oxygen will liberate in the say time more oxygen than is consumed, thus hastening the recovery of the stream. Simultaneously, the fish that require little oxygen such as catfish and carp, are also found. As the dissolved oxygen increases, more types of fish appear.

After recovery, in the **zone of cleaner water**, fish find the stream highly favorable, as the algae support various aquatic insects and other organisms on which fish feed. The water is clear or turbid according to concentration of algae, and may have odor for the same reason.

Throughout the stages of recovery of self-purification, disease organisms are greatly reduced in number because they lack proper food, and experience unfavorable temperatures and pH values of water. However, the water is still dangerous since all disease organisms have not perished.

Q. What does the oxygen sag curve represent? Explain self purification of stream with sag curve.

Ans. The oxygen deficit (D) at the time in a polluted river stream is the difference between the actual DO content of water at that time and the saturation DO content. (Saturation DO means the amount of DO which will be present in water if it is fully saturated with oxygen. The normal saturation DO value for fresh water varies between 7.6mg/l to 14.6mg/l for temperature varying between 00 to 300C) at the water temperature.

Oxygen deficit (D) = Saturation DO - Actual DO

In order to maintain clean condition in a river stream, the oxygen deficit must be will & this can be found out by knowing the rates of deoxygenation & reoxygenation.

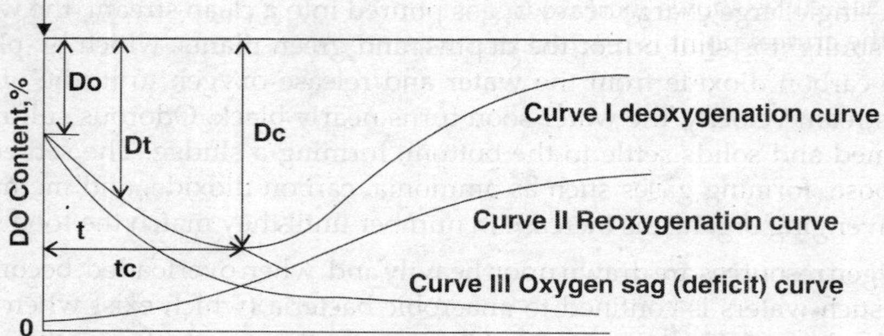

Fig.10: Oxygen sag curve

Curve I - Deoxygenation curve or curve showing depletion of DO in the absence of aeration.

Curve II - Reoxygenation curve (DO content supplied)

Curve III - Oxygen sag curve or oxygen deficit curve (Actual DO Content or oxygen is balance)

Deoxygenation curve

In a polluted stream DO content goes on reducing due to decomposition of organic matter. The rate of deoxygenation depends upon the amount of the organic matter remaining to be oxidized at a given time and as well as on the temperature of the reaction. Hence at a given temperature , the curve showing depletion of DO with time i.e deoxygenation curve is similar to the Ist stage BOD curve.

Reoxygenation curve

In order to counter balance the consumption of DO due to deoxygenation atmosphere supplies oxygen to water and the process is called reoxygenation. The rate at which the oxygen is supplied by the atmosphere to the polluted water depends upon:

(a) The depth of the receiving water (rate is more in a shallow depth)

(b) The condition of the body of the water (rate is more in a running stream than in a quiescent pond)

(c) The saturation deficit or the oxygen deficit (ie. the deficit of DO below the saturation value) and

(d) The temperature of water.

Depending upon these factors, the rate of reoxygenation can also be expressed mathematically and plotted in the form of the curve called reoxygenation curve.

Oxygen Deficit curve (oxygen sag curve): In a running polluted stream exposed to the atmosphere, the deoxygenation as well as reoxygenation go hand in hand. If the deoxygenation is more rapid than the reoxygenation, as oxygen deficit results.

The amount of resultant oxygen deficit can be obtained by adding the deoxygenation & reoxygenation curves. The resultant curve so obtained is called the oxygen sag curve or the oxygen deficit curve. From this curve, the oxygen deficit and oxygen balance isn't he stream after a certain elapse of time can be find out.

It can also be seen that when the deoxygenation rate exceeds the reoxygenation rate, the oxygen sag curve shows increase in deficit of oxygen , but when both the rates become equal , the critical point is reached and then finally when the rate of deoxygenation fall below of reoxygenation, the oxygen deficit goes on decreasing till it becomes zero.

Q. Explain the sources , characteristics & treatment of sugar mill effluent.

Ans. Sources and characteristics of wastes: Waste from the sugar mill house includes the water used as splashes to extract maximum amount of juice & those used to cool the roller bearings. As such the mill house waste contains high BOD due to the presence of sugar & oil from the machineries. The filter cloths used for filtering the juice needs occasional cleaning. The wash waster thus produced though small in volume, contains high BOD & suspended solids.

A large volume of water is required in the barometric condensers of the evaporators & vacuum pan. The water is usually partially or fully circulated after cooling through a spray pond. This cooling water gets polluted as it picks up some organic substances from the vapour of boiling syrup in evaporation & vacuum pans. This polluted water, instead of recirculated is discarded as excess condenser water. These discharges contribute substantially to the waste volume & modulated to BOD in many sugar mills. Additional waste originates due to the leakages & spillages of juice, syrup & molasses in different sections. The periodical washings of the floor through small in volume have got very high BOD. The periodic blow off of the boilers produces another intermittence waste discharge. This waste is high in suspended solids ,low in BOD & usually alkaline. The characteristic of waste water is tabulated below.

Characteristics	Value
pH	4.6-7.1
Total solids	870-3500 mg/lt
BOD	300-200 mg/lt
COD	600-4380 mg/lt
Total suspended solids	220-800 mg/lt
Total Nitrogen	10-40 mg/lt

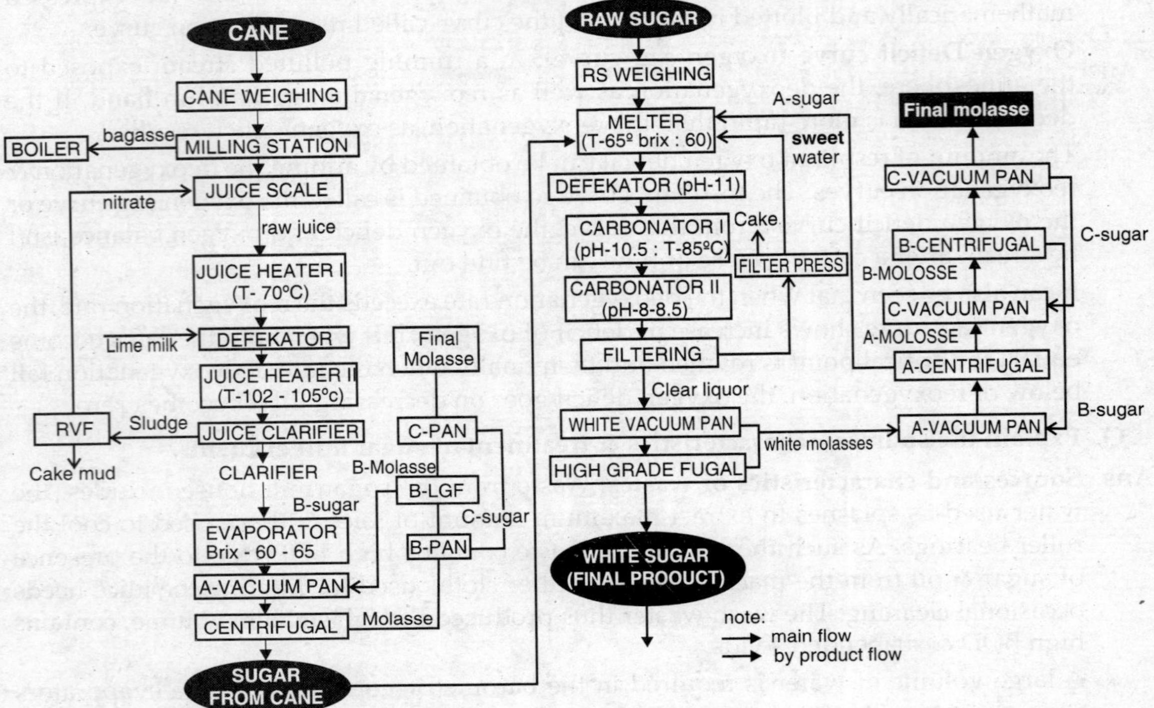

Fig.11: Schematic diagram of sugar processing

Treatment of the wastewater: Like any other industry the pollution low in sugar mills can also be reduced with a better water and material economy practiced in the plant. Judicious use of water in various plant practices & it recycle, wherever practicable, will reduced by recycling cleaning of floors or floor washings using controlled quantity of water will also reduce the volume of waste to certain extent.

The organic load of the waste cam only be reduced by a proper control of the operations. Over loading of the evaporates & the vacuum pans and the extensive boiling of the syrup leads to a loss of sugar through condenser water & this in turn increases both volume & strength of the waste effluent. Disposal of the effluent on land as irrigation water is practiced in many sugar mills, but it is associated with odor problems.

Anaerobic treatment of the effluent using both digesters and lagoons has been found to be very effective & economical. A BOD reduction of both 70- 88% was observed in pilot plant study with an anaerobic digester, where BOD loading was 0.65-1.79 kg/m^3/day with a detention time of 2- 2.4 days at a controlled temperature of 37 ^{0}C.

The effluents of the anaerobic treatment units are found to contain sufficient nutrients (nitrogen & phosphorous) as such further reduction of BOD can be accomplished in aerobic waste stabilization ponds.

Where sufficient land is available a two stage biological treatment with anaerobic lagoons followed by aerobic waste stabilization ponds is recommended for Indian conditions. The mill effluent however is to be pretreated primarily in bar screens & grease trap. The flow sheet for the complete treatment for the sugar mill waste is shown in the fig, it is expected that the BOD reduction in the anaerobic process will be in order of 60% while overall BOD reduction may be in the order of 90%.

Q. Explain the sources and characteristics of tannery wastewater.

Ans. Sources and characteristics of tannery wastewater: The waste water originates from the all operations in the tanning process. The waste may be classified as continues flow water, and intermittent flow waste. Continues flow waste consists of wash water after various process and comprises of large portions of the total waste and are relatively and less polluted then the other one. Spent liquors belonging to soaking, liming and batting, pickling, tanning and finishing operation are discharged intermittently. Although small in volume, they are highly polluted and contain varieties of solute and organic and inorganic substances.

The spent and soaked liquor contains soluble proteins of the hides, dirt and large amount of common salt where salted hides are process. This spent liquor under goes putrefaction very rapidly as it offers a good amount nutrient and favorable environment of bacterial growth. The growth of pathogenic - anthrax bacteria in this waste is also reported.

Flowsheet of tanning industry is shown below.

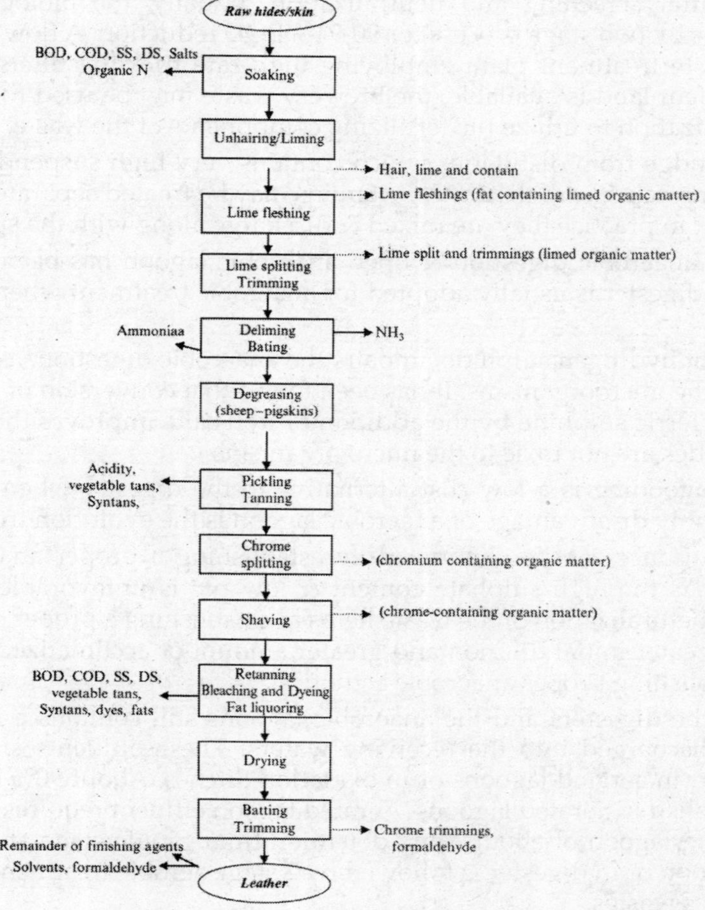

Fig.12: Flowsheet of a tannery

The spent lime liquor contains dissolved and suspended lime, colloidal proteins and their degradation products, sulphides emulsified fatty matters and also carrying a sludge composed of unreacted lime, Calcium sulphide and calcium carbonate. As such, the spent lime liquor as a high alkalinity and moderate BOD and high ammonia nitrogen content.

The spent bate liquor contains high amount of organic and ammonium nitrogen due to the presence of soluble skin proteins and ammonia salts used in bating.

The vegetable ton exact contains tannins and also non tannins. Tannins are of high COD but relatively low BOD value. While non tannins including inorganic salts, organic acids and salts and sugar are of high BOD and COD. The spent vegetable tanning liquor is the strongest individual waste in the vegetable tannin having the highest BOD and very strong dirty brown color.

The spent pickling and chrome tanning waste comprises of as small volume, having a low BOD and contains traces of proteins impurities, sodium chloride and mineral acids and chromium salts.Chromium is highly toxic to the living aquatic creatures.

Q. Discuss the origin and treatment of distillery wastewater.

Ans. Brewery wastes being comparatively less strong can be treated by aerobic biological treatment, after screening and neutralization. Usually, the biological treatment is accomplishes by two stage process for 90-94% BOD reduction. A flow sheet of one such brewery waste treatment plant employing high rate trickling filters is shown in fig. When sufficient land is available, the brewery waste may be used for broad irrigation after neutralization to utilize the fertilizing components of the waste.

The yeast sludge from distilleries which contains very high suspended solids & BOD & in rich in proteins, carbohydrates, vitamins may be treated separately for by product recovery. But in practice they are mixed & discharge along with the spent wash.

Both closed anaerobic digestion & open anaerobic lagoon has been tired in India. A single stage digester is usually adopted for anaerobic treatment when land available is limited.

Production of hydrogen sulphide, impairs the anaerobic digestion, as soluble sulphide are toxic to the microorganisms. It has been found that conversion of soluble sulphides to insoluble ferric sulphide by the addition of iron salts improves the condition as the ferric sulphides are not toxic to the micro organisms.

Anaerobic lagooning is a low cost alternative to the digester when land is available plenty. The only disadvantage of anaerobic lagoon is the evolution from the ponds.

The odor nuisance can be eliminated by establishing a proper anaerobic activity in the lagoon. As the high sulphate content & low pH is unfavorable for the methane fermenters, neturalization of the waste helps in establishing a proper condition for their activity. A greater initial dilution and greater amount of acclimatized sludge may also help in establishing proper anaerobic activities.

Effluent of the digesters and the anaerobic lagoons still contains a high BOD, which cannot be discharged into the receiving waters. These effluents are successfully be treated either in aerated lagoons or in oxidation ditches. About 90% BOD removal can be accomplished in aerated lagoons. Aerated lagoon effluent requires further treatment in a polishing lagoon of about 24hrs detention time. Single stage anaerobic treatment either in lagoon or in digester is followed by 2 stage aerobic treatment is also proposed for distilleries wastes.

Q. Explain the sources and characteristics of waste water from pulp &paper mill.

Ans. The volume & characteristics of the waste depends on the type & manufacturing process adopted & the extent of reuse of water employed in the plant. In the pulp making phase, the chipped cellulose raw materials are digested with different chemicals in one tank under high temperature & pressure. The process thus loosens the cellulose fibers & dissolves the lignin & resin & other non cellulosic materials in the raw material. The kraft process or sulphate process of pulp making uses sodium, sulphate, sodium hydroxide & sodium sulphide as the above mentioned digestive chemicals.

Another process of pulp making known as sulphide process uses magnesium or calcium bisulphate & sulphurous acid as the digestive chemicals. The alkali process uses sodium hydroxide or lime for pulp making.

The spent liquor produced by the above process of digestion is known as Black Liquor. This is not only very rich in lignin content, but also contains a large amount of utilized chemicals. Therefore this black liquor is treated separately for the recovery. So while the entire quantity of the liquor is up hide process makes the colored waste from the section, in the kraft process same is produced due to the leakages, spillages or overflow only from the digester. The cellulosic fiber after being separated from the black liquor is washed & then partially dewatered.

A concentrated wash water may be sent for chemical recovery) in the kraft process. The dilute wash water form the waste water. This volume of waste water is known as Brown stock wash or Unbleached Decter waste.

The washed cellulosic fibers are then sent for the bleaching in 3 stages, where chlorine, caustic & hypochloride are used in successive stages. Waste waters from the first & the last stages are light yellow in color while that from the caustic extraction stage is highly colored. The bleached pulp is then sent for the paper mill. In the paper mill, the pulp is disintegrated & mixed with various filter materials like alum talc, etc and dyes on an along shaped especially pulp is refined in a machine. The refined pulp is then diluted to proper consistency for paper making and passed through a screen to remove lumps or knots. Now this pulp is carried by a travelling belt of fine screen to a series of rolls, where the final product, the paper is produced. The drained water often called as White water forms the waste water from the paper mill section. This waste water contains fine fibers, alum, talc, etc. Usually the fibers are recovered from this waste; the treated liquid is reused for the wet chipping process.

The black liquor of the kraft process concentrated by evaporation and then incineration with the addition of sodium sulphate. The organic like lignin, resin., etc are burnt off & the smell is dissolved in water. The resulting liquid is known as green liquor ; lime is now added to this liquor resulting in the formation of white liquor & lime mud containing chiefly digesting chemicals & is sent for use in digester. The lime mud is calcined (by burning) to form calcium oxide, which is reused to recaustised other green liquors in to the white liquors. Beside those mentioned about, the small volume of waste water is also produced when the bark is removed from the raw wood & the latter is reduced to chips by wet process.

A schematic diagram of pulp & paper industry is shown below.

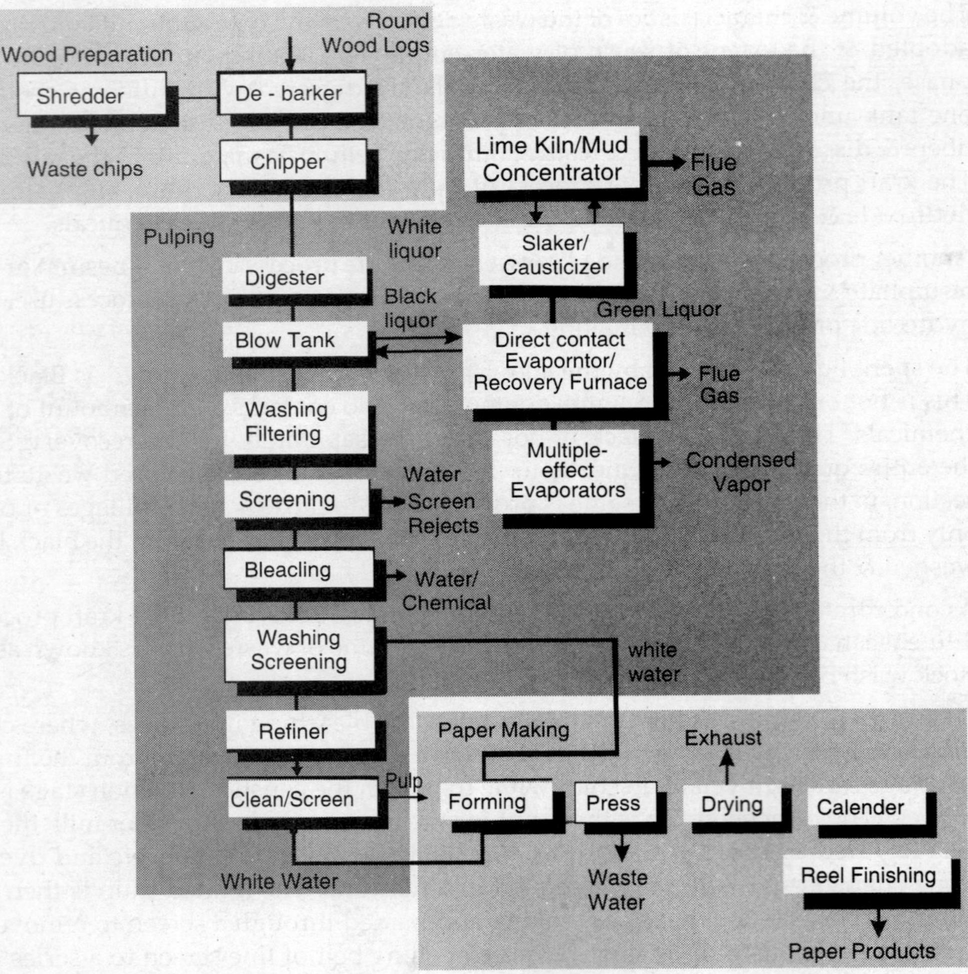

Fig.13: Schematic diagram of a pulp & paper plant

Some very toxic waste materials are also generated during the process of chemical recovery from black liquor. Toxic materials like Dimethyl sulphide, methyl mercaptan etc also comes out with digester gases & forms a colorless waste water after condensation.

Characteristics of pulp & paper mill wastes: The chemical composition of the wastes depends on the size of the plant, manufacturing process and to a great extent on the material economy practiced (by way of recovery of chemicals and fibers) in the plant. In most of the small paper mills in India, the chemical recovery is not practiced due to economical reasons. As such, the pollution potential of the waste of smaller mills is higher than that of the larger mills. Generally, the pulp and paper mills wastes are characterized by very strong color, high BOD, high suspended solids, and a high COD/ BOD ratio.

It may be noticed that too much difference is percent suspended solids contribution from different sections b/w the large and small mills arises due to following reasons.

Large mills produces large amount of lime mud and without being calcined, it is discharged as a waste. This lime mud in the large mills contributes 86.5% of the total suspended solids in the effluent. As such, the percent suspended solids from the digester section; bleaching paper mill section etc. assumes a very low value. On the other hand, no question of lime mud rises in small mill, as it does not have the chemical recovery plant. The waste volume percentage from the digester section is also higher in the case of small mill as the entire quantity of black liquor is wasted in such mills.

Q. Discuss the characteristics of wastes of pharmaceutical industry.

Ans. Pharmaceutical industry produces varied type of products. They range from vitamins, synthetic drugs to antibiotics. The raw materials used are includes both organic and inorganic compounds. Some of the pharmaceutical plants do not generate any liquid effluents, while some others discharge little quantities of strong waste & others let out larger volumes. Due to these wide variations a generalization cannot be drawn on the effluents of pharmaceutical industry.

Most of the antibiotics such as penicillin. Streptomycin, lysine, sulfaquinazoline, nicarbazine & vitamins such as B1, B2, B12 and many steroids are prepared in the fermentation. The most waste produced in the fermentation process is the spent beer liquor. The spent beer liquor is the fermented broth remaining after the recovery of antibiotics and other valuables. It contains large amounts of organic materials, proteins and other nutrients and consequently the BOD of these effluents is abnormally high.

Five main pharmaceutical wastes and their characteristics are as follows:

1. Strong fermentation beers (small in volume but having 4000 to 8000 mg/L BOD)
2. Inorganic solids (waste slurry with little BOD)
3. Washings of floor and equipment (large percentage of total volume and BOD from 600 to 2500 ppm)
4. Chemical waste - solution or solvents which exert a substantial BOD when diluted with other wastes
5. Barometric condenser wastewater - resulting from solids and volatile gases being mixed with condenser wastewater causing 60 to 120 ppm BOD.

The antibiotic wastes impart objectionable odors to stream and inhibit biological population and action. If they are discharged into sewer, they must be properly diluted, otherwise they affect the sewage treatment.

The volume and composition vary from unit to unit. Approximately 1000 to 3000 litres of waste will be discharged per 100 kg of products manufactured. No specific conclusion on the characteristics of the effluents can be drawn. In general, they are either highly acidic or highly alkaline and possess a high BOD and COD. Some of the effluents contain toxic substances like cyanides.

If the wastes are discharged into stream, they deplete the dissolved oxygen immediately. These are corrosive due to their high acidity/alkalinity. Further, some of the substances present in them are toxic to aquatic life.

Q. Discuss the wastes produced by food processing industry (Breweries, wineries etc.)

Ans. While breweries and wineries produce beer and wine respectively, a large no of products are obtained in distilleries. The range of products from distilleries includes industrial alcohols, rectified spirit absolute alcohol, spirit, beverage alcohol etc. But two things are common in all the products mentioned above.

1. All the above products are obtained through the biochemical process of fermentation by yeast using carbohydrates as raw materials and
2. All the products contain ethyl alcohol in different proportions.

In all the industries mentioned above are all characteristically of high BOD and they present a threat to the environment when discharged into the water sources or to the land without treatment.

Making of beer essentially consists of two stages:

1. Preparation of molt from grains like barley.
2. Brewing the barley.

In the malt making the barley grain are steeped (soaked) to bleach out color and then made to sprout under aerobic conditions. The grain malt is then dried and stored after screening the sprouts out.

The malt from the malt house is then transported to the brewing section, where the wart the medium for fermentation is prepared by mixing the course grain malt with hot water and by transforming the starch to sugars boiling in hops. The wort is then inoculated with a prepared suspension of yeast which common the sugar to alcohol.

When the fermentation is complete, the yeast and mall residue is filtered out and finally the beer is carbonated before packing for sale.

As the flavor of the product is of prime importance selection of raw materials & control of process is done accordingly.

Q. Discuss the origin and treatment of dairy waste.

Ans. The liquid waste from a large dairy originates from the following sections or plants. Receiving station bottling plant, cheese plant, butter plant, casein plant, condensing plant, dried milk plant & ice cream plants. Wastes also come from water softening plant & from bottle and washing plants. At the receiving station, the milk is received from the farmers and is emptied into large containers for transport to bottling of other processing plants. The empty is rinsed, washed and sterilized before returning to the farmers. At the bottling plants the raw milk delivered by receiving station is stored. The processing includes cooling, filtration, clarification, pasteurization and bottling.

In the above sections, the liquid wastes originate out of rinse & washings of bottles, cans & equipments & thus contain milk drippings and chemicals used for cleaning containers & equipments.

In a cheese plant, the milk (whole milk or skimmed milk) is pasteurized and cooled and placed in a vat, where a starter (lactic acid producing bacterial culture) and rennet are added. This separates the casein of the milk in the form of curd. The whey is then withdrawn and the curd compressed to allow excess to drain out. Other ingredients are now added and the cheese blocks are cut and packaged for sale. Waste water from this plant includes mainly the discarded whey and wash water used for cleaning vats, equipments floor.

In the creamery process the whole milk is pre heated to about 30^{0}C to separate the cream from the milk. In a butter plant, the cream is pasteurized and may be ripened with a selected acid and a bacterial culture. This is then churned at a temperature of about 7-10^{0}C to produce butter granules. At the proper time the butter ilk is drained out of the churned & the butter is washed & after standardization packaged for sale. Butter milk and wash waters used to clean the churns and small quantity of butter comes out as waste from the butter plants.

The skimmed milk may now be sent for bottling for human consumption or for further processing in the dairy for other products like non fat milk powders. Milk powders are produced by evaporation followed by drying by either roller process or spray process. The dried milk plant wastes consist of chiefly of wash waters used to clean containers & equipments.

The scoured or spoiled milk and some time the skimmed milk are processed to produce caseins used for preparation of some plastics. The process involves the coagulation & precipitation of the casein by the addition of some mineral acids. The wastes from this section include whey, washings and the chemicals used for precipitation. Very large dairies also produce condensed milk & ice creams.

In addition to the wastes from all the above milk processing units, some amount of uncontaminated cooling water comes as waste. These are usually recirculated. The dairy wastes are very often discharged intermittently. The nature & composition of the waste also depends on the types of products produce & the size of the plants.

Due to low COD, BOD ratio the dairy wastes can be treated efficiently by biological processes. Moreover, these wastes contain sufficient nutrients for bacterial growth. But for economical reasons, attempt should be made to reduce the volume & strength of the waste. This can be accomplished by:

1. Prevention of spills, leakages & dropping of milk from the cans.
2. By reducing the amount of water for washes.
3. By segregating the uncontaminated cooling water and recycling the same.
4. By utilizing the buttermilk and whey for the production of dairy bi products of good market value.

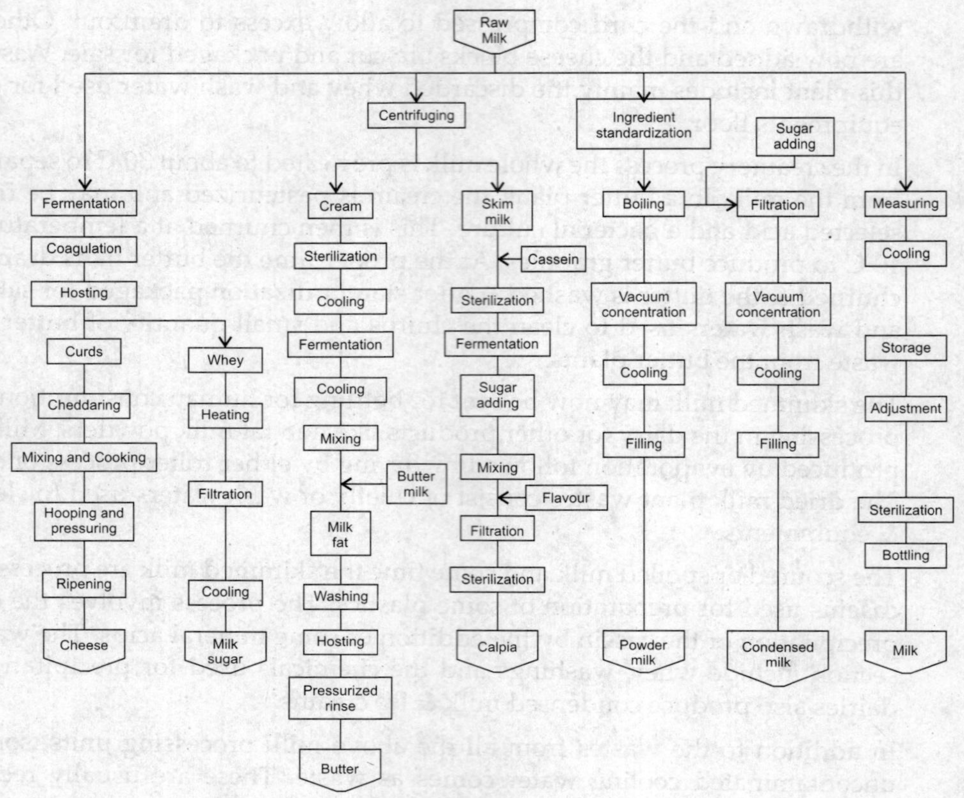

Fig.14: Treatment of dairy waste

Due to the intermittent nature of the waste, it is desirable to provide equalization tank, with or without aeration, before the same is sent for biological treatment. A provision of grease trap is also necessary as a pretreatment to remove fat & other greasy substances from the wastes. A aeration for a day not only prevents the formation of lactic acid, but also reduces the BOD by about 50%.

Both high rate trickling filters & activated sludge plants can be employed very effectively for a complete treatment of the dairy waste. But these convention methods involve much maintenance, skilled methods like oxidation ditch, aerated lagoon, waste stabilization pond etc., can be employed with simpler type of equipments & less maintenance.

Use of the dairy waste for irrigation after primary treatment in an aerated lagoon may be a good answer for the disposal of dairy wastes.

Q. Briefly describe the effect of dairy waste in receiving streams.

Ans. As the dairies are usually situated in rural areas or in small towns, the question of discharging the dairy waste in to the sewers does not arise.

As observed from waste is basically organic in nature. This is also slightly alkaline when fresh. When these wastes are allowed to go into the stream without any treatment, a rapid depletion of DO of the stream occurs along with growth of sewage fungi covering the entire bottom of the stream. The waste is said to carry occasionally, the bacteria responsible for tuberculosis (TB). Though alkaline in fresh condition the milk waste becomes acidic due to the decomposition of lactose into lactic acid under anaerobic condition.

The resulting condition precipitates casein from the waste, which decomposes further into a highly odors black sludge. At certain dilution the dairy waste is found to be toxic to fishes also.

Q. Discuss the effect of untreated/partially treated wastes from cotton textile mill on the receiving stream.

Ans. Effect of textile mill waste on receiving streams/sewers: If the mill waste water is discharged into streams, it causes depletion of DO of the stream. This is due to the settlement of the suspended substances and subsequent decomposition of the same in anaerobic condition. The alkalinity and toxic substances like sulphides & chromium affects the aquatic life and also interferes with the biological treatment processes. Some of the dyes are also found to be toxic. The color often renders the water unfit for use for side. The presence of sulphides makes the waste corrosive particularly to concrete structures. All treatment plants should be planned giving serious consideration for the reduction of waste volume & strength, through process of chemical substitution, chemical recovery & recycling of water.

The pollution load from a textile, mill is dealt with operations like segregation, neutralization, equalization, chemical precipitation, chemical oxidation & biological oxidation. Several chemicals are used to reduce the BOD by chemical coagulation such as alum, ferric sulphate, ferrous sulphate & ferric chloride, lime or H_2SO_4 is used to adjust pH in this process. The dye waste may be economically treated by biological methods prior equalization, neutralization & chemical oxidation.

Q. Discuss the effect of untreated/partially treated wastes from cotton textile mill on the receiving stream.

Ans. If the mill waste water is discharged into streams, it causes depletion of DO of the stream. This is due to the settlement of the suspended substances and subsequent decomposition of the same in anaerobic condition. The alkalinity and toxic substances like sulphides & chromium affects the aquatic life and also interferes with the biological treatment processes. Some of the dyes are also found to be toxic. The color often renders the water unfit for use for side. The presence of sulphides makes the waste corrosive particularly to concrete structures.

All treatment plants should be planned giving serious consideration for the reduction of waste volume & strength, through process of chemical substitution, chemical recovery & recycling of water. The pollution load from a textile, mill is dealt with operations like segregation, neutralization, equalization, chemical ppt, chemical oxidation & biological oxidation. Several chemicals are used to reduce the BOD by chemical coagulation such as alum, ferric sulphate, ferrous sulphate & ferric chloride, lime or H_2SO_4 is used to adjust pH in this process. The dye waste may be economically treated by biological methods prior equalization, neutralization & chemical oxidation.

A Composite waste, when free from toxic substances may be treated as efficiently as domestic sewage, as most of the textile mill wastes contain sufficient nutrients like nitrogen & phosphorous. Trickling filters, activated sludge process & stabilization ponds have been effective in treating textile mill wastes. Extended aeration is found to be very effective in treating strong wastes even without equalization & pretreatment.

Q. Describe the manufacturing process of cotton textile mill and waste generation.

Ans. Manufacturing process: An integrated cotton textile mill produces its own yarn from the raw cotton. Production of yarn from raw cotton includes steps like opening & cleaning picking, carding, and drawing spinning, winding & warping. All these sequences are dry operations and as such do not contribute to the liquid waste of the mill. The entire liquid waste from the textile mills comes from the following operation of slashing (sizing), scouring, desizing, bleaching, mercerizing, dyeing & finishing.

In slashing the yarn is strengthen by loading it with starch or other substances wastes originates from the sections due to spills & floor washings. The substitution of low BOD sizes (such as carboxy methyl cellulose) for the high BOD of the mill effluent by 40 to 90%.

After slashing, the yarn goes for weaving. The prepared cloth now requires scouring & desizing to remove natural impurities and the slashing compounds. Enzymes are usually used in India to hydrolyze the starch, acids may also be used for this purpose. Caustic soda, soda ash, detergents etc. are also used in this section. Bleaching operations use oxidizing chemicals like peroxides & hyper chloride to remove natural coloring material. The section contributes about 10% of the total pollution load.

Mercerizing consists of passing the cloth through 20% caustic soda solution. This process improves the strength elasticity luster & dye affinity. Waste from this section is recycled after sodium hydroxide recovery. Negligible waste which may come out of this section contributes little BOD but a high degree of alkalinity.

Dyeing may be done in various ways, using different types of dyes and chemical classes of dyes include Vat dyes, developing dyes etc. color from the dyes vary widely and although these are not usually toxic, they are treated separately. Thickened dyes are used for probing and subsequent fixation. After fixation of the prints, the fabric is given a thorough wash to remove the unfixed dyes.

The finishing section of the mill imparts various types of chemicals are used for various objectives. These include starches, dextrines, natural & synthetic waxes, synthetics etc. Therefore a composite waste from an integrates cotton textile mill may include the following organic & inorganic substances starch, carboxyl methyl cellulose, sodium hydroxide, detergents, peroxides , hyperchloride dyes & pigments, sodium gums, dextrines, waxes, sulphides, soap etc. Depending on the process & predominant dye used, the characteristics of the mill waste varies widely.

Flow diagram of a cotton textile mill is shown below.

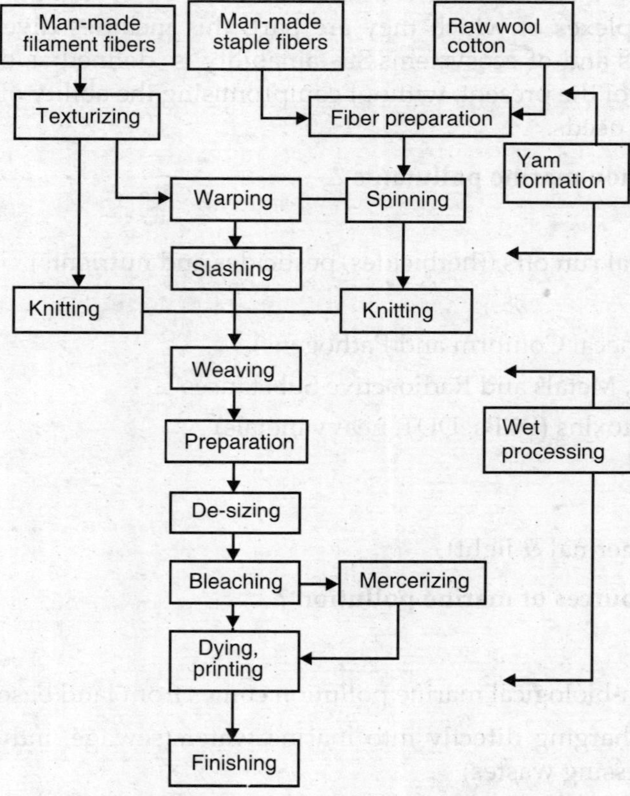

Fig.15: Flow diagram of a cotton textile mill

Q. What is marine pollution ?

Ans. The introduction by man, directly, or indirectly, of substances or energy to the marine environment resulting in deleterious effects such as: hazards to human health, hindrance to marine activities, impairment of the quality of seawater for various uses and reduction of amenities. It does not include natural processes like volcanic eruptions or earthquakes.

Q. Why marine pollution is a concern ?

Ans. Marine pollution means introduction by man of substances or energy into the marine environment (including estuaries) resulting in such deleterious effects as harm to living resources, hazards to human health, hindrance to marine activities including fishing, impairment of quality for use of seawater and reduction of amenities.

We should care about the continuing environmental degradation of our oceans and coastal areas because it is detrimental to human health, economic development, climate and our planet's store of biodiversity. It is interfering with the sustainability of environment and its resources.

Biodiversity or Biological Diversity" means the variability among living organisms from all sources including terrestrial, marine & other aquatic ecosystems and the ecological complexes of which they are part; this includes diversity within species, between species and of ecosystems.Sustainability is defined as the development that meets the need of the present, without compromising the ability of future generation to meet their own needs.

Q. What are common marine pollutants ?

Ans.

- Agricultural run offs ((herbicides, pesticides and nutrients)
- Sediments
- Sewage (Faecal Coliform and Pathogens)
- Chemicals, Metals and Radioactive Substances
- Persistent toxins (PCBs, DDT, heavy metals)
- Oil
- Plastics
- Energy (Thermal & light)

Q. What are the sources of marine pollution ?

Ans. Land sources

- 80% of non-biological marine pollution comes from land based activities
- pipes discharging directly into marine waters(sewage, industrial, chemical and food processing wastes)
- River flows into the sea carry pollutants from the entire catchment area.

From Air

- Global atmospheric inputs to the sea from air discharges.

Oil spills and offshore sources

- Oily discharges from ballast water and bilge water during routine ship operations and illegal dumping of solid waste.
- Designated dumping grounds at sea.
- Accidental spills from Ships carrying hazardous substances, oil, gas etc.

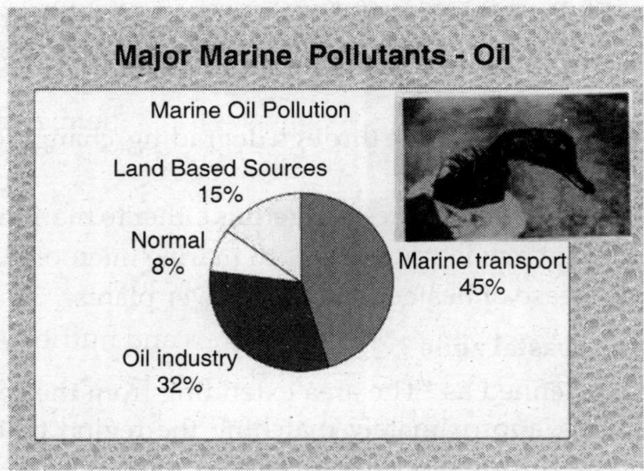

Q. What are the dangers of major marine pollutants ?

Ans.

Major Marine Pollutants - Solid Waste

- A large portion and great danger is nonbiode
- 46,000 prices of floating plastic/sq mile of ocean surface of the N.E U.S. coast.
- Sea turtles mistake plastic bags for jellyfish and die from internal blockages.
- Seals and sea liens starve after being entangled by nets or muzzled by six-pack rings (decomposition time 400 years).
- Plastic debris kill 100,000 marine mammals and 2 million sea birds die annually

Q. What are the effects of marine pollution ?

Ans. Effects are as follows:

- Damages marine life to a large extent, for salt-marsh plants, oil slicks can affect flowering, fruiting and germination.
- Coral reefs
- If liquid oil contaminates a bird's plumage, its water-repellent properties are lost, drown, die
- Drill cuttings dumped on seabed create anoxic conditions & result in the production of toxic sulphides in the bottom sediment thus eliminating the benthic fauna.
- Fish and shellfish production facilities can also be affected by oil slicks. Commercial damage is tainting: imparts an unpleasant flavor to fish and seafood & is detectable even at extremely low levels of contamination.

Q. what are the categories of man made effect on marine environment ?

Ans. Man's effect on the marine environment may be divided into the following general categories:

1. Alteration of bottom substrate through degrading, changing of shoreline structures, and filling.
2. Introduction of toxic substances dangerous either to marine life or to human beings.
3. Release of sewage rich in nutrients with marine microorganisms.
4. Heating and release of heated water by power plants.

Q. How do we define coastal zone ?

Ans. The coastal zone is defined as "The area extending from the coastal plains to the edge of continental shores, approximately matching the region that has been alternatively flooded and exposed during the sea level fluctuation of the late quaternary period".

In the earth's surface, 18% is represented by coastal zone, which provide space for 60% of the world's human population. 70% of the world cities with population exceeding 1.6 million are located in this zone. 90% of the world fish catch is obtained from coastal zone. The coastal zone is only 8% in the hydrosphere, but responsible for 18-33% of the total primary production. The biological wealth of this zone is very high as it serves as feeding, nursery and spawning grounds with rich biodiversity.

Q. What are the Sources of Marine pollution particularly in coastal zone ?

Ans. The main sources of marine pollution are atmosphere, river runoff, agriculture, livestock, urban runoff, automobiles, land clearing, sewage outfall, industrial waste etc. In general the coastal zone receives pollutants from major ways viz., atmosphere, riverine and glacier. Also anthropogenic activities serve as the geological agents by ways of discharging the effluents through piped outfalls, direct dumping, operation of ships etc. Frequently preservatives have been intentionally used assuming that they are relatively harmless to the cultured species. These include antifoulants, of which the broad ecological effect of tributyltin (TBT) is a good example.

Pollution may arise from point source like a single sewer pipe or factory wastewater outfall, or it can arise from a variety of geographic points otherwise called as non point source. In the point source pollution the concentration of the substance or the intensity of the effect (e.g. temperature near cooling water outfalls of power plants) should decline with increasing distance from the point source. The dissipation or decline depends on the nature of substance or factor, water currents and the sedimentary environment and rate of introduction of the substance or factor. In contrast to this, the non-point source effects cannot be attributed to any single spot. Runoff arises due to rain is an example of a non point source. The toxic substance and other fertilizer based nutrients may spread over the coast. However, these sources are more difficult to manage, because the source is widespread over a spectrum of earth surface.

Q. What the indicators of marine pollution ?

Ans. Many organisms serve as indicators of marine pollution. For example, the ecological shift in the succession of phytoplankton communities from diatoms to dinoflagellates due to metal pollution & thermal pollution. The poor hatching of eggs and soft calcium shell of birds and mass mortalities of birds due to failure in the earlier metabolites by the interference of pesticides, large scale mass killing of birds due to casting, asphyxiation & smothering by oil pollution, reproductive failure in mammal are also some of the indicators of severe pollution.

Usually the pollutants exceeding threshold limits combined with environmental variables such as temperature, salinity, pH, dissolved oxygen, hydrogen sulphide etc., give stress to the organisms. In most cases the migratory organisms such as fishes could avoid the polluted area. Whereas the residential or sedentary organisms are exposed to the pollutants experience the stress conditions.

Q. Discuss The Effect Of Pollutants On Marine Organisms ?

Ans. Pollutants that enter into to the oceans are mostly diluted, however the organisms living in the oceans tend to concentrate the pollutants into their body by various mechanisms, like adsorption, absorption, ingestion etc. The concentration of pollutant increases with higher trophic levels, i.e. from primary producers to the tertiary consumer. The threshold limit of organisms to the pollutant concentration increase with increasing trophic levels. Various mode of pollution transport in organisms includes bioconcentration, bioaccumulation , biomagnifications and biodegradation.

Bioconcentration

The bioconcentration is the process by which a contaminant such as oil is directly taken up (i.e., by absorption only) from water and is accumulated to levels greater than those found in the surrounding water.

Bioaccumulation: Marine organisms can accumulate chemical species in amounts far exceeding their surrounding (sea water or sediment) concentrations. The bioaccumulation is the process by which a contaminant is taken up by organisms directly through the physical exposure pathway or through consumption of food containing the contaminated substance.

Bioaccumulation incorporates the concepts of bioconcentration and biomagnification. The term 'concentrator factor' is used to quantify the bioaccumulation and may be defined as the concentration of the chemical species in the organisms or one of its components divided by the seawater or surrounding concentration in the environment from which the organism was living. Organisms may take up the chemical directly from seawater or from their food. Thus, depending upon the composition of the food, concentration factors can vary from one environment to another for a specific organisms with a given chemical.

Biomagnification: Biomagnification is the increase in tissue concentrations of a bioaccumulated chemical substance such as oil as it passes up though the trophic levels.

Biodegradation: Biodegradation is the process of degradation of a contaminant such as oil because of its use as a food a source for certain microorganisms. This process is limited to a great extent by temperature, nutrients and oxygen availability.

Q What are the types of pollutants discharged into marine environment ?

Ans. There are two types of pollutants. They are biodegradable and non-biodegradable pollutants. Most of the materials, which reach the sea, disintegrate either through simple chemical reactions or because of the activities of bacteria and some larger organisms. There are some substances nevertheless which are either extremely stable or else they have a very slow rate of degradation. The biodegradable pollutants are not persistent in the environment (e.g. oil, organic compound in sewage). But the non-biodegradable pollutants are persisting in environment for longer time (e.g. Heavy metals, plastics, nuclear wastes etc.).

In general most of the toxic compounds are considered as Xenobiotics. The term Xenobiotics refers to "bioaccumulative chemical of concern" or "persistent organic pollutants".Some of the compounds are organochlorine compounds ,chlorodanes ,hexachlorobenzene, organophosphorus compounds ,polychlorinated biphenyls (PCBs), butyltin compounds etc.

Q. Discuss the characteristics,sources fate and effect of sewage discharged into ocean/ marine environment.

Ans. Sewage is discharged into the oceans all over the world mostly from urban settlement. Sewage adds to the amount of small particles suspended in the water and contributes large amounts of nutrients.

Characteristics: The raw sewage contains higher concentrations of organic particulate matter, home washings, detergents, small amount of oil, higher concentration of nutrients, toxic heavy metals etc. Also, the sewage contains higher concentration of bacteria and viruses, which includes harmful pathogenic forms. Sewage always has the characteristic of bad odour, irritant and harmful to human & other organisms.

Sources: The main sources of sewage are the coastal outfall located near the cities. The increasing shipping activities also add higher concentration of sewage to the harbour & shipping routes. Moreover many rivers transport sewage from the inner regions of the land.

Fate and effect of sewage: In open coast, the effect of sewage is difficult to detect due to higher mixing up of tidal, wave and current actions. However in semi-enclosed areas like small bays, harbours etc., the effects are devastating such as higher Biochemical Oxygen Demand. The organic load concentration is expressed by means of a quantity called Biochemical Oxygen Demand (BOD) which is the amount of oxygen required by the microorganisms for the decomposition of the organic compounds dissolved in a given volume of sewage.High BOD values indicate high concentrations of organic material and therefore low quantities of dissolved oxygen (DO). Near sewage outflow areas in temperate waters, the benthic invertebrate communities have degraded, Kelp beds have disappeared and diseased fish have become more prevalent. Sewage input in coastal environment normally increases the nutrient concentration of the outfall area. In tropical waters, outflows near coral reefs have caused a bloom of algal species that grows over the coral, and eventually it leads to the death of coral and its associated forms. This some times leads to the change in the ecological succession, that is the coral reef may be replaced by an algal reef.

Q. what are the characteristics,sources,fate and effects of petroleum oil discharged in ocean/marine environment ?

Ans. Oil is discharged in to the ocean in various forms as crude oil and as separate fractions.

Most of the oil fractions are biodegradable.The spilled oil causes more devastation than sewage in the marine environment.

Characteristics of oil: Petroleum oils are very complex mixtures of large numbers of hydrocarbons, containing also small amounts of sulphur, nitrogen, oxygen and different metals combined with the hydrocarbons.

Sources of oil pollution in marine environment

The main ways in which oil enters in to the sea are:

- Natural release
- Tanker operations
- Oil tanker and other ship accidents
- Operation of ships other than tankers
- Offshore oil drilling and production platforms
- Ship-shore oil terminal operation
- Refinery operations
- Discharge of oil products on land and subsequent discharge by runoff

Source of marine oil release:

- 45% is due to marine transportation
- 32% is due to routine loading, discharging and flushing of tanker ships
- 8% is due to natural seeps
- 15% by other means

Fate of oil in the sea: As the oil enters into sea it is exposed to processes, which modify it both physically and chemically. This process is called weathering. As the oil tends to float on the surface of sea, the prevailing wind, wave, tide and current conditions make the oil to spread to a wider spectrum of area from the point of release. The majority of crude oil forms sticky layers on the surface, which prevents free diffusion of gasses, clogs adult organisms feeding structures and decreases the sunlight available for the photosynthesis. Volatile components of an oil spill eventually evaporate into the air, leaving heavier tars behind which forms into tar balls, which fall to the bottom and may be assimilated by bottom organisms or incorporated into sediments.

Weathering: The alteration of physical and chemical properties of discharged oil through a series of natural processes, which begin when the discharge occurs and continue until the oil is removed. Major processes, which contribute to weathering, include evaporation, dissolution, photooxidation and emulsification.

Very little fraction of miscible fraction of oil is dissolved in the seawater and other fractions are immiscible.

Evaporation is the single most important weathering process for the first several days of an oils discharge. Evaporation results in the loss of lighter fractions through volatization from petroleum products (e.g. benzene, naphthalene).

The process whereby oil is incorporated into the medium of concern, usually water in the form of small droplets. Emulsions in water can either be oil-in-water or water-in-oil, formed as a result of wave actions. There is some solution of hydrocarbons in the water and this also is more marked in the more volatile constituents. The oil-in-water leads to the wide dispersion of fine oil droplets in the sea. The water-in-oil composition leads to the formation of semisolid masses known as 'chocolate mousse'.

Tar ball formation: If silt or sediment is present, oil may be absorbed on to it and may fall with it to the seabed. This effect can be seen as tar ball formation, which is mostly observed over the seabed of oil spilled area.

Biodegradation of oil: Some of the oil constituents are naturally attached by bacteria present in the sea. The simpler compounds, the normal paraffins, which have straight chains of carbon atoms, are most easily broken down and consumed. The other higher and complex molecular weight fractions are broken down very slowly.

Effect of oil spill on marine organisms: When there is a major oil spill the degree of effect depends on many factors such as:

- Type oil
- Quantity of oil
- Phsiography of the area
- Weather conditions at the time of spill
- Biodiversity of the area
- Season of the spill
- Previous exposure of the area to oil
- Exposure to other pollutants and
- Treatment of oil spill.

As the oil is of floating type it tend to spread and eventually reach the coastline. In many coastal areas, animals and plant situated between high & low water marks were heavily coated with oil and smothered. The exposure of organism to oil adds foreign smell in its body & tissues and the process is called tainting. The shellfish can be tainted either by deposition of oil on their shell or by ingestion of naturally emulsified oil during normal feeding. The local fish population in the immediate vicinity of oil discharge can acquire a marked taint from oil components dissolved in the water.

Almost all groups of organisms like invertebrates, vertebrates like fishes, reptiles, mammals etc., are affected by oil. When the plumage of a diving birds get oiled, the water repellent properties of the feathers are destroyed, they loose the insulation provided by the feather and its flying capacity. This leads to a condition called hypothermia, subsequently the bird may die. Birds may also ingest the oil. All these cause large scale death of marine birds. It is estimated that between 200,000 and

500,000 marine birds are killed annually by routine releases of oil from tankers.

The tainting can be cleared by rearing the affected organisms in clean water for a considerable period.

In shellfishes, during the process of moulting, the tainted shells are removed. Also, the bird's feathers and some animal's body can be cleaned with detergent. In this type of clean up, the chances of recovery is observed in many cases. Most forms of marine life recover within about five years, as crude oil is not as highly toxic as it is biodegradable.

The polynuclear aromatic hydrocarbons (PAHs) in particular 3, 4 benzopyrene are well known carcinogens (cancer producing). These compounds are present in crude oil in small amounts. The PAH assimilated by marine organisms can be passed upwards through food chain towards man. This will lead to serious concern of human health.

Environmental degradation due to oil: The environmental degradation of oil includes the destruction of algal beds, spreading of oil over rocky shores, incorporation in beach sediment and subsequent loss in scenic beauty.

Q. Discuss the sources &effects of various heavy metals on human beings and marine organisms.

Ans. Normally the metals having the atomic weight more than twenty is considered as heavy metals. The term "heavy metals" is used to denote elements with specific weight higher than those of Iron (Fe) and includes mainly Lead (Pb), Cadmium (Cd) and Chromium (Cr),Zinc (Zn) ,Arsenic (As) ,Mercury (Hg) and Copper(Cu).

Sources of heavy metals: Coal combustion, electric utilities, steel & iron manufacturing, fuel oils, fuel additives, and incineration of urban refuse are the major sources of oceanic & atmospheric contamination by heavy metals. Heavy metal contaminated runoff from the land, rain of pollutants from the air, and fallout from shipwrecks pollute the ocean with dangerous metals.

Effects of heavy metals on organisms: In general, the effects of heavy metals toxicity can be classified into three main categories:

- The blocking of essential functional groups of the biomolecules (e.g. proteins, enzymes).
- The displacement of a metal ion from a biomolecule.
- The modification of the structure of biomolecules in space which is very important for their function.

Some heavy metal pollutants and their effect on human beings & marine organisms is discussed below:

Lead: It affects human central nervous system. A poison by ingestion and moderately irritating. A common air contaminant due to use of bad fuel in the auto industry, now being phased out through the introduction of lead-free petrol; also in air in the vicinity of industrial plants using lead where precautions are not taken. It is a carcinogen of the lungs and kidneys. Lead is flammable in the form of dust when exposed to heat or flame. From the point of view of industrial poisoning, inhalation of lead is much more important than is ingestion. Lead is a cumulative poison. Increasing amounts build up in the body and eventually a point is reached where symptoms and disability occur. Lead can cause irreversible behavioral disturbances, neurological damage and other developmental problems in young children and babies.

Cadmium: This metal is toxic to humans by inhalation and other routes. It can enter through ingestion, intraperitoneal, subcutaneous, intramuscular & intravenous routes. It has no known biochemical or nutritional function. Also cadmium is highly toxic to freshwater & marine organisms. It is bioaccumulative through the food chain. Increased exposure can increase risk of lung cancer. Cadmium was responsible for the "Itai-itai" disease in Japan.

Chromium: This metal exists in two forms, i.e. trivalent and hexavalent. Hexavalent chromium in high dosages has been implicated as the cause of digestive tract cancers, cutaneous & nasal mucous membrane ulcers and dermatitis. Certain chromate salts, i.e. calcium chromate are carcinogenic, at least when inhaled.

Zinc: The Zinc is human skin irritant and produces pulmonary system effects. The difficulty arises from oxidation of zinc fumes prior to inhalation or presence of impurities such as cadmium, antimony, arsenic and lead. Zinc however is a needed micronutrient for humans. It is needed in agriculture. It is used in galvanising industry. The real problem is of its association with heavy metals like lead.

Arsenic: Toxic by subcutaneous, intramuscular & intraperitoneal routes and reported to produce systemic, skin & gastrointestinal effects. It is also a human carcinogen and experimental teratogen.

Mercury: The silvery white liquid metal used in common thermometers, is a potent neurotoxin, capable of causing severe brain damage in developing foetuses and mild tremors & emotional disturbances in exposed adults. Implicated in the occurrence of Minamata disease in Japan , largest source of mercury contamination is through aquatic animals in whom it accumulates and gets bio-transferred to organocompounds, e.g. methyl mercury. Mercury and lead poisoning cause brain damage & behavioral disturbances in children.

Copper: It is also dangerous to marine organisms and it has been used in marine anti-fouling paints.

Q. What are the sources of pesticides in the marine environment and their effects on organisms ?

Ans. Pesticides are toxins (chemical compounds) used in agricultural farm to kill the unwanted pest organisms attaching the cultivated plants. A pesticide is any substance or mixture of substances intended for preventing, destroying, repelling, or mitigating any pest. Pests can be insects, mice and other animals, unwanted plants (weeds), fungi, or microorganisms like bacteria and viruses. Though often misunderstood to refer only to insecticides, the term pesticide also applies to herbicides, fungicides, and various other substances used to control pests. A pesticide is also any substance or mixture of substances intended for use as a plant regulator, defoliant, or desiccant.

There are natural and artificial pesticides. The natural pesticides are biodegradable and are mostly derived from the extracts of plants & animals. However, artificial pesticides are man made and produced extensively only after the industrial revolution. These pesticides belong to four main categories.

SOME COMMON KINDS OF PESTICIDES AND THEIR FEATURES

1. **Organochlorine compounds** eg; DDD- Dichlorodiphenyldichloroe ethane
   - Persistent in the environment or decompose very slowly.
   - Half life is 9 -116 years.
   - Accumulate many organisms.
   - Higher concentrations in higher organisms - mammals like dolphins, whales, tortoise etc.
   - Detected in Antarctic penguins.

2. **Organophosphate esters eg; Parathion**
   - Very toxic
   - Much toxic than previous one.

3. **Carbamates Baygon**
   - Decompose rapidly through hydrolysis.
   - Half-life period approximately one week.
4. **Chlorophenyl-acids 2,4,5-T**
   - Hazardous compounds.
   - Used in Vietnam War for defoliation.

Sources of pesticides in Marine environment: Pesticides are the good examples of non-point source pollutants. These toxic materials used in the highland for agricultural and other purposes end up into the sea either through runoff into streams and rivers or having been transported by the wind (dust and aerial spray).

Depending on the type of the chemical, its partial & vapour pressure, evaporation, condensation and deposition occur at various places of the globe leading to their global partitioning. Because of such conditions these chemicals, which originate from lower latitudes, move to higher latitudes. Depending on various conditions, these chemicals migrate in a series of relatively short jumps termed as "grasshoffer effect". Because of this grasshopper effect, concentrations of such chemicals even in the remote parts of the world (Arctic and Antarctic) may rise even if the use of these chemicals are curtailed in source countries.

Effect of pesticides on organisms: Existence of contamination by these chemicals in the environmental media leads to potential impact on almost all species including human. All the organisms, obtain their first dosage of these chemicals from their food and water, while the mammalian species derive these toxins from theirs mothers through milk. While the plants and lower trophic level organisms also accumulate these chemical in their bodies, the effect of these chemicals on these organisms in not openly seen.

Organochlorines have been found to cause microsomal enzyme induction, accelerating body steroids such as estrogens and also to have an effect on the normal reproductive activities of many experimental animals. It is generally known that 5% egg shell thinning occurs at a concentration of 5 mg/g of DDE in the bird egg. It is also found that concentrations of 20-1000 mg/g of DDE in the liver of birds are considered to pose a threat to individual bird reproduction and therefore on the population as a whole.

Q. What Are The Sources & Effects Of Polychlorinated Biphenyls (Pcbs) ?

Ans. PCBs are a series of technical mixtures consisting of many isomers and compounds that vary from mobile oily liquids to white crystalline solids and hard noncrystalline resins.

Sources of PCBs: PCB's usually come from older electrical equipment like transformers, insulations etc.

Effects of PCBs on organisms: It is toxic by ingestion, inhalation and skin contact. A suspected human carcinogen, affects skin and liver. Usual signs of systemic intoxication include nausea, vomiting, weight loss, jaundice, edema and abdominal pain. Severe liver damage may cause coma and death. PCB's (polychlorinated biphenyls) typically cause reproduction problems in most marine organisms. The reduction in sperm motility in human with increasing PCB concentration was observed indicating the possible effect of these chemicals on the reproductive capability of mammalian species.

Q. What are the sources, fate & effects of radioactive materials in the marine environment?

Ans. The world's oceans have been a sink for radioactive waste from the production of nuclear weapons and electric power since long.

Sources: The sources of radioactivity may be natural and anthropogenic. Radioactive waste enters the ocean from nuclear weapon testing (this is banned by international conventions), the releasing or dumping of wastes from nuclear fuel cycle systems, nuclear accidents (eg,Chernobyl accident in USSR in May 1985), ongoing pollution from smaller but more or less continuous leaks from nuclear power stations or nuclear submarines and carriers etc.

Fate of radioactivity in the marine environment: Dumping of high-level radioactive waste is no longer permitted in the ocean, but dumping of low-level wastes is still permitted. Low-level waste contains less radioactivity per gram than high-level waste. High-level wastes usually have longer half-lives. For example, one common high-level waste that is produced by spent nuclear fuel has a half-life of 24,100 years. It has been suggested that contained nuclear waste should be disposed in the deep sea.

Half life: The time required for half the atoms in a given quantity of a radioactive element to disintegrate is called its half life.

Effects of radioactive materials on organisms: Though little is known about the deep sea environment or the consequences of radioactive materials containment leakage & failure, the effects could be devastating. The bioaccumulation process is not associated with the radioactivity properties of a molecule but with its chemical properties. If, however, a bioacumulated molecule happens to be radioactive, its effects on the organism are much more severe.

Radioactive materials bring about two principal sorts of consequences at the organism level:

- Direct toxicity caused by the ionization atoms and molecules of living materials (and especially water) resulting in the production of strong oxidizing agents, which are harmful to living tissues.
- The mutagen activity: Here radiation is a factor, which causes errors during the DNA transcription leading to genome alteration. These errors, called mutations, are in most cases lethal or else lead to teratogenesis when they occur during reproduction. They can also result in cancer when they take place during the division of somatic cells.

Q. What are the sources, fate & effects of nutrients discharged into marine environment ?

Ans. The nutrients are mainly of nitrogen, phosphorus and silicates. Nutrients are used by plants and animals in a complicated cycle, returned to the environment and then re-used. They are essential for the growth of marine primary producers like phytoplankton, seaweed and seagrass. Too much nutrient in the sea & abnormal algal blooms can upset the balance of plants & animals living near the seabed.

Sources of nutrients: Agriculture in the land, aquaculture practices in the coastal areas and release of sewage as outfall are some of the major sources. Fertilizers used on land are washed into the ocean via rivers and streams. Also the aquaculture farms located near the coastal zone also release the wastewater in to the streams & estuaries or directly to the marine environment. Apart from this the sewage released into the coastal areas also contains higher concentration of nutrients.

Fate of nutrients in the sea: The higher nutrient concentrations are always utilized by phytoplankton. The release of excess nutrients into coastal waters leads to the condition called Eutrophication. Eutrophication leads to higher production of phytoplankton called bloom. The death & decay of phytoplankton leads to higher oxygen consumption, which is detrimental to the organisms.

Effects of nutrients: High nutrient concentrations cause phytoplankton blooms such as, red tides, various yellow & green foams, slimes, and slicks. Although algal blooms are natural, a higher frequency of their occurrence in the past years indicates an unhealthy ecosystem. The toxicity of recent blooms is increasing, which can have a direct effect on the organisms that feed on them. Also, phytoplankton naturally contains DMS (dimethyl sulfoxide), which is released from dead phytoplankton into the atmosphere and can be changed to sulfuric acid, which eventually contribute to acid rain.

Q. What is the sources,fate & effects of thermal pollution of the marine environment ?

Ans. In this type of pollution , the normal temperature of marine source water column is increased by about 7-80C from the ambient temperature.

Source of thermal pollution: Mostly the power generating plants along the ocean coastlines use the marine waters for cooling purposes, which leads to heated water expelled into the marine environment.

Fate of increased temperature in the marine environment: The excess temperature usually gets reduced by various ways. As the water has higher specific heat, it takes prolonged time for cooling unless there is an external mechanism to accelerate this. The dilution by surrounding water, evaporation & precipitation may accelerate cooling within a reasonable time.

Effects of thermal pollution: Thermal pollution seems to only effects the communities immediately adjacent to the discharge. In continuous discharge the exposure time is also continuous, which generally result in community shift that is more heat tolerant species may replace the normal biota. The longtime exposure to the heat conditions may be detrimental to the normal biota. Many studies have been done on the effects of thermal pollution on the marine environment. Thermal discharge is most noted in the tropical areas, where organisms are near their thermal maximum. For example, mangrove trees in a thermal heated bay no longer reproduce and no new seedlings can be found in the lagoon.

Q. What are the sources & effects of solid wastes discharged into the marine environment ?

Ans. Plastic produced in large amount has danger of being discharged as solid waste into the marine environment. Plastic is not biodegradable and therefore effecting the oceans for long periods of time. When compared to other types of pollutants the biological consequences are low.

Source of Solid wastes: Solid wastes are dumped into the sea mainly by the anthropogenic activities. Mostly non biodegradable plastics made materials such as used bottles, cups, plastic bags, plastic fishing nets, metal pieces, abandoned wreckage ships, etc. are dumped into the sea. This is also called Ocean dumping.Some features of plastic dumping in ocean is as follows:

- Six-pack holders will not decompose for 400 years.
- Each square mile of ocean surface off the North East coast of the U.S. has more than 50,000 pieces of plastic floating.
- 100,000 marine mammals and 2 million sea birds die every year after ingesting or being caught in plastic debris.
- Adding ingredients to plastics that would hasten decomposition would add 5-7% to costs, but this price increase is unacceptable to industries.

Usefulness of solid waste: The solid waste on the other hand acts as a substratum for many fouling marine organisms. The initial biofilm formation on a new substratum and subsequent macro fouling is useful in some ways. If the solid waste is not releasing any toxic substances then it can be used build artificial reefs and has been considered as a very good way to increase the environmental heterogeneity (biodiversity) in areas which function as nursery or feeding grounds for important fishing stocks.

Effect on organisms: There have been cases that fish having been trapped inside plastic bags other plastic objects and dying of suffocation. Also some cases have been reported of fish getting entangled in plastic fishing nets which have been torn off from fishing boats and drift in the water column, causing a phenomenon known as "ghost fishing". Sea turtles mistake plastic bags for jellyfish eat and die from internal blockages. Seals and sea lions starve after being muzzled by six-pack rings or entangled by nets.

Birds have become entangled in discarded fishing line & drowned, while lost drift nets are particularly dangerous to marine forms.

Q. What is the effect of sea level rise through global warming.

Ans. The gaseous pollutants (green house gases like corbon-di-oxide, dimethyl sulphoxide, chlorofluorocarbon-CFC etc.) released into the air, have direct implication on the atmosphere. These gaseous pollutants are found to increase the temperature of atmosphere through depletion of ozone layer, which in turn increase the sea surface temperature resulting in melting of ice in the polar regions. Scientists are working to understand how increases in carbon dioxide and other "greenhouse" gases in the atmosphere might change the earth's climate. But so far, complex interactions between oceans, atmosphere, land, and the sun have made precise climatic predictions impossible. Oceans are crucial in shaping climate because they store & move heat around the planet, and they're a major source & storehouse for green house gases (eg, carbon dioxide) that affect Climate.

Effects of Climate change

The climate change may lead to various impacts such as:

- melting of polar ice caps.
- accelerated erosion of beaches and coastal shore areas through raging storms.
- salt water contamination of coastal aquifiers.
- salt water intrusion in to a great distance landward into estuaries and bays.
- destruction of coastal terrestrial vegetation and replacement of salt tolerant species.
- increased frequency of flooding and subsequent submergence of coastal areas.
- Organisms such as coral may bleach and submerge.

If the sea level increased few centimeter, then the existence of many coastal cities & islands will be under severe threat due to flooding and subsequent reduction in living habitat. In most islands, the area is small and the elevation from mean sea level is in few meters. These islands are under severe threat of submergence. Thus concerted efforts must be taken to save the earth from global warming.

Global warming is becoming potential threat to fisheries and biodiversity. The extinction of species has increased from 30,000 in 1970s to 50,000 in 1980s and at this rate 10-20% of the earth's 10 million species of plant & animals would disappear forever. Coastal zones, which provide habitat for 80% of the 13,200 species of marine fish in the world oceans are being affected at alarming rate. Extensive bleaching of corals is spreading fast and there is destruction of valuable fish habitats especially breeding and spawning grounds.

Q. What Initiatives And Actions Are Taken For Marine Pollution Monitoring And Control In India ?

Ans. The Ministry of Environment and Forests under the Environment (Protection) Act, 1986 has declared the coastal stretches as Coastal Regulation Zone (CRZ) for regulating the activities in the CRZ. It declares coastal stretches of seas, bays, estuaries, creeks, rivers and backwater, which are influenced by tidal action (in the landward side) up to 500 meters from the High Tide Line (HTL) and land between Low Tide Line (LTL) and HTL as Coastal Regulating Zone and imposes restrictions on the setting up and expansion of industries operations & processes.

CATEGORIES OF COASTAL REGULATION ZONE IN INDIA

Coastal regulation zone in India has been divided into four categories namely: CRZ-I, CRZ-II, CRZ-III & CRZ-IV.

Description of various Category is given below:

CRZ-I: Areas that are ecologically sensitive and important, such as national parks/ marine parks, sanctuaries, reserve forests, wildlife habitats, mangroves, corals I coral reefs, areas close to breeding and spawning grounds of fish and other marine life, areas of outstanding natural beauty I historical I heritage areas, areas rich in genetic diversity; areas likely to be inundated due to rise in sea level consequent upon global warming and such other areas as may be declared by the Central Government or the concerned authorities at the State I Union Territory level from time to time

Area between the Low Tide Line and the High Tide Line.

CRZ-II: The areas that have already been developed up to or close to the shoreline. For this purpose "developed area" is referred to as that area within the municipal limits or in other legally designated urban areas, which is already substantially build up and which has been provided with drainage and approach roads and other infrastructure facilities, such as water supply and sewerage mains.

CRZ-III: Areas that are relatively undisturbed and those, which do not belong to either Category-I or II. These will include coastal Zone in the rural areas (developed and undeveloped) and also areas within Municipal limits or in other legally designated urban areas, which are not substantially building up.

CRZ-IV: Coastal stretches in the Andaman & Nicobar, Lakshadweep and small island except those designated as CRZ-I, CRZ-II or CRZ-III.

Types of Indian marine environment based on uses: In the coastal zone marine water is subjected to several types of uses. In 1974 the water act (Prevention and control of pollution) was created to keep the coastal water clean and free from pollution, the water bodies area classified according the use. The pollution Control boards at Centre and States have adopted 5 km in to the sea from the highest high tide, as the extent, over which their authority to control waste discharges will extend for the purpose of the water (Prevention and Control of Pollution) Act, 1974.

Coastal waters of India have been used for various activities like fishing, shipping harbour activities, industrial-cooling purposes etc. Depending of the types of uses and activities, water quality criteria have been specified to determine its suitability for a particular purpose. Among the various types of uses there is one use that demands highest level of water quality/purity and that is termed as "designated best use" in that stretch of the coastal segment. Through a survey conducted during 1980-86, all the activities in the Indian coastal areas were identified and based on the "designated best use' concept (i.e., exact or near quality requirement for different uses) the following classifications of seawater has been made.

Based on this, primary water quality criteria have been specified for the following five designated best uses.

Classification of sea water(SW) uses in India

1. Salt pans, Shell fishing, Mariculture and Ecologically Sensitive Zone -**SW-I**
2. Bathing, Contact Water Sports and Commercial fishing- **SW-II**
3. Industrial Cooling, Recreation (non-contact) and Aesthetics -**SW-III**
4. xHarbour- **SW-IV**
5. Navigation and Controlled Waste Disposal -**SW-V**.

Appendix- B ,MCQ On Water Pollution Control Engineering

1. Lagooning process is mainly a means of the
   (a) sludge disposal.
   (b) reduction of excessive flow in sewers.
   (c) biological treatment of wastes.
   (d) none of these.
2. For existence of aquatic life in water, the dissolved oxygen content in it, should not be less than........ ppm.
   (a) 10000 (b) 5
   (c) 500 (d) 1000
3. In water chemical treatment plant, use of chloramines ensures
   (a) disinfection.
   (b) Taste & odour control.
   (c) weed control in reservoirs.
   (d) removal of permanent hardness.
4. Foaming and priming in boiler operation can be reduced by reduction in...... in feed water.
   (a) turbidity (b) color (Hazen)
   (c) total solids (d) all (a), (b) & (c)
5. Scale formation in boiler is controlled by
   (a) preheating of feed water.
   (b) reduction in hardness, silica & alumina in feed water.
   (c) keeping the pH value of feed water just below 7.
   (d) eliminating H_2S in feed water.
6. Which is the most widely used coagulant for the treatment of turbid water?
   (a) Alum (aluminium sulphate)
   (b) Lime
   (c) Ferric chloride
   (d) Sodium aluminate
7. Presence of iron and manganese in water causes
   (a) reduction in its dissolved oxygen content.
   (b) discoloration of bathroom fixtures.
   (c) temporary hardness.
   (d) none of these.
8. Soluble silica present in boiler feed water can be removed by
   (a) coagulation
   (b) filtration
   (c) anion exchanger
   (d) preheating it
9. Corrosion in boilers can be prevented by
   (a) deaeration of feed water.
   (b) enhancing pH of feed water by adding alkali.
   (c) Feeding sodium sulphite or hydrazine phosphate to the boilers, which combines with oxygen and prevents corrosion.
   (d) all (a), (b) and (c).

Answer

1 **(a)**	2 **(b)**	3 **(a)**	4 **(d)**	5 **(b)**
6 **(a)**	7**(b)**	8 **(c)**	9 **(d)**	

10. Tri-sodium phosphate is used in boiler water treatment to reduce
    (a) turbidity
    (b) caustic embrittlement
    (c) suspended silica
    (d) dissolved oxygen
11. Which of the following is the most widely used disinfectant in water treatment?
    (a) Chlorine
    (b) Irradiation of water by ultraviolet light
    (c) Cation exchanger
    (d) Coagulation
12. Replenishment of dissolved oxygen in water stream polluted with industrial waste occurs by
    (a) natural aeration of water stream.
    (b) photosynthetic action of algae.
    (c) both (a) & (b).
    (d) neither (a) nor (b).
13. Pick out the wrong statement.
    (a) Biological oxygen demand (BOD) is a characteristic and not a constituent of water.
    (b) BOD is a measure of the amount of oxygen which will be demanded & used in 5 days by the biological decomposition of the organic matter present in water stream present as food for the living organism.
    (c) BOD is expressed in mg/litre (typically, BOD=2.5 mg/litre for potable water).
    (d) none of these
14. Pick out the correct statement.
    (a) Chemical oxygen demand (COD) is a measure of chemically oxidisable organic matter present in water.
    (b) COD is determined by oxidising the organic matter present in water with potassium dichromate in concentrated. sulphuric acid solution at boiling temperature for specified time.
    (c) COD is related to BOD of a given waste in water but the relationship varies for different wastes. Typically, COD of potable water may be 1-2 mg/litre.
    (d) all (a), (b) and (c).
15. Which of the following is the most lethal water pollutant?
    (a) Phenol and cynide
    (b) Chlorine
    (c) Alkalis
    (d) Suspended solids
16. Presence of soluble organics in polluted water causes
    (a) undesirable plants growth.
    (b) depletion of oxygen.
    (c) fire hazards.
    (d) explosion hazards.
17. Presence of non-biodegradable substances like alkyl benzene sulphonate (ABS) from the detergents in polluted water stream causes
    (a) fire hazards.
    (b) explosion hazards.
    (c) persistent foam.
    (d) depletion of dissolved oxygen.

Answer

10 **(b)**	11 **(a)**	12 **(c)**	13 **(d)**	14 **(d)**
15 **(a)**	16 **(b)**	17 **(c)**		

18. BOD of raw municipal sewage may be in the range of about...... mg/litre.

(a) 1-2 (b) 5-10
(c) 150-300 (d) 2000-3000

19. COD of raw municipal sewage may be in the range of about........ mg/litre.

(a) 1-2 (b) 5-10
(c) 90-120 (d) 1500-2500

20. Disinfection of water is done to destroy pathogenic bacteria and thus prevent water-borne diseases. Disinfection of water may be done by the use of

(a) ozone and iodine.
(b) chlorine or its compounds.
(c) ultraviolet light for irradiation of water.
(d) all (a), (b) & (c).

21. Irradiation of water by ultraviolet light of suitable wavelength is commonly used for disinfection of water in

(a) food industry.
(b) municipal sewage treatment.
(c) petroleum refinery.
(d) iron & steel plant.

22. Deaeration of high pressure boiler feed water is done to reduce

(a) foaming from boilers.
(b) its dissolved oxygen content.
(c) its silica content.
(d) caustic embrittlement.

23. Presence of.......... in water stream are deleterious to aquatic life.

(a) soluble and toxic organics.
(b) suspended solids.
(c) heavy metals and cynides.
(d) all (a), (b) & (c).

24. Presence of nitrogen and phosphorous in waste water discharged into lakes & ponds causes

(a) foaming
(b) odour nuisances
(c) undesirable plant growth
(d) turbidity

25. Presence of volatile compounds like gasoline, oil, alcohol, ether etc. in municipal sewers may cause

(a) explosion
(b) non-biodegradable foam
(c) undesirable plant growth
(d) corrosion

26. Oil and grease present in an emulsified state in waste water discharged from industries can be removed by

(a) biological oxidation.
(b) skimming off.
(c) settling out using chemical reagents.
(d) chlorination.

27. Which is the most practical and economical method for removal of suspended solid matter from polluted water?

(a) Sedimentation
(b) Skimming off
(c) Chlorination
(d) Biological oxidation

Answer

18 **(c)**	19 **(c)**	20 **(d)**	21 **(a)**	22 **(b)**
23 **(d)**	24 **(c)**	25 **(a)**	26 **(d)**	27 **(a)**

28. Which of the following is the most efficient for removal of very finely divided suspended solids and colloidal matter from the polluted water stream?
    (a) Sedimentation tank
    (b) Circular clarifier
    (c) Mechanical flocculation
    (d) Chemical coagulation
29. Pick out the one which is not a chemical coagulant.
    (a) Aluminium sulphate
    (b) Ferrous sulphate
    (c) Hydrated lime
    (d) Chloramine
30. Biological oxidation ponds remove organic matters present in the polluted water by
    (a) using the activities of bacteria and other microorganisms.
    (b) aerobic oxidation.
    (c) both (a) & (b).
    (d) neither (a) nor (b).
31. Bacterial aerobic oxidation of polluted water in biological oxidation ponds is done to purify it. Presence of bacteria helps in
    (a) coagulation and flocculation of colloids.
    (b) oxidation of carbonaceous matter to CO2.
    (c) nitrification or oxidation of ammonia derived from breakdown of nitrogeneous organic matter to the nitrite and eventually to the nitrate.
    (d) all (a), (b) and (c).
32. Polluted water having low BOD are most economically treated in
    (a) sedimentation tanks
    (b) oxidation ponds
    (c) sludge digester
    (d) clarifier
33. Lagoons used for purification of polluted water
    (a) are large shallow artificial lakes also known as clarification lakes, maturation ponds or oxidation ponds.
    (b) use micro-organisms/bacteria in presence of dissolved oxygen.
    (c) gives an excellent final effluent (with 3 to 4 lagoons arranged in series) having suspended solid < 1 mg/litre and BOD = 3.8 mg/litre.
    (d) all (a), (b) & (c).
34. Aerobic biological oxidation ponds used for the purification of polluted water
    (a) destroys/removes pathogen from the sewage.
    (b) is not very effective for non-biodegradable substances (e.g. ABS) containing effluents.
    (c) destroys/removes pathogen much more effectively if the sewage is chlorinated.
    (d) all (a), (b) & (c).
35. Which is the best and the most effective method for the removal of organic contaminant present in the polluted water in very small quantity (say < 200 mg/litre)?
    (a) Lagooning
    (b) Activated carbon adsorption
    (c) Biological oxidation pond
    (d) Chemical coagulation

Answer

28 **(d)**	29 **(b)**	30 **(c)**	31 **(d)**	32 **(b)**
33 **(d)**	34 **(d)**	35 **(b)**		

36. water generated in coke ovens & by-product plant of a steel plant are disposed off by
    (a) quenching of hot coke.
    (b) discharging in the river stream.
    (c) filtration and recycling for cooling coke oven gas.
    (d) none of these.

37. Iron & manganese present in the polluted water is removed by
    (a) simple filtration.
    (b) oxidation followed by settling & filtration.
    (c) chemical coagulation.
    (d) chlorination only.

38. The commonest form of iron & manganese found in ground water as pollutant is in the form of their
    (a) carbonates
    (b) bi-carbonate
    (c) chlorides
    (d) sulphides

39. Iron & manganese present as pollutant in water cannot be removed by
    (a) ion exchange process.
    (b) oxidation followed by settling & filtration.
    (c) lime soda process or manganese zeolite process.
    (d) chlorination.

40. Salt content in sea water is about........... percent.
    (a) 0.5 (b) 1
    (c) 3.5 (d) 10

41. Dose of chlorine for disinfection of water is about..... mg/litre of water.
    (a) 0.01 (b) 0.1
    (c) 0.3 (d) 1

42. What is the major constituents of waste/polluted water discharged from textile, pulp & paper, tanning, distillary, dairy and meat packing industries?
    (a) Radioactive substances
    (b) Natural organic products
    (c) Inorganic pollutants
    (d) None of these

43. Water generated in coke ovens & by-product plant attached to an integrated steel plant containing phenol in concentration of less than 100 mg/litre can be removed by
    (a) chlorination
    (b) treating in biological oxygen pond
    (c) chemical coagulation
    (d) none of these

44. Waste/polluted water discharged from electroplating, blast furnace and coal mining industries contain mainly..... substances.
    (a) radioactive
    (b) organic
    (c) inorganic
    (d) none of these

45. Radioactive substances present in the polluted water stream can be removed by
    (a) biological oxygen treatment.
    (b) coagulation and filtration.
    (c) adsorption in ion exchange materials.
    (d) none of these.

Answer

36 **(a)**	37 **(b)**	38 **(d)**	39 **(d)**	40 **(c)**
41 **(d)**	42 **(b)**	43 **(b)**	44 **(c)**	45 **(c)**

46. The main pollutant in waste water discharged from a petroleum refinery is oil (both in free and emulsified form). Free oil is removed by
   (a) biological oxygen pond.
   (b) aerated lagoons.
   (c) trickling filters.
   (d) gravity separator having oil skimming devices.

47. Suspended solid present in the waste water generated in blast furnace gas cooling and cleaning plant is removed by
   (a) biological oxygen pond.
   (b) radial settling tank (thickener) using coagulant (lime & ferrous sulphate).
   (c) lagoons.
   (d) filtration.

48. The amount of chemical coagulant added for treatment of polluted water .... with increase in temperature of the polluted water to be treated.
   (a) decreases.
   (b) increases.
   (c) remains constant.
   (d) may increase or decrease; depends on the chemical characteristics of polluted water.

49. Presence of bacteria in potable (drinking) water causes
   (a) turbidity
   (b) disease
   (c) bad odour
   (d) bad taste & colour

50. Coal washing waste water containing about 3% suspended solids (comprising of clay, slate, stone etc.) is treated for solid particles removal
   (a) by chemical coagulation.
   (b) in sedimentation tanks equipped with mechanical scrapper.
   (c) in vacuum filter.
   (d) in clarifiers.

51. Coal mines drainage waste water (acidic in nature) results from the earth's water percolating through the voids created in coal bed during mining. This polluted water which either drains out naturally to water courses or are removed before starting the mining is
   (a) neutralised by alkali treatment.
   (b) left as such without any treatment.
   (c) diluted with fresh water to reduce its acidity.
   (d) none of these.

52. The ratio of oxygen available to the oxygen required for stabilisation of sewage is called the
   (a) bacterial stability factor.
   (b) relative stability.
   (c) biological oxygen demand (BOD).
   (d) oxygen ion concentration.

53. In sewage treatment, its sedimentation is speeded up by commonly adding
   (a) hydrochloric acid.
   (b) lime.
   (c) copper sulphate.
   (d) sodium sulphate.

Answer

46 **(d)**	47 **(b)**	48 **(a)**	49 **(b)**	50 **(b)**
51 **(a)**	52 **(b)**	53 **(b)**		

54. Which of the following processes is involved in the biochemical treatment of sewage effluents?
(a) Oxidation (b) Reduction
(c) Dehydration (d) Fermentation

55. Most of the bacteria in sewage are
(a) parasitic (b) saprophytic
(c) anaerobic (d) none of these

56. Solid content in most of the domestic sewage is about............ percent.
(a) 0.001 (b) 0.1
(c) 5 (d) 10

57. Which of the following industries discharge mercury as a pollutant?
(a) Chlor-alkali industry
(b) Tanneries
(c) Beverage plant
(d) Phosphoric acid plant

58. TLV of mercury in potable (drinking) water is about........ ppm.
(a) 0.001 (b) 0.1
(c) 1 (d) 5

59. TLV of lead for public sewer/waste water is about.......... ppm.
(a) 1 (b) 25
(c) 150 (d) 650

60. The destruction of water-borne pathogens is termed as disinfection of water. Which of the following is a water disinfectant?
(a) Chlorine
(b) Alkalis
(c) Benzene hexachloride
(d) Alkyl benzene sulphonate (ABS)

61. The permissible color for domestic water supply is .... ppm.
(a) 1 (b) 20
(c) 100 (d) 1000

62. Presence of nitrates in water in excess of 50 ppm causes
(a) mathenoglobenemia
(b) gastroenteritis
(c) asphyxiation
(d) tooth decay

63. Presence of a certain minimum quantity of flourine is desirable in potable water to prevent
(a) dental cavities
(b) scale formation
(c) water-borne disease
(d) corrosion

64. Presence of .... hardness is responsible for the temporary hardness in water.
(a) carbonate
(b) calcium
(c) choride
(d) sulphate.

65. .... is the process of killing organism in water.
(a) Coagulation
(b) Sterilisation
(c) Disinfection
(d) Sedimentation

66. Growth of .... is promoted by the presence of manganese in water.
(a) files
(b) algae
(c) micro-organisms
(d) mosquitoes

Answer

54 **(a)**	55 **(b)**	56 **(b)**	57 **(a)**	58 **(a)**
59 **(a)**	60 **(a)**	61 **(b)**	62 **(a)**	63 **(a)**
64 **(b)**	65 **(c)**	66 **(c)**		

67. Presence of .... bacteria in water causes disease like typhoid.
    (a) aerobic
    (b) pathogenic
    (c) anaerobic
    (d) non-pathogenic

68. Maximum permissible turbidity in potable water is......... ppm.
    (a) 1 (b) 10
    (c) 250 (d) 1000

69. Turbidity of water is an indication of the presence of
    (a) suspended inorganic matter
    (b) dissolved solids
    (c) floating solids
    (d) dissolved gases

70. Presence of dissolved impurities of .... is responsible for the red brownish color of water.
    (a) carbonates
    (b) bi-carbonates
    (c) iron & manganese
    (d) arsemc.

71. Presence of excess flourine in water causes
    (a) dental cavity
    (b) tooth decay
    (c) fluorosis
    (d) respiratory disease

72. Infective bacteria in water is killed by the....... process.
    (a) sterilisation
    (b) aeration
    (c) disinfection
    (d) none of these.

73. Removal of .... is accomplished by aeration of water.
    (a) dissolved gases
    (b) suspended solids
    (c) dissolved solids
    (d) none of these.

74. The pH value of potable water should be between
    (a) 1 to 1.5 (b) 6.5 to 8
    (c) 13 to 14 (d) 4 to 5

75. A standard test for determination of hardness in water is termed as........ test.
    (a) EDTA (b) electometric
    (c) total count (d) presumptive

76. In a sedimentation tank, the detention period for water ranges from .... hours.
    (a) 2 to 4 (b) 8 to 12
    (c) 16 to 20 (d) 24 to 32

77. In water treatment, alum [$Al_2(SO_4)_3$] is used for the process of
    (a) filtration
    (b) coagulation
    (c) sedimentation
    (d) disinfection.

78. Color test of water is done with an instrument called
    (a) tintometer
    (b) colorimeter
    (c) electro-chemical cell
    (d) turbidimeter.

79. Hydrogen ion concentration in distilled water is
    (a) 10^{-7} (b) 7×10^{-7}
    (c) 10^{7} (d) 7×10^{7}

Answer

67 (b)	68 (b)	69 (a)	70 (c)	71 (c)
72 (a)	73 (a)	74 (b)	75(a)	76 (a)
77 (b)	78 (a)	79 (c)		

80. Removal of .... results from the disinfection of water.
(a) turbidity (b) odour
(c) hardness (d) bacteria.

81. The resistance of water to the passage of light through it is a measure of the
(a) turbidity
(b) color
(c) hardness
(d) dissolved gases.

82. Water filtration rate in a rapid sand filter ranges from .... kilolitres/m^2/hr.
(a) 0.1 to 1 (b) 3 to 6
(c) 10 to 15 (d) 15 to 20

83. Disinfection of water is done to remove
(a) color (b) bad taste
(c) foul odour (d) bacteria

84. Algae growth in water is controlled by
(a) deoxidation (b) chlorination
(c) bleaching (d) aeration.

85. Foul odour and bad taste of water isremoved by treating with
(a) alum
(b) bleaching powder
(c) activated carbon
(d) copper sulphate.

86. In water treatment plant, zeolite process is used to remove the .... of water.
(a) acidity (b) alkalinity
(c) hardness (d) iron & zinc.

87. .... is removed from water by lime-soda process.
(a) Foul smell and taste
(b) Iron and manganese
(c) Temporary hardness
(d) Permanent hardness

88. Maximum permissible residual chlorine in treated water should be .... mg/litre.
(a) 0.001 to 0.01 (b) 0.2 to 0.3
(c) 2 to 3 (d) 5 to 10

89. Fresh domestic sewage is .... in color.
(a) grey (b) dark brown
(c) red (d) black.

90. Brown spots in fabrics will be caused by washing with water containing large amount of
(a) iron (b) zinc
(c) iodine (d) bromine.

91. Water filtration rate in a slow sand filter ranges from .... litres/m^2/hr.
(a) 10 to 20 (b) 100 to 200
(c) 1500 to 2500 (d) 4000 to 5000.

92. Insufficient washing of sand grains in a rapid sand filter causes
(a) air binding
(b) shrinkage of filtering media
(c) mud balls
(d) expansion of filtering media.

93. Fresh sewage is .... in nature.
(a) acidic (b) neutral
(c) alkaline (d) highly acidic.

94. The pH value of oxidised sewage is about
(a) 1.8 (b) 6.2
(c) 7.3 (d) 13.4

Answer

80 **(d)**	81 **(a)**	82 **(b)**	83 **(d)**	84 **(b)**
85 **(c)**	86 **(c)**	87 **(c)**	88 **(b)**	89 **(a)**
90 **(a)**	91 **(b)**	92 **(c)**	93 **(c)**	94 **(c)**

95. ....substances present in sewage are removed in grit chamber during sewage treatment.
(a) Organic (b) Fatty
(c) Inorganic (d) Dissolved

96. The type of bacteria which is active in trickling filter during biological treatment of sewage is the .... bacteria.
(a) anaerobic (b) saphrophytic
(c) anaerobic (d) parasitic.

97. During sewage treatment, the sewage is subjected to .... treatment in Imhoff tank.
(a) filtration
(b) digestion
(c) sedimentation
(d) both (b) and (c).

98. A shallow pond in which the sewage is retained and biologically treated is called
(a) oxidation (b) Imhoff tank
(c) lagoon (d) skimming tank

99. In sewage treatment, the detention period allowed for oxidation ponds ranges from......... weeks.
(a) 1 to 2 (b) 4 to 5
(c) 9 to 10 (d) 15 to 20.

100. The function of skimming tank in sewage treatment is to remove .... substances.
(a) dissolved solid
(b) suspended solid
(c) oil & fatty
(d) gritty & inorganic.

101. Fine grit present in sewage is removed in the .... during sewage treatment.
(a) grit chamber
(b) detritus tank
(c) trickling filter
(d) skimming tank.

102. The main type of sludge gas evolved during sewage treatment in Imhoff tank is
(a) CO_2 (b) CH_4
(c) CO (d) H_2

103. The biological decomposition of organic substances in wastes controlled conditions is called
(a) incineration
(b) biological oxidation
(c) composting
(d) none of these.

104. The most commonly used chemical coagulant in water treatment is
(a) ferrous sulphate
(b) alum
(c) lime
(d) hydrazine

105. Maximum permissible concentration (i.e., TLV) of DDT in public water supply system is......... micro gram (μg)/litre.
(a) 2 (b) 42
(c) 332 (d) 1050

106. TLV of aldrin in public water supply system is about .... μg/litre.
(a) 0.5 (b) 17
(c) 357 (d) 1097

Answer

95 **(c)**	96 **(c)**	97 **(d)**	98 **(a)**	99 **(b)**
100 **(c)**	101 **(b)**	102 **(b)**	103 **(c)**	104 **(b)**
105 **(b)**	106 **(b)**			

107. Inorganic impurities causing water pollution is
    (a) fats
    (b) carbohydrates
    (c) salts of metals
    (d) protein.

108. The detrimental effect on organism and water quality with temperature rise of aquatic system is the reduction of .... of water.
    (a) dissolved oxygen content
    (b) biological oxygen demand
    (c) vapor pressure
    (d) all (a), (b) and (c)

109. Water effluent generated in printing industry is decolourised by
    (a) ion exchange technique.
    (b) reverse osmosis.
    (c) electrolytic decomposition.
    (d) adsorption.

110. Oceans act as sinks for atmospheric gases including carbon dioxide whose concentration in the atmosphere is increased by the
    (a) forestation
    (b) rain
    (c) green house effect
    (d) vegetation

111. Match the deleterious effect of various metallic water pollutants on human health.

List I

(A) Cadmium (B) Cobalt
(C) Zinc (D) Selenium

List II

I. Lung irritation.
II. Bone deformation & growth retardation
III. Liver & kidney damage.
IV. Renal damage.

Code;

	A	B	C	D
(a)	II	I	IV	III
(b)	I	II	III	IV
(c)	IV	III	II	I
(d)	I	II	IV	III

112. Thermal pollution of water increases its toxicity and oxidation of oxygen demanding waste besides favoring bacterial growth. A rise in water temperature by 10°C, doubles the toxic effects of......... present in it.
    (a) coal ash
    (b) potassium cyanide
    (c) ortho-xylene
    (d) none of these.

113. Oilish impurities present the effluent discharged from the electroplating industry is normally not removed by
    (a) chemical coagulation
    (b) floatation & skimming
    (c) centrifugation
    (d) ultrafiltration

114. Dissolved oxygen content in river water is around .... ppm.
    (a) 5 (b) 100
    (c) 250 (d) 500

115. The term Biological Oxygen Demand (BOD) is used in relation to
    (a) potable water
    (b) cooling water
    (c) distilled water
    (d) industrial effluents

Answer

107 **(c)**	108 **(a)**	109 **(c)**	110 **(c)**	111 **(a)**
112 **(b)**	113 **(a)**	114 **(a)**	115 **(d)**	

116. In the context of the chemical process industries, the term BOD is normally associated with the
   (a) characterisation of solid wastes.
   (b) organic concentration in gaseous effluents.
   (c) characterisation of liquid effluents.
   (d) characterisation of boiler feed water.

117. .... method of waste treatment will be recommended, if the ratio of biological oxygen demand (BOD) to chemical oxygen demand (CO D) is 1.
   (a) Thermal (b) Biological
   (c) Physical (d) chemical

118. What is BOD (ppm) of a water sample having initial oxygen concentration of 10 ppm and final oxygen concentration of 4 ppm after keeping it in a closed bottle for 6 days at 25°C?
   (a) 2.5 (b) 6
   (c) 11 (d) 1.5

119. Which of the following is present in maximum amount in acid rain?
   (a) H_2SO_3 (b) H_2SO_4
   (c) HCl (d) HNO_3

120. Organic wastes which can be readily decomposed by micro-organisms are called biodegradable. Pick out the biodegradable pollutant out of the following:
   (a) Pesticides
   (b) Mercuric salts
   (c) Domestic wastes
   (d) Lead compounds

121. Water pollution is not caused by
   (a) oxygen demanding wastes.
   (b) automobile exhaust.
   (c) plant nutrients.
   (d) disease causing agents.

122. If BOD (biological oxygen demand) of river water is high, then it implies that the river is
   (a) polluted with pesticides.
   (b) not at an polluted.
   (c) highly polluted with inorganic chemicals.
   (d) highly polluted with biodegradable organic chemicals.

123. If pH value of rain water is less than.......... , it is termed as acid rain.
   (a) 7 (b) 6.5
   (c) 6 (d) 5.6

124. Pick out the wrong statement about acid rains.
   (a) Europe and north east America is the most affected region of the world by acid rain.
   (b) Lime powder is sprayed on lake waters to neutralise the acidic nature of lake water caused by acid rain.
   (c) Acid rains are more prominent in temperate regions.
   (d) ESP (electrostatic precipitators) are installed in fossil fuel fired thermal power plants to check emission of SO_x& NO_x gases to reduce acid rains.

125. Which of the following gas is generally not present in sewers?
   (a) H_2S (b) CH_4
   (c) CO2 (d) CO

Answer

116 **(c)**	117 **(b)**	118 **(b)**	119 **(b)**	120 **(c)**
121 **(b)**	122 **(d)**	123 **(d)**	124 **(d)**	125 **(d)**

126. When a sheet of paper moistened with lead acetate, held for 5 minutes in a manhole, turns black; then the sewer certainly contains

(a) H_2S (b) CO_2
(c) CH_4 (d) CO

127. Pick out the wrong statement.

(a) Sewage contain about 99.9% water & its specific gravity is slightly more than 1.
(b) Temperature variations affect the viscosity, gas solubility & biological activity of bacteria in sewage.
(c) pH value of fresh sewage is usually more than 7.
(d) Well oxidised sewage contains nitrogen in the form of nitrites & sulfur in the form of sulphites.

128. Which of the following gases is generally not evolved during decomposition of sewage?

(a) H_2S (b) CO_2
(c) CH_4 (d) NH_3

129. Solubility of oxygen in sewage when compared to its solubility in distilled water is about............ percent.

(a) 50 (b) 60
(c) 75 (d) 90

130. Minimum dissolved oxygen (D.O) content prescribed for a river stream to avoid fish kill is........... ppm.

(a) 0.1 (b) 1
(c) 2 (d) 4

131. Pick out the wrong statement.

(a) Pathogenic (disease causing) bacteria enters waste waters primarily from domestic wastes.
(b) Biodegradable organic enters waste waters from both domestic as well as industrial wastes.
(c) Soluble organics in domestic waste waters include carbohydrates, proteins & lipids.
(d) Biological oxygen demand (BOD) of waste water is the oxygen required to oxidise the biologically inactive organic matter.

132. Most of the bacteria in sewage are

(a) anaerobic (b) parasitic
(c) saprophytic (d) pathogenic

133. As compared to fresh river water, sea water contains ,......... oxygen.

(a) 5% more (b) 5% less
(c) 20% more (d) 20% less

134. Dissolved oxygen content in river/ water streams is

(a) minimum at noon.
(b) maximum at noon.
(c) maximum at midnight.
(d) same throughout the day.

135. The natural process under which flowing river water gets cleaned is known as 'self-purification', Dissolved oxygen content may fall to zero. causing anaerobic bio-condition in a river reach called zone of

(a) recovery
(b) degradation
(c) active decomposition
(d) none of these

Answer

126 **(a)**	127 **(d)**	128 **(b)**	129 **(d)**	130 **(d)**
131 **(d)**	132 **(c)**	133 **(d)**	134 **(b)**	135 **(c)**

136. BOD_5 represents 5-days bio-chemical oxygen demand of sewage at a temperature of 20°C; Standard 5 days BOD at 20°C when compared to ultimate BOD, is about...... per cent.
   (a) 21 (b) 38
   (c) 68 (d) 98

137. Pick out the wrong statement
   (a) Microbial utilisation of organics in sewage is first order chemical reaction.
   (b) While testing for COD (chemical oxygen demand) of sewage, organic matter is oxidised by $K_2Cr_2O_7$ in presence of H_2SO_4.
   (c) Between DOD & COD of sewage, the greater value of the two is BOD.
   (d) The average BOD_5 of domestic sewage is about 0.08kg/day/person.

138. Sewage disposal causes formation of sludge banks in sea waters. The most common method of waste water disposal is its
   (a) utilisation for irrigational purposes.
   (b) rapid infiltration.
   (c) dilution in surface water.
   (d) natural evaporation.

139. Complete through treatment should generally be given to sewage, before its disposal in a stream, if the dilution factor available is less than
   (a) 500 (b) 350
   (c) 1250 (d) 150

140. Sewage disposal by dilution is not preferred, where
   (a) the sewage arriving at outfall points is fresh and non-septic.
   (b) there are strong forward currents.
   (c) rivers run with very low flows during summer.
   (d) discharge in water body is very low.

141. Primary treatment of sewage is meant for the removal of
   (a) fine suspended organic matter
   (b) larger suspended matter
   (c) pathogenic (disease causing) bacteria
   (d) dissolved organic matter

142. Match the equipments/techniques used for the removal of unwanted materials during preliminary treatment of sewage
   I. Skimining tanks
      A. Oil At grease
   II. Grit chambers
      B. Floating materials
   III. Screening
      C. Dissolved organic matter
   IV. Reverse osmosis
      D. Sand
   (a) I - A, II - D, III - B, IV - C
   (b) I - A, II - B, III- C, IV - D
   (c) I - D, II - C, III - 13, IV - A
   (d) I - D, 1I - A, III - B, IV - C

143. The sewage treatment units, which work on anaerobic bio-decomposition or organic mutter is
   (a) detritus tank.
   (b) oxidation ponds.
   (c) septic tanks.
   (d) activated sludge plants.

Answer

136 **(c)**	137 **(c)**	138 **(c)**	139 **(d)**	140 **(c)**
141 **(b)**	142 **(a)**	143 **(c)**		

144. The BOD_5 of the effluents obtained from septic tanks is of the order of about............ mg/litre.
   (a) 1 - 5 (b) 20 - 50
   (c) 100 - 200 (d) 1000 - 2000

145. The sewage treatment units, which work on aerobic decomposition of organic matter is
   (a) imhoff tanks.
   (b) sludge digestion tank.
   (c) grit channel.
   (d) trickling filter.

146. The gas which is evolved in sludge digestion tank is mainly composed by
   (a) NH_3 (b) H_2S
   (c) CO_2 (d) CH_4

147. Pick out the wrong statement.
   (a) Sludge digestion tank works on the principle of aerobic decomposition.
   (b) A grit chamber having a higher detention period (about 3 - 4 times) is called detritus tank.
   (c) Chlorine is sometimes used in sewage treatment to help in grease separation.
   (d) A septic tank is both a settling tank as well as digestion tank.

148. Maximum efficiency of BOD removal is achieved in
   (a) aerated lagoon
   (b) oxidation ditch
   (c) oxidation pond
   (d) trickling filter

149. High COD to BOD ratio of an organic waterpollutant indicates
   (a) high biodegradability of the pollutant.
   (b) presence of free toxic material in the pollutant.
   (c) low biodegradability of the pollutant.
   (d) presence of free oxygen for aerobic decomposition.

150. In a conventional activated sludge plant, the oxygen demand is the highest near the inlet end of the aeration tank. Activated sludge is the
   (a) aerated sludge in the aeration unit.
   (b) sludge settled in the humus tank.
   (c) sludge in the secondary tank after aeration and rich in microbial mass.
   (d) sludge in the secondary tank after aeration and rich in nutrients.

151. Composting and lagooning are the methods of sewage disposal. For satisfactory working of a sludge digestion unit, the pH range of digested sludge should be maintained at
   (a) 3 to 5 (b) 6.5 to 8
   (c) 8.5 to 10.5 (d) 11 to 13.5

152. The secondary treatment of sewage is caused by
   (a) chlorine (b) bacteria
   (c) algae (d) coagulants

Answer

144 **(c)**	145 **(d)**	146 **(d)**	147 **(a)**	148 **(b)**
149 **(c)**	150 **(c)**	151 **(b)**	152 **(b)**	

153. Match the detention period adopted for various sewage treatment units.

Sewage treatment units

I. Grit chambers
II. Sedimentation tanks
III. Oxidation ponds
IV. Septic tanks
V. Imhoff tanks

Detention period

A. 1 minute
B. 1- 2 minute
C. 2 - 4 hours
D. 12 - 36 hours
E. 2 - 6 weeks

(a) I - A, II - B, III - C, IV - D, V - E
(b) I- A, II - B, III - E, IV - D, V - C
(c) I - E, II - D, III - C, IV - B, V - A
(d) I - A, II - B, III - C, IV - D, V - E

154. The phenomenon by virtue of which soil is clogged with sewage matter is called sewage

(a) sulking (b) bulking
(c) sickness (d) farming

155. Algae dies out, though fish life may survive, in a river zone, known as the zone of

(a) degradation.
(b) aeration.
(c) active decomposition.
(d) recovery.

156. Imhoff cone is used to measure the total........... content in sewage.

(a) solids
(b) settleable solids
(c) organic solids
(d) inorganic solids

157. The total dissolved solids in drinking water should not exceed............ mg/ litre.

(a) 1 (b) 100
(c) 500 (d) 1000

158. Methamoglobanumia (blue baby disease) is due to presence of excess of.............. in the drinking water.

(a) chlorides
(b) nitrates
(c) dissolved oxygen
(d) PH

159. Flame ionisation detector (FID) is used forthe determination of present in the polluted air.

(a) carbon monoxide (CO)
(b) nitrogen oxides (NO_x)
(c) hydrocarbons (HC)
(d) chlorine

160. Orthotolidine test is used to determine............ present in water.

(a) residual ammonia
(b) nitrates
(c) sulfates
(d) residual chlorine

161. Ether soluble solids in waste water represents its content.

(a) dissolved solids
(b) total solids
(c) grease
(d) none of these

Answer

153 **(b)**	154 **(c)**	155 **(a)**	156 **(b)**	157 (a)
158 **(b)**	159 **(c)**	160 **(d)**	161 **(c)**	

162. Contact stabilisation process is a form of
   (a) stabilisation ponds.
   (b) activated sludge process.
   (c) trickling fillers.
   (d) anaerobic lagoons.

163. Our water ways are not polluted by
   (a) fish farming.
   (b) leaking of old landfills.
   (c) leaking underground gasoline storage tanks.
   (d) dumping used oil on the ground.

164. The presence of high caliform counts in water indicates
   (a) decreased biological oxygen demand (B.O.D)
   (b) contamination by human waste
   (c) hydrocarbon contamination
   (d) phosphorous contamination.

165. Pick out the wrong statement:
   (a) Biotic environment includes decomposers like bacteria and fungi.
   (b) Abiotic environment includes air, water and soil but does not include plants.
   (c) The group of organisms which convert light into food are called autotrophs.
   (d) Ground water mining in coastal area can result into decrease in its toxicity.

166. Presence of which of the following organism is an indicator of sewage pollution?
   (a) Volvox
   (b) Escherichia Coli
   (c) Aerobacter aerogene
   (d) Spirogyra

167. Property of earth to allow water to pass through it is termed as 'perviousness'. What is the capacity of soil to absorb moisture is generally known as?
   (a) Permeability
   (b) Porosity
   (c) Transmissibility
   (d) Infiltration capacity

168. Water borne amoebic dysentery is caused by
   (a) virus (b) protozoa
   (c) bacteria (d) all 'a', 'b'&'c'

169. E. Coli bacteria die in water having pH greater than
   (a) 5.5 (b) 6.5
   (c) 7.5 (d) 9.5.

170. Presence of pathogenic bacteria in drinking water may cause
   (a) cholera (b) dysentery
   (c) typhoid (d) all 'a', 'b' & 'c'

171. Match the maximum permissible concentration of various contaminants in treated water of public water supply for domestic use.

Water contaminants	Maximum permissible concentration, ppm
I. Chloride content	A. 0.35
II. Nitride content	B. 45
III. Nitrate content	C. Nil
IV. Fe & Mn content	D.250

   (a) I - D, II - C, III - B, IV - A
   (b) I - A, II - B, III - C, IV - D
   (c) I - D, II - C, III - A, IV - B
   (d) I - C, II - D, III - A, IV - B

Answer

162 **(b)**	163 **(c)**	164 **(d)**	165 **(d)**	166 **(b)**
167 **(d)**	168 **(b)**	169 **(d)**	170 **(d)**	171 **(a)**

172. Methanoglobinemia or blue baby disease is caused due to the presence of........ air drinking water.
(a) sulfides (b) nitrites
(c) nitrates (d) chlorides

173. Presence of high algal content in water indicates that water is
(a) hard (b) soft
(c) turbidity free (d) acidic.

174. Biochemical Oxygen Demand (BOD) of safe drinking water must be.......... mg/litre.
(a) 0 (b) 3-5
(c) 7-9 (d) < 10

175. To ensure proper growth of teeth of children, the quantity of fluoride used in water supply is....... mg/litre.
(a) 0.1 (b) 1
(c) 5 (d) 10

176. Disinfection of water is done by chlorination for removal of
(a) color (b) turbidity
(c) bacteria (d) odour

177. Pick out the wrong statement about water treatment:
(a) Aeration of water is done for removal of odour.
(b) Treatment of water with alum $[(NH_4)_2SO_4]$ to control algae increases its hardness & alkalinity.
(c) Dissolved carbon dioxide can be removed from water supply main by aeration.
(d) Use of hard water causes scaling of boiler tubes and corrosion & incrustation of pipes.

178. Hard water for public water supply is discarded because, it
(a) contains pathogenic (disease causing) bacteria
(b) consumes more soap
(c) contains lot of turbidity
(d) makes detergents non-biodegrab1e.

179. Most commonly used coagulant for water treatment is
(a) lime
(b) bleaching powder
(c) alum
(d) chlorine

180. Pick out the wrong statement about water treatment:
(a) Efficiency of removing bacteria from water by a slow sand filter is about 98- 99%.
(b) Wash water is required for slow sand filter.
(c) Ozone removes bacteria as well as color & odour from water.
(d) Most ideal disinfectant for drinking water is chlorine.

181. Which of the following is not a chemical used for dechlorination of water?
(a) Chloramine
(b) Sodium thiosulfate
(c) Sulfur dioxide
(d) Sodium sulfate

182. Pick out the wrong statement about sewerage system:
(a) The sewer which receives the discharge from a collecting system and delivers it to the treatment plant is known as trunk sewer.
(b) The rate of flow of sewage is generally assumed less than the rate of water supply.
(c) Fresh sewage is generally highly acidic in nature.
(d) The sewer which resists sulfide corrosion is stone ware pipes.

Answere

172 **(c)**	173 **(d)**	174 **(a)**	175 **(b)**	176 **(c)**
177 **(b)**	178 **(b)**	179 **(c)**	180 **(b)**	181 **(a)**
182 **(c)**				

183. The gas which may cause explosion in sewer pipes is
   (a) CO (b) CO_2
   (c) H_2S (d) CH_4
184. A circular sewer section is preferred because it provides maximum
   (a) economy in construction cost
   (b) hydraulic main depth
   (c) area for a given perimeter
   (d) all 'a', 'b'&'c'.
185. Clogging of sewer pipes is not caused due to
   (a) sitting and low discharge
   (b) domestic wastes thrown in manholes
   (c) greasy & oily matters
   (d) alkalinity of sewage.
186. The non-clog pump which permits solid mater to pass out with the liquid sewage is
   (a) reciprocating pump
   (b) centrifugal pump
   (c) pneumatic ejector
   (d) gear pump
187. Which of the following gases is generally not found in sewers?
   (a) CO (b) CO_2
   (c) H_2S (d) CH_4
188. Pick out the wrong statement about sewage:
   (a) If the flame of a miner's safety lamp in a manhole extinguishes within five minutes, the sewer certainly contains CO_2.
   (b) If paper moistened with lead acetate when placed in a manhole for five minutes. turns black. the sewer certainly contains H_2S.
   (c) Large sewers are cleaned by flushing.
   (d) Asbestors cement sewers are susceptible to corrosion by sulfuric acid.
189. In sewer pipes, a manhole is generally provided at each
   (a) junction
   (b) bend
   (c) change of gradient
   (d) all 'a', 'b'&'c'.
190. Dried sewage after treatment is used as a
   (a) base material for paints
   (b) building material
   (c) chemical for lowering B.O.D
   (d) fertiliser
191. Temperature does not affect the........... in sewage.
   (a) biological activity of bacteria
   (b) solubility of gases
   (c) viscosity
   (d) alkalinity
192. As compared to sea water fresh river water contains...... oxygen.
   (a) 20% more (b) 20% less
   (c) 5% less (d) 50% more
193. Sewage treatment is not required if dilution factor is
   (a) 50 - 100 (b) 150 - 200
   (c) 300 - 350 (d) 500

Answer

183 **(d)**	184 **(d)**	185 **(d)**	186 **(b)**	187 **(a)**
188 **(c)**	189 **(d)**	190 **(d)**	191 **(d)**	192 **(a)**
193 **(d)**				

194. Which of the following crop should not be grown by sewage irrigation?
(a) Banana (b) Cotton
(c) Potato (d) Coconut

195. The coagulant which is generally not used for treating the sewage is
(a) chlorinated copperas
(b) ferric chloride
(c) ferric sulfate
(d) alum.

196. Presence of high algae content in water indicates that water is
(a) having high B.O.D.
(b) having high C.O.D.
(c) free from oxygen.
(d) saturated with oxygen.

197. In sewage having fully oxidised organic matter, nitrogen exists in the form of nitrates. The presence of free ammonia in sewage is detected by
(a) adding H_2SO_4
(b) boiling
(c) adding $KMnO_4$
(d) phenol-disulfuric acid.

198. Dilution method of disposing off sewage is not preferred when the diluting water has
(a) to be added to fresh sewage
(b) high dissolved oxygen content
(c) very high C.O.D.
(d) low dissolved oxygen content

199. If dissolved oxygen concentration falls down to zero in any drainage, it indicates zone of....... in which algae dies out completely.
(a) recovery
(b) degradation
(c) active decomposition
(d) cleaner water

200. Generally, the activity of enzymes is not affected by
(a) temperature
(b) pH value
(c) substrate concentration
(d) none of these,

201. Pick out the wrong statement about sewage:
(a) In sewage, generally B.O.D. is higher than C.O.D.
(b) Fresh sewage becomes stale in 3-4 hours.
(c) Water content of sewage is about 99.9%.
(d) Sewage in oxidation ponds is treated by the action of bacteria & algae both.

202. The digested sludge from septic tanks is removed after a maximum period of....... years.
(a) 1 (b) 3
(c) 5 (d) 7

203. Liquid waste from kitchen, bathrooms & wash basins excluding urinals & water closets is known as 'sullage'. Dry weather flow is the liquid waste generated from
(a) storm run off
(b) industrial effluent
(c) domestic discharge
(d) all 'a', 'b'&'c' together.

Answer

194 **(c)**	195 **(d)**	196 **(d)**	197 **(b)**	198 **(c)**
199 **(c)**	200 **(d)**	201 **(a)**	202 **(b)**	203 **(c)**

204. Decomposition of organic matter is carried out by....... bacteria in septic tanks.
   (a) aerobic (b) anaerobic
   (c) facultative (d) none of these.

205. The gas mainly produced in a sludge digester is
   (a) nitrogen (b) oxygen
   (c) hydrogen (d) methane

206. Aerobic activity is maximum in
   (a) fresh sewage (b) stale sewage
   (c) sewer pipe (d) sewage

207. Fertilising element applied by sewage farming is
   (a) potassium (b) phosphorus
   (c) nitrogen (d) all 'a', 'b'&'c'

208. The term 'elutriation' is related to
   (a) forced sedimentation
   (b) land disposal of sewage
   (c) washing of sludge
   (d) none of these

209. Particle settlement in a sedimentation tank is affected by the
   (a) viscosity of water
   (b) flow velocity
   (c) size, shape & specific gravity of solid
   (d) all 'a', 'b'&'c'.

210. Pick out the wrong statement:
   (a) The percentage ratio of oxygen available to the oxygen required for complete stabilisation of a sewage sample is known as relative stability.
   (b) The effluent of a septic tank is fit for discharge into open drain.
   (c) The process of decomposing the organic matter under controlled anaerobic conditions is called sludge digestion which produces 75% CH_4 containing gases.
   (d) In normal condition, period of sludge digestion is 30 days.

211. For the design of sewers in India, % of sewage discharge is assumed to be........... % of water supplied from water works
   (a) 50 - 60 (b) 75 - 80
   (c) 90 - 95 (d) 100

212. The most suitable section of a sewer in a combined or separate sewerage system is circulate. However, an e.g shaped sewer, when compared to a circular sewer
   (a) is easier to construct
   (b) is more stable
   (c) is economical
   (d) provides better self-cleaning velocity

213. Branch and main sewers are widely made of
   (a) reinforced cement concrete (RCC)
   (b) prestressed cement concrete
   (c) cast iron
   (d) glazed stoneware

Answer

204 **(b)**	205 **(d)**	206 **(a)**	207 **(d)**	208 **(c)**
209 **(d)**	210 **(b)**	211 **(b)**	212 **(d)**	213 **(a)**

214. Match the type of joint used for joining various sewer pipes

Types sewer pipes to be joined

I. Joining smaller sewer pipes of diameter 0.6 metre

II. Joining bigger sewer pipes of diameter 0.6 metre

III. Joining asbestos cement pipes

Preferred joint used

A. Collar joint

B. Simplex joint

C. Spigot and socket joint

(a) I-A, II-B, III-C

(b) I-C, II-B, III-A

(c) I-A, II-C, III-B

(d) I-C, II-A, III-B

215. A pipe installed in house drainage for ventilation purposes is called

(a) waste pipe

(b) soil pipe

(c) vent pipe

(d) anti siphonage pipe

216. A circular sewer section looses most of its merits, when the sewage depth becomes less than..... full.

(a) 1/2 (b) 3/4

(c) 2/3 (d) none of these.

217. B.O.D removal in a oxidation pond may be upto..... percent.

(a) 70 (b) 80

(c) 90 (d) 99

218. When chlorine is added to sewage both at the beginning as well as at the end of the treatment process, the phenomenon is called........ chlorination.

(a) break point (b) post

(c) super (d) split

219. Particles of around one micron (10^{-6} metre) size are best removed by

(a) slow sand filtration

(b) chemical precipitation

(c) chemical coagulation

(d) plain sedimentation

220. Match the detention time in various treatment units.

Treatment units	**Detention time**
A.Primary sedimentation	I. 2 hours
B.Sludge digestion	II. 6 hours
C.Grit chamber	III. 2 minutes
D.Activated sludge	IV. 20 days

(a) A - I, B - IV, C - III, D - II

(b) A - IV, B - III, C - II, D - I

(c) A - I, B - II, C - IV, D - III

(d) A - III, B - IV, C - II, D - I

221. Sewage sickness occurs, when the sewage

(a) clogs the voids of soil due to continuous application of sewage on a piece of land

(b) contains pathogenic organism

(c) enters the water supply system

(d) clogs the sewers due to accumulation of solids

Answer

214 **(d)**	215 **(c)**	216 **(a)**	217 **(c)**	218 **(d)**
219 **(c)**	220 **(a)**	221 **(a)**		

222. Match the property of various sewer pipe materials of construction

Sewer pipe material

A. Steel sewer pipe
B. Concrete sewer pipe
C. Stoneware sewer pipe
D. Cast iron sewer pipe

Property of the pipe material

I. Corrosion resistant to most acids.
II. Corrosion resistance is most natural soils.
III. Unable to withstand high external load.
IV. Unsuitable where soil contains excessive sulfates.

(a) A-III, B-IV, C-I, D-II
(b) A-IV, B-III, C-II, D-I
(c) A-IV, B-III, C-T, D-II
(d) A-Ill, B-IV, C-11, D-I

223. In the oxidation ditch, the excess sludge is taken to
(a) aerobic digester
(b) incinerator
(c) anaerobic digester
(d) drying beds

224. Two main gases liberated from an anaerobic sludge digestion tank are
(a) CH_4&H_2S (b) CH_4& CO_2
(c) NH_3& CO_s (d) NH_3& CH_4

225. Sludge bulking can be controlled by
(a) denitrification
(b) chlorination
(c) aeration
(d) coagulation

226. With rise in temperature, ultimate B.O.D. value of waste water
(a) Increases
(b) decreases
(c) remains constant
(d) doubles for every 10ºC rise

227. A sewer is mainly designed to attain self-cleaning velocity at
(a) average hourly rate of flow
(b) minimum hourly rate of flow
(c) peak hourly rate of flow
(d) sewer running half full

228. Which of the following sewage treatment method requires the largest area for the same discharge?
(a) Oxidation pond
(b) Oxidation ditch
(c) Trickling filter
(d) Anaerobic pond

229. Eutrophication of lakes is primarily due to
(a) multiplication of bacteria
(b) excessive inflow of nutrients
(c) increase in benthic organisms
(d) thermal and density currents.

230. Presence of nitrogen in a waste water sample is due to the decomposition of
(a) fats (b) vitamins
(c) protein (d) carbohydrates.

231. In context of water polluted with sewage BOD signifies
(a) biology of degradation
(b) bacteriological oxygen demand
(c) biological oxygen demand
(d) bio-chemical oxygen demand.

Answer

222 **(a)**	223 **(d)**	224 **(b)**	225 **(b)**	226 **(c)**
227 **(b)**	228 **(a)**	229 **(b)**	230 **(c)**	231 **(d)**

232. Iron may be present in significant quantity in
   (a) irrigation return water
   (b) septic tank effluent
   (c) fresh water from deep tubewells
   (d) rain water.

233. The grown algae are useful in the operation of
   (a) septic tanks
   (b) oxidation ponds
   (c) slow and filters
   (d) sludge digestion lagoons.

234. Secondary treatment of sewage removes
   (a) suspended organic matter
   (b) dissolved and colloidal organic matter
   (c) fats and grease
   (d) pathogenic (disease causing) bacteria

235. Pick out the wrong statement:
   (a) Biochemical oxygen demand of safe drinking water is zero.
   (b) The amount of oxygen consumed by sewage from potassium dichromate is termed as C.O.D.
   (c) The color of fresh domestic sewage is slightly more than l.
   (d) The specific gravity of sewage is slightly more than 1.

236. Most commonly used chemical for controlling the algae is
   (a) alum
   (b) copper sulfate
   (c) bleaching powder
   (d) lime.

237. The stabilisation of sewage in oxidation pond is due to symbiotic action of
   (a) inorganic & organic compounds
   (b) anaerobic & aerobic bacteria
   (c) protozoa & algae
   (d) bacteria & algae

238. Eutrophic condition in lakes are due to excess of
   (a) turbidity
   (b) color
   (c) suspended solid
   (d) algae.

239. Bacteria which use CO_2 as a source of carbon are known as
   (a) heterotrophic
   (b) autotrophic
   (c) anaerobic
   (d) microbacteria.

240. Match the contaminants to their environmental degradation significance.

Water pollutant

A. Suspended solids
B. Dissolved inorganic solids
C. Heavy metals
D. Nutrients e.g, N & P

Environmental significance

I. Causes sludge deposits
II. Interferes with effluent reuse
III. Toxic, may interfere with effluent reuse
IV. Causes eutrophication of lakes

   (a) A - IV, B - III, C - II, D - I
   (b) A - IV, B - II, C - I, D - III
   (c) A - III, B - T, C - II, D - IV
   (d) A - I, B - II, C - III, D - IV

Answer

232 **(c)**	233 **(b)**	234 **(b)**	235 **(c)**	236 **(d)**
237 **(c)**	238 **(d)**	239 **(b)**	240 **(d)**	

241. Velocity control device is an essential component of
   (a) rapid sand filter
   (b) flow equalisation tank
   (c) activated sludge process
   (d) grit chamber

242. Pick out the wrong statement:
   (a) Algal growth is not desirable in waste stabilisation ponds.
   (b) The sewage strength in stabilisation pond is reduced by the principle of algal-bacterial symbiosis.
   (c) About 80% of suspended solids are 1 removed in plain sedimentation of sewage.
   (d) The sewage in a waste stabilisation pond is treated by algae.

243. Municipal sewer is separated from house drainage system by........ trap.
   (a) gully (b) intercepting
   (c) S- (d) P-

244. Septic tank is the most suitable unit for
   (a) hotels (b) hostels
   (c) hospitals (d) all 'a', 'b'&'c'.

245. B.O.D is preferred to C.O.D as an index of sewage concentration, because
   (a) B.O.D represents both nitrogenous & carbonaceous organic matter, while C.O.D may indicate only carbonaceous matter.
   (b) B.O.D relates specifically to putrescible(decomposable) organic matter, which is the most objectionable sewage constituent.
   (c) C.O.D relates to the impurities which can only be removed by chemical treatment which is expensive.
   (d) B.O.D test is easier to perform and gives more reliable results.

246. Re-aeration rate of river does not depend upon the...... of river water.
   (a) depth (b) velocity
   (c) temperature (d) none of these

247. Match the fo1lowing:

Coloumn I

A. Vent pipe
B. Soil pipe
C. Antisiphonage pipe
D. Waste pipe

Coloumn II

I. Carries liquid waste that does not contain human excreta.
II. Carries liquid waste including human excreta.
III. Preserves the water seal of trap through access to atmospheric air.
IV. Provides flow of air to or from drainage system in order to prevent vacuum pressures & excessive pressure and provides escape for fuel gases.

   (a) A - IV, B - III, C - II, D - I
   (b) A - I, B - II, C - IV, D - III
   (c) A - IV, B - II, C - III, D - I
   (d) A - III, B - IV, C - II, D - I

248. Under normal condition of now, an unpolluted river wi1l contain
   (a) more dissolved oxygen in summer than in winter
   (b) less dissolved oxygen in summer than in winter
   (c) same amount of dissolved oxygen in winter and summer
   (d) the least amount of dissolved oxygen during the floods.

Answer

241 **(d)**	242 **(c)**	243 **(b)**	244 **(d)**	245 **(b)**
246 **(d)**	247 **(c)**	248 **(b)**		

249. Correct sequence of unit operations in a sewage treatment plant is

I. Screening
II. Grit removal
III. Secondary sedimentation
IV. Aeration V. Primary sedimentation

(a) I, II, V, IV, III
(b) II, I, III, V, IV
(c) I, II, III, IV, V
(d) II, I, IV, V, I II

250. Pick out the wrong statement:

(a) The normal temperature of sewage is higher than supply water.
(b) Strength of waste water is measured by its pH valve.
(c) The pipe installed in the house drainage to preserve the water seal is ca1led antisiphonage pipe.
(d) The leakage of sewage from sewers into the surrounding is known as infiltration.

251. Match the various sewage treatment processes with the biological agent used

Processes

A. Oxidation ditch
B. Waste stabilisation pond
C. lmholf tank
D. Rotating Biological Contactor (RBC)

Biological agents

I. Facultative bacteria
II. Anaerobic bacteria
III. Aerobic bacteria (suspended culture)
IV. Aerobic bacteria (attached culture)

(a) A - III, B - II, C - I, D - IV
(b) A - IV, B - III, C - II, D - I
(c) A - III, B- I, C - II, D - IV
(d) A - II, B - IV, C - I, D - III

252. The waste stabilisation pond can be

(a) anaerobic (b) aerobic
(c) facultative (d) any of these.

253. Which of the following pairs is not correctly matched about waste water characteristics?

(a) C.O.D - Biodegradability of waste water
(b) B.O.D - strength of sewage
(c) Nitrate - Methemoglobinemia (blue baby disease)
(d) Methane - Product of anaerobic decomposition.

254. The entry of foul smelling gases into the house coming from the sewers can be prevented by providing

(a) water seals for all the fixtures
(b) exhaust fans
(c) water seals for all the fixtures and a vent pipe in the plumbing system
(d) efficient vent pipes in the pumpling system.

255. Which of the following will help to prevent the escape of foul sewer gases from a water closet?

(a) Vent pipe (b) Air gap
(c) Gully trap (d) none of these

Answer

249 **(a)**	250 **(b)**	251 **(c)**	252 **(d)**	253 **(a)**
254 **(c)**	255 **(c)**			

256. Match the sewage impurities removed by various sewage treatment units

Impurities in sewage

A. Large floating matter
B. Suspended inorganic matter
C. Suspended organic matter
D. Dissolved organic matter

Sewage treatment units

I. Trickling filter
II. Primary clarifier
III. Grit chamber
IV. Screens

(a) A-IV, B-III, C-II, D-I
(b) A - IV, B - III, C - I, D - II
(c) A - III, B - IV, C - II, D - I
(d) A - III, B - IV, C - I, D - II

257. Which of the following processes is not involved in self-purification of running water streams?

(a) Dilution (b) Sedimentation
(c) Oxidation (d) Coagulation

258. Waste water solid digestion is done to

(a) separate liquids from solids to facilitate drying
(b) decompose enough of the organic matter to avoid creating a nuissance
(c) decrease the bulk of the waste water solids to facilitate handling
(d) all 'a', 'b'&'c'.

259. Corrosion of concrete sewers are generally caused due to

(a) higher concentration of grit in the sewage
(b) high pH value of the sewage
(c) high velocity of flow of sewage
(d) septic condition of the sewage

260. Pick out the wrong statement:

(a) Sludge digestion in a gobar gas plant is done by anaerobic bacteria.
(b) Permissible B.O.D of sewage effluents when discharged into river is 200mg/litre
(c) BOD value of medium strength domestic sewage could be above 200 mg/litre.
(d) The optimum temperature [or sludge digestion is 25°C.

261. % of activated sludge mixed with the sewage in primary settling tanks is

(a) 5 - 10 (b) 15 - 20
(c) 20 - 40 (d) 40 - 60

262. The first stage of sludge digestion process is

(a) alkaline fermentation
(b) acidic fermentation
(c) acid regression
(d) none of these

263. The sludge does not contain waste water from

(a) kitchen sinks (b) toilets
(c) bath rooms (d) wash basins.

264. Activated sludge process is used to remove

(a) settleable organic matter
(b) soluble organic matter
(c) inorganic matter & toxic chemicals
(d) pathogenic bacteria.

Answer

256 **(a)**	257 **(d)**	258 **(d)**	259 **(d)**	260 **(b)**
261 **(c)**	262 **(b)**	263 **(b)**	264 **(b)**	

265. Pick out the wrong statement
   (a) High C.O.D to B.O.D ratio of waste water indicates its high biodegradability.
   (b) Trickling filter removes dissolved organics from waste water.
   (c) The hospital waste may be disposed off with municipal wastes.
   (d) The most critical characteristics of sewage is Biochemical Oxygen Demand (B.O.D), when it is disposed off in river.

266. Indices of water pollution in case of water streams are dissolved oxygen (D.O) and biochemical oxygen demand (B.O.D). The D.O content in self-purification of water steams is nearly equal to its saturation valve in the zone of
   (a) recovery
   (b) active decomposition
   (c) degradation
   (d) none of these

267. pH value ranges to be maintained for proper functioning of a sludge digestion tank may be
   (a) 3.5 - 4.5 (b) 6.5 - 8.5
   (c) 8.5 - 10.5 (d) < 5

268. Laying of sewer pipes is usually done with the help of
   (a) slight rails and boning rods
   (b) a compass
   (c) a plane table
   (d) a theodolite

269. Match the following:

Coloumn I

A. Water closet
B. Gully trap
C. Cowl
D. Intercepting trap

ColoumnII

I. Prevents gases entering a house from public sewers.
II. Prevents nesting of birds etc.
III. Discharges into soil pipe through floor trap.
IV. Receives human excreta directly.

   (a) A - IV, B - III, C - II, D - I
   (b) A - III, B - II, C - IV, D - I
   (c) A - IV, B - III, C - I, D - II
   (d) A - III, B - IV, C - II, D - I

270. In combined sewer system, storm water are admitted through
   (a) leaping weirs
   (b) intercepting weirs
   (c) inlets
   (d) storm regulators

271. Anaerobic biological treatment of high B.O.D wastes are suggested to
   (a) obtain end products having a fuel value
   (b) yield least amount of sludge
   (c) avoid high requirement of compressed air
   (d) all 'a', 'b'&'c'.

272. Recirculation in activated sludge process is done to
   (a) supply seed materials to the aeration tank
   (b) dilute the incoming sewage
   (c) dampen the effect of flow variation
   (d) operate the plant continuously.

Answer

265 **(d)**	266 **(d)**	267 **(b)**	268 **(a)**	269 **(a)**
270 **(c)**	271 **(d)**	272 **(a)**		

273. Problem encountered by water polluted by excess nutrientsis
   (a) availability of too much oxygen for fish
   (b) decrease in photosynthesis for aquatic plants
   (c) increase in aerobic bacteria
   (d) reduction in water's ability to hold dissolved oxygen.

274. Impact of influx of agricultural waste or mannure on an aquatic ecosystem is
   (a) increase in dissolved oxygen
   (b) reduction in photosynthesis
   (c) excessive growth of algae
   (d) lower biological oxygen demand.

275. Buildings/statue made of....... are particularly more susceptible to damage by acid rain.
   (a) Granite
   (b) Wood
   (c) Marble
   (d) Concrete

276. Which of the following is a hazard of thermal pollution of water?
   (a) Disruption of aquatic life
   (b) Pollution of public drinking water
   (c) Eutrophication
   (d) Increase in algal content.

277. Thermal pollution of aquatic ecosystem
   (a) reduces the ability of water to hold dissolved oxygen.
   (b) reduces the ability of water to hold dissolved CO_2.
   (c) increases the ability of water to hold dissolved CO_2.
   (d) does hot have any impact on dissolved CO_2 or O_2.

278. Pick out the wrong statement about acid rain & its effects:
   (a) Acid rain may leach nutrients from the soil leading to decrease in plant growth.
   (b) Acid rain causes increase in plankton and moss in aquatic ecosystems.
   (c) Acid rain causes increases in sulfur and nitrogen in the soil.
   (d) Acid rain causes nitrogen saturation in aquatic ecosystem.

279. Which of the following pH level of water is harmless to most fish?
   (a) 4 (b) 5
   (c) 6 (d) 8

280. At the mouth of a river, there is an area that is depleted in oxygen. Most likely cause of I this may be
   (a) acid precipitation falling in the ocean
   (b) thermal pollution of water
   (c) fertiliser being carried by river
   (d) overfishing in water

281. Which of the following areas would be most affected by a rise in sea level as a consequence of global warming?
   (a) Grassland (b) Wetland
   (c) Forests (d) Deserts

282. Most major source of ocean pollution is
   (a) run off from pesticides, fertilisers &sewer discharges
   (b) oil spills from ships
   (c) oil exploration & production
   (d) acid rain.

Answer

273 **(b)**	274 **(c)**	275 **(c)**	276 **(a)**	277 **(a)**
278 **(d)**	279 **(d)**	280 **(c)**	281 **(b)**	282 **(a)**

283. Which of the following is a major component in acid rain?

(a) SO_2 (b) SO_3

(c) H_2SO_4 (d) NO_2

284. What consequence of overusing ground water would affect areas that are distant from the point of use?

(a) Saltwater intrusion

(b) Ground water contamination

(c) Aquifer replenishment

(d) Depletion of stream flow.

285. A body of water is considered safe for swimming, if there are no more than 200 calories of coliform bacteria per 100.... of water.

(a) litre (b) gallon

(c) milliliters (d) ppm

286. The eggs and small fry of some aquatic organisms are damaged when a rapidly melting snow pack contains acidic particles, resulting in acidic concentration of 5 to 10 times higher than acidic rainfall. This condition is known as 'acid shock'. Fish may die of acid shock due to too much.

(a) iron (b) aluminium

(c) oxygen (d) lead.

287. The process by which a body of water becomes enriched in dissolved nutrients, stimulating the growth of aquatic plant life and usually resulting in the depletion of dissolved oxygen, is called

(a) algal blooms

(b) eutrophication

(c) ultraviolet oxidation

(d) acid deposition.

288. What impacts do algal blooms have on plants living deeper beneath the water surface?

(a) They supply extra nutrients to deeper growing plants.

(b) They cause deeper growing plants to seek other forms of food.

(c) They kill them

(d) They increase the amount of sunlight to deeper growing plants.

289. Which of the following does not cause water turbidity?

(a) Clay (b) Silt

(c) Plankton (d) None of these

290. Large water projects that slow the flow of rivers may show many economic benefits; but such projects in warm climates can also lead to

(a) valley fever

(b) schistosomiasis

(c) river blindness

(d) malaria.

291. Which of the following water compartments contains the greatest amount of fresh water?

(a) Rivers and streams

(b) Lakes and reservoirs

(c) Ice and snow

(d) Oceans.

292. Thermal pollution in a river

(a) Reduces the O_2 Dissolution

(b) Increases O_2 solubility

(c) Stabilizes O_2 level

(d) None of these

Answer

283 **(c)**	284 **(d)**	285 **(c)**	286 **(b)**	287 **(b)**
288 **(c)**	289 **(d)**	290 **(b)**	291 **(c)**	292 **(a)**

293. chondrococcus columnaris is harmless to fish at 10°C but causes diseases when the temperature of water is
(a) More than 12°C
(b) More than 21°C
(c) Less than 10°C
(d) At 10°C

294. The major effect of discharging heated water from power plants to aquatic ecosystems is
(a) Reducation in DO concentration
(b) Reduction in Biodiversity of the ecosystem
(c) Increase in metabolic of organism
(d) All of above

295. Biochemical Oxygen Demand, (BOD) is a measure of organic material present in water, BOD value less than 5 ppm indicates as water sample to be
(a) rich in dissolved oxygen.
(b) poor in dissolved oxygen.
(c) highly polluted.
(d) not suitable for aquatic life.

296. Sewage containing organic waste should not be disposed in water bodies because it causes major water pollution. Fishes in such a polluted water die because of
(a) Large number of mosquitoes.
(b) Increase in the amount of dissolved oxygen.
(c) Decrease in the amount of dissolved oxygen in water.
(d) Clogging of gills by mud.

297. Which of the following conditions shows the polluted environment.
(a) pH of rain water is 5.6.
(b) amount of carbon dioxide in the atmosphere is 0.03%.
(c) biochemical oxygen demand 10 ppm.
(d) eutrophication.

298. Phosphate containing fertilisers cause water pollution. Addition of such compounds in water bodies causes
(a) enhanced growth of algae.
(b) decreases in amount of dissolved oxygen in water.
(c) deposition of calcium phosphate.
(d) increase in fish population

299. The acid not present in acid rain is
(a) Peroxyacetylnitrate
(b) H_2CO_3
(c) HNO_3
(d) H_2SO_4

300. Assertion (A): The pH of acid rain is less than 5.6.
Reason (R): carbon dioxide present in the atmosphere dissolves in rain water and forms carbonic acid.
(a) Both A and R are correct and R is the correct explanation of A.
(b) Both A and R are correct but R is not the correct explanation of A.
(c) Both A and R are not correct.
(d) A is not correct but R is correct.

301. Assertion (A): Excessive use of chlorinated synthetic pesticides causes soil and water pollution.
Reason (R): Such pesticides are non-biodegradable.
(a) Both A and R are correct and R is the correct explanation of A.
(b) Both A and R are correct but R is not the correct explanation of A.
(c) Both A and R are not correct.
(d) A is not correct but R is correct.

Answer

293 **(b)**	294 **(d)**	295 **(a)**	296 **(a)**	297 **(d)**
298 **(c)**	299 **(a)**	300 **(c)**	301 **(d)**	

302. Assertion (A): If BOD level of water in a reservior is less than 5 ppm it is highly polluted.

Reason (R): High biological oxygen demand means low activity of bacteria in water.

(a) Both A and R are correct and R is the correct explanation of A.

(b) Both A and R are correct but R is not the correct explanation of A.

(c) Both A and R are not correct.

(d) A is not correct but R is correct.

303. The main cause of water pollution is:

(a) untreated sewage, glass bottles and plastic bags

(b) chemical effluent, leaching of pesticides and algae

(c) pesticides, treated sewage and litter

(d) oil spills untreated sewage and chemical effluent

304. Typhoid, cholera and dysentery are diseases that are spread by:

(a) air pollution

(b) water pollution

(c) land/soil pollution

(d) The use of pesticides

305. Biological oxygen demand (BOD) of --- ------ is the least.

(a) Sewage (b) sea water

(c) pure water (d) polluted water

306. Compound of ----------- cause Minamata disease.

(a) mercury (b) cadmium

(c) cobalt (d) fertilizers

307. Due to eutrophication

(a) BOD increases.

(b) BOD decreases.

(c) algae are destroyed.

(d) water becomes less harmful

308. ------------------- is the first step of sewage treatment.

(a) Precipitation

(b) Chlorination

(c) Sedimentation

(d) Aeration

309. BOD is ----------------- in polluted water and -------------- in potable water.

(a) more, less

(b) less, medium

(c) medium, more

(d) less, more

310. Which of the following is responsible for acid rain ?

(a) Hydrogen

(b) Oxygen

(c) Sulphur dioxide

(d) Methane

311. Acid rain is formed due to reaction from the following pair of gases

(a) Methane and ozone

(b) Oxygen and nitrous oxide

(c) methane and sulphur dioxide

(d) Carbon dioxide and sulphur dioxide

312. A major in-stream use of water is for

(a) producing hydroelectric power.

(b) dissolving industrial wastes.

(c) agricultural irrigation.

(d) domestic use.

Answer

302 **(a)**	303 **(d)**	304 **(b)**	305 **(c)**	306 **(a)**
307 **(a)**	308 **(c)**	309 **(a)**	310 **(c)**	311 **(b)**

313. Which of the following are example of Municipal and industrial discharge pipes?
   (a) nonpoint sources of pollution.
   (b) violations of the Clean Water Act.
   (c) point sources of pollution.
   (d) irrigation.

314. The presence of high coliform counts in water indicates
   (a) contamination by human wastes.
   (b) phosphorus contamination.
   (c) decreased biological oxygen demand.
   (d) hydrocarbon contamination.

315. How the biological oxygen demand gets affected with the increased presence of organic matter in water?
   (a) the oxygen demand increases
   (b) the oxygen demand decreases
   (c) the oxygen demand remains unchanged
   (d) None of the above

316. Which of the following is not a major source of groundwater contamination ?
   (a) agricultural products
   (b) landfills
   (c) septic tanks
   (d) all of the above are major sources of groundwater contamination

317. Which of the following is not considered as a part of water use planning ?
   (a) water treatment
   (b) storm sewer drainage
   (c) salinization
   (d) Water use planning considers all of the above issue

318. The stage in which the biological processes are used to purify water in a wastewater treatment plants is called
   (a) secondary sewage treatment
   (b) primary sewage treatment
   (c) wastewater reduction
   (d) biochemical reduction

319. Groundwater mining in coastal areas can result into
   (a) increase in the salinity of groundwater.
   (b) decrease in the toxicity of groundwater.
   (c) decrease in the salinity of groundwater.
   (d) increase in the water table.

320. Standard EDTA (ethylene diamine tetra acetic acid) solution is used to determine the
   (a) hardness in water
   (b) turbidity in water
   (c) dissolved oxygen in water
   (d) residual chlorine in water

321. The water distribution mains are designed for
   (a) maximum daily demand
   (b) maximum hourly demand
   (c) average daily demand
   (d) maximum hourly demand on maximum day

322. The maximum permissible limit for fluoride in drinking water is ---------- mg/litre.
   (a) 0.1 (b) 1.5
   (c) 5 (d) 10

Answer

313 **(c)**	314 **(a)**	315 **(d)**	316 **(d)**	317 **(d)**
318 **(a)**	319 **(a)**	320 **(a)**	321 **(d)**	322 **(b)**

323. The polluted water is one which
   (a) contains pathogenic bacteria
   (b) consists of undesirable substances rendering unfit for drinking and domestic use
   (c) is safe and suitable for drinking and domestic use
   (d) is contaminated

324. If the coliform bacteria is present in a sample of water, then the coliform test to be conducted is i) presumptive coliform test ii) confirmed coliform test iii) completed coliform test. The correct answer is
   (a) only (i)
   (b) both (i) and (ii)
   (c) both (i) and (iii)
   (d) all (i), (ii) and (iii)

325. Alkalinity in water is expressed as milli-grams per litre in terms of equivalent
   (a) calcium carbonate
   (b) magnesium carbonate
   (c) sodium carbonate
   (d) calcium hydroxide

326. The most common cause of acidity in water is
   (a) carbon dioxide
   (b) oxygen
   (c) hydrogen
   (d) nitrogen

327. Ground water is usually free from
   (a) suspended impurities
   (b) dissolved impurities
   (c) both suspended and dissolved impurities
   (d) none of the above

328. The phenolic compounds in public water supply should not be more than ..... ppm.
   (a) 0.1 (b) 0.01
   (c) 0.001 (d) 0.0001

329. If the average daily consumption of a city is 100,000 m^3, the maximum daily consumption on peak hourly demand will be.......................m^3.
   (a) 100000 (b) 150000
   (c) 180000 (d) 270000

330. As compared to shallow-wells, deep wells have
   (a) more depth
   (b) less depth
   (c) more discharge
   (d) less discharge

331. Which of the following causes a decrease in per capita water consumption?
   (a) use of metering system
   (b) good quality of water
   (c) better standard of living of the people
   (d) hotter climate

332. The devices which are installed for drawing water from the sources are called
   (a) aquifers (b) aquaculture
   (c) filters (d) intakes

333. Which of the following is not a water borne disease?
   (a) dysentery (b) cholera
   (c) typhoid (d) malaria

Answer

323 **(b)**	324 **(d)**	325 **(a)**	326 **(a)**	327 **(a)**
328 **(c)**	329 **(d)**	330 **(c)**	331 **(a)**	332 **(d)**
333 **(d)**				

334. Alum as a coagulant is found to be most effective when pH range of water is

(a) 2 to 4 (b) 4 to 6
(c) 6 to 8 (d) 8 to 10

335. The detention period in coagulation tanks is usually kept as

(a) 1 to 2 minutes
(b) 30 to 45 minutes
(c) 2 to 6 hours
(d) 2 to 6 days

336. Air binding phenomena in rapid sand filters occur due to

(a) excessive negative head
(b) mud ball formation
(c) higher turbidity in the effluent
(d) low temperature

337. The chemical most commonly used to increase The speed of sedimentation of sewage is

(a) sulphuric acid
(b) copper sulphate
(c) lime
(d) sodium permanganate

338. The detention period and overflow rate respectively for plain sedimentation as compared to sedimentation with coagulation are generally

(a) less and more
(b) less and less
(c) more and less
(d) more and more

339. If the total hardness of water is greater than its total alkalinity the alkalinity the carbonate hardness will be equal to

(a) total alkalinity
(b) total hardness
(c) total hardness – total alkalinity
(d) non carbonate hardness

340. The effective size of sand particles used in slow sand filters is ….. mm.

(a) 0.25 to 0.35 (b) 0.35 to 0.60
(c) 0.60 to 1.00 (d) 1.00 to 1.80

341. The amount of coagulant needed for coagulation of water increases with i) increase in turbidity of water ii) decrease in turbidity of water iii) increase in temperature of water iv) decrease in temperature of water. The correct answer is

(a) (i) and (ii) (b) (i) and (iv)
(c) (ii) and (iii) (d) (ii) and (iv)

342. In water treatment, rapid gravity filters are adopted to remove

(a) dissolved organic substances
(b) dissolved solids and dissolved gases
(c) floating solids and dissolved inorganic solids
(d) bacteria and colloidal solids

343. The settling velocity of a particle in a sedimentation tank increases, if

(a) particle size is decreased
(b) the surface area of ank is increased
(c) the depth of tank is decrease
(d) none of these

344. Percentage of bacterial load that can be removed from water by the process of plain sedimentation is about

(a) 10 to 20 (b) 50
(c) 75 (d) 95

Answer

334 **(c)**	335 **(c)**	336 **(a)**	337 **(c)**	338 **(c)**
339 **(a)**	340 **(a)**	341 **(b)**	342 **(d)**	343 **(a)**
344 **(a)**				

345. The dissolved oxygen level in natural unpolluted waters at normal temperature is of the order of .... mg/litre.
(a) 1 (b) 10
(c) 100 (d) 1000

346. For a given discharge, the efficiency of sedimentation tank can be increases by
(a) increasing the depth of tank
(b) decreasing the depth of tank
(c) increasing the surface area of tank
(d) decreasing the surface area of tank

347. As compared to rapid sand filters, slow sand filters give i) slower filtration rate ii) higher filtration rate iii) lesser efficiency in removal of bacteria iv) higher efficiency in removal of bacteria. The correct answer is
(a) (i) and (ii) (b) (ii) and (iii)
(c) (i) and (iv) (d) (ii) and (iv)

348. Residual chlorine in water is determined by
(a) starch iodide method
(b) orthotolidine method
(c) both (a) and (b)
(d) none of the above

349. The settling velocity of a particle in a sedimentation tank depends on
(a) depth of tank
(b) surface area of tank
(c) both depth and surface tank
(d) none of the above

350. Assertion A: Slow sand filters and more efficient in removal of a bacteria than rapid filters. Reason R: The sand used in slow sand filters is finer than that in rapid sand filters. Select your answer based on the coding system given below:
(a) Both A and R are true and R is the correct explanation of A.
(b) Both A and R are true but R is not the correct explanation of A.
(c) A is true but R is false.
(d) A is false but R is true.

351. The amount of residual chlorine left in public water supply for safety against pathogenic bacteria is about .... ppm.
(a) 0.01 to 0.05 (b) 0.05 to 0.5
(c) 0.5 to 1.0 (d) 1.0 to 5.0

352. The rate of filtration in slow sand filters in million litres per day per hectare is bout
(a) 50 to 60 (b) 100 t0 150
(c) 500 to 600 (d) 1400 to 1500

353. The overflow rate for the plan sedimentation tanks is about .... litres/hr/m^2.
(a) 500 to 750 (b) 100 to 1250
(c) 1250 to 1500 (d) 1500 to 2000

354. Orthotolidine test in used for determination of
(a) dissolved oxygen
(b) residual chlorine
(c) biochemical oxygen demand
(d) does of coagulant

355. The velocity of flow water in a sedimentation tank is about
(a) 5 to cm/sec.
(b) 15 to 30 cm/sec.
(c) 15 to 30 cm/ minute
(d) 15 to 30 cm/hour

Answser

345 **(b)**	346 **(c)**	347 **(d)**	348 **(c)**	349 **(b)**
350 **(a)**	351 **(c)**	352 **(a)**	353 **(a)**	354 **(c)**
355 **(c)**				

356. On standard slica, the turbidity in drinking water should be limited to …… ppm.
(a) 10 (b) 20
(c) 30 (d) 50

357. The major disadvantage of lime soda process of water softening is that
(a) It is unsuitable for turbid and acidic water.
(b) huge amount of precipitate is formed which creates a disposal problem
(c) the effluent cannot be reduced to zero hardness
(d) it is unsuitable for softening the water of excessive hardness

358. In chlorination, with the rise in temperature of water, death rate of bacteria
(a) increases
(b) decreases
(c) remains unaffected
(d) none of the above

359. The process in which the chlorination is done beyond the break point is known as
(a) prechlorination
(b) post chlorination
(c) super chlorination
(d) break point chlorination

360. Disinfection of water results in
(a) removal of turbidity
(b) removal hardness
(c) killing of disease causing bacteria
(d) complete sterilization

361. Chlorine demand of water is equal to
(a) applied chlorine
(b) residual chloride
(c) sum of applied and residual chlorine
(d) difference of applied and residual chlorine

362. Which of the following chemical compounds can be used for dechlorination of water ?
(a) carbon dioxide
(b) bleaching powder
(c) sulphur dioxide
(d) chloramines

363. The treatment of water with bleaching powder is known as
(a) prechlorination
(b) super chlorination
(c) dechlorination
(d) hypochlorination

364. The suitable layout of a water distribution system for irregularly growing town is
(a) dead end system
(b) grid iron system
(c) radial system
(d) ring system

365. The suitable method for disinfection of swimming pool water is
(a) ultra violet ray's treatment
(b) lime treatment
(c) by using potassium permanganate
(d) clorination

366. The percentage of chlorine in fresh bleaching powder is about….ppm.
(a) 10 to 15 (b) 20 to 25
(c) 30 to 35 (d) 40 to 50

Answer

356 **(a)**	357 **(b)**	358 **(a)**	359 **(c)**	360 **(c)**
361 **(d)**	362 **(c)**	363 **(d)**	364 **(a)**	365 **(a)**
366 **(c)**				

367. Disinfection efficiency is
   (a) reduced at higher pH value of water
   (b) unaffected by pH value of water
   (c) increased at higher pH value of water
   (d) highest at pH value equal to 7

368. Activated carbon is used for..........of water.
   (a) disinfection
   (b) removing hardness
   (c) removing odours
   (d) removing corrosivness

369. The layout of water distribution system in which water flows towards the outer periphery is
   (a) ring system
   (b) dead end system
   (c) radial system
   (d) grid iron system

370. The suitable layout of water distribution system for a city with roads of rectangular pattern is
   (a) grid iron system
   (b) dead end system
   (c) ring system
   (d) radial system

371. Which of the following compounds is widely used for algae control ?
   (a) sodium sulphate
   (b) copper sulphate
   (c) sodium chloride
   (d) calcium chloride

372. The commonly used material for water supply pipes, which has the properties of being strong not easily corroded and long life but is heavy and brittle is
   (a) steel
   (b) cast iron
   (c) copper
   (d) reinforced cement concrete

373. Double filtration is used
   (a) to increase the filtration capacity of show send filters.
   (b) to increase the filtration capacity of rapid send filters.
   (c) for swimming pods, hotels etc.
   (d) all of the above

374. In lime-soda process
   (a) only carbonate hardness is removed
   (b) only non-carbonate hardness is re-moved
   (c) lime reduces the carbonate hardness and soda ash removes the non-carbonate hardness
   (d) lime reduces the non-carbonate hardness and soda ash removes the carbonate hardness

375. The main disadvantage of cement concrete sewers is
   (a) less strength
   (b) difficulty in construction
   (c) difficulty in transportation due to heavy weight
   (d) less life

376. An egg shaped section of sewer
   (a) is economical than circular section.
   (b) provides self cleansing velocity at low discharges.
   (c) is more stable than circular section.
   (d) is easy to construct

Answer

367 **(a)**	368 **(c)**	369 **(c)**	370 **(a)**	371 **(b)**
372 **(b)**	373 **(a)**	374 **(c)**	375 **(c)**	376 **(a)**

377. A sewer that receives the discharge of a number of house sewer is called
   (a) house sewer
   (b) lateral sewer
   (c) intercepting
   (d) submain sewer

378. Which of the following sewer is preferred for combined system of sewage?
   (a) circular sewer
   (b) egg shaped sewer
   (c) rectangular
   (d) none of the above

379. Average rate of water consumption per heap per day as per Indian Standard is ....litres.
   (a) 100 (b) 135
   (c) 165 (d) 200

380. A pipe conveying sewage from plumbing system of a single building to common sewer or point of immediate disposal is called
   (a) house sewer
   (b) lateral sewer
   (c) main sewer
   (d) submain sewer

381. The minimum and maximum diameters (in cms) of sewer should preferably be
   (a) 15 and 100 (b) 15 and 300
   (c) 30 and 450 (d) 60 and 300

382. For a country like India, where rainfall is mainly confined to one season, the suitable sewerage system will be
   (a) Separate system
   (b) Combined system
   (c) Partially combined system
   (d) Partially separate system

383. Sewerage system is usually designed for ....years.
   (a) 10 (b) 25
   (c) 50 (d) 75

384. Most suitable section of sewer in separate sewage system is
   (a) rectangular section
   (b) circular section
   (c) standard form of egg shaped sewer
   (d) modified egg shaped section

385. The specific gravity of sewage is
   (a) much greater than 1
   (b) slightly less than 1
   (c) equal to 1
   (d) slightly greater than 1

386. The suitable system of sanitation for area of distributed rainfall throughout the year less intensity is
   (a) separate system
   (b) combined system
   (c) partially separate system
   (d) partially combined system

387. The slope of sewer should be
   (a) given in the direction of natural slope of ground.
   (b) given in the direction opposite to natural slope of ground
   (c) zero.
   (d) steeper than 1 in 20.

388. The self cleansing velocity for all sewers in India is usually ....m/sec.
   (a) less than 1.0 (b) 1.0 to 1.2
   (c) 1.5 to 2.0 (d) 3.0 to 3.5

Answer

377 **(b)**	378 **(b)**	379 **(b)**	380 **(a)**	381 **(b)**
382 **(a)**	383 **(b)**	384 **(b)**	385 **(d)**	386 **(b)**
387 **(a)**	388 **(b)**			

389. Settling velocity increases with
   (a) specific gravity of solid particles
   (b) size of particles
   (c) depth of tank
   (d) temperature of liquid

390. The type of sewer which is suitable for both combined and separate system is
   (a) circular sewer
   (b) egg shaped sewer
   (c) horse-shoue type sewer
   (d) semi-elliptical sewer

391. Laying of sewers is usually done with the help of
   (a) atheodolite
   (b) a compass
   (c) sight rails and boning rods
   (d) a plane table

392. In a BOD test,.0 ml of raw sewage was diluted to 100 ml and the dissolved oxygen concentration diluted sample at the beginning was 6 ppm andit was 4 ppm at the end of 5 day incubation at 20 ^{0}C. The BOD of raw sewage will be .....ppm.
   (a) 100 (b) 200
   (c) 300 (d) 400

393. Dissolved oxygen in streams is
   (a) maximum at noon
   (b) minimum at noon
   (c) maximum at midnight
   (d) same throughout the day

394. The correct relation between theoretical oxygen demand (TOD), Biochemical oxygen demand (BOD) and chemical oxygen demand (COD) is given by
   (a) TOD>BOD>COD
   (b) TOD>COD>BOD
   (c) BOD>COD>TOD
   (d) COD>BOD>TOD

395. The ratio of 5-day BOD to ultimate BOD is about
   (a) 1/3 (b) 2/3
   (c) 3/4 (d) 1.0

396. The effect of increasing diameter of sewer on the self cleansing velocity is
   (a) to decrease it (b) to increase it
   (c) fluctuation (d) nil

397. The minimum dissolved oxygen which should always be present in water in order to save the aquatic life is..............ppm.
   (a) 1 (b) 4
   (c) 10 (d) 40

398. The characteristics of fresh and septic sewage respectively are
   (a) acidic and alkaline
   (b) alkaline and acidic
   (c) both acidic
   (d) both alkaline

399. Sewage treatment units are normally designed for ....years.
   (a) 5-10 (b) 15-20
   (c) 30-40 (d) 40-50

400. Select the correct statement.
   (a) 5-day BOD is the ultimate BOD.
   (b) 5-day BOD is greater than 4-day BOD keeping other conditions same.
   (c) 5-day BOD is less than 4-day BOD keeping other conditions same.
   (d) BOD does not depend on time.

Answer

389 **(c)**	390 **(b)**	391 **(c)**	392 **(b)**	393 **(a)**
394 **(b)**	395 **(b)**	396 **(b)**	397 **(b)**	398 **(b)**
399 **(b)**	400 **(c)**			

401. Facultative bacteria are able to work in
   (a) presence of oxygen only
   (b) absence of oxygen only
   (c) presence as well as in absence of oxygen
   (d) presence of water

402. The rate of BOD exerted at any time is
   (a) directly proportional of BOD satisfied
   (b) directly proportional of BOD remaining
   (c) inversely proportional to BOD remaining
   (d) none of these

403. Corrosion in concrete sewers is caused by
   (a) septic condition
   (b) dissolved oxygen
   (c) chlorine
   (d) nitrogen

404. The relative stability of a sewage sample, whose dissolved oxygen is same as the total oxygen required satisfy BOD, is
   (a) 1 (b) 100
   (c) 00 (d) zero

405. Sewage treatment units are designed for
   (a) maximum flow only
   (b) minimum flow only
   (c) average flow only
   (d) maximum and minimum flow

406. Standard BOD is measured at
   (a) 20 ^{0}C – 1 day (b) 25 ^{0}C – 3 day
   (c) 20 ^{0}C – 5 day (d) 30 ^{0}C – 5 day

407. Generally the detention period for grid chambers is kept as
   (a) 1 minute (b) 5 minutes
   (c) 2-4 hours (d) 12 hours

408. If Biochemical oxygen demand (BOD) of a town is 20000 kg/day and BOD per capita per day is 0.5 kg, then population equivalent of town is
   (a) 1000 (b) 4000
   (c) 100000 (d) 400000

409. If the sewage contains grease and fatty oils, these are removed in
   (a) grit chambers
   (b) detritus tanks
   (c) skimming tanks
   (d) sediment tank

410 The biochemical treatment of sewage effluents is essential a process of
   (a) oxidation
   (b) dehydration
   (c) reduction
   (d) alkalinization

411. The detention period for oxidation ponds is usually kept as
   (a) 4-8 hours (b) 24 hours
   (c) 10 to 15 days (d) 3 months

412. Sludge volume index is defined as the ratio of the percentage of
   (a) sludge by volume to percentage of suspended solids by weight.
   (b) sludge by volume to percentage of total solids by weight.
   (c) suspended solids by weight.
   (d) total solids by weight to percentage of sludge by volume.

Answer

401 **(c)**	402 **(b)**	403 **(a)**	404 **(b)**	405 **(c)**
406 **(a)**	407 **(d)**	408 **(d)**	409 **(c)**	410 **(a)**
411 **(c)**	412 **(a)**			

413. The maximum efficiency of BOD removal is achieved in
   (a) oxidation
   (b) oxidation ditch
   (c) aerated lagoons
   (d) trickling filters

414. For satisfactory working of a sludge digestion unit, the pH range of digested sludge should be maintained at
   (a) 4.5 to 6.0   (b) 6.5 to 8.0
   (c) 8.5 to 10.0   (d) 10.5 to 12.0

415. The main disadvantage of oxidation pond is that
   (a) large area is required for construction
   (b) maintenance and operation cost are high
   (c) BOD removal is very low
   (d) none of the above

416. The working conditions in Imhoff taks are
   (a) aerobic only
   (b) anaerobic only
   (c) aerobic in lower compartment and anaerobic in upper compartment
   (d) anaerobic in lower compartment and aerobic in upper compartment

417. To divert excessive flow from combined sewers, the most commonly used weir is a
   (a) cipoletti weir
   (b) broadcrested weir
   (c) leaping weir
   (d) submerged weir

418. When there is no recirculation of sewage, then recirculation factor is
   (a) 0
   (b) 1
   (c) infinity
   (d) none of the above

419. In facultative stabilization pond, the sewage is treated by
   (a) aerobic bacteria only.
   (b) algae only.
   (c) dual action of aerobic bacteria and anaerobic bacteria.
   (d) sedimentation.

420. Septic tank is a i) settling tank ii) digestion tank iii) aeration tank. The correct answer is
   (a) only (i)   (b) (i) and (ii)
   (c) (i) and (iii)   (d) only (iii)

421. For the same solid content, if the quantity of sludge with moisture content of 98% is x, then the quantity of sludge with moisture content of 96% will be
   (a) X/4   (b) X/2
   (c) X   (d) 2X

422. For normal sludge, the value of sludge index for Indian conditions is
   (a) 0 to 50   (b) 50 to 150
   (c) 150 to 350   (d) 350 to 500

423. The process of lagooning is primarily a means of
   (a) reducing the excessive flow in sewers
   (b) disposing of sludge
   (c) increasing the capacity of storage reservoirs
   (d) increasing flow of sewage through Imhoff tanks

Answer

413 **(b)**	414 **(b)**	415 **(a)**	416 **(d)**	417 **(c)**
418 **(b)**	419 **(c)**	420 **(b)**	421 **(b)**	422 **(c)**
423 **(b)**				

424. If the average water consumption of a city is 200 liters per capita day, and the population of city is 5,00,000. The maximum hourly demand of water for the maximum day will be .... million litres.
(a) 150 (b) 180
(c) 270 (d) 330

425. The minimum width of a sewer French in mm (millimeter) must be greater than; when D is the external diameter of sewer in mm:
(a) 1.5 D + 100 (b) 1.5 D + 300
(c) 1.5 D + 400 (d) D + 1000

426. Turbidity in water is a physical characteristic of water. It can be easily detectedin a glass of water and are usually objectionable for aesthetic reason, when it concentration is in excess of ....turbidity unit.
(a) 0.1 (b) 1.0
(c) 4.0 (d) 5.0

427. If water having a total hardness of 375 mg/l as $CaCO_3$, the Carbonate Hardness. Is 200 mg/l as $CaCO_3$, will be .... mg/l as $CaCO_3$.
(a) 375 (b) 200
(c) 175 (d) none of these

428. Ground water has to be treated with impurities like hardness, micrograms and low dissolved Oxygen concentration along with different gases. What will be the proper treatment process from the following options:
(a) Sedimentation; Filtration: Coagulation Flocculation Disinfection
(b) Aeration; Sedimentation; Softening; Disinfection
(c) Aeration; Sedimentation; lime soda process; process; Filtration
(d) Screen; Filtration; softening; Disinfection

429. The normal length of a sedimentation tank to remove suspended particles from water should not be more than ________ time of its Width.
(a) 2 (b) 4
(c) 6 (d) 8

430. If the length of settling tank is 5 m, width is 3 m; depth is 1 m of a rectangular tank with discharge (Q) of 1.0 m^3/second; the settling velocity will be ....m/sec.
(a) 0.033 (b) 0.067
(c) 0.167 (d) 0.200

431. Which of the following is an effective alternative to chlorine for disinfecting waste water in a municipal water
(a) Freon
(b) Phosphate
(c) Ammonia
(d) Ozone

432. The presence of fiscal coliform bacteria in a sample of river water suggest which of the following?
(a) The pH of the river is very high.
(b) The water is contaminated with animal waste.
(c) The river is devoid of plant life.
(d) The dissolved oxygen level of the river is high

Answer

424 **(c)**	425 **(b)**	426 **(d)**	427 **(b)**	428 **(b)**
429 **(b)**	430 **(b)**	431 **(d)**	432 **(b)**	

433. If wastewater treatment plant effluent that contains nitrates and phosphates is allowed to flow into a body of water, which of the following may result?
   (a) Chlorination
   (b) Decomposition
   (c) Eutrophication
   (d) Oxygenation

434. As urbanization increases and natural soil surfaces are covered, the groundwater supply is reduced due to
   (a) increased evaporation and transpiration
   (b) decreased surface runoff
   (c) loss of recharge area
   (d) confinement of aquifers

435. Which of the following components of a wastewater treatment plant is designed to facilitate the decomposition of organic material by aerobic microorganisms?
   (a) Bar screen
   (b) Grit-settling tank
   (c) Activated-sludge tank
   (d) Chlorination tank

436. Acid rain is associated with which of the following?
   (a) Formation of the Antarctic ozone hole
   (b) Release of PCB's into the atmosphere
   (c) Damage to tropical rain forests
   (d) The burning of fossil fuels

437. Which of the following would be the strongest evidence In support of the contention that a local area was experiencing acid deposition?
   (a) A sudden die-of all the fish in a local stream
   (b) A gradual increase in the temperature of a local lake
   (c) An increase in the rate of photosynthesis of aquatic plants in a local lake
   (d) An increase in a concentration of soluble heavy metal in a local pond

438. Which of the following is the usual cause of cultural eutrophication in surface water of both developed and developing countries?
   (a) Introduction of cyanobacteria to streamsand river
   (b) Runoff of metal ions in bodies of water
   (c) Runoff of nitrate compounds into bodies of water
   (d) Runoff of herbicides into bodies of water

439. Suppose less water evaporates from the world's oceans. If all other variables remain the same, what will happen to mean global temperature?
   (a) Mean global temperature will increase.
   (b) Mean global temperature will decrease.
   (c) Evaporation from the oceans does not impact mean global temperature.
   (d) Mean global temperature will stay the same.

Answer

433 **(c)**	434 **(c)**	435 **(c)**	436 **(d)**	437 **(b)**
438 **(c)**	439 **(b)**			

440. Which of the following is an example of non point source of pollution?
   (a) the water from the cooling towers of a coal-fired power plant
   (b) the oil from an oil tanker spill off the coast of Alaska
   (c) the effluent from the wastewater-treatment plant
   (d) the water going into the storm drains

441. Which of the following organisms can tolerate low levels of dissolved oxygen in water?
   (a) blackfly larvae and mayfly larvae
   (b) blackfly larvae and sludge worms
   (c) caddis fly and mayfly larvae
   (d) sludge worms and caddis fly larvae

442. You find the following organisms on a stream bottom: mayfly larvae, stonefly larvae, and blackfly larvae. What is the quality of the water?
   (a) greenish
   (b) very cloudy
   (c) severely damaged
   (d) clean water

443. Which of the following is true?
   (a) The sediment bed load impacts only vertebrate life in a stream.
   (b) Sediment bed load does not impact the organisms that live in a stream.
   (c) Streams with a high sediment bed load support a wider variety of organisms than streams with a low sediment bed load.
   (d) Streams with a low sediment bed load support a wider variety of organisms than streams with a high sediment bed load.

444. Sticks and bottles that wash into sewers fit into which category of materials handled at wastewater-treatment plants?
   (a) dissolved inorganic material
   (b) particulate organic material
   (c) debris and grit
   (d) colloidal and dissolved organic material

445. The bar screen at a wastewater-treatment plant breaks. Which phase of treatment is impacted?
   (a) primary treatment
   (b) disinfection
   (c) secondary treatment
   (d) biological nutrient removal

446. Someone adds oxygen to the sludge digester at a wastewater treatment plant. What will happen?
   (a) The digestion of sludge will increase because the bacteria need oxygen.
   (b) The digestion of sludge will decrease because the bacteria cannot live in the presence of oxygen.
   (c) The digestion of sludge will not change.
   (d) The pathogens in the sludge will all die. Clarification will take longer.

Answer

440 **(d)**	441 **(b)**	442 **(d)**	443 **(d)**	444 **(c)**
445 **(a)**	446 **(b)**			

447. Your wastewater treatment plant adds chlorine to the treated water before discharging the effluent to the river. However, many local species of fish are dying because of the presence of this chemical. Which phase of treatment will the plant need to change?
   (a) disinfection
   (b) secondary treatment
   (c) biological nutrient removal
   (d) primary treatment

448. You are a farmer. Which of the following would you most like to have to put on your fields?
   (a) water that remains after treatment is finished
   (b) untreated wastewater
   (c) raw sludge
   (d) dried bio solids

449. Biochemical oxygen demand (BOD) is an important measure of
   (a) oxygen content of water and wastewater
   (b) the oxygen using potential of water and wastewaters
   (c) an organism's natural level of oxygen requirement
   (d) a measure of the biological activity of water and wastewater

450. Which of the following is most likely to be oligotrophic?
   (a) a pond in the middle of a housing development
   (b) a lake surrounded by corn fields
   (c) a pond in the middle of a golf course
   (d) a lake in Superior National Forest

451. The ideal modern sewage system is one in which
   (a) storm water is collected and treated, but sewer water is not.
   (b) individual home owners operate private septic systems on large lots.
   (c) sanitary sewage water and storm water are collected in a single sewer system and treated in a single treatment plant.
   (d) sanitary sewage water is collected separately from storm water and is fully treated to remove all pollutants before the water is put back into the environment.

452. A farmer has a eutrophic pond on her land. Which of the following would work best to reduce the amount of nutrients entering the pond?
   (a) tilling the surrounding fields more often
   (b) planting a field windbreak around the pond
   (c) using more chemical fertilizer on the surrounding fields
   (d) putting manure on the surrounding fields

453. In freshwater systems, ________ is most often the limiting factor, whereas in marine systems the limiting factor is most often ________.
   (a) nitrogen, phosphorus
   (b) potassium, nitrogen
   (c) phosphorus, potassium
   (d) phosphorus, nitrogen

Answer

447 **(a)**	448 **(d)**	449 **(a)**	450 **(d)**	451 **(d)**
452 **(b)**	453 **(d)**			

454. Which of the following is true?
   (a) Warm water and cold water hold the same amount of dissolved oxygen.
   (b) Warm water holds more dissolved oxygen than cold water.
   (c) Cold water holds more dissolved oxygen than warm water.
   (d) The amount of dissolved oxygen that water can hold does not depend on temperature.

455. Which of the following includes all of the others?
   (a) benthic plants
   (b) emergent vegetation
   (c) submerged aquatic vegetation
   (d) pond weeds

456. Which of the following includes all of the others?
   (a) diatoms
   (b) cyanobacteria
   (c) phytoplankton
   (d) green algae

457. The MCL for nitrates in drinking water is 10 ppm. What does this mean?
   (a) You must boil your water for 10 minutes before using it if it contains nitrates.
   (b) Your drinking water should not have more than 10 ppm nitrates.
   (c) You must let your tap run for 10 minutes before using it if the water contains nitrates.
   (d) Your drinking water must have at least 10 ppm nitrates.

458. Biochemical Oxygen Demand, (BOD) is a measure of organic material present in water. BOD value less than 5 ppm indicates a water sample to be
   (a) rich in dissolved oxygen.
   (b) poor in dissolved oxygen.
   (c) highly polluted.
   (d) not suitable for aquatic life.

459. Sewage containing organic waste should not be disposed in water bodies because it causes major water pollution. Fishes in such a polluted water die because of
   (a.) Large number of mosquitoes.
   (b) Increase in the amount of dissolved oxygen.
   (c) Decrease in the amount of dissolved oxygen in water.
   (d) Clogging of gills by mud.

460. The acid that is not present in acid rain is
   (a) H_2CO_3 (b) HNO_3
   (c) H_2SO_4 (d) Oxalic acid

461. Assertion (A): Excessive use of chlorinated synthetic pesticides causes soil and water pollution.

   Reason (R): Such pesticides are non-biodegradable.
   (a) Both A and R are correct and R is the correct explanation of A.
   (b) Both A and R are correct but R is not the correct explanation of A.
   (c) Both A and R are not correct.
   (d) A is not correct but R is correct.

462. **Assertion (A)**: If BOD level of water in a reservoir is less than 5 ppm it is highly polluted.

   Reason (R): High biological oxygen demand means low activity of bacteria in water.
   (a) Both A and R are correct and R is the correct explanation of A.
   (b) Both A and R are correct but R is not the correct explanation of A.
   (c) Both A and R are not correct.
   (d) A is not correct but R is correct.

Answer

454 **(c)**	455 **(a)**	456 **(a)**	457 **(b)**	458 **(a)**
459 **(c)**	460 **(d)**	461 **(a)**	462 **(c)**	